Mineral Nutrients to Mitigate Abiotic Stress in Crops

Plant Biology, Sustainability, and Climate Change

Mineral Nutrients to Mitigate Abiotic Stress in Crops

Edited by

Azamal Husen

Wolaita Sodo University, Wolaita Sodo, South Ethiopia Regional State, Ethiopia; Department of Biotechnology, Graphic Era (Deemed to be University), Dehradun, Uttarakhand, India

Series Editors

Azamal Husen

Muhammad Iqbal

ELSEVIER

ACADEMIC PRESS

An imprint of Elsevier

Academic Press is an imprint of Elsevier
125 London Wall, London EC2Y 5AS, United Kingdom
50 Hampshire Street, 5th Floor, Cambridge, MA 02139, United States

ISBN: 978-0-443-44712-9

For Information on all Academic Press publications
visit our website at https://www.elsevier.com/books-and-journals

Publisher: Andre Wolff
Acquisitions Editor: Chiara Giglio
Editorial Project Manager: Harshita Bathla
Production Project Manager: Fahmida Sultana
Cover Designer: Bakyalakshmi S.

Typeset by MPS Limited, Chennai, India

Contents

Section V Emerging technologies, innovations, and future directions

List of contributors

Zainul Abideen
Sharjah Seed Bank and Herbarium, University of Al Dhaid, Sharjah, United Arab Emirates; Dr. Muhammad Ajmal Khan Institute of Sustainable, Halophyte Utilization, University of Karachi, Karachi, Sindh, Pakistan

Parisa Aghaei-Gharachorlou
Department of Plant Eco-physiology, Faculty of Agriculture, University of Tabriz, Tabriz, Iran

Nazneen Akhtar
Department of Biotechnology and Genetic Engineering, Kohat University of Science & Technology (KUST), Kohat, Pakistan

Morteza Alami-Milani
Department of Plant Eco-physiology, Faculty of Agriculture, University of Tabriz, Tabriz, Iran

Ahmad Ali
Department of Life Science, University of Mumbai, Vidyanagari, Santacruz (East), Mumbai, Maharashtra, India

Hina Alim
Department of Life Science, University of Mumbai, Vidyanagari, Santacruz (East), Mumbai, Maharashtra, India

Hamzeh Amiri
Department of Biology, Lorestan University, Khorramabad, Iran

Anant
Department of Plant Breeding and Genetics, College of Agriculture, Rani Lakshmi Bai Central Agricultural University, Jhansi, Uttar Pradesh, India

Muhammad Arshad
Institute of Environmental Sciences and Engineering (IESE), School of Civil Engineering and Environmental Sciences (SCEE), National University of Sciences and Technology (NUST), Islamabad, Pakistan

Ghorban Badali
Department of Horticulture and Medicinal Plants, Miyaneh Branch, Islamic Azad University, Miyaneh, Iran

Sandeep K. Behera
Department of Life Science, University of Mumbai, Vidyanagari, Santacruz (East), Mumbai, Maharashtra, India

Chedliya Ben Ahmed
Laboratory of Ecosystems and Biodiversity in Arid Areas in Tunisia, Faculty of Sciences of Sfax, University of Sfax, Sfax, Tunisia

Ameni Ben Hassena
Laboratory of Improvement of Olive Productivity and Fruit Trees, Olive Tree Institute of Sfax, University of Sfax, Sfax, Tunisia

Bechir Ben Rouina
Laboratory of Improvement of Olive Productivity and Fruit Trees, Olive Tree Institute of Sfax, University of Sfax, Sfax, Tunisia

Adalberto Benavides-Mendoza
Department of Horticulture, Antonio Narro Autonomous Agrarian University, Saltillo, Coahuila, Mexico; Conahcyt Laboratory of Plant Ecophysiology and Food Safety (LANCEVSA), Antonio Narro Autonomous Agrarian University, Saltillo, Coahuila, Mexico

Ayesha Bibi
Department of Human Nutrition and Dietetics, Women University Mardan, Mardan, Khyber Pakhtunkhwa, Pakistan

Víctor Hugo Carbajal-Vázquez
Department of Plant Physiology, College of Postgraduates in Agricultural Sciences, Montecillo, Texcoco, State of Mexico, Mexico

Noureddine Chaachouay
Agri-Food and Health Laboratory (AFHL), ESEF of Berrechid, Hassan First University, Settat, Morocco; Department of Biotechnology, Graphic Era (Deemed to be University), Dehradun, Uttarakhand, India

Pramita Dan
Department of Biotechnology, St. Xavier's College (Autonomous), Kolkata, West Bengal, India

Zeynep Görkem Doğaroğlu
Department of Environmental Engineering, Mersin University, Mersin, Turkiye

Xiao-Xuan Du
College of Marine and Bioengineering, Yancheng Institute of Technology, Yancheng, Jiangsu, P.R. China

Mamdouh A. Eissa
Department of Soils and Water, Faculty of Agriculture, Assiut University Assiut, Egypt

Nada Elloumi
Laboratory of Water, Energy and Environment, Higher Institute of Biotechnology of Sfax, University of Sfax, Sfax, Tunisia

Fernando Borrego Escalante
Department of Plant Breeding, Antonio Narro Autonomous Agrarian University, Saltillo, Coahuila, Mexico

Salar Farhangi-Abriz
Cotton Research Institute of Iran, Agricultural Research, Education and Extension Organization (AREEO), Gorgan, Iran

Paola Fincheira
Facultad de Ingeniería y Ciencias, Centro de Excelencia en Investigación Biotecnológica Aplicada al Medio Ambiente (CIBAMA), Universidad de La Frontera, Temuco, Región de la Araucanía, Chile

Milagros Melissa Flores Fonseca
Medical Science, University of Colima, Colima, Jalisco, Mexico

Fernando Carlos Gómez-Merino
Department of Plant Physiology, College of Postgraduates in Agricultural Sciences, Montecillo, Texcoco, State of Mexico, Mexico

Iram Gul
Institute of Environmental Sciences and Engineering (IESE), School of Civil Engineering and Environmental Sciences (SCEE), National University of Sciences and Technology (NUST), Islamabad, Pakistan

Jamaa Habchaoui
Laboratory of Botany, Biotechnology and Plant Protection, Department of Biology, Faculty of Sciences, Ibn Tofail University, Kenitra, Morocco; Department of Biotechnology, Graphic Era (Deemed to be University), Dehradun, Uttarakhand, India

Eréndira Esmeralda Hernández-Andrade
Department of Soil Science-Plant Nutrition, College of Postgraduates in Agricultural Sciences, Montecillo, Texcoco, State of Mexico, Mexico

Rosalba Mireya Hernández-Herrera
Department of Botany and Zoology, University Center for Biological and Agricultural Sciences (CUCBA), University of Guadalajara, Zapopan, Jalisco, Mexico

Azamal Husen
Wolaita Sodo University, Wolaita Sodo, South Ethiopia Regional State, Ethiopia; Department of Biotechnology, Graphic Era (Deemed to be University), Dehradun, Uttarakhand, India

Muhammad Imran
Department of Applied Biosciences, College of Agriculture and Life Sciences, Kyungpook National University, Daegu, South Korea

Atia Iqbal
Department of Microbiology and Molecular Genetics, The Women University Multan, Multan, Pakistan

Hina Jabeen
Department of Biotechnology, Women University Mardan, Mardan, Khyber Pakhtunkhwa, Pakistan

Muhammad Jamil
Department of Biotechnology and Genetic Engineering, Kohat University of Science & Technology (KUST), Kohat, Pakistan

Jin Ryeol Jeon
Department of Applied Biosciences, College of Agriculture and Life Sciences, Kyungpook National University, Daegu, South Korea

Sang Mo Kang
Department of Applied Biosciences, College of Agriculture and Life Sciences, Kyungpook National University, Daegu, South Korea

Syeda Irsa Mazhar Kazmi
School of Interdisciplinary Engineering and Sciences (SINES), National University of Sciences and Technology (NUST), Islamabad, Pakistan

Muhammad Aaqil Khan
International Joint Research Laboratory for Global Change Ecology, School of Life Sciences, Henan University, Kaifeng, Henan, P.R. China

Sehresh Khan
Department of Biotechnology and Genetic Engineering, Kohat University of Science & Technology (KUST), Kohat, Pakistan

Prasann Kumar
Department of Agronomy, School of Agriculture, Lovely Professional University, Phagwara, Punjab, India

Swastik Kundu
Department of Biotechnology, St. Xavier's College (Autonomous), Kolkata, West Bengal, India

Pascal Labrousse
Laboratory of Botany and Cryptogamy, Faculty of Pharmacy, University of Limoges, Limoges, France

Konkona Lahiri
Department of Biotechnology, St. Xavier's College (Autonomous), Kolkata, West Bengal, India

Laura Olivia Fuentes Lara
Antonio Narro Autonomous Agrarian University, Department of Animal Nutrition, Saltillo, Coahuila, Mexico

In-Jung Lee
Department of Applied Biosciences, College of Agriculture and Life Sciences, Kyungpook National University, Daegu, South Korea

Ahmad Mahmood
School of Interdisciplinary Engineering and Sciences (SINES), National University of Sciences and Technology (NUST), Islamabad, Pakistan

Lucía Marcial Salvador
Plant Physiology Laboratory, Department of Horticulture, Autonomous Agrarian Antonio Narro University, Saltillo, Coahuila, Mexico

Julia Medrano Macías
Plant Physiology Laboratory, Department of Horticulture, Autonomous Agrarian Antonio Narro University, Saltillo, Coahuila, Mexico

Imed Mezghani
Laboratory of Water, Energy and Environment, Higher Institute of Biotechnology of Sfax, University of Sfax, Sfax, Tunisia

Seyyed Amirreza Mousavi
Department of Plant Eco-physiology, Faculty of Agriculture, University of Tabriz, Tabriz, Iran

Shahnoush Nayeri
Muller Lab, Department of Biology, Pennsylvania State University, University Park, PA, United States

Yong-Sung Park
Department of Applied Biosciences, College of Agriculture and Life Sciences, Kyungpook National University, Daegu, South Korea

Nimisha Patel
Department of Life Science, J C Bose University of Science and Technology, YMCA, Faridabad, Haryana, India

Amit Raj
Department of Agronomy, School of Agriculture, Lovely Professional University, Phagwara, Punjab, India

Ana Cristina Ramírez-Anguiano
Department of Chemistry, University Center for Exact Sciences and Engineering (CUCEI), University of Guadalajara, Guadalajara, Jalisco, Mexico

Luz Leticia Rivera Solís
Plant Physiology Laboratory, Department of Horticulture, Autonomous Agrarian Antonio Narro University, Saltillo, Coahuila, Mexico

Aryadeep Roychoudhury
Discipline of Life Sciences, School of Sciences, Indira Gandhi National Open University, New Delhi, Delhi, India

Shifa Shaffique
Department of Applied Biosciences, College of Agriculture and Life Sciences, Kyungpook National University, Daegu, South Korea

Anis Ali Shah
Department of Botany, University of Education, Lahore, Pakistan

Nadia Sharif
Department of Biotechnology, Women University Mardan, Mardan, Khyber Pakhtunkhwa, Pakistan

Pankaj Sharma
CCS Haryana Agricultural University, Hisar, Haryana, India; University of Illinois, Chicago, IL, United States

Ronanki Srinidhi
Department of Floriculture and Landscaping, College of Horticulture and Forestry, Rani Lakshmi Bai Central Agricultural University, Jhansi, Uttar Pradesh, India

Gayatri A. Talekar
Department of Life Science, University of Mumbai, Vidyanagari, Santacruz (East), Mumbai, Maharashtra, India

Shrejal Tiwari
Department of Floriculture and Landscaping, College of Horticulture and Forestry, Rani Lakshmi Bai Central Agricultural University, Jhansi, Uttar Pradesh, India

Gonzalo Tortella
Facultad de Ingeniería y Ciencias, Centro de Excelencia en Investigación Biotecnológica Aplicada al Medio Ambiente (CIBAMA), Universidad de La Frontera, Temuco, Región de la Araucanía, Chile

Lina Trabelsi
Laboratory of Improvement of Olive Productivity and Fruit Trees, Olive Tree Institute of Sfax, University of Sfax, Sfax, Tunisia

Libia Iris Trejo-Téllez
Department of Soil Science-Plant Nutrition, College of Postgraduates in Agricultural Sciences, Montecillo, Texcoco, State of Mexico, Mexico

Javier Trujillo Rogmas
Department of Applied Biosciences, College of Agriculture and Life Sciences, Kyungpook National University, Daegu, South Korea

Sudhir Kumar Upadhyay
School of Chemical Engineering and Physical Sciences, Lovely Professional University, Phagwara, Punjab, India

Yağmur Uysal
Department of Environmental Engineering, Mersin University, Mersin, Turkiye

Rebeca Casique Valdés
Department of Horticulture, Antonio Narro Autonomous Agrarian University, Saltillo, Coahuila, Mexico

Alejandro Velasco-Ramírez
Department of Agricultural Production, University Center for Biological and Agricultural Sciences (CUCBA), University of Guadalajara, Zapopan, Jalisco, Mexico.

Ana Paulina Velasco-Ramírez
Department of Agricultural Production, University Center for Biological and Agricultural Sciences (CUCBA), University of Guadalajara, Zapopan, Jalisco, Mexico.

Sandra Fabiola Velasco-Ramírez
Department of Chemistry, University Center for Exact Sciences and Engineering (CUCEI), University of Guadalajara, Guadalajara, Jalisco, Mexico.

Weria Weisany
Department of Agriculture Science and Engineering, Science and Research Campus, Islamic Azad University, Tehran, Iran

Ji-In Woo
Department of Applied Biosciences, College of Agriculture and Life Sciences, Kyungpook National University, Daegu, South Korea

Maryam Zakavi
Department of Cellular and Molecular Biology, Faculty of Life Sciences and Biotechnology, Shahid Beheshti University, Tehran, Iran

Neelam Zeb
Department of Biotechnology, Women University Mardan, Mardan, Khyber Pakhtunkhwa, Pakistan

Dan-Dan Zhao
Coastal Agriculture Research Institute, Kyungpook National University, Daegu, Republic of Korea

Mohamed Zouari
Laboratory of Improvement of Olive Productivity and Fruit Trees, Olive Tree Institute of Sfax, University of Sfax, Sfax, Tunisia; Laboratory of Botany and Cryptogamy, Faculty of Pharmacy, University of Limoges, Limoges, France

Nacim Zouari
Laboratory of Biodiversity, Molecules, and Applications (LR22ES02), Higher Institute of Applied Biology of Medenine, University of Gabes, Medenine, Tunisia

Preface

Agriculture is undergoing a paradigm shift in the 21st century. Rapid population growth has intensified demands on food production systems, requiring higher crop yields with minimal inputs. These challenges are further compounded by the increasing frequency, intensity, and unpredictability of climate-change-driven abiotic stresses, including drought, salinity, high temperatures, cold stress, and nutrient deficiencies. Such stresses have direct and profound effects on plant growth, yield, and crop quality. In this rapidly changing context, enhancing crop resilience has emerged as a critical research priority. While innovative approaches to mitigating abiotic stress have gained significant attention, the role of mineral nutrients in addressing these challenges remains unparalleled. Beyond their traditional view as essential components for plant growth, mineral nutrients are now recognized as fundamental regulators of cellular processes and stress response mechanisms that enable plants to adapt to adverse environmental conditions.

This book, part of the *Plant Biology, Sustainability, and Climate Change* series, highlights mineral nutrient management as a potential game changer for sustaining agricultural productivity under environmental stress. Abiotic stresses affect crops at multiple levels, including molecular and cellular processes, whole-plant physiology, and overall growth and yield. These stresses arise from factors such as drought and water limitation, reduced nutrient uptake, ionic and physiological imbalances in saline soils, and extreme temperature conditions. High or low temperatures can disrupt enzyme activity and compromise cell membrane stability, while micronutrient deficiencies further weaken plant resistance. Abiotic stresses may occur independently, but more often they interact, leading to compounded physiological damage. In this context, mineral nutrients—both micronutrients and macronutrients such as nitrogen, calcium, magnesium, phosphorus, sulfur, and potassium—play pivotal roles in plant stress tolerance. These nutrients regulate key processes, including water relations, membrane stability, enzyme activation, photosynthesis, respiration, antioxidant defense, and hormone regulation. Among them, potassium, often referred to as a "stress nutrient," is particularly important for osmotic regulation, stomatal movement, enzyme activation, and maintenance of turgor pressure, thereby enhancing plant resistance to drought and mitigating the adverse effects of salinity. Calcium, while a principal structural component of plant cell walls and membranes, also serves as a crucial secondary messenger that enables rapid cellular responses under stress conditions. Magnesium occupies the central position in the chlorophyll molecule and is therefore essential for photosynthesis, playing key roles in carbon fixation and allocation. Although plants require micronutrients in relatively small amounts, their contributions to abiotic stress tolerance are substantial. Micronutrients such as zinc, copper, iron, manganese, and boron are integral components of numerous enzymes and redox reaction complexes, where they protect plants against oxidative damage. Zinc is involved in enzyme activity and protein biosynthesis, while copper and iron function in electron transport chains and antioxidant defense systems. Manganese plays a critical role in the oxygen-evolving complex of photosystem II and in reactive oxygen species detoxification. Boron contributes to cell wall integrity and membrane stability, particularly under conditions of water deficiency.

Given the diverse applications of mineral nutrients, this book offers an innovative and integrated perspective on nutrient-based strategies for managing abiotic stress in plants. It is designed as a comprehensive resource for scientists, researchers, academicians, and advanced students working in agriculture, horticulture, plant sciences, and biotechnology. This book also serves as a practical reference for practitioners, entrepreneurs, educators, and other stakeholders seeking sustainable strategies to enhance crop resilience in climate change prone regions. This book follows a progressive structure, moving from foundational concepts to advanced applications. Early chapters introduce the nature of abiotic stress in crops and outline fundamental principles of plant mineral nutrition. Subsequent chapters focus on the roles of major and trace minerals in stress tolerance, with particular emphasis on key nutrients such as potassium, calcium, magnesium, phosphorus, sulfur, and iron. Beyond individual nutrient functions, this book addresses broader themes including nutrient imbalances, their physiological and biochemical impacts, and fertility management strategies under stress conditions. Several chapters explore emerging and integrative topics at the forefront of plant stress biology, including nutrient regulation of plant hormones and ethylene biosynthesis, osmotic adjustment, root system architecture, and mineral mediated plant defense mechanisms. This book also covers innovative and future-oriented approaches such as nanofertilizers, nutrient biofortification under stress conditions, and plant-breeding strategies aimed at improving nutrient use efficiency. Collectively, these topics emphasize the development of productive, resource efficient, and environmentally sustainable agricultural systems. The role of beneficial elements in stress management is also examined, alongside future directions in crop nutrition and management. With its strong scientific foundation and practical orientation, this book presents applied strategies in fertilizer technology, soil and crop management, and nutritional interventions for drought- and salinity-affected regions, aiming to improve crop quality while minimizing environmental impact. Overall, this book provides a balanced, forward-looking resource that supports sustainability and climate change adaptability in modern agriculture.

Finally, this book aims to bridge the gap between basic and applied plant sciences by emphasizing mineral nutrition as a central framework for improving crop resilience to abiotic stress. Rather than viewing mineral nutrients solely as inputs for crop management and growth, this book positions them as key regulators of physiological, biochemical, and molecular processes that underpin plant adaptation to adverse environmental conditions. This book encourages the development of new scientific questions, methodologies, and management strategies that leverage the unique roles of mineral nutrients in enhancing stress tolerance and sustaining crop health. By integrating fundamental principles with applied perspectives, it highlights how mineral nutrition can be strategically used to develop crops with improved resistance to environmental stresses. *Mineral Nutrients to Mitigate Abiotic Stress in Crops* is therefore intended as an essential resource for scientists, advanced students, agronomists, and researchers, as well as a practical guide for professionals seeking innovative and sustainable approaches to stress management in agriculture. Through its balanced scientific rigor and applied focus, this book aims to stimulate innovation and support climate-resilient crop production systems.

Azamal Husen

SECTION

I

Foundations of abiotic stress and plant nutrition

CHAPTER

Overview of abiotic stress in crops: challenges and impacts

1

Yağmur Uysal and Zeynep Görkem Doğaroğlu
Department of Environmental Engineering, Mersin University, Mersin, Turkiye

Abbreviations

ABA	abscisic acid
ABC Transporters	ATP-binding cassette transporters
ACC	1-aminocyclopropane-1-carboxylic acid
AMF	arbuscular mycorrhizal fungi
APX	ascorbate peroxidase
ARF	auxin response factor
CAT	catalase
COR	cold-Responsive
CSD1/2	copper/zinc superoxide dismutase 1/2
DAG	diacylglycerol
DREB	dehydration-responsive element binding
ET	ethylene
FAD2/3	fatty acid desaturase 2/3
GPX	glutathione peroxidase
GRF	growth-regulating factor
GST	glutathione-S-transferase
H2O2	hydrogen peroxide
HKT1;5	high affinity potassium transporter 1;5
HSF	heat shock factor
HSP	heat shock protein
HY5	elongated hypocotyl 5
HYL1	hyponastic leaves 1
IP3	inositol-1,4,5-trisphosphate
JA	jasmonic acid
LHC	light-harvesting complex
LEA	late embryogenesis abundant
MDA	malondialdehyde
MAPK	mitogen-activated protein kinase
miRNA	microRNA
ncRNA	noncoding RNA
NHX1	Na^+/H^+ antiporter
P5CS1/2	Δ-1-pyrroline-5-carboxylate synthase 1/2

Mineral Nutrients to Mitigate Abiotic Stress in Crops. DOI: https://doi.org/10.1016/B978-0-443-44712-9.00006-2

PA	phosphatidic acid
PGPR	plant growth-promoting rhizobacteria
Pn	net photosynthetic rate
PSII	photosystem II
RLK	receptor-like kinase
ROS	reactive oxygen species
SA	salicylic acid
SnRK2	SNF1-related protein kinase 2
SOD	superoxide dismutase
SOS1	salt overly sensitive 1
TF	transcription factor
UV-B	ultraviolet-B radiation
UVR8	UV resistance locus 8
VOC	volatile organic compound

1.1 Introduction

Agricultural sustainability is facing increasing challenges due to global climate change, fast population increase, and rising food consumption. By 2050, food needs are predicted to increase by almost 70%, but existing production methods are still inadequate in the face of growing climate-related stress (Falcon et al., 2022). Throughout their lives, plants face a variety of biotic and abiotic challenges. Abiotic elements like drought, salt, extreme temperatures, heavy metals, air pollution, and UV radiation can seriously interfere with water balance, ion homeostasis, and carbon metabolism (Husen, 2021a, 2021b, 2021c, 2023). According to Chowdhary and Songachan (2025), chronic exposure to these disruptions might cause plant mortality as well as growth impairment and metabolic disruption. Abiotic stresses not only reduce crop yield but also diminish nutritional quality, including antioxidant profiles, mineral composition, and phytochemical content. This decline poses a significant food security concern because it affects both production volume and the nutritional value of plant-derived foods (Yuan et al., 2023). Moreover, stress conditions commonly occur in combination rather than individually, producing more severe and complex physiological and molecular disruptions than single stresses. Increased water loss, reduced stomatal conductance, ion toxicity, and elevated oxidative burden collectively impair plant growth (Khan et al., 2025). Under such circumstances, plants temporarily suppress growth and reallocate metabolic resources toward defense, which intensifies biomass and yield losses compared with individual stress exposures.

The reduced performance of modern crop varieties under abiotic stress highlights the need for deeper insights into the biological systems that confer stress resilience. Plant–microbiome interactions, phytohormonal regulation, and reactive oxygen species (ROS) control play central roles in this context. Particularly, Plant

Growth-Promoting Rhizobacteria (PGPR), endophytes, and Arbuscular Mycorrhizal Fungi (AMF) colonizing the rhizosphere support stress tolerance by promoting osmoprotectant accumulation, reducing oxidative pressure, and maintaining ion homeostasis (Azizoglu et al., 2025). In addition, elucidating the genetic and molecular bases of stress tolerance has become a major research focus in recent years. Omics technologies, CRISPR/Cas9-based genome editing, microbial biostimulation, and nanotechnological approaches have emerged as advanced strategies for decoding the multilayered genetic and biochemical networks that regulate stress responses and for developing resilient crop varieties (Zhang and Yang, 2022). Abiotic stresses disrupt plant physiology by inducing water deficit, ionic imbalance, oxidative damage, and photosynthetic impairment, ultimately reducing productivity and nutritional security (Husen, 2024a, 2024b, 2025a, 2025b). Therefore elucidating the molecular, biochemical, and physiological bases of stress adaptation is essential for developing climate-resilient and sustainable agri-food systems. This chapter summarizes plant responses to major abiotic stresses, particularly drought, salinity, and temperature extremes, and highlights emerging approaches, including microbiome-based interventions, omics platforms, and CRISPR-driven genome editing, that enable next-generation stress-tolerant crops.

1.2 Major types of abiotic stresses affecting crops

Every significant abiotic stress factor causes distinct physiological disturbances and sets off distinctive defense mechanisms. The following subsections list the major stress categories along with the integrated response mechanisms for each.

1.2.1 Drought stress: Physiological disruption and integrated resilience mechanisms

Drought is a primary cause of global crop yield reduction, responsible for 20–50% losses in major crops such as maize, wheat, and rice (FAO, 2021). Early physiological responses include turgor loss, stomatal closure, reduced carbon assimilation, and decline in photosystem II (PSII) efficiency (Khan et al., 2025). Drought tolerance is hormonally coordinated by a crosstalk between abscisic acid (ABA) and jasmonic acid (JA). ABA induces stomatal closure to conserve water, while JA reinforces membrane stability, osmoprotectant accumulation, and ROS detoxification (Park et al., 2025). This signaling triggers the accumulation of compatible solutes (proline, glycine betaine, sugars) to maintain osmotic balance. Concurrent morphological adaptations such as reduced leaf area, thicker cuticles, and deeper rooting improve water-use efficiency (Gaoua, 2025).

A major consequence of drought is oxidative stress, marked by elevated levels of H_2O_2, $O_2^{\cdot-}$, and $^1O_2$, which damage lipids, proteins, and DNA. Plants counteract this via enzymatic antioxidants including superoxide dismutase (SOD), catalase

(CAT), ascorbate peroxidase (APX), and glutathione peroxidase (GPX), which collectively facilitate ROS detoxification. In addition to these, nonenzymatic antioxidants (phenolics, flavonoids, carotenoids) provide an additional layer of protection, collectively maintaining redox homeostasis (Mittler et al., 2022).

At the molecular level, drought responses are orchestrated by key transcription factor families. These include DREB, NAC, MYB, bZIP, and AP2/ERF proteins. They induce genes involved in osmotic adjustment and cellular protection, such as those encoding LEA (Late Embryogenesis Abundant) proteins, RD (Responsive to Dehydration) genes, aquaporins, and cell-wall remodeling enzymes, collectively improving tissue flexibility and water retention (Park et al., 2025). Parallel to transcriptional control, post-transcriptional and biochemical adjustments are crucial. MicroRNAs (miRNAs) like miR398, miR159, and miR166 fine-tune antioxidant defenses and water-balance genes, contributing to epigenetic stress memory. At the biochemical level, fatty acid desaturases (FAD2 and FAD3) preserve membrane fluidity by increasing unsaturated lipid content (Tripathi et al., 2025).

Drought also reprograms the rhizosphere microbiome. Severe water deficit reduces microbial diversity, and community recovery after rewatering can be slow or incomplete. In rice, for instance, prolonged drought alters microbial structure for over 50 days after rehydration, favoring drought-resilient taxa such as *Streptomyces* (Metze et al., 2023). In summary, drought is a multifaceted stress that integrates physiological adjustments (water conservation, osmotic regulation), molecular reprogramming (TF- and miRNA-mediated networks), antioxidant defense, and microbiome restructuring. A systems-level understanding of these interconnected mechanisms is essential for breeding and managing drought-resilient crops.

1.2.2 Salt stress: Mechanisms and multilevel adaptation

Salt stress is a rapidly expanding abiotic constraint, threatening global food security. Globally, more than 900 million hectares of land (approximately 1/3 of which is irrigated land for agriculture) are damaged by excessive salt, and the situation is worsening due to climate change and human activity. Particularly in arid and semi-arid regions, salinity severely limits crop productivity and is responsible for nearly 30% of global yield losses (Zhou et al., 2024).

Salinity inflicts damage via a two-phase response: initial osmotic stress that impairs water uptake, followed by ionic toxicity from excessive Na^+ and Cl^- accumulation. This disrupts ion homeostasis, induces stomatal closure, reduces photosynthesis, and triggers K^+ efflux, membrane destabilization, and oxidative effects that are exacerbated when combined with drought. Consequently, plants exhibit reduced turgor, inhibited electron transport, diminished carboxylation efficiency, and altered growth patterns, leading to significant biomass loss (Chauhan et al., 2024).

To mitigate these effects, plants activate a multifaceted defense system. This includes the accumulation of osmoprotectants like proline (via upregulation of

P5CS1/2 genes) and the induction of enzymatic antioxidants (SOD, CAT, APX) to mitigate oxidative damage (Khare et al., 2026). At the molecular level, precise ion transport regulation is critical. Transporters such as HKT1;5, SOS1, and NHX1 work coordinately to exclude Na^+ from the cytosol, a process modulated by signaling molecules like ABA, JA, and molecular hydrogen (H_2) (Rehman et al., 2025; Yang et al., 2024). Furthermore, a microRNA-guided regulatory layer fine-tunes the response, with specific miRNAs (e.g., miR163, miR829) being upregulated or suppressed under stress.

The rhizosphere microbiome is dynamically restructured under salinity. Halophilic bacteria (e.g., *Proteobacteria*) and AMF enhance mineral uptake and ion homeostasis, though community composition shifts with stress intensity. This symbiosis is vital for improving osmoprotectant accumulation and stress resilience (Laiq et al., 2024). Ultimately, salinity tolerance is an integrated trait, involving physiological adjustments, molecular signaling, ion transport, and beneficial microbiome interactions. Membrane lipid peroxidation, indicated by elevated malondialdehyde (MDA), underscores the pervasive cellular damage. Overall, salinity tolerance emerges from interlinked physiological, molecular, and microbial processes, underscoring the need for integrated approaches in developing resilient crop genotypes and sustainable management strategies.

1.2.3 Heat stress: physiological disruption and integrated tolerance mechanisms

Heat stress, intensified by global warming and frequent heatwaves, is a critical abiotic constraint that severely compromises crop yields. Plant growth and crop yields are expected to suffer greatly as a result of the forecast +2°C per decade increase in global temperatures and the depletion of water supplies (Mushtaq et al., 2024). As photosynthetic organisms, plants suffer profound metabolic disruptions when temperatures exceed species-specific thresholds. Sensitive plants exhibit growth retardation above 35°C, while exposure to 40°C–45°C causes membrane destabilization, protein denaturation, and a sharp decline in photosynthetic capacity (Pn). These detrimental effects are often amplified when heat co-occurs with drought or salinity, leading to synergistic damage (Khan et al., 2025).

Cellular and physiological impairment manifests early, primarily as membrane and photosynthetic damage. Elevated temperatures induce lipid peroxidation and ion leakage, compromising membrane integrity. Coupled with dysregulated stomatal conductance and increased transpiration, this disrupts the plant's hydraulic balance, accelerating water loss (Akgul and Aydlnoglu, 2025). The reproductive phase is particularly vulnerable. In cereals like wheat, rice, and maize, heat spikes reduce pollen viability, shorten grain filling, and can cause yield losses exceeding 30%. In horticultural crops, heat stress commonly triggers flower abscission, poor fruit set, and pigment degradation (Zhong et al., 2024).

To counteract these effects, plants deploy a concerted molecular defense system. Heat shock proteins (HSPs), notably HSP70 and HSP90, are central to proteome stability, refolding denatured proteins, and preventing cell death under thermal stress (Tian et al., 2021). Their expression is orchestrated by heat shock factors (HSFs), which transduce temperature signals. Concurrently, plants accumulate osmoprotectants and activate antioxidant systems to mitigate oxidative stress and support enzyme function (Feng et al., 2023). Membrane adaptation through increased sterol and unsaturated fatty acid content helps maintain fluidity and integrity. Phytohormonal networks fine-tune the response: ABA mediates stomatal closure, JA stabilizes membranes, SA alleviates oxidative damage, and ethylene plays a dual role in adaptation versus growth suppression (Shajar et al., 2025).

Microbiome-assisted thermotolerance emerges as a critical biological strategy. PGPR and endophytes enhance antioxidant capacity, upregulate HSP genes, and improve water relations. Furthermore, microbial volatile organic compounds (VOCs) can modulate epigenetic pathways and activate stress-responsive transcription factors, adding a layer of regulatory sophistication (Yuan et al., 2024). In conclusion, thermotolerance is a multidimensional trait integrating membrane stabilization, photosynthetic protection, reproductive success, HSP-mediated proteostasis, hormonal regulation, and beneficial microbiome interactions. A systems-level understanding of these coordinated mechanisms is indispensable for developing crops resilient to the escalating threat of heat stress.

1.2.4 Cold stress: From cellular injury to integrated acclimation

Cold stress comprising chilling (0°C–15°C) and freezing (<0°C) temperatures disrupts cellular homeostasis by reducing membrane fluidity and metabolic activity. Chilling increases membrane viscosity, impairing ion transport and enzyme function, while freezing induces extracellular ice formation, cellular dehydration, and irreversible membrane damage (Jan et al., 2024). Plant sensitivity varies widely: tropical species (e.g., rice, tomato) suffer severe damage during early development, whereas cool-season cereals (e.g., wheat) display moderate tolerance but remain vulnerable at reproductive stages. Common symptoms include leaf chlorosis, curling, growth retardation, and tissue necrosis under prolonged exposure (Ding et al., 2025).

Cold stress primarily targets PSII, leading to chlorophyll degradation, reduced carbon assimilation, and impaired energy production. This damage triggers excessive ROS generation in chloroplasts and mitochondria, resulting in oxidative stress (Liu et al., 2022). To counteract redox imbalance, plants activate key antioxidant enzymes that reinforce the ascorbate-glutathione cycle and enhance ROS detoxification (Yang et al., 2022). Concurrently, cells undergo membrane lipid remodeling by increasing unsaturated fatty acid proportions to maintain fluidity. They also accumulate cryoprotective osmolytes like proline and soluble sugars to lower the cytoplasmic freezing point and limit ice formation. Specialized protective proteins, particularly LEA proteins and dehydrins encoded by COR genes, further stabilize macromolecules and inhibit ice-crystal propagation under chilling stress (Yu et al., 2025). Ultimately, these coordinated physiological and molecular responses are orchestrated by the evolutionarily conserved CBF–COR regulatory module, the central signaling pathway governing cold-inducible gene expression in plants.

Cold acclimation prior to exposure to nonlethal low temperatures (10°C–15°C) primes transcriptional and metabolic pathways, enhancing subsequent freezing tolerance. This adaptive "memory" involves epigenetic modifications such as histone remodeling and DNA methylation, which fine-tune stress-responsive gene expression. Additionally, microRNAs act as key posttranscriptional regulators, targeting transcription factors (e.g., CBF, MYB, bZIP), hydrolases, and oxidoreductases to fine-tune cold adaptation and mediate crosstalk with other stress pathways (Qian et al., 2024).

Agronomic interventions including silicon/potassium supplementation, seed priming, mulching, and inoculation with AMF or PGPR can bolster tolerance by improving water uptake, stabilizing membranes, and strengthening antioxidant defenses. In summary, cold resilience emerges from an integrated network of physiological adjustments (membrane stabilization, osmoprotection), molecular regulation (CBF–COR signaling, miRNA-guided networks), epigenetic priming, and agronomic support. A systems-level understanding of these interconnected mechanisms is essential for developing cold-tolerant cultivars in a changing climate.

1.2.5 Heavy metal stress: Mechanisms of toxicity and integrated tolerance strategies

Heavy metal pollution, driven by industrial discharge, agricultural runoff, and wastewater irrigation, poses a persistent and growing threat to agricultural ecosystems. Toxic metals (e.g., Pb^{2+}, Cd^{2+}, As^{3+}, Cr^{6+}, Ni^{2+}) accumulate in soils, inhibiting plant growth and contaminating the food chain, posing a critical food security and safety concern due to their nonbiodegradable nature and high bioaccumulation potential (Jorjani and Karakaş, 2024).

Metal toxicity initially targets the root system, damaging apical cells, increasing membrane permeability, and impairing water and nutrient uptake. While the cell wall acts as a partial barrier, metals entering the cytosol accumulate in chloroplasts and mitochondria, disrupting their structure and function. Subsequent translocation to shoots inhibits PSII activity, suppresses chlorophyll biosynthesis, and reduces carbon assimilation, leading to visible symptoms such as leaf chlorosis, necrosis, and stunted growth (Yuan et al., 2024). This damage is exacerbated when metals co-occur with salinity, intensifying ionic imbalance and inducing phytochelatin and metallothionein synthesis (Khawula et al., 2025).

A key consequence of metal stress is severe oxidative damage. Metals like Cd and As are potent inducers of ROS, causing lipid peroxidation, protein denaturation, and DNA damage (Zainab et al., 2021). The antioxidant response is uniquely characterized by the prominent role of GSTs and ATP-binding cassette (ABC) transporters in detoxification. The core tolerance mechanism involves a three-step process: (1) metal chelation by phytochelatins or metallothioneins, (2) transport of glutathione-conjugated complexes via GSTs, and (3) vacuolar sequestration mediated by ABC transporters. Osmoprotectants such as proline further aid by stabilizing membranes and mitigating oxidative stress (Mashabela et al., 2023).

At the molecular level, transcription factors orchestrate the expression of genes involved in transport, chelation, and antioxidant defense. Epigenetic regulation including DNA methylation, histone modification, and miRNA activity plays a critical long-term adaptive role. For instance, miR172 in chickpea enhances antioxidant enzyme activities under multimetal stress, while miR2916 in peanut modulates ethylene signaling to improve aluminum tolerance (Ucar et al., 2024).

The impact of heavy metal stress is determined by metal speciation, concentration, soil properties, and plant genotype. Management strategies involve soil amendments (e.g., biochar, silicon), microbial inoculation (PGPR, AMF), and genetic approaches to improve metal exclusion or sequestration. For example, PGPR can immobilize metals by secreting extracellular polymers, forming a protective rhizosphere barrier (Zhang and Li, 2025). In summary, metal tolerance relies on an integrated response combining cellular detoxification (chelation, transport, sequestration), antioxidant defense, transcriptional–epigenetic regulation, and soil–microbiome management. Adopting a systems-level perspective is crucial for developing crops that are both productive in contaminated soils and safe for consumption.

1.2.6 Light and UV-B stress: Photodamage and photoprotection

Excessive light intensity, particularly UV-B radiation (280–315 nm), is a key abiotic stressor that induces photo-oxidative damage in plants. High UV-B directly damages membrane lipids, nucleic acids, and proteins, leading to elevated ROS levels, impairment of PSII, degradation of light-harvesting complexes (LHCs), and a marked decline in carbon fixation efficiency (Chen et al., 2022). While moderate UV-B can synergize with other stresses to stimulate phenolic accumulation, prolonged exposure severely inhibits growth. Interestingly, low-dose UV-B can act as a protective signal, enhancing PSII stability (Fernandes de Oliveira et al., 2024).

A hallmark of UV-B stress is chloroplast deterioration, characterized by thylakoid swelling, disorganized grana, and disrupted lipid homeostasis, which collectively reduce photosynthetic efficiency and lead to photoinhibition and leaf chlorosis (Villano et al., 2025). In response, plants deploy structural and biochemical defenses: increasing leaf thickness, cuticular density, and epidermal compaction, alongside the accumulation of photoprotective compounds like phenolics and anthocyanins (Zheng et al., 2026).

Plants perceive UV-B radiation primarily through the UVR8 (UV Resistance Locus 8) photoreceptor. Upon activation, UVR8 interacts with the central photomorphogenic regulator COP1 (Constitutive Photomorphogenic 1), resulting in the stabilization and accumulation of HY5 (Elongated Hypocotyl 5), which in turn induces genes associated with the phenylpropanoid pathway, antioxidant biosynthesis, and DNA repair. This transcriptional reprogramming enhances the detoxification of reactive species and promotes the repair of UV-induced photodimers. Light-mediated COP1 dynamics also contribute to regulation at the RNA level; Under light conditions, COP1 translocates from the nucleus to the cytoplasm and fine-tunes stress-responsive gene expression by stabilizing the miRNA processing factor HYL1 (Fernandes de Oliveira et al., 2024).

UV-B exposure induces membrane lipid peroxidation, damaging chloroplasts and plasma membranes. In acclimation, plants elevate unsaturated fatty acids to preserve fluidity, a process modulated by ABA, JA, and SA hormones that also regulate stomatal behavior and antioxidant defenses (Shajar et al., 2025). Molecular responses involve oxidative stress TFs (e.g., WRKY, MYB, bZIP) and miRNAs that fine-tune photoprotection. For instance, miR156 suppression alters phytochrome signaling in *Arabidopsis* under specific light (Xie et al., 2017). The stress also reshapes the rhizosphere microbiome; beneficial PGPR and endophytes boost host antioxidant activity and phenolic metabolism via ABA/auxin pathways. Specific light spectra (e.g., blue/red) can further optimize photosynthesis and antioxidant capacity (Villano et al., 2025). Thus adaptation to high light/UV-B integrates direct photoprotection (UVR8, DNA repair), antioxidant defense, membrane remodelling, transcriptional/epigenetic control, and microbiome interactions. Decoding these networks is key to engineering crops resilient to light-induced oxidative stress.

A comparative perspective is necessary to highlight common and unique tolerance mechanisms because of the variety of plant responses to various abiotic stressors. The main physiological, molecular, and microbiome-assisted reactions to each of the main abiotic stresses covered in this chapter are compiled in Table 1.1.

1.3 Integrated biological mechanisms in plant abiotic stress tolerance

The adaptive capacity of plants under abiotic stress is not solely determined by their intrinsic physiological and molecular responses but is also shaped by a multilayered biological system involving the rhizosphere microbiome, natural biostimulants, antioxidant defense networks, and hormonal signaling pathways. Together, these components form an integrated framework that underpins resilience to both acute environmental fluctuations and chronic stress conditions.

Table 1.1 Comparative mechanisms of plant responses to major abiotic stresses.

Plant species	Abiotic stress type	Primary damage	Key physiological responses	Core molecular pathways/TFs	Key citations
Tobacco (**Nicotiana tabacum**)	Drought	Membrane damage, photoinhibition, ROS accumulation	Increased ROS-scavenging enzyme activity, membrane protection, proline accumulation	SOD, APX, CAT, LEA, P5CS	Wang et al. (2023)
Tomato (*Solanum lycopersicum*)	Salinity, Heat	Ion toxicity, protein denaturation, nutrient loss	ABA, JA, SA, glycoalkaloids	ABA, JA, SA, glycoalkaloid biosynthesis genes	Yan et al. (2024)
Mint (*Mentha spicata)*	UV/High Light	DNA/protein/lipid damage, chlorophyll degradation	Increased flavonoids and phenolics, stomatal closure	UVR8, HY5, COP1, WRKY, MYB	Crestani et al. (2023)
Rice *(Oryza sativa*)	Drought, Heat	Growth inhibition, yield loss	Stomatal closure, osmolyte accumulation	ABA, DREB, NAC	Smet et al. (2023)
Citrus (*Citrus macrophylla*)	UV/High Light	DNA/protein/lipid damage, chlorophyll destruction	Flavonoid and phenolic increase, stomatal closure	UVR8, HY5, HYH, COP1, WRKY, MYB	Vives-Peris et al. (2024)
Maize (*Zea mays*)	Cold	Reduced leaf growth, decreased antioxidant activity	Suppression of miR408, decreased CAT and GR, increased APX	miR408, antioxidant enzymes	Akgul and Aydinoglu, (20254)
Arabidopsis thaliana	Salinity	Ion toxicity, osmotic stress	Na^+ exclusion, proline synthesis	DREB, SOS1, NHX1, RD29A/B	Zhang et al. (2023)
Maize (*Zea mays*)	Drought, Salinity, Heat	Reduced photosynthesis, oxidative stress	Antioxidant enzyme activation	NAC, bZIP, MAPK	Pardo et al. (2025)
Wheat (*Triticum aestivum)*	Salinity, Drought	Water loss, reduced turgor, decreased photosynthesis	Stomatal closure, ABA increase, root/shoot ratio change	ABA, DREB, NAC, MYB, bZIP	Di Sario et al. (2025)

1.3.1 Plant–microbe interactions and natural biostimulants

The rhizosphere microbiome is a key regulator of plant responses to drought, salinity, heat, heavy metals, and light stress. PGPR, AMF, and endophytes enhance resilience by improving water/nutrient uptake, stimulating osmoprotectant synthesis, and activating stress-responsive genes. Under water-related stresses, specific PGPR (e.g., *Pseudomonas*, *Bacillus*) modulate osmotic and ionic balance. They upregulate genes like P5CS and TPS1 to boost proline and trehalose accumulation, restrict Na^+ uptake, and improve hydraulic conductivity (Ferioun et al., 2024). AMF colonization expands the root absorptive surface, significantly enhancing water and nutrient acquisition in stressed cereals. Under heat stress, certain endophytes enhance membrane stability by inducing desaturase genes and increasing unsaturated fatty acid content. Multispecies PGPR consortia can also upregulate antioxidant enzymes (APX) and HSPs, reducing oxidative damage (Krishna et al., 2022).

Beyond direct microbial interactions, natural biostimulants such as seaweed extracts enhance phenolic accumulation, photosynthetic stability, and ion homeostasis, thereby mitigating stress severity. PGPR also modulates plant transcription factor networks. For instance, *Paenibacillus polymyxa* CR1 induces RD29A/B expression in wheat under drought, while *Bacillus amyloliquefaciens* MMR04 activates DREB-1E and ERF-1B in pearl millet, strengthening antioxidant defenses (Murali et al., 2021). Despite their potential, the field efficacy of biological strategies can be limited by soil properties, microbial competition, climate variability, and inoculant viability. Advancing region-specific microbial consortia, improved formulations (e.g., microencapsulation), and targeted grower training are essential for reliable, large-scale application.

1.3.2 Modulation of antioxidant defense and hormonal crosstalk

As detailed in Section 2, abiotic stress triggers ROS overproduction in chloroplasts, mitochondria, and peroxisomes. The rhizosphere microbiome acts as a critical biological buffer against this oxidative surge. Beneficial fungi like Trichoderma spp. enhance nutrient uptake and photosynthesis while activating key antioxidant enzymes to bolster ROS scavenging. Similarly, bacterial strains such as *Shewanella putrefaciens* and *Cronobacter dublinensis* elevate SOD and CAT activities, and various PGPR induce the expression of detoxification genes, including GST (Manjunatha et al., 2022).

Phytohormones are central coordinators of abiotic stress responses, with their crosstalk determining the scale and specificity of adaptation. Microbial communities enhance this hormonal network by modulating key regulatory points. A prime example is the production of ACC deaminase by many PGPR, which reduces stress-induced ethylene accumulation, thereby stabilizing root–shoot growth and photosynthetic function (Kumawat et al., 2023). Microorganisms simultaneously modulate ABA, JA, and SA levels, influencing stomatal behavior, antioxidant gene expression, and stress-associated metabolism. JA supports membrane stability and osmoprotectant synthesis under drought and salinity, while SA promotes phenolic production and enhances the glutathione cycle under metal and UV stress. Ethylene acts as a dual regulator, enhancing defense

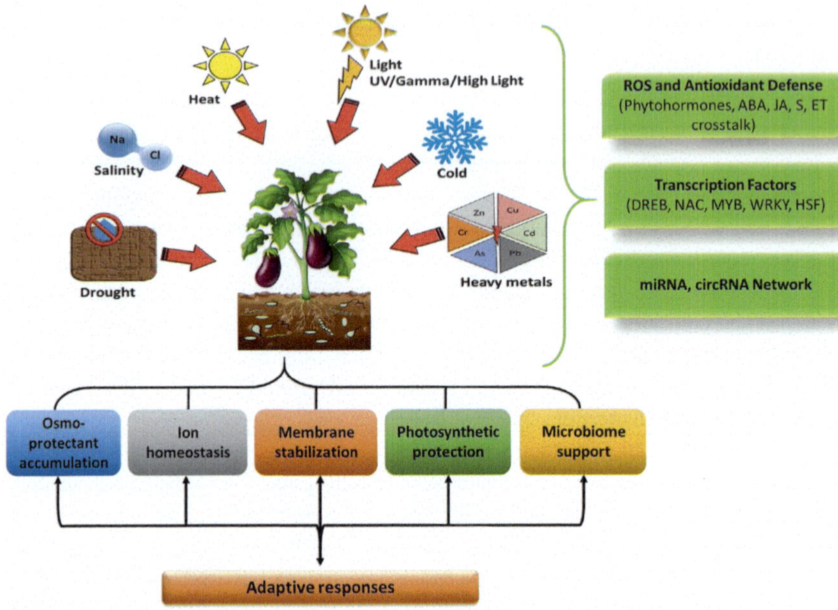

FIGURE 1.1

Overview of plant responses to abiotic stress. Major stresses (drought, salinity, heat, cold, high light, heavy metals) activate hormonal, antioxidant, transcriptional, and RNA-based regulatory pathways that drive osmotic adjustment, ion balance, membrane stability, photosynthetic protection, and microbiome support to enhance plant tolerance.

responses at low concentrations but inhibiting growth when accumulated excessively (Seleiman et al., 2022). In summary, plant–microbiome interactions form an integrated adaptation system that coordinately regulates ion and water balance, antioxidant capacity, and hormonal signaling. This multilayered synergy significantly boosts resilience in complex stress scenarios, establishing microbiome-based approaches as potent, sustainable tools for developing stress-tolerant crops. To synthesize the mechanisms discussed so far, an integrated overview of the major physiological, molecular, and microbiome-mediated layers of abiotic stress tolerance is presented in Fig. 1.1.

1.4 Biotechnological and molecular strategies in abiotic stress tolerance

Abiotic stress tolerance is a complex and multilayered process. Advanced biotechnological tools, building on physiological and molecular responses and microbiome-based adaptations, now play a central role in breeding resilient plants.

Key strategies include engineering of TFs, epigenetic reprogramming, and omics-guided network analysis.

1.4.1 Core signal transduction pathways in abiotic stress response

Plants perceive diverse abiotic stressors through a conserved, multilayered signal transduction network. Initial perception occurs at the plasma membrane through receptor-like kinases (RLKs), mechanosensitive channels, osmosensors, and redox-sensitive proteins. These trigger rapid early signaling events: ROS bursts, cytosolic Ca^{2+} spikes, phospholipid remodeling (e.g., PA, DAG, IP_3), membrane potential shifts, and turgor changes (Zhu, 2016). ROS function dually as damaging agents and essential secondary messengers. Concurrent Ca^{2+} signals are decoded by calmodulins, CDPKs, and CBL–CIPK complexes, which converge on MAPK and SnRK2 kinase cascades. These modules integrate redox, osmotic, and ionic signals, relaying the stress message to the nucleus (Su et al., 2021).

Nuclear TFs including DREB/CBF, NAC, MYB, WRKY, bZIP, and AP2/ERF families reprogram large gene networks controlling osmoprotectant synthesis, antioxidant defense, cell wall remodeling, ion transport, and membrane stabilization. Phytohormonal pathways operate as higher-order regulators: ABA dominates dehydration and salinity responses via stomatal control and SnRK2 activation; JA reinforces membrane stability and ROS detoxification; SA modulates phenolic metabolism under metal/UV stress; and ethylene acts bidirectionally, promoting acclimation at low levels but inhibiting growth when over-accumulated (Ding et al., 2025). While activation dynamics vary by stress type, the core architecture, spanning initial perception, second-messenger signaling (ROS/Ca^{2+}), kinase cascade activation, and TF-mediated transcriptional reprogramming, provides the conserved molecular foundation for acclimation across plant species.

1.4.2 Transcription factor networks and genome editing for stress resilience

Transcription factors serve as master regulatory nodes, converting stress perception into nuclear gene reprogramming. Key families orchestrate adaptive processes such as osmotic adjustment, ion homeostasis, ROS detoxification, and membrane stabilization (Sadau et al., 2024). Under combined stresses, these networks exhibit pronounced plasticity; transcriptomic studies reveal distinct regulation of carbon metabolism, antioxidant defense, and osmoprotectant synthesis pathways compared to single stresses, highlighting their conditional flexibility (Khan et al., 2025).

Hormonal signaling centrally regulates TF activity. ABA activates DREB and AREB TFs to drive osmoprotectant accumulation, while NAC, MYB, and WRKY modules coordinate lignification, root development, and redox balance (Ray et al., 2025). Species-specific studies demonstrate that modules such as MbWRKY50-MbICE1 in *Malus baccata* and FvNAC29-FvMYB44 in *Fragaria vesca* control osmotic and redox responses, while MxFRO4-MxFRO6 in *Malus xiaojinensis* modulates iron uptake and redox homeostasis (Wang et al., 2025; Xu et al., 2025). Collectively, these results illustrate the functional diversity of TFs as drivers of ecological adaptation

HSFs are pivotal for proteostasis under thermal and oxidative stress. HSF1, HSF2, and HSF3 rapidly induce heat shock protein (HSP) genes to maintain protein stability (Cheng et al., 2023). Their activity is modulated by posttranslational modifications, and members like *AtHSFA6b* and *TaHSFC2a* enhance drought tolerance through ABA-mediated pathways (Wenjing et al., 2020). Genome-wide analyses in wheat have identified highly inducible HSF genes, with some showing a 269-fold upregulation within one hour of heat stress, underscoring their potential as breeding targets (Wang et al., 2026).

Auxin response factors (ARFs) modulate growth-defense trade-off regulation under stress. For example, *GmARF16* negatively regulates salt tolerance in soybean, while *CqARF05* overexpression enhances it in quinoa, illustrating their context-dependent roles (Wang et al., 2024). The complexity of TF networks makes them prime targets for genome editing. CRISPR-Cas9 enables precise modifications to develop multistress resilient varieties. Successful examples include editing *ARGOS8* in maize for improved drought tolerance and targeting genes like OsPYL9 and OsSAPK2 in rice to enhance resilience across various stresses. These advances highlight the power of editing tools to refine the critical interface between stress perception and hormonal signaling.

1.4.3 Epigenetic and posttranscriptional regulation: miRNAs and circRNAs

Plants dynamically reprogram their physiology under abiotic stress through posttranscriptional regulation, notably via miRNAs. These small RNAs fine-tune the balance between growth and defense by modulating the expression of target genes involved in heat, cold, drought, salinity, and heavy metal responses. Advances in omics technologies, especially miRNA-seq, have unveiled these complex regulatory layers, highlighting miRNAs as central players in stress adaptation (Xu and Chen, 2023).

The stress-responsive behavior of miRNAs is exemplified by miR398, which exhibits context-dependent regulation. Under heat stress in tomato, miR398 is upregulated, suppressing its CSD1/2 targets and enhancing thermotolerance via modulated ROS signaling. In contrast, under heavy metal stress in hickory, miR398 is downregulated, leading to increased CSD1/2 expression and SOD activity to alleviate oxidative damage (Li et al., 2025). This plasticity allows miR398 to fine-tune redox homeostasis across different stresses. Similarly, cold-responsive miRNAs (e.g., miR156, miR319, miR528) target transcription factors such as AP2/ERF, TCP, and NAC, coordinating gene networks that enhance freezing tolerance (Zhang and Yang, 2022). Beyond miRNAs, circular RNAs (circRNAs) have emerged as key posttranscriptional regulators that function as "miRNA sponges." By sequestering specific miRNAs, circRNAs release their target mRNAs from repression, enabling rapid adjustment of stress-responsive pathways. These competing endogenous RNA (ceRNA) networks regulate critical processes such as ion homeostasis, hormone signaling, and antioxidant defense, facilitating adaptive rewiring under fluctuating conditions (Pan et al., 2025).

Systems biology analyses integrating multiomics data show that plant stress adaptation reflects a coordinated rewiring of molecular networks rather than isolated gene expression changes. Central to this reprogramming is the stress- and tissue-specific modulation of miRNA biogenesis, from MIR transcription to pri- and pre-miRNA processing. Together, miRNAs and circRNAs form regulatory circuits that orchestrate antioxidant defense, hormonal interactions, and ion homeostasis. Their pivotal roles make them valuable targets for biotechnology, including CRISPR/Cas-based editing and epigenetic engineering, to improve crop resilience. A systems-level understanding of these networks is therefore essential for accurately manipulating plant stress tolerance.

1.5 Conclusion

Abiotic stresses, including drought, salinity, temperature extremes, heavy metals, and UV radiation, impose severe physiological, biochemical, and molecular constraints that limit plant productivity under accelerating climate change. To survive, plants must maintain essential functions such as water balance, membrane integrity, ion homeostasis, redox stability, and photosynthetic efficiency amid profound metabolic disruption. This chapter has illustrated that stress resilience arises from an integrated regulatory framework spanning antioxidant defense, phytohormonal crosstalk, transcription factor–mediated reprogramming, miRNA- and circRNA-based posttranscriptional control, and epigenetic memory, a system further reinforced by microbiome-assisted metabolic buffering. Advances in omics technologies, gene editing, biostimulants, and tailored PGPR–AMF consortia now provide robust tools to decode and enhance these multilayered defenses. Ultimately, strengthening abiotic stress tolerance demands a systems biology perspective that can sustain yield stability in increasingly variable environments.

1.6 Future considerations

In order to improve crop resilience under increasingly complicated climatic conditions, future research should concentrate on integrating multiomics platforms, real-time stress phenotyping, and precision engineering of regulatory networks. Important strategic corridors consist of:

- Targeted epigenetic modification in conjunction with thorough mapping of miRNA–circRNA–transcription factor interaction modules to optimize stress-responsive gene networks.
- CRISPR-based precision editing of ion-homeostasis and hormonal signaling pathways in multistress and field-relevant conditions.
- Utilizing AMF consortia, tailored PGPR strains, and sophisticated biostimulant formulations, microbiome engineering maintains plant metabolic stability in a variety of soil types and climates.
- The use of AI-driven decision platforms, remote sensing technologies, and high-resolution phenotyping tools for cropping systems' adaptive management, predictive modeling, and early stress detection.

Designing climate-resilient agricultural systems that can preserve production, nutritional quality, and ecological integrity will require integrating molecular biotechnology with rhizosphere ecology and sustainable agronomy.

AI disclosure

During the preparation of this work, the authors used ChatGPT (OpenAI) in order to language quality, academic phrasing, and manuscript organization. After using this tool, the authors carefully reviewed and edited the content as needed and take full responsibility for the content of the publication. [t

References

Akgul, B., Aydinoglu, F., 2025. Evaluation of zma-miR408 and its target genes function on maize (*Zea mays*) leaf growth response to cold stress by VIGS-based STTM approach. Gene 938, 149161. Available at https://doi.org/10.1016/j.gene.2024.149161.

Azizoglu, U., Argentel-Martínez, L., Peñuelas-Rubio, O., Meriño-Hernández, Y., Rodríguez-Yon, Y., Dell Amico-Rodríguez, J.M., et al., 2025. Arbuscular mycorrhizal fungus *Rhizophagus irregularis* enhances antioxidant system mechanisms in chickpea plants under saline conditions. Plant Growth Regulation 105 (4), 883–894. Available at https://doi.org/10.1007/s10725-025-01315-3.

Chauhan, P., Singh, M., Sharma, A., Singh, M., Chadha, P., Kaur, A., 2024. Halotolerant and plant growth-promoting endophytic fungus *Aspergillus terreus* CR7 alleviates salt stress and exhibits genoprotective effect in *Vigna radiata*. Frontiers in Microbiology 15. Available at https://doi.org/10.3389/fmicb.2024.1336533.

Chen, Z., Dong, Y., Huang, X., 2022. Plant responses to UV-B radiation: signaling, acclimation and stress tolerance. Stress Biology 2 (1). Available at https://doi.org/10.1007/s44154-022-00076-9.

Cheng, Y., Xu, X., Du, R., Liu, Y., Zhang, M., Chen, M., et al., 2023. Heat shock transcription factors (Hsfs) in *Taraxacum kok-saghyz* Rodin: identification, classification, phylogeny and expression analysis. Current Plant Biology 35-36, 100307. Available at https://doi.org/10.1016/j.cpb.2023.100307.

Chowdhary, N.A., Songachan, L.S., 2025. Plant peptides involved in ROS signalling and biotic and abiotic stress responses. International Journal of Peptide Research and Therapeutics 31 (3). Available at https://doi.org/10.1007/s10989-025-10711-4.

Crestani, G., Cunningham, N., Csepregi, K., Badmus, U.O., Jansen, M.A.K., 2023. From stressor to protector, UV-induced abiotic stress resistance. Photochemical & Photobiological Sciences 22 (9), 2189–2204. Available at https://doi.org/10.1007/s43630-023-00441-1.

Ding, F., Fan, X., Tian, R., Wang, M., Sun, Z., 2025. Crosstalk of abscisic acid with other hormones and signaling molecules in tomato cold stress tolerance. Horticulturae 11 (6), 647. Available at https://doi.org/10.3390/horticulturae11060647.

Di Sario, L., Boeri, P., Matus, J.T., Pizzio, G.A., 2025. Plant biostimulants to enhance abiotic stress resilience in crops. International Journal of Molecular Sciences 26 (3), 1129. Available at https://doi.org/10.3390/ijms26031129.

Falcon, W.P., Naylor, R.L., Shankar, N.D., 2022. Rethinking global food demand for 2050. Population and Development Review 48 (4), 921–957. Available at http://doi.org/10.1111/padr.12508.

FAO, 2021. The impact of disasters and crises on agriculture and food security: 2021. FAO, Rome. Available at https://doi.org/10.4060/cb3673en.

Feng, D., Jia, X., Yan, Z., Li, J., Gao, J., Xiao, W., et al., 2023. Underlying mechanisms of exogenous substances involved in alleviating plant heat stress. Plant Stress 10, 100288. Available at https://doi.org/10.1016/j.stress.2023.100288.

Ferioun, M., Zouitane, I., Bouhraoua, S., Elouattassi, Y., Belahcen, D., Errabbani, A., et al., 2024. Applying microbial biostimulants and drought-tolerant genotypes to enhance barley growth and yield under drought stress. Frontiers in Plant Science 15. Available at https://doi.org/10.3389/fpls.2024.1494987

Fernandes de Oliveira, A., Piga, G.K., Najoui, S., Becca, G., Marceddu, S., Rigoldi, M.P., et al., 2024. UV light and adaptive divergence of leaf physiology, anatomy, and ultrastructure drive heat stress tolerance in genetically distant grapevines. Frontiers in Plant Science 15. Available at https://doi.org/10.3389/fpls.2024.1399840.

Gaoua, O., 2025. Integrating physiological, biochemical, and molecular mechanisms reveals insights into drought tolerance in sugar beet (*Beta vulgaris* L.). Discover Plants 2 (1). Available at https://doi.org/10.1007/s44372-025-00416-9.

Husen, A., 2021. Harsh Environment and Plant Resilience: Molecular and Functional Aspects. Springer, Cham. Available at https://doi.org/10.1007/978-3-030-65912-7.

Husen, A., 2021. Plant Performance Under Environmental Stress: Hormones, Biostimulants and Sustainable Plant Growth Management. Springer, Cham. Available at https://doi.org/10.1007/978-3-030-78521-5.

Husen, A., 2023. Plants and Their Interaction to Environmental Pollution: Damage Detection, Adaptation, Tolerance, Physiological and Molecular Responses. Elsevier Inc., Cambridge. Available at https://doi.org/10.1016/C2020-0-03319-0.

Husen, A., 2024. Biostimulants in Plant Protection and Performance. Elsevier Inc., Cambridge. Available at 10.1016/C2022-0-02207-8.

Husen, A., 2024b. Essential Minerals in Plant-Soil System: Coordination, Signalling, and Interaction under Adverse Situations. Elsevier Inc., Cambridge https://doi.org/10.1016/C2022-0-02437-5.

Husen, A., 2025a. Roles of Osmolytes in Changing Environment. Elsevier Inc., Cambridge https://doi.org/10.1016/C2023-0-51382-5.

Husen, A., 2025b. Role of Antioxidants in Mitigating Plant Stress. Elsevier Inc., Cambridge https://doi.org/10.1016/C2023-0-51268-6.

Husen, A., 2021b. The harsh environment and resilient plants: an overview. In: Husen, A. (Ed.), Harsh Environment and Plant Resilience. Springer International Publishing AG, Cham, pp. 1–23 https://doi.org/10.1007/978-3-030-65912-7_1.

Jan, S., Rustgi, S., Barmukh, R., Shikari, A.B., Leske, B., Bekuma, A., et al., 2024. Advances and opportunities in unraveling cold-tolerance mechanisms in the world's primary staple food crops. Plant Genome 17 (1). Available at https://doi.org/10.1002/tpg2.20402.

Jorjani, S., Karakaş, F.P., 2024. Physiological and biochemical responses to heavy metals stress in plants. Journal of Secondary Metabolite 11 (1), 169–190. Available at https://doi.org/10.21448/ijsm.1323494.

Khan, A.A., Wang, Y.F., Akbar, R., Alhoqail, W.A., 2025. Mechanistic insights and future perspectives of drought stress management in staple crops. Frontiers in Plant Science 16. Available at https://doi.org/10.3389/fpls.2025.1547452.

Khan, N.A., Owens, L., Nuñez, M.A., Khan, A.L., 2025. Complexity of combined abiotic stresses to crop plants. Plant Stress 17. Available at https://doi.org/10.1016/j.stress.2025.100926.

Khare, E., Kim, K., Kaistha, S.D., Gupta, G., Kalani, A., 2026. Synergistic alleviation of salt and *Fusarium oxysporum* stresses in *Arabidopsis thaliana* by *Trichoderma cerinum* Gur1 via modulation of the ABA-JA-SA signaling axis. Physiological and Molecular Plant Pathology 141. Available at https://doi.org/10.1016/j.pmpp.2025.103020.

Khawula, S., Daniel, A.I., Nyawo, N., Ndlazi, K., Sibiya, S., Ntshalintshali, S., et al., 2025. Optimizing plant resilience with growth-promoting Rhizobacteria under abiotic and biotic stress conditions. Plant Stress 17. Available at https://doi.org/10.1016/j.stress.2025.100949.

Krishna, R., Jaiswal, D.K., Ansari, W.A., Singh, S., Soumia, P.S., Singh, A.K., et al., 2022. Potential microbial consortium mitigates drought stress in tomato (*Solanum lycopersicum* L.) plant by up-regulating stress-responsive genes and improving fruit yield and soil properties. Journal of Soil Science and Plant Nutrition 22 (4), 4598–4615. Available at https://doi.org/10.1007/s42729-022-00929-2.

Kumawat, K.C., Sharma, B., Nagpal, S., Kumar, A., Tiwari, S., Nair, R.M., 2023. Plant growth-promoting Rhizobacteria: salt stress alleviators to improve crop productivity for sustainable agriculture development. Frontiers in Plant Science 13. Available at https://doi.org/10.3389/fpls.2022.1101862.

Laiq, M., Azeem, M.A., Begum, S., Nazir, N., Ahmad, Z., Ullah, S., et al., 2024. Compositional shifts in rhizobacterial communities of *Brassica napus* under salinity stress. Applied Soil Ecology 195. Available at https://doi.org/10.1016/j.apsoil.2023.105258.

Li, B., Wang, P., Zhao, S., Dong, J., Mao, S., Zhu, X., et al., 2025. Sly-miR398 participates in heat stress tolerance in tomato by modulating ROS accumulation and HSP response. Agronomy 15 (2), 294. Available at https://doi.org/10.3390/agronomy15020294.

Liu, W.C., Song, R.F., Qiu, Y.M., Zheng, S.Q., Li, T.T., Wu, Y., et al., 2022. Sulfenylation of ENOLASE2 facilitates H2O2-conferred freezing tolerance in *Arabidopsis*. Developmental Cell 57 (15), 1883. Available at https://doi.org/10.1016/j.devcel.2022.06.012.

Manjunatha, B.S., Nivetha, N., Krishna, G.K., Elangovan, A., Pushkar, S., Chandrashekar, N., et al., 2022. Plant growth-promoting rhizobacteria *Shewanella putrefaciens* and *Cronobacter dublinensis* enhance drought tolerance of pearl millet by modulating hormones and stress-responsive genes. Physiologia Plantarum 174 (2). Available at https://doi.org/10.1111/ppl.13676.

Mashabela, M.D., Masamba, P., Kappo, A.P., 2023. Applications of metabolomics for the elucidation of abiotic stress tolerance in plants: a special focus on osmotic stress and heavy metal toxicity. Plants 12 (2). Available at https://doi.org/10.3390/plants12020269.

Metze, D., Schnecker, J., Canarini, A., Fuchslueger, L., Koch, B.J., Stone, B.W., et al., 2023. Microbial growth under drought is confined to distinct taxa and modified by potential future climate conditions. Nature Communications 14 (1). Available at https://doi.org/10.1038/s41467-023-41524-y.

Mittler, R., Zandalinas, S.I., Fichman, Y., Van Breusegem, F., 2022. Reactive oxygen species signalling in plant stress responses. Nature Reviews. Molecular Cell Biology 23 (10), 663–679. Available at https://doi.org/10.1038/s41580-022-00499-2.

Murali, M., Singh, S.B., Gowtham, H.G., Shilpa, N., Prasad, M., Aiyaz, M., et al., 2021. Induction of drought tolerance in *Pennisetum glaucum* by ACC deaminase producing PGPR- *Bacillus amyloliquefaciens* through Antioxidant defense system. Microbiological Research 253. Available at https://doi.org/10.1016/j.micres.2021.126891.

Mushtaq, F., Akram, M.H., Usman, M., Mohsin, M., Nawaz, M.S., 2024. Global climate change and its influence on crop production. Journal of Life and Social Sciences 3 (27).

Pan, Y., Peng, Y., Lu, Y., Yao, Y., Yu, W., Li, C., 2025. Research techniques of miRNA and its regulatory network under biotic and abiotic stresses in plants. Plant Stress 18, 101103. Available at https://doi.org/10.1016/j.stress.2025.101103.

Pardo, A.C.H., Pardo, J.D., VanBuren, R., 2025. Stress-responsive transcription factor families are key components of the core abiotic stress response in maize. bioRxiv. Available at https://doi.org/10.1101/2025.02.15.638452.

Park, J., Lee, S.H., Lee, J., Wi, S.H., Seo, T.C., Moon, J.H., et al., 2025. Growing vegetables in a warming world - a review of crop response to drought stress, and strategies to mitigate adverse effects in vegetable production. Frontiers in Plant Science 16. Available at https://doi.org/10.3389/fpls.2025.1561100.

Park, S.R., Jeong, Y., Son, S., 2025. Functions of transcription factor superfamilies in rice immunity. KeAi Communications Co., Cuba Crop Journal 13 (1), 5–22. Available at https://doi.org/10.1016/j.cj.2024.10.006.

Qian, Z., He, L., Li, F., 2024. Understanding cold stress response mechanisms in plants: an overview. Frontiers in Plant Science 15. Available at https://doi.org/10.3389/fpls.2024.1443317.

Ray, M., Burman, S., Meshram, S., 2025. A mini review on plant immune system dynamics: modern insights into biotic and abiotic stress. Phyton-International Journal of Experimental Botany. 94 (8), 2285–2312. Available at https://doi.org/10.32604/phyton.2025.067814.

Rehman, S.ur, Yang, J., Zhang, J., Zhang, L., Hao, X., Song, R., et al., 2025. Salt stress in wheat: a physiological and genetic perspective. Plant Stress 16, 100832. Available at https://doi.org/10.1016/j.stress.2025.100832.

Sadau, S.B., Liu, Z., Ninkuu, V., Guan, L., Sun, X., 2024. DREB transcription factors are crucial regulators of abiotic stress responses in *Gossypium* spp. Plant Stress 11. Available at https://doi.org/10.1016/j.stress.2024.100350.

Seleiman, M.F., Aslam, M.T., Alhammad, B.A., Hassan, M.U., Maqbool, R., Chattha, M.U., et al., 2022. Salinity stress in wheat: effects, mechanisms and management strategies. Phyton-International Journal of Experimental Botany 91 (4), 667–694. Available at https://doi.org/10.32604/phyton.2022.017365.

Shajar, F., Nabi, A.U., Manzoor, A., Saleem, S., Mushtaq, N.U., Manzoor, S., et al., 2025. Membrane-associated mechanisms in plant responses to abiotic stress. Plant Science 360. Available at https://doi.org/10.1016/j.plantsci.2025.112726.

Smet, D., Opdebeeck, H., Vandepoele, K., 2023. Predicting transcriptional responses to heat and drought stress from genomic features using a machine learning approach in rice. Frontiers in Plant Science 14. Available at https://doi.org/10.3389/fpls.2023.1212073.

Su, H., Wang, T., Ju, C., Deng, J., Zhang, T., Li, M., et al., 2021. Abscisic acid signaling negatively regulates nitrate uptake via phosphorylation of NRT1.1 by SnRK2s in Arabidopsis. Journal of Integrative Plant Biology 63 (3), 597–610. Available at https://doi.org/10.1111/jipb.13057.

Tian, F., Hu, X.L., Yao, T., Yang, X., Chen, J.G., Lu, M.Z., et al., 2021. Recent advances in the roles of HSFs and HSPs in heat stress response in woody plants. Frontiers in Plant Science 12. Available at https://doi.org/10.3389/fpls.2021.704905.

Tripathi, S., Dabral, S., Kundu, S., Saini, D.K., Jamal, H., Meena, R.K., et al., 2025. Harnessing the plant-associated microbiome: a sustainable solution for enhancing crop resilience to abiotic stresses and problematic soils. Plant Stress 18. Available at https://doi.org/10.1016/j.stress.2025.101033.

Ucar, S., Yaprak, E., Yigider, E., Kasapoglu, A.G., Oner, B.M., Ilhan, E., et al., 2024. Genome-wide analysis of miR172-mediated response to heavy metal stress in chickpea (*Cicer arietinum* L.): physiological, biochemical, and molecular insights. Plant Biology 24 (1). Available at https://doi.org/10.1186/s12870-024-05786-y.

Villano, F., Balestrini, R., Nerva, L., Chitarra, W., 2025. Harnessing microbes as sun cream against high light stress. New Phytologist 245 (2), 450–457. Available at https://doi.org/10.1111/nph.20206.

Vives-Peris, V., Gómez-Cadenas, A., Pérez-Clemente, R.M., 2024. Proline accumulation and antioxidant response are crucial for citrus tolerance to UV-B light-induced stress. Plant Biology 26 (4), 521–531. Available at https://doi.org/10.1111/plb.13640.

Wang, H., Li, N., Li, H., Zhang, S., Zhang, X., Yan, X., Wang, Z., Yang, Y., Zhang, S., 2023). Overexpression of NtGCN2 Improves DroughtTolerance in Tobacco by Regulating Proline Accumulation, ROS Scavenging Ability, and Stomatal Closure. Plant Physiology and Biochemistry 198, 107665. Available at 10.1016/j.plaphy.2023.107665.

Wang, C., Li, X., Zhuang, Y., Sun, W., Cao, H., Xu, R., et al., 2024. A novel miR160a – GmARF16 – GmMYC2 module determines soybean salt tolerance and adaptation. New Phytologist 241 (5), 2176–2192. Available at https://doi.org/10.1111/nph.19503.

Wang, Q., Wang, X., Zhang, Z., Wang, Z., Chen, W., Hua, X., et al., 2026. Whole-genome identification and expression characterization of the HSFP gene family of Triticum aestivum under heat and drought stress. Plant Science 362, 112780. Available at https://doi.org/10.1016/j.plantsci.2025.112780.

Wang, X., Li, Y., Chen, Z., Li, L., Li, Q., Geng, Z., et al., 2025. MbWRKY50 confers cold and drought tolerance through upregulating antioxidant capacity associated with ROS scavenging. Journal of Plant Physiology 310, 154526. Available at https://doi.org/10.1016/j.jplph.2025.154526.

Wenjing, W., Chen, Q., Singh, P.K., Huang, Y., Pei, D., 2020. CRISPR/Cas9 edited HSFA6a and HSFA6b of *Arabidopsis thaliana* offers ABA and osmotic stress insensitivity by modulation of ROS homeostasis. Plant Signaling and Behavior 15 (12). Available at https://doi.org/10.1080/15592324.2020.1816321.

Xie, Y., Liu, Y., Wang, H., Ma, X., Wang, B., Wu, G., et al., 2017. Phytochrome-interacting factors directly suppress MIR156 expression to enhance shade-avoidance syndrome in *Arabidopsis*. Nature Communications 8 (1). Available at https://doi.org/10.1038/s41467-017-00404-y.

Xu, Y., Chen, X., 2023. microRNA biogenesis and stabilization in plants. Fundamental Research 3 (5), 707–717. Available at https://doi.org/10.1016/j.fmre.2023.02.023.

Xu, Y., Li, Y., Chen, Z., Chen, X., Li, X., Li, W., et al., 2025. *Malus xiaojinensis* MxbHLH30 confers iron homeostasis under iron deficiency in *Arabidopsis*. International Journal of Molecular Sciences 26 (1), 368. Available at https://doi.org/10.3390/ijms26010368.

Yan, W., Sharif, R., Sohail, H., Zhu, Y., Chen, X., Xu, X., 2024. Surviving a double-edged sword: response of horticultural crops to multiple abiotic stressors. International Journal of Molecular Sciences 25 (10), 5199. Available at https://doi.org/10.3390/ijms25105199.

Yang, L., Tian, J., Zhu, M., Yu, B., Sun, Y., 2024. Hydrogen-rich water improvement in root growth in maize exposed to saline stress. Cereal Research Communications 52 (2), 581–590. Available at https://doi.org/10.1007/s42976-023-00413-2.

Yang, X., Luo, Y., Bai, H., Li, X., Tang, S., Liao, X., et al., 2022. DgMYB2 improves cold resistance in chrysanthemum by directly targeting DgGPX1. Horticulture Research 9. Available at https://doi.org/10.1093/hr/uhab028.

Yu, M., Luobu, Z., Zhuoga, D., Wei, X., Tang, Y., 2025. Advances in plant response to low-temperature stress. Plant Growth Regulation 105 (1), 167–185. Available at https://doi.org/10.1007/s10725-024-01253-8.

Yuan, A., Kumar, S.D., Wang, H., Wang, S., Impa, S., Wang, H., et al., 2024. Dynamic interplay among soil nutrients, rhizosphere metabolites, and microbes shape drought and heat stress responses in summer maize. Soil Biology and Biochemistry 191. Available at https://doi.org/10.1016/j.soilbio.2024.109357.

Yuan, L.B., Chen, M.X., Wang, L.N., Sasidharan, R., Voesenek, L.A.C.J., Xiao, S., 2023. Multi-stress resilience in plants recovering from submergence. Plant Biotechnology Journal 21 (3), 466–481. Available at https://doi.org/10.1111/pbi.13944.

Yuan, Z., Cai, S., Yan, C., Rao, S., Cheng, S., Xu, F., et al., 2024. Research progress on the physiological mechanism by which selenium alleviates heavy metal stress in plants: a review. Agronomy 14 (8), 1787. Available at https://doi.org/10.3390/agronomy14081787.

Zainab, N., Amna, Khan, A.A., Azeem, M.A., Ali, B., Wang, T., Shi, F., et al., 2021. PGPR-mediated plant growth attributes and metal extraction ability of *Sesbania sesban* l. In industrially contaminated soils. Agronomy 11 (9). Available at https://doi.org/10.3390/agronomy11091820.

Zhang, F., Yang, J., Zhang, N., Wu, J., Si, H., 2022. Roles of microRNAs in abiotic stress response and characteristics regulation of plant. Frontiers in Plant Science 13. Available at https://doi.org/10.3389/fpls.2022.919243.

Zhang, Y., Zhou, J., Ni, X., Wang, Q., Jia, Y., Xu, X., Wu, H., Fu, P., Wen, H., Guo, Y., Yang, G., 2023). Structural Basis for the Activity Regulation of Salt Overly Sensitive 1 in *Arabidopsis* Salt Tolerance. Nature Plants 9, 1915–1923. Available at 10.1038/s41477-023-01550-6.

Zhang, H., Li, G., 2025. Microbial strategies for enhancing nickel nanoparticle detoxification in plants to mitigate heavy metal stress. Phyton-International Journal of Experimental Botany. 94 (5), 1367–1399. Available at https://doi.org/10.32604/PHYTON.2025.064632.

Zheng, Q., Zhao, H., Duan, M., 2026. Unveiling the dynamic photosynthetic strategies of Heterosigma akashiwo: an interpretable machine learning approach to light and phosphorus stress. Marine Environmental Research 213, 107712. Available at https://doi.org/10.1016/j.marenvres.2025.107712.

Zhong, M., Wang, H., Hao, X., Liu, C., Hao, J., Fan, S., et al., 2024. LsFUL–LsSMU2 module positively controls bolting time in leaf lettuce (*Lactuca sativa* L.) under high temperature. Plant Science 347, 112195. Available at https://doi.org/10.1016/j.plantsci.2024.112195.

Zhou, H., Shi, H., Yang, Y., Feng, X., Chen, X., Xiao, F., et al., 2024. Insights into plant salt stress signaling and tolerance. Journal of Genetics and Genomics 51 (1), 16–34. Available at https://doi.org/10.1016/j.jgg.2023.08.007.

Zhu, J.K., 2016. Abiotic stress signaling and responses in plants. Cell 167 (2), 313–324. Available at https://doi.org/10.1016/j.cell.2016.08.029.

CHAPTER

2 Plant mineral nutrition and nutrient acquisition

Noureddine Chaachouay[1,4], Jamaa Habchaoui[2,4] and Azamal Husen[3,4]

[1]*Agri-Food and Health Laboratory (AFHL), ESEF of Berrechid, Hassan First University, Settat, Morocco,*

[2]*Laboratory of Botany, Biotechnology and Plant Protection, Department of Biology, Faculty of Sciences, Ibn Tofail University, Kenitra, Morocco,*

[3]*Wolaita Sodo University, Wolaita Sodo, South Ethiopia Regional State, Ethiopia,*

[4]*Department of Biotechnology, Graphic Era (Deemed to be University), Dehradun, Uttarakhand, India*

2.1 Introduction

This importance of mineral nutrition in plants is the result of the rich historical experience as well as the wide ecological perspective, which are basic principles that regulate plant growth, yield, and ecosystem processes (Pandey, 2015;Husen. 2024b) Traditionally, mineral nutrition of plants has been studied by early prototypical scientists like de Saussure in the 18th century and 19th century who made seminal observations on the absorption of nutrients by plants as well as the fertility of soil (Mengel et al., 2012). Further developed by technological discoveries, such as hydroponics, a method of growing plants without soil using nutrient-containing solutions, allowed conducting pinpoint experiments with key mineral components and their functions in the development of plants. These developments gave conclusive proof that plants are sensitive to the environment, through which they need specific mineral nutrients in the environment to realize maximum physiological activities (Kathpalia and Bhatla, 2018).

At the heart of mineral nutrition is the idea in the soil–plant–atmosphere continuum to explain the mutualized process of water and mineral nutrients through soil, the plant, and to the atmosphere. Soil acts as the store of mineral nutrients and water, which enable the uptake of roots (Chaachouay et al., 2026; Anbarasan and Ramesh, 2021). However, it is the plant that is the active mechanism of transporting these nutrients to the leaves and other organs, and the atmosphere is the mechanism that helps in water transportation through transpiration that is very significant in transporting nutrients and helping in physiological regulation. This gradient is an essential part of sustaining plant hydration, the acquisition of nutrients, and the gaseous exchange of photosynthesis.

Mineral Nutrients to Mitigate Abiotic Stress in Crops. DOI: https://doi.org/10.1016/B978-0-443-44712-9.00007-4

Mineral nutrition controls important facets of vegetative existence such as the growth rates, biomass production, and reproductive success of plants and has a direct impact on the yield and quality of the agricultural production (Fageria et al., 2010). Besides assisting with the survival of single species of plants, mineral nutrients also help with the larger ecosystem processes of supporting plant communities, regulating microbial activity, and being involved in nutrient cycles that form the foundation of biodiversity and ecosystem stability (Jacoby et al., 2017). Sufficient supply of mineral nutrients increases the strength of the plant against disease and environmental stresses, and this makes the use of optimal nutrient supply of minerals very crucial to sustainable agriculture and the protection of the environment.

The value of the mineral nutrition also helps in that the manner in which plants can change dynamically with their environment based on climate-change conditions, with particular stress of nutrient availability, water stress, and atmospheric variations (Lynch and St.Clair, 2004). This learning forms a basis for enhancing the implementation of crop management systems, breeding nutritionally effective species, and alleviating environmental effects with informed management of soils and nutrients. This chapter is intended to introduce the primary principles of mineral nutrition in the plant, which is primarily due to recent improvements in science. It is concerned with clarifying the absorption, transport, and physiological function of the mineral nutrients and their influence on plant growth, yield, and regulation of the ecology in the soil–plant–atmosphere nexus.

2.2 The essential elements

Arnon and Stout (1939) defined the necessary components of plant growth, which remains the core of recent research in the area of plant nutrition. In their criteria, an element is considered important when it is not redundant since the plant fails to complete its life cycle without it, the purpose a specific and irreplaceable element, and when it has a direct effect on the metabolism of the plant (Bhatla and Kathpalia, 2023). Such strict requirements are meant to distinguish between nutrients that are actually indispensable and those that are possibly absorbed but which do not turn out to be judgmental in normal growth and development.

The main categories of essential elements include macronutrients and micro nutrients, depending on the amount that plants need. The macronutrients include carbon (C), hydrogen (H), oxygen (O), nitrogen (N), phosphorus (P), potassium (K), calcium (Ca), magnesium (Mg), and sulfur (S), which are in great demand owing to the fact that they form structural and functional molecules of plant tissues (Ahmed et al., 2021). These are basic building blocks of some basic physiological functions, such as photosynthesis, energy transduction, and biosynthesis of central biomolecules. Micronutrients, on the other hand, are needed in significantly lower quantities, yet they are also essential. This category includes iron (Fe), manganese

(Mn), boron (B), zinc (Zn), copper (Cu), molybdenum (Mo), chlorine (Cl), and nickel (Ni). In their minimal amounts, micronutrients are also cofactors in enzyme reactions, electron carriers, and maintain cellular processes that cannot be replaced by other means (Hänsch and Mendel, 2009).

Besides the necessary elements, such factors are deemed to be useful, such as silicon (Si), sodium (Na), cobalt (Co), or selenium (Se). These factors fail to meet all the necessary criteria incorporated in essentiality, but can stimulate growth or increase stress-taking or nutrient uptake when specific circumstances are met (Bhatla and Kathpalia, 2023; Nunes da Silva et al., 2022). They play different roles in different plant species and in different environmental conditions, thus aiding in the optimization of plant health and performance in particular agricultural or ecological environments.

2.3 The soil as a nutrient reservoir

Soil is an important bed of nutrients that provide plants with important mineral components to grow and develop (Kathpalia and Bhatla, 2018). It consists of mineral fractions like sand, silt, and clay, which are inorganic, organic matter, which includes decomposed plant/animal remains known as humus, water, air, and various living organisms (Saha et al., 2017; Mansab et al., 2025). These components create a range of nutrient pools, where minerals are stored either in crystalline forms in primary and secondary minerals or that they are adsorbed to the surfaces of clays and organic substances and therefore form a source of plant-relevant nutrients. The active interaction between these pools nourishes the soil solution, which is absorbed by the roots to maintain the metabolism of plants.

Bioavailability of nutrients in soil varies based on various factors that determine the solubility and plants' accessibility to nutrients (Yadav et al., 2012). The pH of the soil is one of the most essential parameters that influence both the chemical speciation and solubility of nutrients and have a specific influence in the rhizosphere, the soil portion directly impacted by the activity of the roots, also known as the rhizosphere effect (Bravin et al., 2012; Sokolova, 2015). Root exudates or microbial activity in this area bring about a change in pH and chemical composition, which can increase food supply. Also, cation exchange capacity (CEC) and anion exchange capacity (AEC) of the soil indicate the capacity of the soil to store and release positive cations (e.g., K^+, Ca^+, etc.) and the negative anions (e.g., NO, PO4, etc.), with a direct effect on nutrient storage and release (Radulov and Berbecea, 2024).

Microbiome of the soil has a central role to play in the cycling of nutrients, decomposition, and general soil fertility (Prasad et al., 2021). Microorganisms process organic matter to produce nutrients like nitrogen, sulfur, phosphorus, and other micronutrients in plant-measurable forms. Microbial interactions also promote biological fixation of nitrogen and phosphorus solubilization that improves the uptake of nutrients (Chaachouay and Husen, 2025; Janati et al., 2021). High soil ecosystem biodiversity leads to enhancement of soil structure, facilitation of nutrient conversion, and may decrease the occurrence of soil-borne pathogens and thus favorable plant nutrition. The chemical properties, soil composition, and biological activity are a complex but still very necessary system of maintaining plant growth due to the regulation of nutrient supply (Delgado and Gómez, 2024). The effective management of soil fertility requires a clear grasp of these factors that will allow the optimum mineral nutrition of plants, or, in other words, the value of agricultural yield and other ecosystem functioning.

2.4 Mechanisms of nutrient acquisition by roots

Mechanisms of nutrient acquisition by roots begin with the transfer of ions across cellular membranes and involve special structures in the form of channels, carriers, and pumps (Dalir, 2023). These membrane transport proteins are involved in the translocation of mineral ions in the soil solution to root cells. Proton ATPases are essential since they actively release protons (H^+ ions) into the root cells so that an electrochemical gradient is maintained across the plasma membrane (Siao et al., 2020). This gradient acts as a driving force and influences the uptake of different nutrient ions passively via channels and actively via cotransporter, which enables plants to absorb essential nutrients, even against a concentration gradient (Fig. 2.1).

The root architecture significantly affects nutrient uptake, especially within a heterogeneous soil where nutrient availability exists in a spatial variation pattern (Li et al., 2014; Mou et al., 2013). Plants vary their root system architecture by varying root length and branching, as well as the development of root hairs to maximize exploration of soil and increase the soil–root contact area. These biochemical modifications on root system architecture, commonly caused by nutrient deficiencies or focal nutrient accents, enrich the ability of roots to graze on immobile nutrients, including phosphorus and micronutrients, and compound uptake effectiveness and total plant growth in nutrient-restricted situations (Maqbool et al., 2022).

Nutrient uptake is further augmented by the symbiotic relations of plants and soil microorganisms. Mycorrhizal relationships, which entail colonization of plant roots by fungi, have a significant positive impact on the uptake of phosphate and various micronutrients because extensions of the root absorptive surface using fungal hyphae increase internalization (Etesami et al., 2021). The fungi move and bring nutrients in the soil that are not accessible to the root, to the plant, increasing nutrient acquisition and availability. Likewise, the rhizobial symbiosis of leguminous plants allows biological fixation of nitrogen, whereby the nitrogen takes the form of various forms that are biologically available and are supplemented by these bacteria in soil nitrogen and are not limited by external fertilizer input (Kheyrodin, 2014).

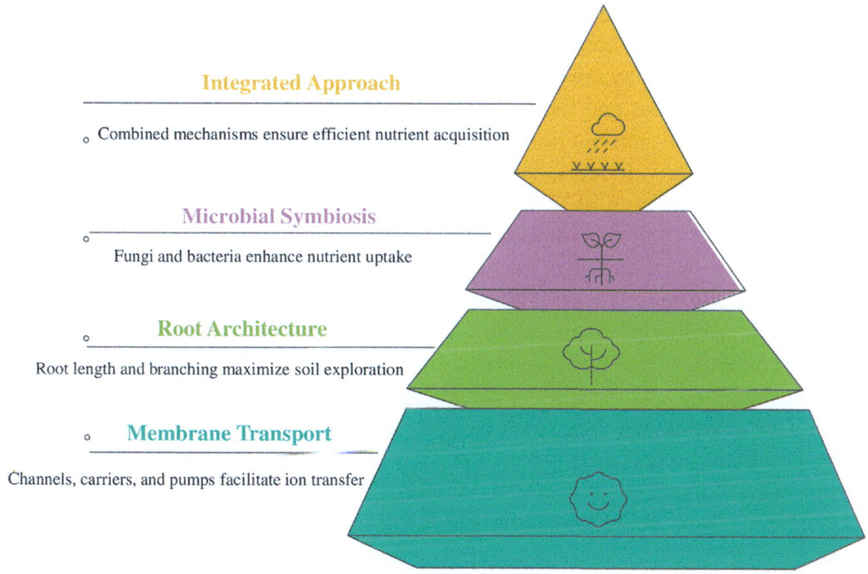

FIGURE 2.1

Mechanisms of nutrient acquisition by roots based on Tang and Riley (2021).

Combined, the membrane transport mechanism, adaptive root architecture, and beneficial microbial symbioses are collectively an integrated approach that plants use to efficiently obtain nutrients in complex and nutrient-variable soil environments (Pallucchini, 2023). The understanding of these mechanisms will give important facts to plant nutrition, soil fertility management, and a sustainable agricultural process, with the purpose of maximizing the utilization of nutrients and crop production.

2.5 Long-distance transport of nutrients

In plants, transport of nutrients over long distances occurs by two major pathways: the apoplastic and symplastic routes (Wolswinkel, 2024). Apoptotic pathway implies cellular transport by way of the cell walls and intercellular spaces without the penetration of the plasma membrane and symplastic cellular transportation by cytoplasm of live cells that are interlinked by plasmodesmata (White and Barton, 2011). The two pathways help in the transport of water and mineral ions in and out of the root cortex into the vascular tissues to enable the transport of nutrients to the xylem to be transferred upward to shoots and leaves (Bollard, 1960). Introduction of nutrients into the xylem is a key process to are long distance transportation needs, which involve selective ion absorption and active transport processes in transporting minerals to the xylem vessels. Proton pumps are commonly utilized in this process, xylem loading, and they create electrochemical concentrations that cause active absorption of nutrient ions. Shoot tissues unloading is used in a bid to facilitate the efficient delivery of nutrients to growing areas like leaves and to growing organs (Rather et al., 2025).

The phloem movement is important in the redistribution of nutrients in the plant. Phloem delivers photosynthates and other nutrients, where they travel out of the source tissues (usually mature leaves) toward sink tissues (roots, fruits, and developing leaves). The elements are grouped into phloem-mobile and phloem-immobile elements depending on whether they can be carried by the phloem or not (Schulz et al., 2001). Older tissues can be remobilized to provide younger tissues with phloem-mobile nutrients, such as nitrogen, phosphorus, and potassium, and phloem-immobile nutrients are more likely to accumulate in older tissues, such as calcium (Chaachouay et al., 2025; Maillard et al., 2015). The source-to-sink relationship controls the mobilization and the distribution of nutrients, which leave the source to go to sinks, so that the growing sink gets proper nutritional supply. In the case of senescence or nutrient deficiency, the nutrients stored in the mature tissues would be frequently transported via the phloem to the active sites of growth. This is an active redistribution of nutrients that is necessary to the development of plants, maximization of use, and adaptation to new environmental conditions.

2.6 Functions of essential mineral elements

Essential mineral elements are important and multiple with important roles in plant growth, development, and physiological functioning (Bhatla and Kathpalia, 2023; Pandey, 2018; Husen, 2024b) Structural roles play a fundamental role and they comprise carbon (C), hydrogen (H) and oxygen (O) which form the backbone of organic molecules carbohydrates, proteins and lipids (Yadav et al., 2012). Calcium (Ca) plays a crucial role in stabilizing and keeping the cell wall integrity of the cell walls, so that cellular structure is not weak and that cells do not become detached in their surrounding matrix. Such structural functions are critical toward the normal growth of the tissues of plants. Potassium plays a central role in energy metabolism, that is, it is a component of adenosine triphosphate, the major intracellular energy currency that makes biochemical reactions possible (Cui and Tcherkez, 2021). Magnesium, which is found at the center of the chlorophyll molecules, is essential in the conversion of light energy when undergoing the process of photosynthesis. The interaction between P and Mg is synergistic, which makes solar energy transduction of chemical energy possible, thus maintaining growth and metabolic processes in plants (Farhat et al., 2016).

Activation and catalysis of enzymes depend on cofactors. Photosynthetic, respiratory, protein-synthesis, and hormone-regulatory enzymes are triggered by a set of divalent and monovalent metal ions, such as potassium, zinc, manganese, copper, and many others. These mineral ions play a significant role in ensuring the efficiency of enzymes and regulation of the metabolic fluxes, and thus the overall plant health and development (MCervela Cardona, 2019; Pandey, 2015) (R.). Also, vital mineral constituents aid in electron transfer and osmotic balance. Iron (Fe) and copper (Cu) are involved in electron transport chains found in mitochondria and chloroplasts, thus helping cells to respire and produce food. Potassium (K^+) and chloride (Cl^-) ions are the primary regulators of the osmotic and charge balance, which control the water uptake, turgor pressure, and ionic balance in cells (Bhatla and Kathpalia, 2023). All these functions are in place so that plants are able to effectively manage energy, nutrients, and water to promote maximum growth.

2.7 Nutrient assimilation and integration into plant metabolism

Nutrient assimilation incorporates the vital mineral-based components in the metabolism by undergoing a sequence of biochemical procedures, the most significant of which is the process of nitrogen assimilation (Ubale et al., 2023). Nitrogen is mainly picked up by plants as nitrate (NO_3^-) and ammonium (NH_4^+) compounds of the soil. The nitrate is broken down to nitrite by nitrate reductase in the cytoplasm and further transformed to ammonium by nitrite reductase in chloroplasts or plastids (Kumari et al., 2022). This ammonium is fast metabolized into amino acids through glutamine synthetase-glutamate synthase (GS-GOGAT) pathway in order to form glutamine and glutamate, which act as carriers of nitrogen in the plant (Zhang et al., 2020). Environmental factors, including light, availability of carbon and nitrogen requirements, strictly regulate the process and optimize its growth, reducing nitrogen wastage (Fig. 2.2).

The first stage of sulfur assimilation starts with the absorption of the sulfate (SO_2^-) in the soil, carried away to the root cells, and reduced under the influence of a set of enzymes to sulfide. The resulting sulfide is then added to organic molecules like cysteine and methionine, which are the essential sulfur-bearing amino acids. The processes of sulfur assimilation are also closely related to the processes of nitrogen metabolism, which contribute to the synthesis of proteins, coenzymes, and vitamins, which are important components of plant structure and functioning. Sulfate reduction and incorporation are adequate to sustain cellular redox balances and also play a role in the defense mechanisms of plants against their harsh environments and their invaders.

Phosphate metabolism involves the uptake of inorganic phosphate and transformation into a wide range of organic molecules, which are critical to energy transduction, signal transduction, and maintenance of structural integrity.

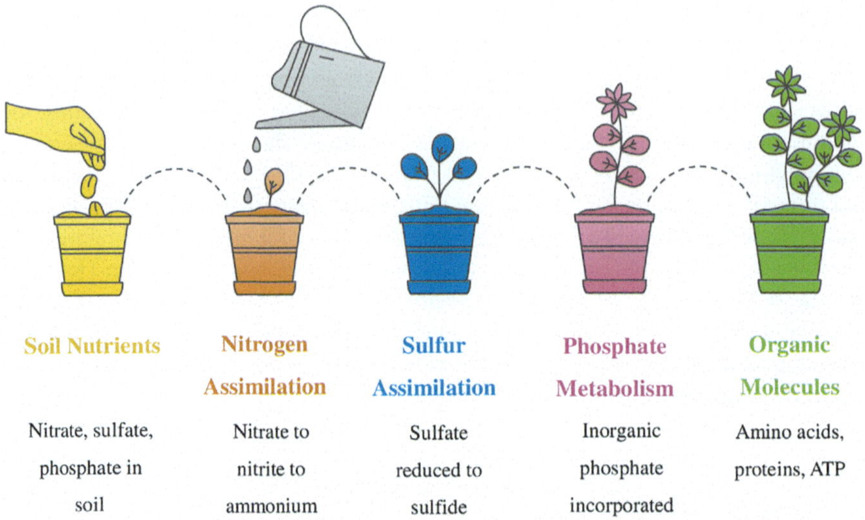

FIGURE 2.2

Nutrient assimilation in plants.

Nucleotides, including ATP and nucleic acids, phospholipids that make up cell membranes, and phosphorylated derivatives of intermediates in central metabolic pathways, all contain phosphate. To maintain metabolic processes and developmental growth, plants have developed regulatory options encompassing adaptation in acquiring and allocating phosphates, especially under low availability, and allocate phosphates in a more accurate manner (White and Barton, 2011). Phosphate plays a critical role in the incorporation of organic compounds to provide energy balance in the cells and generate metabolic reactions in the plant tissues.

2.8 Conclusion and future perspectives

The production of the main tenets related to the mineral nutrition of the plants represents an essential requirement that cannot be overlooked in the development of the long-term agricultural paradigms. The reconciliation of these two, the augmentation of crop yield and responsible environmental stewardship, is the objective of such paradigms. Other strategic measures such as the use of integrated nutrient management include the combination of organic, inorganic, and biological sources of nutrients to improve the use of soil and the high utilization efficiency of nutrients. Integrated nutrient management not only reduces the use of chemical fertilizers but also minimizes the losses of nutrients as well as promotes soil structure and microbial activity, consequently leading to long-term soil sustainability. The processes come with a higher-level knowledge of crop resistance to both biotic and abiotic stresses, producing higher and more consistent yields without disrupting ecosystems. Sustainable nutrient management is also effective in protecting the environment and increasing the economic efficiency of farming enterprises since less money is spent on inputs, and the soil yields more. In prospect, the future trends in plant mineral nutrition are biofortification, nutrient-use efficiency, and climate-resilience approaches. Biofortification aims at the nutritional enrichment of the dietary tissues of plants by raising the amounts of some of the most vital micronutrients with an aim of curbing the eminent challenge of malnutrition around the world. The improvement of plant biology and genetics can be used to create new versions of crops with better nutrient absorption and achievement, which can reduce the use of fertilizers and their impact on the environment. Optimization of nutrient use can make a crucial contribution to the adjustment of the agro systems to the altering weather conditions, which can guarantee the sustainable productivity under the influence of stressors like drought, extreme weather conditions, and soil erosion. Combining new practices with the experience of the agronomic history of the past will result in the creation of an impact on the sphere of agriculture, making it more robust and sustainable. To conclude, a thorough understanding of mineral nutrition and its incorporation into the sustainability environment is a plausible way to overcome the increasing food demand without exhausting natural resources. Continued studies and application of combined nutrition-management policies, crop biofortification, and nutrient-use efficiency indicators are necessary remedial steps toward augmenting food security worldwide, and the health of humans, as well as the reduction of the climate-change consequences. These perspectives denote the pertinence of multidisciplinary cooperation toward sustainable farming.

References

Ahmed, A.B.A., Taha, R.M., Anuar, N., Elias, H., Abdullah, S., Khan, A., et al., 2021. Chapter 6 - Saffron as a natural food colorant and its applications. In: Galanakis, C.M. (Ed.), Saffron. Academic Press, pp. 221–239. Available at https://doi.org/10.1016/B978-0-12-821219-6.00006-3.

Anbarasan, S., Ramesh, S., 2021. The role of plant roots in nutrient uptake and soil health. Plant Science Archives 6 (1), 5–8. Available at https://doi.org/10.51470/psa.2021.6.1.05.

Arnon, D.I., Stout, P.R., 1939. The essentiality of certain elements in minute quantity for plants with special reference to copper. Plant Physiology 14 (2), 371–375. Available at https://doi.org/10.1104/pp.14.2.371.

Bhatla, S.C., Kathpalia, R., 2023. Essential and functional mineral elements. Plant Physiology, Development and Metabolism. Springer Nature Singapore, pp. 25–49 https://doi.org/10.1007/978-981-99-5736-1_2.

Bollard, E.G., 1960. Transport in the Xylem. Annual Review of Plant Physiology 11 (1), 141–166. Available at https://doi.org/10.1146/annurev.pp.11.060160.001041.

Bravin, M.N., Garnier, C., Lenoble, V., Gérard, F., Dudal, Y., Hinsinger, P., 2012. Root-induced changes in pH and dissolved organic matter binding capacity affect copper dynamic speciation in the rhizosphere. Geochimica et Cosmochimica Acta 84, 256–268. Available at https://doi.org/10.1016/j.gca.2012.01.031.

Chaachouay, N., Ansari, M.K.A., Zidane, L., Husen, A., 2026. Uptake, translocation, and transformation of magnetic nanomaterials in plants soil-plant system/ In: Huesen (ed) Magnetic Nanomaterials for Plant and Soil Systems Improvement. Academic Press, London, pp. 53–66. Available at 10.1016/b978-0-443-30200-8.00014-2.

Chaachouay, N., Zidane, L., Husen, A., 2025. Enhancement of seed germination of important crop plant species using carbon-based nanomaterials. In: Husen (ed) Emerging Carbon Nanomaterials for Sustainable Agricultural Practices. Smart nanotechnology. pp. 137–147. Available at 10.1007/978-981-97-5104-4_7.

Chaachouay, N., Husen, A., 2025. Osmolytes and their crosstalk mechanism with plant hormones for higher plant productivity and survival under adverse situations. In: Husen (ed) Roles of Osmolytes in Changing Environment. Academic Press, London, pp. 13–43. Available at 10.1016/b978-0-443-26549-5.00016-3.

Cui, J., Tcherkez, G., 2021. Potassium dependency of enzymes in plant primary metabolism. Plant Physiology and Biochemistry 166, 522–530. Available at https://doi.org/10.1016/j.plaphy.2021.06.017.

Dalir, N., 2023. Nutrients uptake and transport in plants: an overview. In: Singh, V.P., Siddiqui, M.H. (Eds.), Plant Ionomics, 1re éd. Wiley, pp. 180–190 https://doi.org/10.1002/9781119803041.ch10.

Delgado, A., Gómez, J.A., 2024. The soil: physical, chemical, and biological properties. In: Villalobos, F.J., Fereres, E. (Eds.), Principles of Agronomy for Sustainable Agriculture. Springer International Publishing, pp. 15–30 https://doi.org/10.1007/978-3-031-69150-8_2.

Etesami, H., Jeong, B.R., Glick, B.R., 2021. Contribution of arbuscular mycorrhizal fungi, phosphate–solubilizing bacteria, and silicon to P uptake by plant. Frontiers in Plant Science 12. Available at https://doi.org/10.3389/fpls.2021.699618.

Fageria, N.K., Baligar, V.C., Jones, C.A., 2010. Growth and Mineral Nutrition of Field Crops. CRC Press, Boca Raton. Available at https://doi.org/10.1201/b10160.

Farhat, N., Elkhouni, A., Zorrig, W., Smaoui, A., Abdelly, C., Rabhi, M., 2016. Effects of magnesium deficiency on photosynthesis and carbohydrate partitioning, Polish Academy of Sciences, Tunisia. Acta Physiologiae Plantarum 38 (6). Available at https://doi.org/10.1007/s11738-016-2165-z.

Hänsch, R., Mendel, R.R., 2009. Physiological functions of mineral micronutrients (Cu, Zn, Mn, Fe, Ni, Mo, B, Cl). Current Opinion in Plant Biology 12 (3), 259–266. Available at https://doi.org/10.1016/j.pbi.2009.05.006.

Husen, A., 2024. Essential Minerals in Plant-Soil System (Coordination, Signalling, and Interaction under Adverse Situations. Elsevier Inc, USA https://doi.org/10.1016/C2022-0-02437-5.

Jacoby, R., Peukert, M., Succurro, A., Koprivova, A., Kopriva, S., 2017. The role of soil microorganisms in plant mineral nutrition—current knowledge and future directions,

Frontiers Media S.A., Germany. Frontiers in Plant Science 8. Available at https://doi.org/10.3389/fpls.2017.01617.

Janati, W., Benmrid, B., Elhaissoufi, W., Zeroual, Y., Nasielski, J., Bargaz, A., 2021. Will phosphate bio-solubilization stimulate biological nitrogen fixation in grain legumes?, Frontiers Media S.A., Morocco. Frontiers in Agronomy 3. Available at https://doi.org/10.3389/fagro.2021.637196.

Kathpalia, R., Bhatla, S.C., 2018. Plant mineral nutrition. Plant Physiology, Development and Metabolism. Springer Nature Singapore, pp. 37–81 https://doi.org/10.1007/978-981-13-2023-1_2.

Kheyrodin, H., 2014. Plant and soil relationship between fungi. International Journal of Research Studies in Biosciences 2 (9), 42–49.

Kumari, R., Bhatnagar, S., Kalra, C., 2022. Nitrogen assimilation in plants. Advances in Plant Nitrogen Metabolism. CRC Press, pp. 38–54.

MCervela Cardona, L., 2019. Functional studies on the circadian regulation of mitochondrial activity in *Arabidopsis thaliana*. https://ddd.uab.cat/record/240744.

Li, H., Ma, Q., Li, H., Zhang, F., Rengel, Z., Shen, J., 2014. Root morphological responses to localized nutrient supply differ among crop species with contrasting root traits. Plant and Soil 376 (1), 151–163. Available at https://doi.org/10.1007/s11104-013-1965-9.

Lynch, J.P., St.Clair, S.B., 2004. Mineral stress: the missing link in understanding how global climate change will affect plants in real world soils. Field Crops Research 90 (1), 101–115 10.1016/j.fcr.2004.07.008UnitedStates.

Pallucchini, M., 2023. An investigation into the mechanisms underlying the Plant Growth Promoting Properties of G. diazotrophicus [PhD Thesis, University of Nottingham (United Kingdom)]. https://search.proquest.com/openview/ae4e6a258114232f4c2dd74c0b40051a/1?pq-origsite=gscholarandcbl=2026366anddiss=y.

Maillard, A., Diquélou, S., Billard, V., Laîné, P., Garnica, M., Prudent, M., et al., 2015. Leaf mineral nutrient remobilization during leaf senescence and modulation by nutrient deficiency, Frontiers Research Foundation, France. Frontiers in Plant Science 6 (MAY), 1–15. Available at https://doi.org/10.3389/fpls.2015.00317.

Mansab, S., Parveen, K., Nasreen, S., 2025. Understanding soil composition: its effect on plant development. In: Shaaban, M. (Ed.), Soils and Sustainable Agriculture. Springer Nature Switzerland, pp. 27–55 https://doi.org/10.1007/978-3-031-91114-9_3

Maqbool, S., Hassan, M.A., Xia, X., York, L.M., Rasheed, A., He, Z., 2022. Root system architecture in cereals: progress, challenges and perspective, John Wiley and Sons Inc, Pakistan. Plant Journal 110 (1), 23–42. Available at https://doi.org/10.1111/tpj.15669.

Mengel, K., Kirkby, E.A., Kosegarten, H., Appel, T., 2012. Principles of Plant Nutrition. Springer, Dordrecht. Available at https://doi.org/10.1007/978-94-010-1009-2.

Mou, P., Jones, R.H., Tan, Z., Bao, Z., Chen, H., 2013. Morphological and physiological plasticity of plant roots when nutrients are both spatially and temporally heterogeneous. Plant and Soil 364 (1-2), 373–384. Available at https://doi.org/10.1007/s11104-012-1336-y.

Nunes da Silva, M., Machado, J., Osorio, J., Duarte, R., Santos, C.S., 2022. Non-essential elements and their role in sustainable agriculture. Agronomy 12 (4), 888. Available at https://doi.org/10.3390/agronomy12040888.

Pandey, N., 2018. Role of plant nutrients in plant growth and physiology. In: Hasanuzzaman, M., Fujita, M., Oku, H., Nahar, K., Hawrylak-Nowak, B. (Eds.), Plant Nutrients and

Abiotic Stress Tolerance. Springer, Singapore, pp. 51–93 https://doi.org/10.1007/978-981-10-9044-8_2.

Pandey, R., 2015. Mineral nutrition of plants. In: Bahadur, B., Venkat Rajam, M., Sahijram, L., Krishnamurthy, K.V. (Eds.), Plant Biology and Biotechnology. Springer, New Delhi, pp. 499–538 https://doi.org/10.1007/978-81-322-2286-6_20.

Prasad, S., Malav, L.C., Choudhary, J., Kannojiya, S., Kundu, M., Kumar, S., et al., 2021. Soil microbiomes for healthy nutrient recycling. In: Yadav, A.N., Singh, J., Singh, C., Yadav, N. (Eds.), Current Trends in Microbial Biotechnology for Sustainable Agriculture. Springer, Singapore, pp. 1–21 https://doi.org/10.1007/978-981-15-6949-4_1.

Radulov, I., Berbecea, A., 2024. Nutrient Management for Sustainable Soil Fertility. In: Meena, V.S., Bana, R.S., Fagodiya, R.K., Hasanain, M. (Eds.), Sustainable Agroecosystems - Principles and Practices. IntechOpen. Available at 10.5772/intechopen.1006692.

Rather, G.A., Manhas, A., Sharma, N., 2025. Mobility of nutrients and their metabolism in agriculture crops. Agricultural Biotechnology. Apple Academic Press, pp. 75–93.

Saha, J.K., Selladurai, R., Coumar, M.V., Dotaniya, M.L., Kundu, S., Patra, A.K., 2017. Soil and its role in the ecosystem. Soil Pollution - An Emerging Threat to Agriculture. Environmental Chemistry for a Sustainable World, vol 10. Springer, Singapore, pp. 11–36 https://doi.org/10.1007/978-981-10-4274-4_2.

Schulz, A., Thompson, G.A., 2001. Phloem structure and function. Encyclopedia of Life Sciences. Wiley. Available at https://doi.org/10.1002/9780470015902.a0001290.pub2.

Siao, W., Coskun, D., Baluška, F., Kronzucker, H.J., Xu, W., 2020. Root-apex proton fluxes at the centre of soil-stress acclimation. Trends in Plant Science 25 (8), 794–804. Available at https://doi.org/10.1016/j.tplants.2020.03.002.

Sokolova, T.A., 2015. Specificity of soil properties in the rhizosphere: analysis of literature data. Eurasian Soil Science 48 (9), 968–980. Available at https://doi.org/10.1134/s1064229315050099.

Tang, J., Riley, W.J., 2021. On the modeling paradigm of plant root nutrient acquisition, Springer Science and Business Media Deutschland GmbH, United States. Plant and Soil 459 (1-2), 441–451. Available at https://doi.org/10.1007/s11104-020-04798-5.

Ubale, N., Meena, J.K., Shukla, J.P., Hanni, S., Verma, L.K., 2023. Modern Horizons in Horticulture: Emerging Trends and Technologies. Elite Publishing House.

White, R.G., Barton, D.A., 2011. The cytoskeleton in plasmodesmata: a role in intercellular transport? Journal of Experimental Botany 62 (15), 5249–5266. Available at https://doi.org/10.1093/jxb/err227.

Wolswinkel, P., 2024. Long-distance nutrient transport in plants and movement into developing grains. Mineral Nutrition of Crops. CRC Press, pp. 91–113.

Yadav, R.S., Meena, S.C., Patel, S.I., Patel, K.I., Akhtar, M.S., Yadav, B.K., et al., 2012. Bioavailability of soil p for plant nutrition. In: Lichtfouse, E. (Ed.), Farming for Food and Water Security. Sustainable Agriculture Reviews, vol 10. Springer, Dordrecht, pp. 177–200 https://doi.org/10.1007/978-94-007-4500-1_8.

Zhang, J., Lv, J., Xie, J., Gan, Y., Coulter, J.A., Yu, J., et al., 2020. Nitrogen source affects the composition of metabolites in pepper (*Capsicum annuum* L.) and regulates the synthesis of capsaicinoids through the GOGAT-GS pathway, MDPI Multidisciplinary Digital Publishing Institute, China. Foods 9 (2). Available at https://doi.org/10.3390/foods9020150.

SECTION

II

Roles of essential nutrients in stress resilience

CHAPTER

Role of nitrogen in abiotic stress resilience

3

Nazneen Akhtar, Sehresh Khan and Muhammad Jamil

Department of Biotechnology and Genetic Engineering, Kohat University of Science & Technology (KUST), Kohat, Pakistan

3.1 Introduction

The element nitrogen, denoted by the chemical symbol N, is essential to the continuation of life on Earth. Because it facilitates the synthesis of amino acids, which are the building blocks of proteins, it is essential for the health of plants. In turn, proteins are essential for encouraging strong plant growth. Furthermore, nitrogen is a basic building block of nucleic acids, DNA, and RNA, which are necessary for all living things. Plants that lack nitrogen in the soil have limited growth because they are unable to maintain vital nucleic acid structures and manufacture necessary proteins (Zhang et al., 2021). The foundation of the world's food production, nitrogen is essential for plant growth. Studies have shown that nitrogen is the most important component affecting genome size, while phosphorus is less important in this regard. The efficient use of nitrogen results in increases in root diameter, length, branching, and biomass content. Water absorption also improves with the expansion of the root network (Razaq et al., 2017). Although there is roughly 78% nitrogen in the atmosphere, plants are unable to directly utilize this nitrogen. First, either microbes or the production of fertilizer fix this nitrogen. Although the soil contains nitrogen, 98% of it is organic nitrogen, which is indigestible by plants. The conversion of inert nitrogen (N_2) into reactive nitrogen, which plants can absorb and which is what gives cells life, is known as the nitrogen cycle. Before the Haber-Bosch process was created in 1909 for industrial artificial nitrogen fixation, nitrogen was fixed naturally (Stein and Klotz,2016).

Chlorophyll, which is produced by nitrogen, speeds up photosynthesis, provides plants with a greening effect, and improves crop output and quality. By encouraging protein synthesis, stress response enzyme activation, and osmotic control, it also increases plant resilience to abiotic stressors. One of the primary abiotic stresses, salinity, can be moderated by applying the proper amount of nitrogen, which also helps control ion balance, reduce osmotic stress, and lessen competition for water. For instance, sorghum crops' ability to withstand salt was found to be improved by applying 12 g/kg of soil (Elsiddig *et* al., 2023). The crop is greatly impacted by salinity's effect on soil nitrogen availability, which can cause symptoms such as

Mineral Nutrients to Mitigate Abiotic Stress in Crops. DOI: https://doi.org/10.1016/B978-0-443-44712-9.00005-0

chlorosis, slowed development, discoloration, and, in extreme situations, necrosis of older leaves. Small, curled leaves and low-protein grains are also common in several crops. Leaching, or the loss of nitrogen with flowing water, and denitrification, or the transformation of nitrogen from nitrite to gaseous nitrogen as a result of microbial activity, are two processes that cause nitrogen to be lost from soil. Over the past three decades, nitrogen's use has increased because it is an essential nutrient for plant growth, development, and productivity (Sun et al., 2020).

The use of nitrogenous fertilizers has grown by 48% since 2008, when food production surged. Crop output has grown due to greater usage of nitrogenous fertilizers, but environmental pollution is also occurring. Due to soil acidity, excessive nitrogen application also reduces production. Additionally, it lowers the amount of potassium and phosphorus that is available. Reactive nitrogen molecules (ammonia, nitrate, and nitrite) have increased as a result of agricultural activities (mostly fertilizer applications), yet this increase is detrimental to the ecosystem. The timing, accessibility, and movement of nitrogen-related processes, such as nitrification, denitrification, and nitrogen fixation, are also impacted by climate change (Greaver et al., 2016). The second chemical hazard to surface and groundwater is excessive nitrate from synthetic fertilizers. There are four different types of nitrogen in water: nitrate, ammonia, ammonium, and nitrite. Of these, nitrate is the most hazardous. Because it reacts with proteins to cause cancer in adults and blue-baby syndrome in neonates, rising nitrate levels in water are becoming lethal to both aquatic life and humans. The main source of salinity and the previously listed issues is the overuse of nitrogen, which makes it a crucial topic for discussion. The efficiency of nitrogen assimilation in crops will be improved by the harmonization of agronomic, genetic, and biotechnological methods, bringing agricultural systems into line with international demands to protect environmental resources (Stein and Klotz,2016). A high GSH/GSSG ratio, which is required for DHAR (dehydroascorbate reductase) and other GSH-dependent enzymes involved in antioxidant defense mechanisms, was also maintained by increases in NO-mediated glutathione reductase (GR) activity during salinity stress. By altering the physiological, biochemical, and molecular processes of plants, applications of nanoparticles (also known as nano-fertilizers) have shown great promise in improving plants' resistance to environmental stresses such as drought, extreme heat and cold, and salinity (Gojon, 2017; Husen, 2023; Husen, 2025; Siddiqi et al., 2025).

3.2 Abiotic stress from salinity

Desert playas are flat, dry areas at the bottom of undrained desert basins where water evaporates. In the spring, when these areas are submerged in water, the concentrations of salt rise from 7000 to 14,000 mM NaCl. Due to its impact on agricultural productivity through the suppression of photosynthesis, salinity is a significant abiotic stressor in arid and semi-arid regions of the world. In this instance, the potassium ion concentration is lower and the sodium ion concentration is higher (Soliman et al., 2020).

Presently, salinity affects 20% of the world's geographical area. Salt stress is predicted to cause 30% of agricultural land to disappear over the next 25 years, with a 50% potential loss by 2050. Different soluble salts, such as sodium sulfate and sodium chloride, are found in various fields depending on the geography and environmental factors. Depending on the source of nitrogen, salinity has different effects on the uptake of different nutrients. For example, plants become more sensitive when ammonia is applied as a fertilizer. Salinity has a major impact on major cereal crops like rice, wheat, maize, and sorghum, as well as equally significant cash crops like cotton and sugarcane. Salinity currently affects 20% of the world's land area. Salt stress is thought to be responsible for 30% of agricultural land loss in the next 25 years, with a 50% loss potential by 2050. Sodium sulfate and sodium chloride are two examples of the various soluble salts that are present in various fields depending on the geography and environmental circumstances (Elsiddig et al., 2023). Depending on the source of nitrogen, salinity has different effects on the intake of different nutrients. For example, plants become more sensitive when ammonia is used as a fertilizer. Salinity has a significant impact on both vital commercial crops like cotton and sugarcane as well as major grain crops like rice, wheat, maize, and sorghum. According to molecular research, salts inhibit plant growth in three different ways. The concentrations of plant toxins (ion toxicity due to the accumulation of $Na+$ and $Cl-$ ions) are increased by osmotic stress (water is drawn out of the cell due to higher salt concentrations), ionic stress, and oxidative stress caused by reactive oxygen species (ROS). Ion, nutritional, and hormone imbalances all disrupt photosynthetic activity, but the most significant is the disruption in water uptake brought on by salinity (Hussain et al., 2019). As salinity rises, soil's inorganic nitrogen content falls. Nitrogen fertilizer promotes N_2O emission from saline soils because nitrogen is present in large amounts as a substrate for soil nitrification and denitrification. The relationship between salinity and different nitrogen dosages was investigated in an experiment on *Crocus sativus* L. Accordingly, 50 kg of N/ha reduced the negative effects of salt, while 100 kg of N/ha increased the negative effects on the crop. The detrimental effects of salt can be measured using the number of leaves, length of leaves, weight of dry shoots, number of corms, and weight of corms (Hashemi et al., no.date).

3.3 Significance of nitrogen

3.3.1 In plants

The only energy source for life on Earth is the Sun. Chlorophyll in plants, algae, and cyanobacteria absorbs sunlight, particularly blue and red light. These organisms use photosynthesis to transform carbon dioxide, water, and sunlight into chemical energy, which is then converted into glucose. The thylakoid membrane of the chloroplast contains chlorophyll, which is present in the green portions of plants. The magnesium ion in chlorophyll is surrounded by four nitrogen atoms, creating a complicated structure. This is frequently called a tetrapyrrole ring or porphyrin ring (Pareek et al., 2017). The amount of chlorophyll is determined by the structure of the chloroplasts, which is influenced by the plants' access to nitrogen. The photosynthetic area and chlorophyll content decreased when the effect of nitrogen was studied on maize crops using a low quantity of nitrogen fertilizer. Nitrogen is the primary component that regulates the dynamics of many ecosystems on Earth. Compared to other nutrients, wheat consumes nitrogen more often. When it is lacking, the leaf's surface area and photosynthetic area diminish, which also lessens the plant's greenness (Bojović and Marković, 2009). Compared to other nutrients, wheat consumes nitrogen more often. When it is lacking, the leaf's surface area and photosynthetic area diminish, which also lessens the plant's greenness. When 138 kg/ha of fertilizer was applied, the leaf's photosynthetic area and pigment PS II both increased. The sugar beet produced the most chlorophyll and the highest productivity when nitrogen was applied at a rate of 120 kg/ha (Bianco et al., 2022).

When nitrogen was applied at a rate of 120 kg/ha, sugar beet produced the most chlorophyll and the maximum productivity. By fostering root growth, changing pH, boosting microbial activity, and making these fertilizers more soluble, high nitrogen concentrations enhance plants' uptake of potassium, phosphorus, iron, calcium, magnesium, and copper. Grain production in China has increased by 70% since 1980 as a result of the usage of nitrogenous fertilizers. Cell walls' composition and structure are also influenced by nitrogen (Rivai et al., 2021).

In the case of oats, nitrogen is also in charge of vegetative growth and reproduction. Nitrogen application raised the amount of chlorophyll in saline soils, which improved photosynthesis and oat crop output. The reduction of salinity through nitrogen use was credited with increasing grain concentration (quantity or density of grains in crops such as wheat). While excessive nitrogen treatment increases soil salinity, raising nitrogen concentration up to a certain point lowers soil salinity (Jiang et al., 2023). Although N fertilization may limit the plant's ability to fix atmospheric N, it seems to have a positive impact on lowering alfalfa's salt stress and maintaining the production of shoot biomass. When 12 g N kg^{-1} of soil was treated, N fertilizer proved effective in reducing the negative effects of NaCl and Na_2SO_4 in sorghum (Ali et al., 2023).

3.3.2 The dual nature of nitrogen: industrial wonder and vital nutrient

Since humans cannot directly use elemental nitrogen, they must take in other forms, such as NO_2^- and NO_3^-, with NO_3^- coming from green plants and unintentionally drinking tainted water. While some neurotransmitters and peptide hormones are composed of amino acids, which are the building blocks of proteins, nitrogen accounts for 16% of the structure of proteins. DNA and RNA are composed of carbon, phosphorus, and nitrogen. Purines, which have three nitrogens, and pyrimidines, which have five, are the two primary forms of nucleotides (Melino et al., 2018). Children require 35% protein for growth, compared to 15% for adults. Understanding metabolism, consumption, and health implications is crucial given the growing need for nitrogen. Adults' bodies maintain a nitrogen balance and neither store nor accumulate protein. But growth, reproduction, illness, or injury increase the body's need for nitrogen, creating a positive balance. Conversely, when protein use exceeds intake or catabolism exceeds consumption, a negative nitrogen balance results (Rivai et al., 2021).

However, the body's need for nitrogen rises with growth, reproduction, disease, or damage, leading to a positive balance. On the other hand, a negative nitrogen balance happens when protein use surpasses intake or catabolism surpasses consumption. As an essential messenger molecule, nitric oxide promotes physiological processes such as immunological regulation, platelet activity, and blood vessel dilatation. Like carbohydrates, proteins contribute 2%–5% of the body's energy, or 4 calories per gram (Lavrentieva and Dudaryonok, 2022). In addition to its usage in the pharmaceutical industry for the preservation of blood, eggs, and sperm, nitrogen has numerous other applications. Nitrous oxide has an anesthetic effect as well. Nitrogen is a crucial element that stops oxidation during the metal manufacturing process. It can be applied as a blanketing agent as well. Its low temperature makes it useful in the food business as well, where it is used to freeze and chill food. Additionally, it lessens food deterioration and facilitates long-distance food transportation. Additionally, unstable organic compounds are eliminated from the process using nitrogen (Hashemi et al., 2023). Fig. 3.1 illustrates the key transformations and roles of nitrogen (N) in the environment and biological systems.

3.4 Nitrogen cycle and fixation

There are two main sources of nitrogen on Earth: the atmosphere, which contains nitrogen in the form of molecules (N_2), and the reactive nitrogen pool, which is found in biological molecules. The nitrogen cycle is the interconversion of nitrogen between these two pools. The nitrogen cycle comprises nitrogen fixation, mineralization, immobilization, nitrification, and denitrification as its five main stages. Nitrogen Fixation is the process by which molecular nitrogen is transformed into biological nitrogen (Takai, 2019). In addition to being carried out by free-living and symbiotic microorganisms, lightning can also be used to artificially fix nitrogen. With the aid of the nitrogenase enzyme, prokaryotes carry out biological nitrogen fixation by converting atmospheric nitrogen molecules into ammonia. Beijerinck made the discovery of this method in 1901. These prokaryotes include free-living bacteria like Azotobacter, which inhabits soil, Cyanobacteria, which inhabit water, and

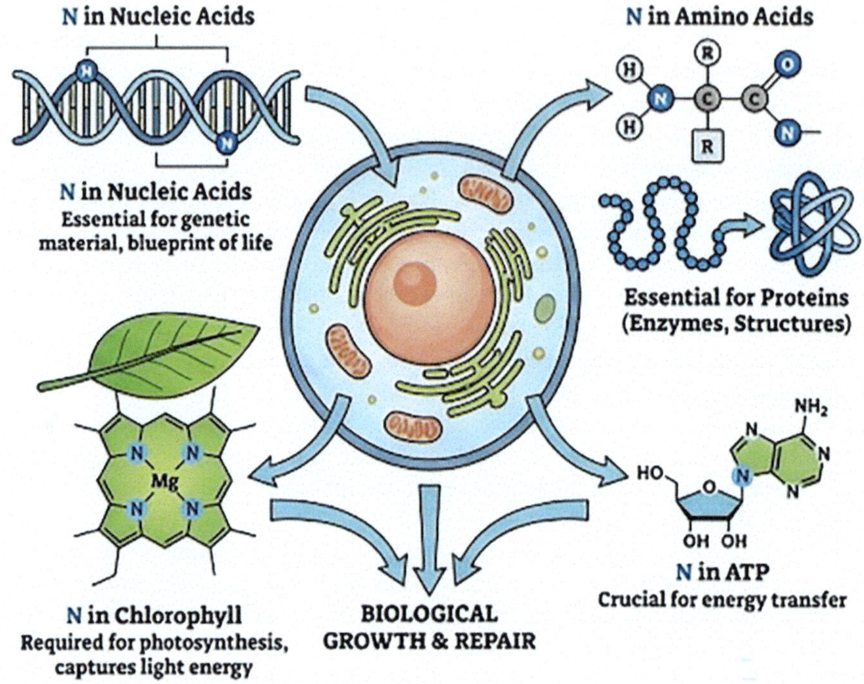

FIGURE 3.1

Roles of nitrogen (N) in the environment and biological systems.

Azospirillum and *Rhizobium*, which collaborate with legumes. The nitrogenase enzyme contains molybdenum, iron, and vanadium (Bueno Batista and Dixon, 2019).

Two nitrogen atoms joined by a triple covalent bond form the nitrogen molecule. Nitrogenase catalyzes the addition of three hydrogen atoms to every nitrogen atom. Because the nitrogenase enzyme can only function in an anaerobic environment, leg Hb is a pigment found in leguminous plants. It scavenges oxygen. For nitrogenase to function, an anaerobic environment must be created, and all oxygen must be consumed. Nitrogen is repaired using hydrogen donors. Hydrogen donors include pyruvic acid, glucose, sucrose, and numerous other substances. Energy is produced using ATP. The generated ammonia molecules must be bound by another carbon-containing substance (Ali et al., 2023). In the center of each mature nodule are billions of microorganisms that fix nitrogen. White or gray nodules are immature and do not fix nitrogen. The nitrogen is being fixed when nodules have a pink or reddish hue. Old nodules that aren't fixing nitrogen turn green in color. Plant roots can form an intracellular or intercellular relationship with cyanobacteria. The rate at which cyanobacteria and heterotrophic organisms fix nitrogen depends on the availability of micronutrients and the quality of carbon, respectively. On land, nitrogen fixation is slowed down by a low phosphorus supply (Rivai et al., 2021).

In nonsymbiotic relationships, bacteria fix nitrogen by breaking down their own cells and using carbon and energy from the surroundings. This mechanism is powered by the energy found in plant roots, specifically in the form of carbohydrates. Tillage, acidic soils, intercropping, fungicides, insecticides, nitrogen fertilizer, temperature, and moisture are some of the variables that affect biological nitrogen fixation. Aerobic organisms such as Methanosarcina and Clostridium are examples of free-living heterotrophs; heterotrophic microaerophiles include Frankia, Azospirillum, and Bradyrhizobium; and aerobic heterotrophic free-living organisms include Derxia and Azobacter (Melino et al., 2018). Microaerophilic Herbaspirillium and endophytic Acetobacter are examples of root-associated species, while Bradyrhizobium and Frankia form symbiotic relationships. Microaerophilic organisms such as Rhodospirillum and Bradyrhizobium, aerobic free-living autotrophs like Cyanobacteria, and symbiotic autotrophs such as Anabaena azalea and Cyanobacteria are examples of free-living autotrophs. Lightning can fix atmospheric nitrogen by combining nitrogen and oxygen to create NO, which subsequently combines with rainwater to make nitric or nitrous acid, which is then transformed into nitrates for plant uptake (Mahmud et al., 2020).

In the case of abiotic nitrogen fixation in soil, photochemical fixation can fix ten T g of nitrogen per year in the desert. Sometimes, more than half of the nitrogen available to the ecosystem comes from organic nitrogen, which is used in photochemical nitrogen fixation to create ammonium. Photochemical nitrogen fixation is essential because sunlight can fix organic nitrogen, which microbes cannot always convert into ammonium (Hashemi et al., 2023). Nitrification is the second stage of the nitrogen cycle, where ammonium is converted to nitrite via oxidation. Other nitrifying bacteria, such as Nitrobacter, further convert nitrite ions to nitrate ions. Plants can easily absorb both nitrite and nitrate, which are types of nitrogen. Microorganisms are responsible for the final stage of the nitrogen cycle, which involves converting nitrate into molecular nitrogen (Takai, 2019). Fig. 3.2 shows the transformation of nitrogen through various forms, making it available to and removed from the biosphere

3.5 What are the benefits, significance, and advancements of nitrogen use efficiency?

Excessive fertilizer use not only increases plant growth and yield but also raises production costs and damages the environment because nitrogen is necessary for crop production. This emphasizes how crucial it is that users understand nitrogen

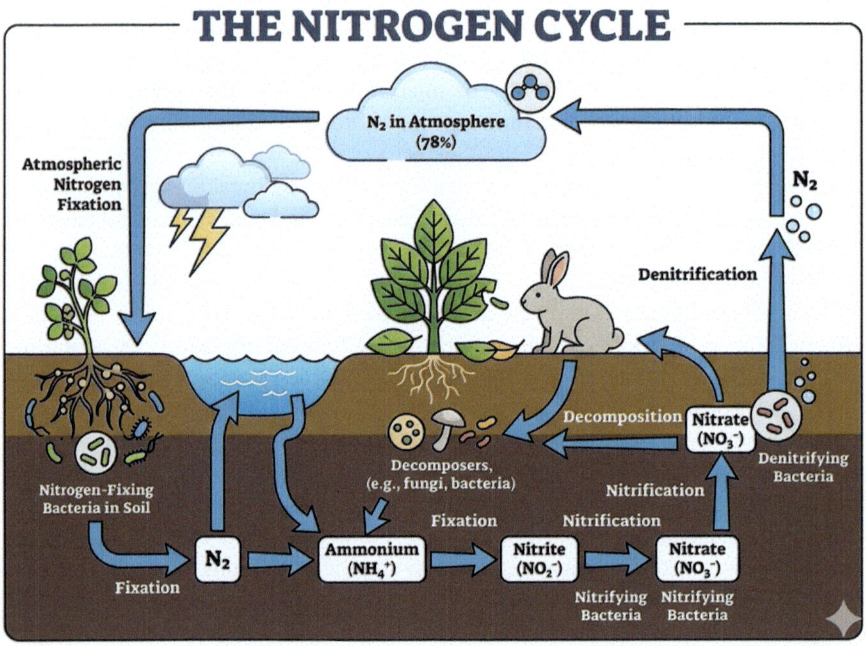

FIGURE 3.2

This diagram illustrates the nitrogen cycle, a critical biogeochemical process for sustaining life.

use efficiency and cultivate cultivars with higher nitrogen use efficiency. Because only 30%–50% of applied nitrogen is used by plants and the remainder is wasted, nitrogen use efficiency is frequently used to estimate the conversion of nitrogen inputs into agricultural products and to assess the risk of nitrogen losses to the wider environment. This is dependent upon factors such as nitrogen uptake, assimilation, and mobilization (Suri et al., 2021). Agronomists and genetic researchers have different definitions of nitrogen usage efficiency (NUE). The relationship between the amount of nitrogen that plants can use or absorb and the amount of grain produced per unit of nitrogen is known as NUE, according to agronomists. According to genetic researchers, NUE is associated with the quantity of genes necessary for increased nitrogen use efficiency, such as those that produce glutamine (Hashemi et al., 2023). Many plants absorb some of the nitrate through their roots, but most of it is transported to the shoot, where nitrate reductase in the cytoplasm first converts it to nitrite. The cytoplasmic and plastid-based glutamine synthetase (GS) and nitrite reductase subsequently converted it into ammonium. As a result, glutamine is essential for increasing NUE. By altering the expression of genes involved in nitrate transport, the Indica subspecies of rice was genetically modified to increase NUE through increased nitrate absorption. China employed the same genetic alterations to raise the rice cultivar's NUE (Rivai et al., 2021).

It can also be defined as the amount of nitrogen that is either stored by plants or released into the atmosphere as nitrous oxide. Because they store the most nitrogen, leguminous crops have a higher NUE; however, it's unclear if they release nitrogen into the atmosphere in trace amounts or at all. The amount of nitrogen that plants use and retain after harvesting is generally referred to as nitrogen utilization efficiency. Nitrogen usage efficiency is the total of the efficiency of using and absorbing nitrogen, including the efficiency of remobilization and absorption (Ali et al., 2023). For cereal crops, the global nitrogen utilization efficiency remained at 30%. Improvements in NUE aid in controlling the overuse of nitrogen. With better crop management, an estimate for England's fields shows that NUE can rise but will remain below 75%. The efficiency of nitrogen consumption varied depending on whether ammonium and nitrates were present. It was demonstrated that when ammonium levels were high, NUE decreased. Compared to wealthy countries, China's nitrogen use efficiency is 10% to 20% lower. Significant NUE has been demonstrated in microbes even when reduced nitrogen is present.

Out of the 83 Mt of nitrogen applied to cereal crops in prior years, Southeast Asia used 71%, and North East Asia used 32%. Nitrogen recovery is influenced by crop type, fertilizer type, cultural methods, and environmental factors. NUE is lower when farmers grow their crops in traditional ways. Furthermore, it is reduced when plant requirements and application timing are not well synchronized (typically applied sooner that is lost due to temperature or any other abiotic stress). NUE is also decreased by the nutrient's applications being balanced (Shahzad et al., 2019). NUE can be raised by fertilizers using a slow-release method. In Pakistan, it has been demonstrated that potassium fertilizer increases NUE. Cultural practices can be improved to increase NUE. Utilizing Nitrogen In detecting leaves, mycorrhizal fungus can boost efficiency, which is reliant on nitrogen transfer. NUE can also be raised by using genetically engineered varieties. Urea, which is covered with sulfur to boost its efficacy, is the most commonly used fertilizer in the world. To improve the efficiency of nitrogen use, granular urea can be substituted with a Urea Triazone solution, which dissolves or releases nitrogen gradually (Suri et al., 2021).

Applying nutrients and water directly to the plant's rhizosphere increases the NUE. A broad and vigorous rhizosphere also has a higher NUE, as rice tests have shown. We can employ cultivars with high NUE to protect the environment by analyzing their genetic makeup. The dynamics of soil nitrogen and the availability of nitrogen for plants are influenced by the extent of soil disturbance brought about by different tillage techniques. The conservation tillage (less or no tillage) approach typically results in higher NUE overall, while the link between the two varies across studies. Cotton yields were higher in conservation tillage plots than in conventional tillage plots with an ideal nitrogen treatment, according to a lengthy (10-year) study carried out in the southern United States (Shahzad et al., 2019). Slow-release nitrogen fertilizers may reduce nitrogen leaching and denitrification losses in addition to enhancing the timing of nitrogen release and uptake in response to crop needs. Similarly, the NUE has been enhanced using coated N sources like crotobylidene diurea and isobutylidene diurea, which are slow-release synthetic urea-based fertilizers. In winter wheat, N fertilizer injection, band application, and placement beneath the seeds during planting produced higher NUE and lower NH_3 volatilization as compared to broadcasting. Because they reduce pollution and raise crop yields, the previously described NUE-boosting components are crucial to sustainable agriculture. To get NUE, the plant's intrinsic capacity must be enhanced and chosen in a way that promotes effective uptake, utilization, and yield generation under circumstances where N supply is scarce or nonexistent (Melino et al., 2018).

3.6 Nitrogen symptoms and variations

3.6.1 Nitrogen deficiency

Reduced nitrogen levels in plants result in lower levels of chlorophyll and enzyme degradation. Protein breakdown causes changes in enzyme activity, which leads to leaf senescence. The amounts of Rubisco activase, endase, ATP synthesis, glutamine synthesis, and other enzymes decreased as a result of the reduced nitrogen. When leaves attain full senescence, the amount of mRNA dramatically decreases after rising in the early stages of senescence. Reduced agricultural plant development, primarily due to low nitrogen levels, affects overall crop output because chlorophyll breakdown causes yellowing of the leaves (Shahzad et al., 2019). This lack of nitrogen reduces the amount of protein in the seed and prevents cell growth, which first manifests as decreased yellowing of the leaves and, in severe cases, may cause leaf drop. Additionally, early maturity caused by a nitrogen shortage reduces yield even more. During vegetative growth, the leaf's surface area shrinks. Low fertilization rates may result in seed yields that fall short of expectations. There has been a 30% drop in seed production. It was discovered that the concentration and inflorescence of chlorophyll were decreased when photosynthetic equipment lacked nitrogen (Ali et al., 2023). The main metabolism of plants is reduced when there is insufficient nitrogen in the soil. Because there is less nitrogen available, transpiration and photosynthesis go more slowly. This has an effect on plant growth. Low nitrogen levels promoted blooming in Arabidopsis. Yield losses are most when the plant is nitrogen-deficient during the tailoring stage, but a lack of nitrogen has negligible effect on output during the heading and seedling stages (Hashemi et al., 2023).

3.6.2 Nitrogen excess

Plants suffer, and the ecosystem is disturbed when nitrogen levels are higher than necessary. It affects plant growth, development, and composition in addition to seed quality. By reducing the quantity of flavonoids and phenolics, increased nitrogen consumption lowers antioxidant capability. Overuse of nitrate increases plant biomass, but overuse of ammonium has a negative impact on plant growth. By turning leaves dark green, nitrogen toxicity extends the vegetative development cycle (Suri et al., 2021). The plant grows taller as the stem remains thin, which ultimately results in lodging. The plant's water use efficiency decreases due to nitrogen poisoning, which results in stunted development and burning of the leaves. Lesions appear on the leaves, stems, and roots in addition to the downward rolling of the leaves. Particle disintegration and the creation of hazardous substances known as carcinogens and allergens are caused by nitrogen toxicity (Hashemi et al., 2023).

3.7 Nitrogen and cereal crops

3.7.1 Nitrogen use efficiency in cereals

The largest quantity of arable land in the world is used to cultivate cereal crops. Furthermore, maize, rice, and wheat account for roughly 90% of grain output. The ecology is at risk because 60% of the 94 million tons of nitrogen that are applied to cereal crops annually are lost. In cereal crops, a plant's nitrogen utilization efficiency is determined by how well it absorbs nitrogen and how much of it it can use. During harvest, NUE can also be calculated. Moreover, higher nitrogen application lowers NUE. It can be enhanced by crop rotation and the availability of extra macronutrients and micronutrients. Nano-fertilizers can also be used to increase NUE (Salim and Raza, 2020).

3.7.2 Nitrogen deficiency in wheat

Wheat is one of the most significant staple crops, providing people with a lot of carbohydrates and good proteins (10%–15%). As more people move into cities and new industries are established, there is an increasing demand for this staple meal. Dietary fiber and vitamin B are also found in wheat. Older leaves first show signs of nitrogen stress by changing from green to light green before turning yellow due to a lack of chlorophyll. The entire field turns yellow due to nitrogen stress (Wang et al., 2021). Older leaves first show signs of nitrogen stress by changing from green to light green before turning yellow due to a lack of chlorophyll. The entire field looks yellow due to nitrogen stress. A lack of nitrogen in wheat reduces the synthesis of chloroplasts, which contain the photosynthetic pigment chlorophyll. This directly affects photosynthesis. Plant growth is slowed, crop yield is reduced, and leaf size is reduced, all of which interfere with the plant's physiology. Grain quality is affected when the amount of protein and amino acids decreases; the quantity of spikes and spikelets decreases as the supply of nitrogen rises and the demand for nitrogen for wheat harvests rises; grain weight also decreases; crop yield is lost due to decreased NUE and low reproductive development; nitrogen poisoning causes the wheat plants to lose their resistance and exhibit excessive vegetative growth; weed problems are more common in these types of fields (Melino et al., 2018).

3.7.3 Nitrogen deficiency in rice

One of the three major food crops is rice. About half of the world's population relies on rice for their diet. In terms of cultivable land, rice comes in second. Asia produces 90 percent of the world's rice. Ninety-eight percent of the world's rice production comes from Asia, Latin America, and Africa combined. When nitrogen demands rise, such as during the tailoring stage, rice experiences a nitrogen deficit. Furthermore, older leaves begin to change color from the tip, which causes the entire leaf to turn yellow. Furthermore, leaves may perish in extreme cases of N deprivation. The leaves' size is also impacted, as they get shorter and narrower, which disrupts the photosynthetic area and turns the leaves yellow. The enhanced foliage of rice plants under nitrogen toxicity draws insects and intensifies disease assault. Lodging occurs as the plant achieves maturity, causing the leaves to fold and form patches. Plants develop more slowly in the absence of photosynthesis, which depends on the quantity of chlorophyll. The structure of the chloroplast, which includes chlorophyll, is destroyed by nitrogen poisoning in rice plants (Khan et al., 2021). This figure (Fig. 3.3) presents a concise overview of the nitrogen deficiency in rice plant.

3.8 Nitrogen deficiency in maize

Maize is the third most grown crop in the world, behind wheat and rice. The Queen of Cereals takes some time to mature. It can be eaten by people and fed to animals as a feeder crop. It can be used for a variety of purposes, including making flour for baking, consuming it after roasting, and having a high fructose and amino acid content. We get a variety of nutrients from maize, as will be discussed below. Necrosis at the ends of older leaves and other nitrogen-deficient leaves that turn pale green are symptoms of the illness (Singh et al., 2020). We get a variety of nutrients from maize, as will be discussed below. Nitrogen stress led plants to develop more slowly, which decreased yield since nitrogen is an essential component of the cell protoplast. There were also more plants that were barren. The lack of nitrogen has a comparable effect on the biomass of the plant's shoots. Nitrogen-deficient maize leaves showed high spectrum reflection in a spectral and true color response experiment, which was found to be caused by a loss of chlorophyll (Adhikari et al., 2021).

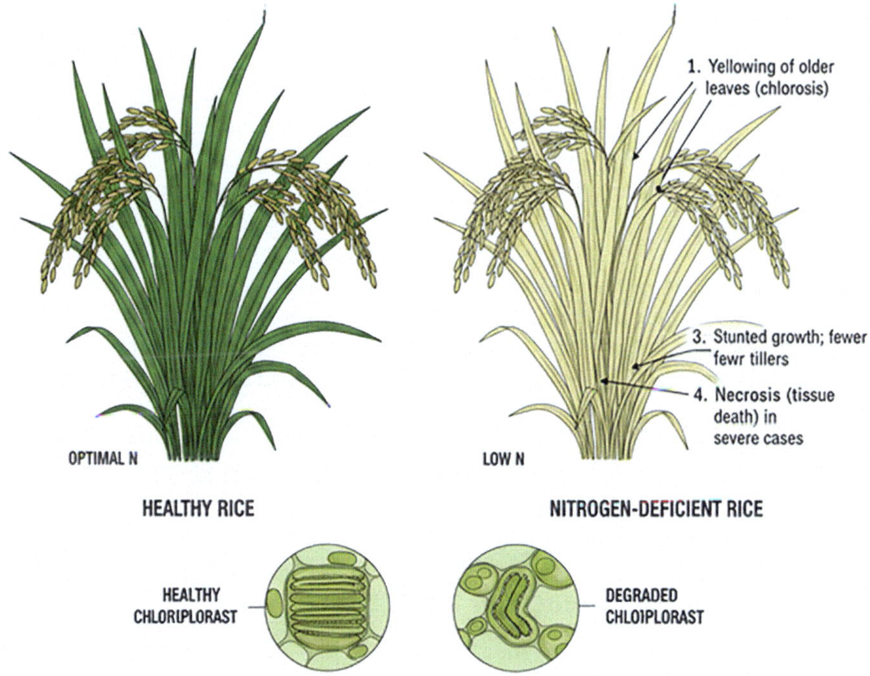

FIGURE 3.3

This figure presents a concise overview of the nitrogen (N) deficiency in rice plant.

Since nitrogen is a component of the cell protoplast and is crucial to plant growth, nitrogen stress causes plants to grow more slowly, which reduces output. Additionally, there were more barren plants. The plant's shoot biomass is similarly impacted by the nitrogen shortage. When maize leaves were evaluated in a spectral and true color response experiment, nitrogen-deficient leaves had strong spectral reflection, which was determined to be caused by a loss of chlorophyll. Scarlet eggplant growth is restricted by nutritional deficiencies, which change the efficiency of macronutrient absorption and utilization, resulting in lower dry matter output, plant height, stem diameter, leaf number, and leaf area (Ali et al., 2023). Insufficient nitrogen impairs the production of photosynthetic pigments, hinders cell division, and interferes with cell elongation in subsequent leaf growth. This in turn has an impact on the absorption of light and the metabolism of magnesium, which is essential for chlorophyll molecules. The rate of transpiration and the effectiveness of absorbing and utilizing other nutrients are similarly reduced by a nitrogen shortage. Furthermore, excessive nitrogen application reduces the pH of the soil, which can hinder crop growth and diminish yield (Salim and Raza, 2020).

3.9 Nitrogen deficiency in barley

A lack of nitrogen disturbs the plant because it is essential for the formation of barley. Nitrogen stress first manifests in older leaves, which start to change color at the tip. The lower foliage takes on a light green hue. When under continuous stress, the delicate green tint of older leaves turns white, while other leaves start to change from green. Finally, it lowers protein levels as well as yield. Nitrogen stress has been demonstrated to cause a 50% reduction in photosynthesis in barley. Additionally, it prevents shoots from growing (Salim and Raza, 2020). Plants under nitrogen stress produce more carbohydrates, proteins, and amino acids even when their rate of development slows. It has been observed that the leaves are smaller than those of plants that receive enough nitrogen. Plant growth is impacted by nitrogen poisoning, which also increases pathogenic and nonpathogenic stresses. Leaf yellowing is caused by excessive ammonia consumption, and in severely affected types, yield may decrease by up to 60%. Sometimes it can be fatal to the plants. Additionally, shoots and roots may droop (Adhikari et al., 2021).

3.10 Nitrogen deficiency in sorghum

After the four aforementioned crops, sorghum is the fifth most important crop, providing food for almost 750 million people. It thrives in hot settings and requires a temperature of 25°C for optimal growth. When water is scarce, it curls its leaves to conserve it. Grown on 44 million hectares in over 100 countries, sweet sorghum has a protein content of 17%. Because of its low production costs and ability to produce ethanol, it is a financially viable option. When sorghum plants are stressed by nitrogen, their growth slows, their leaves become chlorotic, and sometimes they die. It decreases the number of leaves and shortens the roots (Ahmad et al., 2022).

3.11 Nutritional and nitrogen fixation in legumes

The Leguminosae family, which has 800 genera, is the third largest family of flowering plants. Some members of this family known as pulses are categorized as grain crops while others are weeds. Legumes have two to three times the protein of cereal crops due to their high nutritious value. They are high in dietary fiber, carbs, fatty acids, vitamins, minerals, and folic acid in addition to protein. They are essential to millions of people's ability to survive. Around 121.5 million hectares and 82.4 million hectares of the world's cultivable land are used for the production of pulses and soybeans, respectively. Grain legumes are popular due to their higher protein or oil content (Schwember et al., 2019). With 15% of all pulse production, chickpeas rank as the third most significant crop. It is utilized to feed livestock and serves as an essential source of protein for small farmers. Cowpeas increase soil fertility and have a high nutritional value. Soybeans contain 20%–22% oil and 40%–42% protein. More than 60% of vegetable proteins come from soybeans. Lentils are grown all over the world because they provide the highest concentrations of essential minerals, vitamin B1, and amino acids. Its main manufacturers are Canada, Turkey, and India (Singh et al., 2020).

3.11.1 Nitrogen deficiency in legumes (chickpeas, cowpeas, lentils, beans)

Because of the absence of nitrogen, the plant's overall growth is limited and appears pale yellow. Upper-middle leaves cup and older leaves turn yellow, which might occasionally cause them to die. A lack of nitrogen in the late season lowers biomass, which in turn lowers lentil crop production. It is shown that lentils have a maximum nitrogen use efficiency of 110.1%. Early in the season, nitrogen deficit is seen during nodule growth, and late in the season, it can be caused by a number of biotic or abiotic reasons. Younger leaves are unharmed but older leaves show the first signs. The plant's predominant color in mild conditions is light green. The top three leaves are clearly yellowing (Schwember et al., 2019). The quantity of nodules and the dry weight of fava beans are reduced when nitrogen is used excessively. Furthermore, because nitrogen fertilizer is harmful flavonoid secretion is decreased in subterranean areas. In one experiment, the flavonoid content of a crop with nitrogen was 13.2% lower than that of a crop with nitrogen deficiency. Excess nitrogen has an impact on nodule growth and nitrogenase activity in soybeans. N toxicity reduces the amount of food that is traveling in the direction of the nodules. Lastly, there is a decrease in both nitrogen fixation and nitrogen use efficiency (Liu et al., 2019). Table 3.1 shows the nitrogen deficiency and their effect on different plants.

3.12 Vegetables and nitrogen

The basic process of photosynthesis depends on nitrogen, and any change in its quantity can interfere with this crucial activity. Although nitrogen fertilizer is essential for plant growth its application can result in nitrogen oxide and dinitrogen oxide emissions, which lowers crops' nitrogen use efficiency. Vegetable nitrogen use efficiency ranges from 5% to 67.3% for various crops, with an increase noted when manure and industrial nitrogen are mixed. Because they are picked before they reach the reproductive stage leafy plants like spinach and lettuce require nitrogen throughout their vegetative growth phase in order to build their leaf size quickly (Schwember et al., 2019). Smaller leaves could be the result of not applying nitrogen to spinach during its early growth stage. On the other hand, tuber vegetables need the same amount of nitrogen in both the vegetative and tuber stages. For the best growth and development, cucurbits need nitrogen levels between 110 kg and less than 220 kg per acre. While potatoes show maximum productivity with nitrogen

Table 3.1 Nitrogen deficiency and their effect on different plants.

Nitrogen deficiency plant	Effects on plants	Key references
Sorghum	N is highly mobile, leading to initial light green color and early senescence of older leaves as the plant mobilizes N to younger tissues	Zhao et al. (2005)
Arabidopsis plant	Confirms that reduced chlorophyll results in a pale yellow color (chlorosis), and older leaves turn completely yellow	Huang et al. (2020)
Oryza Sative L.	Highlights the progression from pale green to yellowing, starting from older leaves, and eventually leading to more serious symptoms like necrosis and the stems acquiring purple hues	Chen et al. (2024)
Maize	Slow growth and uniform yellowing of older leaves are the first and most common symptoms.	Amoah et al. (2025)
Wheat	N deficiency significantly reduced leaf area (LA) and chlorophyll (Chl) content, resulting in lower biomass production (dry weight).	Ali (2023)

application rates between 100 and 150 kg per hectare, onions grow and yield at their best with 100 kg per hectare (Wang et al., 2021). Radishes need 2.15 kg of nitrogen per 1000 kg of fleshy leaves for optimal growth, whereas tomatoes need 240 kg of nitrogen for better growth, development, and output. Vegetables are essential for human growth because they contain minerals, vitamins, plant chemicals, and dietary fiber. Additionally, they reduce the risk of diabetes, heart disease, and certain types of cancer while also improving vision. Crucifers have 50% dietary fiber and 40% nonstarch components by dry weight, making onions the family's largest source of dietary fiber. Members of the Alliaceae family contain 5% sulfur by dry weight. In First World countries, tomatoes are a source of potassium and help protect the skin, and potatoes provide a lot of carbs (they contain 7% starch when cooked and 13% when chilled) and important amino acids (Singh et al., 2020).

3.12.1 Nitrogen deficiency and toxicity in vegetable crops (cucurbits, onion, potato, radish, tomato)

While tomatoes require 240 kg of nitrogen for improved growth, development, and yield, radishes require 2.15 kg of nitrogen per 1000 kg of fleshy leaves for optimal growth. Because they include nutritional fiber, vitamins, plant compounds, and minerals, vegetables are vital for human growth. They also improve vision and lower the risk of diabetes, heart disease, and some types of cancer. Onions are the family's biggest source of dietary fiber, with crucifers having 50% dietary fiber and 40%

nonstarch components by dry weight. Alliaceae members have 5% sulfur by dry weight. While tomatoes require 240 kg of nitrogen for improved growth, development, and yield, radishes require 2.15 kg of nitrogen per 1000 kg of fleshy leaves for optimal growth. Because they include nutritional fiber, vitamins, plant compounds, and minerals, vegetables are vital for human growth. Additionally, they reduce the risk of diabetes, heart disease, and certain types of cancer while also improving vision. Crucifers have 50% dietary fiber and 40% nonstarch components by dry weight, making onions the family's largest source of dietary fiber. Members of the Alliaceae family contain 5% sulfur by dry weight. Potatoes supply a lot of carbohydrates (they contain 7% starch when cooked and 13% when cooled), essential amino acids, and potassium, which helps protect the skin in First World countries (Schwember et al., 2019). Notably, radishes have elevated anthocyanin levels under nitrogen stress, which is a sign of physiological reactions to nutrient deficits, and cotyledons can die from a nitrogen shortage. Likewise, in tomatoes, nitrogen stress causes the nitrogen content of the leaves to drop, which lowers the amount of nitrogen in the tissues and, in turn, lowers the amount of biomass produced by the plant. Leaves vary in hue from green to less green. Tomato plants' metabolic processes are disrupted by nitrogen stress. Ammonium poisoning alters the PH between cells, throws off the osmosis balance, and prevents the plant from getting nutrients. It causes the plant's calcium and magnesium levels to drop (Ahmad et al., 2022).

The primary component of photosynthesis, chlorophyll production, is reduced when ammonium is poisonous. As a result, plants develop more slowly because photosynthesis is slowed down and less food is produced. The plant's underground portions and leaves also exhibit signs of necrosis and yellowing. The lowest and middle leaves have a curvature. Additionally, fruits get smaller. Onions that are exposed to nitrogen toxicity have weakened immune systems and are more susceptible to biotic and abiotic stress. Overuse of nitrogen delays onion development, resulting in two centers at the bulb. A reduction in yield is the result of nitrogen poisoning (Schwember et al., 2019). When too much nitrogen is used, plant growth first increases and then continues to decline. Additionally, there is a decrease in nitrogen use efficiency. The radish's ammonium toxicity results in a reduction in photosynthesis. Additionally, it slows down the rate of transpiration. There is less buildup of dry materials in the case of subterranean portions. However, the chloroplasts' elevated ammonium content prevents materials from passing through and influences the plants' enzymatic activity. Nitrogen poisoning results in a reduction of the plant's resistance to biotic stress. It alters the amount of vitamins in fruits, impacts secondary metabolites of plants, and degrades fruit quality (Adhikari et al., 2021).

3.13 Nitrogen and fruits

Fruit growth and development are greatly influenced by nitrogen because it is a part of the cell wall and is necessary for plant metabolism. The edible part of a plant is referred to as "fruit" in biology. Fruits are good for people because they include dietary fiber, which reduces cholesterol. Because they provide a high concentration of vitamins, minerals, and plant chemicals, they are thought to be protective. Fruits include 61%–89.1% water, 0.5%–1.1% protein, 4.4%–34.8% fat, 4.4%–34.8% sugar, 2%–14.8% dietary fiber, and trace amounts to 3% carbohydrate. Citrus, bananas, and apples have protein contents of 0.17%, 0.18%, and 0.15%, respectively. About carbohydrates (Ahmad et al., 2022).

3.13.1 Nitrogen deficiency in fruits (apple, banana, citrus, grapes, mango)

When apples lack nitrogen, their leaves become thinner and their color changes steadily without streaks or blotches. Moreover, symptoms that initially manifest on older leaves before progressing to the higher leaves may cause necrosis in more severe cases. Terminal shoots are shorter, even though they are visible. In terms of fruit development, apples deficient in nitrogen yield fewer, smaller, and earlier-ripening fruits. Furthermore, plants deficient in nitrogen produce fruits with reduced nutritional content and quality (Singh et al., 2020). A lack of nitrogen causes the amount of chlorophyll in leaves to drop, which causes the leaves to lose their green hue. Lower leaves first exhibit symptoms, and in more severe cases, upper leaves turn pink. The stem thins out and forms a pseudo-stem. Petioles are becoming smaller, and leaves have a shorter lifespan. There is also a reduction in plant growth. Initially affecting the lower leaves, the symptoms of the shortage spread to the upper leaves, which turn light green to yellow. New leaves are more thin and smaller. Furthermore, older leaves turn yellow prior to shedding, and green striations show up on the leaves. It also causes fruits to turn pale in hue. Furthermore, older leaves turn yellow prior to shedding, and green striations show up on the leaves. It also causes fruits to turn pale in hue. A decrease in growth may cause the nitrogen-deficient plants to stay smaller. Occasionally, a portion of the leaf stays green while the veins become yellow. While other plant components display crimson or pink hues, the leaves transition from green to light green and then yellow. There is a decrease in shoot growth. Fruits reduce the amount of nitrogen accessible to yeast (Hashemi et al., 2023).

This occurs in leaves as a result of low protein and plastid breakdown, which reduces chlorophyll levels and results in chlorosis. Plant vegetative growth has decreased. Like other fruits, this one has inadequacies that start with older leaves and go on to younger ones. In addition, in severe conditions, the entire plant becomes yellow. Reduced chlorophyll content inhibits vegetative growth, which in turn inhibits plant growth. Finally, the production of mangoes is declining (Suri et al., 2021). During nitrogen toxicity, bananas form short pseudo stems that bend when bearing fruit, and fruits with low calcium levels are produced as a result of high nitrogen consumption. Reduced Nitrogen Use Efficiency and higher soil pH coincide with this later-season delay in the commencement of reproductive growth.

Nitrogen poisoning in citrus causes decreased enzyme activity, weakened cell membranes, and impaired root conduction of nitrates toward the leaves. Furthermore, nitrogen toxicity-induced increased vegetative growth reduces the time needed for reproductive growth, which eventually results in lower yield (Khan et al., 2021).

3.14 Nitrogen and climate

The nitrogen cycle is significantly influenced by human activity, and different types of nitrogen affect the climate, which in turn leads to environmental degradation. While some kinds of nitrogen cause warming, others have cooling effects. One important type of nitrogen that is essential for cooling the environment is nitrogen. Like ozone, which absorbs solar radiation N_2 gas strengthens the ozone layer, retaining more sunlight and producing cooling effects. On the other hand, as more rays reach the Earth, nitrous oxide, a chemical that destroys the ozone layer causes a warming effect. Nitrous oxide and nitrogen are both categorized as greenhouse gases. By creating algal blooms upsetting aquatic life and degrading our soil too much nitrogen can pollute the air and water. As a result, ammonia, nitrous oxide, nitrogen dioxide, and nitric oxide are among the gases released as a result of nitrogen's toxicity and are the cause of pollution (Fremantle et al., 2022).

3.14.1 Effects on the atmosphere: ammonia and nitrous oxide emissions

Ammonia is an unpleasant-smelling gas that is hazardous for the environment and people because it irritates the skin and eyes, and in high concentrations, it messes up our lungs and airways. Additionally, it results in death, while in the case of the eyes it damages them permanently. In addition to causing heart problems, ammonia can occasionally trigger heart attacks. The synthesis of certain enzymes such as aconitase, NADPH-oxidase, ribonucleotide reductase, and other oxidoreductases is inhibited by oxides of nitrogen (NO). Elevated NO_2 levels can seriously harm the lungs and prolong and worsen common viral infections (Adhikari et al., 2021).

Recent estimates indicate that NO_2 pollution causes 4.0 (1.8–5.2) million new cases of asthma in children annually, or 13% (range; 6%–16%) of all asthma cases worldwide. In terms of the environment, it is deposited in the air, soil, and lakes. It makes the lake water more acidic, which can kill aquatic life. Then, problems may arise from the release of ammonia into the soil. Any kind of nitrogen emission from the soil results from a disturbance in the carbon-to-nitrogen ratio. There are two methods for ammonia to enter the soil: wet deposition and dry deposition (Schwember et al., 2019). When dry deposition occurs, ammonia mixes with atmospheric dust particles and is pulled to the ground by gravity. Ammonia and precipitation combine to form wet deposition, where the agricultural sector is responsible for up to 70% of all anthropogenic emissions. This is mainly due to the overuse of fertilizers, particularly nitrogen, and livestock waste or excreta (Singh et al., 2020).

3.14.2 Nitrogen's impact on the aquatic environment

Since they dissolve in water, ammonium, nitrate, nitrite, and ammonia can seep into rivers, lakes, and groundwater. Nitrate can have a number of negative health impacts on people after they drink this water, however ammonium is usually thought to be nontoxic. When a child is more than three months old, drinking water tainted with nitrate can cause birth defects, chest cancer, and other issues. The remaining tainted water is then put into the lakes, endangering aquatic life. Aquatic plants and algae grow and produce more because the water has more nitrogen in the form of nitrate. Additionally, there is more competition in the aquatic habitat for food and other necessities (Fremantle et al., 2022). Algal bloom is the term for the situation where the growth of algae is so accelerated that it covers the water's surface. Additionally, because the algae bloom consumes a large portion of the aquatic ecosystem's available food and minerals and functions as a translucent object, light cannot penetrate the water. Some poisonous blooms release toxic substances in addition to these regular blooms. These pollutants have detrimental impacts on the land ecology in addition to harming aquatic life. By entering the human body through fish, these hazardous substances that humans eat can cause a variety of issues and occasionally even death (Khan et al., 2021). By devouring fish impacted by the algal bloom or drinking tainted water, they also harm other creatures, such as dogs, birds, etc. When an algal bloom dies, the amount of carbon dioxide in the water rises, and the amount of oxygen available decreases, which is lethal to aquatic life. Aquatic animals are impacted not just by algal blooms but also by the direct introduction of various kinds of nitrogen. This affects the creatures' way of life in a number of ways, including reducing the amount of food they can eat and making calcium carbonate, which is required to make some species' skeletons, more acidic (Ahmad et al., 2022).

3.14.3 Impact on biodiversity and the carbon cycle

One of the main causes of the carbon cycle disruption is the overuse of nitrogen. In order to help bacteria break down and release carbon dioxide, nitrogen is essential. While plants that cannot withstand high nitrogen levels die, those that do flourish. Additionally, the imbalance has a detrimental effect on a number of species, making some more vulnerable to drought and causing them to grow poorly under less drought stress. Overdosing on nitrogen prevents it from being absorbed by the plants and from leaking down with the water. As a result, it gets incorporated into the soil and undergoes several reactions that lower the pH of the soil and produce acidification (Ahmad et al., 2022). Plants have many problems absorbing that aluminum because of its toxicity. In addition to internally interfering with respiration, metabolism, cell division, and many other physiological processes, it affects the plant's leaves, roots, and shoots. It also prevents nutrients like phosphorus, magnesium, and others from being absorbed. Nitrogen toxicity causes soil salinity, which modifies plant morphology, physiology, and molecular structures, thereby affecting plant growth. It also modifies the morphology of the soil (Suri et al., 2021). The length and strength of roots are reduced when excessive fertilizers are introduced to the soil. In a similar vein, they are poisonous and even lethal to soil microbes. Soil erosion is supported by both of these factors. Compacted soil is held together by a variety of processes, including plant roots and soil microbes. Therefore nutrient-rich topsoil that flows with water into rivers and oceans can be removed by wind, water, snow, and animals. We refer to this as soil erosion. As a result, the soil behind is unsuitable for plant growth, and the resultant plants show signs of various nutritional deficiencies and poor health (Khan et al., 2021).

Similarly, when the compact layers of ground are held together by no microbes and have short roots. Topsoil will then flow readily, and because it lacks minerals. The weak soil that remains will not be conducive to plant growth. The leaching of other minerals is another issue associated with too much nitrogen. When nitrate leaches down, it takes those minerals with it because it mixes with other minerals like calcium. It aids in the addition of harmful metals to plants as well. Toxic metals like lead and cadmium change forms when the pH is decreased. Furthermore, these forms become accessible to plants and have the potential to be lethal. Additionally, it lowers the soil's organic carbon content (Ahmad et al., 2022). Rainwater turns acidic as a result of nitrogen being released into the atmosphere. Furthermore, plants become more acidic when it rains, which can result in a number of issues, including the body of the plant losing minerals and possibly dying. Not only do plants suffer from the effects of acid rain, but microbial communities are also impacted. The ecosystems are in balance. The process by which bacteria transform nitrogen into nitrogen gas and nitrogen oxide is known as denitrification (Suri et al., 2021).

3.14.4 Effects of climate change on the nitrogen cycle

Human activities such as the overuse of nitrogenous fertilizers, deforestation, construction sites, burning of fossil fuels, pharmaceutical businesses, and the cultivation of leguminous crops are disrupting the nitrogen cycle. It has been discovered that burning fossil fuels adds nitrogen to the atmosphere by producing nitrous oxide gas. Thus acid rain, which is detrimental to plant health, is a result of this nitrous oxide gas. Over the past three centuries, there have been disruptions to the nitrogen cycle. One of the main causes of this nitrogen cycle disruption is the burning of fossil fuels, which releases reactive nitrogen into the atmosphere (Fremantle et al., 2022). The nitrogen cycle is being upset by human activities such as the usage of nitrogenous fertilizers, deforestation, building sites, burning fossil fuels, pharmaceutical companies, and the production of leguminous crops. By generating nitrous oxide gas, burning fossil fuels has been found to contribute nitrogen to the atmosphere. As a result, this nitrous oxide gas causes acid rain, which is bad for plants. The nitrogen cycle has been disturbed during the last three centuries. Fossil fuel combustion, which releases reactive nitrogen into the atmosphere, is one of the primary reasons for this disturbance of the nitrogen cycle (Ahmad et al., 2022).

These medications' waste is dumped into water bodies, where nitrifying bacteria are found. The nitrifying bacteria that are present during the nitrification process are disturbed by these additional compounds. According to reports, the nitrification process eliminates dangerous chemical substances. Since a wide range of bacteria are involved in the ammonification process, nitrification uses compounds that are more vulnerable to damage. While some medicinal compounds have no effect on ammonia production, others disrupt ammonification. The production of NH_3- is impacted when the nystatin and pharmacy complex is present in the water in concentrations of 100 milligrams per kilogram (Singh et al., 2020). Similarly, the production of NH_3- is decreased when the amount of erythromycin is 20 mg per kilogram of water. The presence of benzyl penicillin, however, had no effect. Pharmaceutical sector chemicals have less of an impact on denitrification, but some kinds of chemicals can stop the process entirely. For instance, the denitrification process is totally stopped when 250 mg of acetaminophen is present in one liter of water. Anammox aids in the fixation of nitrogen. Although there are five varieties of these bacteria, they are all susceptible to abiotic stressors and contaminants, such as those found in medications (Liu et al., 2019).

3.15 Conclusion

This chapter underscores the critical, dual role of nitrogen (N) metabolism in mediating plant growth and tolerance to major abiotic stressors, particularly salinity. We established that N availability operates along a narrow optimal window, where appropriate supplementation can synergistically mitigate stress-induced damage (e.g., chlorophyll loss), while over-application severely exacerbates it. By detailing the physiological roles of N in chloroplasts, specific toxicity/deficiency symptoms across diverse crop types, and the nuanced impact of different N sources on toxic ion balance (Na vs. C), this chapter illuminates the complexity of N management under stress. Ultimately, achieving sustainable crop productivity and effectively combating global soil salinity requires substantial, multi-faceted improvements in NUE. Future efforts must focus on developing superior N-consuming crop varieties, selecting low-loss fertilizer sources, and employing precise application strategies to harmonize plant nutrition with environmental resilience and health.

Reference

Adhikari, K., Bhandari, S., Aryal, K., Mahato, M., Shrestha, J., 2021. Effect of different levels of nitrogen on growth and yield of hybrid maize (*Zea mays* L.) varieties. Journal of Agriculture and Natural Resources 4 (2), 48–62. Available at https://doi.org/10.3126/janr.v4i2.33656.

Ahmad, I., Zhu, G., Zhou, G., Song, X., Hussein Ibrahim, M.E., Ibrahim Salih, E.G., 2022. Effect of N on growth, antioxidant capacity, and chlorophyll content of sorghum. Agronomy 12 (2), 501. Available at https://doi.org/10.3390/agronomy12020501.

Ali, A.M., 2023. Inducing nitrogen deficiency at early growth stages of wheat favors high yield and nitrogen recovery efficiency. Journal of Plant Nutrition 46 (7), 1368–1376. Available at https://doi.org/10.1080/01904167.2022.2061992.

Amoah, J.N., Keitel, C., Kaiser, B.N., 2025. Nitrogen deficiency impacts growth and modulates carbon metabolism in maize. Planta 262 (4), 94.

Bianco, G.V., Sacchetti, A., Grande, M., D'Orazio, A., Milella, A., Bruno, G., 2022. Effective hole conductivity in nitrogen-doped CVD-graphene by singlet oxygen treatment under photoactivation conditions. Scientific Reports 12 (1). Available at https://doi.org/10.1038/s41598-022-12696-2.

Bojović, B., Marković, A., 2009. Correlation between nitrogen and chlorophyll content in wheat (*Triticum aestivum* L.). Kragujevac Journal of Science 31, 69–74.

Bueno Batista, M., Dixon, R., 2019. Manipulating nitrogen regulation in diazotrophic bacteria for agronomic benefit. Biochemical Society Transactions 47 (2), 603–614. Available at https://doi.org/10.1042/bst20180342.

Chen, L.H., Xu, M., Cheng, Z., Yang, L.T., 2024. Effects of nitrogen deficiency on the photosynthesis, chlorophyll a fluorescence, antioxidant system, and sulfur compounds in *Oryza sativa*. International Journal of Molecular Sciences 25 (19). Available at https://doi.org/10.3390/ijms251910409.

Elsiddig, A.M.I., Zhou, G., Zhu, G., Nimir, N.E.A., Suliman, M.S.E., Ibrahim, M.E.H., et al., 2023. Nitrogen fertilizer promoting salt tolerance of two sorghum varieties under different salt compositions. Chilean Journal of Agricultural Research 83 (1), 3–13. Available at https://doi.org/10.4067/S0718-58392023000100003.

Fremantle, K., De Vito, L., Williams, B., Longhurst, J., 2022. A framework for assessing the contribution of UK local authorities toward compliance with the EU ambient air quality directive. Journal of Environmental Planning and Management 65 (2), 356–373. Available at https://doi.org/10.1080/09640568.2021.1888698.

Gojon, A., 2017. Nitrogen nutrition in plants: rapid progress and new challenges. Journal of Experimental Botany 68 (10), 2457–2462. Available at https://doi.org/10.1093/jxb/erx171.

Greaver, T.L., Clark, C.M., Compton, J.E., Vallano, D., Talhelm, A.F., Weaver, C.P., et al., 2016. Key ecological responses to nitrogen are altered by climate change. Nature Climate Change 6 (9), 836–843. Available at https://doi.org/10.1038/nclimate3088.

Hashemi, S.E., Madahhosseini, S., Pirasteh-Anosheh, H., Sedaghati, E., Race, M., 2023. The role of nitrogen in inducing salt stress tolerance in *Crocus sativus L.*: assessment based on plant growth and ions distribution in leaves. Sustainability 15 (1), 567. Available at https://doi.org/10.3390/su15010567.

Huang, C.H., Singh, G.P., Park, S.H., Chua, N.H., Ram, R.J., Park, B.S., 2020. Early diagnosis and management of nitrogen deficiency in plants utilizing Raman spectroscopy. Frontiers in Plant Science 11. Available at https://doi.org/10.3389/fpls.2020.00663.

Husen, A, 2023. Nanomaterials and Nanocomposites Exposures to Plants (Response, Interaction, Phytotoxicity and Defense Mechanisms). Springer, Singapore. Available at https://doi.org/10.1007/978–981-99–2419-6.

Husen, A, 2025. Nanobiotechnology for Abiotic Stress Adaptation and Mitigation in Agricultural Crops. Springer, Singapore. Available at https://doi.org/10.1007/978–981-96–9298-9.

Hussain, S., Shaukat, M., Ashraf, M., Zhu, C., Jin, Q., Zhang, J., 2019. Salinity stress in arid and semi-arid climates: effects and management in field crops. In: Hussain, S. (Ed.), Climate Change and Agriculture. IntechOpen. Available at 10.5772/intechopen.87982.

Jiang, J., Liu, W., Yang, Z., Ma, J., Zhang, C., Zhang, X., 2023. Effects of irrigation and nitrogen on growth and water productivity of maize grown in saline soil. Archives of Agronomy and Soil Science 69 (8).

Khan, Z., Nauman Khan, M., Luo, T., Zhang, K., Zhu, K., Rana, M.S., et al., 2021. Compensation of high nitrogen toxicity and nitrogen deficiency with biochar amendment through enhancement of soil fertility and nitrogen use efficiency promoted rice growth and yield. GCB Bioenergy 13 (11), 1765–1784. Available at https://doi.org/10.1111/gcbb.12884.

Lavrentieva, N.N., Dudaryonok, A.S., 2022. Nitrogen dioxide line shift coefficients induced by air pressure. Federation Molecular Physics 120 (9). Available at https://doi.org/10.1080/00268976.2022.2052370.

Liu, Y., Yin, X., Xiao, J., Tang, L., Zheng, Y., 2019. Interactive influences of intercropping by nitrogen on flavonoid exudation and nodulation in faba bean. Scientific Reports 9 (1), 2045–2322. Available at https://doi.org/10.1038/s41598-019-41146-9.

Mahmud, K., Makaju, S., Ibrahim, R., Missaoui, A., 2020. Current progress in nitrogen fixing plants and microbiome research. Plants 9 (1). Available at https://doi.org/10.3390/plants9010097.

Melino, V.J., Casartelli, A., George, J., Rupasinghe, T., Roessner, U., Okamoto, M., et al., 2018. RNA catabolites contribute to the nitrogen pool and support growth recovery of wheat. Frontiers in Plant Science. Available at https://doi.org/10.3389/fpls.2018.01539.

Pareek, S., Sagar, N.A., Sharma, S., Kumar, V., Agarwal, T., González-Aguilar, G.A., 2017. Chlorophylls: chemistry and biological functions. Fruit and Vegetable Phytochemicals: Chemistry and Human Health. John Wiley & Sons, Ltd, pp. 269–284.

Razaq, M., Zhang, P., Shen, H.L., Salahuddin, 2017. Influence of nitrogen and phosphorous on the growth and root morphology of *Acer mono*. PLoS One 12 (2). Available at https://doi.org/10.1371/journal.pone.0171321.

Rivai, R.R., Miyamoto, T., Awano, T., Takada, R., Tobimatsu, Y., Umezawa, T., et al., 2021. Nitrogen deficiency results in changes to cell wall composition of sorghum seedlings Scientific Reports 11 (1). Available at https://doi.org/10.1038/s41598-021-02570-y.

Salim, N., Raza, A., 2020. Nutrient use efficiency (NUE) for sustainable wheat production: a review. Journal of Plant Nutrition 43 (2), 297–315. Available at https://doi.org/10.1080/01904167.2019.1676907.

Schwember, A.R., Schulze, J., del Pozo, A., Cabeza, R.A., 2019. Regulation of symbiotic nitrogen fixation in legume root nodules. Plants 8 (9), 333. Available at https://doi.org/10.3390/plants8090333.

Shahzad, A.N., Qureshi, M.K., Wakeel, A., Misselbrook, T., 2019. Crop production in Pakistan and low nitrogen use efficiencies. Nature Sustainability 2 (12), 1106–1114. Available at https://doi.org/10.1038/s41893-019-0429-5.

Siddiqi, K.S., Husen, A., Zahra, N., Moheman, A., 2025. Harnessing silicon nanoparticles and various forms of silicon for enhanced plant growth performance under salinity stress: application and mechanism. Discover Nano 20 (1). Available at https://doi.org/10.1186/s11671-025-04270-2.

Singh, A.K., Singh, A.K., Singh, J.P., 2020. Boron in crop production from soil to plant system: a review. Archives of Agriculture and Environmental Science 5 (2), 218–222. Available at https://doi.org/10.26832/24566632.2020.0502020.

Soliman, M., Elkelish, A., Souad, T., Alhaithloul, H., Farooq, M., 2020. Brassinosteroid seed priming with nitrogen supplementation improves salt tolerance in soybean. Physiology and Molecular Biology of Plants 26 (3), 501–511. Available at https://doi.org/10.1007/s12298-020-00765-7.

Stein, L.Y., Klotz, M.G., 2016. The nitrogen cycle. Current Biology: CB 26, 94–98.

Sun, J., Li, W., Li, C., Chang, W., Zhang, S., Zeng, Y., et al., 2020. Effect of different rates of nitrogen fertilization on crop yield, soil properties and leaf physiological attributes in banana under subtropical regions of China. Frontiers in Plant Science 11. Available at https://doi.org/10.3389/fpls.2020.613760.

Suri, N., Zhang, Y., Gieg, L.M., Ryan, M.C., 2021. Denitrification biokinetics: towards optimization for industrial applications sustain. Frontiers in Microbiology 12. Available at https://doi.org/10.3389/fmicb.2021.610389.

Takai, K., 2019. The nitrogen cycle: a large, fast, and mystifying cycle. Microbes and Environments 34 (3), 223–225.

Wang, Y., Wang, D., Tao, Z., Yang, Y., Gao, Z., Zhao, G., et al., 2021. Impacts of nitrogen deficiency on wheat (*Triticum aestivum* L.) grain during the medium filling stage: transcriptomic and metabolomic comparisons. Frontiers in Plant Science 12. Available at https://doi.org/10.3389/fpls.2021.674433.

Zhang, H., Zhao, Q., Wang, Z., Wang, L., Li, X., Fan, Z., et al., 2021. Effects of nitrogen fertilizer on photosynthetic characteristics, biomass, and yield of wheat under different shading conditions. Agronomy 11 (10), 1989. Available at https://doi.org/10.3390/agronomy11101989.

Zhao, D., Reddy, K.R., Kakani, V.G., Reddy, V.R., 2005. Nitrogen deficiency effects on plant growth, leaf photosynthesis, and hyperspectral reflectance properties of sorghum. Journal of Agronomy 22 (4), 391–403. Available at https://doi.org/10.1016/j.eja.2004.06.005

CHAPTER

4 Role of phosphorus in resistance to abiotic stresses

Hamzeh Amiri

Department of Biology, Lorestan University, Khorramabad, Iran

4.1 Introduction

Phosphorus (P) is an indispensable macronutrient fundamental to all life on Earth. In plants, phosphorus is a structural component of key biomolecules, including nucleic acids (DNA and RNA), phospholipids that form cellular membranes, and the energy currency adenosine triphosphate (ATP) (Veneklaas et al., 2012). Beyond these basic physiological roles, phosphorus is increasingly recognized as a central regulator of sophisticated metabolic pathways and signal transduction networks that govern plant growth, development, and responses to the environment. However, the availability of soluble inorganic phosphate (Pi), the primary form taken up by plant roots, is severely limited in most natural and agricultural soils. This limitation arises from its propensity to form insoluble complexes with cations such as aluminum, iron, and calcium in acidic and alkaline soils, respectively (Shen et al., 2011). Consequently, phosphorus deficiency is one of the most critical constraints on global agricultural productivity.

Compounding the challenge of innate soil phosphorus scarcity, modern agriculture faces the escalating threat of abiotic stresses driven by climate change and environmental degradation. Factors such as drought, salinity, extreme temperatures, and heavy metal contamination in soils are becoming more frequent and severe, posing a monumental threat to global food security (Fahad et al., 2017). These stresses disrupt cellular homeostasis, inhibit photosynthesis, impair nutrient acquisition, and can ultimately lead to significant yield losses. Traditionally, plant responses to nutrient deficiency and abiotic stresses have been studied in isolation. However, a growing body of evidence reveals intricate and synergistic crosstalk between nutrient signaling pathways and stress response mechanisms (Mohammadi et al., 2021).

It is within this context that the role of phosphorus in plant resistance to abiotic stresses emerges as a critical area of research. Phosphorus status profoundly influences a plant's ability to perceive, signal, and mount an effective defense against environmental adversities. For instance, an adequate phosphorus supply is known to promote a deeper and more extensive root system, which is a primary

Mineral Nutrients to Mitigate Abiotic Stress in Crops. DOI: https://doi.org/10.1016/B978-0-443-44712-9.00004-9

adaptive trait for overcoming both water and nutrient deficits (López-Bucio et al., 2003). Furthermore, phosphorus is integral to the antioxidant system, helping to mitigate the oxidative damage caused by reactive oxygen species (ROS), which are generated at elevated levels under most abiotic stresses (Tewari et al., 2020). The phosphorus signaling network, orchestrated by key regulators like the PHOSPHATE STARVATION RESPONSE (PHR) transcription factors, directly interacts with stress hormone pathways, such as those mediated by abscisic acid (ABA), to coordinate a tailored response to combined challenges (Rouached et al., 2010; Wang et al., 2025).

This chapter aims to synthesize current knowledge on the multifaceted role of phosphorus in enhancing plant resilience to major abiotic stresses. We will explore the physiological, biochemical, and molecular mechanisms through which phosphorus nutrition modulates plant responses to drought, salinity, temperature extremes, and heavy metal toxicity. Finally, we will discuss practical agronomic and biotechnological strategies aimed at improving phosphorus use efficiency under stress conditions, highlighting the potential for integrated approaches to develop more resilient crops for a sustainable future.

4.2 Phosphorus metabolism and homeostasis under optimal conditions

A comprehensive understanding of how phosphorus (P) contributes to abiotic stress resistance first requires an examination of its metabolism and the sophisticated homeostatic mechanisms that govern its acquisition and utilization under optimal conditions. Phosphorus homeostasis is a dynamic process involving coordinated uptake from the soil, assimilation into organic compounds, internal recycling, and remobilization, all tightly regulated by a complex molecular signaling network.

4.2.1 Uptake and assimilation: phosphate transporters and root architecture

The primary form of P absorbed by plant roots is inorganic phosphate (Pi), which is often present in soil solutions at micromolar concentrations despite its abundance in total soil P. To overcome this limitation, plants have evolved a two-pronged strategy: (1) the development of a root system architecture (RSA) that maximizes soil exploration and (2) the deployment of high-affinity transport systems (Wang et al., 2025).

Under sufficient P availability, root architecture is typically characterized by a primary root with longer lateral branches, optimizing exploration of the topsoil where P tends to accumulate. The uptake of Pi across the plasma membrane of root epidermal and cortical cells is mediated by a family of proton-coupled phosphate transporters (PHTs). These are categorized into high-affinity and low-affinity transporters, allowing plants to acquire Pi across a wide range of soil concentrations (Nussaume et al., 2011). The high-affinity PHT1 family is particularly crucial for primary Pi acquisition from the soil. Once inside the root cell, Pi is rapidly assimilated into organic compounds. The primary entry point into metabolic pathways is through phosphorylation, catalyzed by various kinases using ATP. This assimilated P is incorporated into crucial molecules like nucleic acids (DNA, RNA), phospholipids for membranes, and coenzymes essential for energy transfer and redox reactions (Plaxton and Lambers, 2015; Versaw and Garcia, 2017).

4.2.2 Internal recycling and remobilization

Phosphorus homeostasis is not solely dependent on external acquisition; efficient internal recycling is paramount. Plants are highly proficient at remobilizing P from older, senescing tissues to younger, growing organs such as new leaves, roots, and developing seeds. This process is vital for maintaining metabolic activity in meristematic regions when external supply is limited. This internal recycling is facilitated by a suite of enzymes that hydrolyze organic P esters, releasing free Pi for reuse. Key among these are acid phosphatases (APases) and ribonucleases (RNases). Intracellular and secreted APases play a critical role in scavenging Pi from a wide range of organic P compounds, both within the plant apoplast and in the rhizosphere. Similarly, RNases catalyze the breakdown of RNA, a significant intracellular P pool, liberating nucleotides that can be further degraded to release Pi (Plaxton and Tran, 2011). This constant cycle of hydrolysis and reassimilation ensures that P is allocated to the most metabolically active sinks within the plant, demonstrating a high degree of metabolic efficiency (Nussaume and Kanno, 2024).

4.2.3 The P signaling network: phosphate starvation response transcription factors and SPX modules

The coordination of P acquisition, assimilation, and recycling is governed by a central regulatory system. At the heart of this system in Arabidopsis and other plants are the PHOSPHATE STARVATION RESPONSE 1 (PHR1) transcription factors and their functional orthologues in other species (e.g., OsPHRs in rice). PHR1 acts as a master regulator, binding to the P1BS (PHR1-binding sequence) motif in the promoters of a vast array of phosphate starvation-induced (PSI) genes (Bustos et al., 2010). These PSI genes include *PHT1* phosphate transporters, *APase* genes, and genes involved in root architecture modification.

Under P-replete conditions, the activity of PHR1 is suppressed by a family of SPX proteins (named after SYG1/Pho81/XPR1). SPX proteins act as intracellular P sensors. When cellular Pi levels are high, SPX proteins bind to PHR1, preventing it from activating its target genes. This interaction is mediated by inositol polyphosphate signaling molecules (e.g., IP6), whose levels correlate with P status. As cellular Pi levels decline, the synthesis of these inositol polyphosphates decreases, leading to the dissociation of the SPX-PHR1 complex. This releases PHR1, allowing it to dimerize, enter the nucleus, and activate the expression of PSI genes to orchestrate the P starvation response (PSR) (Wild et al., 2016; Wang et al., 2025, 2018). This SPX-PHR module represents an elegant feedback loop, ensuring that the energetically costly processes of enhanced P acquisition and recycling are only activated when necessary, thereby maintaining P homeostasis under optimal and fluctuating conditions. This foundational understanding of P metabolism and its strict regulation provides the essential context for comprehending how these processes are disrupted, co-opted, and integrated into broader signaling networks when plants face abiotic stresses.

4.3 The interplay between phosphorus and abiotic stress signaling

The plant's response to environmental challenges is not a simple linear pathway but a complex, integrated network where different signaling cascades converge and interact. The phosphorus status of a plant serves as a critical modulator within this network, influencing how abiotic stresses are perceived and the efficacy of the ensuing response.

4.3.1 Convergent pathways: integration with stress hormones

A key mechanism for signal integration occurs through the interplay between the central P regulator, PHR1, and stress hormones, particularly abscisic acid (ABA). ABA is a primary hormone mediating plant responses to drought, salinity, and cold. Research has revealed a significant overlap between the transcriptomes of P-deficient plants and ABA-treated plants, suggesting shared regulatory nodes (Lei et al., 2011).

PHR1 has been shown to directly bind to the promoters of several ABA-responsive genes, indicating that it can regulate aspects of the ABA response. Conversely, ABA application can influence the expression of PHT transporters and induce root architectural changes reminiscent of the P starvation response (Kanno et al., 2016). This reciprocal regulation allows the plant to synchronize its nutrient acquisition strategy with its water-status needs. For instance, under impending drought stress, the activation of PHR-targeted genes can preemptively enhance root foraging capacity, an adaptive trait beneficial for both water and P capture. Similarly, jasmonic acid (JA) and ethylene signaling pathways, which are activated in response to various stresses including wounding and heavy metals, also interact with P signaling, often in an antagonistic manner to fine-tune resource allocation between growth and defense (Khan et al., 2016; Zhou et al., 2021).

4.3.2 Reactive oxygen species: a double-edged sword modulated by P status

The generation of reactive oxygen species (ROS) such as hydrogen peroxide (H_2O_2) and superoxide (O_2^-) is a common consequence of most abiotic stresses, leading to oxidative damage to lipids, proteins, and DNA. However, ROS also function as important signaling molecules that activate defense pathways. The cellular redox state is therefore tightly regulated by antioxidant systems, many of which are dependent on P availability (Al-Qahtani et al., 2025).

Adequate P nutrition is crucial for maintaining the antioxidant capacity of the plant. P is a key component of NADPH, the reducing power required for the ascorbate-glutathione cycle, a major ROS-scavenging pathway. Furthermore, P sufficiency supports the synthesis of antioxidant enzymes like superoxide dismutase (SOD) and catalase (CAT) (Tewari et al., 2020). Under P deficiency, the compromised antioxidant system can lead to a dangerous accumulation of ROS under subsequent stress, exacerbating cellular damage. Interestingly, the P starvation response itself can generate a mild oxidative burst, which may act as a priming signal, preconditioning the plant's antioxidant machinery for more severe stress later (Shin et al., 2006). This dual role underscores how P status dictates whether ROS act as destructive molecules or protective secondary messengers (Al-Qahtani et al., 2025).

4.3.3 The role of P in osmoprotectant synthesis

Osmotic adjustment is a fundamental strategy for maintaining cell turgor and protecting cellular structures under drought and salinity stress. This is achieved through the accumulation of compatible solutes, or osmoprotectants, such as proline, glycine betaine, and soluble sugars. The biosynthesis of these compounds is highly energy-intensive, requiring substantial amounts of ATP and phosphorylated intermediates.

Phosphorus availability directly governs the capacity for osmoprotectant synthesis. The carbon skeletons for these molecules are derived from photosynthetic carbon metabolism and glycolysis, both processes replete with phosphorylated compounds (e.g., glucose-6-phosphate, glyceraldehyde-3-phosphate) and dependent on ATP and NADPH. Under optimal P nutrition, plants can readily generate the necessary energy and carbon precursors to accumulate osmoprotectants to cytotoxic levels, thereby stabilizing proteins and membranes, and maintaining osmotic balance (Garg et al., 2004). In contrast, P-deficient plants have a limited energy budget and impaired carbon metabolism, constraining their ability to synthesize these critical defense compounds and rendering them more susceptible to osmotic stress.

In conclusion, the interplay between P and abiotic stress signaling is profound and multifaceted. P status influences the core hormonal signaling networks, modulates the delicate balance of ROS signaling and scavenging, and provides the energetic and metabolic foundation for crucial osmotic adjustments. Understanding this crosstalk is essential for developing strategies to enhance crop resilience to multiple concurrent stresses.

4.4 Phosphorus-mediated resistance to specific abiotic stresses

Abiotic stresses, such as drought, salinity, heat, and pollution, represent the primary constraints on global agricultural productivity, threatening food security (Shaffique et al., 2025; Kumar et al., 2025; Husen, 2023, 2024, 2025). Plants have evolved a suite of physiological, biochemical, and molecular adaptations to cope with these challenges. Among the factors influencing these adaptive responses, plant nutrition, particularly phosphorus nutrition, is paramount.

Phosphorus is absorbed by plants primarily as inorganic phosphate ($H_2PO_4^-$ or HPO_4^{2-}), which is often immobilized in soils, making it one of the most limiting nutrients for crop production (Veneklaas et al., 2012). While the essentiality of P for basic metabolism is well-established, its role as a mediator of stress tolerance is a dynamic area of research. This paper explores the specific mechanisms through which phosphorus confers resistance to three major abiotic stresses: drought, salinity, and heavy metal toxicity (Fig. 4.1).

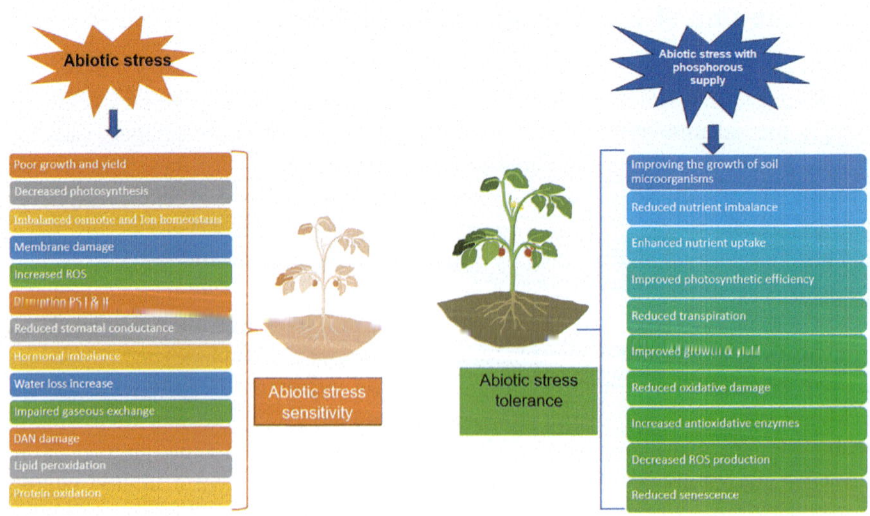

FIGURE 4.1

Effects of abiotic stresses on plants in the presence and absence of phosphorus supply.

4.4.1 Phosphorus-mediated resistance to drought stress

Drought stress represents one of the most significant abiotic constraints on global agricultural productivity, severely limiting plant growth and yield outcomes across diverse cropping systems. Phosphorus (P), as an essential macronutrient, plays a pivotal role in enhancing plant drought resistance through multiple interconnected physiological and biochemical mechanisms that operate at various organizational levels within the plant system. The fundamental importance of phosphorus in energy transfer processes, particularly through adenosine triphosphate (ATP) molecules, establishes its critical role in maintaining metabolic functions during water-deficit conditions. Adequate phosphorus availability significantly improves root system architecture by promoting deeper root penetration, enhancing lateral root development, and increasing root hair density, thereby substantially improving the plant's capacity for water acquisition from deeper soil layers that retain moisture during drought periods (López-Bucio et al., 2002). This morphological adaptation represents a primary defense mechanism against water scarcity, enabling plants to maintain hydraulic continuity with diminishing soil water resources. Beyond structural modifications, phosphorus nutrition directly influences stomatal regulation through its essential role in ATP-dependent ion pumps that govern guard cell functioning, allowing for more precise control of stomatal aperture and consequently improving water-use efficiency while maintaining photosynthetic activity under water-limited conditions (Bechtaoui et al., 2021; Waratch et al., 2011). The metabolic functions of phosphorus extend to osmotic adjustment mechanisms where it facilitates the accumulation of compatible solutes including proline, glycine betaine, and various soluble sugars that collectively help maintain cell turgor pressure and protect cellular structures against dehydration-induced damage (Garg et al., 2004). Furthermore, phosphorus enhances the antioxidant defense system by upregulating the activity of crucial enzymes including superoxide dismutase, catalase, and peroxidase, thereby effectively reducing oxidative damage caused by drought-induced reactive oxygen species accumulation (Parecido et al., 2021). The integration of phosphorus nutrition with phytohormonal signaling pathways, particularly those involving abscisic acid, demonstrates the sophisticated regulatory role phosphorus plays in coordinating the plant's comprehensive drought response strategy (Li et al., 2022). These multifaceted mechanisms collectively underscore how optimal phosphorus nutrition provides a comprehensive approach to drought resistance, encompassing improved water acquisition, enhanced water conservation, maintained metabolic functionality, and reduced oxidative damage (Fig. 4.1). The strategic management of phosphorus nutrition therefore represents a crucial component in developing sustainable agricultural practices for drought-prone environments, offering significant potential for enhancing crop resilience and food security in the face of increasing climate variability and water scarcity challenges.

4.4.2 Phosphorus-mediated resistance to salt stress

Salt stress represents a major abiotic constraint affecting agricultural productivity worldwide, particularly in arid and semi-arid regions where soil salinity continues to expand due to irrigation practices and climate change. Phosphorus (P) nutrition has emerged as a crucial factor in mitigating the adverse effects of salt stress on plant growth and development through multiple physiological, biochemical, and molecular mechanisms. The role of phosphorus in enhancing salt tolerance begins with its influence on ion homeostasis, where adequate phosphorus availability helps maintain the optimal K^+/Na^+ ratio by reducing sodium uptake and enhancing potassium absorption through regulation of ion transporters and channels (Chakraborty et al., 2016; Mohamed et al., 2021). This ionic balance is crucial for maintaining enzymatic activity and cellular functions under saline conditions. Phosphorus also contributes significantly to osmotic adjustment by facilitating the accumulation of compatible solutes such as proline, glycine betaine, and soluble sugars, which help maintain cell turgor and protect cellular structures against osmotic stress induced by high salt concentrations (Abbas et al., 2021). Furthermore, phosphorus nutrition enhances the antioxidant defense system by upregulating the activity of key enzymes including superoxide dismutase (SOD), catalase (CAT), peroxidase (POD), and ascorbate peroxidase (APX), thereby effectively scavenging reactive oxygen species (ROS) and minimizing oxidative damage to cellular components (Liu et al., 2021). The protective role of phosphorus extends to photosynthetic processes, where it helps maintain chlorophyll content, photosynthetic rate, and stomatal conductance under saline conditions by supporting ATP synthesis and energy metabolism crucial for photosynthetic machinery (Taliman et al., 2019). Additionally, phosphorus influences phytohormonal regulation, particularly abscisic acid (ABA) metabolism and signaling, which plays a central role in coordinating plant responses to salt stress (Loudari et al., 2022; Zheng et al., 2023). The integration of phosphorus nutrition with other physiological processes including water relations, nutrient uptake, and gene expression related to salt tolerance further demonstrates its comprehensive role in enhancing plant resilience to salinity. These multifaceted mechanisms collectively highlight the importance of optimal phosphorus nutrition in developing sustainable agricultural strategies for salt-affected soils, offering significant potential for improving crop productivity and food security in salinity-prone regions (Fig. 4.1).

4.4.3 Phosphorus-mediated resistance to heat stress

Heat stress has emerged as a critical abiotic factor limiting crop productivity worldwide, particularly in the context of global climate change and increasing temperature extremes. Phosphorus (P) nutrition plays a multifaceted role in enhancing plant thermotolerance through various physiological, biochemical, and molecular mechanisms that collectively mitigate the detrimental effects of elevated temperatures. The fundamental involvement of phosphorus in energy metabolism, primarily through adenosine triphosphate (ATP) molecules, establishes its crucial function in maintaining cellular homeostasis under heat stress conditions. Adequate phosphorus availability supports photosynthetic stability by protecting the photosystem II (PSII) complex from heat-induced damage, maintaining chlorophyll content, and enhancing the electron transport rate, thereby ensuring continued carbon assimilation under high-temperature conditions (Bechtaoui et al., 2021; Waraich et al., 2012). Phosphorus nutrition further contributes to membrane stability by maintaining the integrity of phospholipid bilayers and reducing membrane fluidity, which is essential for preventing electrolyte leakage and maintaining cellular compartmentalization under heat stress (Wang et al., 2021). The role of phosphorus in antioxidant defense mechanisms is particularly significant, as it enhances the activities of key enzymes including superoxide dismutase (SOD), catalase (CAT), peroxidase (POD), and ascorbate peroxidase (APX), thereby effectively scavenging reactive oxygen species (ROS) and minimizing oxidative damage to cellular components (Parecido et al., 2021). Additionally, phosphorus facilitates the accumulation of compatible solutes such as proline, glycine betaine, and soluble sugars, which contribute to osmotic adjustment and protein stabilization under high-temperature conditions (Garg et al., 2020). The involvement of phosphorus in heat shock protein (HSP) synthesis and signaling pathways demonstrates its regulatory role in thermotolerance mechanisms, as adequate phosphorus status enhances the expression of HSP genes that function as molecular chaperones to prevent protein denaturation and facilitate refolding of damaged proteins (Li et al., 2022). Phosphorus nutrition also influences phytohormonal balance, particularly the synthesis and signaling of abscisic acid (ABA) and cytokinins, which are integral to coordinating plant responses to heat stress through stomatal regulation and growth maintenance (Khan et al., 2023). The integration of these mechanisms highlights how optimal phosphorus nutrition provides a comprehensive approach to heat stress mitigation, encompassing improved photosynthetic performance, enhanced membrane stability, reduced oxidative damage, and better protein protection (Table 4.1). These collective benefits underscore the importance of strategic phosphorus management in developing climate-resilient agricultural systems, offering significant potential for sustaining crop productivity in heat-prone environments and contributing to global food security under changing climatic conditions.

4.4.4 Phosphorus-mediated resistance to heavy metal stress

Heavy metal contamination in agricultural soils has become a significant environmental concern worldwide, posing serious threats to crop productivity and food safety. Phosphorus nutrition plays a crucial role in mitigating heavy metal toxicity in plants through multiple interconnected mechanisms that operate at physiological, biochemical, and molecular levels. The protective effects of phosphorus begin in the rhizosphere, where phosphate ions can immobilize heavy metals through

Table 4.1 The roles of phosphorus in plant tolerance to abiotic stresses

Abiotic stress	Role of phosphorus	Key mechanisms	References
Drought stress	Improves water-use efficiency and maintains cellular function under water deficit.	– **Root system development**: P promotes deeper and more extensive root systems, allowing access to deeper soil water. – **Osmotic adjustment**: P-containing compounds (e.g., phytic acid) contribute to osmotic adjustment, maintaining turgor pressure. – **Energy metabolism**: Adequate ATP is crucial for energy-intensive stress responses, including the synthesis of osmoprotectants and stress proteins. – **Stomatal regulation:** P is involved in the signal transduction pathways that control stomatal closure.	Garg et al. (2004); He and Dijkstra (2014); Yang et al. (2024)
Salinity stress	Mitigates ionic toxicity and osmotic stress caused by high Na^+ and Cl^- concentrations.	– **Ion homeostasis:** Enhances the selectivity of K^+ over Na^+ uptake, maintaining a favorable K^+/Na^+ ratio, crucial for enzyme activation. – **Osmoprotection:** Supports the synthesis of organic osmolytes (e.g., proline, glycine betaine). – **Membrane stability:** Phospholipids are key components of cell membranes; P sufficiency maintains membrane integrity against ionic assault. – **Photosynthetic protection:** Supports photosynthetic phosphorylation (ATP for Calvin cycle) under stress.	Talbi et al. (2015); Ashraf et al. (2013)
Cold/ Chilling stress	Maintains membrane fluidity and energy status during low-temperature stress.	– **Membrane lipid composition:** P is a component of phospholipids. Adequate P helps maintain unsaturated fatty acids in membranes, preventing solidification. – **Energy charge:** ATP from P metabolism provides energy for metabolic heat production and synthesis of cryoprotective compounds. – **Antioxidant system:** Activates antioxidant enzymes (e.g., superoxide dismutase, catalase) to scavenge reactive oxygen species (ROS) generated under cold stress.	Jan et al. (2020), Hurry et al. (2000)

(Continued)

Table 4.1 The roles of phosphorus in plant tolerance to abiotic stresses (*Continued*)

Abiotic stress	Role of phosphorus	Key mechanisms	References
Heat stress	Protects photosynthetic apparatus and protein structures from denaturation.	– **Photosynthesis:** Stabilizes the photosynthetic machinery by supporting the regeneration of ribulose-1,5-bisphosphate (RuBP). – **Protein stability:** ATP-dependent chaperone proteins, which refold denatured proteins, require P. – **ROS detoxification:** Fuels the antioxidant system to mitigate oxidative damage caused by high temperatures.	Sattar et al. (2020), Bechtaoui et al. (2021), Mohammadi et al. (2021), Waraich et al. (2012)
Heavy metal stress	Reduces metal bioavailability and mitigates oxidative damage.	– **Precipitation/Immobilization:** In the rhizosphere, phosphate ions can precipitate with heavy metals (e.g., Pb, Zn, Cd), reducing their uptake. – **Chelation:** P nutrition influences the synthesis of metal-chelating compounds like phytochelatins and metallothioneins. – **Compartmentalization:** ATP-dependent transporters are involved in sequestering toxic metals into vacuoles.	Wu et al. (2017), Liu et al. (2021), Shahid et al. (2013)

precipitation and complexation reactions, forming insoluble metal-phosphate compounds that reduce metal bioavailability and uptake by plant roots (Liu et al., 2021; Wu et al., 2017). This precipitation mechanism is particularly effective for metals such as cadmium (Cd), lead (Pb), and zinc (Zn), significantly decreasing their translocation to aerial plant parts. Within plant tissues, phosphorus enhances the synthesis of metal-chelating compounds including phytochelatins, metallothioneins, and organic acids, which sequester heavy metals in less sensitive cellular compartments and prevent their interaction with vital metabolic processes (Chen et al., 2020). Phosphorus nutrition also strengthens the antioxidant defense system by upregulating the activities of key enzymes such as superoxide dismutase (SOD), catalase (CAT), peroxidase (POD), and ascorbate peroxidase (APX), thereby effectively scavenging reactive oxygen species (ROS) generated by heavy metal stress and minimizing oxidative damage to cellular components (Ma et al., 2021). Furthermore, adequate phosphorus availability improves membrane stability through the maintenance of phospholipid bilayers and reduces electrolyte leakage, which helps preserve cellular integrity under metal stress conditions (Song et al., 2022).

The role of phosphorus in nutrient homeostasis is particularly important, as it competes with heavy metals for uptake transporters and reduces metal accumulation in plants while maintaining essential nutrient balance (Sörengård et al., 2023). Phosphorus also influences gene expression related to heavy metal detoxification, enhancing the transcription of genes involved in metal transport, chelation, and sequestration mechanisms (Lin et al., 2022). (Table 4.1). Additionally, phosphorus-mediated improvements in root system architecture and mycorrhizal associations enhance plant ability to tolerate metal stress by improving nutrient acquisition and creating physical barriers against metal uptake(Soliman et al., 2023). These comprehensive mechanisms demonstrate that optimal phosphorus nutrition provides an effective strategy for reducing heavy metal toxicity in plants, offering significant potential for phytoremediation applications and safe crop production in metal-contaminated soils. The integration of phosphorus management with other agronomic practices represents a promising approach for mitigating heavy metal stress in agricultural systems, contributing to improved food security and environmental sustainability.

4.5 Strategies for enhancing phosphorus use efficiency under stress

Phosphorus (P) is an essential macronutrient that plays a critical role in numerous plant physiological processes including energy transfer through ATP, nucleic acid synthesis, membrane formation, and photosynthesis (Veneklaas et al., 2012). Despite its abundance in many soils, phosphorus availability to plants is often limited due to its tendency to form insoluble complexes with soil components, particularly in alkaline and acidic soils (Hinsinger, 2001). This problem is significantly exacerbated under various environmental stress conditions such as drought, salinity, extreme temperatures, and heavy metal contamination, which further reduce phosphorus solubility, mobility, and root acquisition capacity (Wang and Lambers, 2020). (Table 4.1).

Plants have evolved sophisticated adaptive mechanisms to enhance phosphorus use efficiency (PUE) under stress conditions. Morphological modifications include enhanced root growth, increased root hair density and length, and development of specialized root structures like proteoid roots that improve soil exploration and phosphorus acquisition (Lambers et al., 2006). Physiological adaptations involve the exudation of organic acids (citrate, malate, oxalate), protons, and enzymes (phosphatases, phytases) that solubilize fixed phosphorus (Wang and Lambers, 2020). Molecular responses include upregulation of high-affinity phosphate transporters (PHT family) and activation of phosphate starvation-induced (PSI) genes regulated by the PHR1 transcription factor (Shen et al., 2011).

Agronomic strategies to improve PUE under stress include precision fertilization techniques such as band application and foliar feeding, soil amendments with organic matter and biochar, conservation agriculture practices including reduced tillage and cover cropping, and improved water management strategies (Richardson et al., 2011). Microbial interventions offer promising approaches through phosphorus-solubilizing microorganisms (PSMs) including bacteria (Pseudomonas, Bacillus) and fungi (Aspergillus, Penicillium) that can convert insoluble phosphorus forms into plant-available phosphate, and arbuscular mycorrhizal fungi (AMF) that form symbiotic associations with plant roots, extending their hyphal networks to access phosphorus beyond the root depletion zone (Richardson et al., 2011).

Genetic improvement and breeding strategies provide sustainable solutions through conventional breeding approaches, molecular breeding techniques including marker-assisted selection, and genetic engineering approaches that have targeted key genes involved in phosphorus acquisition, with recent advances in gene editing technologies particularly CRISPR-Cas9, offering precise tools for developing crops with optimized PUE traits (Veneklaas et al., 2012). Integrated approaches that combine multiple strategies show the greatest potential for enhancing PUE under stress conditions through the combination of stress-tolerant genotypes with improved soil management practices, appropriate microbial inoculants, and precision nutrient management (Richardson et al., 2011).

Future research should focus on understanding the complex interactions between different stress factors and phosphorus nutrition, developing field-ready microbial consortia, and creating integrated management packages tailored to specific cropping systems and environmental conditions to ensure sustainable agricultural productivity in the face of increasing environmental challenges and limited phosphorus resources (Veneklaas et al., 2012; Wang and Lambers, 2020).

4.6 Conclusion and future perspectives

This chapter has synthesized the critical role of phosphorus (P) nutrition in mediating plant responses to abiotic stresses, establishing that P homeostasis is integral to stress resilience rather than merely a component of growth. The evidence presented demonstrates that phosphorus status orchestrates a wide array of physiological and molecular adaptations that enhance plant survival under drought, salinity, heavy metal toxicity, and temperature extremes. The mechanisms include morphological modifications in root architecture that improve soil exploration, biochemical responses involving exudation of organic acids and phosphatases that enhance P availability, and metabolic adjustments that optimize internal P utilization through lipid remodeling and enhanced recycling efficiency. Furthermore, adequate P nutrition strengthens plant defense systems by supporting antioxidant activity, osmotic adjustment, and membrane stability. The synergistic relationship between P and arbuscular mycorrhizal fungi further expands the plant's capacity to withstand environmental challenges through enhanced nutrient and water acquisition. These findings collectively position phosphorus management as a strategic approach for improving crop resilience in suboptimal environments.

Looking forward, several key research directions emerge as critical for advancing this field. First, there is a need to elucidate the precise molecular signaling networks that integrate P starvation responses with abiotic stress signaling pathways, particularly the crosstalk between P regulatory elements and stress-responsive hormones. Second, the potential of the soil microbiome remains underexploited; developing tailored synthetic microbial communities that enhance P solubilization while priming plant stress responses represents a promising frontier. Third, advancing precision breeding approaches through integrated phenotyping and genomics will accelerate the development of cultivars with enhanced P-use efficiency under stress conditions. Fourth, research must address the complex interactions between multiple concurrent stresses and P nutrition to better predict plant performance in real-world scenarios. Finally, developing climate-smart P management strategies that account for changing environmental conditions will be essential for sustainable agriculture. These future efforts will require interdisciplinary approaches combining molecular biology, microbiology, breeding, and agronomy to fully harness the potential of phosphorus in enhancing crop resistance to abiotic stresses, thereby contributing to global food security under increasingly challenging environmental conditions.

References

Abbas, T., Balal, R.M., Shahid, M.A., Pervez, M.A., Ayyub, C.M., Aqueel, M.A., et al., 2021. Silicon-induced alleviation of NaCl toxicity in okra (*Abelmoschus esculentus*) is associated with enhanced photosynthesis, osmoprotectants and antioxidant metabolism. Acta Physiologiae Plantarum 43 (7), 1–15. Available at https://doi.org/10.1007/s11738-021-03262-w.

Al-Qahtani, W.H., Abdel-Maksoud, M.A., Saleh, I.A., Zomot, N., Almunqedhi, B.M., Kataya, A.R.A., et al., 2025. Effects of phosphorus sources on arsenic stress mitigation in wheat via proline and antioxidant pathways, Springer, Saudi Arabia. Journal of Plant Growth Regulation 44 (2), 821–838. Available at https://doi.org/10.1007/s00344-024-11490-w.

Ashraf, M., Shahbaz, M., Ali, Q., 2013. Drought-induced modulation in growth and mineral nutrients in canola (*Brassica napus* L.). Pakistan Journal of Botany 45 (1), 93–98.

Bechtaoui, N., Rabiu, M.K., Raklami, A., Oufdou, K., Hafidi, M., Jemo, M., 2021. Phosphate-dependent regulation of growth and stresses management in plants, Frontiers Media S.A., Morocco. Frontiers in Plant Science 12. Available at https://doi.org/10.3389/fpls.2021.679916.

Bustos, R., Castrillo, G., Linhares, F., Puga, M.I., Rubio, V., Pérez-Pérez, J., et al., 2010. A central regulatory system largely controls transcriptional activation and repression responses to phosphate starvation in *Arabidopsis*. PLoS Genetics 6 (9), e1001102. Available at https://doi.org/10.1371/journal.pgen.1001102.

Chakraborty, K., Bose, J., Shabala, L., Shabala, S., 2016. Difference in root K + retention ability and reduced sensitivity of K + -permeable channels to reactive oxygen species confer differential salt tolerance in three *Brassica* species. Journal of Experimental Botany 67 (15), 4611–4625. Available at https://doi.org/10.1093/jxb/erw236.

Chen, M., Shu, S., Li, J., Lv, X., Dong, F., Jiang, G., 2020. Activating palladium nanoparticles via a Mott-Schottky heterojunction in electrocatalytic hydrodechlorination reaction. Journal of Hazardous Materials 389, 121876. Available at https://doi.org/10.1016/j.jhazmat.2019.121876.

Fahad, S., Bajwa, A.A., Nazir, U., Anjum, S.A., Farooq, A., Zohaib, A., et al., 2017. Crop production under drought and heat stress: plant responses and management options, Frontiers Media S.A., China. Frontiers in Plant Science 8. Available at https://doi.org/10.3389/fpls.2017.01147.

Garg, B.K., Burman, U., Kathju, S., 2004. The influence of phosphorus nutrition on the physiological response of moth bean genotypes to drought. Journal of Plant Nutrition and Soil Science 167 (4), 503–508. Available at https://doi.org/10.1002/jpln.200320368.

Garg, V., Kühn, C., Garg, N., Singla, P., Bhandari, P., 2020. Phosphorus nutrition and arbuscular mycorrhizal symbiosis regulate osmotic adjustment and antioxidant metabolism in chickpea under heat stress. Plant Physiology and Biochemistry 151, 284–291. Available at https://doi.org/10.1016/j.plaphy.2020.03.023.

He, M., Dijkstra, F.A., 2014. Drought effect on plant nitrogen and phosphorus: a meta-analysis. New Phytologist 204 (4), 924–931. Available at https://doi.org/10.1111/nph.12952.

Hinsinger, P., 2001. Bioavailability of soil inorganic P in the rhizosphere as affected by root-induced chemical changes: a review. Plant and Soil 237 (2), 173–195. Available at https://doi.org/10.1023/a:1013351617532.

Hurry, V.M., Strand, A., Tobiæson, M., Gardeström, P., Öquist, G., 2000. Cold hardening of spring and winter wheat and rape results in differential effects on growth, carbon metabolism, and carbohydrate content, American Society of Plant Biologists, Australia. Plant Physiology 109 (2), 697–706. Available at https://doi.org/10.1104/pp.109.2.697.

Husen, A., 2023. Plants and Their Interaction to Environmental Pollution: Damage Detection, Adaptation, Tolerance, Physiological and Molecular Responses. Elsevier. Available at https://doi.org/10.1016/C2020-0-03319-0.

Husen, A., 2024. Essential Minerals in Plant Soil System: Coordination, Signalling, and Interaction under Adverse Situations. Elsevier. Available at https://doi.org/10.1016/C2022-0-02437-5.

Husen, A., 2025. Role of Antioxidants in Mitigating Plant Stress. Academic Press. Available at https://doi.org/10.1016/C2023-0-51268-6.

Jan, M.F., Abbas, G., Arif, M., Akhtar, K., 2020. Phosphorus nutrition improves chilling tolerance in maize. Journal of Plant Growth Regulation 39, 753–762.

Kanno, S., Arrighi, J.F., Chiarenza, S., Bayle, V., Berthomé, R., Péret, B., et al., 2016. A novel role for the root cap in phosphate uptake and homeostasis. eLife 5, 2016. Available at https://doi.org/10.7554/eLife.14577.

Khan, F., Siddique, A.B., Shabala, S., Zhou, M., Zhao, C., 2023. Phosphorus plays key roles in regulating plants' physiological responses to abiotic stresses. Plants 12 (15), 2861. Available at https://doi.org/10.3390/plants12152861.

Khan, G.A., Vogiatzaki, E., Glauser, G., Poirier, Y., 2016. Phosphate deficiency induces the jasmonate pathway and enhances resistance to insect herbivory, American Society of

Plant Biologists, Switzerland. Plant Physiology 171 (1), 632–644. Available at https://doi.org/10.1104/pp.16.00278.

Kumar, A., Choudhary, A., Kaur, H., Singh, K., Guha, S., Choudhary, D.R., et al., 2025. Exploring the role of silicon in enhancing sustainable plant growth, defense system, environmental stress mitigation and management. Discover Applied Sciences 7 (5). Available at https://doi.org/10.1007/s42452-025-06866-w.

Lambers, H., Shane, M.W., Cramer, M.D., Pearse, S.J., Veneklaas, E.J., 2006. Root structure and functioning for efficient acquisition of phosphorus: matching morphological and physiological traits. Annals of Botany 98 (4), 693–713. Available at https://doi.org/10.1093/aob/mcl114.

Lei, M., Zhu, C., Liu, Y., Karthikeyan, A.S., Bressan, R.A., Raghothama, K.G., et al., 2011. Ethylene signalling is involved in regulation of phosphate starvation-induced gene expression and production of acid phosphatases and anthocyanin in *Arabidopsis*. New Phytologist 189 (4), 1084–1095. Available at https://doi.org/10.1111/j.1469-8137.2010.03555.x.

Li, X., Zhang, Y., Yang, J., 2022. Phosphorus supplementation enhances heat shock protein expression and thermotolerance in tomato plants. Scientia Horticulturae 295, 110881. Available at https://doi.org/10.1016/j.scienta.2021.110881,.

Li, Z., Xu, C., Wang, J., Li, J., 2022. Phosphorus nutrition and arbuscular mycorrhizal symbiosis regulate phytohormone patterns of tomato under drought stress. Journal of Plant Growth Regulation 41 (5), 3031–3043. Available at https://doi.org/10.1007/s00344-021-10492-2.

Lin, J., Xiao, Y., Liu, Y., Lei, Y., Cai, Y., Liang, Q., et al., 2022. Leachate from plastic food packaging induced reproductive and neurobehavioral toxicity in zebrafish. Ecotoxicology and Environmental Safety 231, 113189. Available at https://doi.org/10.1016/j.ecoenv.2022.113189.

Liu, Z., Tang, J.C., Ren, X., Schaeffer, S.M., 2021. Effects of phosphorus modified nZVI-biochar composite on emission of greenhouse gases and changes of microbial community in soil. Environmental Pollution 274, 116483. Available at https://doi.org/10.1016/j.envpol.2021.116483.

López-Bucio, J., Cruz-Ramírez, A., Herrera-Estrella, L., 2003. The role of nutrient availability in regulating root architecture, Elsevier Ltd, Mexico. Current Opinion in Plant Biology 6 (3), 280–287. Available at https://doi.org/10.1016/S1369-5266(03)00035-9.

López-Bucio, J., Hernández-Abreu, E., Sánchez-Calderón, L., Nieto-Jacobo, M.F., Simpson, J., Herrera-Estrella, L., 2002. Phosphate availability alters architecture and causes changes in hormone sensitivity in the *Arabidopsis* root system, American Society of Plant Biologists, Mexico. Plant Physiology 129 (1), 244–256. Available at https://doi.org/10.1104/pp.010934.

Loudari, A., Mayane, A., Zeroual, Y., Colinet, G., Oukarroum, A., 2022. Photosynthetic performance and nutrient uptake under salt stress: differential responses of wheat plants to contrasting phosphorus forms and rates, Frontiers Media S.A., Morocco. Frontiers in Plant Science 13. Available at https://doi.org/10.3389/fpls.2022.1038672.

Ma, Y., Sun, M./, Liang, Q., Wang, F., Lin, L., Li, T., et al., 2021. The relationship between long-term exposure to PM2.5 and hypertension in women: A meta-analysis. Ecotoxicology and Environmental Safety 208, 111492. Available at https://doi.org/10.1016/j.ecoenv.2020.111492.

Mohamed, H.I., El-Sayed, A.A., Rady, M.M., Caruso, G., Sekara, A., Abdelhamid, M.T., 2021. Coupling effects of phosphorus fertilization source and rate on growth and ion accumulation of common bean under salinity stress. PeerJ 9, e11463. Available at https://doi.org/10.7717/peerj.11463.

Mohammadi, M.A., Cheng, Y., Aslam, M., Jakada, B.H., Wai, M.H., Ye, K., et al., 2021. ROS and oxidative response systems in plants under biotic and abiotic stresses: revisiting the crucial role of phosphite triggered plants defense response, Frontiers Media S.A., China. Frontiers in Microbiology 12. Available at https://doi.org/10.3389/fmicb.2021.631318.

Nussaume, L., Kanno, S., 2024. Reviewing impacts of biotic and abiotic stresses on the regulation of phosphate homeostasis in plants. Journal of Plant Research 137 (3), 297–306. Available at https://doi.org/10.1007/s10265-024-01533-4.

Nussaume, L., Kanno, S., Javot, H., Marin, E., Pochon, N., Ayadi, A., et al., 2011. Thibaud, Phosphate import in plants: focus on the PHT1 transporters, Frontiers Research Foundation, France. Frontiers in Plant Science 2. Available at https://doi.org/10.3389/fpls.2011.00083.

Parecido, R.J., Soratto, R.P., Perdoná, M.J., Gitari, H.I., Dognani, V., Anderson, R., et al., 2021. Liming method and rate effects on soil acidity and arabica coffee nutrition, growth, and yield. Journal of Soil Science and Plant Nutrition 21, 2613–2625. Available at https://doi.org/10.1007/s42729-021-00550-9.

Plaxton, W.C., Tran, H.T., 2011. Metabolic adaptations of phosphate-starved plants, American Society of Plant Biologists, Canada. Plant Physiology 156 (3), 1006–1015. Available at https://doi.org/10.1104/pp.111.175281.

Plaxton, W.C., Lambers, H., 2015. Phosphorus: Back to the Roots. In: Plaxton, W.C., Lambers, H. (Eds.), Annual Plant Reviews Volume 48. John Wiley & Sons, Ltd, pp 1–36. Available at https://doi.org/10.1002/9781118958841.ch1.

Richardson, A.E., Lynch, J.P., Ryan, P.R., Delhaize, E., Smith, F.A., Smith, S.E., et al., 2011. Plant and microbial strategies to improve the phosphorus efficiency of agriculture. Plant and Soil 349 (1-2), 121–156. Available at https://doi.org/10.1007/s11104-011-0950-4.

Rouached, H., Arpat, A.B., Poirier, Y., 2010. Regulation of phosphate starvation responses in plants: signaling players and cross-talks. Molecular Plant 3 (2), 288–299. Available at https://doi.org/10.1093/mp/ssp120.

Sattar, A., Sher, A., Ijaz, M., Ul-Allah, S., Rizwan, M.S., Hussain, M., et al., 2020. Terminal drought and heat stress alter physiological and biochemical attributes in flag leaf of bread wheat. PLoS One 15 (5). Available at https://doi.org/10.1371/journal.pone.0232974.

Shaffique, S., Injamum-Ul-Hoque, M., Husen, A., kang, S.-M., Lee, I.-J., 2025. Revolutionizing heat stress tolerance in *Glycine max*: exploring the latest advances in microbial application. Plant Stress 15, 100725. Available at https://doi.org/10.1016/j.stress.2024.100725.

Shahid, M., Austruy, A., Echevarria, G., Arshad, M., Sanaullah, M., Aslam, M., et al., 2013. EDTA-enhanced phytoremediation of heavy metals: a review. Soil and Sediment Contamination: An International Journal 23 (4), 389–416. Available at https://doi.org/10.1080/15320383.2014.831029.

Shen, J., Yuan, L., Zhang, J., Li, H., Bai, Z., Chen, X., et al., 2011. Phosphorus dynamics: from soil to plant. Plant Physiology 156 (3), 997–1005. Available at https://doi.org/10.1104/pp.111.175232.

Shin, H., Shin, H.S., Chen, R., Harrison, M.J., 2006. Loss of At4 function impacts phosphate distribution between the roots and the shoots during phosphate starvation. Plant Journal 45 (5), 712–726. Available at https://doi.org/10.1111/j.1365-313X.2005.02629.x.

Soliman, M.H., Alnusairi, G.S.H., Abdullah Khan, A., Alnusaire, T.S., Fakhr, M.A., Abdulmajeed, A.M., et al., 2023. Biochar and Selenium Nanoparticles Induce Water Transporter Genes for Sustaining Carbon Assimilation and Grain Production in Salt-Stressed Wheat. Journal of Plant Growth Regulation 42 (3), 1522–1543. Available at https://doi.org/10.1007/s00344-022-10636-y.

Song, J., Zou, X., Liu, P., Cardoso, J.A., Schultze-Kraft, R., Liu, G., et al., 2022. Differential expressions and enzymatic properties of malate dehydrogenases in response to nutrient and metal stresses in Stylosanthes guianensis. Plant Physiology and Biochemistry 170, 325–337. Available at https://doi.org/10.1016/j.plaphy.2021.12.012.

Sörengård, M., Bergström, S., McCleaf, P., Wiberg, K.I., Ahrens, L., 2023. Corrigendum:"Long-distance transport of per- and polyfluoroalkyl substances (PFAS) in a Swedish drinking water aquifer" [Environ. Pollut. 311(2022) 119981]. Environmental Pollution 316, 120588. Available at 10.1016/j.envpol.2022.120588.

Talbi, S., Rojas, J.A., Sahrawy, M., Rodríguez-Serrano, M., Cárdenas, K.E., Debouba, M., et al., 2015. Effect of drought on growth, photosynthesis and total antioxidant capacity of the saharan plant *Oudeneya africana*. Environmental and Experimental Botany 111, 114–126.

Taliman, N.A., Dong, Q., Echigo, K., Raboy, V., Saneoka, H., 2019. Effect of phosphorus fertilization on the growth, photosynthesis, nitrogen fixation, mineral accumulation, seed yield, and seed quality of a low-phytate soybean line. Plants 8 (5), 119. Available at https://doi.org/10.3390/plants8050119.

Tewari, R.K., Kumar, P., Sharma, P.N., 2020. Morphology and oxidative physiology of boron-deficient mulberry plants. Tree Physiology 27 (4), 581–590. Available at https://doi.org/10.1093/treephys/27.4.581,.

Veneklaas, E.J., Lambers, H., Bragg, J., Finnegan, P.M., Lovelock, C.E., Plaxton, W.C., et al., 2012. Opportunities for improving phosphorus-use efficiency in crop plants. New Phytologist 195 (2), 306–320. Available at https://doi.org/10.1111/j.1469-8137.2012.04190.x.

Versaw, W.K., Garcia, L.R., 2017. Intracellular transport and compartmentation of phosphate in plants. Current Opinion in Plant Biology 39, 25–30. Available at https://doi.org/10.1016/j.pbi.2017.04.015.

Wang, L., Li, H., Suo, Y., Han, W., Diao, S., Mai, Y., et al., 2021. Effects of plant growth regulators, soil moisture contents, and carbon/nitrogen ratios on sex differentiation in persimmon (Diospyros kaki Thunb.) flowers. Journal of Plant Growth Regulation 40 (3), 1121–1138. Available at https://doi.org/10.1007/s00344-020-10170-9.

Wang, L., Ju, C., Han, C., Yu, Z., Bai, M.Y., Wang, C., 2025. The interaction of nutrient uptake with biotic and abiotic stresses in plants, John Wiley and Sons Inc, China. Journal of Integrative Plant Biology 67 (3), 455–487. Available at https://doi.org/10.1111/jipb.13827.

Wang, X., Li, Z., Li, Q., Hu, Z., 2025. Alleviation of plant abiotic stress: mechanistic insights into emerging applications of phosphate-solubilizing microorganisms in agriculture. Plants 14 (10), 1558.

Wang, Y., Lambers, H., 2020. Root-released organic anions in response to low phosphorus availability: recent progress and challenges. Journal of Experimental Botany 71 (22), 6565–6575.

Wang, Z., Ruan, W., Shi, J., Zhang, L., Xiang, D., Yang, C., et al., 2018. Rice SPX1 and SPX2 inhibit phosphate starvation responses through interacting with PHR2 in a phosphate-dependent manner. Proceedings of the National Academy of Sciences 111 (41), 14953–14958. Available at https://doi.org/10.1073/pnas.1404680111.

Waraich, E.A., Ahmad, R., Halim, A., Aziz, T., 2012. Alleviation of temperature stress by nutrient management in crop plants: a review. Journal of Soil Science and Plant Nutrition 12 (2), 221–244. Available at https://doi.org/10.4067/s0718-95162012000200003.

Wild, R., Gerasimaite, R., Jung, J.-Y., Truffault, V., Pavlovic, I., Schmidt, A., et al., 2016. Control of eukaryotic phosphate homeostasis by inositol polyphosphate sensor domains. Science (New York, N.Y.) 352 (6288), 986–990. Available at https://doi.org/10.1126/science.aad9858.

Wu, W., Wu, J., Liu, X., Chen, X., Wu, Y., Yu, S., 2017. Inorganic phosphorus fertilizer ameliorates maize growth by reducing metal uptake, improving soil enzyme activity and microbial community structure. Ecotoxicology and Environmental Safety 143, 322–329. Available at https://doi.org/10.1016/j.ecoenv.2017.05.039.

Yang, S.Y., Lin, W.Y., Hsiao, Y.M., Chiou, T.J., 2024. Milestones in understanding transport, sensing, and signaling of the plant nutrient phosphorus, American Society of Plant Biologists, Taiwan. The Plant Cell 36 (5), 1504–1523. Available at https://doi.org/10.1093/plcell/koad326.

Zheng, M., Deng, M., Zhou, Y., Liu, R., Liu, Y., Wang, H., et al., 2023. Multifaceted effects of difenoconazole in tomato fruit ripening: Physiology, flavour and nutritional quality. Plant Physiology and Biochemistry 194, 223–235. Available at https://doi.org/10.1016/j.plaphy.2022.11.015.

Zhou, J., Hu, Q., Xiao, X., Yao, D., Ge, S., Ye, J., et al., 2021. Mechanism of phosphate sensing and signaling revealed by rice SPX1-PHR2 complex structure. Nature Communications 12 (1). Available at https://doi.org/10.1038/s41467-021-27391-5.

CHAPTER

Potassium in enhancing stress tolerance

5

Mamdouh A. Eissa

Department of Soils and Water, Faculty of Agriculture, Assiut University Assiut, Egypt

5.1 Introduction

Potassium (K) is one of the three primary macronutrients required in large quantities by plants, along with nitrogen (N) and phosphorus (P). Unlike N and P, potassium does not become a structural component of organic molecules but instead functions predominantly as a regulator of numerous physiological and biochemical processes essential for plant growth and productivity. It is indispensable for maintaining ionic balance, regulating osmotic pressure, and supporting enzyme activation, photosynthesis, and carbohydrate metabolism (Liao et al., 2023; Hossain et al., 2024). Potassium plays a central role in water and nutrient transport within the plant, supporting cell turgor, expansion, and overall metabolic efficiency (Liao et al., 2023; Easwaran et al., 2024). It activates over 60 enzymes that govern critical processes such as protein synthesis, starch formation, and energy metabolism (Liao et al., 2023; Sarikhani and Ebrahimi, 2024). Adequate potassium availability enhances root growth, improves nitrogen and water uptake, and increases cellulose and protein accumulation, ultimately translating into higher yields and improved crop quality (Easwaran et al., 2024; Kaur et al., 2022). Beyond growth, potassium strengthens plants' resilience to abiotic and biotic stresses, including drought, cold, and pathogen attacks, by improving physiological stability and enhancing disease resistance (Easwaran et al., 2024; Chen et al., 2021). In addition, potassium contributes to grain filling, kernel weight, and fiber properties in cotton, which are vital for overall crop marketability (Easwaran et al., 2024; Kaur et al., 2022). Its role in stomatal regulation is also crucial, ensuring efficient photosynthesis and transpiration under fluctuating environmental conditions (Sarikhani and Ebrahimi, 2024; Liao et al., 2023). Despite its importance, the availability of potassium in soils is often limited, as most of it exists in insoluble mineral forms such as feldspars, micas, and illites (Ramírez-Cariño et al., 2023; Chen et al., 2021). Only a small portion is readily available to plants, and this limitation has prompted interest in sustainable strategies such as the use of potassium-solubilizing bacteria to enhance bioavailability (Ramírez-Cariño et al., 2023; Chen et al., 2021). Given its multifaceted roles in plant nutrition, productivity, and stress adaptation, potassium remains a cornerstone element for

Mineral Nutrients to Mitigate Abiotic Stress in Crops. DOI: https://doi.org/10.1016/B978-0-443-44712-9.00009-8

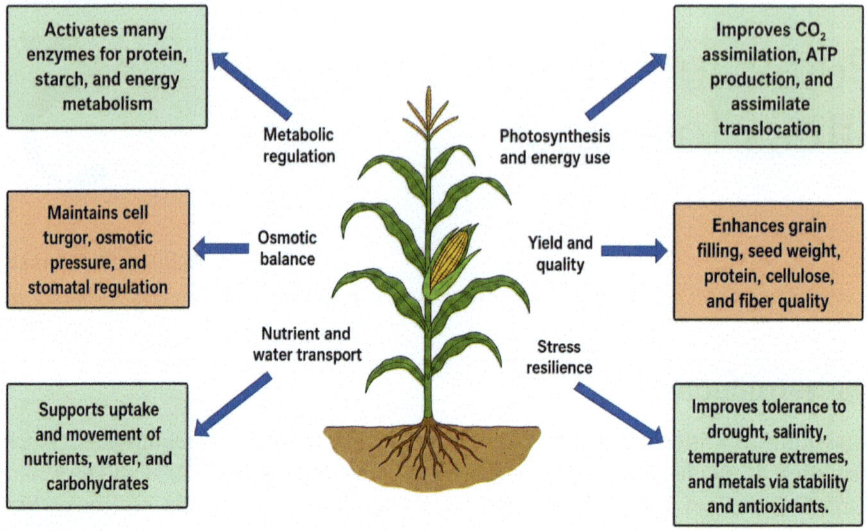

FIGURE 5.1

Role of potassium in plant nutrition.

optimizing crop performance and ensuring global food security (Fig. 5.1). Drought, salinity, temperature extremes, pollution, and heavy metals are major abiotic stressors that significantly reduce global crop productivity, threatening food security worldwide (Husen, 2022, 2025; Shaffique et al., 2025; Kumar et al., 2025; Ansari et al., 2025; Tadayon et al., 2025; Husen, 2024). These stressors can act individually or in combination, often leading to more severe impacts when they co-occur, a phenomenon referred to as "multifactorial stress combination" (Pascual et al., 2023). Salinity affects about 800 million hectares of land globally, particularly in arid and semi-arid regions, restricting crop productivity (Ansari et al., 2025; Tadayon et al., 2025). Drought is becoming more frequent and severe due to climate change, limiting water availability and plant growth, especially in water-scarce regions (Ansari et al., 2025, Tadayon et al., 2025). Temperature extremes, including heat waves and cold snaps, are causing more frequent and intense environmental stresses, further reducing crop yields and quality (Manorma et al., 2024; Ansari et al., 2025).

Heavy metals in soils, often from industrial pollution, not only limit crop productivity but also pose serious risks to human health through the food chain (Pascual et al., 2023; Ansari et al., 2025). The combined effects of these abiotic stresses result in billions of dollars in agricultural losses annually and have significant repercussions for food prices, land use, and economic stability (Safdar et al., 2022; Manorma et al., 2024). High management costs for these stresses can be prohibitive for many farmers, particularly in developing regions, exacerbating poverty and food insecurity (Manorma et al., 2024). These stresses impair plant physiology through multiple mechanisms (Husen, 2021a, 2021b). Salinity induces osmotic stress, ion toxicity, nutrient imbalances, and water deprivation, all of which reduce growth and photosynthesis (Ansari et al., 2025; Tadayon et al., 2025). Drought and high temperatures limit water and nutrient absorption, disrupt physiological processes, and lower fruit quality (Ansari et al., 2025; Tadayon et al., 2025). Heavy metals cause toxicity in plants, further reducing growth and yield, and can accumulate in edible plant parts, posing health risks (Pascual et al., 2023; Ansari et al., 2025). Climate change exacerbates these stresses by increasing their frequency and severity, making them more challenging to manage and intensifying their impact on global agriculture (Ansari et al., 2025; Tadayon et al., 2025). Changes in precipitation patterns, increased evapotranspiration, and rising global temperatures are worsening drought and salinity issues (Ansari et al., 2025; Tadayon et al., 2025).

These challenges are global, affecting both developed and developing countries, and are particularly acute in arid and semi-arid regions (Ansari et al., 2025; Tadayon et al., 2025). Salinity affects over 20 % of irrigated lands and may impact up to 50 % by 2050 due to industrial and agricultural irrigation practices (Hopmans et al., 2021; Singh, 2021).

5.2 Plants grown under drought stress

5.2.1 Adverse impacts of drought stress on plant productivity

Drought stress is a major environmental constraint that severely limits plant growth and crop yield worldwide. It impairs physiological and biochemical processes, primarily by inhibiting photosynthesis. Under water-deficit conditions, stomatal closure restricts CO_2 uptake and gas exchange, reducing carbon assimilation and energy production, which directly limits growth and biomass accumulation (Javaid et al., 2024; Benavides-Mendoza et al., 2024). Water scarcity also decreases nutrient absorption and transport, disrupting hormonal balance, including abscisic acid and cytokinins, and altering cell membrane selectivity, further constraining plant development (Javaid et al., 2024; Benavides-Mendoza et al., 2024). At the cellular level, drought causes dehydration, reduces turgor pressure, and inhibits cell elongation, resulting in stunted growth and delayed development (Ayaz et al., 2015; Javaid et al., 2024).

Morphologically, drought limits leaf expansion, root growth, and overall biomass accumulation, while also modifying root exudation patterns (Furtak and Wolińska, 2023; Pantha et al., 2024). Reproductive structures are highly sensitive to water deficit; flower size, number, pollen viability, and nectar quality decline, leading to fewer and lower-quality reproductive organs (Amin Jafari et al., 2024; Al Tawaha et al., 2025). Drought can also delay seed germination, seedling emergence, and establishment, which are crucial for crop productivity (Benavides-Mendoza et al., 2024; Amin Jafari et al., 2024). Stress during reproductive stages such as anthesis and grain filling can reduce grain number, grain weight, and overall yield by more than 40% under severe conditions, while moderate drought may cause 10%–15% losses (Pantha et al., 2024; Al Tawaha et al., 2025). Drought also affects product quality, causing fruit shriveling, rice cracking, and decreased seed viability, and increases production and harvesting costs, posing risks to food security and farmer livelihoods (Ayaz et al., 2015; Furtak and Wolińska, 2023).

Sensitivity to drought varies across developmental stages, with the reproductive phase being the most vulnerable. Nonetheless, all growth stages from seedling emergence to stem elongation can be affected, and the severity of impact depends on the intensity, timing, and duration of drought events (Pantha et al., 2024; Ishfaq et al., 2024). Combined stresses, such as drought accompanied by heat waves, exacerbate yield losses by shortening the plant life cycle and altering seed composition (Furtak and Wolińska, 2023). Some plants develop adaptive physiological, morphological, and molecular mechanisms to tolerate drought; however, these responses are often species-specific and may be insufficient under prolonged or severe stress (Pantha et al., 2024; Al Tawaha et al., 2025). Repeated exposure to drought may induce stress memory, enhancing responsiveness to future events, although protection against extreme conditions remains limited (Pantha et al., 2024).

Drought is one of the most limiting abiotic stresses affecting crop productivity worldwide. Potassium (K), being the most abundant cation in plant cells, plays a central role in enhancing drought tolerance by maintaining physiological and biochemical stability under water-deficit conditions (Chowdhury et al., 2024; Thakur et al., 2025). Potassium's functions during drought are multifaceted, encompassing osmotic regulation, stomatal control, antioxidant defense, root development, and yield maintenance.

5.2.2 Potassium and osmotic adjustment in drought-stressed plants

One of the key drought-resistance mechanisms facilitated by potassium is osmotic adjustment. Adequate K levels lower the osmotic potential of plant cells, allowing them to retain water and maintain turgor pressure, which is critical for sustaining cell expansion, stomatal function, and photosynthesis under stress (Xu et al., 2021; Thakur et al., 2025). Plants store K^+ in vacuoles to improve water uptake efficiently, a more energy-conserving strategy than synthesizing organic solutes (Bellaloui et al., 2013; Thakur et al., 2025). Potassium activates numerous enzymes involved in photosynthesis, protein synthesis, glycolysis, and carbohydrate transport, supporting plant growth under drought (Chowdhury et al., 2024; Singh and Lone, 2021). It also maintains electrical potential gradients across membranes, essential for nutrient uptake and metabolic regulation (Chowdhury et al., 2024; Singh and Lone, 2021).

5.2.3 Potassium and stomatal control under drought

Potassium is central to stomatal movement, controlling guard cell turgor. Under drought, K^+ accumulation in guard cells enables precise stomatal closure, reducing transpiration and water loss while maintaining CO_2 uptake for photosynthesis (Xu et al., 2021; Thakur et al., 2025). K-deficient plants exhibit delayed stomatal closure, leading to excessive water loss and reduced drought tolerance (Cakmak, 2005; Chen et al., 2017) (). K fertilization decreased stomatal conductance by 30–40% under drought, and water-use efficiency improved by up to 25% in K-supplemented plants (Thakur et al., 2025; Singh and Lone, 2021).

5.2.4 Potassium-mediated antioxidant defense and ROS detoxification in crops under drought stress

Drought induces reactive oxygen species (ROS) accumulation, causing oxidative damage to lipids, proteins, and DNA. Potassium enhances antioxidant enzyme activities—including superoxide dismutase (SOD), catalase (CAT), and ascorbate peroxidase (APX)—to minimize oxidative stress (Rashki et al., 2022; Chowdhury et al., 2024). In tomato, K treatment increased SOD activity by 42%, CAT by 36%, and APX by 29%, while in wheat, lipid peroxidation was reduced by up to 40% (Hasanuzzaman et al., 2018). K-deficient plants show increased ROS and are more susceptible to drought-induced damage (Rashki et al., 2022; Chowdhury et al., 2024).

5.2.5 Potassium effects on root development and nutrient uptake in crops under drought stress

Potassium (K) is a critical macronutrient that profoundly influences root development in plants. Its functions span the regulation of growth, metabolic activity, water relations, nutrient uptake, and stress resilience. Potassium directly regulates root growth and shapes root system architecture. Deficiency in potassium leads to poorly developed roots, slower growth rates, and reduced seed and yield production. Adequate potassium supply enhances root length, density, and overall vigor, improving plant resilience and productivity (Toribio et al., 2025; Singh et al., 2025). It maintains cell turgor, which is essential for cell expansion and root elongation, allowing roots to explore soil efficiently for water and nutrients (Shankar et al., 2013, Toribio et al., 2025). Potassium also activates numerous enzymes involved in metabolism, including those regulating root growth and nutrient uptake, ensuring optimal root function (Singh et al., 2025; Khan et al., 2023). By regulating osmotic pressure within root cells, potassium facilitates water uptake, particularly under drought or water-limited conditions, and enhances aquaporin activity for improved water transport (Yang et al., 2023a; Toribio et al., 2025). Adequate potassium also improves the uptake of other nutrients and promotes organic acid exudation, which increases nutrient availability in the rhizosphere. Interactions with potassium-solubilizing microorganisms further enhance root growth and nutrient acquisition (Jiang et al., 2023; Toribio et al., 2025). Potassium contributes to root resilience under abiotic stresses such as drought and biotic stresses such as diseases, supporting overall root health and function (Toribio et al., 2025; Singh et al., 2025). Insufficient potassium results in stunted root systems, reduced proliferation, and compromised root functionality, making plants more susceptible to environmental stresses and diseases (Kour et al., 2020; Singh et al., 2025). Potassium-deficient roots also lead to lower crop yields, smaller seeds, and decreased quality, particularly in root and tuber crops (Jiang et al., 2023; Singh et al., 2025).

5.2.6 Potassium-mediated growth and yield under drought stress

Potassium (K) is a crucial macronutrient that plays an essential role in enhancing crop yield under drought stress conditions. Table 5.1 surmises the roles of potassium (K) in mitigating drought stress in plants. It is involved in a range of physiological and biochemical processes, including osmotic adjustment, stomatal regulation, enzyme activation, and the maintenance of turgor pressure at low leaf water potentials. Through these functions, potassium optimizes water use efficiency and reduces membrane damage, thereby contributing to plant resilience under water-deficit conditions (Aksu and Altay, 2020; Bhardwaj et al., 2025). Several studies have demonstrated that potassium application significantly improves crop yield under drought. In maize, the application of 75 kg K ha^{-1} resulted in higher biological and grain yield components (Ali et al., 2019), while wheat yield was enhanced when potassium fertilization was combined with salicylic acid (Tian et al., 2020; Munsif et al., 2022). Similar yield improvements have also been observed in sunflower and sugar beet under potassium application during drought stress (Aksu and Altay, 2020; Shehzad et al., 2020). These effects are largely attributed to potassium's role in maintaining leaf water status, enhancing osmotic adjustment, and upregulating antioxidant defense systems, which collectively improve drought tolerance in crops such as sunflower and wheat (Shehzad et al., 2020; Sun et al., 2025). Moreover, potassium facilitates carbohydrate partitioning into stems, which is vital for sustaining grain filling under drought conditions (Cakmak and Rengel, 2024). The extent of potassium's effectiveness, however, can vary across crop genotypes. For example, barley lines with higher potassium retention in flag leaves exhibited superior drought tolerance and delayed leaf senescence (Hosseini et al., 2016). Likewise, genotypic variability in cocoa yield responses to drought stress has been linked to potassium fertilization (Adet et al., 2024). The beneficial effects of potassium are also influenced by application rates, as exemplified in maize, where 75 kg K ha^{-1} proved optimal, while higher rates led to a slight decline in yield (Ali et al., 2019). Similarly, potassium application at 80 mg/kg soil improved growth and yield in rape subjected to drought conditions (Zhu et al., 2021). Table 5.2 show the optimal potassium application rates and associated benefits under drought stress.

Table 5.1 Key roles of potassium (K) in mitigating drought stress in plants.

Function	Mechanism/Effect	References
Osmotic adjustment and water retention	Maintains cell turgor and water potential; lowers osmotic potential by storing K^+ in vacuoles; improves water uptake and retention; energy-efficient osmoregulation compared to organic solutes. Enhances relative water content (15%–30% higher under drought).	Thakur et al. (2025), Xu et al. (2021)
Stomatal regulation and water-use efficiency	Controls guard cell turgor; facilitates precise stomatal closure under drought; reduces transpiration while maintaining CO_2 uptake; improves water-use efficiency by up to 25% in some crops.	Thakur et al. (2025), Singh and Lone (2021)
Enzyme activation and metabolic stability	Activates enzymes involved in photosynthesis, protein synthesis, glycolysis, and carbohydrate translocation; maintains electrical potential gradients across membranes; stabilizes metabolism under water deficit.	Singh and Lone (2021), Chowdhury et al. (2024)
Antioxidant defense and ROS detoxification	Enhances activity of SOD, CAT, and APX; reduces reactive oxygen species accumulation; protects chloroplasts, membranes, and proteins from oxidative damage; lowers lipid peroxidation by up to 40%.	Rashki et al. (2022), Chowdhury et al. (2024)
Root development and nutrient uptake	Promotes deeper and more branched roots; increases root length density and dry mass; improves K^+ acquisition under low soil K conditions; enhances water and nutrient uptake from subsoil.	Thakur et al. (2025), Chowdhury et al. (2024)
Yield and growth maintenance	Sustains photosynthesis, carbohydrate metabolism, and overall growth; mitigates drought-induced reductions in yield; improves grain filling, kernel weight, and yield stability under water-limited conditions (up to 30%–40% yield improvement in some crops).	Rashki et al. (2022), Chowdhury et al. (2024)

5.3 Salinity-induced limitations on plant growth and yield

5.3.1 Salinity stress in plants

Saline soils occupy a vast area worldwide, with the global extent reported at 1.381 billion hectares, accounting for 10.7% of the total land area (FAO, 2024). Similarly,

Table 5.2 Optimal potassium application rates and associated benefits under drought stress.

Crop	Optimal K application	Key benefits under drought stress	References
Maize	75 kg ha^{-1}	Increased biological and grain yield components	Ali et al. (2019)
Wheat	100 kg ha^{-1}	Improved yield, nutrient uptake, and antioxidant profile	Munsif et al. (2022)
Sunflower	11.48 g L^{-1}	Enhanced leaf gas exchange, antioxidative defense	Shehzad et al. (2020)
Sugar Beet	10–80 mg kg^{-1}	Increased relative water content, reduced membrane damage	Aksu and Altay (2020)
Rape	80 mg kg^{-1} soil	Improved growth, yield, and seed quality	Zhu et al. (2021)

the majority of Earth's water resources are saline; estimates indicate that approximately 96%–97.5% of the global water supply is salty, primarily stored in oceans and seas (Ren et al., 2021). Salt stress, also referred to as salinity stress, is one of the most critical abiotic factors limiting plant growth, development, and productivity. It arises when plants are exposed to excessive concentrations of soluble salts, predominantly sodium (Na^+) and chloride (Cl^-), in soils or irrigation water (Rekaby et al., 2021; Gao et al., 2024; Pandit et al., 2024; Umer et al., 2024). The adverse effects of salt stress on plants are diverse, encompassing osmotic stress, ion toxicity, and metabolic disturbances. Elevated salt concentrations in the rhizosphere reduce the soil water potential, restricting water uptake by roots and leading to water-deficit conditions. This osmotic stress is typically the earliest physiological response to salinity, resulting in reduced root and shoot elongation, decreased biomass accumulation, and overall stunted plant development (Ding et al., 2022; Liu et al., 2022; Pandit et al., 2024).

In addition to osmotic imbalance, the excessive accumulation of sodium (Na^+) and chloride (Cl^-) ions disrupts nutrient homeostasis and triggers ion toxicity. Sodium can displace calcium (Ca^{2+}) from the plasma membrane, impairing membrane stability and further suppressing growth (Rekaby et al., 2021; Gao et al., 2024; Zhang et al., 2025). Such ionic disturbances interfere with the uptake of essential nutrients, aggravating nutrient deficiencies and compromising plant productivity (Ding et al., 2022; Pandit et al., 2024; Umer et al., 2024). Salt stress also leads to significant physiological and metabolic impairments. It reduces photosynthetic efficiency by inducing stomatal closure, damaging chloroplast structures, and impairing electron transport processes (Zhang et al., 2025; Gao et al., 2024). Enzyme activities linked to carbon assimilation are also diminished, further limiting plant energy metabolism. Moreover, salinity promotes the overproduction of reactive oxygen species (ROS), which causes oxidative stress and cellular injury (Rekaby et al., 2021; Dzinyela et al., 2023; Chen et al., 2025). These physiological changes manifest in altered plant growth and development.

Salt stress commonly reduces leaf size, modifies leaf morphology, and accelerates senescence under prolonged conditions (Guo et al., 2024; Yang et al., 2025). Reproductive stages, including seed germination, flowering, and fruiting, are particularly sensitive, often resulting in reduced yields and poor crop quality (Ding et al., 2022; Gao et al., 2024). At the tissue and ecosystem level, the root system is highly vulnerable, with reductions in root length, surface area, and volume under salinity stress (Zhang et al., 2025). Furthermore, salt stress alters the composition and diversity of microbial communities in the rhizosphere and phyllosphere, which in turn can influence nutrient cycling and plant–microbe interactions (Yang et al., 2025). The severity of salt stress depends on multiple factors, including plant species, growth stage, and environmental conditions. Younger plants and sensitive species are more susceptible, while stress intensity increases under unfavorable climatic conditions such as high temperatures and low humidity (Rehman et al., 2025; Umer et al., 2024). Additionally, both the concentration and duration of salinity exposure are critical, with higher levels and longer exposure times causing more pronounced damage (Abd El-Wahed et al., 2023; Guo et al., 2024; Umer et al., 2024).

Salt stress is a major factor reducing crop yields and overall productivity, posing a global threat to agriculture and food security (Yang et al., 2023b; Shengcai et al., 2025). High salt concentrations inhibit photosynthesis and protein synthesis, leading to reduced plant growth and yield (Yin et al., 2019; Quan et al., 2025). Salt stress also induces the production of reactive oxygen species (ROS), causing oxidative stress and cellular damage (Dzinyela et al., 2023; Shengcai et al., 2025). Excess salts further cause ionic toxicity, particularly from sodium and chloride ions, and disrupt the uptake of essential nutrients such as potassium (K^+) and calcium (Ca^{2+}), leading to nutrient imbalances (Yin et al., 2019; Shengcai et al., 2025). In addition, increased soil salinity raises osmotic pressure, making it harder for plants to absorb water, which limits growth and yield (Kaur and Nayyar, 2015; Mohamed et al., 2021). Growth and development are significantly impaired under salt stress. Seed germination, plant height, and root length are suppressed, leading to poor plant establishment (Ayub et al., 2020; Kaur and Nayyar, 2015). Growth delays and developmental restrictions ultimately result in yield loss (Zhang et al., 2021; Quan et al., 2025). Root and shoot growth, tillering, and panicle branching are also reduced, all of which are critical for yield formation (Ayub et al., 2020; Quan et al., 2025). At the molecular level, salt stress triggers changes in gene expression, including those associated with ABA signaling and pathways regulating salt tolerance. The ability of crop varieties to activate such genes determines their performance under saline conditions (Zhang et al., 2021; Quan et al., 2025). Yield losses due to salinity represent a serious concern for global food security, particularly as the salinization of arable land continues to rise (Dzinyela et al., 2023; Li et al., 2011). Developing and cultivating salt-tolerant crop varieties is therefore essential, although progress remains constrained by the limited number of genes with practical breeding value (Abd El-Wahed et al., 2023; Zhang et al., 2021; Quan et al., 2025).

Salt stress directly reduces crop yields and overall productivity, posing a significant threat to global agriculture and food security (Gao et al., 2024; Pandit et al., 2024; Umer et al., 2024). Potassium is essential for plant adaptation to salt stress by maintaining ion balance, supporting physiological processes, and mitigating the negative effects of sodium toxicity. Ensuring adequate potassium supply or enhancing its uptake can significantly improve plant resilience under saline conditions (Miao et al., 2024; Han et al., 2025).

5.3.2 Role of potassium in maintaining ion homeostasis under salinity stress

Potassium helps maintain a high potassium-to-sodium (K/Na) ratio, which is crucial for plant growth under salt stress. A low K/Na ratio negatively affects enzyme activity, photosynthesis, turgor maintenance, stomatal movement, and root membrane integrity. Under salt stress, plants often lose K due to competition with Na for uptake and transport, leading to K deficiency and aggravated stress symptoms (Yan et al., 2021; Alam et al., 2022). Table 5.3 shows the impacts of potassium in mitigating salt stress in plants. Endophytic fungi can support plants by reducing Na uptake, increasing Na efflux, and sequestering Na into vacuoles, partly through regulating ion transporters and related genes (Abd El-Wahed et al., 2023; Zhang et al., 2024a; Xie et al., 2025).

Table 5.3 Role of potassium in mitigating salt stress in plants.

Effect	Mechanism/Outcome	References
Increased antioxidant enzyme activity	Enhances SOD, CAT, and APX activity to scavenge ROS	Che et al. (2022), Ditta and Ullah (2023), Chaudhary et al. (2024)
Maintains K^+/Na^+ ratio	Promotes selective K^+ uptake and reduces Na^+ absorption	Degl'innocenti et al. (2009), Ditta and Ullah (2023), Han et al. (2025)
Reduces ROS production	Lowers NAD(P)H oxidase activity and supports photosynthetic efficiency	Degl'innocenti et al. (2009)
Improves growth and stress tolerance	Supports metabolism, chlorophyll stability, and photosynthesis	Abdel-Haliem et al. (2017), Han et al. (2025)
Supports osmotic regulation	Maintains cell turgor, prevents leaf senescence, and preserves cellular function	Abdel-Haliem et al. (2017), Chen et al. (2021), Che et al. (2022)

Reducing Na^+ uptake and increasing Na^+ efflux (removal from cells), sequestering Na^+ into vacuoles to lower cytoplasmic Na^+ levels, and enhancing the expression of potassium transporters and channels (e.g., HKT1, KAT1, KAT2) are key mechanisms that improve K^+ uptake and retention (Zhang et al., 2024a; Xie et al., 2025). Additionally, improved K^+/Na^+ homeostasis is achieved through the regulation of genes such as *NHX*, *SOS1*, and *CNGC15*(Abd El-Wahed et al., 2023; Xie et al., 2025; Ma et al., 2025).

5.3.3 Alleviating salt stress symptoms through potassium nutrition

Potassium supplementation increases the K/Na ratio in plant tissues, a central mechanism for alleviating salt stress. In corn, increased K content and K/Na ratio in leaf sap were directly linked to reduced salt stress symptoms (Hameed et al., 2024). Similarly, exogenous application of potassium can significantly enhance K concentration in salt-stressed plants, reversing part of the K loss (Zulfiqar et al., 2023). Salt stress causes K deficiency by promoting Na^+ competition for transport proteins and by causing K leakage due to membrane damage. This leads to further nutrient deficiency symptoms and impairs plant growth (Yan et al., 2021; Hiya et al., 2024). Maintaining adequate potassium levels helps counteract these deficiencies, supporting normal metabolic and physiological functions under stress (Hiya et al., 2024; Wang et al., 2021). Potassium is essential for enzyme activation, regulation of osmotic pressure, and stomatal movement. Under salt stress, K^+ helps maintain cellular homeostasis, supports photosynthesis, and regulates water status in plant tissues (Zou et al., 2021; Moghaddam et al., 2023). Adequate K^+ levels prevent cellular damage by supporting membrane integrity and reducing oxidative stress, which are common consequences of salt stress (Moghaddam et al., 2023; Hiya et al., 2024).

5.3.4 Potassium in supporting physiological and metabolic functions under salinity stress

Potassium is essential for chlorophyll synthesis, photosynthesis, and metabolic activity. Under salt stress, it sustains photosynthetic efficiency by promoting ATP synthesis, accelerating carbon assimilation, and preventing electron transport chain overreduction. These processes improve growth and stress tolerance (Miao et al., 2024; Han et al., 2025). In addition, K enhances antioxidant enzyme activity and proline accumulation, further contributing to stress resilience (Han et al., 2025). Potassium application enhances the activity of key antioxidant enzymes, including superoxide dismutase (SOD), catalase (CAT), and ascorbate peroxidase (APX), which are otherwise suppressed under salt stress. This enhancement strengthens plant defense against reactive oxygen species (ROS), reducing cellular damage and improving tolerance to high salinity (Degl'innocenti et al., 2009; Che et al., 2022; Ditta and Ullah, 2023; Chaudhary et al., 2024). Adequate potassium nutrition can also lower ROS production by reducing NAD(P)H oxidase activity and maintaining efficient photosynthetic electron transport, thereby mitigating oxidative stress. This effect is particularly important because salt stress typically increases ROS generation, which can damage cellular components if not properly managed (Degl'innocenti et al., 2009).

5.3.5 Role of potassium in gene expression and stress response under salt stress

Potassium can modulate the expression of stress-responsive genes and pathways linked to plant hormone regulation, strengthening the plant's ability to cope with salinity (Miao et al., 2024). Potassium influences gene regulation under salt stress by modulating the expression of potassium transporter and channel genes, both directly and via miRNA-mediated pathways. This regulation is essential for maintaining ion homeostasis, particularly the Na^+/K^+ ratio, and supports broader stress adaptation mechanisms in plants (Shobbar et al., 2023; Zhou et al., 2023; Meng et al., 2020; Zhang et al., 2024a). Salt stress alters the expression of genes involved in potassium transport. Genes encoding potassium transporters (e.g., BpKAT, BpHAK5) are highly induced in leaves under salt stress, indicating upregulation to maintain K^+ homeostasis. In certain mutants (bpsos1), these genes show differential regulation, with some upregulated and others downregulated compared to controls, suggesting a complex regulatory network responsive to salt stress. Potassium channel genes (BpAKTs) are generally expressed at lower levels under salt stress and are further downregulated in mutants, highlighting the sensitivity of K^+ channel gene expression to salt-induced cellular changes (Zhang et al., 2024b). Salt-responsive microRNAs (miRNAs) target genes involved in potassium transport. miRNAs regulate genes such as the potassium transporter KUP3p and the potassium channel akt2-like under salt stress, affecting ion homeostasis and plant adaptation to salinity. This posttranscriptional regulation by miRNAs is part of the plant's strategy to restore ion balance and enhance salt tolerance (Shobbar et al., 2023). Salt stress also increases the expression of genes related to potassium transport and homeostasis, as well as those involved in cell division, amino acid metabolism, sucrose synthesis, photosynthetic activity, and ABA signaling. These changes are particularly evident in salt-tolerant plant varieties, supporting the role of potassium-related gene regulation in the overall stress response (Meng et al., 2020).

5.3.6 Effects of potassium on crop yield under salt stress

Potassium (K) plays a pivotal role in mitigating the adverse effects of salt stress on crop yield. In soybean (*Glycine max*), K application improved growth and biomass under salinity by enhancing relative water content, chlorophyll content, membrane stability, leaf area, photosynthetic rate, and reducing sodium accumulation, indicating its role in maintaining ionic balance and physiological performance (Parveen et al., 2025). In wheat (*Triticum aestivum*), co-application of K and γ-aminobutyric acid (GABA) enhanced growth and yield traits by improving defense mechanisms, nitrogen and carbon assimilation, and photosynthesis-related processes, highlighting the synergistic effects of K with other compounds (Kumari et al., 2023). In peanut (*Arachis hypogaea*), K fertilization at recommended doses mitigated salt stress during critical growth stages, increasing pod mass, the number of marketable pods, and overall yield, emphasizing the importance of application timing and dosage (Abreu et al., 2024). In sugarcane (*Saccharum officinarum*), K application improved growth, cane yield, and juice quality under salinity by reducing sodium uptake and enhancing K concentration in plant tissues, demonstrating its role in both yield and quality improvement (Ashraf et al., 2012). Field experiments in wheat (*Triticum aestivum*) showed that K enhanced antioxidant enzyme activity, chlorophyll content, photosynthesis rate, and yield components under salt stress, reflecting its contribution to physiological resilience (Bybordi, 2015). In tomato (*Solanum lycopersicum*), K increased growth and yield, particularly in salt-tolerant genotypes, and improved fruit quality characteristics under salinity (Amjad et al., 2014). In rice (*Oryza sativa*), optimal K levels significantly increased grain yield and reduced soil salinity, enhancing salt resistance (Trinh et al., 2024). In cotton (*Gossypium hirsutum*), K application raised the K^+/Na^+ ratio, improved photosynthesis, and enhanced carbohydrate metabolism, resulting in higher yields (Ju et al., 2021). Similarly, in mungbean (*Vigna radiata*), potassium nitrate improved growth, photosynthetic pigments, and antioxidant enzyme activities, contributing to increased yield under salt stress (Ibrahimi et al., 2025).

The optimum dose of potassium under salt stress varies considerably among crop species. In rice (*Oryza sativa*), application of 50 kg K_2O ha^{-1} under saline field conditions resulted in the highest grain yield and improved K^+/Na^+ ratios (Trinh et al., 2024). A supra-optimal dose of 150 kg K_2O ha^{-1} enhanced growth, nutrient uptake, and economic yield in sugarcane (*Saccharum officinarum*) under salinity (Ashraf et al., 2008). Application of 300 kg K ha^{-1} effectively alleviated salinity-induced stress and enhanced growth and physiological traits in maize (*Zea mays*) (Abbasi et al., 2015)

5.4 Physiological and metabolic responses of plants to heat stress

5.4.1 Effects of heat stress on plant growth and yield

Heat stress refers to exposure to elevated temperatures above a critical threshold that causes irreversible damage to plants at morphological, biochemical, and physiological levels (Du et al., 2019; Lovio-Fragoso et al., 2023). It is a major abiotic stress that negatively influences plant growth, development, reproduction, and yield, ultimately reducing agricultural productivity (Szádeczky-Kardoss et al., 2022; Londoño Vélez et al., 2022; Lovio-Fragoso et al., 2023). Heat stress impairs seed germination, seedling growth, photosynthetic activity, protein stability, and enzymatic reactions. It also disrupts pollen development and pollination, leading to reduced fertility. Visible symptoms include leaf scorching, early senescence, rolling and drying, fruit discoloration, and tissue necrosis. At the cellular level, heat stress causes protein denaturation, membrane damage, and elevated production of reactive oxygen species (Szádeczky-Kardoss et al., 2022; Londoño Vélez et al., 2022; Lovio-Fragoso et al., 2023).

Plants respond through basal and acquired thermotolerance, supported by protective mechanisms such as the synthesis of heat-shock proteins and the activation of antioxidant defense systems. Heat acclimation, in which prior exposure to moderate stress enhances subsequent tolerance, further contributes to adaptation (Szádeczky-Kardoss et al., 2022; Londoño Vélez et al., 2022; Lovio-Fragoso et al., 2023). Heat stress reduces both yield and quality of crops, particularly when stress coincides with reproductive stages. The extent of damage depends on stress intensity, duration, developmental stage, and species sensitivity (Szádeczky-Kardoss et al., 2022; Londoño Vélez et al., 2022; Lovio-Fragoso et al., 2023).

Heat stress has profound and multifaceted impacts on plant growth, influencing physiological, biochemical, molecular, and developmental processes. Germination and early growth are particularly vulnerable, with reduced germination rates, retarded shoot and root growth, and poor seedling establishment being common outcomes (Tushabe et al., 2023; Kashyap et al., 2024; Ying et al., 2025). Seedling development is highly sensitive, and severe heat stress can cause chlorosis, etiolation, and plant death (Du et al., 2019; Kashyap et al., 2024). At the physiological level, heat stress disrupts photosynthesis and respiration, reducing energy availability for plant development, while also interfering with evapotranspiration and decreasing relative water content, which results in wilting and senescence (Molina-Bravo et al., 2011; Ying et al., 2025). At the cellular level, heat stress compromises membrane stability and enhances oxidative stress, primarily through the overproduction of reactive oxygen species (ROS). This leads to protein denaturation, enzyme inactivation, and damage to DNA, lipids, and other macromolecules, further accelerating senescence and yield decline (Khalid et al., 2025; Ying et al., 2025).

Morphological alterations are also evident, including reduced shoot dry mass, decreased relative growth rate, leaf scorching, early senescence, abscission, and fruit discoloration (Molina-Bravo et al., 2011; Kashyap et al., 2024). Elevated temperatures often accelerate crop development, shortening the duration of key metabolic processes and thereby reducing final yields (Pandey and Choudhary, 2019). Reproductive development is especially sensitive to heat, as pollen fertility, ovule quality, and floral organ development are negatively affected. These disruptions lead to reduced seed mass, fewer flowers, and altered flowering times, ultimately diminishing reproductive success and yield across crops, particularly when stress coincides with reproductive phases (Tushabe et al., 2023; Kashyap et al., 2024). On a biochemical and genetic level, heat stress alters hormone and metabolite levels, disrupting normal signaling pathways, while also inducing changes in gene expression. The upregulation of microRNAs, stress-responsive genes, and heat-shock proteins (HSPs) provides partial protection but is often insufficient to counteract the cumulative damage caused by prolonged or intense heat exposure (Hemantaranjan et al., 2018; Khalid et al., 2025).

Potassium plays a vital role in enhancing plant resilience to heat stress by regulating stomatal conductance and maintaining water balance, thereby improving cellular hydration and preventing heat-induced damage (El-Sappah et al., 2022; Sharma et al., 2024). Acting as an osmolyte, potassium protects cellular structures from dehydration and oxidative stress, while simultaneously strengthening the antioxidant defense system (Sharma et al., 2024). It also enhances photosynthetic efficiency and metabolic activity, which supports plant vigor and ensures sustained growth under elevated temperatures (El-Sappah et al., 2022; Sharma et al., 2024). Through these combined mechanisms, potassium contributes significantly to improving plant tolerance and adaptation to heat stress.

5.4.2 Potassium mitigation of heat stress in plants

Potassium (K) plays a pivotal role in enhancing plant resilience to heat stress through multiple physiological, biochemical, and molecular mechanisms. One of its primary functions is osmotic regulation, where potassium helps maintain turgor pressure and water uptake under heat stress. This is particularly important because elevated temperatures often disrupt water relations and reduce mineral availability, making potassium supply essential for sustaining growth (El-Sappah et al., 2022; Sharma et al., 2024). In addition to its osmotic role, potassium contributes to cellular protection by activating antioxidant enzymes that mitigate the damaging effects of heat-induced reactive oxygen species (ROS). This action delays leaf senescence and protects plant tissues from oxidative damage during high temperatures (Sharma et al., 2024).

Another important function is the stabilization of cell membranes, which are highly vulnerable to heat-induced injury, adequate potassium maintains membrane integrity and ensures the continuity of normal metabolic processes (El-Sappah et al., 2022). Potassium also enhances chlorophyll synthesis and supports photosynthetic activity, enabling plants to sustain carbon assimilation and delay heat-induced leaf senescence. These benefits extend to improved nutrient uptake and activation of various enzymes, which are crucial for metabolism, particularly when heat stress impairs the absorption of essential elements (Khan et al., 2020; El-Sappah et al., 2022). On a molecular scale, potassium availability influences the expression of stress-responsive genes and transporters, including high-affinity potassium carriers and transcription factors that strengthen plant adaptation to elevated temperatures (Khan et al., 2020). Table 5.4 explains potassium-mediated mechanisms for heat stress mitigation in plants.

Potassium mitigates heat stress in plants by supporting water balance, activating antioxidant defenses, stabilizing membranes, maintaining photosynthesis,

Table 5.4 Potassium-mediated mechanisms for heat stress mitigation in plants.

Mechanism	Effect under heat stress	References
Osmotic regulation	Maintains water uptake and turgor pressure	El-Sappah et al. (2022), Sharma et al. (2024)
Antioxidant enzyme activation	Reduces oxidative damage and delays senescence	Sharma et al. (2024)
Membrane stability	Preserves cell integrity and physiological function	El-Sappah et al. (2022)
Chlorophyll synthesis and photosynthesis	Supports photosynthesis and delays leaf aging	Sharma et al. (2024)
Nutrient uptake and enzyme activity	Sustains metabolism and nutrient absorption	El-Sappah et al. (2022), Khan et al. (2020)
Gene expression regulation	Enhances stress-responsive pathways and K^+ transport	Khan et al. (2020)

improving nutrient uptake, and enhancing stress-responsive gene expression. These combined effects help plants survive and maintain productivity during periods of high temperature (Khan et al., 2020; El-Sappah et al., 2022; Sharma et al., 2024). Collectively, these mechanisms underline potassium's indispensable role in mitigating heat stress, highlighting its contribution to maintaining plant vigor, productivity, and survival under challenging climatic conditions.

5.5 Cold stress-induced changes in growth and crop yield

5.5.1 Effects of cold stress on plant growth and yield

Cold stress in plants refers to the adverse effects of low temperatures on growth, development, physiology, and overall productivity. It represents a major abiotic stress that significantly restricts crop yields and determines the geographical distribution of plant species. Cold stress is generally classified into two categories: chilling stress, which occurs at temperatures between 0°C and 20°C (sometimes defined as 0°C–15°C), and freezing stress, which occurs at temperatures below 0°C (Ritonga and Chen, 2020; Manasa S et al., 2022). Cold stress in plants is a major environmental challenge that disrupts growth, development, and productivity through a combination of physiological, biochemical, and molecular changes. Plants have evolved various mechanisms to sense, respond, and adapt to cold stress, but it remains a significant factor limiting agricultural output and plant distribution worldwide (Ritonga and Chen, 2020; Manasa S et al., 2022; Ali et al., 2024).

Cold stress is a major environmental constraint that disrupts plant growth, development, and productivity. It impairs photosynthesis, accelerates chlorophyll degradation, and inhibits key metabolic reactions, leading to stunted growth and reduced vigor (Dev et al., 2025; Zhang et al., 2020). Membrane rigidification, suppression of nutrient uptake, and hormonal imbalances, including altered levels of abscisic acid and jasmonic acid, further limit growth (Dev et al., 2025; Zhang et al., 2020). The inhibition of cell division and elongation, together with reduced photosynthetic efficiency, directly compromises developmental processes (Dev et al., 2025; Zhang et al., 2020). Cold stress also reduces crop yield and quality through cellular injury, impaired photosynthesis, and metabolic disruption (Dev et al., 2025; Zhang et al., 2020). Both chilling (0°C–15°C) and freezing (<0°C) stress reduce the productivity of major crops such as rice, wheat, and maize, while also shortening growing seasons and restricting geographical distribution (Dev et al., 2025; Zhang et al., 2020).

At the biochemical level, cold stress promotes excessive accumulation of reactive oxygen species (ROS), including hydrogen peroxide and superoxide, which damage lipids, proteins, and nucleic acids. Activities of antioxidant enzymes such as superoxide dismutase (SOD), peroxidase (POD), and catalase (CAT) are often suppressed, disturbing ROS homeostasis. To compensate, plants enhance antioxidant defenses and redox regulation, while exogenous application of minerals such as potassium and zinc, or compounds like melatonin, further strengthens antioxidant activity and tolerance (Dev et al., 2025; Zhang et al., 2020). Cold acclimation, characterized by increased antioxidase activity and the production of protective proteins, enhances cellular stability and survival under prolonged cold (Huliaieva et al., 2025; Dev et al., 2025). At the molecular level, cold stress is sensed at the plasma membrane, activating calcium influx, MAPK cascades, and hormone-regulated signaling pathways that reprogram gene expression (Dev et al., 2025; Zhang et al., 2020). The ICE1–CBF–COR signaling cascade is particularly crucial, as it upregulates cold-responsive genes and improves tolerance (Qian et al., 2025; Meng et al., 2025). Cold-tolerant species typically maintain greater membrane fluidity, robust antioxidant systems, and stronger expression of resistance genes, allowing them to mitigate damage and recover growth under low-temperature stress (Dev et al., 2025; Huliaieva et al., 2025; Zhang et al., 2020). Potassium mitigates cold stress by acting as an osmolyte, lowering the freezing point of cell sap, and increasing frost resistance. It stabilizes membranes, enhances antioxidant defenses, supports water and nutrient transport, and promotes the accumulation of sugars and protective compounds during cold acclimatization.

5.5.2 Role of potassium in alleviating cold stress

Potassium (K) plays a vital role in enhancing plant tolerance to cold stress, improving crop yield, and activating antioxidant defenses. Adequate potassium improves plant survival under chilling and freezing conditions by stimulating antioxidant enzymes that maintain membrane integrity, whereas potassium deficiency increases oxidative damage and reduces biomass and survival (El-Mahdy et al., 2018; Saadati et al., 2021; Salimi et al., 2024). High potassium levels lower the osmotic potential of cells, reducing cytoplasmic freezing and preventing dehydration, thereby mitigating cold-induced injury (Salimi et al., 2024). Cytosolic potassium is also essential for activating enzymes associated with frost resistance, further supporting cold tolerance (Saadati et al., 2021).

Potassium supply enhances growth and yield in crops such as ginseng, potato, pepper, tomato, eggplant, and maize by reducing the severity of chilling damage and sustaining photosynthesis and stomatal conductance under low-temperature stress (Salimi et al., 2024; Sharma et al., 2024). Moreover, potassium enhances the activities of key antioxidant enzymes, including superoxide dismutase, catalase, ascorbate peroxidase, and peroxidase, effectively scavenging reactive oxygen species and reducing oxidative damage (El-Mahdy et al., 2018; Saadati et al., 2021; Salimi et al., 2024; Sharma et al., 2024). Sufficient potassium supply also promotes the accumulation of secondary metabolites, such as ginsenosides in ginseng, which contribute to improved cold tolerance and antioxidant capacity (El-Mahdy et al., 2018; Saadati et al., 2021; Sharma et al., 2024).

Potassium (K) plays a critical role in enhancing the antioxidant defense system of plants under cold stress. Adequate potassium improves plant survival by stimulating antioxidant enzymes, including superoxide dismutase (SOD), catalase (CAT), ascorbate peroxidase (APX), and glutathione reductase (GR), which scavenge reactive oxygen species (ROS) generated during low-temperature stress (El-Mahdy et al., 2018; Saadati et al., 2021; Azari et al., 2023; Zulfiqar and Ashraf, 2022; Chrysargyris and Tzortzakis, 2025). Exogenous potassium application, such as KNO_3, has been shown to enhance cold tolerance in ginseng by upregulating both antioxidant systems and secondary metabolite-related genes (El-Mahdy et al., 2018; Saadati et al., 2021). Potassium deficiency under cold stress increases oxidative damage, reduces plant survival, and lowers biomass. Adequate potassium supply minimizes ROS accumulation, reduces lipid peroxidation, and maintains membrane integrity (El-Mahdy et al., 2018; Tiwari et al., 2020; Saadati et al., 2021; Azari et al., 2023; Chrysargyris and Tzortzakis, 2025). In crops such as carnation, potato, tomato, pepper, and eggplant, potassium fertilization alleviates chilling-induced oxidative stress and associated damage (Tiwari et al., 2020). Table 5.5. shows the role of potassium in antioxidant defense and cold stress tolerance in plants.

Potassium also enhances the expression of genes involved in secondary metabolite pathways, contributing to improved cold tolerance and antioxidant capacity (El-Mahdy et al., 2018; Saadati et al., 2021; Chrysargyris and Tzortzakis, 2025). Maintenance of proper cytosolic potassium levels is essential for optimal activity of frost-resistance and antioxidant enzymes, preventing oxidative stress-induced lipid membrane damage (Saadati et al., 2021; Azari et al., 2023; Chrysargyris and Tzortzakis, 2025). While adequate potassium enhances antioxidant defense, excessive potassium can induce oxidative stress, as indicated by increased malondialdehyde (MDA) levels at very high K concentrations. Therefore maintaining optimal potassium levels is crucial to balance antioxidant activity and prevent stress (Chrysargyris and Tzortzakis, 2025).

Studies have shown that potassium application can lead to significant improvements in plant growth and yield under cold stress conditions. For instance, in coconut seedlings, potassium application helped maintain nutrient and osmotic regulation, promoted ROS removal, and reduced cold damage, leading to better growth and yield

Table 5.5 Summary of potassium's role in antioxidant defense and cold stress tolerance in plant.

Effect of potassium	Mechanism	References
Stimulates antioxidant enzymes	Increases activity of SOD, CAT, APX, GR; upregulates secondary metabolite genes	El-Mahdy et al. (2018), Saadati et al. (2021), Azari et al. (2023), Zulfiqar and Ashraf (2022), Chrysargyris and Tzortzakis (2025)
Reduces oxidative damage	Lowers ROS, prevents lipid peroxidation, maintains membrane integrity	El-Mahdy et al. (2018), Saadati et al. (2021), Azari et al. (2023), Chrysargyris and Tzortzakis (2025), Tiwari et al. (2020)
Enhances cold tolerance	Improves survival, biomass, and overall stress resistance	El-Mahdy et al. (2018), Saadati et al. (2021), Tiwari et al. (2020), Sattar et al. (2019)
Dose-dependent response	Excess K may increase oxidative stress (MDA); optimal K enhances antioxidant defense	Chrysargyris and Tzortzakis (2025)

(Tian et al., 2020; Lu et al., 2024). Similarly, in citrus plants, potassium application improved photosynthesis and carbohydrate metabolism, enhancing cold tolerance and yield (Iqbal et al., 2024). A study on alfalfa reported that 200 kg ha^{-1} K_2O effectively enhances cold resistance, helping to maintain cellular integrity and osmotic balance under low-temperature stress (Sun et al., 2023).

5.6 Plant responses to metal toxicity

5.6.1 Source of heavy metals in plants and soil

Toxic metals in agriculture originate from both natural and anthropogenic sources, with the latter being the dominant contributor to soil and crop contamination. Natural processes such as weathering of rocks, volcanic eruptions, and atmospheric dust emissions release metals into the environment, although their impact is relatively minor compared with industrial and agricultural activities (Bakkali et al., 2009; Mng'ong'o et al., 2021). Anthropogenic inputs, on the other hand, are widespread and severe. Industrial operations including mining, smelting, steel production, and waste incineration release large quantities of cadmium (Cd), lead (Pb), zinc (Zn), and copper (Cu), while emissions from coal-fired plants and industrial discharges add to atmospheric deposition of toxic metals (Palacios and Capdevila, 2013; Park and Choi, 2021). In agriculture, phosphate-based fertilizers are a significant source of arsenic (As), cadmium (Cd), chromium (Cr), and lead (Pb), while pesticides and herbicides, particularly older formulations containing metal compounds, further contribute to long-term soil contamination (Bakkali et al., 2009; Park and Choi,

2021; Hossain et al., 2022). Similarly, livestock feed additives and growth promoters introduce trace metals that accumulate in manure and subsequently enter agricultural soils when applied as fertilizer (Mng'ong'o et al., 2021; Akoury et al., 2023). Waste and by-products such as sewage sludge, wastewater, and animal manures are also major sources of cadmium, lead, and zinc contamination, while urban and industrial runoff deposits nickel (Ni), zinc (Zn), and lead (Pb) onto nearby croplands (Eissa and Negim, 2018; Almaroai and Eissa, 2020). Atmospheric deposition from industrial emissions and vehicular exhaust further exacerbates soil metal loading (Bakkali et al., 2009; Jahan and Rautela, 2022; Park and Choi, 2021).

5.6.2 Metal toxicity in plants

Metal toxicity in plants refers to the harmful effects caused by excessive accumulation of metals in plant tissues, which disrupts physiological, biochemical, and metabolic processes (Eissa, 2014; Eissa, 2016; Eissa et al., 2016). It impairs photosynthesis, lipid metabolism, nucleic acid biosynthesis, and protein synthesis, ultimately leading to stunted growth and reduced biomass production. Visual symptoms include reduced root elongation, chlorosis of young leaves, leaf rolling, necrosis, wilting, and premature senescence (Ma et al., 2020; Deepa et al., 2023; Kotnala et al., 2025). At the physiological and biochemical level, toxic metals disturb ionic homeostasis and enzyme activities, negatively affecting nutrient uptake, water balance, and metabolic functions. They often trigger the excessive generation of reactive oxygen species (ROS), causing oxidative stress that damages nucleic acids, proteins, and lipids. Such oxidative damage can impair photosynthesis, disrupt nutrient transport, and lead to genotoxic effects (Guédard et al., 2012; Tipayno, 2024; Kotnala et al., 2025).

At the cellular and molecular scale, metal toxicity interferes with cell proliferation, cytoskeletal organization, and enzymatic activity, which can result in programmed cell death or necrosis (Eissa and Almaroai, 2019). The stability of microtubules, essential for morphogenesis, is compromised, and the homeostasis of cytoplasmic metal ions is disrupted, inhibiting vital enzymes. These alterations collectively weaken cellular integrity and metabolic regulation (Rossini-Oliva et al., 2018; Ma et al., 2020; Kotnala et al., 2025). Both essential micronutrients such as iron, manganese, zinc, and copper, when present in excess, and nonessential elements including aluminum, mercury, arsenic, lead, and cadmium can induce toxicity. The problem is further aggravated by anthropogenic activities such as mining, smelting, and industrial pollution, which elevate metal concentrations in soil, water, and air, thereby intensifying toxicity risks for crops and ecosystems (Kar and Raichaudhuri, 2021; Czajka and Nkongolo, 2023; Kotnala et al., 2025).

5.6.3 Effects of metal toxicity on plant growth and crop yield

Cadmium and chromium are among the most detrimental heavy metals to crop growth and productivity. In maize, cadmium exposure drastically reduces root and shoot elongation, chlorophyll content, grain yield, and protein accumulation, with severe reductions of up to 65% in root length and 32% in shoot length at high concentrations (Rizvi and Khan, 2019). The combined toxicity of aluminum and chromium further impairs physiological performance by reducing photosynthetic efficiency, promoting oxidative stress, and enhancing lipid peroxidation, which collectively limit biomass and yield (Anjum et al., 2016). In tomato, excessive copper and zinc concentrations trigger oxidative stress, stimulating proline and antioxidant production but ultimately suppressing growth and development, particularly under elevated metal loads (Badiaa et al., 2020). Nickel toxicity further exacerbates yield losses by reducing plant height, biomass accumulation, and chlorophyll content (Zeeshan et al., 2020).

In wheat, cadmium stress—especially when combined with drought—significantly decreases dry weight and chlorophyll concentration, intensifying oxidative damage. Nevertheless, the application of zinc oxide nanoparticles and biochar has been shown to alleviate these adverse effects, improving biomass production and reducing cadmium uptake by plants (Bashir et al., 2021). Collectively, these findings highlight those toxic metals severely reduce crop yields by disrupting physiological, biochemical, and metabolic pathways. Mitigation approaches, such as the integration of soil amendments and nutrient management strategies, including potassium application, provide promising solutions to enhance crop resilience, protect food quality, and ensure agricultural sustainability under conditions of heavy metal contamination (Francini et al., 2022).

5.6.4 Potassium increases metal toxicity tolerance in plants

Potassium (K) plays a central role in mitigating the adverse effects of heavy metal toxicity in plants and can be regarded as a crucial nutrient for enhancing plant tolerance under such stress conditions (Eissa and Almaroai, 2019; Francini et al., 2022). Potassium is effective in increasing plant tolerance to metal toxicity through multiple physiological and biochemical mechanisms, including reducing oxidative stress, restricting heavy metal accumulation, and enhancing antioxidant defenses (Eissa and Almaroai, 2019; Francini et al., 2022). Its application is beneficial for mitigating the negative impacts of heavy metals on plant growth and productivity (Yaashikaa et al., 2022; Koleva et al., 2022; Ahmed et al., 2023; Han et al., 2025). Recent evidence demonstrates that potassium application reduces physiological impairments caused by toxic metal accumulation and supports photosynthetic efficiency, metabolic stability, and overall plant growth. For example, the combined application of potassium and silicon was shown to significantly alleviate ferrous ion (Fe^{2+}) toxicity in rice, leading to improved mesophyll performance and reduced damage to photosynthetic and metabolic processes (Ahmed et al., 2023).

Several mechanisms have been proposed to explain the protective role of potassium against heavy metal stress. One important mechanism is the modulation of cell wall biosynthesis, which enhances the retention of heavy metals within the cell wall, thereby limiting their mobility and toxicity within the cytoplasm (Eissa and Almaroai, 2019; Han et al., 2025). In addition, potassium helps maintain ionic and osmotic balance, regulates protein synthesis, and activates key enzymes, all of which contribute to reduced oxidative stress under metal toxicity (Mousa et al., 2022; Yaashikaa et al., 2022). Potassium also exerts antagonistic effects on certain metals, such as cadmium (Cd), restricting their transport and accumulation in plant tissues and aiding in detoxification (Koleva et al., 2022).

Another critical function of potassium is its ability to enhance antioxidant defense systems. Elevated potassium levels have been reported to increase the activity of antioxidant enzymes, such as superoxide dismutase (SOD), catalase (CAT), and peroxidases, thereby reducing oxidative damage as indicated by lower levels of lipid peroxidation markers like malondialdehyde (MDA) (Koleva et al., 2022). Furthermore, potassium supplementation improves the uptake of essential nutrients under heavy metal stress, which strengthens plant metabolic processes and enhances stress tolerance (Mousa et al., 2022). The physiological and agronomic benefits of potassium application under heavy metal stress are evident in improved chlorophyll stability, enhanced photosynthate production, and increased growth and yield of crops (Koleva et al., 2022). Thus potassium fertilization represents an important strategy for plant adaptation and resilience to multiple abiotic stresses, including heavy metal toxicity (Han et al., 2025).

Potassium plays a central role in photosynthesis, being essential for chlorophyll synthesis, enzymatic activation in carbon fixation, and overall photosynthetic efficiency. Adequate potassium supply enhances leaf area development, increases CO_2 assimilation, and improves plant productivity under both normal and stress conditions (Rashki et al., 2022; Zulfiqar et al., 2022; Han et al., 2025). Furthermore, potassium application stabilizes chlorophyll architecture, thereby ensuring efficient photosynthate production and mitigating the negative effects of metal-induced stress (Koleva et al., 2022). In contrast, potassium deficiency reduces photosynthetic activity and increases respiration, rendering plants more vulnerable to environmental stresses (Rashki et al., 2022).

Potassium (K) is essential for enhancing crop yield and stress tolerance, including under heavy metal stress. It regulates turgor, membrane potential, osmotic balance, and enzyme activation, supporting plant resilience (Batool et al., 2020). Potassium application improves growth, nutrient uptake, and yield in crops such as maize, wheat, and tomato (Batool et al., 2020). In maize, K increases salinity tolerance, grain yield, and essential nutrient content (Zn and Fe), while reducing cadmium (Cd) accumulation in postphosphate mining soils (Batool et al., 2020; Ruthrof et al., 2021). In wheat, it mitigates salinity stress, enhancing growth, photosynthetic pigments, enzyme activity, and yield, with split applications proving more effective than basal fertilization (El-Lethy et al., 2013; Jatoi et al., 2024). In tomato, intercropping with *Solanum nigrum* or combining K with chitosan and nanosilver increases K uptake, improves yield, and strengthens resistance to heavy metal stress (Krupa-Małkiewicz and Ochmian, 2024; Li et al., 2019). Overall, potassium enhances nutrient acquisition, osmotic regulation, and reduces heavy metal translocation, promoting plant health and productivity, especially when applied at optimal doses or combined with complementary treatments (Fig. 5.2 A & B).

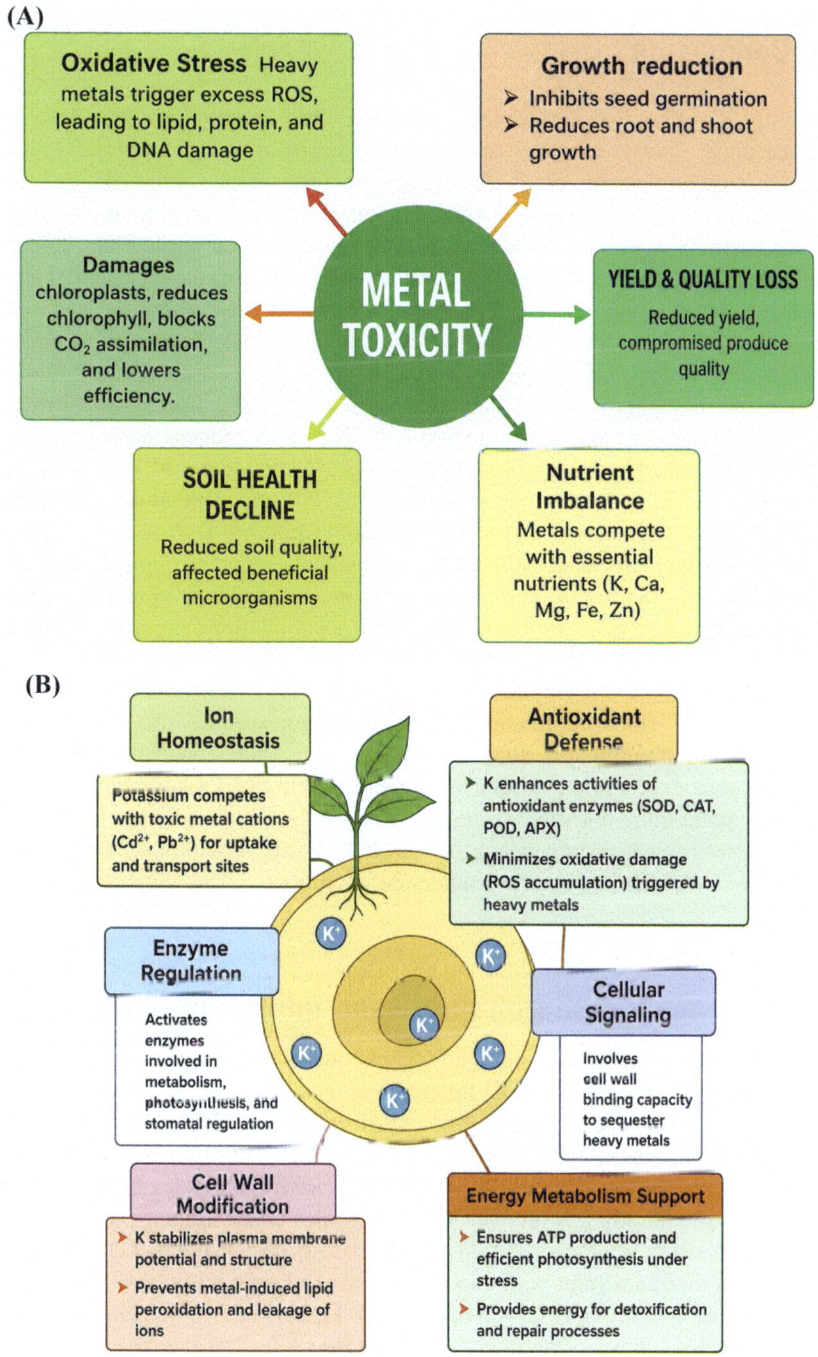

FIGURE 5.2

Potassium impacts on metal toxicity tolerance in plants.

5.7 Conclusion

Potassium (K) is indispensable for sustaining crop growth, productivity, and resilience under diverse abiotic stresses. Unlike nitrogen and phosphorus, K does not enter structural compounds but regulates fundamental physiological and biochemical processes, including osmotic adjustment, enzyme activation, photosynthesis, carbohydrate metabolism, stomatal regulation, and antioxidant defense. Across all major stresses—drought, salinity, heat, cold, and heavy metal toxicity—adequate potassium supply strengthens plant stability, supports metabolic efficiency, and safeguards yield. Under drought conditions, potassium improves relative water content, enhances stomatal control, activates antioxidant defenses, promotes root development, and increases yield stability by up to 30%–40%. In saline environments, it maintains a favorable K^+/Na^+ ratio, reduces sodium toxicity, enhances antioxidant activity, and regulates key stress-responsive genes and transporters. For heat stress, potassium ensures osmotic regulation, membrane stability, delayed senescence, and sustained photosynthesis. During cold stress, it lowers cytoplasmic freezing, stimulates antioxidant enzymes, upregulates secondary metabolite pathways, and improves crop survival and biomass. Against heavy metal toxicity, potassium restricts metal accumulation, strengthens antioxidant defenses, improves photosynthetic efficiency, and supports nutrient uptake. Collectively, these findings emphasize that potassium is not only a nutrient for plant growth but also a critical regulator of stress adaptation mechanisms. Its functions extend from maintaining ion homeostasis and activating enzymatic systems to modulating gene expression and improving yield stability under adverse conditions. Optimizing potassium nutrition is therefore a fundamental strategy to mitigate abiotic stresses, enhance crop performance, and ensure agricultural sustainability under increasingly challenging environmental conditions.

5.8 Challenges, research gaps, and future directions

Despite the clear role of potassium in alleviating abiotic stresses, several challenges and gaps remain. Most soil potassium exists in insoluble mineral forms such as feldspars, micas, and illites, leaving only a small fraction readily available for plant uptake. This limitation reduces fertilizer efficiency, particularly under drought and saline conditions, and poses additional challenges for farmers in developing regions where input costs are high. Moreover, crop genotypes differ in their potassium use efficiency and stress tolerance, making it difficult to design universal application strategies. Another constraint is the uncertainty around optimal potassium dosages, as while rates such as 75 kg K ha^{-1} in maize or 50 kg K_2O ha^{-1} in rice improve yield, supra-optimal doses have sometimes led to yield declines. At the molecular level, although potassium has been shown to regulate transporters, channels, and stress-responsive genes, the complexity of these pathways is not yet fully understood.

Looking forward, sustainable approaches must be prioritized to address these gaps. Potassium-solubilizing bacteria and beneficial microorganisms offer a promising avenue to increase K bioavailability in soils naturally low in available potassium. Advances in molecular biology and genomics can further unravel the roles of transporters (e.g., HKT1, KAT1, KAT2, NHX, SOS1) and salt-responsive microRNAs in stress tolerance, guiding breeding programs for genotypes with improved potassium efficiency. Refining crop- and environment-specific K application strategies, including timing and dosage, will also be crucial to maximize benefits while preventing yield penalties from over-application. Integrating potassium management with complementary nutrient and stress-mitigation strategies provides another opportunity to strengthen resilience. Addressing these challenges through targeted research and innovation will ensure more effective potassium use, enhance crop performance under multifactorial stresses, and contribute to sustainable agricultural productivity and global food security.

References

Abbasi, G.H., Akhtar, J., Ahmad, R., Jamil, M., Anwar-Ul-Haq, M., Ali, S., et al., 2015. Potassium application mitigates salt stress differentially at different growth stages in tolerant and sensitive maize hybrids. Plant Growth Regulation 76 (1), 111–125. Available at https://doi.org/10.1007/s10725-015-0050-1.

Abd El-Wahed, M.H., Eissa, M.A., Almasoudi, N.M., Abo-Elyousr, K.A.M., 2023. Macronutrient-rich biochar induces boron nanoparticles in improving the salt tolerance of pomegranate (*Punica granatum* L.) in arid degraded soils. Scientia Horticulturae 313. Available at https://doi.org/10.1016/j.scienta.2023.111908.

Abdel-Haliem, M.E.F., Hegazy, Hegazy S., Hassan, N.S., Naguib, D.M., 2017. Effect of silica ions and nano silica on rice plants under salinity stress. Ecological Engineering 99, 282–289. Available at https://doi.org/10.1016/j.ecoleng.2016.11.060.

Abreu, F.D.S., Viana, T.V.A., De Sousa, G.G., Baldé, B., De Lacerda, C.F., Goes, G.F., et al., 2024. Salt stress and potassium fertilization on the agronomic performance of peanut crop [Estresse salino e adubação potássica no desempenho agronômico da cultura do amendoim]. Revista Caatinga 37, e11996. Available at https://doi.org/10.1590/1983-21252024v3711996rc.

Adet, L., Rozendaal, D.M.A., Tapi, A., Zuidema, P.A., Vaast, P., Anten, N.P.R., 2024. Negative effects of water deficit on cocoa tree yield are partially mitigated by irrigation and potassium application. Agricultural Water Management 296. Available at https://doi.org/10.1016/j.agwat.2024.108789.

Ahmed, S.F., Biswas, A., Ullah, H., Himanshu, S.K., Tisarum, R., Cha-Um, S., et al., 2023. Interactive effects of silicon and potassium on photosynthesis and physio-biochemical traits of rice (*Oryza sativa* L.) leaf mesophyll under ferrous iron toxicity. Plant Stress 10, 100203. Available at https://doi.org/10.1016/j.stress.2023.100203.

Akoury, E., Mansour, N., Reda, G.A., Dimassi, H., Karam, L., Alwan, N., et al., 2023. Toxic metals in packed rice: effects of size, type, origin, packing season, and storage duration. Journal of Food Composition and Analysis 115, 104920. Available at https://doi.org/10.1016/j.jfca.2022.104920.

Aksu, G., Altay, H., 2020. The effects of potassium applications on drought stress in sugar beet. Sugar Tech 22 (6), 1092–1102. Available at https://doi.org/10.1007/s12355-020-00851-w.

Al Tawaha, A.R.M., Al-Tawaha, A.R., Qotb, M.A., Karnwal, A., Nesterova, N., Singh, A., et al., 2025. Chapter 15 - Resilience in the face of drought: strategies for optimizing plant resilience and enhancing nutrient transfer. In: Etesami, H., Chen, Y. (Eds.), Sustainable Agriculture under Drought Stress: Integrated Soil, Water and Nutrient Management. Academic Press, pp. 221–236. Available at 10.1016/b978-0-443-23956-4.00015-6.

Alam, H., Zamin, M., Adnan, M., Ahmad, N., Nawaz, T., Saud, S., et al., 2022. Evaluating the resistance mechanism of *Atriplex leucoclada* (Orache) to salt and water stress: a potential crop for biosaline agriculture. Frontiers in Plant Science 13. Available at https://doi.org/10.3389/fpls.2022.948736.

Ali, A., Kant, K., Kaur, N., Gupta, S., Jindal, P., Gill, S.S., et al., 2024. Salicylic acid: homeostasis, signalling and phytohormone crosstalk in plants under environmental challenges. South African Journal of Botany 169, 314–335. Available at https://doi.org/10.1016/j.sajb.2024.04.012.

Ali, S., Inamullah, Arif, M., Ali, M., Iqbal, M.O., Munsif, F., Khan, A., 2019. Maize productivity as influenced by potassium under reduced irrigation regimes. Sarhad Journal of Agriculture 35 (1), 171–181. Available at https://doi.org/10.17582/journal.sja/2019/35.1.171.181.

Almaroai, Y.A., Eissa, M.A., 2020. Effect of biochar on yield and quality of tomato grown on a metal-contaminated soil. Scientia Horticulturae 265. Available at https://doi.org/10.1016/j.scienta.2020.109210.

Amin Jafari, M.A., Naderidarbaghshahi, M.R., Soleymani, A., Nasiri, B.M., 2024. Sunflower grain yield and oil content affected by zinc fertilization and genotype in drought stress conditions. Journal of Trace Elements and Minerals 9, 100169. Available at https://doi.org/10.1016/j.jtemin.2024.100169.

Amjad, M., Akhtar, J., Anwar-Ul-Haq, M., Imran, S., Jacobsen, S.E., 2014. Soil and foliar application of potassium enhances fruit yield and quality of tomato under salinity. Turkish Journal of Biology 38 (2), 208–218. Available at https://doi.org/10.3906/biy-1305-54.

Anjum, S.A., Ashraf, U., Khan, I., Tanveer, M., Saleem, M.F., Wang, L., 2016. Aluminum and chromium toxicity in maize: implications for agronomic attributes, net photosynthesis, physio-biochemical oscillations, and metal accumulation in different plant parts. Water, Air, and Soil Pollution 227 (9). Available at https://doi.org/10.1007/s11270-016-3013-x.

Ansari, W.A., Farah, M.A., Kumar, S.C., Mir, M.I., Shahid, M., Al-Anazi, K.M., et al., 2025. Harnessing *Achromobacter xylosoxidans* SPSB-5 for enhanced P-solubilization, biotic/abiotic stress tolerance, and improved growth of *Brassica juncea* (L.) seedlings in saline environment. Physiological and Molecular Plant Pathology 137, 102615. Available at https://doi.org/10.1016/j.pmpp.2025.102615.

Ashraf, M., Afzal, M., Ahmad, R., Maqsood, M.A., Shahzad, S.M., Tahir, M.A., et al., 2012. Growth response of the salt-sensitive and the salt-tolerant sugarcane genotypes to potassium nutrition under salt stress. Archives of Agronomy and Soil Science 58 (4), 385–398. Available at https://doi.org/10.1080/03650340.2010.529609.

Ashraf, M.Y., Hussain, F., Akhter, J., Gul, A., Ross, M., Ebert, G., 2008. Effect of different sources and rates of nitrogen and supra optimal level of potassium fertilization on growth, yield and nutrient uptake by sugarcane grown under saline conditions. Pakistan Journal of Botany 40 (4), 1521–1531.

Ayaz, M., Ahmad, R., Shahzad, M., Khan, N., Shah, M.M., Khan, S.A., 2015. Drought stress stunt tomato plant growth and up-regulate expression of SlAREB, SlNCED3, and SlERF024 genes. Scientia Horticulturae 195, 48–55. Available at https://doi.org/10.1016/j.scienta.2015.08.025.

Ayub, M.A., Ahmad, H.R., Ali, M., Rizwan, M., Ali, S., Zia Ur Rehman, M., et al., 2020. Chapter 3 - Salinity and its tolerance strategies in plants. In: Tripathi, D.K., Singh, V.P., Chauhan, D.K. (Eds.), Plant Life Under Changing Environment: Responses and Management. Academic Press, pp. 47–76. Available at https://doi.org/10.1016/b978-0-12-818204-8.00003-5.

Azari, S.J., Sorooshzadeh, A., Nabati, J., Oskoueian, E., 2023. Relationship between fertilization and planting depths on antioxidant activity in saffron (*Crocus sativus* L). Industrial Crops and Products 191, 116004. Available at https://doi.org/10.1016/j.indcrop.2022.116004.

Badiaa, O., Yssaad, H.A.R., Topcuoglu, B., 2020. Effect of heavy metals (Copper and zinc) on proline, polyphenols and flavonoids content of tomato (*Lycopersicon esculentum* Mill.). Plant Archives, Algeria Plant Archives 20 (1), 2125–2137.

Bakkali, K., Martos, N.R., Souhail, B., Ballesteros, E., 2009. Characterization of trace metals in vegetables by graphite furnace atomic absorption spectrometry after closed vessel microwave digestion. Food Chemistry 116 (2), 590–594. Available at https://doi.org/10.1016/j.foodchem.2009.03.010.

Bashir, A., Ur Rehman, M.Z., Hussaini, K.M., Adrees, M., Qayyum, M.F., Sayal, A.U., et al., 2021. Combined use of zinc nanoparticles and co-composted biochar enhanced wheat growth and decreased Cd concentration in grains under Cd and drought stress: a field study. Environmental Technology and Innovation 23. Available at https://doi.org/10.1016/j.eti.2021.101518.

Batool, M., Bashir, M.A., Qayyoum, M.A., Hussain, A., Hameed, S.A., Alghanem, S.M., et al., 2020. Effect of potassium application on the growth and Fe/Zn use efficiency of salt-tolerant and sensitive maize genotypes in response to salinity. Fresenius Environmental Bulletin 29 (5), 3778–3789.

Bellaloui, N., Gillen, A.M., Mengistu, A., Kebede, H., Fisher, D.K., Smith, J.R., et al., 2013. Responses of nitrogen metabolism and seed nutrition to drought stress in soybean genotypes differing in slow-wilting phenotype. Frontiers in Plant Science 4. Available at https://doi.org/10.3389/fpls.2013.00498.

Benavides-Mendoza, A., Naz, M., González-Morales, S., López-Sánchez, R.C., Robledo-Olivo, A., 2024. Role of Beneficial Elements Signaling and Metabolic Performance in Plants Under Drought Stress. Elsevier BV, pp. 265–301. Available at https://doi.org/10.1016/b978-0-443-16082-0.00008-4.

Bhardwaj, S., Kapoor, B., Kapoor, D., Thakur, U., Dolma, Y., Raza, A., 2025. Manifold roles of potassium in mediating drought tolerance in plants and its underlying mechanisms. Plant Science 351, 112337. Available at https://doi.org/10.1016/j.plantsci.2024.112337.

Bybordi, A., 2015. Influence of exogenous application of silicon and potassium on physiological responses, yield, and yield components of salt-stressed wheat. Communications in Soil Science and Plant Analysis 46 (1), 109–122. Available at https://doi.org/10.1080/00103624.2014.956936.

Cakmak, I., 2005. The role of potassium in alleviating detrimental effects of abiotic stresses in plants. Journal of Plant Nutrition and Soil Science 168 (4), 521–530. Available at https://doi.org/10.1002/jpln.200420485.

Cakmak, I., Rengel, Z., 2024. Humboldt review: potassium may mitigate drought stress by increasing stem carbohydrates and their mobilization into grains. Journal of Plant Physiology 303, 154325. Available at https://doi.org/10.1016/j.jplph.2024.154325.

Chaudhary, M.T., Majeed, S., Rana, I.A., Ali, Z., Jia, Y., Du, X., et al., 2024. Impact of salinity stress on cotton and opportunities for improvement through conventional and biotechnological approaches. BMC Plant Biology 24 (1). Available at https://doi.org/10.1186/s12870-023-04558-4.

Che, Y., Fan, D., Wang, Z., Xu, N., Zhang, H., Sun, G., et al., 2022. Potassium mitigates salt-stress impacts on photosynthesis by alleviation of the proton diffusion potential in thylakoids. Environmental and Experimental Botany 194, 104708. Available at https://doi.org/10.1016/j.envexpbot.2021.104708.

Chen, F., Shen, Z., Shi, R., Zhang, X., Zhang, H., Li, W., et al., 2025. Carbon dots-mediated plant adaptive responses to abiotic stress. Materials Research Bulletin 182, 113137. Available at https://doi.org/10.1016/j.materresbull.2024.113137.

Chen, G., Liu, C., Gao, Z., Zhang, Y., Jiang, H., Zhu, L., et al., 2017. Oshak1, a high-affinity potassium transporter, positively regulates responses to drought stress in rice. Frontiers in Plant Science 8. Available at https://doi.org/10.3389/fpls.2017.01885.

Chen, Y., Li, R., Ge, J., Liu, J., Wang, W., Xu, M., et al., 2021. Exogenous melatonin confers enhanced salinity tolerance in rice by blocking the ROS burst and improving Na+/K+ homeostasis. Environmental and Experimental Botany 189, 104530. Available at https://doi.org/10.1016/j.envexpbot.2021.104530.

Chowdhury, M.S.N., Hossain Sani, M.N., Siddique, A.B., Hossain, M.S., Yong, J.W.H., 2024. Synergistic effects of biochar and potassium co-application on growth, physiological attributes, and antioxidant defense mechanisms of wheat under water deficit conditions. Plant Stress 12, 100452. Available at https://doi.org/10.1016/j.stress.2024.100452.

Chrysargyris, A., Tzortzakis, N., 2025. Optimizing nitrogen, phosphorus, and potassium requirements to improve *Origanum dubium* Boiss. growth, nutrient and water use efficiency, essential oil yield and composition. Industrial Crops and Products 224, 120291. Available at https://doi.org/10.1016/j.indcrop.2024.120291.

Czajka, K.M., Nkongolo, K.K., 2023. Whole genome expression analysis of trembling aspen (*Populus tremuloides*) genotypes exposed to increasing concentrations of nickel. Plant Gene 34, 100416. Available at https://doi.org/10.1016/j.plgene.2023.100416.

Deepa, Sharaya, R., Nehra, A., Agarwala, N., Khan, N.A., Tuteja, N., et al., 2023. Biostimulants in the alleviation of metal toxicity: an overview. Biostimulants in Alleviation of Metal Toxicity in Plants: Emerging Trends and Opportunities. Elsevier, India, pp. 1–19. Available at https://doi.org/10.1016/B978-0-323-99600-6.00017-7.

Degl'innocenti, E., Hafsi, C., Guidi, L., Navari-Izzo, F., 2009. The effect of salinity on photosynthetic activity in potassium-deficient barley species. Journal of Plant Physiology 166 (18), 1968–1981. Available at https://doi.org/10.1016/j.jplph.2009.06.013.

Dev, W., Sultana, F., Li, H., Hu, D., Peng, Z., He, S., et al., 2025. Molecular mechanisms of cold stress response in cotton: transcriptional reprogramming and genetic strategies for tolerance. Plant Science 352, 112390. Available at https://doi.org/10.1016/j.plantsci.2025.112390.

Ding, Z., Zhao, F., Zhu, Z., Ali, E.F., Shaheen, S.M., Rinklebe, J., et al., 2022. Green nanosilica enhanced the salt-tolerance defenses and yield of Williams banana: a field trial for using saline water in low fertile arid soil. Environmental and Experimental Botany 197. Available at https://doi.org/10.1016/j.envexpbot.2022.104843.

Ditta, A., Ullah, N., 2023. Perspectives of Using Plant Growth-Promoting Rhizobacteria Under Salinity Stress for Sustainable Crop Production. Elsevier BV, pp. 231–247. Available at https://doi.org/10.1016/b978-0-323-89871-3.00024-0.

Du, X., Zhu, X., Yang, Y., Wang, Y., Arens, P., Liu, H., Liu, J.H., et al., 2019. De novo transcriptome analysis of *Viola* × *wittrockiana* exposed to high temperature stress. PLoS One 14 (9), e0222344. Available at https://doi.org/10.1371/journal.pone.0222344.

Dzinyela, R., Alhassan, A.R., Suglo, P., Movahedi, A., 2023. Advanced study of functional proteins involved in salt stress regulatory pathways in plants. South African Journal of Botany 159, 425–438. Available at https://doi.org/10.1016/j.sajb.2023.06.029.

Easwaran, C., Christopher, S.R., Moorthy, G., Mohan, P., Marimuthu, R., Koothan, V., et al., 2024. Nano hybrid fertilizers: a review on the state of the art in sustainable agriculture. Science of the Total Environment 929, 172533. Available at https://doi.org/10.1016/j.scitotenv.2024.172533.

Eissa, M.A., Ahmed, E.M., Reichman, S.M., Goss, M., 2016. Production of the forage halophyte *Atriplex amnicola* in metal-contaminated soils. Soil Use and Management 32 (3), 350–356. Available at https://doi.org/10.1111/sum.12275.

Eissa, M.A., 2014. Performance of river saltbush (*Atriplex amnicola*) grown on contaminated soils as affected by organic fertilization. International Digital Organization for Scientific Information, Egypt World Applied Sciences Journal 30 (12), 1877–1881. Available at https://doi.org/10.5829/idosi.wasj.2014.30.12.19.

Eissa, M.A., 2016. Phosphate and organic amendments for safe production of okra from metal-contaminated soils. Agronomy Journal 108 (2), 540–547. Available at https://doi.org/10.2134/agronj2015.0460.

Eissa, M.A., Almaroai, Y.A., 2019. Phytoremediation capacity of some forage plants grown on a metals-contaminated soil. Soil and Sediment Contamination 28 (6), 569–581. Available at https://doi.org/10.1080/15320383.2019.1634674.

Eissa, M.A., Negim, O.E., 2018. Heavy metals uptake and translocation by lettuce and spinach grown on a metal-contaminated soil. Journal of Soil Science and Plant Nutrition 18 (4), 1097–1107. Available at https://doi.org/10.4067/S0718-95162018005003101.

El-Lethy, S.R., Abdelhamid, M.T., Reda, F., 2013. Effect of potassium application on wheat (*Triticum aestivum* L.) cultivars grown under salinity stress. World Applied Sciences Journal 26 (7), 840–850. Available at https://doi.org/10.5829/idosi.wasj.2013.26.07.13527.

El Mahdy, M.T., Youssef, M., Eissa, M.A., 2018. Impact of in vitro cold stress on two banana genotypes based on physio-biochemical Evaluation. South African Journal of Botany 119, 219–225. Available at https://doi.org/10.1016/j.sajb.2018.09.014.

El-Sappah, A.H., Rather, S.A., Wani, S.H., Elrys, A.S., Bilal, M., Huang, Q., et al., 2022. Heat stress-mediated constraints in Maize (*Zea mays*) production: challenges and solutions. Frontiers in Plant Science 13. Available at https://doi.org/10.3389/fpls.2022.879366.

FAO, 2024. Global status of salt-affected soils – Main report. FAO, Rome. Available at https://doi.org/10.4060/cd3044en.

Francini, A., Toscano, S., Romano, D., Ferrante, A., 2022. An overview of potassium in abiotic stress: emphasis on potassium transporters and molecular mechanism. Role of

Potassium in Abiotic Stress. Springer Nature, Italy, pp. 249–262. Available at https://doi.org/10.1007/978-981-16-4461-0_12.

Furtak, K., Wolińska, A., 2023. The impact of extreme weather events as a consequence of climate change on the soil moisture and on the quality of the soil environment and agriculture – a review. CATENA 231, 107378. Available at https://doi.org/10.1016/j.catena.2023.107378.

Gao, G., Yan, L., Tong, K., Yu, H., Lu, M., Wang, L., et al., 2024. The potential and prospects of modified biochar for comprehensive management of salt-affected soils and plants: a critical review. Science of the Total Environment 912, 169618. Available at https://doi.org/10.1016/j.scitotenv.2023.169618.

Guédard, M.L., Faure, O., Bessoule, J.-J., 2012. Early changes in the fatty acid composition of photosynthetic membrane lipids from Populus nigra grown on a metallurgical landfill. Chemosphere 88 (6), 693–698. Available at https://doi.org/10.1016/j.chemosphere.2012.03.079.

Guo, L., Mao, X., Chen, Y., Li, L., Hu, Y., Zhang, H., et al., 2024. Transcriptome analysis reveals salt stress-related genes in *Rhododendron simii* and RsWRKY40 is referred to salt tolerance. Environmental and Experimental Botany 220. Available at https://doi.org/10.1016/j.envexpbot.2024.105678.

Hameed, R., Abbas, A., Li, G., Shahani, A.A.A., Roha, B., Du, D., 2024. Biochar as a soil amendment for saline soils reclamation: mechanisms and efficacy. Biochar Production for Green Economy: Agricultural and Environmental Perspectives. Elsevier, China, pp. 205–225. Available at https://doi.org/10.1016/B978-0-443-15506-2.00005-5.

Han, X., Zhou, Y., Feng, X., Wang, Y., Zhang, H., 2025. Potassium uptake function of LbKT1 and LbSKOR from *Lycium barbarum* and their influence on the arbuscular mycorrhizal symbiosis. Plant Science 359, 112587. Available at https://doi.org/10.1016/j.plantsci.2025.112587.

Hasanuzzaman, M., Bhuyan, M., Nahar, K., Hossain, M., Mahmud, J., Hossen, M., et al., 2018. Potassium: a vital regulator of plant responses and tolerance to abiotic stresses. Agronomy 8 (3), 31. Available at https://doi.org/10.3390/agronomy8030031.

Hemantaranjan, A., Malik, C.P., Bhanu, A.N., 2018. Physiology of heat stress and tolerance mechanisms - an overview. The Journal of Plant Science Research 34 (1), 51–64. Available at https://doi.org/10.32381/jpsr.2018.34.01.7.

Hiya, H.J., Nakashima, Y., Takeuchi, A., Nakamura, T., Nakamura, Y., Murata, Y., et al., 2024. Outward-rectifying potassium channels GORK and SKOR function in regulation of root growth under salt stress in *Arabidopsis thaliana*. Journal of Plant Physiology 302, 154322. Available at https://doi.org/10.1016/j.jplph.2024.154322.

Hopmans, J.W., Qureshi, A.S., Kisekka, I., Munns, R., Grattan, S.R., Rengasamy, P., et al., 2021. Critical knowledge gaps and research priorities in global soil salinity. Advances in Agronomy 169, 1–191. Available at https://doi.org/10.1016/bs.agron.2021.03.001.

Hossain, J., Azam, M.G., Gaber, A., Aftab, T., Hossain, A., 2022. CHAPTER 18 - Cytotoxicity of metal/metalloids' pollution in plants. In: Aftab, T., Hakeem, K. (Eds.), Metals Metalloids Soil Plant Water Systems: Phytophysiology and Remediation Techniques. Academic Press, pp. 371–394. Available at https://doi.org/10.1016/b978-0-323-91675-2.00017-2.

Hossain, M.I., Khaleque, M.A., Ali, M.R., Bacchu, M.S., Hossain, M.S., Shahed, S.M.F., et al., 2024. Development of electrochemical sensors for quick detection of environmental (soil and water) NPK ions. RSC Advances 14 (13), 9137–9158. Available at https://doi.org/10.1039/d4ra00034j.

Hosseini, S.A., Hajirezaei, M.R., Seiler, C., Sreenivasulu, N., Von Wirén, N., 2016. A potential role of flag leaf potassium in conferring tolerance to drought-induced leaf senescence in barley. Frontiers in Plant Science 7. Available at https://doi.org/10.3389/fpls.2016.00206.

Huliaieva, H., Tokovenko, I., Kharchuk, M., Bohdan, M., Pasichnyk, L., 2025. The influence of iodine nano citrates on juvenile wheat plants at phytopathogenic infection and cooling stress. Plant Nano Biology 13, 100175. Available at https://doi.org/10.1016/j.plana.2025.100175.

Husen, A., 2023. Plants and Their Interaction to Environmental Pollution: Damage Detection, Adaptation, Tolerance, Physiological and Molecular Responses. Elsevier Inc., Ethiopia, pp. 1–449. Available at https://doi.org/10.1016/C2020-0-03319-0.

Husen, A., 2024. Essential Minerals in Plant-Soil System: Coordination, Signalling, and Interaction under Adverse Situations. Elsevier Inc, India, pp. 1–463. Available at https://doi.org/10.1016/C2022-0-02437-5.

Husen, A., 2025. Role of Antioxidants in Mitigating Plant Stress. Elsevier Inc, Ethiopia, pp. 1–442. Available at https://doi.org/10.1016/C2023-0-51268-6.

Husen, A., 2021. Harsh Environment and Plant Resilience: Molecular and Functional Aspects. Springer Nature, Switzerland AG. Available at https://doi.org/10.1007/978-3-030-65912-7.

Husen, A., 2021. The harsh environment and resilient plants: an overview. In: Husen, A. (Ed.), Harsh Environment and Plant Resilience. Springer, Cham, pp. 1–23. Available at https://doi.org/10.1007/978-3-030-65912-7_1.

Ibrahimi, A., Yadav, K., Jakhar, S., 2025. Mitigating salinity stress in mung bean [*Vigna radiate* (L.) R. Wilczek] using potassium nitrate. Universitatea din Oradea, India Analele Universitatii din Oradea, Fascicula Biologie 32 (1), 7–15.

Iqbal, S., Balal, R.M., Seleiman, M.F., Mattia, M., Chater, J.M., Shahid, M.A., 2024. Silicon and potassium-induced modulations in leaf carbohydrate metabolism confer freezing tolerance in Satsuma mandarin. Silicon 16 (12), 5135–5146. Available at https://doi.org/10.1007/s12633-024-03070-0.

Ishfaq, N., Ahmad Waraich, E., Ahmad, M., Hussain, S., Zulfiqar, U., Din, K.U., et al., 2024. Mitigating drought-induced oxidative stress in wheat (*Triticum aestivum* L.) through foliar application of sulfhydryl thiourea. Scientific Reports 14 (1). Available at https://doi.org/10.1038/s41598-024-66506-y.

Jahan, S., Rautela, S., 2022. Chapter 16 - Role of phytohormones in mitigating the harmful impacts of hazardous and trace materials on agriculture crops. In: Naeem, M. (Ed.), Hazardous and Trace Materials in Soil and Plants: Sources, Effects and Management. Academic Press, pp. 231–246. Available at https://doi.org/10.1016/b978-0-323-91632-5.00001-x.

Jatoi, T.A., Jatoi, W.A., Memon, S., Rajper, A.A., Ilyas, M., 2024. Impact of basal and split potassium application on yield parameters of two wheat varieties. Pakistan Journal of Scientific and Industrial Research Series B: Biological Sciences 67 (1), 86–90.

Javaid, A., Munir, N., Abideen, Z., Siddiqui, Z.S., Yong, J.W.H., 2024. The role of natural and synthetic zeolites as soil amendments for mitigating the negative impacts of abiotic stresses to improve agricultural resilience. Plant Stress 14, 100627. Available at https://doi.org/10.1016/j.stress.2024.100627.

Jiang, M., Lan, S., Peng, M., Wang, Z., Zhuang, L., 2023. The diversity of *Ferula* species and environmental factors on metabolite composition using untargeted metabolomics. Food Bioscience 56, 103075. Available at https://doi.org/10.1016/j.fbio.2023.103075.

Ju, F., Pang, J., Huo, Y., Zhu, J., Yu, K., Sun, L., et al., 2021. Potassium application alleviates the negative effects of salt stress on cotton (*Gossypium hirsutum* L.) yield by improving the ionic homeostasis, photosynthetic capacity and carbohydrate metabolism of the leaf subtending the cotton boll. Field Crops Research 272, 108288. Available at https://doi.org/10.1016/j.fcr.2021.108288.

Kar, M.M., Raichaudhuri, A., 2021. Role of microRNAs in mediating biotic and abiotic stress in plants. Plant Gene 26, 100277. Available at https://doi.org/10.1016/j.plgene.2021.100277.

Kashyap, S., Biswal, B., Bhakuni, K., Ali, G., Bhattacharjee, S., Yadav, M.R., et al., 2024. Amelioration of abiotic stresses in forage crop production using microbial stimulants: an overview. Microbial Biostimulants for Plant Growth and Abiotic Stress Amelioration. Elsevier, India, pp. 397–456. Available at https://doi.org/10.1016/B978-0-443-13318-3.00019-7.

Kaur, S., Nayyar, H., 2015. Selenium fertilization to salt-stressed mungbean (*Vigna radiata* L. Wilczek) plants reduces sodium uptake, improves reproductive function, pod set and seed yield. Scientia Horticulturae 197, 304–317. Available at https://doi.org/10.1016/j.scienta.2015.09.048.

Kaur, T., Kour, D., Yadav, A.N., 2022. 1 - Trends of agricultural microbiology for sustainable crops production and economy: An introduction. In: Soni, R. (Ed.), Trends of Applied Microbiology for Sustainable Economy. Academic Press, pp. 1–44. Available at https://doi.org/10.1016/b978-0-323-91595-3.00003-3.

Khalid, S., Amanullah, Tawaha, A.R.M.A., Nadia, Gunal, H., Al-Tawaha, A.R., et al., 2025. Rizwan Rashid, Impacts of Climate Change and Drought Stress on Plant Metabolome. Elsevier BV, pp. 105–113. Available at https://doi.org/10.1016/b978-0-443-23956-4.00009-0.

Khan, M.A., Asaf, S., Latif Khan, A., Jan, R., Kang, S.-M., Kim, K.-M., et al., 2020. Extending thermotolerance to tomato seedlings by inoculation with SA1 isolate of *Bacillus cereus* and comparison with exogenous humic acid application. PLoS One 15 (4), e0232228. Available at https://doi.org/10.1371/journal.pone.0232228.

Khan, A., Panthari, D., Sharma, R., Punetha, A., Singh, A., Upadhayay, V., 2023. Chapter 6 - Biofertilizers: a microbial-assisted strategy to improve plant growth and soil health. In: Pandey, S.C. (Ed.), Advanced Microbial Techniques in Agriculture, Environment, and Health Management. Academic Press, pp. 97–118. Available at https://doi.org/10.1016/B978-0-323-91643-1.00007-7.

Koleva, L., Umar, A., Yasin, N.A., Shah, A.A., Siddiqui, M.H., Alamri, S., et al., 2022. Iron oxide and silicon nanoparticles modulate mineral nutrient homeostasis and metabolism in cadmium-stressed *Phaseolus vulgaris*. Frontiers in Plant Science 13. Available at https://doi.org/10.3389/fpls.2022.806781.

Kotnala, S., Tiwari, S., Nayak, A., Bhushan, B., Chandra, S., Medeiros, C.R., et al., 2025. Impact of heavy metal toxicity on the human health and environment. Science of the Total Environment 987, 179785. Available at https://doi.org/10.1016/j.scitotenv.2025.179785.

Kour, D., Rana, K.L., Yadav, A.N., Yadav, N., Kumar, M., Kumar, V., et al., 2020. Microbial biofertilizers: bioresources and eco-friendly technologies for agricultural and environmental sustainability. Biocatalysis and Agricultural Biotechnology 23, 101487. Available at https://doi.org/10.1016/j.bcab.2019.101487.

Krupa-Małkiewicz, M., Ochmian, I., 2024. Sustainable approaches to alleviate heavy metal stress in tomatoes: exploring the role of chitosan and nanosilver. Agronomy 14 (11), 2477. Available at https://doi.org/10.3390/agronomy14112477.

Kumar, A., Choudhary, A., Kaur, H., Singh, K., Guha, S., Choudhary, D.R., et al., 2025. Exploring the role of silicon in enhancing sustainable plant growth, defense system, environmental stress mitigation and management. Discover Applied Sciences 7 (5). Available at https://doi.org/10.1007/s42452-025-06866-w.

Kumari, S., Nazir, F., Jain, K., Khan, M.I.R., 2023. GABA and potassium modulates defence systems, assimilation of nitrogen and carbon, and yield traits under salt stress in wheat. Journal of Plant Growth Regulation 42 (10), 6721–6740. Available at https://doi.org/10.1007/s00344-023-10992-3.

Li, D., Zhang, Y., Hu, X., Shen, X., Ma, L., Su, Z., et al., 2011. Transcriptional profiling of *Medicago truncatula* under salt stress identified a novel CBF transcription factor MtCBF4 that plays an important role in abiotic stress responses. BMC Plant Biology 11 (1). Available at https://doi.org/10.1186/1471-2229-11-109.

Li, R., Sun, G., Tang, Y., 2019. Effect of intercropping of *Solanum nigrum* L., tomato and eggplant on potassium content under cadmium stress. IOP Conference Series: Earth and Environmental Science 233 (4), 042024. Available at https://doi.org/10.1088/1755-1315/233/4/042024.

Liao, Z., Fan, J., Lai, Z., Bai, Z., Wang, H., Cheng, M., et al., 2023. Response network and regulatory measures of plant-soil-rhizosphere environment to drought stress. Advances in Agronomy 180, 93–196. Available at https://doi.org/10.1016/bs.agron.2023.03.002.

Liu, H., Liu, Y., Xu, N., Sun, Y., Li, Q., Yue, L., et al., 2022. Chrysanthemum × grandiflora leaf and root transcript profiling in response to salinity stress. BMC Plant Biology 22 (1). Available at https://doi.org/10.1186/s12870-022-03612-x.

Londoño Vélez, V., Alquraish, F., Tarbiyyah, I., Rafique, F., Mao, D., Chodasiewicz, M., 2022. Landscape of biomolecular condensates in heat stress responses. Frontiers in Plant Science 13. Available at https://doi.org/10.3389/fpls.2022.1032045.

Lovio-Fragoso, J.P., De Jesús-Campos, D., Razo-Mendivil, F.G., García-Coronado, H., Domínguez-Rosas, E., Trillo-Hernández, E.A., et al., 2023. Chapter 8 - Desert plant transcriptomics and adaptation to abiotic stress. Transcriptome Profiling: Progress and Prospects. Academic Press, pp. 199–256. Available at https://doi.org/10.1016/b978-0-323-91810-7.00013-3.

Lu, L., Wang, Y., Sayed, M.A., Iqbal, A., Yang, Y., 2024. Exploring the physiological and molecular mechanisms by which potassium regulates low-temperature tolerance of coconut (*Cocos nucifera* L.) seedlings. Agronomy 14 (12), 2983. Available at https://doi.org/10.3390/agronomy14122983.

Ma, Q.-J., Sun, M.-H., Lu, J., Hu, D.-G., Kang, H., You, C.-X., et al., 2020. Phosphorylation of a malate transporter promotes malate excretion and reduces cadmium uptake in apple. Journal of Experimental Botany 71 (12), 3437–3449. Available at https://doi.org/10.1093/jxb/eraa121.

Manasa S, L., Panigrahy, M., Panigrahi, K.C.S., Rout, G.R., 2022. Overview of cold stress regulation in plants. Botanical Review 88 (3), 359–387. Available at https://doi.org/10.1007/s12229-021-09267-x.

Manorma, K., Devi, S., Sharma, M., Chand, S., Rana, K., Kesta, K., et al., 2024. Microbial biostimulants in the amelioration of climate change. Microbial Biostimulants for Plant Growth and Abiotic Stress Amelioration. Elsevier, India, pp. 253–282. Available at https://doi.org/10.1016/B978-0-443-13318-3.00006-9.

Meng, F., Ma, M., Li, S., Liang, P., Liang, Y., Shi, H., et al., 2025. Genome-wide identification of light-harvesting chlorophyll a/b-binding (LHC) gene family in tomato and functional analysis of SlLhcb1.11 and SlELIP1 under cold stress. Genomics 117 (2), 111022. Available at https://doi.org/10.1016/j.ygeno.2025.111022.

Meng, X., Liu, S., Dong, T., Xu, T., Ma, D., Pan, S., et al., 2020. Comparative transcriptome and proteome analysis of salt-tolerant and salt-sensitive sweet potato and overexpression of IbNAC7 confers salt tolerance in arabidopsis. Frontiers in Plant Science 11. Available at https://doi.org/10.3389/fpls.2020.572540.

Miao, S., Wang, X., Song, Y., Zhang, Y., Li, X., Che, W., et al., 2024. Optimal agronomic measures combined with biochar increased rice yield through enhancing nitrogen use efficiency in soda saline-alkali fields. European Journal of Agronomy 161, 127365. Available at https://doi.org/10.1016/j.eja.2024.127365.

Mng'ong'o, M., Munishi, L.K., Ndakidemi, P.A., Blake, W., Comber, S., Hutchinson, T.H., 2021. Toxic metals in East African agro-ecosystems: key risks for sustainable food production. Journal of Environmental Management 294, 112973. Available at https://doi.org/10.1016/j.jenvman.2021.112973.

Moghaddam, A., Larijani, H.R., Oveysi, M., Moghaddam, H.R.T., Nasri, M., 2023. Alleviating the adverse effects of salinity stress on *Salicornia persica* using sodium nitroprusside and potassium nitrate. BMC Plant Biology 23 (1). Available at https://doi.org/10.1186/s12870-023-04179-x.

Mohamed, H.I., Sajyan, T.K., Shaalan, R., Bejjani, R., Sassine, Y.N., Basit, A., 2021. Plant-Mediated Copper Nanoparticles for Agri-Ecosystem Applications Agri-Waste and Microbes for Production of Sustainable Nanomaterials. Elsevier, Egypt, pp. 79–120. Available at https://doi.org/10.1016/B978-0-12-823575-1.00025-1.

Molina-Bravo, R., Arellano, C., Sosinski, B.R., Fernandez, G.E., 2011. A protocol to assess heat tolerance in a segregating population of raspberry using chlorophyll fluorescence. Scientia Horticulturae 130 (3), 524–530. Available at https://doi.org/10.1016/j.scienta.2011.07.022.

Mousa, M.A.A., Abo-Elyousr, K.A.M., Ibrahim, O.H.M., Alshareef, N.O., Eissa, M.A., 2022. Shrimp-waste-derived biochar induces metal toxicity tolerance of wastewater-irrigated Quinoa (*Chenopodium quinoa*). Agriculture 12 (11), 1748. Available at https://doi.org/10.3390/agriculture12111748.

Munsif, F., Farooq, U., Arif, M., Shah, T., Jehangir, M., Zaheer, S., et al., 2022. Potassium and salicylic acid function synergistically to promote the drought resilience through upregulation of antioxidant profile for enhancing potassium use efficiency and wheat yield. Annals of Applied Biology 180 (2), 273–282. Available at https://doi.org/10.1111/aab.12731.

Palacios, Ò., Capdevila, M., 2013. 3.04 - Toxicology (Pb, Hg, Cd, As, Al, Cr, and Others). Comprehensive Inorganic Chemistry II: From Elements to Applications. Elsevier BV, pp. 51–63. Available at https://doi.org/10.1016/b978-0-08-097774-4.00304-1.

Pandey, B., Choudhary, K.K., 2019. Chapter 9 - Air Pollution: Role in Climate Change and Its Impact on Crop Plants. In: Choudhary, K.K., Kumar, A., Singh, A.K. (Eds.), Climate Change and Agricultural Ecosystems: Current Challenges and Adaptation. Woodhead Publishing, pp. 211–247. Available at https://doi.org/10.1016/B978-0-12-816483-9.00009-8.

Pandit, K., Chandni, Kaur, S., Kumar, M., Bhardwaj, R., Kaur, S., 2024. Salinity stress: impact on plant growth. Advances in Food Security and Sustainability 9, 145–160. Available at https://doi.org/10.1016/bs.af2s.2024.07.002.

Pantha, S., Kilian, B., Özkan, H., Zeibig, F., Frei, M., 2024. Physiological and biochemical changes induced by drought stress during the stem elongation and anthesis stages in the *Triticum* genus. Environmental and Experimental Botany 228, 106047. Available at https://doi.org/10.1016/j.envexpbot.2024.106047.

Park, M.-K., Choi, S.-D., 2021. Monitoring and risk assessment of arsenic species and metals in the Taehwa River in Ulsan, the largest industrial city in South Korea. Marine Pollution Bulletin 172, 112862. Available at https://doi.org/10.1016/j.marpolbul.2021.112862.

Parveen, Din, M.U., Anwar-Ul-Haq, M., Li, D., Aamer, M., Al-Khayri, J.M., et al., 2025. Exogenous application of potassium mitigates salt stress and enhances soybean growth through improved photosynthetic efficiency, antioxidant activities, and ionic homeostasis. Applied Ecology and Environmental Research 23 (4), 6359–6372. Available at https://doi.org/10.15666/aeer/2304_63596372.

Pascual, L.S., Mittler, R., Sinha, R., Peláez-Vico, M.Á., López-Climent, M.F., Vives-Peris, V., et al., 2023. Jasmonic acid is required for tomato acclimation to multifactorial stress combination. Environmental and Experimental Botany 213, 105425. Available at https://doi.org/10.1016/j.envexpbot.2023.105425.

Qian, Z., Zhao, C., Wan, H., Rao, X., Luo, Z., He, L., et al., 2025. Genome-wide identification of DREB1 transcription factors in *Erianthus fulvus* and the functional role of EfDREB1C in regulating cold tolerance of transgenic *Arabidopsis* and sugarcane. International Journal of Biological Macromolecules 317 (Pt 1), 144859. Available at https://doi.org/10.1016/j.ijbiomac.2025.144859.

Quan, R., Wang, J., Qin, H., Chen, L., Xiao, D., Zhao, Z., et al., 2025. Improving grain yield and salt tolerance by optimizing plant height with beneficial haplotypes in rice (*Oryza sativa*). Journal of Advanced Research 76, 17–32. Available at https://doi.org/10.1016/j.jare.2024.12.007.

Ramírez-Cariño, H.F., Morales, I., Guadarrama-Mendoza, P.C., González-Terreros, E., Martínez-Gutiérrez, G.A., Dunlap, C.A., et al., 2023. Biofertilizing effect of putative plant growth promoting rhizobacteria in vitro and in tomatillo seedlings (*Physalis ixocarpa* Brot). Scientia Horticulturae 308, 111567. Available at https://doi.org/10.1016/j.scienta.2022.111567.

Rashki, P., Piri, H., Khamari, E., 2022. Effect of potassium fertilization on roselle yield and yield components as well as IWUE under deficit irrigation regimes. South African Journal of Botany 148, 21–32. Available at https://doi.org/10.1016/j.sajb.2022.03.048.

Rehman, B., Khalifa, W.M.A., Mirza, C.R., Butt, T.A., Manan, N., Sajjad, A., et al., 2025. Salt tolerance potential of native plant species and halophilic bacteria from Bahadur Khel and Khewra, Pakistan. Kuwait Journal of Science 52 (3), 100428. Available at https://doi.org/10.1016/j.kjs.2025.100428.

Rekaby, S.A., Awad, M., Majrashi, A., Ali, E.F., Eissa, M.A., 2021. Corn cob-derived biochar improves the growth of saline-irrigated quinoa in different orders of egyptian soils. Horticulturae 7 (8). Available at https://doi.org/10.3390/horticulturae7080221.

Ren, F., Yang, G., Li, W., He, X., Gao, Y., Tian, L., et al., 2021. Yield-compatible salinity level for growing cotton (*Gossypium hirsutum* L.) under mulched drip irrigation using saline water. Agricultural Water Management 250, 106859. Available at https://doi.org/10.1016/j.agwat.2021.106859.

Ritonga, F.N., Chen, S., 2020. Physiological and molecular mechanism involved in cold stress tolerance in plants. Plants 9 (5), 560. Available at https://doi.org/10.3390/plants9050560.

Rizvi, A., Khan, M.S., 2019. Heavy metal-mediated toxicity to maize: oxidative damage, antioxidant defence response and metal distribution in plant organs. International Journal of Environmental Science and Technology 16 (8), 4873–4886. Available at https://doi.org/10.1007/s13762-018-1916-3.

Rossini-Oliva, S., Abreu, M.M., Leidi, E.O., 2018. A review of hazardous elements tolerance in a metallophyte model species: *Erica andevalensis*. Geoderma 319, 43–51. Available at https://doi.org/10.1016/j.geoderma.2017.12.035.

Ruthrof, K.X., Steel, E., Yates, R., Skinner, P., Ballard, N., De Prato, L., et al., 2021. Responses to controlled release potassium fertilisers in agriculture following phosphate mining. Soil Research 59 (7), 727–736. Available at https://doi.org/10.1071/SR20309.

Saadati, S., Baninasab, B., Mobli, M., Gholami, M., 2021. Foliar application of potassium to improve the freezing tolerance of olive leaves by increasing some osmolite compounds and antioxidant activity. Scientia Horticulturae 276, 109765. Available at https://doi.org/10.1016/j.scienta.2020.109765.

Safdar, M., Kim, W., Park, S., Gwon, Y., Kim, Y.-O., Kim, J., 2022. Engineering plants with carbon nanotubes: a sustainable agriculture approach. Journal of Nanobiotechnology 20 (1). Available at https://doi.org/10.1186/s12951-022-01483-w.

Salimi, I., Khoshgoftarmanesh, A.H., Markarian, S., Maibodi, S.A.M.M., 2024. Response of strawberry (*Fragaria ananassa* L.) to chilling and potassium supply from inorganic and amino acid-complexed sources. Scientia Horticulturae 325, 112655. Available at https://doi.org/10.1016/j.scienta.2023.112655.

Sarikhani, M.R., Ebrahimi, M., 2024. 10 - Potassium releasing bacteria (KRB): role of KRB in horticultural crops. In: Rakshit, A., Meena, V.S., Fraceto, L.F., Parihar, M., Mendonza, A.B., Singh, H.B. (Eds.), Bio-Inoculants in Horticultural Crops: Advances in Bio-Inoculant Sciences, Volume 3. Woodhead Publishing, pp. 175–195. Available at https://doi.org/10.1016/b978-0-323-96005-2.00004-0.

Sattar, A., Naveed, M., Ali, M., Zahir, Z.A., Nadeem, S.M., Yaseen, M., et al., 2019. Perspectives of potassium solubilizing microbes in sustainable food production system: a review. Applied Soil Ecology 133, 146–159. Available at https://doi.org/10.1016/j.apsoil.2018.09.012.

Shaffique, S., Injamum-Ul-Hoque, M., Husen, A., Kang, S.-M., Lee, I.-J., 2025. Revolutionizing heat stress tolerance in *Glycine max*: exploring the latest advances in microbial application. Plant Stress 15, 100725. Available at https://doi.org/10.1016/j.stress.2024.100725.

Shankar, A., Singh, A., Kanwar, P., Srivastava, A.K., Pandey, A., Suprasanna, P., et al., 2013. Gene expression analysis of rice seedling under potassium deprivation reveals major changes in metabolism and signaling components. PLoS One 8 (7), e70321. Available at https://doi.org/10.1371/journal.pone.0070321.

Sharma, M.M.M., Kapoor, D., Rohilla, R., Loyal, A., Sharma, P., Husen, A., 2024. Role of Essential Minerals and Rhizosphere in Plant under Changing Environmental Conditions. Elsevier BV, pp. 235–252. Available at https://doi.org/10.1016/b978-0-443-16082-0.00002-3.

Shehzad, M.A., Nawaz, F., Ahmad, F., Ahmad, N., Masood, S., 2020. Protective effect of potassium and chitosan supply on growth, physiological processes and antioxidative machinery in sunflower (*Helianthus annuus* L.) under drought stress. Ecotoxicology and Environmental Safety 187, 109841. Available at https://doi.org/10.1016/j.ecoenv.2019.109841.

Shengcai, L., Tianliu, H., Yingxiao, H., Weijia, L., Mengbo, Y., Rui, Z., et al., 2025. CaNAC014 transcription factor enhances salt stress tolerance in pepper by regulating the activity of the CaSOS1 promoter. Plant Stress 18, 101025. Available at https://doi.org/10.1016/j.stress.2025.101025.

Shobbar, Z.S., Amirbakhtiar, N., Mirdar Mansuri, R., Loni, F., Akbari, A., Sasaninezhad, M., 2023. Small RNAs involved in salt stress tolerance of food crops. Plant Small RNA in Food Crops. Elsevier, Iran, pp. 295–346. Available at https://doi.org/10.1016/B978-0-323-91722-3.00003-8.

Singh, A., 2021. Soil salinization management for sustainable development: a review. Journal of Environmental Management 277, 111383. Available at https://doi.org/10.1016/j.jenvman.2020.111383.

Singh, S., Shraboni, F.A., Islam, A.K.M.A., Bhanu, A.N., 2025. Genetic engineering of rhizosphere microbes towards crop improvement. Advances in Botanical Research 116, 235–282. Available at https://doi.org/10.1016/bs.abr.2025.04.001.

Singh, V., Lone, R.A., 2021. High-affinity K+ transporters and their functions in plants. Cation Transporters in Plants. Elsevier, Canada, pp. 49–61. Available at https://doi.org/10.1016/B978-0-323-85790-1.00009-9.

Sun, C., Xu, L., Ullah, A., Liu, X., Sun, J., Jiang, D., et al., 2025. Drought priming combined with additional potassium application enhances wheat tolerance to post-anthesis drought via osmotic adjustment and leaf hydraulic conductance. Plant Physiology and Biochemistry 227, 109984. Available at https://doi.org/10.1016/j.plaphy.2025.109984.

Sun, M., Zhang, Y., Xia, Q., Wang, X., Zhang, Q., Liu, T., et al., 2023. Effects of potassium fertilizer types and amounts on nitrogen protective substances in alfalfa root crown and cold resistance under low temperature stress. Editorial Department of Chinese Journal of Grassland, China Chinese Journal of Grassland 45 (3), 78–86. Available at https://doi.org/10.16742/j.zgcdxb.20210436.

Szádeczky-Kardoss, I., Szaker, H.M., Verma, R., Darkó, É., Pettkó-Szandtner, A., Silhavy, D., et al., 2022. Elongation factor TFIIS is essential for heat stress adaptation in plants. Nucleic Acids Research 50 (4), 1927–1950. Available at https://doi.org/10.1093/nar/gkac020.

Tadayon, M.S., Mousavi, S.M., Hosseini, S.M., Sadeghi, S., 2025. Optimizing pistachio yield and efficiency: evaluating artificial mulch and shade nets for enhanced drought and salinity resilience. European Journal of Agronomy 168, 127655. Available at https://doi.org/10.1016/j.eja.2025.127655.

Thakur, P.P., Jadhav, P.V., Dasgupta, D., Mote, G., Bhagat, R., Kale, P.B., et al., 2025. Mineral nutrient acquisition under stress: sensing, signaling, and transportation. Plant Science 360, 112702. Available at https://doi.org/10.1016/j.plantsci.2025.112702.

Tian, C., Zhou, X., Liu, Q., Peng, J., Zhang, Z., Song, H., et al., 2020. Increasing yield, quality and profitability of winter oilseed rape (*Brassica napus*) under combinations of nutrient levels in fertiliser and planting density. Crop and Pasture Science 71 (12), 1010–1019. Available at https://doi.org/10.1071/CP20328.

Tipayno, S., 2024. Application of microorganisms for cross-protection against biotic and metal-induced oxidative stresses in plants. Beneficial Microbes for Sustainable Agriculture Under Stress Conditions. Available at https://doi.org/10.1016/B978-0-443-13193-6.00011-7.

Tiwari, S., Patel, A., Singh, M., Prasad, S., 2020. Regulation of temperature stress in plants. Plant Life Under Changing Environment. Available at https://doi.org/10.1016/B978-0-12-818204-8.00002-3.

Toribio, A.J., Suárez-Estrella, F., Jurado, M.M., López-González, J.A., Martínez-Gallardo, M.R., Estrella-González, M.J., et al., 2025. Boosting plant welfare and rhizospheric health through the application of phosphorus and potassium-solubilizing fungi from compost and vermicompost. Biocatalysis and Agricultural Biotechnology 68, 103741. Available at https://doi.org/10.1016/j.bcab.2025.103741.

Trinh, T.S., Hoang, T.T.H., Vu, T.M., Tran, T.H.S., Tran, V.T., Alharby, H.F., 2024. Potassium nutrition in rice (*Oryza sativa* L.) for higher yield with lower sodium uptake under saline conditions of Central Vietnam. Indian Journal of Agricultural Research 58 (5), 775–781. Available at https://doi.org/10.18805/IJARe.AF-849.

Tushabe, D., Altmann, F., Koehler, E., Woods, S., Rosbakh, S., 2023. Negative effects of high-temperature stress on gametophyte performance and their consequences for seed reproduction in wild plants. Environmental and Experimental Botany 216, 105532. Available at https://doi.org/10.1016/j.envexpbot.2023.105532.

Umer, S., Akram, N.A., Bukhari, I., Abdel Latef, A.A.H., 2024. Biochar for alleviation of salinity stress in plants. Biochar in Mitigating Abiotic Stress in Plants. Elsevier, Pakistan, pp. 173–192. Available at https://doi.org/10.1016/B978-0-443-24137-6.00011-2.

Wang, Y., Zeng, F.R., Wang, Y., Xu, S., Chen, Z.H., 2021. Potassium transporters and their evolution in plants under salt stress. Cation Transporters in Plants. Elsevier, China, pp. 63–83. Available at https://doi.org/10.1016/B978-0-323-85790-1.00022-1.

Xie, Z., Meng, Y., Zhao, M., Liu, D., Bi, W., 2025. Bioremediation potential of rhizosphere microbes. Advances in Botanical Research 116, 157–194. Available at https://doi.org/10.1016/bs.abr.2025.03.003.

Xu, F.-J., Zhang, A.-Y., Yu, Y.-Y., Sun, K., Tang, M.-J., Zhang, W., et al., 2021. Soil legacy of arbuscular mycorrhizal fungus *Gigaspora margarita*: the potassium-sequestering glomalin improves peanut (*Arachis hypogaea*) drought resistance and pod yield. Microbiological Research 249, 126774. Available at https://doi.org/10.1016/j.micres.2021.126774.

Yaashikaa, P.R., Kumar, P.S., Jeevanantham, S., Saravanan, R., 2022. A review on bioremediation approach for heavy metal detoxification and accumulation in plants. Environmental Pollution 301, 119035. Available at https://doi.org/10.1016/j.envpol.2022.119035.

Yan, G., Fan, X., Zheng, W., Gao, Z., Yin, C., Li, T., et al., 2021. Silicon alleviates salt stress-induced potassium deficiency by promoting potassium uptake and translocation in rice (*Oryza sativa* L). Journal of Plant Physiology 258-259, 153379. Available at https://doi.org/10.1016/j.jplph.2021.153379.

Yang, C., Zhang, W., Gu, H., Liu, A., Guo, Q., Chen, Y., et al., 2023a. Field, plant, to leaf: a meta-analysis on crop water use efficiency response to potassium fertilization. Journal of Hydrology 621, 129578. Available at https://doi.org/10.1016/j.jhydrol.2023.129578.

Yang, H., Liu, J., Ma, M., Tan, Z., Zhang, K., Sun, R., et al., 2025. Leaf development and its interaction with phyllospheric microorganisms: impacts on plant stress responses. Plant Stress 16, 100843. Available at https://doi.org/10.1016/j.stress.2025.100843.

Yang, S., Wang, J., Tang, Z., Li, Y., Zhang, J., Guo, F., et al., 2023b. Calcium/calmodulin modulates salt responses by binding a novel interacting protein SAMS1 in peanut (*Arachis hypogaea* L.). The Crop Journal 11 (1), 21–32. Available at https://doi.org/10.1016/j.cj.2022.06.007.

Yin, J., Jia, J., Lian, Z., Hu, Y., Guo, J., Huo, H., et al., 2019. Silicon enhances the salt tolerance of cucumber through increasing polyamine accumulation and decreasing oxidative damage. Ecotoxicology and Environmental Safety 169, 8–17. Available at https://doi.org/10.1016/j.ecoenv.2018.10.105.

Ying, J., Balasubramanian, B., Park, S., Anand, A., Meyyazhagan, A., Pappusamy, M., et al., 2025. Green nanoparticles in agriculture: enhancing crop growth and stress tolerance. Plant Stress 18, 101017. Available at https://doi.org/10.1016/j.stress.2025.101017.

Zeeshan, M., Ahmad, W., Hussain, F., Ahamd, W., Numan, M., Shah, M., et al., 2020. Phytostabalization of the heavy metals in the soil with biochar applications, the impact on chlorophyll, carotene, soil fertility and tomato crop yield. Journal of Cleaner Production 255, 120318. Available at https://doi.org/10.1016/j.jclepro.2020.120318.

Zhang, C., Meng, Y., Zhao, M., Wang, M., Wang, C., Dong, J., et al., 2024a. Advances and mechanisms of fungal symbionts in improving the salt tolerance of crops. Plant Science 349, 112261. Available at https://doi.org/10.1016/j.plantsci.2024.112261.

Zhang, M., Wu, M., Xu, T., Cao, J., Zhang, Z., Zhang, T., et al., 2024b. A putative Na+/H+ antiporter BpSOS1 contributes to salt tolerance in birch. Plant Science 346, 112181. Available at https://doi.org/10.1016/j.plantsci.2024.112181.

Zhang, X., Liu, P., Qing, C., Yang, C., Shen, Y., Ma, L., 2021. Comparative transcriptome analyses of maize seedling root responses to salt stress. PeerJ 9, e10765. Available at https://doi.org/10.7717/peerj.10765.

Zhang, Y., Wan, S., Xing, B., Peng, C., Zhu, J., Shao, Q., et al., 2025. An HD-Zip transcription factor ArHDZ22 regulates plant height and decreases salt tolerance in *Anoectochilus roxburghii*. Industrial Crops and Products 223, 120251. Available at https://doi.org/10.1016/j.indcrop.2024.120251.

Zhang, Z., Xiao, W., Qiu, J., Xin, Y., Liu, Q., Chen, H., et al., 2020. Nystose regulates the response of rice roots to cold stress via multiple signaling pathways: a comparative proteomics analysis. PLoS One 15 (9), e0238381. Available at https://doi.org/10.1371/journal.pone.0238381.

Zhou, Y., Zhang, Z., Zhao, X., Liu, L., Tang, Q., Fu, J., et al., 2023. Receptor-like cytoplasmic kinase STK confers salt tolerance in rice. Rice 16 (1). Available at https://doi.org/10.1186/s12284-023-00637-0.

Zhu, B., Xu, Q.W., Ma, S.M., Liu, B.Y., Duan, M.C., Wang, L.C., 2021. Effects of potassium fertilizer rate on growth, seed quality and potassium use efficiency in *Brassica napus* under drought stress. Journal of Plant Nutrition and Fertilizers 27 (6), 1016–1026. Available at https://doi.org/10.11674/zwyf.20561.

Zou, P., Yang, X., Yuan, Y., Jing, C., Cao, J., Wang, Y., et al., 2021. Purification and characterization of a fucoidan from the brown algae *Macrocystis pyrifera* and the activity of enhancing salt-stress tolerance of wheat seedlings. International Journal of Biological Macromolecules 180, 547–558. Available at https://doi.org/10.1016/j.ijbiomac.2021.03.039.

Zulfiqar, F., Moosa, A., Ferrante, A., Nafees, M., Darras, A., Nazir, M.M., et al., 2023. Exogenous foliar application of melatonin mitigates salt induced oxidative stress and promotes growth in *Gerbera jamosonii*. South African Journal of Botany 161, 678–684. Available at https://doi.org/10.1016/j.sajb.2023.08.055.

Zulfiqar, F., Nafees, M., Darras, A., Shaukat, N., Chen, J., Ferrante, A., et al., 2022. Pre-harvest potassium foliar application improves yield, vase life and overall postharvest quality of cut gladiolus inflorescences. Postharvest Biology and Technology 192, 112027. Available at https://doi.org/10.1016/j.postharvbio.2022.112027.

Zulfiqar, F., Ashraf, M., 2022. Antioxidants as modulators of arsenic-induced oxidative stress tolerance in plants: an overview. Journal of Hazardous Materials 427, 127891. Available at https://doi.org/10.1016/j.jhazmat.2021.127891.

CHAPTER

Calcium: membrane stability and stress signaling in plants

6

Konkona Lahiri[1], Swastik Kundu[1], Pramita Dan[1] and Aryadeep Roychoudhury[2]

[1]*Department of Biotechnology, St. Xavier's College (Autonomous), Kolkata, West Bengal, India,*
[2]*Discipline of Life Sciences, School of Sciences, Indira Gandhi National Open University, New Delhi, Delhi, India*

6.1 Introduction

Calcium ions (Ca^{2+}) serve as crucial intracellular second messengers, playing indispensable roles in plant growth, development, and stress responses by modulating downstream cellular activities such as gene expression and metabolic functions (Tuteja and Mahajan, 2014; Lu et al., 2025). The dynamic regulation of cytosolic calcium concentrations, often referred to as "calcium signatures," enables plants to perceive and respond to a wide array of environmental challenges, including fluctuations in light intensity, temperature, and nutrient availability (Hashimoto and Kudla, 2011; Jing et al., 2024; Wdowiak et al., 2024). These signatures, characterized by specific spatiotemporal patterns of Ca^{2+} transients, are critical for encoding information from diverse stimuli into precise physiological responses (Negi et al., 2023; Hong-Bo et al., 2008). Indeed, environmental stresses such as heat, salinity, and heavy metal toxicity trigger a rapid increase in cytosolic Ca^{2+}, acting as a critical alert signal (Park and Shin, 2022). This elevation in cytosolic Ca^{2+} serves as a universal signal, initiating a cascade of downstream events that enable plants to adapt to or resist various adverse conditions (Li et al., 2022).

The postgenomic era has significantly advanced the understanding of calcium signaling machinery, revealing a complex network of sensors, responders, and transporters, crucial for stress perception and developmental cues (Ghosh et al., 2021). As a ubiquitous intracellular second messenger, Ca^{2+} acts as a vital relay hub for numerous biological messages, orchestrating signal transduction networks across all eukaryotic organisms (Patra et al., 2021; Tuteja and Mahajan, 2014). In plants, calcium is particularly essential for mineral nutrition and overall cellular physiology, highlighting the need to understand its metabolism and diverse functions (Wdowiak et al., 2024; Wang et al., 2023). These functions encompass not only abiotic stress responses, such as those to cold or drought, but also biotic interactions, where specific Ca^{2+} signatures are generated to combat pathogen

Mineral Nutrients to Mitigate Abiotic Stress in Crops. DOI: https://doi.org/10.1016/B978-0-443-44712-9.00010-4

infections (Park and Shin, 2022; Mori et al., 2018; Wang et al., 2024). External environmental cues, including changes in soil mineral content and the presence of harmful elements, directly influence calcium dynamics, underscoring its role in plant adaptation to diverse ecological niches (Wang et al., 2023).

6.2 Cellular calcium homeostasis

6.2.1 Basal cytosolic Ca^{2+} concentration

Maintaining basal cytosolic Ca^{2+} concentrations at remarkably low levels, typically around 100 nM, is crucial for plant cells, contrasting sharply with the millimolar concentrations found in extracellular spaces and internal stores like the endoplasmic reticulum and vacuoles (Lu et al., 2025). This steep electrochemical gradient across membranes is meticulously maintained by a sophisticated interplay of Ca^{2+} channels, pumps, and exchangers, which govern both Ca^{2+} influx and efflux to prevent cytotoxicity and facilitate precise signaling events (Patra et al., 2021; Tuteja and Mahajan, 2014).

6.2.2 Calcium storage compartments

6.2.2.1 *Vacuole*

The vacuole, serving as the primary Ca^{2+} storage compartment, plays a critical role in sequestering excess cytosolic Ca^{2+}, thereby preventing cellular toxicity and contributing to the maintenance of low basal cytosolic Ca^{2+} concentrations necessary for signaling fidelity (Wdowiak et al., 2024). This sequestration is primarily mediated by Ca^{2+}-ATPases and Ca^{2+}/H^+ antiporters located on the tonoplast, which actively transport Ca^{2+} into the vacuolar lumen against its concentration gradient (Dodd et al., 2010).

6.2.2.2 *Endoplasmic reticulum*

The endoplasmic reticulum also functions as a vital internal Ca^{2+} store, intricately involved in protein folding and modification processes, with its Ca^{2+} concentration typically maintained in the micromolar to millimolar range, allowing for rapid release during stress (Park and Shin, 2022; Li et al., 2022).

6.2.2.3 *Apoplast*

The apoplast, comprising the cell walls and intercellular spaces, acts as a significant extracellular reservoir of Ca^{2+}, with concentrations often reaching millimolar levels (Wdowiak et al., 2024). This external Ca^{2+} pool is readily available for influx into the cytosol, contributing to the generation of Ca^{2+} signals in response to various environmental stimuli (Michal Johnson et al., 2014).

6.2.3 Maintenance of steep Ca^{2+} gradients

The precise regulation of Ca^{2+} gradients is of paramount importance for the integrity of Ca^{2+} signaling pathways, as even subtle perturbations can disrupt downstream cellular responses and lead to impaired plant development or stress acclimation (Jing et al., 2024; Wang et al., 2024). This strict spatiotemporal regulation of Ca^{2+} uptake, efflux, and buffering within the cytosol is critical for its multifaceted roles as both a nutrient and a signaling molecule (Wilkins et al., 2016). However, due to its ability to interact readily with proteins, membranes, and organic acids, calcium can become a toxic cellular compound at elevated concentrations, potentially leading to protein and nucleic acid aggregation, phosphate precipitation, and compromised lipid membrane integrity (Bose et al., 2011). Consequently, plants have developed sophisticated Ca^{2+} efflux mechanisms to maintain consistently low cytosolic free calcium levels by actively expelling Ca^{2+} out of the cell or sequestering it into intracellular organelles (Bose et al., 2011).

6.2.4 Ca^{2+} channels, pumps, and exchangers

The establishment and maintenance of steep Ca^{2+} gradients across cellular membranes are orchestrated by a sophisticated interplay of Ca^{2+} channels, pumps, and exchangers, which facilitate both rapid influx and precise efflux of Ca^{2+} ions (Tuteja and Mahajan, 2014). This dynamic equilibrium is crucial for maintaining low cytosolic Ca^{2+} concentrations under resting conditions, which are typically 10,000 times lower than in the extracellular space or intracellular stores (Wdowiak et al., 2024; Lu et al., 2025). This substantial gradient enables rapid Ca^{2+} influx upon stimulation, thereby mediating diverse physiological processes in plants, including responses to light, touch, pathogens, hormones, and various abiotic stresses (Tuteja and Mahajan, 2014). Calcium ions are essential for plant survival, given that plants, being sessile organisms, must adapt to the soil conditions in which they germinate (Wdowiak et al., 2024). The dynamic regulation of cytosolic Ca^{2+} concentration (Ca^{2+}cyt) is essential for plants to transduce environmental signals into specific cellular responses (Mori et al., 2018). This intricate Ca^{2+} communication across membranes involves a diverse array of Ca^{2+}-permeable channels and transporters that receive environmental cues and transduce them into cellular information (Park and Shin, 2022).

6.3 Calcium signatures

6.3.1 Definition and properties of Ca^{2+} signatures

Stress is any environmental condition hinder the plant from growing with its optimum genetic potential. Plants are subjected to many types of fluctuations in the environment. Animals being able to move can flee away from these environmental fluctuations. However, plants, being sessile, have to combat these stress conditions in the physical environment. Plants adjust to these fluctuations in nature as they are plastic in terms of metabolism and development. However, acute or chronic changes in the environmental conditions can adversely affect the growth and development of the plant. Plants have evolved a host of strategies, like different signaling pathways, to withstand these severe conditions. Ca^{2+} ions are one of the most universal secondary messengers in plant cells (Ranty et al., 2016). In plants, environmental cues such as drought, temperature extremes, salt stress, heavy metal toxicity, or biotic stress like herbivore attack or infections are perceived by the receptors on the plasma membrane or in organelles. These receptors trigger an influx of Ca^{2+} in the cytosol. Normally, cytosolic Ca^{2+} is kept at extremely low levels (~10^{-7} M). Outside the cell and inside the ER, Ca^{2+} concentration is much higher (~10^{-3} M). This huge difference means that when Ca^{2+} channels open, Ca^{2+} floods into the cytosol, raising its concentration 10–20 times. This sudden spike activates Ca^{2+}-responsive proteins, leading to metabolic changes and stress-responsive gene expression(Xu et al., 2022). Cytosolic Ca^{2+} levels are kept low in resting cells by different pumps and ion exchangers of the cell membrane, and the ER pumps Ca^{2+} out of the cytosol using ATP. As Ca^{2+} levels in cytosol are normally kept extremely low, even a small influx from internal stores (vacuoles or ER) or external space (apoplast) represents a large and significant signal. When a stress signal is perceived, these channels open briefly, allowing influx of Ca^{2+} into the cytosol from outside or internal stores. This results in a rapid, transient rise called a calcium spike or calcium signature (Ranty et al., 2016). The shape of the Ca^{2+} spike (with respect to amplitude, frequency, and duration) encodes information about the type of stress. Cold, salt, and drought each generate distinct Ca^{2+} spikes known as stress-specific Ca^{2+} signatures.

6.3.2 Involvement of calcium in varied abiotic stresses

6.3.2.1 *Drought stress*

Drought is a severe form of abiotic stress that greatly impairs plant growth and development. Water is absorbed from the soil and is delivered to different organs of the plant by the vascular system. As water is lost from leaves by transpiration, a water deficit is created; water will flow into the leaves to replenish the loss only if water is available in the soil. However, if the soil starts drying up, a deficit is created between the supply from roots and demand from leaves (Peiter et al., 2005). Abscisic acid (ABA) is found to raise the levels of intracellular calcium. It has been reported that oscillations in intracellular calcium occur against several stimuli in plants. Guard cells in plants are able to translate artificially imposed calcium with specific frequencies and amplitudes, required for stomatal closure during drought stress (Whalley and Knight, 2013). Specific Ca^{2+} elevation patterns are necessary in controlling stomatal closure response and the final steady state of stomatal aperture (Shao et al., 2008).

6.3.2.2 Salinity stress

High amounts of salt in soil are one of the common agricultural concerns throughout the globe. Saline soil severely impairs plant growth. Sodic soils can change the soil structure and decrease soil porosity and water permeability. Plants can tolerate a certain concentration of salts, beyond which they start showing symptoms of growth inhibition. Halophytes are salt-loving plants that can grow as well as sustain themselves in saline conditions. On the contrary, glycophytes (literally "sweet plants") cannot tolerate salt to the same extent as halophytes. Highly saline soil prevents roots from taking up water due to a decreased solute potential (solute potential is always less than water potential). Moreover, when these ions accumulate to a great extent in cells, they can directly affect protein synthesis and enzyme activity, thereby inhibiting metabolism. Plants have devised ways by which they can avoid salt stress. To ensure tolerance, plants reduce the uptake of Na^+ and stimulate the entry of K^+ ions into cytosol or sequester Na^+ ions into vacuoles. Once sequestered, these ions are no longer available for interference in other metabolic activities. Cytosolic Na^+ can be sequestered into different intracellular organelles like ER and Golgi. During Na^+ ion toxicity, intracellular calcium concentration increases from its basal level, which is read by a special sensor called SOS3, discussed later. As soon as salt stress is sensed, a short-term, stable oscillating change in intracellular calcium is observed, which triggers downstream response machinery. A distinct calcium oscillation was elicited in cells when treated with NaCl. However, this varies with the duration of exposure to stress as well as the intensity of stress imposed, the plant species, and the type of cell concerned (Kader and Lindberg, 2014).

6.3.2.3 Freezing stress and heat stress

Plants grow at an optimum temperature of 25°C–35°C when all enzymes needed for growth and metabolism function well. However, when they are subjected to low temperature like 10°C–15°C, chilling injury can occur, which may slow down growth. Sudden exposure to temperature close to 0°C may lead to cold shock, and this might lead to discoloration of leaves, inhibition of protein synthesis, and loss of enzyme activity. The plasmalemma is made up of a lipid bilayer comprising phospholipids and proteins. The plasma membrane, with a high percentage of saturated fatty acids, may undergo solidification in chilling-sensitive plants, reducing the activity of membrane transporters and pumps and impairing solute transport. As cold stress is perceived, definite changes in calcium signatures with respect to amplitude, frequency, and duration are observed. Low temperature causes distinct changes in the intracellular calcium concentrations (Yuan et al., 2018).

High temperature beyond the optimum can severely impair plant growth. It can cause denaturation of proteins, loss of enzyme activity, and other biochemical and molecular aberrations. Changes in climate and global warming are serious threats to plant growth and crop yield. However, plants have devised ways by which they combat the emerging heat stress. In plants, many calcium channels act as thermosensors that directly modulate calcium signaling when exposed to heat. Exposure to high temperature peaks calcium levels in cells, which is decoded by different calcium-sensing effector proteins like calcium-binding proteins and calcium-permeable channels. These transmit the signal downstream to elicit specific cellular responses.

6.3.2.4 UV-B radiation

UV-B radiation (~280–315 nm) is a biologically active, powerful part of the electromagnetic spectrum coming from sunlight. UV-B radiation causes damage to cellular components, reduced photosynthetic efficiency, and impaired growth and development. It has been seen that when the microalgae *Dunaliella salina* was exposed to UV-B, there was a surge in the level of reactive oxygen species (ROS) due to activation of an enzyme, NADPH oxidase. This generated ROS causes oxidative damage to biomolecules, impairing cell metabolism and sustenance. ROS inhibitors like 3-(3,4-dichlorophenyl)-1,1-dimethylurea (DCMU) decelerate death induced by UV-B, suggesting the fact that ROS accumulation triggers cell death. Moreover, Ca^{2+} has a significant role in response to UV-B irradiation. As the amount of ROS in the cells increases, it leads to high activity of Ca^{2+} ATPase pumps of the membrane and a surge in the levels of intracellular calcium concentration. Addition of exogenously added calcium (in the form of $CaCl_2$) did not worsen the conditions, but intracellular free calcium did (Zhang et al., 2017). Experiments in *Rhododendron* have shown that UV-B upregulates Ca^{2+} to activate protein phosphorylation cascades that control phenolic acid biosynthesis. Accumulation of such phenolic compounds through Ca-dependent activation of transcriptional regulators helps regulate ROS scavenging enzyme activity (Zhou et al., 2024a). Therefore Ca signatures, on perception of UV-B, crosstalk with ROS scavenging machinery to ROS.

6.4 Decoding calcium signatures

6.4.1 Ca^{2+} binding proteins

- **Calmodulins**

  Calmodulin (CaM) is a universal Ca^{2+} sensor found in nearly all eukaryotic cells. CaM is shaped like a dumbbell with two globular ends attached by an α helix. Each end has a couple of closely associated EF hands. These structures interact via antiparallel β sheets, and this becomes the basis of cooperative binding. CaM is the best characterized Ca^{2+}-responsive protein, being 148 amino acids in length. When Ca^{2+} binds, calmodulin exposes its two hydrophobic surfaces, one in each globular domain. It changes its shape and wraps around its target protein. This conformational change activates the target protein directly or indirectly. Target sites for CaM are called Calmodulin-Binding Domains (CaMBDs), which have a stretch of 12–30 positively charged, amphiphilic amino acids. These CaMBPs have major roles to play in metabolism, transcription, phosphorylation, and many unknown functions (Bouché et al., 2005).

- **Calmodulin-like proteins**
  CaM-like proteins (CMLs) are the biggest group of Ca^{2+} relay sensors in plants. Arabidopsis itself has around 50 CMLs. The expression of these CMLs is induced by different abiotic and biotic stresses (Ranty et al., 2016). They are comprised mostly of EF-hand calcium-binding domains and are more or less similar to CaMs. However, CMLs have one to six EF hands. Different biophysical studies show that CML binds Ca^{2+} and hence acts as a vital calcium sensor. All EF hands in CMLs cannot bind calcium. Studies have shown that only one in three EF hands in AtCML14 binds Ca^{2+}, while only two or three EF-hand motifs were able to bind Ca^{2+} out of the total four EF hands, each in AtCML15 and AtCML16 (Ogunrinde et al., 2017). This suggests that CMLs get activated at different calcium concentrations when compared to CaMs.
- **Calcineurin B-like proteins**
  Calcineurin B-like (CBLs) proteins are another class of Ca^{2+} sensors that recognize calcium signals in cells. Calcineurin is a calcium-dependent protein phosphatase that has a catalytic subunit (Calcineurin A) and a Ca^{2+}-binding regulatory subunit (Calcineurin B) (Sanyal et al., 2020). Although the sequences of these subunits are highly conserved from yeast to mammals, no such homolog has been found in plants till now. *CBL* genes were found for the first time in Arabidopsis, which showed a folding pattern very similar to calcineurin B. It is interesting to find that unlike Calcineurin B, which is usually associated with calcineurin A phosphatase in yeast and mammals, CBL was found to be associated with a different family of kinases called CBL-interacting protein kinases (CIPK) (Tang et al., 2020). Whenever calcium spikes in cytosol, calcium binds CBL, which activates CIPK. Now as CBL activates CIPK, CIPK phosphorylates CBL at some conserved Ser/Thr residues on its C-terminal, enhancing its activity (Yadav et al., 2018). The activated CBL–CIPK complex relays the signal and brings about changes in downstream targets.
- **Calcium-dependent protein kinases**
  Calcium-dependent protein kinases (CDPKs) are a family of calcium sensors that not only sense the calcium signatures but also respond within a single protein to translate the calcium signal into phosphorylation events (Boudsocq et al., 2010). They have an important role to play in mitigating plant abiotic stress (Asano et al., 2014). CDPKs have been found to possess an N-terminal variable domain, a Ser/Thr kinase domain, an autoinhibitory junction region, and a regulatory calmodulin domain (Ranty et al., 2016). The regulatory domain is arranged in two lobes, each comprising four EF-hand calcium-binding motifs with differences in calcium-binding affinities. When *CDPK* genes were over-expressed, it was found to contribute to better tolerance to drought stress, heat stress, and salinity stress. CDPKs have been found to phosphorylate a host of

proteins like NADPH oxidase or ascorbate peroxidase, enzymes of secondary metabolite biosynthesis, or transcription factors.

6.4.2 Activation of downstream components

- **Protein kinases**
  Protein kinases are vital regulators of any signaling pathway. Phosphorylation by protein kinases can activate a downstream protein or serve as a stepping stone in protein-protein interactions. The mitogen activated protein kinase (MAPK) cascade, sucrose nonfermenting (SNF1) related protein kinase 2, and calcium-dependent protein kinases (CDPK) are important to modulate stress response pathways in plants (Chen et al., 2021). *Arabidopsis* has been found to have around 1000 such protein kinases. Sucrose nonfermenting (SNF1) related protein kinase 2 (SnRK2s) proteins are prominent players in dehydration stress response. Abscisic acid accumulates during drought stress, which causes stomata to close and activate downstream stress-responsive genes. ABA activates SnRK2.2, SnRK2.3, and SnRK2.6 in *Arabidopsis* against several abiotic stresses like dehydration stress. The MAPK cascade comprises three important kinases: MAP Kinases (MAPK/MPK), MAP Kinase Kinases (MAPKK/MKK), and MAP Kinase Kinase Kinases (MAPKKK/MAP3Ks). Studies help us to understand that the MAPK cascade is associated closely with drought stress and ABA signaling, for instance, CAT1 expression and peroxide generation are regulated by MKK1-MPK6 in the ABA-dependent signaling pathway. MAPKKK18-MAPKK3 regulates drought stress tolerance (Chen et al., 2021). The salt overly sensitive (SOS) pathway actively participates in drought and salinity stress response. It consists of the Na^+/H^+ antiporter called SOS1, SnRK3.11/SOS2/CIPK24, and SOS3/CBL4. SOS3 gets activated at elevated calcium levels. SOS3 activates SOS2. Activated SOS2 further activates SOS1 at the plasma membrane, helping in the efflux of Na^+ ions (Chen et al., 2021).
- **Transcription factors**
  Induction of downstream stress-related genes takes place during transcription, where activation of particular transcription factors leads to the expression of desired genes. ERF transcription factors are found to bind to the dehydration-responsive element (DRE), responsible for the expression of drought- and freezing-stress-related genes. When the ABA signaling pathway is activated during drought stress, ABA is recognized by Pyrabactin Resistance 1 (PYR), which inhibits PP2Cs and activates SnRK2 kinases. These activate ABA-responsive element binding factors (ABFs), viz., basic leucine zipper (bZIP) transcription factors that carry out the expression of ABA-responsive genes. The activated ABFs can also activate transcription of other factors like AP2/ERFs, which also regulate the expression of stress-responsive genes. Jasmonic acid participates in the developmental modification under stress conditions. Under

abiotic stress, JA interacts with a JA receptor called CORONATINE INSENSITIVE 1 (COI1), which leads to 26S proteasomal destruction of JA signaling repressive proteins called JASMONATE ZIM-DOMAIN (JAZ) proteins. MYC2 is the vital transcription factor responsible for jasmonic acid-related gene expression. However, when jasmonic acid is absent, MYC2 is suppressed by JAZ. MYC2 is characterized by a basic helix-loop-helix (bHLH) domain and is found to promote the expression of dehydration-responsive genes like *rd22* (Yoon et al., 2020). Expression of WRKY transcription factors is dependent on different stress signals and salicylic acid levels. AtWRKY53 is essential to mitigate oxidative stress response (Jiang et al., 2017). MEKK1 interacts with WRKY53 to regulate the expression of CAT1/CAT2/CAT3. WRKY transcription factors like GhWRKY25 in *Gossypium hirsutum* and VvWRKY11 in *Arabidopsis* facilitate drought stress response (Phukan et al., 2016).

- **Ion transporters**

  Ion transporters, mainly those that are involved in ion exclusion, are vital for protecting plants from salinity stress. Prolonged salt stress can disrupt homeostasis in plants and inhibit normal growth. Efficient efflux systems in the plasma membrane, like the H^+/Na^+ antiporters, remove potentially toxic Na^+ from the cytosol to the apoplast (Malakar and Chattopadhyay, 2021). In Arabidopsis, a Na^+/H^+ exchanger called SOS1 (Salt Overlay Sensitive) is present on the plasmalemma and extrudes excess Na^+ ions from root epidermal cells to the soil-root interface. The function of SOS1 is regulated by SOS2/CIPK24 Ser/Thr protein kinase in *Arabidopsis*. As Ca^{2+} levels increase in cells, the signals are perceived by SOS3 (a Ca^{2+} sensor on the plasma membrane) and activate SOS2, which forms a complex with SCaBP8/CBL10 and regulates the activity of SOS1. Active transport is a necessity for excluding Na^+ from the cytosol. Na^+ ions are sequestered in vacuoles by the vacuolar Na^+/H^+ antiporters and H+ ATPase pumps. There are eight *NHX* gene families in *Arabidopsis*. The plasma membrane-localized P-type H^+ ATPase pumps are responsible for huge pH and potential gradients across the plasma membrane, while the V-type H^+ ATPase pumps and H^+-PPase generate a gradient of pH and membrane potential across the vacuolar tonoplast (Brini and Masmoudi, 2012). These efficiently help in the sequestration or compartmentalization of Na^+ in vacuoles, protecting the plants from excess sodicity.

- **Metabolic enzymes**

  Different environmental abiotic stresses can lead to the production of ROS like peroxide, superoxide, and hydroxide, which cause oxidative damage to an array of biomolecules in plants. One of the enzymatic antioxidants called superoxide dismutase (SOD) regulates the amount of formed superoxide and H_2O_2 and decreases the risk of formation of hydroxyl radical. SOD activity increases significantly during drought stress, as in *Euphorbia esula*, maize, and wheat. Another enzyme called ascorbate peroxidase (APX) detoxifies H_2O_2 via the ascorbate glutathione pathway, where ascorbic acid is utilized to reduce H_2O_2 to water and monodehydroascorbate. In *APX1* mutants, H_2O_2 accumulation causes stomatal closure (Koyro et al., 2012).

6.4.3 Translating signatures into physiological responses

As we discussed earlier, the shape of the calcium spike with respect to amplitude and frequency varies from stress to stress. Each stress is defined by its characteristic calcium signature, which needs to be translated or decoded so that specific downstream actions can be elicited. A host of different calcium decoders like the CDPKs, CBL–CIPK modules, CaM and CMLs, sense calcium signatures and relay the signal to downstream targets. Salinity stress activates the SOS pathway as it produces a distinct calcium spike upon exposure to salt. Similarly, low temperature generates a calcium spike, which is decoded by CaM or CMLs that further activate CAMTA transcription factors to express *COR* genes. Stress-induced calcium influx activates respiratory burst oxidase homolog (RBOH) upon binding to calcium that generates ROS. This further activates MAPK and antioxidant enzyme machinery to mitigate stress conditions. Thus translation of stress-specific spatio-temporal patterns of Ca signatures is highly important for downstream responses to be elicited for the plants to combat the stress situation.

6.5 Calcium-mediated stress signaling pathways

6.5.1 Signaling under abiotic stress

- **Drought stress**
  As leaf turgor falls, the stomata close to reduce transpiration and conserve water. When turgor is lost, the leaf cells collapse and the leaf wilts. ABA binds to PYR/PYL/PCAR receptors that lead to the inhibition of PP2Cs (a negative regulator). This activates SnRK2 kinases that trigger downstream responses like the closing of stomata that inhibits water loss by stomatal transpiration. SnRK2 activates OST1 (OPEN STOMATA1), which transmits the signal downstream (Chen et al., 2021). ABA also induces ROS production that serves as a second messenger. ABA-induced calcium signaling can activate the AtCBL1/9-AtCIPK26 module, leading to the phosphorylation of effector proteins such as RBOH, resulting in the generation of ROS (Postiglione and Muday, 2020). ROS activates Ca^{2+} channels in the plasma membrane (Pei et al., 2000). Treating guard cells with ABA induces a spike in the inositol 1,4,5-trisphosphate (IP3) levels due to high phospholipase C (PLC) activity. This IP3 causes a rise in the cytosolic Ca^{2+} levels. Using neomycin and U73122, which are the inhibitors of phosphoinositide signaling, it was observed that ABA-induced Ca^{2+} elevations were suppressed (Staxén et al., 1999). Microinjection of cyclic ADP-ribose (cADPR) uplifts calcium levels without external calcium, meaning the calcium

ions were released from vacuoles. Using a cADPR antagonist like nicotinamide reduced the ABA-induced Ca^{2+} increase (Leckie et al., 1998). Cytosolic Ca^{2+} levels increased from both the arms—the ROS-mediated and IP3-cADPR branches. This inhibits AKT1/KAT1 channels, which facilitates preventing K^+ ion influx into the guard cells. Activated OST1 activates SLAC1 (Slow Anion Channel Associated 1), an S-type anion channel, from where chloride and malate ion efflux take place (Chen et al., 2021). ABA-induced Ca^{2+} increase inhibits the plasmalemma H^+ pump and makes the cytosol alkaline. It also depolarizes the membrane. This activates K^+ out channels, leading to the efflux of K^+ ions. Vacuolar channels like TPC1 (two-pore channel 1) release K^+ and Cl^- ions into the cytosol, leading to the shrinkage of guard cells, and hence stomatal closure.

Arabidopsis reduced hyperosmolarity-induced Ca^{2+} increase (AtOSCA1) encodes a plasma membrane calcium-permeable channel responsible for hyperosmolarity-induced transient elevation in Ca^{2+} (Yuan et al., 2014). When cell turgor is lost due to hypo-osmotic conditions, this activates certain osmosensitive ion channels in plants. Structural biology-based studies support that these OSCA family members have membrane tension-responsive gating mechanisms (Jojoa-Cruz et al., 2018). As OSCA1 gets activated, Ca^{2+} ion entry takes place from the apoplast that produces the required calcium signatures, serving as second messengers for downstream signaling pathways. This suggests that OSCA1 is central to the plant osmotic stress Ca^{2+} influx system required to mitigate water deficit or hyperosmotic stress (Yuan et al., 2014).

- **Salt stress**

  Na^+ ATPase or Na^+/K^+ ATPase pumps are not much characterized in plants. Plants rely only on H^+-ATPases or H^+-pyrophosphatases. As soon as the saline stress is sensed, there is a spike in the cytosolic Ca^{2+} levels. The amplitude and frequency of the calcium spike in the cytosol provide specific information about the nature of saline stress to plants (Kader and Lindberg, 2014). The increase in calcium levels in the cytosol is perceived by the calmodulin domain on CPKs (calcium-dependent protein kinases) that activate the downstream gene (Ranty et al., 2016). Ion exclusion takes place by the SOS pathway, where SOS3/CBL4 (Calcineurin B-like protein 4) translates salinity stress-mediated cytoplasmic Ca^{2+} signals. SOS3 cooperates with SOS2/CIPK24 (CBL interacting protein kinase 24) to induce phosphorylation and activation of SOS1/NHX7, a plasma membrane Na^+/H^+ transporter (Kader and Lindberg, 2014). This pumps excess Na^+ out of the cell, thereby reducing cytosolic Na^+ toxicity (Zhu, 2002). In Arabidopsis, activation of SOS1 is achieved by the activation of MAPK6 (mitogen activated protein kinase 6). SOS2 facilitates the activity of vacuolar V-type H^+ ATPase that leads to Na^+ sequestration into vacuoles.

- **Heat stress**

  Heat stress causes fluidization of the plasma membrane, causing an inflow of calcium ions into the cytoplasm. Sensing of heat stress takes place via CNGCs that elevate the calcium levels of cytosol when cAMP or cGMP binds

(Swarbreck et al., 2013). High temperature elicits one of the earliest measurable increases in cytosolic Ca^{2+}, which activates Ca^{2+}-binding effectors like CDPKs, CaMs, and CMLs that engage the heat shock response at multiple levels. Arabidopsis CaM isoforms CaM1/4, CaM2/3/5. CaM 6 and CaM 7 bind differently to CNGC6, and this binding is dependent on Ca^{2+}. When CaM2/3/5/7 were knocked out, there was an increase in calcium concentration inside the cell under heat stress. However, no such effect was observed in CaM1/4/6 knockouts (Niu et al., 2020). It was observed that nitric oxide (NO) has a role to play in heat stress mitigation, as it stimulates the heat stress transcription factors and accumulation of heat shock proteins (HSPs), suggesting that NO acts upstream of AtCaM3 in the signaling pathway. Overexpression of the *CaM3* gene in cucumber plants showed enhanced heat stress tolerance and provided protection against ROS and damage to the photosynthetic apparatus. In Arabidopsis, heat stress prominently increased the CML12 and CML24 mRNAs. Both CaMs and CMLs play central roles in mitigating heat stress in plants. Ca^{2+}-dependent protein kinases (CDPKs) phosphorylate as well as regulate chaperone systems, including HSP pathways, and modulate ROS scavenging enzymes to limit oxidative damage that accompanies heat stress. Inactivation of the autoinhibitory domain of CDPK1 leads to improved thermotolerance, while CPK28 in tomato phosphorylates ascorbate peroxidase (APX), helping mitigate ROS alongside heat stress tolerance. The synthesis of HSPs helps plants cope with heat stress better. Translation of these HSPs increases dramatically during heat stress, while the synthesis of other proteins may decrease. Certain sequences called heat shock elements (HSEs) are bound by heat shock transcription factors (HSF) that cause the transcription of HSP mRNA. These HSPs act as chaperones that help in protein folding and prevent protein aggregation during heat stress (Kang et al., 2024).

- **Cold stress**

  Low temperatures may lead to the expression of *cold-regulated (COR)* genes like *RD29A* (responsive to desiccation 29A) and *COR15a* that help in cold acclimation. The DRE (dehydration responsive element) is bound by a factor called CBF1 (C-repeat Binding Factors), whose expression is induced by cold acclimation, causing the expression of *COR* genes. Upon sustained or sudden exposure to low temperatures, multiple Ca^{2+} channels like CNGCs (cyclic nucleotide-gated channels) and MCA (Mid1 complementary activity) type mechano-sensitive channels and complexes like COLD1-RGA1 (in *Oryza sativa*) are activated, causing a quick but short-lived increase in Ca^{2+} levels in the cytosol. In *Arabidopsis*, cold stress is perceived by the sensor called COLD1 (chilling-tolerance divergence 1), which activates cold-responsive Ca^{2+} channels called MCA1 and MCA2. This increases the cytosolic Ca^{2+} levels. The expression of AtSRC2 facilitates ROS production via AtRBOHF. These ROS may lead to increased Ca^{2+} levels as positive feedback. CPKs relay the signal downstream to activate the MAPK cascade (Fig. 6.1). A key transcriptional regulator called the CAMTA family (calmodulin-binding transcriptional

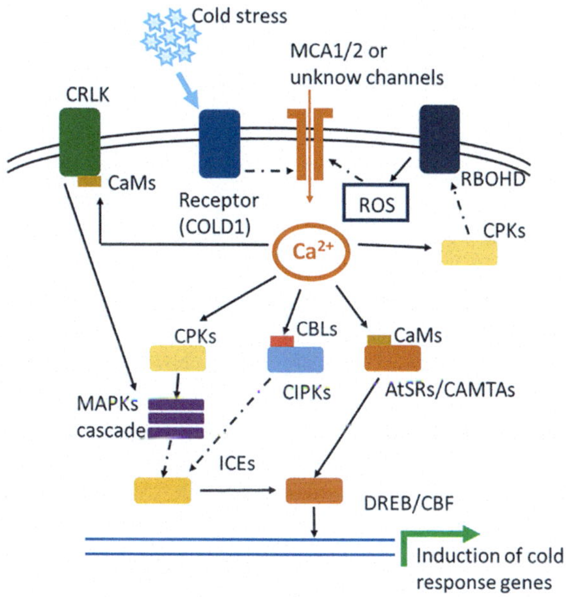

FIGURE 6.1

Mitigation of freezing stress by calcium-based signaling machinery in *Arabidopsis*.

The image has been adopted from Yuan, P., Yang, T., Poovaiah, B.W., 2018. Calcium signaling-mediated plant response to cold stress. International Journal of Molecular Sciences. 19 (12), 3896 https://doi.org/10.3390/ijms19123896 under the Creative Commons Attribution license (http://creativecommons.org/licenses/by/4.0/).

activators) acts downstream of the Ca²/CaM complex to regulate the CBF/DREB1 pathway that upregulates *COR* genes (Yuan et al., 2018).

- **Heavy metal stress**

  Heavy metal contamination arising from different natural or anthropogenic sources, like mining, industrial discharges, and many others, tends to pollute the surrounding soil. Accumulation of these heavy metals, like cadmium, mercury, lead, arsenic, chromium, etc., occurs in plants when they take them up through water via roots. This leads to heavy metal toxicity impairing photosynthesis, growth, and metabolism in plants. Heavy metal stress causes the production of ROS like superoxide, peroxide, and hydroxides, which can lead to oxidative stress. Cd is found mostly in chemical fertilizers and pesticides. Cd inhibits the uptake of major macro elements like Mg, Ca, K, and P. This also leads to disruption in chromosomal segregations and DNA fragmentation. Mercury (Hg) toxicity can affect mitochondrial and chloroplast activity. There are visible lesions on the affected areas and obstruction to water flow in the vascular bundles of plants. Chromium (Cr) causes chlorosis in leaves and growth retardation in affected plants. It also inhibits ROS scavenging enzymes. Lead (Pb) inhibits germination, disrupts the ultrastructure of chloroplasts, and blocks the C3 or Calvin Cycle (Ghori et al., 2019). The activity of calcium channels gets altered on exposure to heavy metals, and there is a surge of calcium in the cells. Calcium binds to calmodulin that regulates the uptake of heavy metals. The high calcium levels also elevate NO levels that convert toxic oxygen radicals to H_2O_2 and oxygen, acting as an antioxidant. The H_2O_2 produced may activate G-protein-coupled receptor signaling (GPCR) and phospholipid signaling pathways. Activation of the MAPK pathway by the phospholipid pathway phosphorylates downstream transcription factors, which are responsible for the expression of stress-responsive genes (Ghori et al., 2019). Arsenite uptake in plants is carried out by NIPs, mainly NIP1, that allow arsenic (III) to enter the cells. CPK31 interacts with NIP1 that modulates absorption of As. *cpk31* mutants show much lower As(III) toxicity and increased resistance to As in comparison to wild-type plants. This means that CPK31 phosphorylates NIP1, promoting As uptake (Ji et al., 2017). NRAMP6-like metal transporters are involved in the uptake of Cd^{2+}. During Cd toxicity, intracellular calcium levels increase, which activates CPK21/CPK23, interacting with and phosphorylating NRAMP6. This reduces the uptake of Cd in plants (Zhang et al., 2023).

The table (Table 6.1) represents the calcium-dependent transcription factors induced under abiotic stress conditions.

6.5.2 Signaling under biotic stress

When a pathogen invades a plant, PRRs (pathogen recognition receptors) on the plant cell surface identify the PAMPs (pathogen-associated molecular patterns) on the pathogen surface. This initial immune response triggered in the host plant is called PAMP-triggered immunity (PTI). Pathogens, on the contrary, secrete elicitor molecules that try to suppress the plant PTI and initiate pathogenesis. However, the host plant has specific resistance genes called *R* genes that detect the presence of the elicitors and trigger a much more robust and effective immune response called effector-triggered immunity (ETI). This often leads to regulated cell death at the infection site called hypersensitive response (HR) that prevents the spread of infection into other organs of the plant. Each of these different layers of immunity in plants differs in its calcium signature, which are sensed by a host of different calcium sensors like CaM, CMLs, or CDPKs, for instance, different calmodulin-binding transcription factors associate with plant defense responses. CBP60g (Calmodulin-Binding Protein 60g), in *Arabidopsis thaliana*, that regulates the production of salicylic acid in a positive way, activates expression of ICS1 (Isochorismate Synthase 1), a major enzyme of the salicylic acid biosynthetic pathway. This strengthens the defense response and resistance. Overexpression of CMLs (SCaM-4/-5 in tobacco) is found to increase resistance against a broad array of virulent pathogens. CDPKs on Ca^{2+} binding undergo a conformational change,

Table 6.1 Calcium-dependent transcription factors to mitigate different abiotic stress conditions.

Transcription factor involved	Concerned sensor	Type of stress	Downstream target genes	Role	Key citations
CAMTA family	Ca^{2+}/CaM	Freezing stress	*CBF*, *COR* genes	Cold acclimation	Yuan et al. (2018)
ABFs (bZIP)	SnRK2 (Ca^{2+}-dependent ABA signaling)	Drought	ABA-responsive genes	Stomatal regulation	Yoon et al. (2020)
WRKYs	Ca^{2+}-MAPK signaling	Drought, oxidative stress	*CAT* genes	ROS scavenging	Jiang et al. (2017)
MYC2	Ca^{2+}/CaM	Drought stress	*rd22* and other dehydration-responsive genes	Stomatal regulation	Yoon et al. (2020)

activate their kinase activity, and phosphorylate the downstream targets (Aldon et al., 2018). Pathogen attack leads to the production of NO, a second messenger, which modulates disease resistance via HR. Ca^{2+}/CaM is necessary for the expression of AtNOS1 (nitric oxide synthase) in *Arabidopsis*. Moreover, accelerated production of ROS in response to pathogen invasion and its subsequent accumulation is another prerequisite for eliciting HR. An increase in cytosolic calcium is responsible for the generation of H_2O_2 by NAD kinase (a CaMBP). A subset of CDPKs is also found to phosphorylate NADPH oxidase (RBOHD), meaning that it has a role in ROS-mediated signal transduction. Thus both NO and ROS contribute to the activation of HR in plants (Aldon et al., 2018).

6.5.3 Calcium interactions with other signaling networks

- **Reactive oxygen species**
  Perception of different stress stimuli leads to a surge of intracellular calcium levels, giving rise to varied calcium signatures. RBOH is activated by various protein kinases like CPK5 and CIPK. Activation of RBOHs produces ROS. The ROS burst involves the production of superoxide by RBOH. H_2O_2 is generated from superoxide by the action of SOD. HPCA1, a plasma membrane-associated leucine-rich repeat (LRR) receptor-like kinase (RLK), is an important receptor for H_2O_2. This leads to the activation of different calcium channels and transporters (induced by H_2O_2) in guard cells, causing the stomata to close during drought stress. The ability of phosphorylation by HPCA1 is found low in the absence of H_2O_2, and therefore the calcium channels are closed. Therefore HPCA1 links Ca^{2+} and ROS accumulation in cells in mitigating stress (Ravi et al., 2023). This crosstalk between Ca^{2+} and ROS is used in long-distance signal transmission. Accumulation of nonprotoplast ROS causes Ca^{2+} release from adjacent cells. Thereafter, another Ca^{2+}-dependent protein kinase gets activated, and the signal gets transmitted over long distances (Li et al., 2022).
- **Phytohormones (ABA, JA and SA)**
  Abscisic acid induces calcium release from internal calcium stores. During water stress, ABA accumulates in leaves, activates the phospholipase C pathway, and breaks down IP3, which activates intracellular calcium in guard cells. ABA can also induce intracellular calcium by H_2O_2, causing plasmalemma hyperpolarization and activating plasma membrane hyperpolarization-activated calcium channels (HACCs) (Li et al., 2022). ABA binds to PYR/PYL/PCAR receptors that lead to the inhibition of PP2Cs (a negative regulator) like ABI1. This activates SnRK2 kinases that trigger downstream responses. SnRK2 activates OST1, which transmits the signal downstream (Chen et al., 2021). ABA also induces ROS production that serves as a second messenger (Pei et al., 2000). ROS activates Ca^{2+} channels in the plasma membrane. ABA-induced Ca^{2+} increase inhibits the proton pump on plasmalemma that makes the cytosol alkaline and depolarizes the membrane. This activates K^+ out channels,

removing K^+ ions from the cytosol. Guard cells undergo shrinkage, and the stoma closes.

When a plant recognizes a pathogen, there is a surge in calcium concentration in the cell that binds CaM. This Ca-CaM complex interacts with CAMTA3 or AtSR1. When calcium levels are low, AtSR1 binds to the promoter of EDS1 (Enhanced Disease Susceptibility 1), a regulator of salicylic acid (SA) biosynthesis, and prevents its expression when not required, keeping SA levels at a minimum. However, during an infection, Ca-CaM binds to AtSR1. This activates EDS1 and allows the expression of SA biosynthetic genes. SA accumulation leads to Systemic Acquired Resistance (SAR) in parts far away from the site of infection, which confers a long-distance, versatile resistance to pathogens. This leads to our understanding that AtSR1 negatively regulates SAR (Du et al., 2009).

Under abiotic stress, JA interacts with a JA receptor called COI1, which leads to 26 S proteasomal degradation of JA signaling repressors called JAZ proteins. These JAZ proteins suppress MYC2 when JA is not available. As discussed earlier, MYC2 carries out transcription of jasmonic acid-responsive genes (Yoon et al., 2020). In *Arabidopsis*, when JA was exogenously applied, it induced a rise in calcium spike in the cytosol. Ca binds with CaM, which regulates the downstream JA signaling. In a different study, it has been shown that *Arabidopsis thaliana* Cyclic nucleotide-gated channel 2 (AtCNGC2) on the plasma membrane is involved in JA-induced Ca^{2+} influx (Lu et al., 2016). Fig. 6.2 represents the crosstalk of calcium with ROS and phytohormone signaling pathways.

6.6 Restoration of calcium homeostasis

An essential plant nutrient is calcium, and it plays a crucial structural role in the maintenance of cell wall and membrane integrity. Calcium can easily interact with proteins, membranes, and organic acids through its ability to form different coordination bonds (from six to nine), which results in a high affinity for carboxylate oxygen, rapid binding kinetics, and complex geometries. In plants, calcium homeostasis restoration is critically important because Ca^{2+} acts like a central second messenger in virtually all environmental and developmental signaling pathways, including responses to drought, salinity, cold, pathogen attack, and mechanical stress. Stimulus-induced Ca^{2+} transients encode information through their amplitude, frequency, and spatial distribution, forming stimulus-specific "calcium signatures" (Wdowiak et al., 2024).

The vacuole is important for its role as the major calcium store. The vacuoles of the plant contain millimolar concentrations of Ca^{2+}, whereas during resting conditions the cytosol contains submicromolar levels of Ca^{2+}. Ca^{2+} is released into the cytosol from the vacuole and mediated by voltage-gated Ca^{2+} channels, cyclic ADP-ribose receptor channels, and inositol 1,4,5-triphosphate receptor channels. A low Ca^{2+} concentration in the cytosol near the vacuole is believed to

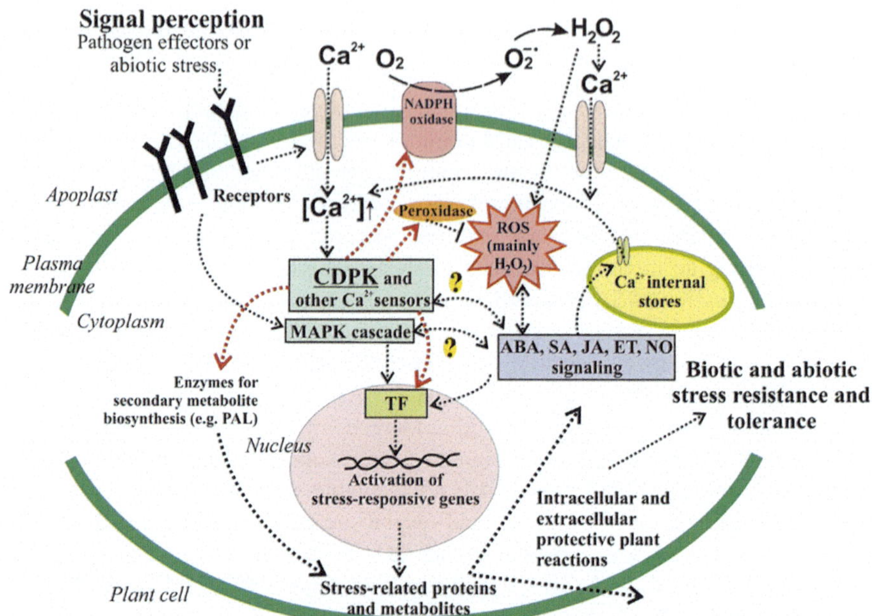

FIGURE 6.2

Crosstalk of different calcium sensors in plants with ROS and hormone signaling pathways to mitigate abiotic and biotic stress conditions.

The image has been adopted from (Kiselev and Dubrovina, 2025) under the Creative Commons Attribution license (http://creativecommons.org/licenses/by-nc-nd/4.0/).

be achieved by high-affinity Ca^{2+}-ATPases and low-affinity H^+/Ca^{2+} antiporters, which are involved in the removal of Ca^{2+} from the cytosol (Hirschi, 2001; Sanders et al., 2002).

Rapid restoration of basal cytosolic Ca^{2+} levels by Ca^{2+}-ATPases, Ca^{2+}/H^+ exchangers, and sequestration into the vacuole, endoplasmic reticulum, and other organelles is essential to terminate signals accurately and prevent aberrant activation of downstream pathways. Failure to re-establish calcium homeostasis leads to prolonged Ca^{2+} elevation, which disrupts enzyme activities, impairs membrane integrity, interferes with cytoskeletal dynamics, and compromises photosynthesis and growth. Moreover, sustained Ca^{2+} imbalance can trigger oxidative stress and programmed cell death, particularly under abiotic and biotic stress conditions (Liu et al., 2011). Thus efficient restoration of calcium homeostasis ensures signal fidelity, stress tolerance, developmental plasticity, and overall plant fitness in fluctuating environments. Under normal conditions the calcium levels inside different components of the cell are kept in the range of 1–10 mM in the axoplasm and around 100–200 nM in the cytoplasm either through transportation outside the cell or pumping into the vacuoles, where the concentration is kept at 0.2–10 mM (Ettinger et al., 1999); the mechanisms will be discussed further in this section.

6.6.1 Mechanisms of Ca^{2+} clearance

There are various efflux systems to remove the excess Ca ions from the cell. We shall discuss about two such mechanisms, namely ATP-dependent Ca^{2+} pumps and Ca^{2+}/H^+ exchangers. The contribution of Ca^{2+}-ATPase pumps and Ca^{2+}/H^+ exchangers in shaping $[Ca^{2+}]_{cytosolic}$ signatures is modeled by using a four-component model (plasma- and endomembrane-based Ca^{2+} permeable channels and efflux systems), taking into account the cytosolic Ca^{2+} buffering. It is concluded that physiologically relevant variations in the activity of Ca^{2+}-ATPase pumps and Ca^{2+}/H^+ exchangers are sufficient to fully describe all the $[Ca^{2+}]_{cytosolic}$ signatures in response to environmental stimuli, emphasizing the crucial role these active efflux systems play in plant adaptive responses to the environment.

- **ATP-dependent Ca^{2+} pumps**
  The course of biological evolution was predetermined when nucleosides became the elementary units of genetic code so that ATP emerged as a universal energy substrate. These events defined the molecular canvas of living cells and thus shaped major signaling cascades. In this interaction, cells used what was readily available to them; purines and inorganic phosphates were readily available, so that ATP became central for energy metabolism (Galimov, 2009). Pyrophosphate bonds, phosphorylated nucleosides, and ATP are thought to have emerged during the prebiotic period and were foundational to early organic evolution, including the origin of RNA and DNA. Although the precise source of organic phosphates remains uncertain, hypotheses range from endogenous prebiotic synthesis to extraterrestrial delivery, supported by the detection of phosphates in meteorites (Ponnamperuma et al., 1963). Nucleotide triphosphates were therefore likely available at the earliest stages of biological evolution, first sustaining an RNA world and later supporting DNA-based information storage and translation. The evolutionary association between ATP and Ca^{2+} may reflect physicochemical conditions of the primordial oceans, where Ca^{2+} availability could have shaped early cellular environments (Bryant et al., 2013). Changes in ocean chemistry, particularly fluctuations in Ca^{2+} concentration, have been proposed as selective pressures driving the evolution of sophisticated ion homeostasis mechanisms, although this view remains debated. Nonetheless, these early geochemical constraints provide an important backdrop for understanding the central role of Ca^{2+} regulation in modern biological systems (Eigen, 1971; Kazmierczak et al., 2013).

Arabidopsis thaliana has 10 ACA-type Ca^{2+} pumps that pump Ca^{2+} out from the cytosol to the other compartments. ACA1, ACA2, and ACA7 are localized on the ER; ACA4 and ACA11 are localized on the vacuole (Geisler et al., 2000; Lee and Seo, 2021), while ACA8, ACA9, ACA10, ACA12, and ACA13 are localized on the plasma membrane (PM) (Bonza et al., 2000; dit Frey et al., 2012). ACA proteins have been shown to regulate pollen growth, rosette growth, and immune responses in *Arabidopsis thaliana.* The loss-of-function (LOF) mutant of the ER-localized ACAs aca1, aca2, and aca7 had reduced rosette growth in *Arabidopsis.* The LOF mutants of PM-localized ACAs, such as *aca10* in the No-0 background or *aca8* and *aca10* in the Col-0 background, showed reduced growth of rosette and inflorescence. The LOF of ACA13 or ACA12 did not lead to overt defects but enhanced the defects of ACA8 and ACA10, indicating a relatively minor role of ACA12 and ACA13 among the PM-localized ACAs (Li et al., 2023).

Plasma membrane Ca^{2+}-ATPases (PM Ca^{2+}-ATPases) play a central role in Ca^{2+} clearance by actively exporting Ca^{2+} from the cytosol to maintain low resting cytosolic Ca^{2+} levels following signaling events. In *Arabidopsis thaliana*, At-ACA8 and At-ACA9 are bona fide PM-localized Ca^{2+}-ATPases, while At-ACA10 is a closely related isoform with overlapping functional characteristics. Expression analyses indicate that At-ACA8 and At-ACA10 are broadly expressed across plant organs, whereas At-ACA9 shows strong flower-specific expression, suggesting functional specialization among PM Ca^{2+}-ATPase isoforms. All three isoforms are detectable in young seedlings, although At-ACA9 transcripts accumulate at substantially lower levels compared with At-ACA8 and At-ACA10 (Bonza et al., 2000).

In *Arabidopsis thaliana*, fourteen Ca^{2+}-ATPases have been identified and are classified into two major families based on protein structure and regulatory mechanisms: the autoinhibited Ca^{2+}-ATPases (ACAs) and the endoplasmic reticulum–type Ca^{2+}-ATPases (ECAs). ACA-type Ca^{2+}-ATPases share closest homology with calmodulin (CaM)-stimulated plasma membrane Ca^{2+}-ATPases (PMCAs) found in animal cells. While animal PMCAs are exclusively localized to the plasma membrane, plant ACAs exhibit a broader subcellular distribution and are found not only at the plasma membrane but also within endomembrane systems, including the endoplasmic reticulum, vacuole, plastids, and the prevacuolar compartment (Baxter et al., 2003).

A defining feature of plant ACAs is the presence of a calmodulin-regulated autoinhibitory domain located at the N-terminus, which contrasts with the C-terminal regulatory domains characteristic of animal PMCAs. This structural distinction underpins differences in regulatory control and allows plant ACAs to respond flexibly to diverse Ca^{2+} signaling contexts. Based on amino acid sequence similarity and predicted subcellular localization, ACAs are further subdivided into four subfamilies, reflecting functional specialization within the family. Among these, ACA4 and ACA11 cluster within the same subfamily (Baxter et al., 2003).

A key regulator of stress and developmental signaling is the phytohormone abscisic acid (ABA), which rapidly and specifically induces the expression of At-

ACA8 and At-ACA9, but not At-ACA10, in young seedlings. ABA treatment also increases At-ACA8 protein abundance at the plasma membrane, providing strong evidence that PM Ca^{2+}-ATPases contribute directly to ABA-regulated Ca^{2+} signaling pathways. These findings highlight a mechanism by which Ca^{2+} clearance at the plasma membrane is dynamically up-regulated to terminate ABA-induced Ca^{2+} transients and restore cytosolic Ca^{2+} homeostasis. Collectively, these observations establish plasma membrane Ca^{2+}-ATPases as key components of Ca^{2+} clearance machinery that integrate hormonal, developmental, and environmental signals to shape plant growth and architecture (Cerana et al., 2006).

6.6.1.1 Ca^{2+}/H^+ exchangers

CAX is a member of the CaCA (Ca^{2+}/cation antiporter) superfamily, and they mediate the transport of various cations like calcium and magnesium. These transmembrane proteins consist of 11 transmembrane domains arranged in two pseudo-symmetric repeats. CAXs are involved in a wide range of essential physiological processes, including Ca^{2+} signaling, ion homeostasis regulation, and responses to environmental stress (Sarkar et al., 2024). CAX family members are distributed across various domains of life forms, e.g., bacteria, fungi, algae, and higher plants. Previous studies classified the CAX family into three major clades: Type I, encompassing plant, fungal, and bacterial members; Type II, found in lower vertebrates; and Type III, primarily composed of bacterial members. The advent of next-generation sequencing technologies has led to the identification of an increasing number of CAXs in plants (Shigaki et al., 2006).

In contrast to lower plants like algae, which typically have a low copy number of CAXs, higher plants exhibit significant expansion and diversification of the CAX family (Pittman et al., 2016). Whole-genome duplication (WGD) events are considered to be a major driving force behind the expansion of gene families, and this phenomenon can be used to explain the presence of plant CAXs in every stratum of life. One explanation is via the ancient WGD events in the common ancestor of land plants. Gene family expansion typically increases gene copy number, enabling functional diversification and contributing to the adaptive evolution of plant species. Species with varying life histories and ecological niches often exhibit distinct patterns of gene family expansion and contraction, allowing them to respond effectively to diverse environmental pressures (Blanc and Wolfe, 2004; Zheng et al., 2021; Panchy et al., 2016).

The Zygnematophyceae, which are considered the closest living relative of embryophyte land plants, provide further insight into the evolutionary origins of plant Ca^{2+} clearance systems. The presence of a single CAX protein *in Mesotaenium endlicherianum*, phylogenetically positioned near early bryophyte CAXs, suggests that ancestral CAX-mediated Ca^{2+} sequestration mechanisms were already established prior to land colonization. Subsequent gene duplication events during bryophyte diversification likely enhanced the capacity for vacuolar Ca^{2+} storage and buffering, thereby improving the precision and robustness of Ca^{2+} signal termination (Li et al., 2018).

As terrestrial plants diversified, independent expansions of the *CAX* gene family occurred in both ferns and seed plants, despite their early evolutionary divergence. This parallel expansion underscores the strong selective pressure to refine Ca^{2+} clearance systems in land plants. Increased CAX diversity would have enabled specialization toward distinct tissues, developmental stages, or stress conditions, allowing fine-tuned regulation of cytosolic Ca^{2+} levels (Cheng et al., 2019). Such specialization is critical for maintaining signal specificity while preventing prolonged Ca^{2+} elevation that could disrupt enzyme activity, membrane integrity, and overall cellular function.

In *Arabidopsis thaliana*, members of the CAX family exhibit distinct biochemical properties and tissue-specific expression patterns that collectively contribute to efficient cytosolic Ca^{2+} clearance. AtCAX1 is characterized by a relatively low affinity for Ca^{2+} but a high transport capacity, enabling rapid sequestration of excess cytosolic Ca^{2+} into intracellular compartments, primarily the vacuole. This property makes AtCAX1 particularly important for restoring basal cytosolic Ca^{2+} levels following signaling events. Loss-of-function *atcax1* mutants display heightened sensitivity to elevated Ca^{2+} and reduced Ca^{2+} accumulation in leaf tissues compared with wild-type plants, underscoring its role in Ca^{2+} clearance and storage in aerial organs (Cheng et al., 2005). In contrast, AtCAX2 displays preferential expression in vascular and floral tissues, highlighting its involvement in long-distance Ca^{2+} transport and localized Ca^{2+} clearance within specialized tissues. By facilitating vacuolar sequestration of Ca^{2+} in these regions, AtCAX2 likely contributes to the regulation of cytosolic Ca^{2+} transients associated with vascular signaling and reproductive development. AtCAX5 is similarly enriched in floral organs, and comparable expression patterns observed across multiple plant species suggest a conserved function for CAX-mediated Ca^{2+} clearance in reproductive tissues (Edmond et al., 2009).

Expansion of the *CAX* gene family has facilitated functional diversification within Ca^{2+} clearance pathways. While AtCAX1 and AtCAX3 function primarily as dedicated Ca^{2+} transporters involved in cytosolic Ca^{2+} removal, AtCAX2 and AtCAX4 exhibit broader substrate specificity, mediating the vacuolar sequestration of additional cations such as Cd^{2+}, Mn^{2+}, Zn^{2+}, and Na^{+}. This dual role integrates Ca^{2+} clearance with general mineral homeostasis, allowing plants to simultaneously terminate Ca^{2+} signals and detoxify excess metal ions (Pittman et al., 2004; Edmond et al., 2009).

Together, these findings illustrate how tissue-specific expression and functional specialization of CAX transporters underpin effective Ca^{2+} clearance in plants.

Under basal physiological conditions, plants employ a tightly regulated Ca^{2+}–CBL–CIPK–CAX signaling cascade to couple cytosolic Ca^{2+} entry with its rapid sequestration into the vacuole, thereby maintaining resting cytosolic Ca^{2+} concentration. In this pathway, stimulus-induced Ca^{2+} influx is sensed by calcineurin B-like (CBL) proteins, which recruit and activate CBL-interacting protein kinases (CIPKs) (Catalá et al., 2003). Activated CIPKs subsequently phosphorylate and stimulate CAX transporters, enhancing vacuolar Ca^{2+} uptake. This mechanism constitutes a Ca^{2+}-dependent negative feedback loop that ensures efficient Ca^{2+} clearance and prevents prolonged cytosolic Ca^{2+} elevation, enabling plants to adapt dynamically to the variable Ca^{2+} levels encountered in natural soil.

Beyond their role in basal Ca^{2+} homeostasis, CAX1 and CAX3 are also integrated into plant innate immune signaling pathways. Through this integration, CAX-mediated Ca^{2+} sequestration contributes to shaping the amplitude and duration of immune-associated Ca^{2+} transients and ensures appropriate signal termination once defense responses are initiated (He et al., 2022).

Disruption of CAX-dependent Ca^{2+} clearance mechanisms results in sustained elevation of cytosolic Ca^{2+} levels. Such prolonged Ca^{2+} accumulation is sufficient to activate SA-dependent immune pathways constitutively, leading to autoimmune phenotypes and growth inhibition. Thus CAX-mediated vacuolar sequestration represents a critical regulatory node in plant Ca^{2+} clearance, safeguarding signal specificity, while maintaining cellular and physiological homeostasis. Two principal mechanisms have been identified that operate under distinct physiological contexts to ensure tight control of CAX-mediated Ca^{2+} sequestration into the vacuole. These kinases subsequently stimulate CAX transporters, promoting vacuolar Ca^{2+} sequestration and restoring cytosolic Ca^{2+} to resting levels (Cheng et al., 2005). During pathogen challenge, a distinct regulatory module operates to prevent sustained cytosolic Ca^{2+} accumulation following immune activation. In this context, receptor-like kinases (RLKs) and downstream receptor-like cytoplasmic kinases (RLCKs) activate CAX1 and CAX3 independently of cytosolic Ca^{2+} levels. This RLK–RLCK–CAX module promotes vacuolar Ca^{2+} sequestration during pattern-triggered immunity, thereby limiting prolonged Ca^{2+} elevation after defense-associated Ca^{2+} influx. Unlike the Ca^{2+}–CBL–CIPK–CAX cascade, this mechanism represents a Ca^{2+}-independent strategy for cytosolic Ca^{2+} scavenging, ensuring timely signal termination even under conditions of intense immune signaling (Zhou et al., 2024b). A key future challenge will be to determine how these pathways are integrated and coordinated to activate CAX1/3 with temporal and spatial precision, thereby ensuring accurate control of cytosolic Ca^{2+} dynamics across diverse physiological states (Fig. 6.3).

6.6.2 Importance for signal reset and future responsiveness

Efficient restoration of resting cytosolic Ca^{2+} levels is essential not only for terminating an ongoing signaling event, but also for ensuring that cells remain competent to perceive and respond to subsequent stimuli. Failure to fully reset cytosolic Ca^{2+} to basal levels compromises the fidelity of these signals, as residual Ca^{2+} can obscure stimulus-specific Ca^{2+} signatures and lead to inappropriate or constitutive activation of downstream pathways. In the absence of efficient Ca^{2+} clearance, prolonged cytosolic Ca^{2+} elevation can result in signal desensitization,

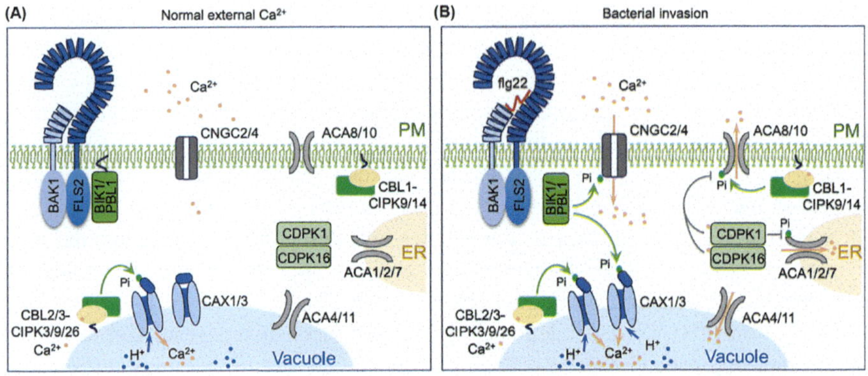

FIGURE 6.3

Model of plant Ca^{2+} signaling during normal conditions and bacterial invasion. (A) Under basal conditions, extracellular Ca^{2+} enters the cytoplasm through plasma-membrane-localized CNGC2/4 and related channels, causing a transient rise in cytosolic Ca^{2+} levels. This increase is sensed by the Ca^{2+} sensors CBL2/3, which promote sequestration of excess Ca^{2+} into the vacuole. (B) Upon bacterial infection, the PRR complex FLS2–BAK1 perceives flagellin (flg22), leading to the activation of BIK1/PBL1. These kinases stimulate Ca^{2+} influx via CNGC2/4 and Ca^{2+} efflux via CAX1/3. The coordinated regulation of Ca^{2+} entry and removal shapes the amplitude and duration of cytosolic Ca^{2+} signals required for effective immune responses. Additional Ca^{2+} pumps, including tonoplast-localized ACA4/11, ER-localized ACA1/2/7, and plasma-membrane-localized ACA8/10, help restore Ca^{2+} homeostasis, with CIPK- and CDPK-mediated phosphorylation, further fine-tuning the ACA activity.

The image has been adopted from Zhou, J., He, Y., Meng, X., 2024. Phosphorylation of CAX transporters controls Ca^{2+}2+ homeostasis. Plant Communications. 5 (9), 101042. https://doi.org/10.1016/j.xplc.2024.101042 under the Creative Commons Attribution license (http://creativecommons.org/licenses/by-nc-nd/4.0/).

aberrant crosstalk between pathways, and, in extreme cases, activation of stress or cell death programs.

Genetic evidence from *Arabidopsis thaliana* demonstrates that efficient Ca^{2+} clearance by plasma membrane Ca^{2+}-ATPases (ACAs) is essential for both reproductive development and immune homeostasis (Schiøtt et al., 2004). Loss-of-function (LOF) mutations in ACA9, or combined LOF of ACA1, ACA2, and ACA7 in pollen, result in markedly reduced pollen tube growth, highlighting the requirement for precise cytosolic Ca^{2+} regulation during polarized cell expansion. Similarly, LOF of ACA13 in papilla cells compromises pollen–stigma interactions, further underscoring the importance of ACA-mediated Ca^{2+} efflux in reproductive signaling. Beyond development, the disruption of ACA-dependent Ca^{2+} clearance has profound consequences for immune regulation (Yang et al., 2017). Multiple

ACA mutant combinations, including ACA1 ACA2 ACA7, ACA4 ACA11, and ACA8 ACA10, exhibit constitutive activation of defense pathways in the absence of pathogen challenge, manifesting as autoimmune responses and enhanced resistance to bacterial pathogens. In several cases, these phenotypes are accompanied by SA-dependent programmed cell death in leaf tissues, indicating that failure to reset cytosolic Ca^{2+} levels is sufficient to trigger sustained immune signaling. Growth defects observed in ACA mutants, such as reduced rosette expansion and compact inflorescence architecture, are largely attributable to these autoimmune responses rather than to primary developmental abnormalities. Notably, inhibition of SA biosynthesis or downstream immune signaling substantially rescues growth in *aca10 (cif1), aca1 aca2 aca7*, and *aca4 aca11* mutants. These findings demonstrate that ACA-mediated Ca^{2+} efflux is critical for terminating Ca^{2+} signals and preventing inappropriate activation of immune pathways (Yang et al., 2017). Thus the inability to perform a signal reset can lead to increased toxicity in the cells due to the excess calcium ions. Additionally, in the process of plant development, lack of ACA and CAX transporters leads to the sterility of the pollen tube, so that no fertilization would occur. This is detrimental, as the plant life cycle would be impaired.

6.7 Role of calcium in membrane stability

Calcium has long been recognized as a key structural and signalling component in plant stress responses, particularly through its roles in membrane stabilization, cell wall integrity, and Ca^{2+}-dependent signaling cascades. Under moderate stress conditions, Ca^{2+} can enhance membrane stability by shielding negatively charged phospholipid headgroups, reducing ion leakage, and maintaining membrane selectivity. In several plant species, supplemental Ca^{2+} has been shown to alleviate salt-induced membrane damage by limiting Na^{+} influx, preserving lipid organization, and supporting the function of Ca^{2+}-dependent transporters and signaling pathways. However, evidence from cowpea (*Vigna unguiculata*) exposed to severe salinity (75 mM NaCl) illustrated that Ca^{2+}-mediated membrane protection is context dependent. Despite a wide range of supplemental Ca^{2+} concentrations and forms ($CaCl_2$ and $CaSO_4$), calcium addition failed to restore membrane integrity or counteract lipid compositional changes. Notably, increased Ca^{2+} availability even intensified fatty acid saturation in leaf membranes, suggesting that excessive Ca^{2+} under high salinity may exacerbate membrane rigidification rather than confer protection (Guimarães et al., 2011). However, it is important to note that Cl^{-} toxicity is species dependent. In several other plant species, increased Cl^{-} levels did not result in significant cellular damage, and Ca^{2+} supplementation partially mitigated the inhibitory effects of NaCl salinity. Furthermore, the beneficial effects of Ca^{2+} on cowpea growth under saline conditions appear to be strongly modulated by environmental context. Positive effects of Ca^{2+} supplementation were observed only when plants were grown under conditions of high evaporative demand, indicating that Ca^{2+}-mediated stress mitigation depends not only on ionic composition, but also on external climatic factors that influence transpiration and ion fluxes. In addition to lipid depletion, salt stress is commonly associated with enhanced lipid peroxidation, as reflected by increased levels of malondialdehyde (MDA), a widely used indicator of oxidative damage to membrane fatty acids (Chakravarthy and Giorgio, 1992).

6.7.1 Calcium in membrane integrity and fluidity

Calcium is not only an essential factor that serves as a second messenger in plant stress responses, but also a vital component in maintaining membrane integrity and fluidity. Calcium crosslinks negatively charged carboxyl groups of de-esterified pectin (by pectin methyl esterase) in the middle lamella that cements one cell with another (Thor, 2019). High calcium concentration makes the cell wall more rigid and sturdy. Calcium stabilizes membranes via interaction with phospholipids, and this provides structural integrity to cell membranes. It has been observed that calcium maintains the membrane permeability. Cells grown in low calcium but in the presence of the metal chelator EDTA showed ion leakage. In soybean and maize, the roots failed to absorb solutes when low calcium prevailed. Thus it was found that a $[Ca^{2+}]$ between 0.1 and 1.0 mM was required to maintain the membrane integrity in plants. It is also reported that calcium through crosslinking pectin units and cementing one cell with the other retards abscission (detachment of parts) or senescence. Reduced calcium levels were found to create lesions in the nuclear envelope, plasma membrane, and tonoplasts, thereby hampering membrane stability. These studies provide us with a comprehensive idea about the role of calcium in maintaining membrane integrity in plants (Hepler, 2005).

Plant cell membranes are sensitive to temperature changes. Different thermosensors present on the cell membrane sense these external fluctuations in temperature. Plasmalemma is composed of amphipathic phospholipids, supporting membrane fluidity alterations. At elevated temperature, when the level of unsaturation in the membrane lipids of *Physcomitrella patens* was decreased, a corresponding increase in membrane rigidity was observed, which explained the inward flow of calcium during heat stress. So, temperature shifts transiently increase the fluidity of cell membranes, opening the calcium channel, thereby improving plant stress tolerance (Saidi et al., 2014). Membrane lipid 'mesomorphism' gives the lipid bilayer a property to take different physical conformations in accordance with different temperature conditions. Plasma membrane fluidity responds fast and independently of cellular metabolism (Cano-Ramirez et al., 2021). Ca^{2+} ions are found to reduce the membrane disruptions made by caffeic acid methyl ester (CAME) and caffeic acid (CAF), as phenolic acids are potential metal ion chelating agents. UV–Vis spectroscopic studies confirm that CAME and CAF never bind to calcium when they are present in solution.

Molecular dynamic simulation showed that calcium increases membrane thickness and hence lipid packaging, indicating that calcium reduces the ability of CAME and CAF to disrupt lipid bilayers (Hossain et al., 2022). This has a direct impact on calcium on lipid packaging and ion–lipid interactions, thereby determining membrane stability and fluidity.

6.7.2 Regulation of membrane transport and permeability

Membrane transport regulation is extensively controlled by posttranslational modifications, particularly reversible protein phosphorylation. Many protein kinases that regulate membrane transport are themselves activated by calcium, providing an additional layer of signal integration. Calcium-dependent protein kinases constitute a large family of kinases that respond directly to Ca^{2+} via EF-hand motifs and frequently associate with membranes through lipid modifications. These kinases play key roles in regulating ion channels and transporters, including those involved in stomatal function, gas exchange, and water balance, thereby linking calcium signaling to physiological adaptation. A second major calcium-decoding system in plants is the calcineurin B-like protein family. These small EF-hand calcium sensors selectively interact with a corresponding family of protein kinases, forming the CBL–CIPK signaling network. This network functions as a central coordinator of ion transport, integrating calcium signals with membrane transport activity across multiple cellular compartments. Through selective pairing between sensors and kinases, the CBL–CIPK system enables stimulus-specific regulation of diverse transport processes, ensuring appropriate ionic responses to environmental cues. While significant progress has been made in understanding calcium-decoding mechanisms, key challenges remain in identifying the transport systems that generate calcium signals and in elucidating how calcium signaling networks integrate responses to multiple concurrent environmental stresses under natural and agricultural conditions (Kleist and Luan, 2016).

6.8 Integrated model of Ca^{2+}-based stress signaling

There are mainly two types of stress encountered by a plant, viz., abiotic and biotic. Both abiotic and biotic stresses converge on Ca^{2+}-based signaling networks, with stress-specific Ca^{2+} signatures, enabling discrimination between environmental cues and biological threats. Integration of these signals allows plants to balance growth, defense, and survival. Within an integrated signaling framework, Ca^{2+} dynamics cannot be viewed as arising solely from plasma membrane influx. Instead, cytosolic Ca^{2+} signatures are shaped by coordinated interactions between Ca^{2+} influx channels and organelle-localized Ca^{2+} efflux systems, including pumps and exchangers associated with the endoplasmic reticulum, vacuole, and other compartments. This spatial coordination ensures precise control of Ca^{2+} amplitude, duration, and oscillatory behavior, which is essential for signal specificity (Kleist and Luan, 2016).

Advances in molecular genetics and live-cell Ca^{2+} imaging have substantially enhanced our capacity to dissect stress-sensing mechanisms. In particular, Ca^{2+} imaging–based forward genetic screens have emerged as powerful tools for identifying novel components of stress perception and Ca^{2+} signaling. Fluorescence-based Ca^{2+} indicators, including single-fluorophore sensors (e.g., GCaMP and GECO variants) and ratiometric FRET-based indicators such as yellow cameleons, now permit visualization of Ca^{2+} dynamics at cellular and tissue scales. Continued refinement of these tools is expected to drive major discoveries in plant stress signaling (Mehra and Bennett, 2022).

Importantly, plants in natural environment are rarely exposed to single stresses in isolation. Accumulating evidence indicates that certain Ca^{2+} sensors, particularly calcium-dependent protein kinases (CPKs) and calmodulin-like proteins (CMLs), function as central signaling hubs, capable of integrating Ca^{2+} signals generated by both biotic and abiotic stresses. Using AtCPK3 as a representative example, it is possible to conceptualize how a single Ca^{2+} sensor can selectively drive different adaptive responses, depending on the nature of the stimulus. Beyond the generation of stimulus-specific Ca^{2+} signatures, signaling specificity is likely achieved through the spatial and temporal co-localization of Ca^{2+} sensors with their downstream targets and interacting partner proteins. In guard cells, AtCPK3 is activated by Ca^{2+} influx through nonselective Ca^{2+} channels in response to abiotic stimuli such as drought. Once activated, AtCPK3 phosphorylates S-type anion channels, leading to stomatal closure and reduced water loss. In this context, AtCPK3 operates within a signaling module optimized for rapid abiotic stress adaptation. In mesophyll cells, AtCPK3 participates in more complex signaling scenarios, as these cells are exposed to diverse biotic challenges that elicit distinct outcomes (Huang et al., 2019).

Herbivore attack by *Spodoptera littoralis*, for example, activates glutamate receptor-like channels, inducing Ca^{2+} influx. The resulting Ca^{2+} transients activate AtCPK3, which then phosphorylates the transcription factor HsfB2a, promoting the expression of defensin genes involved in herbivore defense. In contrast, exposure to fumonisin B1 or long-chain bases (FB1/LCBs), which trigger programmed cell death (PCD), generates cytosolic Ca^{2+} elevations that are independent of glutamate receptor activity, as ionotropic glutamate channel inhibitors fail to block these responses. This indicates that distinct Ca^{2+} channels are responsible for shaping the Ca^{2+} signatures associated with PCD. These observations suggest that the functional switching of AtCPK3 is not determined solely by Ca^{2+} concentration but also by the origin of Ca^{2+} influx and the identity of the channels involved (Huang et al., 2019).

AtCPK21 interacts with the anion channel SLAH3 in an abscisic acid (ABA)-dependent manner within plasma membrane microdomains of guard cells. This interaction is regulated by a protein scaffold involving the protein phosphatase ABI1 and an ABA receptor from the RCAR1/PYR/PYL family, which together fine-tune AtCPK21 activity and downstream signaling. Similar scaffold-based organization is well established in animal systems, where Ca^{2+} channels cluster within membrane microdomains and associate with dedicated downstream signaling complexes (Huang et al., 2019).

6.8.1 From stimulus → Ca^{2+} signature → decoding → response

In this section, the entire mechanism will be discussed in two parts: the stimulus being abiotic or biotic. However, regardless of the varying nature of the stimulus, the secondary messenger is in all cases calcium. There are various mechanisms used to change the calcium signatures for the ultimate recognition of the type of stimulus and thus the downstream effect of the same. Stress-specific sensing mechanisms initiate Ca^{2+} influx, while coordinated contributions from plasma membrane channels, organellar Ca^{2+} release, and ROS-mediated feedback shape cytosolic Ca^{2+} signatures. Integration of these signals enables plants to discriminate among stresses, activate appropriate responses, and restore homeostasis, underscoring the central role of Ca^{2+} dynamics in environmental adaptation.

6.8.1.1 Abiotic stimulus

- **Salt stress**
  Soil salinization is a major abiotic stress that severely impairs seed germination, plant growth, and crop productivity (Husen, 2021; Zahra et al., 2024; Husen et al., 2019; Zhao et al., 2024). Unlike animals, sodium is not an essential element for most plants, and excessive accumulation of Na^+ and Cl^- ions induces both osmotic and ionic stress. Exposure to high salinity triggers large, low-affinity Na^+ influxes mediated by HKT-type transporters and nonselective cation channels (NSCCs), including glutamate receptor-like channels (GLRs) and cyclic nucleotide-gated channels (CNGCs). A critical advance in salt stress sensing was the identification of the *monocation-induced [Ca^{2+}] increase 1 (moca1)* mutant in *Arabidopsis thaliana*, which exhibits impaired Na^+-induced cytosolic Ca^{2+} elevation and hypersensitivity to salt stress. MOCA1 encodes a glucuronosyltransferase responsible for the conversion of inositol phosphorylceramide (IPC) to glycosylinositol phosphorylceramide (GIPC), a negatively charged sphingolipid located in the outer leaflet of the plasma membrane. Loss of MOCA1 reduces GIPC levels and increases IPC accumulation, thereby decreasing membrane Na^+-binding capacity. Direct binding of Na^+ to GIPCs is proposed to initiate membrane depolarization and activate Ca^{2+} influx pathways, providing a molecular basis for Na^+ perception in plants. However, the identity of the Ca^{2+} channels downstream of GIPCs remains unresolved (Tartaglio et al., 2017). Downstream of Ca^{2+} influx, vacuolar Ca^{2+} release mediated by the two-pore channel TPC1 contributes to Ca^{2+} wave propagation, as evidenced by altered Ca^{2+} dynamics in *tpc1* mutants. Additional regulators of salt-induced Ca^{2+} signaling include the glycosyltransferase QUA1, which modulates cytosolic Ca^{2+} levels under salinity stress. Plasma membrane-localized annexins also

participate in salt sensing; AtANN1 and AtANN4 are required for NaCl-induced Ca^{2+} influx in root epidermal cells, and loss-of-function mutants show attenuated Ca^{2+} responses. Furthermore, CNGC10 negatively regulates salt tolerance, suggesting complex regulatory roles for CNGCs in salinity responses.

The plant cell wall also contributes to salt stress perception. The receptor-like kinase FERONIA (FER) is implicated in sensing salt-induced alterations in cell wall integrity, as *fer* mutants display reduced Ca^{2+} signaling and increased salt sensitivity. FER functions together with cell wall-associated leucine-rich repeat extensins (LRX3, LRX4, and LRX5) and secreted RALF peptides to mediate salt stress signaling, although the initial sensing mechanism remains unclear (Zhao et al., 2024).

6.8.1.1.1 Temperature stress

Temperature extremes represent major environmental constraints on plant growth and productivity. Cold stress reduces membrane fluidity, destabilizes protein complexes, and impairs enzyme activity, necessitating rapid perception and signal transduction to activate protective responses. A rapid and transient increase in cytosolic Ca^{2+} is among the earliest events following cold exposure, occurring within seconds and often displaying oscillatory behavior upon repeated stimulation. Unlike animals, plants lack canonical transient receptor potential (TRP) channels, and the molecular basis of thermosensation remains incompletely understood. In rice, the transmembrane protein COLD1 represents the first identified cold sensor. COLD1 interacts with the G-protein subunit RGA1 to enhance GTPase activity and induce cold-dependent Ca^{2+} influx, thereby promoting chilling tolerance (Qian et al., 2024). However, whether COLD1 directly functions as a temperature-regulated ion channel remains unknown.

Changes in membrane fluidity induced by cold or heat may also be sensed by membrane-localized proteins such as receptor-like kinases (RLKs) and ion channels. Several kinases, including CRLK1 and CRLK2, positively regulate cold tolerance, whereas cold-responsive protein kinase 1 (CRPK1) acts as a negative regulator (Yang et al., 2017).

6.8.1.1.2 Oxidative stress

Reactive oxygen species (ROS), particularly hydrogen peroxide (H_2O_2), function as key signaling molecules in plant development and stress responses (Waszczak et al., 2018). Extracellular H_2O_2 acts as a signal regulating root growth, pollen tube elongation, stomatal movement, and responses to abiotic and biotic stresses. Guard Cell Hydrogen Peroxide-Resistant 1 (GHR1) was identified as a regulator of H_2O_2-dependent ABA signaling and stomatal closure. While extracellular H_2O_2 can enter cells via aquaporins, the mechanisms underlying its perception at the cell surface were unclear until the identification of HPCA1. Using Ca^{2+} imaging–based forward genetic screens, HPCA1 was identified as a leucine-rich repeat receptor kinase that functions as an extracellular H_2O_2 sensor (Feng et al., 2023). HPCA1 is activated through covalent modification of extracellular cysteine residues, triggering autophosphorylation and Ca^{2+} influx. *hpca1* mutants exhibit impaired H_2O_2- and ABA-induced Ca^{2+} signaling and stomatal responses, demonstrating the central role of HPCA1 in oxidative stress perception.

ROS and Ca^{2+} signaling are tightly coupled through feedback mechanisms. Ca^{2+} binds EF-hand motifs of the NADPH oxidase RBOHD, promoting ROS production, which in turn activates the Ca^{2+} channels to amplify signaling. Additionally, ROS stimulates the vacuolar channel TPC1, releasing Ca^{2+} from the internal stores and facilitating long-distance Ca^{2+} wave propagation (Evans and Blackwell, 2015). Recent evidence suggests that MSL3 functions downstream of HPCA1, contributing to systemic ROS–Ca^{2+} signaling and stress acclimation (Mittler et al., 2022). Collectively, salt, temperature, and oxidative stresses exemplify how diverse environmental cues converge on Ca^{2+}-based signaling networks.

6.8.1.2 Biotic stimulus

- **Ca^{2+} signaling in herbivore-induced stress responses**
  In addition to microbial elicitors, herbivore-associated molecular patterns (HAMPs) present in insect oral secretions or oviposition fluids also induce Ca^{2+} signaling. Using fluorescent Ca^{2+} probes and aequorin-expressing soybean cells, herbivory by *Spodoptera littoralis* larvae was shown to elicit localized Ca^{2+} increases at leaf margins, extending 30–200 μm from the wound site (Maffei et al., 2004). Components of larval regurgitate, such as linolenoyl-L-glutamine and volicitin, similarly induced Ca^{2+} transients in soybean cells. These Ca^{2+} signals were closely associated with early membrane depolarization and hyperpolarization events, which are known to activate downstream defense pathways involving system in synthesis, ROS production, and JA signaling (Arimura et al., 2000). These studies established Ca^{2+} transients as early integrators of mechanical damage and chemical cues during herbivore attack.

The expression of the Ca^{2+}-sensing aequorin system in transgenic soybean (*Glycine max*) suspension cells and the effects of herbivore-wounding in lima bean leaves showed increasing amounts of H_2O_2 that correlate with a higher $[Ca^{2+}]_{cytosolic}$(Maffei et al., 2004). The same responses were found in the living fossil *Gingko biloba* leaves damaged by *S. littoralis* and in the fern *Pteris vittata* (Imbiscuso et al., 2009). Changes of $[Ca^{2+}]$ fluxes and H_2O_2 were detected in a tobacco variety (G140) infested by the tobacco aphid *M. persicae*. Even in this case, H_2O_2 accumulation depended on Ca^{2+} influx, which was both increased in a long period of aphid feeding (Ren et al., 2014). ROS signals primarily act by inducing the opening of calcium-permeable channels in the membrane. This is applicable to both abiotic and biotic stimuli.

6.8.2 Systems-level view of plant stress tolerance

System-level view of the plant stress tolerance includes extensive interaction amongst the different signaling components, viz., Ca^{2+} ions, ROS, and phytohormones. Ca^{2+} ions and their interaction with ROS have been discussed in the previous sections. At the system level, plants perceive environmental and biological challenges through diverse sensors that converge on a limited set of core signaling hubs, among which Ca^{2+}, ROS, phytohormones, and electrical signals play dominant roles. Within this framework, Ca^{2+} signaling functions as a central integrator that links stress perception to downstream responses. Local stress perception can trigger Ca^{2+} and ROS waves that propagate through vascular tissues, enabling distal organs to preemptively activate defense or acclimation responses (Mori and Schroeder, 2004). This systemic coordination allows plants to balance localized damage control with global resource allocation, ensuring that growth, reproduction, and defense are optimally prioritized under fluctuating environmental conditions.

6.9 Conclusion

Calcium is recognized as an essential regulator of plant growth, serving as both a building nutrient and an incredibly adaptable second messenger that controls how plants react to a variety of signals from both their internal and external environment. It is apparent from this chapter that the ability of plants to perceive, decode and react appropriately to abiotic and biotic stresses is largely dependent on accurate spatiotemporal modifications in the concentration of Ca^{2+} in the cytosol, translated as stimulus-specific Ca^{2+} signatures that permit plants to decode external signals, such as those generated during salinity, drought, heavy metal toxicity, heat, and pathogen attack, into highly accurate responses. One of the hallmarks of Ca^{2+} signaling is the existence of a large electrochemical potential difference across cellular membranes, which enables a rapid burst of Ca^{2+} entry into the cell in response to any stress and prevents a significant presence of Ca^{2+} in the cell under normal conditions. Plasma membrane channels, endomembrane compartments, and the apoplast are responsible for producing stress-responsive Ca^{2+} signatures. Ca^{2+} sensors such as calmodulins, calmodulin-like proteins, CBLs, and CDPKs function as molecular decoders downstream of Ca^{2+} signatures, which translate the signals into kinases, transcription factors, ion channels, and enzymes. These decoding activities are responsible for important adaptive cellular responses such as stomatal control, ion homeostasis, antioxidant defenses, immune response, and growth regulation. Another aspect is the rapid reestablishment of Ca^{2+} homeostasis. ATP-Ca^{2+} pumps, along with Ca^{2+}/H^+ exchanger proteins, especially the ACA-type Ca^{2+}-ATPases and the CAX transporters, play a crucial part in Ca^{2+} signal termination by either exporting Ca^{2+} in the cytoplasm or sequestering Ca^{2+} ions in the vacuole and endoplasmic reticulum. The evolution of these Ca^{2+} homeostatic mechanisms in terrestrial plants indicates the adaptive pressure on the plant to maintain Ca^{2+} signal integrity as well as prevent the cytotoxic build-up of Ca^{2+}. Impaired homeostatic recovery often results in the inhibition of organogenesis. Additionally, Ca^{2+} signaling is thoroughly interwoven into the larger complex of signaling pathways, including phytohormones like ABA, JA, and SA. Such pervasive interaction enables plants to manage local and systemic responses to stress in addition to managing growth and defense. The complex interplay between Ca^{2+} entry into cells and faster Ca^{2+} signaling clearance brings into play a highly reliable, yet dynamic mechanism that enables plants to successfully thrive and survive amidst challenging living conditions. In conclusion, Ca^{2+}-mediated membrane stability and stress signals offer a highly conserved and evolutionarily optimized process that maintains plant plasticity and robustness. The elucidation of plant molecular components that control Ca^{2+} dynamics and Ca^{2+} homeostasis can improve fundamental knowledge of plant biology and has great significance and potential for plant biotechnology aimed at strengthening plant stress tolerance against adverse conditions of climate change and environmental deterioration.

References

Aldon, D., Mbengue, M., Mazars, C., Galaud, J.-P., 2018. Calcium signalling in plant biotic interactions. International Journal of Molecular Sciences 19 (3), 665. Available at https://doi.org/10.3390/ijms19030665.

Arimura, G.-i, Ozawa, R., Shimoda, T., Nishioka, T., Boland, W., Takabayashi, J., 2000. Herbivory-induced volatiles elicit defence genes in lima bean leaves. Nature 406 (6795), 512–515. Available at https://doi.org/10.1038/35020072.

Asano, T., Hayashi, N., Kikuchi, S., Ohsugi, R., 2014. CDPK-mediated abiotic stress signaling. Plant Signaling & Behavior 7 (7), 817–821. Available at https://doi.org/10.4161/psb.20351.

Baxter, I., Tchieu, J., Sussman, M.R., Boutry, M., Palmgren, M.G., Gribskov, M., et al., 2003. Genomic comparison of P-type ATPase ion pumps in *Arabidopsis* and rice. Plant Physiology 132 (2), 618–628. Available at https://doi.org/10.1104/pp.103.021923.

Blanc, G., Wolfe, K.H., 2004. Widespread paleopolyploidy in model plant species inferred from age distributions of duplicate genes[W]. The Plant Cell 16 (7), 1667–1678. Available at https://doi.org/10.1105/tpc.021345.

Bonza, M.C., Morandini, P., Luoni, L., Geisler, M., Palmgren, M.G., De Michelis, M.I., 2000. At-ACA8 encodes a plasma membrane-localized calcium-ATPase of *Arabidopsis* with a calmodulin-binding domain at the N terminus. Plant Physiology 123 (4), 1495–1505. Available at https://doi.org/10.1104/pp.123.4.1495.

Bose, J., Pottosin, I.I., Shabala, S.S., Palmgren, M.G., Shabala, S., 2011. Calcium efflux systems in stress signaling and adaptation in plants. Frontiers in Plant Science 2. Available at https://doi.org/10.3389/fpls.2011.00085.

Bouché, N., Yellin, A., Snedden, W.A., Fromm, H., 2005. Plant-specific calmodulin-binding proteins. Annual Review of Plant Biology 56 (1), 435–466. Available at https://doi.org/10.1146/annurev.arplant.56.032604.144224.

Boudsocq, M., Willmann, M.R., McCormack, M., Lee, H., Shan, L., He, P., et al., 2010. Differential innate immune signalling via Ca^{2+} sensor protein kinases. Nature 464 (7287), 418–422. Available at https://doi.org/10.1038/nature08794.

Brini, F., Masmoudi, K., 2012. Ion transporters and abiotic stress tolerance in plants. International Scholarly Research Notices 2012, 927436. Available at https://doi.org/10.5402/2012/927436.

Bryant, D.E., Greenfield, D., Walshaw, R.D., Johnson, B.R.G., Herschy, B., Smith, C., et al., 2013. Hydrothermal modification of the Sikhote-Alin iron meteorite under low pH geothermal environments. A plausibly prebiotic route to activated phosphorus on the early Earth. Geochimica et Cosmochimica Acta 109, 90–112. Available at https://doi.org/10.1016/j.gca.2012.12.043.

Cano-Ramirez, D.L., Carmona-Salazar, L., Morales-Cedillo, F., Ramírez-Salcedo, J., Cahoon, E.B., Gavilanes-Ruíz, M., 2021. Plasma membrane fluidity: an environment thermal detector in plants. Cells 10 (10). Available at https://doi.org/10.3390/cells10102778.

Catalá, R., Santos, E., Alonso, J.M., Ecker, J.R., Martínez-Zapater, J.M., Salinas, J., 2003. Mutations in the Ca^{2+}/H+ transporter CAX1 increase CBF/DREB1 expression and the cold-acclimation response in *Arabidopsis*. The Plant Cell 15 (12), 2940–2951. Available at https://doi.org/10.1105/tpc.015248.

Cerana, M., Bonza, M.C., Harris, R., Sanders, D., de Michelis, M.I., 2006. Abscisic acid stimulates the expression of two isoforms of plasma membrane Ca^{2+}-ATPase in *Arabidopsis thaliana* seedlings. Plant Biology 8 (5), 572–578. Available at https://doi.org/10.1055/s-2006-924111.

Chakravarthy, S.R., Giorgio, T.D., 1992. Shear stress-facilitated calcium ion transport across lipid bilayers. Biochimica et Biophysica Acta (BBA) - Biomembranes 1112 (2), 197–204. Available at https://doi.org/10.1016/0005-2736(92)90392-y.

Chen, X., Ding, Y., Yang, Y., Song, C., Wang, B., Yang, S., et al., 2021. Protein kinases in plant responses to drought, salt, and cold stress. Journal of Integrative Plant Biology 63 (1), 53–78. Available at https://doi.org/10.1111/jipb.13061.

Cheng, N.H., Pittman, J.K., Shigaki, T., Lachmansingh, J., LeClere, S., Lahner, B., et al., 2005. Functional association of *Arabidopsis* CAX1 and CAX3 is required for normal growth and ion homeostasis. Plant Physiology 138 (4), 2048–2060. Available at https://doi.org/10.1104/pp.105.061218.

Cheng, S., Xian, W., Fu, Y., Marin, B., Keller, J., Wu, T., et al., 2019. Genomes of subaerial zygnematophyceae provide insights into land plant evolution. Cell 179 (5), 1057. Available at https://doi.org/10.1016/j.cell.2019.10.019.

dit Frey, N.F., Mbengue, M., Kwaaitaal, M., Nitsch, L., Altenbach, D., Häweker, H., et al., 2012. Plasma membrane calcium ATPases are important components of receptor-mediated signaling in plant immune responses and development. Plant Physiology 159 (2), 798–809. Available at https://doi.org/10.1104/pp.111.192575.

Dodd, A.N., Kudla, J., Sanders, D., 2010. The language of calcium signaling. Annual Review of Plant Biology 61, 593–620. Available at https://doi.org/10.1146/annurev-arplant-070109-104628.

Du, L., Ali, G.S., Simons, K.A., Hou, J., Yang, T., Reddy, A.S.N., et al., 2009. Ca^{2+}/calmodulin regulates salicylic-acid-mediated plant immunity. Nature 457 (7233), 1154–1158. Available at https://doi.org/10.1038/nature07612.

Edmond, C., Shigaki, T., Ewert, S., Nelson, M.D., Connorton, J.M., Chalova, V., et al., 2009. Comparative analysis of CAX2-like cation transporters indicates functional and regulatory diversity. Biochemical Journal 418 (1), 145–154. Available at https://doi.org/10.1042/BJ20081814.

Eigen, M., 1971. Selforganization of matter and the evolution of biological macromolecules. Die Naturwissenschaften 58 (10), 465–523. Available at https://doi.org/10.1007/bf00623322.

Ettinger, W.F., Clear, A.M., Fanning, K.J., Peck, M.L., 1999. Identification of a Ca^{2+}/H+ antiport in the plant chloroplast thylakoid membrane. Plant Physiology 119 (4), 1379–1385. Available at https://doi.org/10.1104/pp.119.4.1379.

Evans, R.C., Blackwell, K.T., 2015. Calcium: amplitude, duration, or location? Biological Bulletin 228 (1), 75–83. Available at https://doi.org/10.1086/BBLv228n1p75.

Feng, D., Wang, X., Gao, J., Zhang, C., Liu, H., Liu, P., et al., 2023. Exogenous calcium: Its mechanisms and research advances involved in plant stress tolerance. Frontiers in Plant Science 14. Available at https://doi.org/10.3389/fpls.2023.1143963.

Galimov, E.M., 2009. Concept of sustained ordering and an ATP-related mechanism of life's origin. International Journal of Molecular Sciences 10 (5), 2019–2030. Available at https://doi.org/10.3390/ijms10052019.

Geisler, M., Frangne, N., Gomès, E., Martinoia, E., Palmgren, M.G., 2000. The ACA4 gene of *Arabidopsis* encodes a vacuolar membrane calcium pump that improves salt tolerance in yeast. Plant Physiology 124 (4), 1814–1827. Available at https://doi.org/10.1104/pp.124.4.1814.

Ghori, N.-H., Ghori, T., Hayat, M.Q., Imadi, S.R., Gul, A., Altay, V., et al., 2019. Heavy metal stress and responses in plants. International Journal of Environmental Science and Technology 16 (3), 1807–1828. Available at https://doi.org/10.1007/s13762-019-02215-8.

Ghosh, S., Bheri, M., Pandey, G.K., 2021. Delineating calcium signaling machinery in plants: tapping the potential through functional genomics. Current Genomics 22 (6), 404–439. Available at https://doi.org/10.2174/1389202922666211130143328.

Guimarães, F.V.A., de Lacerda, C.F., Marques, E.C., de Miranda, M.R.A., de Abreu, C.E.B., Prisco, J.T., et al., 2011. Calcium can moderate changes on membrane structure and lipid composition in cowpea plants under salt stress. Plant Growth Regulation 65 (1), 55–63. Available at https://doi.org/10.1007/s10725-011-9574-1.

Hashimoto, K., Kudla, J., 2011. Calcium decoding mechanisms in plants. Biochimie 93 (12), 2054–2059. Available at https://doi.org/10.1016/j.biochi.2011.05.019.

He, Z., Webster, S., He, S.Y., 2022. Growth–defense trade-offs in plants. Current Biology 32 (12), R634. Available at https://doi.org/10.1016/j.cub.2022.04.070.

Hepler, P.K., 2005. Calcium: a central regulator of plant growth and development. The Plant Cell 17 (8), 2142–2155. Available at https://doi.org/10.1105/tpc.105.032508.

Hirschi, K., 2001. Vacuolar H+/Ca^{2+} transport: who's directing the traffic? Trends in Plant Science 6 (3), 100–104. Available at https://doi.org/10.1016/s1360-1385(00)01863-x.

Hong-Bo, S., Li-Ye, C., Ming-An, S., 2008. Calcium as a versatile plant signal transducer under soil water stress. Bioessays: News and Reviews in Molecular, Cellular and Developmental Biology 30 (7), 634–641. Available at https://doi.org/10.1002/bies.20770.

Hossain, S.I., Seppelt, M., Nguyen, N., Stokes, C., Deplazes, E., 2022. The role of ion-lipid interactions and lipid packing in transient defects caused by phenolic compounds. Biophysical Journal 121 (18), 3520–3532. Available at https://doi.org/10.1016/j.bpj.2022.08.001.

Huang, S., Waadt, R., Nuhkat, M., Kollist, H., Hedrich, R., Roelfsema, M.R.G., 2019. Calcium signals in guard cells enhance the efficiency by which abscisic acid triggers stomatal closure. New Phytologist 224 (1), 177–187. Available at https://doi.org/10.1111/nph.15985.

Husen, A., 2021. The Harsh environment and resilient plants: an overview. In: Husen, A. (Ed.), Harsh Environment and Plant Resilience. Springer International Publishing AG, Gewerbestrasse 11, 6330 Cham, pp. 1–23. Available at https://doi.org/10.1007/978-3-030-65912-7_1.

Husen, A., Iqbal, M., Khanum, N., Aref, I.M., Sohrab, S.S., Meshresa, G., 2019. Modulation of salt-stress tolerance of niger (*Guizotia abyssinica*), an oilseed plant, by application of salicylic acid. Journal of Environmental Biology 40, 94–104.

Imbiscuso, G., Trotta, A., Maffei, M., Bossi, S., 2009. Herbivory induces a ROS burst and the release of volatile organic compounds in the fern *Pteris vittata* L. Journal of Plant Interactions 4 (1), 15–22. Available at https://doi.org/10.1080/17429140802387739.

Ji, R., Zhou, L., Liu, J., Wang, Y., Yang, L., Zheng, Q., et al., 2017. Calcium-dependent protein kinase CPK31 interacts with arsenic transporter AtNIP1;1 and regulates arsenite uptake in *Arabidopsis thaliana*. PLoS One 12 (3), e0173681. Available at https://doi.org/10.1371/journal.pone.0173681.

Jiang, J., Ma, S., Ye, N., Jiang, M., Cao, J., Zhang, J., 2017. WRKY transcription factors in plant responses to stresses. Journal of Integrative Plant Biology 59 (2), 86–101. Available at https://doi.org/10.1111/jipb.12513.

Jing, T., Li, J., He, Y., Shankar, A., Saxena, A., Tiwari, A., et al., 2024. Role of calcium nutrition in plant physiology: advances in research and insights into acidic soil conditions - a comprehensive review. Plant Physiology and Biochemistry 210. Available at https://doi.org/10.1016/j.plaphy.2024.108602.

Jojoa-Cruz, S., Saotome, K., Murthy, S.E., Tsui, C.C.A., Sansom, M.S.P., Patapoutian, A., et al., 2018. Cryo-EM structure of the mechanically activated ion channel OSCA1.2. eLife 7. Available at https://doi.org/10.7554/elife.41845.

Kader, M.A., Lindberg, S., 2014. Cytosolic calcium and pH signaling in plants under salinity stress. Plant Signaling & Behavior 5 (3), 233–238. Available at https://doi.org/10.4161/psb.5.3.10740.

Kang, X., Zhao, L., Liu, X., 2024. Calcium signaling and the response to heat shock in crop plants. International Journal of Molecular Sciences 25 (1), 324. Available at https://doi.org/10.3390/ijms25010324.

Kazmierczak, J., Kempe, S., Kremer, B., 2013. Calcium in the early evolution of living systems: a biohistorical approach. Current Organic Chemistry 17 (16), 1738–1750. Available at https://doi.org/10.2174/13852728113179990081.

Kleist, T.J., Luan, S., 2016. Constant change: dynamic regulation of membrane transport by calcium signalling networks keeps plants in tune with their environment. Plant Cell and Environment 39 (3), 467–481. Available at https://doi.org/10.1111/pce.12599.

Koyro, H.W., Ahmad, P., Geissler, N., 2012. Abiotic stress responses in plants: an overview. Environmental Adaptations and Stress Tolerance of Plants in the Era of Climate Change. Springer New York, Germany, pp. 1–28. Available at http://doi.org/10.1007/978-1-4614-0815-4, https://doi.org/10.1007/978-1-4614-0815-4_1.

Leckie, C.P., Mcainsh, M.R., Allen, G.J., Sanders, D., Hetherington, A.M., 1998. Abscisic acid-induced stomatal closure mediated by cyclic ADP-ribose. Proceedings of the National Academy of Sciences of the United States of America 95 (26), 15837–15842. Available at https://doi.org/10.1073/pnas.95.26.15837.

Lee, H.J., Seo, P.J., 2021. Ca^{2+}talyzing initial responses to environmental stresses. Trends in Plant Science 26 (8), 849–870. Available at https://doi.org/10.1016/j.tplants.2021.02.007.

Li, F.W., Brouwer, P., Carretero-Paulet, L., Cheng, S., De Vries, J., Delaux, P.M., et al., 2018. Fern genomes elucidate land plant evolution and cyanobacterial symbioses. Nature Plants 4 (7), 460–472. Available at https://doi.org/10.1038/s41477-018-0188-8.

Li, Y., Liu, Y., Jin, L., Peng, R., 2022. Crosstalk between Ca^{2+} and other regulators assists plants in responding to abiotic stress. Plants 11 (10), 1351. Available at https://doi.org/10.3390/plants11101351.

Li, Z., Harper, J.F., Weigand, C., Hua, J., 2023. Resting cytosol Ca^{2+} level maintained by Ca^{2+} pumps affects environmental responses in *Arabidopsis*. Plant Physiology 191 (4), 2534–2550. Available at https://doi.org/10.1093/plphys/kiad047.

Liu, T.Y., Aung, K., Tseng, C.Y., Chang, T.Y., Chen, Y.S., Chiou, T.J., 2011. Vacuolar ca^{2+}/H+ transport activity is required for systemic phosphate homeostasis involving shoot-to-root signaling in arabidopsis. Plant Physiology 156 (3), 1176–1189. Available at https://doi.org/10.1104/pp.111.175257.

Lu, M., Zhang, Y., Tang, S., Pan, J., Yu, Y., Han, J., et al., 2016. AtCNGC2 is involved in jasmonic acid-induced calcium mobilization. Journal of Experimental Botany 67 (3), 809–819. Available at https://doi.org/10.1093/jxb/erv500.

Lu, S., Sun, Y., Ma, L., Luan, S., Yang, G., 2025. Advancing insights into calcium homeostasis and signaling in plant growth and resilience. Science Bulletin 70 (2), 125–127. Available at https://doi.org/10.1016/j.scib.2024.09.031.

Maffei, M., Bossi, S., Spiteller, D., Mithöfer, A., Boland, W., 2004. Effects of feeding Spodoptera littoralis on Lima bean leaves. I. Membrane potentials, intracellular calcium variations, oral secretions, and regurgitate components. Plant Physiology 134 (4), 1752–1762. Available at https://doi.org/10.1104/pp.103.034165.

Malakar, P., Chattopadhyay, D., 2021. Adaptation of plants to salt stress: the role of the ion transporters. Journal of Plant Biochemistry and Biotechnology 30 (4), 668–683. Available at https://doi.org/10.1007/s13562-021-00741-6.

Mehra, P., Bennett, M.J., 2022. A novel Ca^{2+} sensor switch for elevated salt tolerance in plants. Developmental Cell 57 (17), 2045–2047. Available at https://doi.org/10.1016/j.devcel.2022.08.006.

Michal Johnson, J., Reichelt, M., Vadassery, J., Gershenzon, J., Oelmüller, R., 2014. An *Arabidopsis* mutant impaired in intracellular calcium elevation is sensitive to biotic and abiotic stress. BMC Plant Biology 14 (1). Available at https://doi.org/10.1186/1471-2229-14-162.

Mittler, R., Zandalinas, S.I., Fichman, Y., Van Breusegem, F., 2022. Reactive oxygen species signalling in plant stress responses. Nature Reviews. Molecular Cell Biology 23 (10), 663–679. Available at https://doi.org/10.1038/s41580-022-00499-2.

Mori, I.C., Schroeder, J.I., 2004. Reactive oxygen species activation of plant Ca^{2+} channels. A signaling mechanism in polar growth, hormone transduction, stress signaling, and hypothetically mechanotransduction. Plant Physiology 135 (2), 702–708. Available at https://doi.org/10.1104/pp.104.042069.

Mori, K., Renhu, N., Naito, M., Nakamura, A., Shiba, H., Yamamoto, T., et al., 2018. Ca^{2+}-permeable mechanosensitive channels MCA1 and MCA2 mediate cold-induced cytosolic

Ca^{2+} increase and cold tolerance in *Arabidopsis*. Scientific Reports 8 (1). Available at https://doi.org/10.1038/s41598-017-17483-y.

Negi, N.P., Prakash, G., Narwal, P., Panwar, R., Kumar, D., Chaudhry, B., et al., 2023. The calcium connection: exploring the intricacies of calcium signaling in plant-microbe interactions. Frontiers in Plant Science 14. Available at https://doi.org/10.3389/fpls.2023.1248648.

Niu, W.-T., Han, X.-W., Wei, S.-S., Shang, Z.-L., Wang, J., Yang, D.-W., et al., 2020. *Arabidopsis* cyclic nucleotide-gated channel 6 is negatively modulated by multiple calmodulin isoforms during heat shock. Journal of Experimental Botany 71 (1), 90–104. Available at https://doi.org/10.1093/jxb/erz445.

Ogunrinde, A., Munro, K., Davidson, A., Ubaid, M., Snedden, W.A., 2017. *Arabidopsis* calmodulin-like proteins, CML15 and CML16 possess biochemical properties distinct from calmodulin and show non-overlapping tissue expression patterns. Frontiers in Plant Science 8. Available at https://doi.org/10.3389/fpls.2017.02175.

Panchy, N., Lehti-Shiu, M., Shiu, S.-H., 2016. Evolution of gene duplication in plants. Plant Physiology 171 (4), 2294–2316. Available at https://doi.org/10.1104/pp.16.00523.

Park, C.-J., Shin, R., 2022. Calcium channels and transporters: roles in response to biotic and abiotic stresses. Frontiers in Plant Science 13. Available at https://doi.org/10.3389/fpls.2022.964059.

Patra, N., Hariharan, S., Gain, H., Maiti, M.K., Das, A., Banerjee, J., 2021. TypiCal but DeliCate Ca++re: dissecting the essence of calcium signaling network as a robust response coordinator of versatile abiotic and biotic stimuli in plants. Frontiers in Plant Science 12. Available at https://doi.org/10.3389/fpls.2021.752246.

Pei, Z.-M., Murata, Y., Benning, G., Thomine, S., Klüsener, B., Allen, G.J., et al., 2000. Calcium channels activated by hydrogen peroxide mediate abscisic acid signalling in guard cells. Nature 406 (6797), 731–734. Available at https://doi.org/10.1038/35021067.

Peiter, E., Maathuis, F.J.M., Mills, L.N., Knight, H., Pelloux, J., Hetherington, A.M., et al., 2005. The vacuolar Ca^{2+}-activated channel TPC1 regulates germination and stomatal movement. Nature 434 (7031), 404–408. Available at https://doi.org/10.1038/nature03381.

Phukan, U.J., Jeena, G.S., Shukla, R.K., 2016. WRKY transcription factors: molecular regulation and stress responses in plants. Frontiers in Plant Science 7 (2016). Available at https://doi.org/10.3389/fpls.2016.00760.

Pittman, J.K., Shigaki, T., Marshall, J.L., Morris, J.L., Cheng, N.H., Hirschi, K.D., 2004. Functional and regulatory analysis of the *Arabidopsis thaliana* CAX2 cation transporter. Plant Molecular Biology 56 (6), 959–971. Available at https://doi.org/10.1007/s11103-004-6446-3.

Pittman, J.K., Hirschi, K.D., Weber, A., 2016. CAX -ing a wide net: Cation/H + transporters in metal remediation and abiotic stress signalling. Plant Biology 18 (5), 741–749. Available at https://doi.org/10.1111/plb.12460.

Ponnamperuma, C., Sagan, C., Mariner, R., 1963. Synthesis of adenosine triphosphate under possible primitive earth conditions. Nature 199 (4890), 222–226. Available at https://doi.org/10.1038/199222a0.

Postiglione, A.E., Muday, G.K., 2020. The role of ROS homeostasis in ABA-induced guard cell signaling. Frontiers in Plant Science 11. Available at https://doi.org/10.3389/fpls.2020.00968.

Qian, Z., He, L., Li, F., 2024. Understanding cold stress response mechanisms in plants: an overview. Frontiers in Plant Science 15. Available at https://doi.org/10.3389/fpls.2024.1443317.

Ranty, B., Aldon, D., Cotelle, V., Galaud, J.P., Thuleau, P., Mazars, C., 2016. Calcium sensors as key hubs in plant responses to biotic and abiotic stresses. Frontiers in Plant Science 7 (2016). Available at https://doi.org/10.3389/fpls.2016.00327.

Ravi, B., Foyer, C.H., Pandey, G.K., 2023. The integration of reactive oxygen species (ROS) and calcium signalling in abiotic stress responses. Plant Cell and Environment 46 (7), 1985–2006. Available at https://doi.org/10.1111/pce.14596.

Ren, G., Wang, X., Chen, D., Wang, X., Liu, X., 2014. Effects of aphids Myzus persicae on the changes of Ca^{2+} and H_2O_2 flux and enzyme activities in tobacco. Journal of Plant Interactions 9 (1), 883–888. Available at https://doi.org/10.1080/17429145.2014.982221.

Saidi, Y., Peter, M., Finka, A., Cicekli, C., Vigh, L., Goloubinoff, P., 2014. Membrane lipid composition affects plant heat sensing and modulates Ca^{2+} -dependent heat shock response. Plant Signaling & Behavior 5 (12), 1530–1533. Available at https://doi.org/10.4161/psb.5.12.13163.

Sanders, D., Pelloux, J., Brownlee, C., Harper, J.F., 2002. Calcium at the crossroads of signaling. The Plant Cell 14 (suppl 1), S401. Available at https://doi.org/10.1105/tpc.002899.

Sanyal, S.K., Mahiwal, S., Nambiar, D.M., Pandey, G.K., 2020. CBL–CIPK module-mediated phosphoregulation: facts and hypothesis. The Biochemical Journal 477 (5), 853–871. Available at https://doi.org/10.1042/BCJ20190339.

Sarkar, S., Rhein, H.S., Pittman, J.K., Hirschi, K.D., 2024. A dominant-negative *Arabidopsis* cation exchanger 1 (CAX1): N-terminal autoinhibition and membrane topology. Plant Journal 119 (6), 2982–2999. Available at https://doi.org/10.1111/tpj.16966.

Schiøtt, M., Romanowsky, S.M., Bækgaard, L., Jakobsen, M.K., Palmgren, M.G., Harper, J.F., 2004. A plant plasma membrane Ca^{2+} pump is required for normal pollen tube growth and fertilization. Proceedings of the National Academy of Sciences 101 (25), 9502–9507. Available at https://doi.org/10.1073/pnas.0401542101.

Shao, H.-B., Song, W.-Y., Chu, L.-Y., 2008. Advances of calcium signals involved in plant anti-drought. Comptes Rendus Biologies 331 (8), 587–596. Available at https://doi.org/10.1016/j.crvi.2008.03.012 ISSN 1631-0691.

Shigaki, T., Rees, I., Nakhleh, L., Hirschi, K.D., 2006. Identification of three distinct phylogenetic groups of CAX cation/proton antiporters. Journal of Molecular Evolution 63 (6), 815–825. Available at https://doi.org/10.1007/s00239-006-0048-4.

Staxén, I., Pical, C., Montgomery, L.T., Gray, J.E., Hetherington, A.M., Mcainsh, M.R., 1999. Abscisic acid induces oscillations in guard-cell cytosolic free calcium that involve phosphoinositide-specific phospholipase C. Proceedings of the National Academy of Sciences of the United States of America 96 (4), 1779–1784. Available at https://doi.org/10.1073/pnas.96.4.1779.

Swarbreck, S.M., Colaço, R., Davies, J.M., 2013. Plant calcium-permeable channels. Plant Physiology 163 (2), 514–522. Available at https://doi.org/10.1104/pp.113.220855.

Tang, R.-J., et al., 2020. The CBL–CIPK calcium signaling network: unified paradigm from 20 years of discoveries trends in plant science. Trends in Plant Science 25 (6), 604–617.

Tartaglio, V., Rennie, E.A., Cahoon, R., Wang, G., Baidoo, E., Mortimer, J.C., et al., 2017. Glycosylation of inositol phosphorylceramide sphingolipids is required for normal growth and reproduction in *Arabidopsis*. Plant Journal 89 (2), 278–290. Available at https://doi.org/10.1111/tpj.13382.

Thor, K., 2019. Calcium—nutrient and messenger. Frontiers in Plant Science 10. Available at https://doi.org/10.3389/fpls.2019.00440.

Tuteja, N., Mahajan, S., 2014. Calcium signaling network in plants. Plant Signaling & Behavior 2 (2), 79–85. Available at https://doi.org/10.4161/psb.2.2.4176.

Wang, T., Chen, X., Ju, C., Wang, C., 2023. Calcium signaling in plant mineral nutrition: from uptake to transport. Plant Communications 4 (6), 100678. Available at https://doi.org/10.1016/j.xplc.2023.100678.

Wang, C., Tang, R.-J., Kou, S., Xu, X., Lu, Y., Rauscher, K., et al., 2024. Mechanisms of calcium homeostasis orchestrate plant growth and immunity. Nature 627 (8003), 382–388. Available at https://doi.org/10.1038/s41586-024-07100-0.

Waszczak, C., Carmody, M., Kangasjärvi, J., 2018. Reactive oxygen species in plant signaling. Annual Review of Plant Biology 69 (1), 209–236. Available at https://doi.org/10.1146/annurev-arplant-042817-040322.

Wdowiak, A., Podgórska, A., Szal, B., 2024. Calcium in plants: an important element of cell physiology and structure, signaling, and stress responses. Acta Physiologiae Plantarum 46 (12). Available at https://doi.org/10.1007/s11738-024-03733-w.

Whalley, H.J., Knight, M.R., 2013. Calcium signatures are decoded by plants to give specific gene responses. New Phytologist 197 (3), 690–693. Available at https://doi.org/10.1111/nph.12087.

Wilkins, K.A., Matthus, E., Swarbreck, S.M., Davies, J.M., 2016. Calcium-mediated abiotic stress signaling in roots. Frontiers in Plant Science 7 (2016). Available at https://doi.org/10.3389/fpls.2016.01296.

Xu, T., Niu, J., Jiang, Z., 2022. Sensing mechanisms: calcium signaling mediated abiotic stress in plants. Frontiers in Plant Science 13. Available at https://doi.org/10.3389/fpls.2022.925863.

Yadav, A.K., Jha, S.K., Sanyal, S.K., Luan, S., Pandey, G.K., 2018. *Arabidopsis* calcineurin B-like proteins differentially regulate phosphorylation activity of CBL-interacting protein kinase 9. The Biochemical Journal 475 (16), 2621–2636. Available at https://doi.org/10.1042/BCJ20180372.

Yang, D.-L., Shi, Z., Bao, Y., Yan, J., Yang, Z., Yu, H., et al., 2017. Calcium pumps and interacting BON1 protein modulate calcium signature, stomatal closure, and plant immunity. Plant Physiology 175 (1), 424–437. Available at https://doi.org/10.1104/pp.17.00495.

Yoon, Y., Seo, D.H., Shin, H., Kim, H.J., Kim, C.M., Jang, G., 2020. The role of stress-responsive transcription factors in modulating abiotic stress tolerance in plants. Agronomy 10 (6). Available at https://doi.org/10.3390/agronomy10060788.

Yuan, F., Yang, H., Xue, Y., Kong, D., Ye, R., Li, C., et al., 2014. OSCA1 mediates osmotic-stress-evoked Ca^{2+} increases vital for osmosensing in *Arabidopsis*. Nature 514 (7522), 367–371. Available at https://doi.org/10.1038/nature13593.

Yuan, P., Yang, T., Poovaiah, B.W., 2018. Calcium signaling-mediated plant response to cold stress. International Journal of Molecular Sciences 19 (12), 3896. Available at https://doi.org/10.3390/ijms19123896.

Zahra, N., Kausar, A., Sharma, P., Wabaidur, S.M., Husen, A., 2024. Serotonin application to Guizotia abyssinica improves growth, physiological functions, antioxidant system and alleviates the deleterious effects induced by salt stress. Plant Biosystems - An International Journal Dealing with all Aspects of Plant Biology 158 (5), 1006–1014. Available at https://doi.org/10.1080/11263504.2024.2384972.

Zhang, X., Tang, X., Wang, M., Zhang, W., Zhou, B., Wang, Y., 2017. ROS and calcium signaling mediated pathways involved in stress responses of the marine microalgae Dunaliella salina to enhanced UV-B radiation. Journal of Photochemistry and Photobiology B: Biology 173, 360–367. Available at https://doi.org/10.1016/j.jphotobiol.2017.05.038.

Zhang, Y., Wang, Z., Liu, Y., Zhang, T., Liu, J., You, Z., et al., 2023. Plasma membrane-associated calcium signaling modulates cadmium transport. New Phytologist 238 (1), 313–331. Available at https://doi.org/10.1111/nph.18698.

Zhao, X., Lin, S., Yu, S., Zhang, Y., Su, L., Geng, L., et al., 2024. Exogenous calcium enhances the physiological status and photosynthetic capacity of rose under drought stress. Horticultural Plant Journal 10 (3), 853–865. Available at https://doi.org/10.1016/j.hpj.2023.01.010.

Zheng, Y., Wang, L.B., Sun, S.F., Liu, S.Y., Liu, M.J., Lin, J., 2021. Phylogenetic and ion-response analyses reveal a relationship between gene expansion and functional divergence in the Ca^{2+}/cation antiporter family in Angiosperms. Plant Molecular Biology 105 (3), 303–320. Available at https://doi.org/10.1007/s11103-020-01088-7.

Zhou, X., Gong, F., Cao, K., Xu, H., Zhou, X., 2024a. Calcium signaling regulates the accumulation of phenolic acids in response to UV-B stress in Rhododendron chrysanthum Pall. Plant Cell Reports 43 (9). Available at https://doi.org/10.1007/s00299-024-03308-6.

Zhou, J., He, Y., Meng, X., 2024b. Phosphorylation of CAX transporters controls Ca^{2+} homeostasis. Plant Communications 5 (9), 101042. Available at https://doi.org/10.1016/j.xplc.2024.101042.

Zhu, J.-K., 2002. Salt and drought stress signal transduction in plants. Annual Review of Plant Biology 53 (1), 247–273. Available at https://doi.org/10.1146/annurev.arplant.53.091401.143329.

CHAPTER

7

Magnesium's role in photosynthesis and stress resistance

Muhammad Imran[1], Dan-Dan Zhao[2], Shifa Shaffique[1], Ji-In Woo[1], Xiao-Xuan Du[3], Jin Ryeol Jeon[1], Yong-Sung Park[1], Anis Ali Shah[4] and In-Jung Lee[1]

[1]*Department of Applied Biosciences, College of Agriculture and Life Sciences, Kyungpook National University, Daegu, South Korea,*
[2]*Coastal Agriculture Research Institute, Kyungpook National University, Daegu, Republic of Korea,*
[3]*College of Marine and Bioengineering, Yancheng Institute of Technology, Yancheng, Jiangsu, P.R. China,*
[4]*Department of Botany, University of Education, Lahore, Pakistan*

Abbreviations

Mg^{2+}	magnesium ion
ATP	adenosine triphosphate
ADP	adenosine diphosphate
RuBisCO	ribulose-1,5-bisphosphate carboxylase/oxygenase
CBB cycle	Calvin–Benson–Bassham cycle
PSI	photosystem I
PSII	photosystem II
WOC	water-oxidizing complex
LHC	light-harvesting complex
MGT	magnesium transporter
MRS2	magnesium resistance suppressor 2
NIPA	nuclear interactor of ARF and GBF1
CNGC	cyclic nucleotide-gated channel
TOC	translocon at the outer chloroplast membrane
TIC	translocon at the inner chloroplast membrane
TPC	two-pore channel
ROS	reactive oxygen species
PR	proteins (pathogenesis-related proteins)
miRNA	microRNA
bHLH	basic helix-–loop–helix
WRKY	WRKY-type transcription factor family
MYB	MYB-type transcription factor family
IoT	internet of things
INM	integrated nutrient management

Mineral Nutrients to Mitigate Abiotic Stress in Crops. DOI: https://doi.org/10.1016/B978-0-443-44712-9.00001-3

7.1 Introduction

Magnesium (Mg) makes its appearance as approximately 10%–15% of the total Mg content in leaf tissues and is the second most abundant cation in the chloroplast and fourth most abundant cation in plant cells (Chen et al., 2018; Gerendás and Führs, 2013). Beyond its structural function as the primary atom in chlorophyll molecules carrying out photosynthesis and converting sunlight energy into chemical energy is Mg's profound implication for plant life health. This vital macronutrient is the cofactor of more than 300 enzymes, participates in protein synthesis, maintains a cell's pH balance, and provides for stress-tolerant (Guo et al., 2016). Since Mg insufficiency impacts crop yields globally, especially in intensively managed agricultural systems where soil depletion and unbalanced fertilization practices have become more prevalent, the importance of Mg in agriculture cannot be overlooked (Chen et al., 2018). These issues have been made worse by climate change, as rising temperatures, changed precipitation patterns, and an increase in the frequency of extreme weather events have made plants more stressed and raised the need for proper Mg nutrition (Ahmed et al., 2023).

Recent discoveries of the intricate regulatory networks controlling Mg homeostasis and its interaction with other physiological processes have been made possible by recent developments in biotechnology, genomics, and metabolomics (Kobayashi et al., 2013). Mg's are involved in almost every part of photosynthetic metabolism, from carbon fixation and photoprotection to light harvesting and electron transport (Adaskaveg and Blanco-Ulate, 2023). Furthermore, new research indicates that Mg availability affects that how plants react to a range of stressors, such as pathogen attack, drought, salt, and temperature extremes (Guo et al., 2016). Similarly, for protein synthesis, Mg is essential with specific requirements for mRNA binding, ribosome assembly, and peptidyl transfer reactions (Guo et al., 2016; Yamagami et al., 2021). For the structural integrity and functional activity of Mg, ribosomes contain numerous binding sites. These Mg ions help stabilize the complex three-dimensional structure of ribosomal RNA and facilitate the proper positioning of tRNA molecules during translation. Moreover, the biosynthesis of plant hormones also involves Mg-dependent steps, particularly in the production of cytokinins and gibberellins. Such as Mg-dependent enzymes for the modification of tRNA-derived precursors are crucial for cytokinin biosynthesis, while Mg-dependent oxidases and hydroxylases are involved in gibberellin biosynthesis. These interactions between Mg and hormone biosynthesis provide a mechanism through which mineral nutrition can influence plant development and responses to environmental conditions. Furthermore, in glycolysis, key enzymes such as hexokinase, phosphoglucose isomerase, and pyruvate kinase require Mg for their proper function (Chen et al., 2023; Guo et al., 2016). These enzymes catalyze essential steps in glucose metabolism, and their Mg requirements link carbohydrate metabolism to mineral nutrition status, such as hexokinase, which catalyzes the first step of glycolysis and requires Mg for the binding and activation of ATP, while pyruvate kinase requires Mg for both substrate binding and catalytic activity.

Understanding the molecular pathways that underpin Mg's role in photosynthesis and stress tolerance is crucial for developing strategies to enhance crop productivity and resilience. The purpose of this chapter is to summarize and provide a comprehensive overview of the multifaceted roles of Mg in plant biology, with a focus on photosynthetic activities and mechanisms for stress tolerance, aiming to synthesize current research findings on Mg's critical role in photosynthetic efficiency and to explores how Mg mediates plant responses to a wide range of abiotic and biotic stressors, particularly in the context of climate change.

7.2 Mg distribution and homeostasis in plants

In plants, Mg homeostasis is maintained via a complex web of subcellular compartmentalization techniques, regulatory systems, and transport proteins. There is considerable diversity across species and tissue types in the total Mg concentration of plant tissues, which normally ranges from 1.5 to 3.5 mg g^{-1} dry weight (Assunção et al., 2022). Mg is carefully regulated and its distribution throughout cellular compartments, with the chloroplast retaining the largest quantities (20–25 mM), followed by the vacuole (1–5 mM) and the cytoplasm (0.25–1 mM) (Guo et al., 2023). In order to maintain Mg homeostasis, complex transport systems that function at several levels, from subcellular distribution to long-distance transport and root uptake, are needed. MGT (Magnesium Transporter) proteins and other members of the CorA family of Mg transporters are essential for the uptake of Mg from the soil and its passage across cellular membranes (Ahmed et al., 2023; Li et al., 2016). These transporters are regulated by intricate processes that include transcriptional, posttranscriptional, and posttranslational regulation, and they display tissue-specific expression patterns (Li et al., 2016). Recent research has identified many important regulators of Mg homeostasis, such as the NIPA (Nuclear Interactor of ARF and GBF1) proteins, which are involved in Mg transport across the Golgi membrane, and the MRS2 (magnesium resistance suppressor 2) family of proteins, which act as Mg channels in the inner membrane of the mitochondria (Ishijima, 2021). Mg transport, depending on the external Mg content, both high-affinity and low-affinity transport mechanisms are involved in the uptake of Mg from the soil solution at the root level.

Mg travels long distances mostly through the xylem, where it associates with organic acids like citrate and malate to move as a divalent cation (Guo et al., 2023). In the xylem, certain transporter proteins found in the leaf mesophyll cells and root pericycle are involved in the loading and unloading of Mg. The MGT and CNGC (Cyclic Nucleotide-Gated Channel) family's genes have been found to encode these transporters by recent molecular research (Jarratt-Barnham et al., 2021). The Mg transport through phloem is very important, especially when it comes to the transfer of Mg from source to sink tissues. In the phloem, the regulation of Mg transport is impacted by the plant developmental stage, the surrounding environment, and its general nutritional state (Marschner, 2011; Tang et al., 2022). Although the molecular mechanisms behind Mg phloem loading and unloading are not fully known, they seem to involve transporter families that are comparable to those engaged in xylem transport. Furthermore, Mg's physiological functions depend on its subcellular distribution, where different transport pathways allow each compartment to maintain a unique concentration of Mg. As the major location for photosynthesis, the chloroplast keeps the cell's Mg concentrations at their highest. Although the precise mechanisms are yet unknown, the TIC (translocon at the inner chloroplast membrane) and TOC (Translocon at the Outer Chloroplast membrane) complexes are responsible for the import of Mg into the chloroplast (Rochaix, 2022). Within the chloroplast, in response to light conditions and metabolic demands, Mg is distributed between the stroma and the thylakoid system. Upon illumination, the stromal Mg concentration increases significantly by about 3–5 mM, which is essential for the activation of key photosynthetic enzymes (Hermans et al., 2013; Tränkner et al., 2018). In many plant species, 60%–90% of the total cellular Mg is stored in the vacuole, which acts as the main Mg storage space (Gao et al., 2015). Maintaining cytoplasmic Mg homeostasis and serving as a reserve for mobilization during times of high demand or restricted external supply depend on vacuolar Mg storage. Certain transporters and channels, such as those in the MRS2 family and perhaps some TPC (Two-Pore Channel) proteins, are involved in the movement of Mg across the tonoplast (Chen et al., 2018; Upadhyay, 2021).

7.3 Mg's central role in photosynthesis

7.3.1 Chlorophyll structure and functions

Mg is well recognized for its function in photosynthesis due to its pivotal position in the chlorophyll molecule, where it is coordinated by four nitrogen atoms in the porphyrin ring structure (Tränkner et al., 2018). Chlorophyll spectroscopic features depend on the coordination of Mg, which also determines how well it absorbs light and enables it to capture photons for photosynthetic energy conversion. Additional ligands, typically water molecules or amino acid residues from related proteins, stabilize the Mg–nitrogen coordination in chlorophyll by adjusting the chromophore's electronic properties (Zhu et al., 2025). One Mg atom is included in each of the various types of chlorophyll molecules, which include higher plants' chlorophyll a and chlorophyll b. The amount of chlorophyll in leaves usually amounts to 2%–3% of the total Mg content, indicating its important function while it only makes a minor portion of the plant's overall Mg needs (Sato et al., 2018). One of the most important regulatory processes in the production of chlorophyll is the insertion of Mg into protoporphyrin IX, which is mediated by Mg chelatase, a multisubunit enzyme complex (Yao et al., 2023). Chlorophyll function is fundamentally

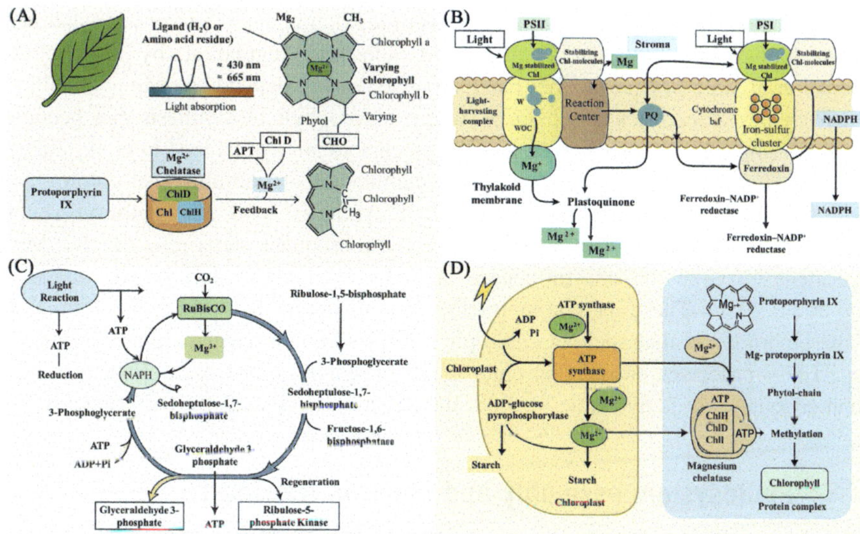

FIGURE 7.1

Overview of magnesium (Mg^{2+}) involvement in (A) the structure and biosynthesis of chlorophyll, where Mg^{2+} is chelated by Mg chelatase (ChlH, ChlD, and ChlI) to form Mg-protoporphyrin IX and produce chlorophyll a and b. (B) Photosystem assembly and electron transport in the thylakoid membrane, Mg^{2+} stabilizes chlorophyll molecules within photosystem II (PSII) and photosystem I (PSI), facilitating light absorption, electron transport. (C) The Calvin Benson cycle uses ATP and NADPH generated during light reactions, and Mg^{2+} works as a cofactor for RuBisCO and kinases. (D) ATP synthesis, where Mg^{2+} regulates ATP synthase activity in chloroplasts, influencing ATP production and starch biosynthesis.

dependent on the spectroscopic features that the Mg center confers (Zhang et al., 2023). The electronic transitions that cause light absorption are influenced by the metal center, with chlorophyll a exhibiting significant absorption peaks at roughly 430 and 665 nm, whereas chlorophyll b absorbs at its maximum at 453 and 642 nm (Fig. 7.1A). A wide range of visible spectrum wavelengths can be effectively captured due to these fluctuations in absorption spectra, which are partially caused by changes in the chemical environment surrounding the Mg center.

7.3.2 Chlorophyll biosynthesis

Mg is a cofactor for the biosynthesis of chlorophyll, which involves a complex series of enzymatic reactions, and is regulated by Mg availability (Liu et al., 2025). The insertion of Mg into protoporphyrin IX, catalyzed by Mg chelatase, is a key regulatory step in chlorophyll biosynthesis. This enzyme complex consists of three subunits (ChlI, ChlH, and ChlD) and requires ATP and Mg for activity (Yang et al., 2023).

The Mg chelatase is subject to multiple levels of regulation, including posttranslational modifications, transcriptional control, and allosteric regulation by metabolites (Farmer et al., 2019). On the other hand, enzyme activity depends on Mg availability, which creates a feedback mechanism, where chlorophyll biosynthesis can become limited during Mg deficiency even if other precursors are available (Liu et al., 2025). This regulatory mechanism prevents the accumulation of nonfunctional chlorophyll precursors and helps coordinate chlorophyll synthesis with Mg nutrition status (Fig. 7.1A). In chlorophyll biosynthesis, many subsequent steps also involve Mg-dependent enzymes or are influenced by Mg availability during Mg insertion (Liu et al., 2025). These include the methylation reactions that modify the formation of the phytol chain that anchors chlorophyll to protein complexes and the porphyrin ring. These processes ensure that chlorophyll biosynthesis proceeds efficiently only when adequate Mg is available for both the structural and catalytic requirements.

7.3.3 Photosystem assembly and electron transport

Mg is necessary for the proper formation and operation of photosystem complexes in addition to its structural involvement in chlorophyll. Mg is essential for the structural stability and ideal electron transport activity of photosystem I (PSI) and photosystem II (PSII) (Tränkner et al, 2018). Mg ions participate in PSII's stabilization of the water-oxidizing complex (WOC), where they support the preservation of the appropriate geometric configuration of the manganese cluster in charge of oxygen evolution and water oxidation (Fig. 7.1B). The structure of light-harvesting antenna complexes (LHCs), in which Mg-containing chlorophyll molecules are organized precisely to promote efficient energy transfer to reaction centers, is another example of how Mg plays a role in photosystem assembly (Yang et al., 2025). These chlorophyll molecules are surrounded by protein environments that offer particular binding sites for the Mg center and related ligands, resulting in the best possible electronic coupling for energy transfer activities (Nagchowdhury et al., 2024; Yang et al., 2025). For both photosystems to function at their best, electron transport chains need the right amounts of Mg. Mg ions support the correct construction and folding of electron transport proteins, such as iron-sulfur clusters and cytochrome complexes (Hermans et al., 2013; Wang et al., 2023). Mg stromal content, which rises with light, is especially crucial for controlling electron transport rates and avoiding photo-inhibitory damage in high light.

7.3.4 Calvin–Benson–Bassham cycle and carbon fixation

Mg plays a pivotal role as a cofactor in the Calvin–Benson–Bassham (CBB) cycle, the central pathway of photosynthetic carbon fixation (Jensen et al., 2017). Among the enzymes in this cycle, ribulose-1,5-bisphosphate carboxylase/oxygenase (RuBisCO) is the most significant Mg-dependent enzyme, as it catalyzes the first step of incorporating atmospheric CO_2 into organic molecules (Barnhart, 2024; Hazra et al., 2015). Mg is indispensable for RuBisCO's structural integrity and catalytic function, since the enzyme possesses multiple Mg-binding sites critical for substrate orientation and stabilization during the reaction (Fig. 7.1C). Without adequate Mg, RuBisCO cannot maintain its active conformation or efficiently bind ribulose-1,5-bisphosphate (RuBP) and CO_2 (Zhang, 2024).

The mechanism of RuBisCO activation further illustrates the central role of Mg. Before the enzyme can function, a carbamate group must form between CO_2 and a conserved lysine residue at the active site. Mg then coordinates this carbamate, along with nearby water molecules, to stabilize the complex and allow catalysis to proceed (Wolf et al., 2019). This Mg-carbamate complex is therefore a prerequisite for enzymatic activity, making Mg availability a direct regulator of RuBisCO's activation state. When Mg supply is limited, the rate of CO_2 fixation declines markedly, leading to suppressed photosynthetic carbon assimilation and overall plant productivity (Bloom and Lancaster, 2018). In addition to RuBisCO, Mg also supports the function of other enzymes within the CBB cycle, emphasizing its broader regulatory significance. Key examples include sedoheptulose-1,7-bisphosphatase, ribulose-5-phosphate kinase, and fructose-1,6-bisphosphatase (Fig. 7.1C) (Bloom and Lancaster, 2018; Bracher et al., 2017). These enzymes catalyze essential reactions that regenerate RuBP, thereby sustaining the continuous turnover of the cycle. Their strict dependence on Mg highlights how carbon fixation is not only initiated but also maintained by sufficient Mg availability. Even a moderate change in Mg availability can alter their efficiency, disrupting the metabolic balance of the cycle (Ishijima, 2021). Moreover, the CBB cycle is further shaped by the influence of Mg and its dynamic regulation within the chloroplast stroma. During the darkness, the Mg concentrations range from 1 to 2 mM, while during the light, they rise to 3–5 mM due to ion transport across the chloroplast envelope (Shabala and Hariadi, 2005). This light-dependent process ensures that significant Mg is available when photosynthetic requirements are highest, thereby aligning enzyme activity with the availability of energy and reducing power from the light reactions. In this way, between light-driven processes and carbon fixation, Mg mobilization provides a crucial link (Shabala and Hariadi, 2005; Ye et al., 2019). In summary, Mg is indispensable for both the initiation and regulation of photosynthetic carbon assimilation. By enabling stabilizing enzyme-substrate complexes, RuBisCO activation, and supporting the activity of additional CBB cycle enzymes, Mg emerges as a central determinant of photosynthetic performance (Ye et al., 2019). Deficiency in this element compromises multiple enzymatic steps and ultimately disrupts the entire cycle, underscoring Mg's fundamental importance for sustaining plant growth and productivity.

7.3.5 ATP synthesis and chlorophyll biosynthesis pathway

Mg is essential for chloroplast energy metabolism and ATP production, and is necessary for the maximum activity of ATP synthase, the enzyme that produces ATP during photophosphorylation (Zhu et al., 2025). ADP and inorganic phosphate bind to the enzyme with the help of Mg ions, which speeds up the phosphorylation process that yields ATP. Thus the energy state of the chloroplast and the cell's ability to perform biosynthetic processes are directly impacted by the availability of Mg (Wang et al., 2020). Under different light conditions, Mg's control over ATP synthase is very crucial (Shabala and Hariadi, 2005). Increased ATP demand for CO_2 fixation and other biochemical processes during high-light intensity periods necessitates optimal ATP synthase activity, which is dependent on sufficient Mg availability (Yamori and Shikanai, 2016). On the other hand, Mg availability may become a limiting factor for ATP production in low light levels, which could decrease photosynthetic efficiency. Other enzymes involved in chloroplast energy metabolism, such as ADP-glucose pyrophosphorylase, which catalyzes the initial committed step in starch production, also depend on Mg for their proper function (Fig. 7.1D). Starch buildup is linked to Mg levels since this enzyme needs Mg for both its catalytic activity and allosteric control. Thereby, integration of Mg requirements across multiple aspects of energy metabolism ensures that photosynthetic carbon assimilation and energy storage are coordinated with mineral nutrition status (He et al., 2024).

Furthermore, many of the enzymatic activities involved in the formation of chlorophyll depend on Mg as a cofactor or are controlled by its availability (Tränkner et al., 2018). Mg chelatase catalyzes the insertion of Mg into protoporphyrin IX, which is a crucial regulatory step in the production of chlorophyll. The three subunits that make up this enzyme complex, ChlH, ChlD, and ChlI, need ATP and Mg to function (Fig. 7.1D) (Sato et al., 2018; Yao et al., 2023). Several levels of regulation affect Mg chelatase, including allosteric regulation by metabolites, transcriptional control, and posttranslational modifications (Farmer et al., 2019). Because the enzyme's activity is dependent on the availability of Mg, a Mg deficit can restrict the production of chlorophyll even in the presence of other precursors (Farhat et al., 2016). By avoiding the buildup of nonfunctional chlorophyll precursors, this regulatory mechanism aids in coordinating chlorophyll synthesis with Mg dietary status. The availability of Mg affects or involves Mg-dependent enzymes in a number of subsequent steps in chlorophyll production after Mg insertion (Tian et al., 2021). These include the creation of the phytol chain, which binds chlorophyll to protein complexes, and the methylation processes that alter the porphyrin ring. These procedures work together to guarantee that chlorophyll production only progresses effectively when there is sufficient Mg available to meet the structural and catalytic requirements.

7.4 Mg stress resistance mechanisms

7.4.1 Abiotic stress tolerance

Mg is involved in several aspects of how plants react to abiotic stressors, such as HMs toxicity, drought, salt, and temperature fluctuations (Guo, 2017). The processes that underlie Mg-mediated stress tolerance include indirect control of stress-responsive pathways as well as direct protective effects. Sufficient Mg intake supports the maintenance of photosynthetic efficiency, osmotic adjustment, and antioxidant capability during drought, all of which enhance drought tolerance (Ahmed et al., 2023; Jamali Jaghdani et al., 2021). Mg's effects on stomatal control and water consumption efficiency are partially linked to its involvement in drought tolerance (Tränkner et al., 2016). Reduced drought tolerance and greater water loss result from Mg deficiency's impairment of stomatal closure responses (Fig. 7.2A). Conversely, adequate Mg absorption promotes healthy stomatal activity, which helps plants maintain water balance during dry spells. This effect is mediated by the function of Mg in the metabolism of guard cells and its impact on potassium transport, which is necessary for stomatal motions (Ahmed et al., 2023). Based on recent research, Mg

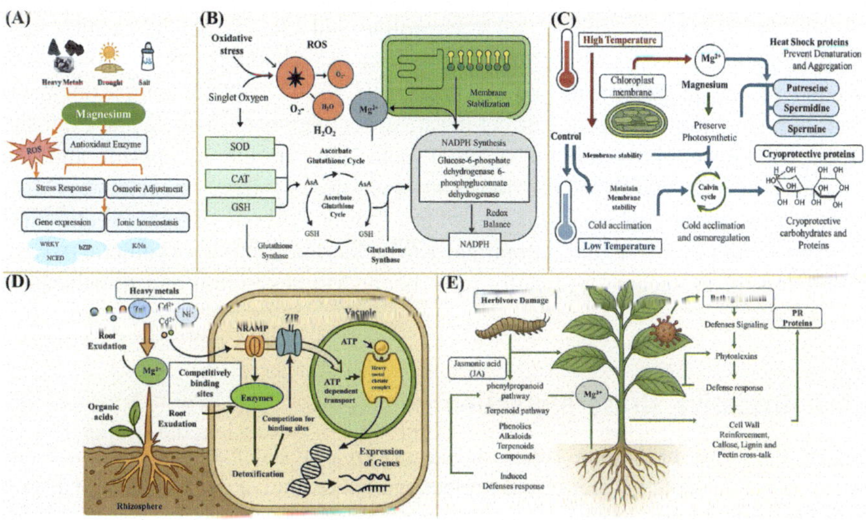

FIGURE 7.2

Schematic representation of magnesium (Mg^{2+}) mediated (A) Mg^{2+} mitigates the effects of abiotic stress by activating antioxidant enzymes, reducing reactive oxygen species (ROS), and stress-responsive gene expression (WRKY, bZIP, NCED, K^+/Na^+). (B) Mg^{2+} mitigates oxidative stress, Mg^{2+} contributes to ROS detoxification by the ascorbate-glutathione cycle, stabilizes membranes, and supports NADPH synthesis for redox balance. (C) Mg^{2+} mitigates temperature stress by promoting heat shock proteins, polyamines, and cryoprotective proteins, thereby maintaining membrane stability and Calvin cycle efficiency. (D) Mg^{2+} mitigates metal stress by alleviating toxicity through root exudation, competition for binding sites, and enhanced enzymatic detoxification and vacuolar sequestration. (E) Mg^{2+} mitigates biotic stress by regulating jasmonic acid-mediated defense pathways, promoting the synthesis of phenolic, alkaloid, terpenoid compounds, and reinforcing cell walls to induce pathogenesis-related (PR) proteins.

deficiency dramatically lowers the expression of genes that respond to drought, such as those that encode transcription factors known as dehydration-responsive element-binding (DREB) (Chaudhry et al., 2021).

Beyond stomatal regulation, Mg and water use efficiency are related in ways that affect hydraulic conductivity and root architecture (Farhat et al., 2016). Plants that have enough Mg tend to grow larger, more branched root systems, which improves their ability to explore different soil volumes in search of water. Mg also affects whole-plant hydraulic conductance during drought by influencing the production and function of aquaporins, membrane proteins that help move water across cellular membranes (Tränkner et al., 2016). By controlling osmotic adjustment, safeguarding the photosynthetic apparatus, and preserving the K/Na balance, adequate Mg intake also increases resistance to salt stress. In saline contexts, salt and Mg compete for binding sites on enzymes and cellular structures to maintain enzyme activity (Fig. 7.2A). Additionally, Mg is necessary for the synthesis of appropriate solutes such as proline and glycine betaine, which help plants maintain osmotic balance under salt stress (Roychoudhury et al., 2020). The function of Mg in preserving ionic homeostasis within cellular compartments is strongly related to its protective effect in salt stress tolerance (Shabala and Pottosin, 2014). Moreover, the retention of vital potassium in the cytoplasm and the sequestration of harmful sodium ions into vacuoles are both impacted by Mg's effects on the activity of ion channels and transporters in both plasma and tonoplast membranes. In order to maintain optimal K/Mg/Na ratios in photosynthetic tissues, salt-tolerant plant species frequently have better Mg absorption and translocation mechanisms

7.4.2 Oxidative stress protection

One of the most promising roles of Mg in stress resistance is to enhance the function of antioxidant systems and protect against oxidative damage. The majority of biotic and abiotic stressors include oxidative stress, which is caused by the generation of reactive oxygen species (ROS) that can harm cellular constituents (Zheltova et al., 2016). Mg protects against oxidative stress in a number of ways, such as by stabilizing cellular membranes and preserving the action of antioxidant enzymes (Zheltova et al., 2016). Mg is necessary for the maximum activity of several important antioxidant enzymes, such as glutathione reductase, catalase, and superoxide dismutase (Fig. 7.2B). These enzyme activities are directly influenced by Mg and are responsible for scavenging ROS and maintaining cellular redox balance (Chen et al., 2018; Morabitoa et al., 2019). Under low Mg availability, the antioxidant capacity is reduced, which leads to enhanced susceptibility to oxidative damage and impaired stress tolerance. In plants, the availability of Mg is especially important for the ascorbate-glutathione cycle, a vital antioxidant system. Several Mg-dependent enzymes are involved in this cycle, which detoxifies hydrogen peroxide and keeps ascorbate and glutathione in their reduced forms (Zhang, 2024).

This cycle is disrupted by Mg shortages, which result in the buildup of oxidized versions of these antioxidants and a decreased ability to neutralize ROS. Furthermore, as glutathione synthase requires Mg as a cofactor, Mg affects the manufacture of glutathione itself (Noctor et al., 2012).

The regulation of NADPH synthesis, which is necessary to preserve cellular redox balance, is also related to Mg and oxidative stress. Mg is essential for the action of two important enzymes in the oxidative pentose phosphate pathway that produce NADPH: glucose-6-phosphate dehydrogenase and 6-phosphogluconate dehydrogenase (Hanau and Helliwell, 2022). Since this reducing power is required to support other antioxidant activities and regenerate reduced glutathione, Mg is indirectly necessary for prolonged antioxidant capacity under stress (Aydemir and Ulusu, 2020; Hanau and Helliwell, 2022). In situations of oxidative stress, Mg is also crucial for preserving the integrity of cellular membranes. The metal ion inhibits lipid peroxidation and preserves membrane function by stabilizing proteins and phospholipids in the membrane. This protective mechanism is especially crucial in chloroplasts since intense sunlight can cause the creation of ROS, including singlet oxygen, which can harm photosynthetic membranes (Fig. 7.2B) (Morais et al., 2017). Multiple mechanisms, such as direct interaction with phospholipid head groups, modification of membrane-associated enzymes, and control of lipid composition, support Mg's protective effect on membranes (Ota et al., 2019). In order to increase membrane stability and decrease permeability to hazardous substances, Mg ions create bridges between negatively charged phospholipids. This structural role of Mg becomes particularly important under stress conditions, which promote membrane destabilization, such as combined heat and drought stress or exposure to reactive oxygen species.

7.4.3 Temperature stress responses

Plant metabolism and needs for Mg are greatly impacted by temperature stress, including both extremes of high and low temperatures. Due to higher metabolic rates and increased demands for stress-reduction strategies, high temperature stress raises the demand for Mg (Shao et al., 2021). Adequate Mg supplementation can improve the effects of heat stress on protein folding, enzyme function, and membrane stability. Stabilization of chloroplast membranes and preservation of photosynthetic electron transport are two of Mg's protective actions against heat stress (Fig. 7.2C) (Shao et al., 2021). Mg usually separates from chlorophyll and other cellular constituents under heat stress, necessitating active processes to preserve Mg homeostasis (Boaretto et al., 2020). When exposed to extreme temperatures, plants that receive enough Mg nourishment maintain their photosynthetic potential better and sustain less oxidative damage. Mg is essential for the production and operation of heat shock proteins (HSPs), which work as molecular chaperones to stop protein denaturation and aggregation during heat stress (Silva et al., 2017). The synthesis of HSPs necessitates transcription and translation machinery that is dependent on Mg, and the availability of Mg directly affects the ATPase activity of several HSPs. Research has indicated that plants lacking Mg have lower levels of HSP accumulation, which leads to more severe heat-induced damage to cellular membranes and proteins (Senbayram et al., 2015).

Mg and heat stress interact by modifying polyamine metabolism, which stabilizes membranes and scavenges free radicals to promote thermal tolerance (Luo et al., 2020). Polyamine production involves enzymes that are dependent on Mg, and under heat stress, a sufficient Mg diet encourages the buildup of polyamines such as putrescine, spermidine, and spermine. These substances aid in the preservation of cellular integrity and promote the healing process following heat-induced harm (Iwadate et al., 2023). Mg-dependent systems are also involved in cold stress responses, especially when it comes to membrane stability and cold acclimatization processes. Adequate Mg levels can help partially avoid membrane damage caused by low temperatures that alter membrane fluidity (Akçay et al., 2025). Cryoprotective substances and antifreeze proteins are produced during cold acclimation, and many of these require enzymes that are dependent on Mg (Jahed et al., 2023). During cold stress, Mg deficiency intensifies the symptoms of chilling injury, such as enhanced photoinhibition, lipid peroxidation, and electrolyte leakage. Due to decreased enzymatic activity and compromised electron transport, the photosynthetic apparatus is susceptible to photo-oxidative damage at low temperatures, making the element's function in preserving chloroplast structure even more crucial (Zhu et al., 2025). Mg helps maintain a balance between light absorption and carbon fixation capacity by promoting the activity of Calvin cycle enzymes at temperatures below ideal levels (Fig. 7.2C). Because Mg is necessary for the synthesis of cryoprotective carbohydrates and the remodeling of membrane lipid composition, its availability has a significant impact on cold acclimation processes (Zhao et al., 2024). Plants undergoing cold acclimation with adequate Mg nutrition show enhanced capacity for osmoregulation through sugar accumulation and improved membrane stability through altered lipid saturation levels, both of which contribute to freezing tolerance.

7.4.4 Heavy metal stress mitigation

Heavy metals (HMs) toxicity is a concerning environmental stress that reduces crop production in contaminated soils. Mg contributes to the reduction of HMs stress in a variety of ways, such as by supporting detoxification processes, maintaining vital enzyme activity, and competitively inhibiting HMsuptake (Lu et al., 2021). With regard to the toxicity of cadmium, lead, and aluminum, Mg's preventive actions against HMs stress are especially well-established. Mg protects against HMs stress by competing for binding sites on enzymes, transporters, and other parts of the cell (Ahmed et al., 2023; Lu et al., 2021). Mg can be relocated from enzyme active sites through heavy metals like cadmium, which may lead to metabolic disturbance and enzyme inactivation. Adequate Mg consumption helps sustain enzyme performance by competing with heavy metals for binding sites and assisting in the repair of damaged enzymes. At the molecular level, heavy metals and Mg compete for binding to membrane transporters, specifically those that are members of the NRAMP (natural resistance-associated macrophage protein) and ZIP (ZRT/IRT-like protein) families (Fig. 7.2D) (Kanwal et al., 2024). An increased Mg supply can lower the intake of heavy metals by overloading these transport channels and preventing harmful metals from entering cells (Singh et al., 2016). This competitive interaction is significantly effective for divalent cations such as zinc, cadmium, and nickel, due to their similar transport paths to Mg (Singh et al., 2016). The mobility and speciation of heavy metals in the rhizosphere are influenced by Mg intake, which impacts the metals' bioavailability to plants. The Mg status can influence root-induced pH changes and exudation of organic acids, which can alter the solubility and thus the uptake of heavy metals. Therefore plants that receive significant Mg nutrition often show modified root exudation patterns that reduce HMs bioavailability in the soil.

Additionally, the synthesis of these compounds requires Mg-dependent enzymes, which sequester heavy metals and reduce their toxicity. During HMs efflux and compartmentalization, Mg provides the energy required for it, which is important for maintaining cellular homeostasis under HMs stress conditions (Rekha et al., 2021). For HMs tolerance, Mg's function in promoting ATP-dependent HMs transport into vacuoles is crucial, such as ABC transporters and other ATP-binding cassette proteins that pump heavy metal-chelate complexes into vacuoles and sequester them away from delicate metabolic processes (Fig. 7.2D) (Dahuja et al., 2021). These activities take a significant amount of energy, and a Mg deficit can affect the ability of vacuolar sequestration, which can result in higher cytoplasmic concentrations of heavy metals and increased toxicity (Huang et al., 2019; Tang et al., 2022). Additionally, through its impact on signaling pathways and transcription factor activity, Mg affects the expression of genes that respond to HMs stress. Under sufficient Mg intake, many genes related to HMs detoxification and tolerance are activated, indicating that Mg status modulates the plant's HMs stress response system (Rengel et al., 2015). This comprises genes that produce antioxidant enzymes, chelating agents, and metal transporters, all of which work together to improve the plant's ability to withstand HMs pollution.

7.4.5 Biotic stress resistance

Mg helps plants withstand biotic stresses, such as pathogen attack and herbivore damage, while its functions in abiotic stress tolerance have received more attention. Both induced and preformed defenses are involved in the pathways that underlie Mg-mediated biotic stress resistance. The synthesis of several secondary metabolites, such as phenolic compounds, alkaloids, and terpenoids, that are involved in plant defense, depends on Mg (Ali et al., 2024). Mg's impacts on cell wall integrity and structure contribute to its function in infection resistance. Mg helps cell walls' pectin molecules cross-link, which increases their mechanical resistance to pathogen penetration (Hassan et al., 2021). Mg is also necessary for the production of lignin and other substances that reinforce cell walls and act as barriers to prevent the entry of pathogens. Callose, lignin, and other reinforcing components are deposited at the sites of attempted disease entry as part of the Mg-dependent process of cell wall alteration during pathogen attack (Fig. 7.2E). Mg is necessary for the best activity of the enzymes that cause these changes, such as callose synthases and peroxidases that are involved in lignin polymerization. Plants lacking Mg are more vulnerable to pathogen colonization because they frequently show delayed or insufficient cell wall reinforcement responses (Ahmed et al., 2021). The availability of Mg affects the production of antimicrobial phytoalexins, which are low-molecular-weight secondary metabolites with direct antibacterial activity (Ahmed et al., 2021; Ahuja et al., 2012). Mg-dependent enzymes, especially those in the phenylpropanoid and terpenoid pathways, are involved in many of the biosynthetic pathways that result in the formation of phytoalexin. After a pathogen challenge, plants that receive enough Mg nutrition exhibit stronger and faster phytoalexin accumulation, which improves disease resistance.

Adequate Mg diet is also necessary for induced defense responses, especially for the production of defense chemicals and pathogenesis-related (PR) proteins (Bouain et al., 2018). Many defensive signaling pathway enzymes, such as protein kinases and phosphatases that facilitate signal transmission, depend on Mg to function (Bouain et al., 2018). Thus the plant's capacity to mount efficient defense responses against biotic stressors can be influenced by the availability of Mg. Additionally, Mg level affects the systemic acquired resistance (SAR) response, which offers sustained defense against a wide range of infections (Yamagami et al., 2021). Mobile signals that prime distant tissues for improved defense responses are produced and translocated during SAR (Bernat and Disney, 2015). The creation and perception of these signals, such as salicylic acid (SA) and related chemicals, depend on Mg. According to studies, plants lacking Mg have poorer SAR signaling, which lowers their defenses against infections by secondary pathogens (Cakmak and Yazici, 2010). Both direct and indirect defense mechanisms are involved in Mg's role in herbivore resistance. Toxins and feeding deterrents are examples of direct defenses, and many of these are produced via pathways that depend on Mg (Fig. 7.2E). Mg-dependent enzymes in the terpenoid and phenylpropanoid pathways are necessary for the manufacture of the volatile organic chemicals that are released as part of indirect defenses and draw in herbivores' natural enemies (Honarmandrad et al., 2022). Another crucial component of biotic stress resistance is the interplay between Mg feeding and the jasmonic acid (JA) signaling system, which is essential for wound responses and herbivore defense (Wasternack and Hause, 2013). Mg affects tissues' susceptibility to the hormone as well as the manufacture of jasmonic acid. The expression of JA-responsive genes implicated in herbivore defense, such as proteinase inhibitors and enzymes involved in the synthesis of toxic secondary metabolites, is increased by an adequate Mg diet (Dar et al., 2015).

7.5 Molecular mechanisms and genetic regulation

The molecular basis of Mg transport in plants involves multiple families of transport proteins, each with distinct characteristics and regulatory mechanisms. The main high-affinity Mg uptake mechanism is represented by the MGT (Mg Transport) family, whose members are found in plasma membranes, chloroplast membranes, and other cellular compartments (Chen et al., 2018). In response to Mg levels and environmental factors, these proteins display tissue-specific expression patterns and are regulated by intricate transcriptional and posttranscriptional processes. Recent studies have identified that Mg-responsive elements and binding sites for transcription factors involved in nutritional homeostasis are among the important regulatory elements found in MGT gene promoters (Wang et al., 2024). Under conditions of Mg deprivation, the expression of MGT genes is typically elevated, adhering to a typical homeostatic response pattern (Fig. 7.3A) 114, (Hermans et al., 2013). The control, however, entails intricate interactions with other signaling pathways linked to development, environmental stress, and nutrient status rather than being merely inverse to Mg availability (Verbruggen and Hermans, 2013; Hermans et al., 2013). Phosphorylation, ubiquitination, and trafficking processes that regulate protein activity and localization are all part of the posttranslational regulation of Mg transporters (Kobayashi et al., 2013). When it comes to regulating transporter function in response to signaling molecules and cellular energy status, protein kinases and phosphatases are crucial. The subcellular

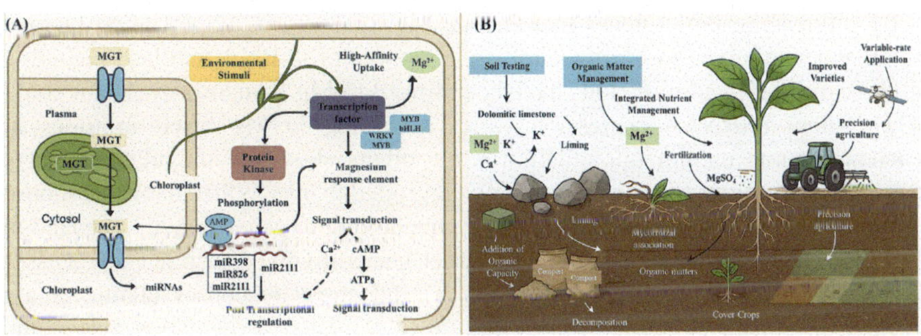

FIGURE 7.3

representation of and agronomic strategies regulating magnesium (Mg^{2+}) uptake, signaling, and management in plants. (A) Molecular machanism of Mg^{2+} transport by magnesium transporters (MGTs) in the plasma membrane, cytosol, chloroplasts, and environmental stimuli activate transcription factors (MYB, WRKY, bHLH), protein kinases, triggering phosphorylation cascades and secondary messengers (Ca^{2+}, cAMP, ATP), and modulate posttranscriptional regulation by MicroRNAs (miR398, miR826, miR2111). (B) Mg^{2+} sustainable management, soil and crop-based management approaches, include soil testing, liming with dolomitic limestone, organic matter incorporation, and integrated nutrient management using $MgSO_4$ fertilizers.

localization of transporters is also dynamically regulated, with some proteins showing stimulus-dependent trafficking between different membrane compartments.

Furthermore, there are several levels of control involved in the transcriptional regulation of Mg homeostasis, such as responses to environmental stimuli, developmental regulation, and tissue-specific expression patterns (Guo et al., 2016). Members of the WRKY, MYB, and bHLH (basic Helix-Loop-Helix) transcription factor families have all been linked to Mg homeostasis. These transcription factors coordinate the expression of several genes involved in the transport and use of Mg by identifying particular DNA sequences in the promoters of genes that respond to Mg (Zuo et al., 2023). Complex regulatory hierarchies incorporating both positive and negative feedback processes have been discovered through the identification of Mg-responsive transcriptional networks (Hermans et al., 2013). While genes encoding Mg-dependent enzymes may be downregulated to preserve available Mg for vital processes, the expression of genes encoding Mg transporters is typically elevated during conditions of Mg deficit. This coordinate regulation helps maintain cellular Mg homeostasis while prioritizing critical physiological processes (Fig. 7.3A). Numerous miRNA families exhibit responsiveness to Mg level, making microRNAs (miRNAs) a crucial layer of posttranscriptional regulation in Mg homeostasis (Guo et al., 2016; Verbruggen and Hermans, 2013; Hermans et al., 2013). These tiny regulatory RNAs can cause translational inhibition or mRNA destruction when they attach to complementary regions in target mRNAs.

By influencing target genes that encode Mg transporters, transcription factors, and enzymes involved in Mg metabolism, a number of miRNA families, including miR398, miR826, and miR2111, have been linked to Mg homeostasis (Islam et al., 2022). These miRNAs' expression varies according to Mg level; in times of deficiency, some exhibit upregulation while others show downregulation. Many facets of Mg homeostasis can be precisely controlled thanks to this distinct response (Islam et al., 2022; Kobayashi et al., 2013). In addition to genes directly engaged in Mg transport, Mg-responsive miRNAs target genes involved in related processes like photosynthesis, chlorophyll production, and stress responses (Jamla et al., 2023). The broader targeting implies that miRNA-mediated regulation aids in the coordination of Mg homeostasis with other physiological processes, guaranteeing that the responses of cells to Mg status are appropriately incorporated into the metabolism and development of the entire plant. At the molecular level, our understanding of the sensing and signaling pathways that identify Mg status and trigger homeostatic reactions is growing (Guo, 2017). Although no particular Mg sensor in plants has been found with certainty, a number of candidates have been put out because of their capacity to bind Mg and their participation in signaling cascades that are responsive to Mg. Mg-binding proteins, RNA structures, and enzyme systems that are responsive to Mg availability are examples of these possible sensors. Signaling pathways involved in Mg homeostasis appear to involve calcium (Ca) signaling, protein phosphorylation cascades, and possibly cyclic nucleotide signaling (Tian et al., 2021). Since both ions can affect the activity of Ca-dependent protein kinases and other signaling enzymes, the relationship between Mg and Ca signaling is very significant (Fig. 7.3A). Thus variations in Mg levels inside cells can set off signaling cascades that eventually alter metabolic activity and gene expression. Crosstalk across various signaling pathways, such as those responsive to light, hormones, and environmental stressors, is necessary for the integration of Mg status signals with other physiological processes (Li et al., 2023; Verbruggen and Hermans, 2013; Hermans et al., 2013). This integration guarantees that reactions to Mg status are suitable for the developmental stage and environmental conditions of the plant, and that Mg homeostasis is synchronized with overall plant physiology.

7.6 Sustainable management strategies

Understanding and maximizing soil Mg availability through strategies that preserve or improve soils' natural ability to produce Mg is the first step toward sustainable Mg management. This approach places a strong emphasis on the use of Mg-containing substances that offer both short-term and long-term advantages, as well as the significance of soil testing and organic matter management. Since it can act as a source or a sink for Mg depending on the processes of mineralization and decomposition, soil organic matter is essential to the retention and cycling of Mg (Chaudhry et al., 2021). Environmental factors, crop needs, and soil conditions all influence the selection of suitable Mg sources (Farhat et al., 2016). Dolomitic limestone is especially well suited for long-term soil development initiatives since it corrects soil acidity and offers a slow-release source of Mg (Fig. 7.3B). Although Mg sulfate, also known as Epsom salt, offers easily accessible Mg and sulfur, it has little effect and might cause soil acidification if used frequently without pH control. Sustainable production systems require integrated nutrient management strategies that take Mg into account in addition to other minerals (Marschner, 2011). Fertilizer programs must carefully balance the antagonistic interaction between Mg and potassium, especially in high-yielding systems where potassium requirements are considerable. Similarly, the interplay between Mg and Ca must be considered in liming programs, with dolomitic limestone being favored over calcitic limestone in Mg-deficient soils.

Additionally, selecting and developing crop varieties with enhanced Mg absorption, transport, and utilization capabilities is the primary objective of plant-based techniques for improving Mg efficiency (Qu et al., 2023). These tactics include biotechnological methods that insert or alter genes involved in Mg homeostasis, as well as traditional breeding methods that take advantage of natural genetic variation. A sustainable method of controlling Mg nutrition while lowering fertilizer inputs and environmental effects is provided by the creation of cultivars that are Mg efficient. A useful strategy for increasing Mg efficiency in perennial crops is rootstock selection; certain rootstocks exhibit better Mg absorption and transport properties (Kuzin and Solovchenko, 2021). Variations in root shape, transporter expression levels, and rhizosphere change are the mechanisms that underlie the impacts of rootstock on Mg nutrition (Fig. 7.3B). In an effort to find genotypes with enhanced traits, screening procedures for rootstock Mg efficiency usually assess performance under both adequate and deficient Mg circumstances (Qu et al., 2023). By increasing the effective root surface area and facilitating access to soil Mg reserves, mycorrhizal interactions can dramatically improve Mg intake (Boaretto et al., 2020). Plant species, management techniques, and soil conditions all affect how effective mycorrhizal connections are; low-fertility soils usually show the biggest benefits. Practices that encourage positive mycorrhizal connections, such as reduced tillage, the addition of organic matter, and suitable crop rotations, should be taken into account by sustainable management systems.

Similar to this, site-specific application techniques based on crop needs, environmental factors, and soil variability can be optimized with precision agricultural technologies (da Silva et al., 2023). Variable-rate application equipment makes it possible to precisely supply Mg fertilizers based on geographical variations within fields, while GPS-guided soil sampling permits detailed mapping of soil Mg levels. These technologies have the potential to greatly increase Mg management's economic and environmental efficiency. Satellite photography and drone-based sensors are examples of remote sensing technologies that hold promise for tracking crop Mg levels and identifying signs of deficiencies before they become noticeable (Wang et al., 2020). A Mg deficit can be detected early with the use of spectral indicators based on photosynthetic efficiency and chlorophyll concentration, enabling prompt remedial intervention. A thorough method of precision Mg management is offered by the integration of yield mapping, soil data, and remote sensing data. Decision support systems can optimize Mg management recommendations for particular field conditions and production goals by integrating weather data, soil information, crop models, and economic aspects (Jones et al., 2017). These systems can provide tailored recommendations that optimize production and profitability while reducing environmental consequences by taking into account the intricate relationships between Mg nutrition, environmental factors, and other management considerations.

Additionally, the integrated nutrient management (INM) systems offer a comprehensive approach to Mg nutrition by integrating various sources and tactics to maximize sustainability over the long term as well as short-term performance (Wu and Ma, 2015). These systems usually combine crop management techniques, biological inputs, organic amendments, and mineral fertilizers to provide synergistic effects that improve nutrient availability and usage efficiency. Understanding how various components interact and how they affect soil-plant systems is essential to the success of INM systems. Compost, manure, and crop residues are examples of organic amendments that support Mg nutrition through a variety of means, including direct nutrient supply, higher microbial activity, improved soil structure, and increased cation exchange capacity (Fig. 7.3B) (Assunção et al., 2022; Wang et al., 2020). Depending on the source materials and processing techniques, organic amendments'

Mg concentration and release characteristics might vary greatly, necessitating careful management and selection to provide the intended results. In balanced feeding regimens, the slow-release properties of organic Mg sources work in tandem with fast-acting mineral fertilizers. By absorbing Mg from deeper soil layers, reducing leaching losses, and adding organic matter that improves Mg retention, cover crop systems can support the cycling of Mg through a number of mechanisms (Blanco-Canqui and Ruis, 2018). Through residue decomposition, certain cover crop species, especially deep-rooted legumes, can retrieve Mg from subsurface layers and transfer it to surface horizons. In cropping systems, the choice of suitable cover crop species and management techniques can have a big impact on Mg cycling.

7.7 Future research directions

To determine the regulatory networks regulating Mg homeostasis, future studies can increasingly employ systems-level techniques that include genomes, transcriptomics, proteomics, and metabolomics. Novel genes and pathways involved in Mg signaling can be found using multi-omics analysis. Since knowing these pathways would allow for targeted regulation of Mg homeostasis in crops, identifying Mg sensors and signaling components continues to be a top priority. Synthetic biology, in conjunction with sophisticated molecular and structural methods, will be essential to these discoveries. Unprecedented resolution of intercellular signaling and cell-type-specific Mg responses is promised by single-cell technology. Furthermore, through changes in temperature, CO_2, and precipitation regimes, climate change poses new problems for Mg nutrition by affecting soil availability and plant demand. In order to develop adaptation methods, such as stress-tolerant and nutrient-efficient crops, future research must assess how these changes impact Mg cycling. Optimizing Mg use efficiency while preserving yields and soil health is necessary for sustainable intensification of agriculture. Nutritional quality and efficiency, particularly Mg bio-fortification, should be the focus of integrated breeding strategies. IoT sensors, artificial intelligence, and machine learning are examples of digital and precision agriculture technologies that will improve real-time nutrient monitoring and management. Blockchain-based solutions have the potential to increase fertilizer use's sustainability and transparency. Addressing Mg management at various scales requires interdisciplinary cooperation amongst ecologists, social scientists, agronomists, and plant scientists.

7.8 Conclusions

Mg plays vital roles in plant physiology, extending beyond its function in pH regulation, stress response, and chlorophyll to enzyme activation. Its importance in photosynthesis and metabolism is highlighted by integrated molecular and physiological research, which connects Mg transport to energy balance and carbon fixation. The significance of Mg in preserving cellular homeostasis in dynamic situations is highlighted by new discoveries about its transport and compartmental control. Through osmotic balance, antioxidant activity, and enzyme stability, an adequate Mg diet improves tolerance to temperature extremes, salinity, and drought. Thus effective Mg management promotes resilience and yield, which helps to achieve sustainable intensification. Despite the prevalence of Mg deficiency, balanced fertilization and precision management can greatly increase output and financial returns; in systems with deficiencies, benefit-cost ratios range from 2:1 to 4:1. Future research must concentrate on creating crop varieties that are Mg efficient, maintaining good yields while using fewer inputs. These advancements will be accelerated by combining molecular breeding with conventional agronomy. Research is still needed since there are still unanswered questions about Mg sensing, signaling, and interactions with environmental stresses. For responsive treatment, enhanced, quick diagnostic techniques for Mg status using spectral, fluorescence, or machine learning approaches are essential. It is also crucial to have an extensive understanding of the Mg cycle and how it interacts with organic matter and soil biota. There is a nutritional problem with the declining Mg content of basic crops, which closely links Mg management to human health and food security. The more general objectives of water quality, soil and ecosystem health, and cyclical resource usage must all be in line with sustainable Mg plans. In conclusion, developing Mg research necessitates combining digital technologies, interdisciplinary cooperation, and molecular techniques to develop resilient, resource-efficient agricultural systems that can maintain productivity in the face of climate change.

References

Adaskaveg, J.A., Blanco-Ulate, B., 2023. Targeting ripening regulators to develop fruit with high quality and extended shelf life. Current Opinion in Biotechnology 79, 102872. Available at https://doi.org/10.1016/j.copbio.2022.102872.

Ahmed, N., Zhang, B., Bozdar, B., Chachar, S., Rai, M., Li, J., et al., 2023. The power of magnesium: unlocking the potential for increased yield, quality, and stress tolerance of horticultural crops. Frontiers in Plant Science 14. Available at https://doi.org/10.3389/fpls.2023.1285512.

Ahmed, T., Noman, M., Luo, J., Muhammad, S., Shahid, M., Ali, M.A., et al., 2021. Bioengineered chitosan-magnesium nanocomposite: a novel agricultural antimicrobial agent against *Acidovorax oryzae* and *Rhizoctonia solani* for sustainable rice production. International Journal of Biological Macromolecules 168, 834–845. Available at https://doi.org/10.1016/j.ijbiomac.2020.11.148.

Ahuja, I., Kissen, R., Bones, A.M., 2012. Phytoalexins in defense against pathogens. Trends in Plant Science 17 (2), 73–90. Available at https://doi.org/10.1016/j.tplants.2011.11.002.

Akçay, M., Yazıcılar, B., Kassa, S.B., Ilhan, D., Shadıdızajı, A., Bezırganoglu, I., 2025. Synergistic effects of salicylic acid and magnesium oxide nanoparticles on cold stress and miRNA expression in alfalfa (*Medicago sativa* L.) genotypes. Plant Cell, Tissue and Organ Culture (PCTOC 161 (3). Available at https://doi.org/10.1007/s11240-025-03097-0.

Ali, S., Ulhassan, Z., Shahbaz, H., Kaleem, Z., Yousaf, M.A., Ali, S., et al., 2024. Application of magnesium oxide nanoparticles as a novel sustainable approach to enhance crop tolerance to abiotic and biotic stresses. Environmental Science: Nano 11 (8), 3250–3267. Available at https://doi.org/10.1039/d4en00417e.

Assunção, A.G.L., Cakmak, I., Clemens, S., González-Guerrero, M., Nawrocki, A., Thomine, S., et al., 2022. Micronutrient homeostasis in plants for more sustainable agriculture and healthier human nutrition. Journal of Experimental Botany 73 (6), 1789–1799. Available at https://doi.org/10.1093/jxb/erac014.

Aydemir, D., Ulusu, N.N., 2020. The possible role of the glucose-6-phosphate dehydrogenase enzyme deficiency in the polyneuropathies. Journal of Basic and Clinical Health Sciences 4 (3), 212–217.

Barnhart, M.H., 2024. The Phenotypic and Transcriptomic Response of Wild and Cultivated Sunflower (*Helianthus annuus* L.) to Several Climate Change Related Abiotic StressesUniversity of Georgia.

Bernat, V., Disney, M.D., 2015. RNA structures as mediators of neurological diseases and as drug targets. Neuron 87 (1), 28–46. Available at https://doi.org/10.1016/j.neuron.2015.06.012.

Blanco-Canqui, H., Ruis, S.J., 2018. No-tillage and soil physical environment. Geoderma 326, 164–200. Available at https://doi.org/10.1016/j.geoderma.2018.03.011.

Bloom, A.J., Lancaster, K.M., 2018. Manganese binding to Rubisco could drive a photorespiratory pathway that increases the energy efficiency of photosynthesis. Nature Plants 4 (7), 414–422. Available at https://doi.org/10.1038/s41477-018-0191-0.

Boaretto, R.M., Hippler, F.W.R., Ferreira, G.A., Azevedo, R.A., Quaggio, J.A., Mattos, D., 2020. The possible role of extra magnesium and nitrogen supply to alleviate stress caused by high irradiation and temperature in lemon trees. Plant and Soil 457 (1-2), 57–70. Available at https://doi.org/10.1007/s11104-020-04597-y.

Bouain, N., Satbhai, S.B., Korte, A., Saenchai, C., Desbrosses, G., Berthomieu, P., et al., 2018. Natural allelic variation of the AZI1 gene controls root growth under zinc-limiting condition. PLoS Genetics 14 (4), e1007304. Available at https://doi.org/10.1371/journal.pgen.1007304.

Bracher, A., Whitney, S.M., Hartl, F.U., Hayer-Hartl, M., 2017. Biogenesis and metabolic maintenance of rubisco. Annual Review of Plant Biology 68, 29–60. Available at https://doi.org/10.1146/annurev-arplant-043015-111633.

Cakmak, I., Yazici, A.M., 2010. Magnesium: a forgotten element in crop production. Better crops 94 (2), 23–25.

Chaudhry, A.H., Nayab, S., Hussain, S.B., Ali, M., Pan, Z., 2021. Current understandings on magnesium deficiency and future outlooks for sustainable agriculture. International Journal of Molecular Sciences 22 (4), 1–18. Available at https://doi.org/10.3390/ijms22041819.

Chen, S., Zou, Y., Song, C., Cao, K., Cai, K., Wu, Y., et al., 2023. The role of glycolytic metabolic pathways in cardiovascular disease and potential therapeutic approaches. Basic Research in Cardiology 118 (1). Available at https://doi.org/10.1007/s00395-023-01018-w.

Chen, Z.C., Peng, W.T., Li, J., Liao, H., 2018. Functional dissection and transport mechanism of magnesium in plants. Seminars in Cell and Developmental Biology 74, 142–152. Available at https://doi.org/10.1016/j.semcdb.2017.08.005.

da Silva, A., Coolong, T., Neilsen, D., Neilsen, G., Simonne, E., Ozores-Hampton, M., et al., 2023. Instant Insights: Irrigation Management in Horticultural Production. Burleigh Dodds Science Publishing.

Dahuja, A., Kumar, R.R., Sakhare, A., Watts, A., Singh, B., Goswami, S., et al., 2021. Role of ATP-binding cassette transporters in maintaining plant homeostasis under abiotic and biotic stresses. Physiologia Plantarum 171 (4), 785–801. Available at https://doi.org/10.1111/ppl.13302.

Dar, T.A., Uddin, M., Khan, M.M.A., Hakeem, K.R., Jaleel, H., 2015. Jasmonates counter plant stress: a review. Environmental and Experimental Botany 115, 49–57. Available at https://doi.org/10.1016/j.envexpbot.2015.02.010.

Farhat, N., Elkhouni, A., Zorrig, W., Smaoui, A., Abdelly, C., Rabhi, M., 2016. Effects of magnesium deficiency on photosynthesis and carbohydrate partitioning. Acta Physiologiae Plantarum 38 (6). Available at https://doi.org/10.1007/s11738-016-2165-z.

Farmer, D.A., Brindley, A.A., Hitchcock, A., Jackson, P.J., Johnson, B., Dickman, M.J., et al., 2019. The ChlD subunit links the motor and porphyrin binding subunits of magnesium chelatase. Biochemical Journal 476 (13), 1875–1887. Available at https://doi.org/10.1042/BCJ20190095.

Gao, C., Zhao, Q., Jiang, L., 2015. Vacuoles protect plants from high magnesium stress. Proceedings of the National Academy of Sciences 112 (10), 2931–2932. Available at https://doi.org/10.1073/pnas.1501318112.

Gerendás, J., Führs, H., 2013. The significance of magnesium for crop quality. Plant and Soil 368 (1-2), 101–128. Available at https://doi.org/10.1007/s11104-012-1555-2.

Guo, J., Jiao, Y., Wang, Y., Hu, W., Jia, Y., Huang, Z., et al., 2023. Regulation of magnesium and calcium homeostasis in citrus seedlings under varying magnesium supply. Plant Physiology and Biochemistry 204. Available at https://doi.org/10.1016/j.plaphy.2023.108146.

Guo, W., 2017. Magnesium homeostasis mechanisms and magnesium use efficiency in plants. Plant Macronutrient Use Efficiency: Molecular and Genomic Perspectives in Crop Plants. Elsevier, China, pp. 197–213.

Guo, W., Nazim, H., Liang, Z., Yang, D., 2016. Magnesium deficiency in plants: an urgent problem. the Crop Journal 4 (2), 83–91. Available at https://doi.org/10.1016/j.cj.2015.11.003.

Hanau, S., Helliwell, J.R., 2022. 6-Phosphogluconate dehydrogenase and its crystal structures. Structural Biology and Crystallization Communications 78 (3), 96–112.

Hassan, S.E.D., Fouda, A., Saied, E., Farag, M.M.S., Eid, A.M., Barghoth, M.G., et al., 2021. Rhizopus oryzae-mediated green synthesis of magnesium oxide nanoparticles (Mgo-nps): a promising tool for antimicrobial, mosquitocidal action, and tanning effluent treatment. Journal of Fungi 7 (5). Available at https://doi.org/10.3390/jof7050372.

Hazra, S., Henderson, J.N., Liles, K., Hilton, M.T., Wachter, R.M., 2015. Regulation of ribulose-1,5-bisphosphate carboxylase/oxygenase (Rubisco) activase: Product inhibition, cooperativity, and magnesium activation. Journal of Biological Chemistry 290 (40), 24222–24236. Available at https://doi.org/10.1074/jbc.M115.651745.

He, Z., Shang, X., Zhang, T., Yun, J., 2024. Effect of calcium and magnesium on starch synthesis in maize kernels and its physiological driving mechanism. Frontiers in Plant Science 14. Available at https://doi.org/10.3389/fpls.2023.1332517.

Hermans, C., Conn, S.J., Chen, J., Xiao, Q., Verbruggen, N., 2013. An update on magnesium homeostasis mechanisms in plants. Metallomics 5 (9), 1170–1183. Available at https://doi.org/10.1039/c3mt20223b.

Honarmandrad, Z., Kucharska, K., Gębicki, J., 2022. Processing of biomass prior to hydrogen fermentation and post-fermentative broth management. Molecules (Basel, Switzerland) 27 (21), 7658. Available at https://doi.org/10.3390/molecules27217658.

Huang, J.H., Xu, J., Ye, X., Luo, T.Y., Ren, L.H., Fan, G.C., et al., 2019. Magnesium deficiency affects secondary lignification of the vascular system in *Citrus sinensis* seedlings. Trees - Structure and Function 33 (1), 171–182. Available at https://doi.org/10.1007/s00468-018-1766-0.

Ishijima, S., 2021. Magnesium and Mg^{2+} transport proteins in cells. Journal of Biological Macromolecules 21 (2), 55–67. Available at https://doi.org/10.14533/jbm.21.55.

Islam, W., Tauqeer, A., Waheed, A., Zeng, F., 2022. MicroRNA mediated plant responses to nutrient stress. International Journal of Molecular Sciences 23 (5), 2562. Available at https://doi.org/10.3390/ijms23052562.

Iwadate, Y., Golubeva, Y.A., Slauch, J.M., 2023. Cation homeostasis: coordinate regulation of polyamine and magnesium levels in *Salmonella*. mBio 14 (1). Available at https://doi.org/10.1128/mbio.02698-22.

Jahed, K.R., Saini, A.K., Sherif, S.M., 2023. Coping with the cold: unveiling cryoprotectants, molecular signaling pathways, and strategies for cold stress resilience. Frontiers in Plant Science 14. Available at https://doi.org/10.3389/fpls.2023.1246093.

Jamali Jaghdani, S., Jahns, P., Tränkner, M., 2021. The impact of magnesium deficiency on photosynthesis and photoprotection in *Spinacia oleracea*. Plant Stress 2, 100040. Available at https://doi.org/10.1016/j.stress.2021.100040.

Jamla, M., Joshi, S., Patil, S., Tripathi, B.N., Kumar, V., 2023. MicroRNAs modulating nutrient homeostasis: a sustainable approach for developing biofortified crops. Protoplasma 260 (1), 5–19. Available at https://doi.org/10.1007/s00709-022-01775-w.

Jarratt-Barnham, E., Wang, L., Ning, Y., Davies, J.M., 2021. The complex story of plant cyclic nucleotide-gated channels. International Journal of Molecular Sciences 22 (2), 1–26. Available at https://doi.org/10.3390/ijms22020874.

Jensen, E., Clément, R., Maberly, S.C., Gontero, B., 2017. Regulation of the Calvin–Benson–Bassham cycle in the enigmatic diatoms: biochemical and evolutionary variations on an original theme. Philosophical Transactions of the Royal Society B. Biological Sciences 372 (1728), 20160401. Available at https://doi.org/10.1098/rstb.2016.0401.

Jones, J.W., Antle, J.M., Basso, B., Boote, K.J., Conant, R.T., Foster, I., et al., 2017. Toward a new generation of agricultural system data, models, and knowledge products: State of agricultural systems science. Agricultural Systems 155, 269–288. Available at https://doi.org/10.1016/j.agsy.2016.09.021.

Kanwal, F., Riaz, A., Ali, S., Zhang, G., 2024. NRAMPs and manganese: magic keys to reduce cadmium toxicity and accumulation in plants. Science of the Total Environment 921, 171005. Available at https://doi.org/10.1016/j.scitotenv.2024.171005.

Kobayashi, N.I., Saito, T., Iwata, N., Ohmae, Y., Iwata, R., Tanoi, K., et al., 2013. Leaf senescence in rice due to magnesium deficiency mediated defect in transpiration rate before sugar accumulation and chlorosis. Physiologia Plantarum 148 (4), 490–501. Available at https://doi.org/10.1111/ppl.12003.

Kuzin, A., Solovchenko, A., 2021. Essential role of potassium in apple and its implications for management of orchard fertilization. Plants 10 (12), 2624. Available at https://doi.org/10.3390/plants10122624.

Li, H., Du, H., Huang, K., Chen, X., Liu, T., Gao, S., et al., 2016. Identification, and functional and expression analyses of the CorA/MRS2/MGT-type magnesium transporter family in maize. Plant and Cell Physiology 57 (6), 1153–1168. Available at https://doi.org/10.1093/pcp/pcw064.

Li, J., Muneer, M.A., Sun, A., Guo, Q., Wang, Y., Huang, Z., et al., 2023. Magnesium application improves the morphology, nutrients uptake, photosynthetic traits, and quality of tobacco (*Nicotiana tabacum* L.) under cold stress. Frontiers in Plant Science 14. Available at https://doi.org/10.3389/fpls.2023.1078128.

Liu, X., Wang, Y., Hu, C., Dong, Z., Moussa, M.G., Huang, K., et al., 2025. Magnesium enhances peel coloration in citrus by regulating carotenoid and chlorophyll contents and biosynthetic gene expression. Scientia Horticulturae 343, 114088. Available at https://doi.org/10.1016/j.scienta.2025.114088.

Lu, M., Liang, Y., Lakshmanan, P., Guan, X., Liu, D., Chen, X., 2021. Magnesium application reduced heavy metal-associated health risks and improved nutritional quality of field-grown Chinese cabbage. Environmental Pollution 289, 117881. Available at https://doi.org/10.1016/j.envpol.2021.117881.

Luo, L., Li, Z., Tang, M.Y., Cheng, B.Z., Zeng, W.H., Peng, Y., et al., 2020. Metabolic regulation of polyamines and γ-aminobutyric acid in relation to spermidine-induced heat tolerance in white clover. Plant Biology 22 (5), 794–804. Available at https://doi.org/10.1111/plb.13139.

Marschner, H., 2011. Marschner's Mineral Nutrition of Higher Plants. Academic Press.

Morabitoa, R., Remigante, A., Marino, A., 2019. Protective role of magnesium against oxidative stress on SO. Cellular Physiology and Biochemistry 52 (6), 1292–1308.

Morais, J.B.S., Severo, J.S., Santos, L.Rd, de Sousa Melo, S.R., de Oliveira Santos, R., de Oliveira, A.R.S., et al., 2017. Role of magnesium in oxidative stress in individuals with obesity. Biological Trace Element Research 176 (1), 20–26. Available at https://doi.org/10.1007/s12011-016-0793-1.

Nagchowdhury, P., Krishna, M., Patra, C.R., 2024. Biomedical applications of magnesium oxide nanoparticles. Nanomaterials for Biomedical and Bioengineering Applications. Springer Nature, India, pp. 97–132.

Noctor, G., Mhamdi, A., Chaouch, S., Han, Y., Neukermans, J., Marquez-Garcia, B., et al., 2012. Glutathione in plants: an integrated overview. Plant, Cell and Environment 35 (2), 454–484. Available at https://doi.org/10.1111/j.1365-3040.2011.02400.x.

Ota, M., Imada, K., Sasaki, K., Kajihara, H., Sakai, S., Ito, S., 2019. MgO-induced defence against bacterial wilt disease in *Arabidopsis thaliana*. Plant Pathology 68 (2), 323–333. Available at https://doi.org/10.1111/ppa.12939.

Qu, S., Li, H., Zhang, X., Gao, J., Ma, R., Ma, L., et al., 2023. Effects of magnesium imbalance on root growth and nutrient absorption in different genotypes of vegetable crops. Plants 12 (20), 3518. Available at https://doi.org/10.3390/plants12203518.

Rekha, K., Usha, B., Keeran, N.S., 2021. Role of ABC transporters and other vacuolar transporters during heavy metal stress in plants. Metal and Nutrient Transporters in Abiotic Stress. Elsevier, United States, pp. 55–76.

Rengel, Z., Bose, J., Chen, Q., Tripathi, B.N., 2015. Magnesium alleviates plant toxicity of aluminium and heavy metals. Crop and Pasture Science 66 (12), 1298–1307. Available at https://doi.org/10.1071/CP15284.

Rochaix, J.D., 2022. Chloroplast protein import machinery and quality control. FEBS Journal 289 (22), 6908–6918. Available at https://doi.org/10.1111/febs.16464.

Roychoudhury, A., Kumar Tripathi, D., Roychoudhury, A., 2020. Role of proline and glycine betaine in overcoming abiotic stresses. Protective Chemical Agents in the Amelioration of Plant Abiotic Stress: Biochemical and Molecular Perspectives. Wiley, pp. 1–23.

Sato, T., Shimoda, Y., Matsuda, K., Tanaka, A., Ito, H., 2018. Mg-dechelation of chlorophyll a by Stay-Green activates chlorophyll b degradation through expressing Non-Yellow Coloring 1 in *Arabidopsis thaliana*. Journal of Plant Physiology 222, 94–102. Available at https://doi.org/10.1016/j.jplph.2018.01.010.

Senbayram, M., Gransee, A., Wahle, V., Thiel, H., 2015. Role of magnesium fertilisers in agriculture: plant-soil continuum. Crop and Pasture Science 66 (12), 1219–1229. Available at https://doi.org/10.1071/CP15104.

Shabala, S., Hariadi, Y., 2005. Effects of magnesium availability on the activity of plasma membrane ion transporters and light-induced responses from broad bean leaf mesophyll. Planta 221 (1), 56–65. Available at https://doi.org/10.1007/s00425-004-1425-0.

Shabala, S., Pottosin, I., 2014. Regulation of potassium transport in plants under hostile conditions: implications for abiotic and biotic stress tolerance. Physiologia Plantarum 151 (3), 257–279. Available at https://doi.org/10.1111/ppl.12165.

Shao, Y., Li, S., Gao, L., Sun, C., Hu, J., Ullah, A., et al., 2021. Magnesium application promotes rubisco activation and contributes to high-temperature stress alleviation in wheat during the grain filling. Frontiers in Plant Science 12. Available at https://doi.org/10.3389/fpls.2021.675582.

Silva, D.Md, Souza, K.R.Dd, Vilas Boas, L.V., Alves, Y.S., Alves, J.D., 2017. The effect of magnesium nutrition on the antioxidant response of coffee seedlings under heat stress. Scientia Horticulturae 224, 115–125. Available at https://doi.org/10.1016/j.scienta.2017.04.029.

Singh, V., Tripathi, B.N., Sharma, V., 2016. Interaction of Mg with heavy metals (Cu, Cd) in T. aestivum with special reference to oxidative and proline metabolism. Journal of Plant Research 129 (3), 487–497. Available at https://doi.org/10.1007/s10265-015-0767-y.

Tang, R.J., Yang, Y., Yan, Y.W., Mao, D.D., Yuan, H.M., Wang, C., et al., 2022. Two transporters mobilize magnesium from vacuolar stores to enable plant acclimation to magnesium deficiency. Plant Physiology 190 (2), 1307–1320. Available at https://doi.org/10.1093/plphys/kiac330.

Tian, X.Y., He, D.D., Bai, S., Zeng, W.Z., Wang, Z., Wang, M., et al., 2021. Physiological and molecular advances in magnesium nutrition of plants. Plant and Soil 468 (1-2). Available at https://doi.org/10.1007/s11104-021-05139-w.

Tränkner, M., Jákli, B., Tavakol, E., Geilfus, C.M., Cakmak, I., Dittert, K., et al., 2016. Magnesium deficiency decreases biomass water-use efficiency and increases leaf water-use efficiency and oxidative stress in barley plants. Plant and Soil 406 (1-2), 409–423. Available at https://doi.org/10.1007/s11104-016-2886-1.

Tränkner, M., Tavakol, E., Jákli, B., 2018. Functioning of potassium and magnesium in photosynthesis, photosynthate translocation and photoprotection. Physiologia Plantarum 163 (3), 414–431. Available at https://doi.org/10.1111/ppl.12747.

Upadhyay, S.K., 2021. Cation Transporters in Plants. Academic Press.

Verbruggen, N., Hermans, C., 2013. Physiological and molecular responses to magnesium nutritional imbalance in plants. Plant and Soil 368 (1-2), 87–99. Available at https://doi.org/10.1007/s11104-013-1589-0.

Wang, L., Zhong, Y., Liu, J., Ma, R., Miao, Y., Chen, W., et al., 2023. Pigment biosynthesis and molecular genetics of fruit color in pepper. Plants 12 (11). Available at https://doi.org/10.3390/plants12112156.

Wang, R., Dong, X., Gao, Y., Hao, F., Zhang, H., Lin, G., 2024. Molecular mechanism of exogenous magnesium in regulating cation homeostasis in roots of peanut seedlings under salt stress. Agronomy 14 (4), 724. Available at https://doi.org/10.3390/agronomy14040724.

Wang, Z., Hassan, M.U., Nadeem, F., Wu, L., Zhang, F., Li, X., 2020. Magnesium fertilization improves crop yield in most production systems: a meta-analysis. Frontiers in Plant Science 10. Available at https://doi.org/10.3389/fpls.2019.01727.

Wasternack, C., Hause, B., 2013. Jasmonates: biosynthesis, perception, signal transduction and action in plant stress response, growth and development. An update to the 2007 review in Annals of Botany. Annals of Botany 111 (6), 1021–1058. Available at https://doi.org/10.1093/aob/mct067.

Wolf, J., Straten, S., Pitann, B., Mühling, K.H., 2019. Foliar Magnesium supply increases the abundance of RuBisCO of Mg-deficient maize plants. Journal of Applied Botany and Food Quality 92, 274–280. Available at https://doi.org/10.5073/JABFQ.2019.092.038.

Wu, W., Ma, B., 2015. Integrated nutrient management (INM) for sustaining crop productivity and reducing environmental impact: a review. Science of the Total Environment 512-513, 415–427. Available at https://doi.org/10.1016/j.scitotenv.2014.12.101.

Yamagami, R., Sieg, J.P., Bevilacqua, P.C., 2021. Functional roles of chelated magnesium Ions in RNA folding and function. Biochemistry 60 (31), 2374–2386. Available at https://doi.org/10.1021/acs.biochem.1c00012.

Yamori, W., Shikanai, T., 2016. Physiological functions of cyclic electron transport around photosystem I in sustaining photosynthesis and plant growth. Annual Review of Plant Biology 67 (1), 81–106. Available at https://doi.org/10.1146/annurev-arplant-043015-112002.

Yang, X., Cai, J., Xue, J., Luo, X., Zhu, W., Xiao, X., et al., 2023. Magnesium chelatase subunit D is not only required for chlorophyll biosynthesis and photosynthesis, but also affecting starch accumulation in Manihot esculenta Crantz. BMC Plant Biology 23 (1). Available at https://doi.org/10.1186/s12870-023-04224-9.

Yang, Z., Li, F., Shen, S., Wang, X., Nihmot Ibrahim, A., Zheng, H., et al., 2025. Natural chlorophyll: a review of analysis methods, health benefits, and stabilization strategies. Critical Reviews in Food Science and Nutrition 65 (16), 3069–3083. Available at https://doi.org/10.1080/10408398.2024.2356259.

Yao, Y., Zhang, H., Guo, R., Fan, J., Liu, S., Liao, J., et al., 2023. Physiological, cytological, and transcriptomic analysis of magnesium protoporphyrin IX methyltransferase mutant reveal complex genetic regulatory network linking chlorophyll synthesis and chloroplast development in rice. Plants 12 (21), 3785. Available at https://doi.org/10.3390/plants12213785.

Ye, X., Chen, X.F., Deng, C.L., Yang, L.T., Lai, N.W., Guo, J.X., et al., 2019. Magnesium-deficiency effects on pigments, photosynthesis and photosynthetic electron transport of leaves, and nutrients of leaf blades and veins in citrus sinensis seedlings. Plants 8 (10). Available at https://doi.org/10.3390/plants8100389.

Zhang, C., Ma, C., Zhu, L., Yao, M., 2023. Simultaneous determination of protoporphyrin IX and magnesium protoporphyrin IX in *Arabidopsis thaliana* and Camellia sinensis using

UPLC-MS/MS. Plant Methods 19 (1). Available at https://doi.org/10.1186/s13007-023-01008-y.

Zhang, W. (2024) Exploring novel ligands interacting with RuBisCO. University of Canterbury. https://ir.canterbury.ac.nz/server/api/core/bitstreams/7fd3b3c5-f4b1-480f-a3a9-2573723893b6/content

Zhao, Y., Li, S., Wu, J., Liu, H., Wang, P., Xu, L., 2024. Insights into membrane lipids modification in barley leaves as an adaptation mechanism to cold stress. Plant Growth Regulation 103 (2), 369–388. Available at https://doi.org/10.1007/s10725-023-01114-w.

Zheltova, A.A., Kharitonova, M.V., Iezhitsa, I.N., Spasov, A.A., 2016. Magnesium deficiency and oxidative stress: an update. Federation BioMedicine (Taiwan) 6 (4), 8–14. Available at https://doi.org/10.7603/s40681-016-0020-6.

Zhu, Z., Zhang, H., Tian, H., Chai, G., Muhammad, R., Wang, Q., et al., 2025. Comprehensive analysis of the effects on photosynthesis and energy balance in tomato leaves under magnesium deficiency. Plant Physiology and Biochemistry 222, 109671. Available at https://doi.org/10.1016/j.plaphy.2025.109671.

Zuo, Z.F., Lee, H.Y., Kang, H.G., 2023. Basic helix-loop-helix transcription factors: regulators for plant growth development and abiotic stress responses. International Journal of Molecular Sciences 24 (2). Available at https://doi.org/10.3390/ijms24021419.

CHAPTER

8

Sulfur and plant stress responses

Neelam Zeb[1], Nadia Sharif[1], Hina Jabeen[1], Ayesha Bibi[2] and Zainul Abideen[3,4]

[1]*Department of Biotechnology, Women University Mardan, Mardan, Khyber Pakhtunkhwa, Pakistan,* [2]*Department of Human Nutrition and Dietetics, Women University Mardan, Mardan, Khyber Pakhtunkhwa, Pakistan,* [3]*Sharjah Seed Bank and Herbarium, University of Al Dhaid, Sharjah, United Arab Emirates,* [4]*Dr. Muhammad Ajmal Khan Institute of Sustainable, Halophyte Utilization, University of Karachi, Karachi, Sindh, Pakistan*

Abbreviations

ACC	1, amino-cyclopropane carboxylate
ABA	abscisic acid
APS	adenosine 5'-phospho-sulfate
APR	adenosine 5'-phosphosulphate reductase
AUX	auxin
CAM	calmodulin-like proteins
Ca^{+2}	calcium ions
CDPKs	Ca+2-dependent kinases
CO_2	carbon dioxide
CRLs	calcineurin-like B proteins
ET	ethylene
IAA	indole-3-acetic acid
K	potassium
Met	methionine
N	nitrogen
NAD+	nicotinamide adenine diphosphate (oxidized)
P	phosphorous
PDH	pyruvate dehydrogenase
ROS	reactive oxygen species
SAT	serine acetyl transferase
SAM	S-adenosyl methionine
S	sulphur
S^{-2}	sulphide
SO_2	sulphur dioxide
SO_3^{-2}	sulfites
SO_4^{-2}	sulphates
Trx	thioredoxins

Mineral Nutrients to Mitigate Abiotic Stress in Crops. DOI: https://doi.org/10.1016/B978-0-443-44712-9.00011-6

8.1 Introduction

The population is growing continuously with food requirements and demands productivity enhancement. Therefore, the requirement for food production and yield is a major challenge that can be improved by ensuring food security. The common factor affecting plants' growth and productivity can be the environment stress and pathogen attacks (Adichirattle et al., 2024). Plant stress can be determined by any damage or internal disturbance that alters optimum plant physiology due to unfavorable changes and environmental factors. The abiotic stress made by environmental factors like temperature, heavy metals, pollution, salinity, drought etc., which are most challenging threats to all economically important plants and adversely affects the molecular, physiological and biochemicals plant processes, that results in the loss of about 50% plant vegetation (Islam and Mohammad, 2020; Saleem et al., 2021a,b; Husen, 2021a,b,c, 2023, 2024, 2025). Biological agents, including insects, nematodes, mites, and microbial pathogens (bacteria, fungi and viruses) are the source of the biotic stress which utilizes the plants nutrients and affecting plant health and by disturbing the internal system by genetic manipulations, hormonal imbalance, disorder in metabolic activities and other physiological conditions and leads to the oxidative stress by producing reactive oxygen species (Raza et al., 2022). In the complex system, the plants created a range of defense strategies to deal with biotic and abiotic factors in order to acclimatize, grow, and survive in stressful conditions (Pieterse et al., 2009).

8.1.1 Abiotic stress in plants

The stress caused by abiotic factors in any atmosphere is known as abiotic stress and undesirably affects the gene expression, growth and development, metabolism, and other biological processes. Following are the main factors intricate in abiotic stress in plants.

a. Temperature
b. Drought
c. Salinity
d. Metals
e. Nutrients
f. Flooding (Zhang et al., 2023).

Various transduction signals are involved in stress responses in plants. The temperature and drought stress factors usually occur together and are interrelated. Similarly, osmotic pressure is enhanced by the elevated soil salt content and drought, which leads to the damage in plants through regulation of organic and inorganic metabolites. The damage caused by abiotic stress includes the generation and accretion of reactive oxygen species (ROS), plasma membrane and cell membranes disruption (Zhang et al., 2023). Moreover, the occurrence of certain metals particularly heavy metals can lead to the toxicity and nutrient scarcity, which also results in activation stress in plants. The plant stress responses can cause two types of changes;

i. Nonadaptive changes, caused by protein misfolding and cell wall damage
ii. Adaptive changes, causing enhancement in plants' resistance.

In short, the plant cell response is a complex and multiple-level process involving various biological processes such as cell signaling, transcription of stress genes, translation of cellular protein, processing, modification, and adaptation (Zhang et al., 2023).

8.1.2 Biotic stress in plants

The plants are also adversely exaggerated by biotic factors like bacteria, viruses, pests, and weeds that can cause serious threat to the plant development and yield. These biological agents can damage the roots, stem, leaves, and fruits of the plants. Pathogenic microbes (bacteria, viruses, and fungi) can cause various infectious diseases in plants that can inhibit photosynthesis and thus reduce crop yield. The infectious diseases can be transmitted from one plant to another thus can destroy the whole field. Similarly, the weed plants are parasitic plants and also produce biotic stress by utilizing nutrients and water, and also compete for space and light, which badly affects the plant development and productivity (Dixit et al., 2024).

8.2 Stress signaling in plants

The stress (both biotic and abiotic) is perceived by the plants through receptors and consequently the signal activates downstream transduction and generates Ca^{+2}, inositol phosphates, and other reactive oxygen species (ROS) as second messengers, which further amplify the signals (Jain et al., 2018). Stress messenger impart significant roles in the activation of stress genes through different mechanisms such as inositol phosphates, changes the level of $Ca^{+\ 2}$ level with in the cell, which is further sensed by Ca^{+2} sensors or Ca^{+2} binding proteins, which includes CAM (calmodulin-like proteins), CRLs (calcineurin-like B proteins), and CDPKs (Ca^{+2}-dependent kinases). In the absence of signal, these proteins are in pr-enzyme form, but modulation in Ca^{+2} level changes the confirmation of second messengers which initiate the phosphorylation cascade and thus activate the transcription factors and stress-responsive genes. Through this strategy, the plants can tolerate unfavorable conditions and help the plant to survive either individually or synergistically (Raina et al., 2021). The stress-associated genes after activation lead to the production of growth hormones (ABA, ethylene, and salicylic acid), which further amplify the initial signal. These growth hormones also start the signaling pathway (second round) in which the same pathway can be initiated again, and also different pathways can be started by utilizing different components. There are certain additional components known as accessory components that are not directly involved in signaling nevertheless participate in the modulation or assembly of signaling molecules such as protein modifiers and enzymes involved in myristoylation, methylation, ubiquitination, and glycosylation (Gallino et al., 2018).

8.3 Nutritional requirements of plants

The plant systems are dependent on the macroelements as nutritional requirements for the growth, survival, and yield. There are seventeen (17) essential elements utilized by plants as nutrients, out of which seven are present in higher concentration, therefore, considered as macro-elements (Nazar et al., 2011). The nutrients that have higher concentration (1–10 mg) per gram dry weight of plants are known as macro-elements which are essential in biological processes. Among the essential macroelements, calcium (Ca), potassium (K), and nitrogen (N) are nontoxic to plants in lower to moderate concentrations, while sulphur (S) and phosphorus (P) are nontoxic in lower toxicity but have detrimental effects on plants if present in higher concentrations in soils. The macroelements present in soil in different concentrations can also lead to the abiotic stress within the plants and contribute as environmental pollutants (Sutar et al., 2017).

8.3.1 Sulfur as a major nutritional element

Among all macroelements, sulfur is an essential element present in soil and required by the plants along with other living organisms and considered as fourth significant element after nitrogen, phosphorus and potassium. Adding sulfur as fertilizer to the crops continuously increasing as the estimated cost on sulfur containing fertilizers was 4.52 billion US dollars, which is probable to reach 7.03 billion US dollars in 2026 (Zenda et al., 2021). Sulfur plays key part in plant physiology, development and metabolic processes, and structural components. Sulfur is the major component of several amino acids (Cysteine, glutathione, S-adenosyl methionine and methionine), vitamins (thiamine and biotin), plants pigments (chlorophyll), phytochelatins, and co-enzyme A. It is also taking part in the synthesis of proteins and enzymes by forming disulfide bonds, regulating enzymatic reactions such as reduction–oxidation reactions and preventing oxidative stress through the derivatives of glutathione. Moreover, different species of plants are dependent on sulfur at various developmental stages such as higher concentration of sulfur are required for seed development and plant growth (Abadie and Tcherkez, 2019). Furthermore, the iron and sulfur-containing proteins (Fe-S) are playing important roles in various metabolic activities and other biological processes including photosynthesis, respiration and energy production, and protection against light. Plants accumulate sulfur in the form of inorganic sulfates (SO_4^{-2}), and utilized in the development of cysteine and methionine in seeds and roots functions. The sulfur and sulphur-comprising substances are also convoluted in the

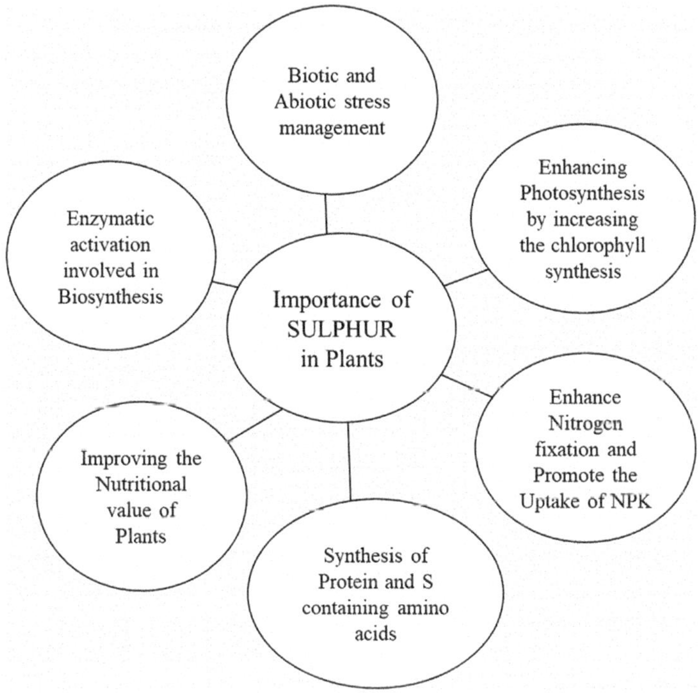

FIGURE 8.1

Sulfur role in plants growth, physiology, and development.

Table 8.1 Requirement of sulphur in different types of crops.

Sr. no.	Crop type	Sulfur requirement for producing 1-ton seeds (kg)	References
1.	Cereals	3–4 kg	Abou Seeda et al. (2020)
2.	Legume	8 kg	
3.	Oil plants	12 kg	

control and management of biotic and abiotic stress, signaling, and metabolic rate. Sulfur in the form of glutathione (Thiol) metabolite is also playing an important role in plants defense against herbivores, microbial pathogens, and pests even in very small concentrations (millimolar) (Shafiee et al., 2020). The specific roles of sulfur in plants are given in Fig. 8.1.

Sulfur is also playing key role in enhancing the oil contents in oil crops, which requires comparatively more sulfur than cereal grains and for producing 1 ton of seed, different concentration of sulfur is required by different plants as given in Table 8.1

The requirements of sulfur for increasing oil contents in plants are almost equivalent to the phosphorus. Moreover, during crop rotations, sulfur uptake is usually high; however, the removal residual crops can lead to the depletion of sulfur, which can be fulfilled by adding fertilizers (Abou Seeda et al., 2020).

8.3.2 Sources of sulphur in soil

Sulfur is naturally found in soil in various forms. The organic matter present in the soil keep 95% of the total sulfur content, and this produces sulfates after mineralization, which will be assimilated in the plants from soil. Sulfur is present in both organic and inorganic forms (Shah et al., 2022).

8.3.2.1 *Inorganic sulphur*

The sulfur is contemporaneous in the inorganic form in soil in diverse concentrations such as ammonium sulfate (24%), thiosulphate (26%), ammonium phosphosulfate (13%), magnesium sulfate (22%), potassium-magnesium sulfate (23%), iron sulfate or pyrates (53%), calcium sulfate (17%), magnesium phosphate (18%), single superphosphate (12%–14%) and sulfur in elemental form (85%–100%). The sulfur is also added to the soil as fertilizers at different stages of vegetation, before and after sowing the seeds, at vegetation and flowering stages, foliage stage depending upon the plant's species and developmental stages. Moreover, the soil around industrial areas is rich source of sulfur and also is in atmosphere as SO_2, which after dissolving in rainwater also reaches the soil. The pesticides also contain small amount of sulfur, which is added in the soil. Furthermore, the chemical fertilizers also contain sulfur along with phosphorus and potassium (Shah et al., 2022).

8.3.2.2 *Organic sulphur sources*

The organic sulfur constitutes about 98% of total sulfur present in soil as sulfates, sulfides, and elemental sulfur, which contains a mixture of soil microbes, animal manure, and plant residues, and are inaccessible to plants as such. The organic sulfur can be utilized after its transformation through mineralization, oxidation, mobilization, immobilization, and reduction to simple form (Kopriva et al., 2019).

8.4 Uptake and assimilation of sulphur by plants

Sulfur, SO_4^{-2} is absorbed by the plants from the soil through sulfate transporters *SULTR1* and *SULTR2* present in the root cells and later dispersed to all plant parts according to the requirement and accessibility of the sulfur, where it is utilized (Khan et al., 2014). Moreover, the SO_2 gas (atmospheric sulfur) is directly

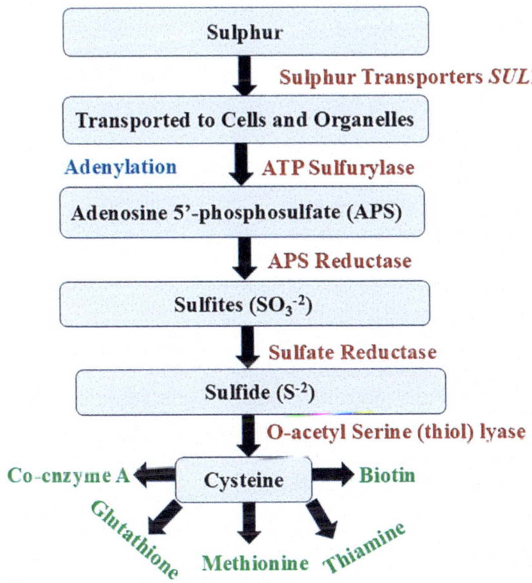

FIGURE 8.2

The assimilation of sulphur in plant cells and conversion to cysteine and other products.

consumed by the leaves and aerial parts of the plant. The common source of sulfur in soil is SO_4^{-2}, which is present naturally in soil and also added in the form of fertilizers, and its uptake and assimilation is regulated by transcription factors. The SO_4^{-2} (anionic or divalent) is assimilated by the roots from soil, and afterwards transported throughout the plants through transmembrane SO_4^{-2} transporters to the xylem and phloem. During the process of sulfur assimilation, cysteine is produced in the plants, which is then used for protein synthesis and also acts as a cause for the production of additional amino acids (methionine) and different metabolites (Koprivova and Kopriva, 2014). The assimilation of sulfur and its conversion to cysteine is given in Fig. 8.2.

8.5 Functions of sulphur-containing compounds in plants

The sulfur is also an important part of various biological components in plants, which include the sulfur-containing vitamins (thiamine and biotin) and amino acids (cysteine and methionine), and various other metabolites (Glutathione and Glucosinolates) (Fuentes-Lara et al., 2019). The importance of sulfur-containing molecules in plants is explained in Tables 8.2 and 8.3.

Table 8.2 Importance of organic sulfur in plants function and metabolic reactions.

Organic sulfur	Plants functions	References
Cysteine	Enzymatic Reactions, stabilization of proteins structure and folding, present in protein-synthesizing compartment, generation of methionine, phytochelatins, Fe-S clusters, and glutathione, heavy metals chelation, cyanides detoxification in mitochondria, and providing plant immunity.	Damodaran (2017), Birke et al. (2012), Romero et al. (2014)
Methionine	Involved in cellular metabolism, S-adenosyl methionine's regulatory roles, the start of mRNA translation, precursor for numerous metabolites including polyamine, thiamine, and ethylene, act as stress protectant.	Amir and Hacham (2015), Hesse et al. (2004)
Thiamine diphosphate	It is vitamin B1, serves as a cofactor for metabolic pathway enzymes and offers defense against biotic and abiotic stress	Rapala-Kozik (2011)
Biotin	It is vitamin B7, water soluble, serves as enzyme cofactor in metabolic processes.	Yusof (2019)
Glutathione	Tripeptide of glutamate, cysteine, and glycine involved in the storage of reduced sulfur, regulates metabolic functions such as cell differentiation and death, detoxification of toxins and xenobiotics, protection in contrast to biotic and abiotic stress by maintaining redox environment and scavenging ROS, acts as antioxidants and substrate for antioxidant enzymes.	Hasanuzzaman et al. (2017), Sabetta et al. (2017), Cohen et al. (2020)
Glucosinolate	Secondary metabolites, precursors of mustard oil, provide defense against pathogens, weeds, and biofumigants.	Sønderby et al. (2010), Jeschke et al. (2019), Capaldi et al. (2015)

8.6 Deficiency of sulphur in plants

In case of sulfur deficit, the older leaves appear yellow, indicating the signs of chlorosis due to the unavailability of sulfur, proving that sulfur cannot reach older leaves when present in limiting amount in plants and soil. However, the young leaves appear green. Sulfur deficiency is also associated with reduction in CO_2 assimilation rates and photosynthetic capability, thus affecting the metabolic activities of plants. During metabolic processes, various reactive oxygen species (ROS), mainly during nutrient deprivation, which can be controlled through

Table 8.3 The global estimated loss of different types of crops due to heat and climate change over a period of 1961–2014.

Sr. no.	Crops	Percentage loss
1.	Cereal crops	1.4%
2.	Oilseed crops	0.5%
3.	Pulses	0.6%
4.	Fruits	0.2%
5.	Vegetables	0.09%

scavenging enzymes, increased photosynthesis rate, and antioxidative system of plants. Thus the generation of ROS like superoxides and hydrogen peroxide in chloroplast (main source) also enhances at a lower rate of photosynthesis and carbon fixation (Aarabi et al., 2016). The deficiency of sulfur can be fulfilled by using the sulfur in the form of fungicides and insecticides, which alleviates the sulfur content in the soil (Sharma et al., 2024).

8.7 Role of sulfur in phytohormone regulation

Sulfur is also imparting a significant role in the biosynthesis of phytohormones and their derivatives, and in turn the phytohormones are involved in the metabolism of sulfur. This nutrient (S) is involved in the synthesis of ethylene (ET), which is then utilized in the methionine metabolism (Capaldi et al., 2015). The biosynthetic pathway for the synthesis of ethylene from methionine (Met) and then converting again to Met is given in Fig. 8.3

The deficiency of sulfate (SO_4^{2-}) in plants leads to the reduction of S-adenosyl methionine (SAM), which also results in the inhibition of production of ethylene, cysteine, and abscisic acid (ABA) biosynthesis, because of the abiotic stress tolerance is also reduced. Moreover, the sulfur is also required in the biosynthesis of auxin (AUX) and indole-3-acetic acid (IAA), therefore the deficiency of sulfur can affect the biosynthesis of all these phytohormones, which can affect the plant development and yield (Hasanuzzaman et al., 2018).

8.8 Role of sulfur in plant stress responses

8.8.1 Abiotic stress

8.8.1.1 Temperature and drought

Temperature is the key factor in maintaining various plant functions including plants development, growth, germination of seeds, photosynthesis, respiration,

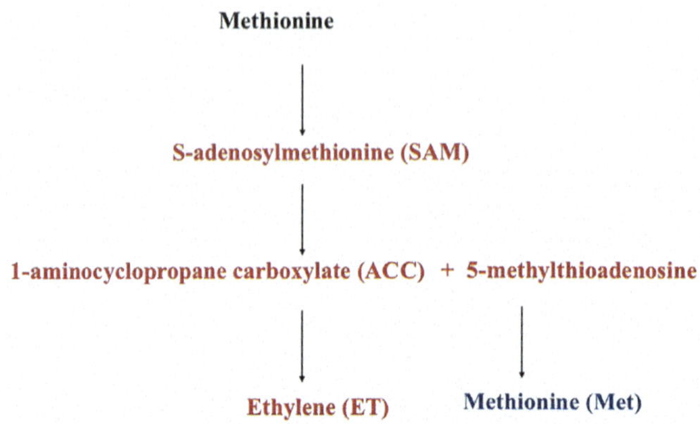

FIGURE. 8.3

Biosynthesis of ET from methionine.

transpiration, and grains formation. But for all these processes, optimum temperature is required; a slight increase in temperature can cause plants to experience heat stress, which may have damaging effects. Heat stress is the rise of temperature from optimum or normal temperature. The climate change is occurring drastically due to anthropogenic activities, and the estimated rise in temperature by the end of 21st century is expected to increase (**1.8°C–5.8°C rise**), consequently it will cause heat stress, which considered as the key issue affecting badly the plants yield and growth. In addition to heat stress, climate change will elevate the risk of drought due to which the rate of transpiration also increases (Mehrabi and Ramankutty, 2017). Survey was conducted for the period of 1961–2014 in order to determine the loss of crops globally due to heat and drought and estimated loss is presented in Table 8.3.

Moreover, the heat stress is also responsible for impaired pollination due to lowering the fertility of pollen grains, which results in the formation of embryo (Alghabari et al., 2015; Alghabari et al., 2016). In a study, it was stated that about 1% decrease in the yield of Zey mays kernel was observed at the temperature above 30°C, which is the optimum or threshold value (Ihsan et al., 2019). Similarly, in *Oryza sativa*, the productivity was reduced due to increase in temperature at night over high daytime temperature (Fahad et al., 2018).

Furthermore, the heat stress can also adversely affect various metabolic pathways, as a result of which the levels of primary and secondary metabolites are also lowered, which trigger the plants to activate defense mechanisms. Increase in temperature also cause elevation of ROS and other free radicles, which cause disruption of cellular structure and functions by inhibiting the enzymatic activities and protein synthesis. Moreover, high temperatures can also affect the rate of photosynthesis by reducing the amount of chlorophyll in chloroplasts. For example, the temperature above threshold value caused imbalance and chlorophyll loss in the

Chlorophyll "a" and "b" ratio in the Triticum. aestivum and other field crops (Shanmugam et al., 2013).

The sulfur-containing amino acid, thiamine, is present in grains and all green vegetables, which is involved in the TCA (tricarboxylic acid cycle), amino acid biosynthesis, and glycolytic pathway. It acts as a cofactor, which increases the catalytic efficiency of two enzyme complexes, **α-ketoglutarate dehydrogenase and pyruvate dehydrogenase (PDH), which** are involved in providing tolerance against heat. Due to a rise in temperature, the reoxidation of dihydrolipoamide moiety occurs by utilizing **NAD+** by upregulating the PDH at the last stage of the conversion of pyruvate to acetyl-CoA. The amount of thiamine usually increases during heat stress for the generation of free radicals, and also improves the expression of antioxidant enzymes, which produce the lenience beside heat stress in plants. Experimentally, it was proved that sulfur is involved in antioxidative metabolic pathways at elevated temperature (Mobin et al., 2017).

Moreover, at high temperature, the concentration of intracellular ammonium ions concentrations increased due to the breakdown of proteins. But under heat stress, the amount of asparagine increases to reduce the ammonium toxicity. The methionine also plays important role by increasing plants' ability to withstand heat stress and is also involved in the mRNA translation (Du et al., 2011). Another sulfur-containing protein is "Thioredoxins" (Trx), which are very heat stable and found in almost all organisms, which protects the mitochondria from oxidative stress. The Thioredoxin type C proteins increase the heat tolerance in *Arabidopsis thaliana* through holdase chaperone proteins (redox-dependent) (Chae et al., 2013). Similarly, the glucosinolates level also increased in *Brassica* in the response of increasing temperature (Del Carmen Martínez-Ballesta et al., 2013).

The drought causes abiotic stress due to shortage of water, which causes ROS to be produced due to oxidative stress and is also responsible for osmotic stress within the plants (Salehi-Lisar and Bakhshayeshan-Agdam, 2016). Glutathioneplays a crucial part in ROS detoxification by maintaining the redox condition in cells and is oxidized to GSSG under heat and drought conditions. It also upregulates the level of serine acetyl transferase (SAT) and Adenosine 5'-phosphosulphate reductase (APR) enzymes during scarcity of water in plant cells (Queval et al., 2011).

8.8.2 Salinity stress

The other most common abiotic factor affecting plants to larger extent is salinity or salt content of soil where plant is present, and it can affect about 7% of land area, while 33% irrigated land globally (Wani, 2023). The salinity stress is continuously elevating and therefore, is the major disquiet that needs to be addressed. The salinization in global cultivable lands has reached about 10%, whereas it is anticipated to reach roughly 50% by the year 2050. Salinity can be classified into two types,

1. Primary salinity that results from the long-term buildup of salts in the soil or surface water such as sodium, calcium, magnesium, sulfates, and carbonates that develop either naturally or due to anthropogenic activities.
2. Secondary soil salinity is caused due to the disturbance of balance between the water supplied to plants for irrigation or through rainfall and water utilized by the plants in the form of transpiration (Rasheed et al., 2022).

Plants that grow and develop at different stages can be adversely affected by high salt content in soil. For example, the first stage of seed germination is highly subtle to salinity and can inhibit the plant growth. The plant's rate of photosynthesis can also be affected by salinity stress (Jahan et al., 2019). The uptake of nutrients (Ca, P, N, S, and K) by the transporters in roots plasma membranes and metabolism is also affected by salinity stress, which generates the competition between ions directly or can also be responsible for reduction of osmotic potential indirectly, thus minimizing the nutrients level in the roots. Salinity stress can also produce reactive oxygen species (ROS) like hydrogen peroxide, oxygen, and hydroxyl radicals, which cause disruption of macromolecules and cellular components like lipid, carbohydrates, and proteins (Jahan et al., 2020; Sehar et al., 2019).

Various methionine-derived compounds including spermine, putrescine, spermidine, and cadaverine enhances the tolerance against salinity. The sulfur-containing vitamins B1 (Thiamine) and B7 (Biotin), are involved in various metabolic pathways as cofactors. The sulfur methyl methionine is utilized during the early phases of development and germination of *A. thaliana*, and produces salt tolerance in plants (Das and Roychoudhury, 2014).

8.8.3 Cold stress

The cold stress is associated with extremely low temperature and that results in causing two types of injuries in plants.

a. Chilling injury; due to chilling stress instigated by temperature that exceeds the freezing point.
b. Freezing injury; due to freezing stress instigated by temperature below the freezing point (Ritonga and Chen, 2020).

Both cold and chilling temperature induce oxidative and osmotic pressure in plants. Sulfur can generate the tolerance against the cold temperature. For example, the treatment of plant seeds with SO_2 gas in different concentrations from 0 to 100 μM, increases the cold tolerance in plants. Moreover, the cold tolerance has been increased in the seed of A. thaliana, when exposed to 50 μM SO_2, which upregulated the expression of two genes AtCAT3 and AtPOD, that reduce the concentration of ROS in the cells thus protect the plants from the damaging effect of H_2O_2 (Ma, 2021).

8.8.4 Heavy metal stress

Because of the advancement of modern agricultural and industrial techniques, the heavy metals concentration has been raised drastically and has become a major pollutant causing abiotic stress. Heavy metals can affect the plants by inhibiting seed germination, disrupting of cellular metabolism, reducing plant growth, transpiration, and photosynthesis, and unbalancing the homeostasis through ROS. Heavy metals can be toxic for human when entered into their bodies through the food chain (Ghori *et al.*, 2019; Li et al., 2022).

Heavy metals and metalloids are classified into two types that are inorganic in nature.

a. Essential heavy metals, which includes Ni (nickel), Cu (copper), and Zn (zinc), are involved in various physiological processes, but can cause abiotic stress when present in higher concentration.
b. Nonessential heavy metals, which includes As (arsenic), Hg (mercury), Pb (lead), Cd (cadmium), are required in minute concentration and are toxic if present even in low dose (Xiao et al., 2023; Cao et al., 2023).

Sulfur plays A very significant role in reducing the heavy metals toxicity. When there is a high concentration of Cd, the synthesis of amino acids (methionine, lysin and histidine) that contain sulfur increases in the roots of rice plants, which possess chelation activity, binds with the heavy metals, thus reduces its toxicity. These amino acids also stimulate the growth of rhizosphere bacteria that utilize the heavy metals in the soil and thus make them unavailable for the plants to be absorbed by the roots. Moreover, the sulfur bacteria in the soil convert the heavy metals into insoluble form (precipitates of sulfides), which prevents their entry into the plants. The phytochelatins possess sulfhydryl groups, that shows affinity for different heavy metals including Pd, Cu, Hg, Zn, Sn, Ar, Se, Au, Cd, Ag, etc. Similarly, glutathione also showed chelating potential in plants and can reduce heavy metal stress in plants (Yu et al., 2019).

8.8.5 Biotic stress

Sulfur also possesses antimicrobial activity and can protect the plants from pathogenic diseases. Sulfur possesses antifungal activity, and it was discovered at the end of 1980s in Scotland, Germany, and other Western European Countries, when the Clean Air Acts removed the atmospheric sulfur deposits, and that leads to sulfur deficiency, which results in increasing the fungal diseases in plants. But when inorganic sulfate salts were applied to the agricultural crops, the fungal diseases were significantly reduced, indicating their role in controlling infections. In a study, the tolerance against *Pyrenopeziza brassicae* in oilseed rape, *Rhizoctonia solani* in potato tubers, and *Uncinula necator* in grapes by applying sulfate soil (Künstler et al., 2020).

The sulfur as SO_2 is involved in creating resistance against pathogens and various other biotic factors. For example, the resistance against bacterial pathogen *Botrytis cinerea* was increased in *Arabidopsis* plantS after exposure to SO_2 (Xue and Yi, 2018).

8.9 Conclusion

The climate change is responsible for abiotic stress in plants, and the major influences include temperature, salinity, drought, heavy metals, cold, etc., which are responsible for reactive oxygen species generation in plants that cause oxidative and osmotic stress in plants. Moreover, the abiotic stress factors can also affect the metabolic pathways and respiration, photosynthesis, transpiration, and other physiological processes, embryo formation, and seed germination, which can affect the yield. The biotic factors include bacteria, fungi, and viral strains that can cause various types of infections in plants. Sulfur, which the major nutritional element, plays vital part in controlling the biotic and abiotic stress conditions in plants. The sulfur keeping amino acids (methionine, cysteine), secondary metabolites (glutathione), and SO_2, are involved in detoxification of ROS, maintaining the redox status, boosting the immunity, chelating the heavy metals, and thus protectING the plants from damage. Therefore, the sulfur should be available in sufficient amountS to the plants to improve the growth and yield, which can be added through fertilizers.

8.10 Future considerations

The investigations on the impact of sulfur on plant development, growth, and metabolism are elaborated in this chapter, and the mechanisms through which the plants cope with the biotic and abiotic factors have also been explained, but the following are a few future recommendations to overcome the sulfur deficiency:

1. The organic matter in the form of compost or crop residues can be added to the soil, which can provide the sulfur in the long term as it can be the source of slow-releasing sulfate.
2. The soil should be tested and upon requirement the sulfur should be added along with N, P, K, etc., as balanced fertilizers.

References

Aarabi, F., Kusajima, M., Tohge, T., Konishi, T., Gigolashvili, T., Takamune, M., et al., 2016. Sulfur deficiency–induced repressor proteins optimize glucosinolate biosynthesis in plants. Science Advances 2 (10). Available at https://doi.org/10.1126/sciadv.1601087.

Abadie, C., Tcherkez, G., 2019. Plant sulphur metabolism is stimulated by photorespiration. Communications Biology 2 (1). Available at https://doi.org/10.1038/s42003-019-0616-y.

Abou Seeda, M., Abou El-Nour, E., Yassen, A., Gad Mervat, M., Zaghloul, S., 2020. Importance of sulfur and its roles in plants physiology: a review. Current Science International. 9, 198–231.

Adichirattle, N., Munir, M., Thomas, B., 2024. Stress response crosstalk in plants - a review. Pakistan Journal of Botany 56 (5), 2049–2057. Available at https://doi.org/10.30848/PJB2024-5(40).

Alghabari, F., Ihsan, M.Z., Hussain, S., Aishia, G., Daur, I., 2015. Effect of Rht alleles on wheat grain yield and quality under high temperature and drought stress during booting and anthesis. Environmental Science and Pollution Research 22 (20), 15506–15515. Available at https://doi.org/10.1007/s11356-015-4724-z.

Alghabari, F., Ihsan, M.Z., Khaliq, A., Hussain, S., Daur, I., Fahad, S., et al., 2016. Gibberellin-sensitive Rht alleles confer tolerance to heat and drought stresses in wheat at booting stage. Journal of Cereal Science 70, 72–78. Available at https://doi.org/10.1016/j.jcs.2016.05.016.

Amir, R., Hacham, Y., 2015. Methionine Metabolism in Plants Sulfur: A Missing Link between Soils, Crops, and Nutrition. Wiley, Israel, pp. 251–279. Available at http://doi.org/10.2134/agronmonogr50.c16.

Birke, H., Haas, F.H., De Kok, L.J., Balk, J., Wirtz, M., Hell, R., 2012. Cysteine biosynthesis, in concert with a novel mechanism, contributes to sulfide detoxification in mitochondria of *Arabidopsis thaliana*. Biochemical Journal 445 (2), 275–283. Available at https://doi.org/10.1042/BJ20120038.

Cao, Y., Ma, C., Yu, H., Tan, Q., Dhankher, O.P., White, J.C., et al., 2023. The role of sulfur nutrition in plant response to metal(loid) stress: facilitating biofortification and phytoremediation. Journal of Hazardous Materials 443. Available at https://doi.org/10.1016/j.jhazmat.2022.130283.

Capaldi, F.R., Gratão, P.L., Reis, A.R., Lima, L.W., Azevedo, R.A., 2015. Sulfur metabolism and stress defense responses in plants. Tropical Plant Biology 8 (3-4), 60–73. Available at https://doi.org/10.1007/s12042-015-9152-1.

Del Carmen Martínez-Ballesta, M., Moreno, D., Carvajal, M., 2013. The physiological importance of glucosinolates on plant response to abiotic stress in *Brassica*. International Journal of Molecular Sciences 14 (6), 11607–11625. Available at https://doi.org/10.3390/ijms140611607.

Chae, H.B., Moon, J.C., Shin, M.R., Chi, Y.H., Jung, Y.J., Lee, S.Y., et al., 2013. Thioredoxin reductase type C (NTRC) orchestrates enhanced thermotolerance to *Arabidopsis* by its redox-dependent holdase chaperone function. Molecular Plant 6 (2), 323–336. Available at https://doi.org/10.1093/mp/sss105.

Cohen, A., Hacham, Y., Welfe, Y., Khatib, S., Avice, J.C., Amir, R., 2020. Evidence of a significant role of glutathione reductase in the sulfur assimilation pathway. Plant Journal 102 (2), 246–261. Available at https://doi.org/10.1111/tpj.14621.

Damodaran, S., 2017. Amino acids, peptides, and proteins. Fennema's Food Chemistry. CRC Press, Boca Raton, pp. 235–356. Available at 10.1201/9781315372914.

Das, K., Roychoudhury, A., 2014. Reactive oxygen species (ROS) and response of antioxidants as ROS-scavengers during environmental stress in plants. Frontiers in Environmental Science 2. Available at https://doi.org/10.3389/fenvs.2014.00053.

Dixit, S., Sivalingam, P.N., Baskaran, R.K.M., Senthil-Kumar, M., Ghosh, P.K., 2024. Plant responses to concurrent abiotic and biotic stress: unravelling physiological and morphological mechanisms. Plant Physiology Reports 29 (1), 6–17. Available at https://doi.org/10.1007/s40502-023-00766-0.

Du, H., Wang, Z., Yu, W., Liu, Y., Huang, B., 2011. Differential metabolic responses of perennial grass *Cynodon transvaalensis* × *Cynodon dactylon* (C 4) and *Poa pratensis* (C 3) to heat stress. Physiologia Plantarum 141 (3), 251–264. Available at https://doi.org/10.1111/j.1399-3054.2010.01432.x.

Fahad, S., Ihsan, M.Z., Khaliq, A., Daur, I., Saud, S., Alzamanan, S., et al., 2018. Consequences of high temperature under changing climate optima for rice pollen characteristics-concepts and perspectives. Archives of Agronomy and Soil Science 64 (11), 1473–1488. Available at https://doi.org/10.1080/03650340.2018.1443213.

Fuentes-Lara, L.O., Medrano-Macías, J., Pérez-Labrada, F., Rivas-Martínez, E.N., García-Enciso, E.L., González-Morales, S., et al., 2019. From elemental sulfur to hydrogen sulfide in agricultural soils and plants. Molecules (Basel, Switzerland) 24 (12). Available at https://doi.org/10.3390/molecules24122282.

Gallino, J.P., Ruibal, C., Casaretto, E., Fleitas, A.L., Bonnecarrère, V., Borsani, O., et al., 2018. A dehydration-induced eukaryotic translation initiation factor iso4G identified in a slow wilting soybean cultivar enhances abiotic stress tolerance in *Arabidopsis*. Frontiers in Plant Science 9. Available at https://doi.org/10.3389/fpls.2018.00262.

Ghori, N.-H., Ghori, T., Hayat, M.Q., Imadi, S.R., Gul, A., Altay, V., et al., 2019. Heavy metal stress and responses in plants. International Journal of Environmental Science and Technology 16 (3), 1807–1828. Available at https://doi.org/10.1007/s13762-019-02215-8.

Hasanuzzaman, M., Nahar, K., Anee, T.I., Fujita, M., 2017. Glutathione in plants: biosynthesis and physiological role in environmental stress tolerance. Physiology and Molecular Biology of Plants 23 (2), 249–268. Available at https://doi.org/10.1007/s12298-017-0422-2.

Hasanuzzaman, M., Bhuyan, M.H.M.B., Mahmud, J.A., Nahar, K., Mohsin, S.M., Parvin, K., et al., 2018. Interaction of sulfur with phytohormones and signaling molecules in conferring abiotic stress tolerance to plants. Plant Signaling & Behavior 13 (5), e1477905. Available at https://doi.org/10.1080/15592324.2018.1477905.

Hesse, H., Kreft, O., Maimann, S., Zeh, M., Hoefgen, R., 2004. Current understanding of the regulation of methionine biosynthesis in plants. Journal of Experimental Botany 55 (404), 1799–1808. Available at https://doi.org/10.1093/jxb/erh139.

Husen, A., 2021a. Harsh Environment and Plant Resilience (Molecular and Functional Aspects). Springer Nature Switzerland AG, Cham https://doi.org/10.1007/978-3-030-65912-7.

Husen, A., 2021b. The harsh environment and resilient plants: an overview. In: Husen, A. (Ed.), Harsh Environment and Plant Resilience. Springer International Publishing AG, Cham, pp. 1–23 https://doi.org/10.1007/978-3-030-65912-7_1.

Husen, A., 2021c. Plant Performance Under Environmental Stress (Hormones, Biostimulants and Sustainable Plant Growth Management). Springer Nature Switzerland AG, Cham https://doi.org/10.1007/978-3-030-78521-5.

Husen, A., 2023. Plants and Their Interaction to Environmental Pollution (Damage Detection, Adaptation, Tolerance, Physiological and Molecular Responses). Elsevier Inc, Cambridge, MA https://doi.org/10.1016/C2020-0-03319-0.

Husen, A., 2024. Essential Minerals in Plant-Soil System (Coordination, Signalling, and Interaction under Adverse Situations. Elsevier Inc, Cambridge, MA https://doi.org/10.1016/C2022-0-02437-5.

Husen, A., 2025. Role of Antioxidants in Mitigating Plant Stress. Elsevier Inc, Cambridge, MA https://doi.org/10.1016/C2023-0-51268-6.

Ihsan, M.Z., Daur, I., Alghabari, F., Alzamanan, S., Rizwan, S., Ahmad, M., et al., 2019. Heat stress and plant development: role of sulphur metabolites and management strategies. Acta Agriculturae Scandinavica Section B: Soil and Plant Science 69 (4), 332–342. Available at https://doi.org/10.1080/09064710.2019.1569715.

Islam, S., Mohammad, F., 2020. Triacontanol as a dynamic growth regulator for plants under diverse environmental conditions. Physiology and Molecular Biology of Plants 26 (5), 871–883. Available at https://doi.org/10.1007/s12298-020-00815-0.

Jahan, B., AlAjmi, M.F., Rehman, M.T., Khan, N.A., 2020. Treatment of nitric oxide supplemented with nitrogen and sulfur regulates photosynthetic performance and stomatal behavior in mustard under salt stress. Physiologia Plantarum 168 (2), 490–510. Available at https://doi.org/10.1111/ppl.13056.

Jahan, B., Sehar, Z., Masood, A., Anjum, N.A., Khan, M.I.R., Khan, N.A., 2019. Sulfur availability potentiates phytohormones-mediated action in plants. Plant Signaling Molecules: Role and Regulation under Stressful Environments. Elsevier, India, pp. 287–301. Available at 10.1016/B978-0-12-816451-8.00017-4.

Jain, M., Nagar, P., Goel, P., Singh, A.K., Kumari, S., Mustafiz, A., 2018. Second messengers: central regulators in plant abiotic stress response abiotic stress-mediated sensing and. Signaling in Plants: An Omics Perspective. Springer, Singapore, pp. 47–94. Available at 10.1007/978-981-10-7479-0_2.

Jeschke, V., Weber, K., Moore, S.S., Burow, M., 2019. Coordination of glucosinolate biosynthesis and turnover under different nutrient conditions. Frontiers in Plant Science 10. Available at https://doi.org/10.3389/fpls.2019.01560.

Khan, N., Khan, M., Asgher, M., Fatma, M., Masood, A., Syeed, S., 2014. Salinity tolerance in plants: revisiting the role of sulfur metabolites. Journal of Plant Biochemistry Physiology 2 10.4172.

Kopriva, S., Malagoli, M., Takahashi, H., 2019. Sulfur nutrition: impacts on plant development, metabolism, and stress responses. Journal of Experimental Botany 70 (16), 4069–4073. Available at https://doi.org/10.1093/jxb/erz319.

Koprivova, A., Kopriva, S., 2014. Molecular mechanisms of regulation of sulfate assimilation: first steps on a long road. Frontiers in Plant Science 5. Available at https://doi.org/10.3389/fpls.2014.00589 October.

Künstler, A., Gullner, G., Ádám, A.L., Kolozsváriné Kolozsváriné Nagy, J., Király, L., 2020. The versatile roles of sulfur-containing biomolecules in plant defense—a road to disease resistance. Plants 9 (12), 1705. Available at https://doi.org/10.3390/plants9121705.

Li, Z.G., Li, X.E., Chen, H.Y., 2022. Sulfur Dioxide: an emerging signaling molecule in plants. Frontiers in Plant Science 13. Available at https://doi.org/10.3389/fpls.2022.891626.

Ma, L., 2021. Mechanism of sulfur dioxide enhancing cold stress adaptability of *Arabidopsis thaliana*. Master Thesis, Shanxi University, Taiyuan, pp. 1-72 (in Chinese).

Mehrabi, Z., Ramankutty, N., 2017. The cost of heat waves and droughts for global crop production. bioRxiv. Available at https://www.biorxiv.org. https://doi.org/10.1101/188151.

Mobin, M., Khan, M.N., Abbas, Z.K., Ansari, H.R., Al-Mutairi, K.A., 2017. Significance of sulfur in heat stressed cluster bean (*Cymopsis tetragonoloba* L. Taub) genotypes: responses of growth, sugar and antioxidative metabolism. Archives of Agronomy and Soil Science 63 (2), 288–295. Available at https://doi.org/10.1080/03650340.2016.1196811.

Nazar, R., Iqbal, N., Masood, A., Syeed, S., Khan, N.A., 2011. Understanding the significance of sulfur in improving salinity tolerance in plants. Environmental and Experimental Botany 70 (2-3), 80–87. Available at https://doi.org/10.1016/j.envexpbot.2010.09.011.

Pieterse, C.M.J., Leon-Reyes, A., Van Der Ent, S., Van Wees, S.C.M., 2009. Networking by small-molecule hormones in plant immunity. Nature Chemical Biology 5 (5), 308–316. Available at https://doi.org/10.1038/nchembio.164.

Queval, G., Jaillard, D., Zechmann, B., Noctor, G., 2011. Increased intracellular H_2O_2 availability preferentially drives glutathione accumulation in vacuoles and chloroplasts. Plant, Cell & Environment 34 (1), 21–32. Available at https://doi.org/10.1111/j.1365-3040.2010.02222.x.

Raina, M., Kisku, A.V., Joon, S., Kumar, S., Kumar, D., 2021. Calmodulin and calmodulin-like Ca^{2+} binding proteins as molecular players of abiotic stress response in plants. Calcium Transport Elements in Plants. Elsevier, India, pp. 231–248. Available at http://doi.org/10.1016/B978-0-12-821792-4.00001-1.

Rapala-Kozik, M., 2011. Vitamin B1 (thiamine): a cofactor for enzymes involved in the main metabolic pathways and an environmental stress protectant. Advances in Botanical Research. Elsevier.

Rasheed, F., Anjum, N.A., Masood, A., Sofo, A., Khan, N.A., 2022. The key roles of salicylic acid and sulfur in plant salinity stress tolerance. Journal of Plant Growth Regulation 41 (5), 1891–1904. Available at https://doi.org/10.1007/s00344-020-10257-3.

Raza, A., Salehi, H., Rahman, M.A., Zahid, Z., Madadkar Haghjou, M., Najafi-Kakavand, S., et al., 2022. Plant hormones and neurotransmitter interactions mediate antioxidant defenses under induced oxidative stress in plants. Frontiers in Plant Science 13. Available at https://doi.org/10.3389/fpls.2022.961872.

Ritonga, F.N., Chen, S., 2020. Physiological and molecular mechanism involved in cold stress tolerance in plants. Plants 9 (5), 560. Available at https://doi.org/10.3390/plants9050560.

Romero, L.C., Aroca, M.A., Laureano-Marín, A.M., Moreno, I., García, I., Gotor, C., 2014. Cysteine and cysteine-related signaling pathways in *Arabidopsis thaliana*. Molecular Plant 7 (2), 264–276. Available at https://doi.org/10.1093/mp/sst168.

Sabetta, W., Paradiso, A., Paciolla, C., de Pinto, M.C., 2017. Chemistry, biosynthesis, and antioxidative function of glutathione in plants. Glutathione in Plant Growth, Development, and Stress Tolerance. Springer International Publishing, Italy, pp. 1–27. Available at http://doi.org/10.1007/978-3-319-66682-2_1.

Saleem, M., Fariduddin, Q., Castroverde, C.D.M., 2021a. Salicylic acid: a key regulator of redox signalling and plant immunity. Plant Physiology and Biochemistry 168, 381–397. Available at https://doi.org/10.1016/j.plaphy.2021.10.011.

Saleem, M., Fariduddin, Q., Janda, T., 2021b. Multifaceted role of salicylic acid in combating cold stress in plants: a review. Journal of Plant Growth Regulation 40 (2), 464–485. Available at https://doi.org/10.1007/s00344-020-10152-x.

Salehi-Lisar, S.Y., Bakhshayeshan-Agdam, H., 2016. Drought stress in plants: causes, consequences, and tolerance. Drought Stress Tolerance in Plants, Vol 1: Physiology and

Biochemistry. Springer International Publishing, Iran, pp. 1–16. Available at 10.1007/978-3-319-28899-4_1.

Sehar, Z., Masood, A., Khan, N.A., 2019. Nitric oxide reverses glucose-mediated photosynthetic repression in wheat (*Triticum aestivum* L.) under salt stress. Environmental and Experimental Botany 161, 277–289. Available at https://doi.org/10.1016/j.envexpbot.2019.01.010.

Shafiee, A.S., Amini, D.M., Rezazadeh, A., Naji, A.M., 2020. Evaluation of sulfur and foliar application of Zn and Fe on yield and biochemical factors of cumin (Cuminum cyminum L.) under irrigation regimes. Journal of Herbmed Pharmacology 9(2):161-170 9 (2), 161–170. Available at 10.34172/jhp.2020.21.

Shah, S.H., Islam, S., Mohammad, F., 2022. Sulphur as a dynamic mineral element for plants: a review. Journal of Soil Science and Plant Nutrition 22 (2), 2118–2143. Available at https://doi.org/10.1007/s42729-022-00798-9.

Shanmugam, S., Kjaer, K.H., Ottosen, C.-O., Rosenqvist, E., Kumari Sharma, D., Wollenweber, B., 2013. The alleviating effect of elevated CO $_2$ on heat stress susceptibility of two wheat (*Triticum aestivum* L.) cultivars. Journal of Agronomy and Crop Science 199 (5), 340–350. Available at https://doi.org/10.1111/jac.12023.

Sharma, R.K., Cox, M.S., Oglesby, C., Dhillon, J.S., 2024. Revisiting the role of sulfur in crop production: a narrative review. Journal of Agriculture and Food Research 15. Available at https://doi.org/10.1016/j.jafr.2024.101013.

Sutar, R.K., Pujar, A.M., Kumar, B.A., Hebsur, N., 2017. Sulphur nutrition in maize-a critical review. International Journal of Pure & Applied Bioscience 5, 1582–1596.

Sønderby, I.E., Geu-Flores, F., Halkier, B.A., 2010. Biosynthesis of glucosinolates - gene discovery and beyond. Trends in Plant Science 15 (5), 283–290. Available at https://doi.org/10.1016/j.tplants.2010.02.005.

Wani, S.H., 2023. Mechanisms of abiotic stress responses and tolerance in plants: physiological, biochemical and molecular interventions, volume II. Frontiers Media SA.

Xiao, X., Li, W., Jin, M., Zhang, L., Qin, L., Geng, W., 2023. Responses and tolerance mechanisms of microalgae to heavy metal stress: a review. Marine Environmental Research 183, 105805. Available at https://doi.org/10.1016/j.marenvres.2022.105805.

Xue, M., Yi, H., 2018. Enhanced *Arabidopsis* disease resistance against *Botrytis cinerea* induced by sulfur dioxide. Ecotoxicology and Environmental Safety 147, 523–529. Available at https://doi.org/10.1016/j.ecoenv.2017.09.011.

Yusof, Z.N.B., 2019. Thiamine and its role in protection against stress in plants (enhancement in thiamine content for nutritional quality improvement). Nutritional Quality Improvement in Plants. Concepts and Strategies in Plant Sciences. Springer Science and Business Media LLC, Cham, pp. 177–186. Available at 10.1007/978-3-319-95354-0_7.

Yu, G., Ma, J., Jiang, P., Li, J., Gao, J., Qiao, et al., 2019. The mechanism of plant resistance to heavy metal. IOP Conference Series: Earth and Environmental Science. IOP Publishing, pp. 052004.

Zenda, T., Liu, S., Dong, A., Duan, H., 2021. Revisiting sulphur—the once neglected nutrient: it's roles in plant growth, metabolism, stress tolerance and crop production. Agriculture 11 (7), 626. Available at https://doi.org/10.3390/agriculture11070626.

Zhang, Y., Xu, J., Li, R., Ge, Y., Li, Y., Li, R., 2023. Plants' response to abiotic stress: mechanisms and strategies. International Journal of Molecular Sciences 24 (13), 10915. Available at https://doi.org/10.3390/ijms241310915.

CHAPTER

Heat stress: role of potassium and calcium

9

Shrejal Tiwari[1], Anant[2], Ronanki Srinidhi[1], Pankaj Sharma[3,4] and Azamal Husen[5,6]

[1]*Department of Floriculture and Landscaping, College of Horticulture and Forestry, Rani Lakshmi Bai Central Agricultural University, Jhansi, Uttar Pradesh, India,*
[2]*Department of Plant Breeding and Genetics, College of Agriculture, Rani Lakshmi Bai Central Agricultural University, Jhansi, Uttar Pradesh, India,*
[3]*CCS Haryana Agricultural University, Hisar, Haryana, India,*
[4]*University of Illinois, Chicago, IL, United States,*
[5]*Wolaita Sodo University, Wolaita Sodo, South Ethiopia Regional State, Ethiopia,*
[6]*Department of Biotechnology, Graphic Era (Deemed to be University), Dehradun, Uttarakhand, India*

Abbreviations

Ca^{2+}	Calcium
K^+	Potassium
ROS	Reactive oxygen species
SOD	Superoxide dismutase
GPX	Glutathione reductase
POX	peroxidase
GP	Guaiacol peroxidase
APX	Ascorbate peroxidase
CAT	Catalase
GST	glutathione S transferase
O_2^-	Superoxide anions
H_2O_2	Hydrogen peroxide
C	Carbon
N	Nitrogen
Fe	Iron
B	Boron
Mg^{2+}	Magnesium
Mn^{2+}	Manganese
NO_3^-	Nitrate
NH_4^+	Ammonium
LATS	low-affinity transport system
HATS	high-affinity transport system
NRT2	Nitrate transporter 2

Mineral Nutrients to Mitigate Abiotic Stress in Crops. DOI: https://doi.org/10.1016/B978-0-443-44712-9.00008-6

AMTs	Ammonium transporter genes
HACCs	Hyperpolarization- activated Ca^{2+}channels
DACCS	Depolarization- activated Ca^{2+} channels
VICCs	Voltage–independent Ca channels
CML	CaM-like protein
CaM	Calmodulin
CAMTAs	Calmodulin-binding transcription activators
GST	glutathione S transferase
MDHAR	Mono de-hydro ascorbate reductase
CIPKs	CBL-interacting protein kinases
SA	salicylic acid
ABA	abscisic acid
PAs	polyamines
JA	jasmonic acid
ETH	ethylene
WUE	water use efficiency
KIN1	kinase
gs	stomatal conductance
E	transpiration rate
Pn	Chl a fluorescence, and photosynthetic rate
CO_2	Carbon dioxide
ELISA	Enzyme-linked immunosorbent assay
NRT1	nitrate transporter 1
LATS	low-affinity transport system
NRT2	nitrate transporter 2
AMTs	ammonium transporter genes
PEST	Polyproline-threonine-glutamic-acid-serine
AQ	Anthraquinone
OST1	Open stomata 1 kinase

9.1 Introduction

Heat stress is known to be one of the most intense abiotic factors restricting crop productivity as well as threatening global food security (Tovar et al., 2022). This stress occurs when temperatures surpass the optimal physiological range of crop plants. This resulted in a series of stress events leading to disrupted cellular homeostasis (Sawar et al., 2019). High temperatures cause a sudden increase in evapotranspiration rates due to high osmotic potential. This results in unbalanced ion homeostasis, enhanced reactive oxygen species (ROS) accumulation, and increased ionic imbalances. Together, these affect the major biochemical events (AbdElgawad et al., 2016; Zhang et al., 2022). The stability of the cell membrane is disrupted by lipid peroxides. The light reactions as well as the assimilation of carbon during the photosynthesis process are negatively affected. The activity of major enzymes is reduced. The major metabolic pathways essential for plant growth are hindered (Zhang et al., 2022; Shaffique et al., 2025). In addition, the effect of high stress temperatures disrupts the mitochondrial respiration and energy metabolism, hormone regulation, and the uptake and transport of essential mineral nutrients, thereby contributing to the diminished performance and efficacy of plants (AbdElgawad et al., 2016; Sawar et al., 2019; Shaffique et al., 2025). This integrated and collective consequence of the above-shown differences is reflected as diminished biomass, flower, and fruit abortion, and the diminished quality and quantity of grains in different plants like cereals, legumes, and other horticulturals. This recent body of research also points out that the extended exposure to temperatures above the optimum level together accelerates the onset of senescence and disrupts the protein structures, thereby contributing to stress vulnerability at the cellular and whole-plant level (Zhang et al., 2022; Shaffique et al., 2025). In general, various plants detect abiotic stress via specific sensing mechanisms, which activate intracellular signaling cascades resulting in the production of a variety of ROS and the induction of antioxidant defense systems that reduce stress-mediated damage. Such signaling pathways activate various protective reactions that involve the induction of stress-responsive enzymes and metabolites, all of which together contribute to minimizing cellular damage and permit plants to continue growing and reproducing under poor conditions. While these adaptive reactions provide some protection, they are often inadequate to surmount the negative effects of prolonged or intense abiotic stress entirely. Fahad et al. (2017) strongly supported the importance of comprehending the physiological and biochemical processes that enable plants to resist abiotic stresses like heat stress, high light, and water deficit, owing to the phenomenon of frequent occurrence of high temperatures and environmentally stressful conditions. Tiwari et al. (2023) supported the opinion of experts like Fahad et al. (2017) and mentioned the same aspect, adding that the community of agronomists and plant biologists has been interested in understanding stress protection strategies in plants and, in this regard, trying to develop feasible means to counter stress effects introduced in plants through adverse environmental conditions.

An extensive amount of research work has already proven that calcium and potassium ions, especially when added exogenously, are crucially involved in controlling plant growth, development, and functional efficiency. These macro-mineral nutrients have been found to make a significant contribution to developing plant resistance by conferring a capability to counter adverse environmental conditions (Poovaiah and Reddy, 1993; Sanders et al., 2002, Harper et al., 2004; Reddy and Reddy, 2004; Bouché et al., 2005; Hepler, 2005; Shabala and Pottosin, 2010; Marschner, 2011; Waraich et al., 2012; Wang and Wu, 2013; Tripathi et al., 2014). While calcium acts as a biomolecular structural and signaling molecule that stabilizes cell membranes and regulates stress-induced signal transduction, potassium is essentially involved in functions such as osmoregulation, activation of enzymes, and stomatal regulation, which are of utmost importance during stressful conditions to maintain the balance of plant functions. Among the various techniques that plants utilize to increase their thermotolerance, the regulation and utilization of minerals, specifically the ions Ca^{2+} and K^+, have been established as an important index that contributes to the survival and productivity of plants under high temperature stress. Recently, researchers suggested that the optimal use of minerals has an imperative contribution to the elevated tolerance of plants to high temperature stress through the maintenance of high photochemical efficiency and the regulation of antioxidant defense systems to navigate the high temperature stress-induced injury (Cakmak, 2005; Tiwari et al., 2023).

9.1.1 Heat stress challenges in modern agriculture

Heat stress has a profound effect on agricultural productivity for various ranges of crops and agroclimatic zones. Rising heat waves affect not only warm-season crops like cotton, rice, and wheat but also cool-season crops such as potato, which grow in limited fashion when summer temperatures consistently exceed 40°C. Heat stress impacts various plant parts, leading to scorching, sunburn symptoms on leaves, scorching on twigs, premature senescence and abscission on leaves, and marked inhibition of shoot and root growth. Heat stress can result in reduced plant vigor and, ultimately, diminished agricultural productivity (Waraich et al., 2012). Crop-specific reactions emphasize even more the severity of the effects caused by heat stress. In melons, above-optimal temperature conditions prevent seed germination and the establishment of young seedlings, reduce female flower production, affect fertilization, and, accordingly, hamper fruit setting and sugar accumulation. In sugarcane, high temperature stress severely limits internode growth, especially the first internode, and could cause early plant death. Sugarcane genotypes grown under continuous high-temperature stress tend to produce short internodes, a high number of tillers, early senescence, and reduced total biomass production, thus confirming that such a crop is highly responsive to high-temperature environments (Suzuki and Masuda, 1961). Thermosensitivity is very high during the reproductive phase in rice, with anthesis and fertilization, along with microsporogenesis at the booting stage, being the most sensitive developmental stages. Prasad et al. (2023) showed that spikelet sterility in rice increases significantly with heat stress due to impaired anther dehiscence, reduced pollen shedding, poor viability and germination of pollen on the stigma, and restricted elongation of the pollen tube, each factor collectively impeding successful fertilization. These reproductive failures, therefore almost directly relate to heavy yield losses under conditions of high temperatures. At the cellular and molecular levels, the effect of high temperatures involves protein denaturation, destabilization of nucleic acids, and impairment of the function of critical enzymes associated with photosynthesis, respiration, and key biosynthetic pathways. Additionally, heat shock stimulates the production of reactive oxygen species, such as superoxide anions and hydrogen peroxide, that cause oxidative damage to cellular membranes, proteins, and genomic DNA (Anderson et al., 2020). Even though plants are always producing ROS due to normal metabolic events, which proceed in chloroplasts, mitochondria, and peroxisomes, excessive ROS accumulation under heat shock conditions overwhelms the antioxidant defense systems and extends cellular injury (Del Río et al., 2006; Navrot et al., 2007). The superoxide radicals undergo a cascade of chemical reactions and are quickly converted into other active oxygen species, hydrogen peroxide, and hydroxyl radicals. These ROS include superoxide, hydrogen peroxide, singlet oxygen, hydroperoxyl radicals, hydroxyl radicals, and a variety of organic peroxides and radicals, which are highly reactive and may cause serious damage to cellular structure. They readily attack integral biomolecules such as proteins, lipids, carbohydrates, and nucleic acids, which leads to oxidative damage, membrane lipid peroxidation, enzyme inactivation, genetic instability, and, if not efficiently detoxified, the death of cells. When plants experience high temperatures and other stresses, ROS build up and can severely damage plant cells. This accumulation is a key reason for reduced crop yields globally, since high ROS levels during stress events contribute significantly to the loss of productivity (Mittler, 2006; Apel and Hirt, 2004; Tuteja and Mahajan, 2007; Tuteja et al., 2010; Khan et al., 2017) as shown in Fig. 9.1

The impact of heat stress on agriculture is very drastic, with the reproductive phase being highly sensitive to increased temperatures. Heat stress causes a drastic impairment of the reproductive organs by promoting pollen sterility, ovule malfunction, and reduced seed formation, ultimately leading to reduced productivity

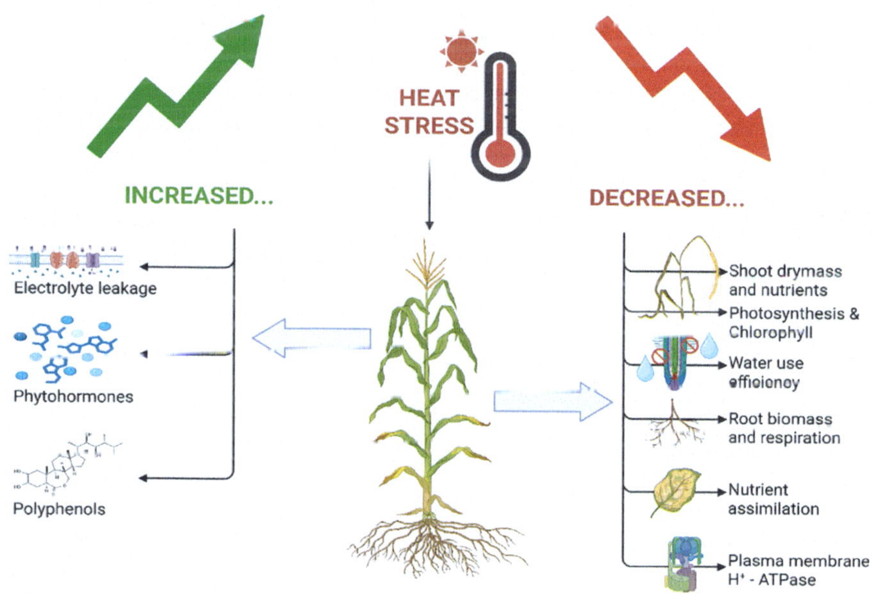

FIGURE 9.1

Effect of heat stress on the plant system.

and plant yield. Even brief periods of increased temperatures during the key developmental stages like flowering or anthesis bring about irreparable changes to plant morphometrics, physiological processes, and seed composition. Moreover, plant heat stress is often simultaneous with other abiotic stresses like drought, bringing about an accumulation of physiological limitations that bring about increased cell damage.

9.2 Heat stress decreases plant growth by affecting root mass and photosynthesis

The response of aerial and subterranean plant organs to heat stress depends largely on experimental exposure conditions, which vary significantly across studies. Generally, root growth is more sensitive to high temperatures than shoot growth due to the root system's lower optimal temperature range, while a decrease in root mass may cause a drop in shoot mass. For instance, in the aeroponic culture of salad rocket plants (*Eruca sativa*), a 5°C increase in root-zone temperature from 20°C to 25°C reduced shoot and root fresh weight with decreased net photosynthesis. High soil temperature is more detrimental to root and shoot growth than high air temperature; it is the exposure of just the roots to high temperature that can inhibit shoot growth. This has been confirmed by previous studies on noncrop species as well, where differential temperatures were imposed on roots and shoots with exposure to different air and soil temperatures. After six days of acute heat stress, tomato plants recapitulated these findings. Severe heat stress (42°C) reduced both root and shoot biomass as well as the root:shoot ratio. A higher temperature would reduce the root:shoot ratio by restricting root development and changing the architecture of the root system. Remarkably, a number of studies have verified that heat stress has a detrimental effect on both root architecture and root mass, ultimately affecting the ratio of lateral roots to root biomass. Consequently, these alterations in root architecture also have an adverse effect on the absorption of nutrients and water. Photosynthesis is particularly vulnerable to high temperatures. Heat stress impairs the photosynthetic apparatus by denaturing chloroplast proteins and uncoupling the electron transport chain, leading to reduced carbon assimilation. Notably, the extent of this damage correlates directly with the intensity of the stress. Surprisingly, net photosynthesis rose in tomato plants subjected to 6 days of moderate heat stress (35°C); however, it fell under severe heat stress (42°C) due to reduced stomatal conductance. Moderate or severe heat stress damages the photosynthetic enzymes and proteins, which primarily affect photochemical reactions in the thylakoid lamellae and carbon metabolism in the stroma of chloroplasts.

The photosynthetic apparatus, photosystems, pigments, photosynthesis-related enzyme activities, and gas exchange could all be directly harmed by heat stress, which would inhibit different redox and metabolic reactions as well as the entire process of photosynthesis. Numerous studies on the effects of heat stress on photosynthesis, whether short-term or long-term, describe alterations in the structure or anatomy of leaves as well as modifications in the way the photosynthetic apparatus functions. For instance, Shen et al. (2017) examined the effects of heat stress on photosynthesis by comparing the leaf anatomy of several cultivars of Glenn Dale Azalea (Rhododendron hybridum) grown at 38°C/30°C (heat stress) and 25°C/17°C (control) for six days. In Rhododendron cultivar Liu Qiu Hong, heat stress dramatically increased stomatal length (54.3%), in cultivar "Lan Yin," only a 10.8% increase was observed. These heat stress effects on photosynthesis were confirmed by the reduction in size and percentage of stomatal aperture, which slowed transpirational water loss. Although moderate heat stress suppresses photosynthetic rates, it causes less permanent damage to PSII compared to severe stress. However, any reduction in photosynthesis cascades into nutrient acquisition issues. Photosynthesis provides the carbon skeletons necessary for nitrogen uptake and assimilation. Therefore heat stress limits plant growth not only by reducing carbon fixation but also by uncoupling the regulation of nutrient homeostasis.

9.2.1 Effects of heat stress on nutrient content in plants

Development of a root system and sufficient mineral nutrient availability are fundamentally associated with plant growth and development. Plant nutrient uptake is a complex process influenced by several variables such as rhizospheric environment, root system, and nutrient transport rates. When different plant organs such as the entire plant (root + shoot) or just the roots are exposed to heat stress, the effects on nutrient concentration change. In *Chenopodium*, heat stress imposed during the critical anthesis stage (11 days at 30°C–35°C) led to a marked decline in the accumulation of As, Cd, Rb, and Sr within the main panicles. Remarkably, heat stress treatment of *Chenopodium* roots resulted in a 30% increase in element concentration, while heat stress exposure of the entire plant reduced seed nutrient concentration by 12%, which had an impact on yield. In tomato, acute and chronic heat stress lowered root Fe, N, and C levels, while acute stress (42°C) also reduced shoot N. Conversely, moderate stress (35°C) increased boron uptake. Such nutrient responses vary significantly by species, element, and stress protocol. For example, 37°C exposure reduced Fe and Zn in *Lens culinaris*, whereas heat stress during grain filling had opposing effects on wheat genotypes: reducing tissue Fe in bread wheat while increasing it in durum wheat. Table 9.1 gives an outline regarding the effects of heat stress on various crops.

9.2.2 Reduced nutrient uptake during heat stress due to damage to uptake proteins

Two important factors in root nutrient uptake are nutrient demand and nutrient use efficiency. The ability of crops to absorb and use nutrients for maximum yields is

Table 9.1 Effects of calcium and potassium on plant responses to heat stress.

Type of Ca and K used	Method of application	Concentration	Crops	Plant response against heat stress	Key reference
$CaCl_2$	Foliar spraying	20 mM	*Phalaenopsis aphrodite*	Increased antioxidant enzymatic (SOD, POD, CAT, proline) to combat ROS spp. Combating heat stress	Yang and Yang (2021)
$CaCl_2$	Foliar spraying	5 mM	*Glycine max*	Reduced lipid peroxidation while increasing enzymatic activities and sugar accumulation	Naeem et al. (2019)
$CaCl_2$	Foliar spraying	20 mM	*Nicotiana tabacum*	Improved stomatal conductance	Tan et al. (2011)
$CaCl_2$	Foliar spraying	20 mM	*Camellia sinensis*	Upregulation of 299 DEG, while downregulating almost 624 genes related to translation and signal transduction	Wang et al. (2019)
$CaCl_2$	Foliar spraying	100 μM	*Rosa rugosa*	Regulating SOD, POD, and photosynthesis	Wang et al. (2022)
$CaCl_2$	Foliar spraying	10 mM	*Triticum aestivum*	Reduced MDA content and increased in the antioxidant enzymatic system of plant	Goswami et al. (2015)
$CaCl_2$ and KCl	Foliar spraying	250, 500 mg/l	*Pithecellobium dulce*	Increased total chrolophyll and carotenoid plus reduced electrolyte leakage and ABA production	Abdallah et al. (2019)
$CaCl_2$ and $CaNO_3$	Foliar spraying	1 mM	*Solanum lycopersicum* cv. 'Luktho	Improved CO_2 assimilation rate than control and better functioning of Photosystem II	Sakhonwasee et al. (2017)
$CaCl_2$	Foliar spraying	10 mM	*Rhododendron* × *pulchrum*	Regulating leaf anatomical structure (epidermis thickness, stomatal aperture) protects photosystems I and II	Shen et al. (2022)
$CaCl_2$	Foliar spraying	15 mM	*Dianthus caryophyllus*	Upregulation of photosynthesis-related pathways, including PSII proteins	Zhang et al. (2022)

K_2SO_4	Foliar spraying	2%	*Triticum aestivum* cv Anaj-17	Improved chlorophyll content, reduced cellular damage, and improved SOD, POD, CAT production	Sarwar et al. (2022)
KNO_3	Foliar spraying	15–60 g/L	*Triticum aestivum*	Improved chlorophyll content and morphological attributes	Shahid et al. (2019), Du et al. (2019)
KNO_3	Foliar spraying	1 %	*Triticum aestivum*	Boost in yield	Singh et al. (2025)
K_2SO_4	Injecting into trunk	48%	*Phoenix dactylifera*	Imporved absorption of mineral nutrient and better fruit quality	Elsayd et al. (2018)
$CaCl_2$	Mircoplants priming	40 mM	*Fragaria* x *ananassa*	Superior morphology (less leaf damage rate) and higher total chlorophyll content, lower H_2O_2 levels	Wang et al. (2022)

known as nutrient use efficiency. It involves three main processes in plants: nutrient uptake, assimilation, and utilization. The impact of temperature on plant–nutrient interactions in terms of nutrient uptake in a changing environment, however, is comparatively poorly understood. The activity of nutrient-uptake proteins in plants mediates the uptake of the majority of mineral elements. The concentration of uptake proteins per unit root and the rate of protein function determine the activities of nutrient-uptake proteins. For instance, plants use three distinct transporter proteins to absorb nitrogen as either ammonium (NH_4^+) or nitrate (NO_3^-). The low-affinity transport system (LATS) and the high-affinity transport system (HATS), two kinetically different nitrate uptake systems, regulate the root's uptake of nitrate (NO_3^-). The nitrate transporter 1 (NRT1) family encodes the low-affinity transport system (LATS), while the nitrate transporter 2 (NRT2) family encodes HAT. Similar to this, plants only use ammonium transporter genes (AMTs) to absorb ammonium.

Plant uptake of phosphorus has been extensively studied, leading to the identification of five Pi transporter families (PHT1–5) that mediate uptake under P-limiting conditions. For instance, the primary proteins facilitating Iron (Fe) uptake are FRO1–7 and IRT1. Likewise, under Boron (B) deficiency, uptake is predominantly managed by the BOR1 and NIP5;1 transporter. Zinc regulation also involves transporter modulation; plants that received foliar zinc treatment expressed more Zn-uptake proteins (ZIPs). Similarly, nitrate transporters show distinct localization patterns. For example, NRT1–5 is located on the plasma membrane of the root pericycle near the xylem, facilitating long-distance transport of nitrate from the root to shoot, while AtNRT1-1 is expressed in epidermis, cortex, and endodermis of mature roots and scavenges nitrate under low nitrogen availability. This precise localization highlights the strict regulation governing macronutrient and micronutrient uptake in response to environmental cues. A similar mechanism has been reported for the uptake of the micronutrient boron. In B-deficient environments, boric acid channels and borate exporters [NIP5;1 and BOR1, respectively] allow for boron uptake and translocation throughout the entire plant. Once it reaches a sufficient level, it triggers the downregulation of BOR1 and NIP5;1 through proteolysis via endocytosis and mRNA degradation, respectively.

Remarkably, the majority of research on plant nutrient uptake was limited to altering nutrient concentrations to levels of toxicity or deficiency, while the state of nutrient uptake proteins under abiotic stress is subjected to very few studies. Protein-specific antibodies were used in Western blots or ELISA (enzyme-linked immunosorbent assay) to measure the relative amounts of the major nutrient-uptake proteins (NRT1, NRT2, AMT1, PHT1, BOR1, NIP5); (1 and FRO1) per unit total root protein. In order to target conserved domains of specific nutrient-uptake proteins across species, polyclonal antibodies were created. Only one of the seven aforementioned studies found that heat stress alone altered nutrient-uptake protein. The effects of high temperatures on proteins involved in the uptake of particular nutrients may be the cause of changes in the concentration of nutrients in plants.

To meet the global food demands, crop production must increase substantially including nutritional management, optimizing the status of key macronutrients like calcium and potassium, which represents a practical and cost-effective approach to improving plant heat tolerance. Understanding the specific mechanisms through which these minerals enhance thermotolerance is therefore critical for developing targeted crop management strategies.

9.3 Calcium and potassium in thermotolerance

9.3.1 Signaling of Ca^{2+} under abiotic stress

Calcium is the most essential secondary messenger in most pathways activated from stress signals like drought, salinity, cold, heat, and heavy metals (Singh, 2020; Verma et al., 2024). The calcium signaling machineries in stressed plants are not yet fully understood. For instance, osmotic adjustment involves the induction of changes in many metabolic pathways that respond to water, drought conditions, hyperosmotic stress, high salt levels, and low temperatures (Singh, 2020). Along with these, the expression of many genes that together lead to the assembly of several proteins actively participating in signal transduction pathways through which plants resist the stress condition.

9.3.1.1 *Channels to transport Ca in cell*

The three main types of Ca channels for transport that have been discovered and classified in plants are: (1) Hyperpolarization-activated Ca^{2+} channels (HACCs). (2) Ca^{2+} channels activated by depolarization (DACCS). (3) Voltage-independent calcium channels (VICCs) (Demidchik et al., 2018). Furthermore, stress-induced ROS are required to control and regulate calcium discharge through the plasma membrane. Calcium plays an important role in osmotic stress, and it is also evident that mechanosensitive Ca^{2+} channels responding to mechanical stimuli, such as MSL2 and MSL3, perform a protective function of osmotic homeostasis (Zhang et al., 2016). The gene MCA limits Ca^{2+} concentrations under hypoosmotic stress; upregulation of MCA would permit increased Ca accumulation in the cell that modulates ROS levels within. Another gene reported to be involved in osmo-sensing is OSCA1. The expression of OSCA1 increased after osmotic stress, causing increases in cytosolic Ca^{2+} levels in Arabidopsis. Loss of function in the OSCA1 gene impaired Ca^{2+} transport and downregulated transpiration mechanism (Yuan et al., 2017).

9.3.1.2 *Calcium sensing protein*

Abiotic stress raises calcium concentrations in the cytosol, including salts, heat, cold, and drought. Elevated cystolic calcium is interpreted by sensor proteins, triggering a signal transduction cascade, while sequestration into vacuoles and organelles restores homeostasis, which works in conjunction with downstream effector proteins that interpret the perceived signals. Calcium sensor proteins in plants include calmodulin (CaM) and calmodulin-like protein (CML), calcineurin-B-like (CBL), calcium-dependent protein kinases (CDPKs, known as CPKs in Arabidopsis) (Tong et al., 2021; Raina et al., 2021). These operate as a complex relay protein network that transfers the stress signal to induce various responses in plants under abiotic stressors (Mohanta et al., 2018). CaMs/CMLs and CBLs presumably serve as sensor relays, regulating targets downstream. CaMs influence transcription by regulating calmodulin-binding transcription activators (CAMTAs), whereas CBLs interact with CIPKs (serine/threonine protein kinases) to regulate their activity (Virdi et al., 2015).

9.3.1.3 *Calmodulin and calmodulin-like protein*

The major calcium-sensing protein, CaM, consists of 149 amino acids (Iqbal et al., 2020). The 29 amino-acid section referred to as the EF hand is also a name for this helix-loop-helix motif. CaM possesses a total of four EF hands, where the E helix is from the first 1–10 amino acids and the F helix is from 19 to 29 amino acids, while the core region (amino acids 10–21) of the EF-hand domain constitutes the core region of the domain binding to Ca^{2+} (Iqbal et al., 2020; Ranjan et al., 2025). For activity, CaMs and CML proteins recruit other active kinases. Regulation of cellular response was assessed by Thoday-Kennedy et al. (2015) involving the potential of a protein that interacts with the calcineurin-B. CaMs have a long, flexible central helix that separates two globular lobes at each end. CaM-Ca^{2+} complexes activate a plethora of transcription factors (TFs) that have roles in various molecular, physiological, and biochemical processes in the cell. CAMTA transcription factors are the ones that are calcium-induced and calcium-modulated. This category shows a narrower specification and includes TFs active in calcium-induced signaling that are evidently very highly conserved. Many studies have underlined the involvement of Arabidopsis CAMTA3, also known as SR1, which controls a number of biotic and abiotic stress response processes. Members of CAMTA transcription factors significantly affect the plant development in response to both biotic and abiotic stresses. These transcription factors might be differentially expressed against a variety of stimuli across plant species. Based on their promoter analysis, PeCAMTAs exhibit many cis-acting regions probably connected to drought stress, while experiments in drought stress response showed a similarly high expression of CAMTA. TaCAMTA was found to predominantly influence drought stress responses at the seedling stage, with TaCAMTA1b-B being critical in regulating responses to water deficit during the nursery stage (Wang et al., 2022). According to Meenakshi et al. (2022), the antioxidant enzymes APX, CAT, GST, SOD, and mono de-hydro ascorbate reductase (MDHAR) showed increased activity in the CAMTA-overexpressing lines. Lowering the stress markers TBARS and H_2O_2 translated to enhanced survival of Cicer arietinum against drought- and salinity-induced shocks.

9.3.1.4 *Calcineurin-B-like*

Plant cells have two known intracellular Ca^{2+} sensor proteins, CaM1 and CBL. Phosphorylation cascades could not be initiated by them since, lacking a kinase domain, CaM/CML/CBL must interact with specific protein kinases that phosphorylate substrate proteins, affecting gene expression downstream. CBL interacts with the CBL-interacting protein kinase (CIPK) family of serine/threonine protein kinases (Tang et al., 2020; Xiao et al., 2022). According to La Verde et al. (2018) , CML proteins containing EF-hand domains are unique to plants and some protists. Each of the EF-hand motifs has Ca^{2+} binding sites at residues 1 (X), 3 (Y), 5 (Z), 7 (Y), 9 (X), and 12 (Z) (Su et al., 2020). A rise in cytosolic Ca^{2+} in response to hormone signals such as ABA likely causes CBLs to undergo conformational changes in order to recruit the CBL-CIPK module, and this again activates RBOHF-CIPK pathway activity to enhance ROS generation, therefore stomatal closure under dehydration caused by several abiotic stresses, including salt. Leading to activation of SOS3, and SOS2 (CIPK24) or CBL4, which form the physiological basis of salt tolerance pathway, or SOS pathway t. Relationship of SOS2 and NDPK2 (nucleotide diphosphate kinase 2) in salt-stressed plants regulates CAT2 and CAT3 activities, lowers ROS formation, and confers tolerance against salt stress.

CBL and CIPK complexes act as mediators of calcium intracellular signals for downstream defense gene modulation under different stimuli. CBLs bind calcium ions and interact with CIPKs through a highly conserved core structure consisting of four elongation factor hand (EF-hand) domains (Ma et al., 2023a, 2023b; Xiao et al., 2022). In addition, CBLs would not be phosphorylated by CIPKs unless a PFPF motif at the C-terminus is present (Ma et al., 2023a, 2023b). CIPK proteins are serine/threonine kinases of the SnRK3 subfamily, which have a regulatory domain at the C-terminus that inhibits them. The regulatory domain at its C-terminus comprises a unique NAF/FISL motif and a protein–phosphatase interaction motif termed PPI, while the N-terminal catalytic kinase domain contains an ATP-binding site and also an activation loop. The conserved NAF/FISL motif interacts with the Ca^{2+} -bound CBLs for the activation of the calcium-dependent target CIPKs. Also, in some CIPKs, the PPI motif is utilized to target certain members of PP2 (Arab et al., 2023) studied 20 AlCIPK genes and 6 AlCBL genes, and they concluded that the relationship between AlCBL genes and AlCIPK is tissue-specific, and different relationships are likely to occur in root and leaf tissues.

These relationships suggest that AlCIPK3.1-AlCBL4.1, AlCIPK1.2-AlCBL4.4, and AlCBL10 may coexpress in the root and leaf tissues. AlCBL10 may also coexpress with AlCIPK5, AlCIPK26, and AlCIPK12.3 in root tissue. A total of 8 CBLs and 17n CIPKs from cold-tolerant satsuma mandarin were studied by Xiao et al. (2022). They identified that cold and Xcc infection signals were transduced through CBL6-CIPK8, and CBL6/CBL8-CIPK14 was only cold sensitive in citrus. Transgenic Arabidopsis overexpressing CuCIPK16 exhibited increased cold tolerance, associated with elevated POD activity and diminished level of MDA.

Multiple Ca^{2+} -binding proteins function as the primary information transducers recognizing cellular changes in Ca^{2+} concentration.

9.3.1.5 Ca^{2+}-dependent protein kinases

According to Mohanta et al. (2018) and Atif et al. (2019), CPKs are considered versatile regulators of plants toward abiotic stress. In 1984, the first plant CPKs were found in Pisum sativum (Hetherington and Trewavas, 1984). CPKs play major roles in plant growth, development, and signal transduction. Structurally, there are four functional domains that constitute CDPKs or CPKs: protein kinase domain, variable N-terminal domain (VNTD), C-terminal regulatory calmodulin-like domain (CBD), and AJ (Singh and Reddy, 2017; Atif et al., 2019). The VNTD often contains myristoylation and palmitoylation sites that facilitate membrane association and are upstream of polyproline-threonine-glutamic-acid-serine (PEST) sequences, targeting protein for rapid proteolytic degradation. The CBD on the other hand are four loops of EF-hand which each has 20 amino acids that directly bind to calcium ions . Upon calcium binding to these hands, the autoinhibitory junction (AJ)is released from kinase active sites, thereby relieving inhibition and allowing CPK substrate to phosphorylate (Yip Delormel and Boudsocq, 2019; Nehra et al., 2022).

Most proteomic studies indicated that intercellular activity of CPKs is calcium-dependent auto- phosphorylation of serine and threonine residues. There are two basic categories of CDPKs: inactive and active. During the inactive state, AJ stays bound at the active site of the PKD domain and stops the kinase from working, preventing substrate phosphorylation. However, when Ca^{+2} ions attach themselves to EF lobes of CBD force AJ to release from the active site of the PKD domain and thereby activating the kinase (Klimecka and Muszyńska, 2007). While the recent research has investigated the involvement of the VNTD in substrate specificity (Asai et al., 2013). Functional studies of CPK highlight diverse roles, according to Veremeichik et al. (2019), overexpression of AtCPK1 in soybean cells leads to accumulation of phytoalexins. Similarly, heterologous expression of AtCPK1 in cells of Rubia cordifolia has been shown to enhance anthraquinone (AQ) production (Shkryl et al., 2011). In some cases, CDPKs such as HbCPK9/15 were associated with latex biosynthesis, while VaCPK20 and VaCPK29 increased the content of resveratrol in cells of Vitisa murensis (Xiao et al., 2017). CsCDPK20 and CsCDPK26 in the tea plant stimulate heat-stress-responsive genes and manage proper heat-stress signaling (Wang et al., 2019). Furthermore, some of these CDKs, AtCPK10/AtCPK30/AtCPK32, are known to phosphorylate NIN-LIKEPROTEIN transcription factors to regulate transcriptional changes in response to the availability of nitrate (Dekomah et al., 2022). Triple mutants of these genes exhibit impaired cotyledon greening and expansion and reduced expression of nitrate-responsive genes (Liu and Tang, 2023). Additionally, CPKs participate in a range of metabolic and developmental processes, including ion transport (example CPK13), stem elongation and secondary growth (AtCPK28), pollen tube growth, latex biosynthesis, higher biomass accumulation, wounding and herbivory attack, germination and seedling growth, and early maturity (Xiao et al., 2017; Aleynova et al., 2017).

9.3.1.6 Induction of stress-related genes

Several marker genes including RD29A, RD29B, KIN1/2, RD22, DREB1A, and DREB2A acts as key indicator of stresses (drought, salinity, heat, and cold), whereas RD29B was primarily stimulated by osmotic stress. Ndathe et al. (2022). KIN (cold-inducible) genes respond to drought, cold, and salt stress (Tahtiharju et al., 1997), while DREB1A (or CBF3) is induced by cold, and DREB2B is induced by osmotic stress (Liu et al., 1998). Since DREB genes are expressed earlier than RD/KIN genes, they serve as markers for early-stage stress responses, while RD/KIN genes indicate late-stage responses.

In a plant cell, the resting cytosolic levels of Ca range between 100 and 200 nmol (Jezek and Blatt, 2017). Under stress, cytosolic Ca^{2+} levels rise transiently. Because high cytosolic calcium can be cytotoxic (due to negative charge interactions), calcium homeostasis is strictly regulated via transport systems that pump Ca^{2+} back into organelles or the apoplast (Demidchik et al., 2018).

9.3.1.7 Role of stress hormones with calcium to the plant under abiotic stress

Calcium deficiency is rare in the soil but might develop under soils whose pH is strongly acidic (Mukherjee, 2022). Where soils are endowed with adequate amount of Ca^{2+} in their habitat, shoot generally contains about 0.15% of its dry weight in Ca^{2+} concentration (Marschner, 2011) Ca^{2+} also plays the role of a messenger in higher plants interacting with plant hormones - salicylic acid (SA), abscisic acid (ABA), polyamines (PAs), jasmonic acid (JA), and ethylene (ETH) in combating the deleterious effects of abiotic stress. According to the reports, Ca^{2+} signaling is particularly prominent in stomatal regulation. ABA enhances cytosolic calcium levels in guard cells via Ca^{2+} influx across the plasma membrane triggering the signaling cascade in guard cells; this induces stomatal closure lowering water loss, which may impact plants' water use efficiency (WUE). Similarly, the signaling pathways of CDPK dependence are activated by JA to cause stomatal closure (Shi et al., 2018; Kolupaev et al., 2023).

Transcriptomics studies have revealed that CaM regulates CAMTA, which is made up of a group of transcription factors that include members of bZIP, MYB, WRKY, and NAC families that regulate different types of plant defense reactions to stress conditions (Popescu et al., 2007). Besides this, Ca^{2+} is necessary for JA synthesis. The Ca^{2+} channel blocker-$LaCl_3$ inhibited the expression of the genes associated with JA production in potato plants under stress conditions (Yang et al., 2022). Research shows that an increase in cold stress raises cytosolic Ca^{2+} in Arabidopsis cells, thus activating the gene expression of JA-producing enzymes and JA signaling proteins (Kim et al., 2024). The number of investigations regarding the roles played by Ca^{2+} ions in JA signal transduction is enormous. Consequently, it was proven that in A. thaliana, Ca^{2+} enters the cytosol from the apoplast in response to JA through cyclic nucleotide-regulated channel 2 (AtCNGC2) (Lu et al., 2022). Following the activation of Ca^{2+} mobilization from intracellular Ca^{2+} depots. JA activates adenylate cyclase through an unidentified pathway, which leads to an increase in cytosolic Ca^{2+}. Ca^{2+} binding to CaM and AtCNGC2 suppression results in reversibility in this signaling process (Kolupaev et al., 2023). It has been concluded based on Khan et al. (2017) that the Ca^{2+} sensor CML regulates MYC2 and has a distinct yet critical function in mediating signaling from JAs during abiotic stress. Ca^{2+} /CaM advanced the participation of AhSAMS1 (Polyamines synthesis gene) in ion homeostasis by regulation of ROS, whereas CaM regulated methylation induced by AhSAMS1, suggesting a new strategy to improve the resistance of plants to salt stress (Yang et al., 2022). Application of exogenous Ca^{2+} with plant hormones can be applied by either root application or seed priming or foliar spray, which may provide better tolerance in plants to abiotic stress; it also helps curb the generation of ROS species and makes the plants efficient in photosynthesis and nitrogen assimilation (Li and Howell, 2021; Naeem et al., 2019).

Plants use various strategies to optimize water and nutrient uptake under salinity stress in response to high osmotic potential of the rhizosphere. Plants have specialized sensors and signaling to detect and respond to salt stress, triggering the signaling pathways closely linked to Ca^{2+} sensing. The influx of Na^{+} through voltage-insensitive monovalent cation channels (VIC), nonselective cation channels (NSCCs), and Na^{+}/H^{+} antiporters causes damage to the structure of cell walls and activates a Ca^{2+} channel through an unknown pa^{2+}thway (Feng et al., 2023; Rajput et al., 2021; Zhang et al., 2019). Prior findings suggested a major role for the Na^{+}/H^{+} (NHX) transporter in salt tolerance, where phenotypic analysis revealed tolerance reduction in plants having shortened NHX, while overexpression of the transporter led to enhanced salt stress resistance (Li and Howell, 2021). Ca^{2+} was found, in fact, to be really important by Lynch et al. (1989) through observation that intracellular concentrations of Ca^{2+} underwent a sharp increase within a few seconds of exposure to salinity shock in maize root protoplasts. The SOS signal transduction cascade is triggered by an oscillation in calcium in response to Na^{+} detection in the root. The SOS pathway consists of three key SOS proteins: SOS3 (also referred to as SCaBPs or CBL4), SOS2 (serine/threonine protein kinase), and SOS1, a plasma membrane Na^{+}/H^{+} antiporter (Kang et al., 2024). As Ca^{2+} sensor proteins, SOS3 and SCaBP8 interpret the Ca^{2+} signal brought on by salt stress, draw SOS2 to the membrane of the plasma membrane, and activate its kinase activity (Zhao et al., 2021). The harmful accumulation of Na^{+} is then prevented when SOS2 phosphorylates and activates SOS1, enabling it to export Na^{+} ions from cytosol to apoplast (Feki et al., 2017). In addition to the SOS pathway, there exists a grouping of Ca^{2+}-dependent membrane-binding proteins, called AtAN NEXINs, which were suggested to mediate Ca^{2+} signaling invoked by both ROS and NaCl (Bachani et al., 2022). It is proposed that AtANN4 mediates initial responses in salt-stressed plants through a Ca^{2+} signal (Ma et al., 2022). Upon bioinformatics investigation, an

AtNCL gene from the Arabidopsis source was found; this gene codes for a plasma membrane-localized protein (PmE1665455) with NCX-like function and calcium-binding capability. Interestingly, the CBL-CIPK signaling network regulates both Na^+ levels under NHX control and SOS function.

Heat stress adversely affects plant physiology and yield by accelerating more water molecules to evaporate from the leaves of plants, resulting in desiccation and low turgor pressure. Symptoms include wilting, scorching of leaf margins, and possibly death of cells. To minimize water loss, plants alter the regulation of stomatal activity, causing stomatal closure in order to reduce water loss; however, this closure ensures that gases necessary for photosynthesis will not be exchanged and, therefore the amount of CO_2 absorbed is reduced with further retarded growth (Balfagón et al., 2020).

Cellular membrane stability and protein synthesis are affected adversely by heat stress due to adverse high temperature. Most proteins denatured under these conditions become ineffective. The fluidization of cellular membranes compromises selective permeability, leading to the disruption of various cellular functions. The effect of heat stress varies from one species to other, with some species exhibiting more tolerance than others. In addition, previously experiencing sublethal heat stress may enhance a plant's ability to withstand other heat episodes. Several techniques may be employed to reduce the impacts of heat stress, including irrigation, shading, mulching, fogging systems, and the use of heat-tolerant plant types or genetically modified organisms (Chaudhary et al., 2022; Choudhury et al., 2017). Extreme heat episodes can also be minimized through modifications of crop management practices and scheduling of activities. According to Siddiqui et al. (2017), plants respond to heat stress by activating their antioxidant enzymes (SOD, POD, and CAT) to minimize damage caused by ROS. Heat shock proteins such as HSF7, HSP70.1, and HSP70.11 are also induced, which enhance the assembly of photosynthates and alleviate heat-sensitive damage to RUBISCO enzyme (Xu et al., 2022). Intensive investigations have been conducted on the mechanisms of Ca^{2+} action in the heat stress response in various plant species like Ipomoea batatas (Tang et al., 2023), Cucumis sativus (Ma et al., 2023a,b), *Zea mays* (Li and Howell, 2021), Hordeum vulgare (Poonia et al., 2020), and Solanum lycopersicum (Siddiqui et al., 2017).

Cold stress, consisting of chilling (<20°C) and freezing (<0°C) temperatures, disrupts cellular homeostasis and hampers plants' development, growth, and crop yield (Raza et al., 2024). Sensors, such as Chilling Tolerance Divergence 1 (COLD1), detect membrane rigidity changes and trigger Ca^{2+} influx (Mori et al., 2018; Shi et al., 2018). These Ca^{2+} sensors decode the Ca^{2+} signals and initiate downstream signaling events such as phosphorylation, transcriptional reprogramming, activation of MAPK cascades, and accumulation of ROS or NO in concert with other signaling components (). Furthermore, low temperature conditions induce the expression of certain cold-regulated (COR) genes, including responsive to desiccation29 (RD29A), COR15a, kinase (KIN1), and RsERF40 to protect cell membranes from freeze-induced damage (Ritonga and Chen, 2020).

9.4 K+ as a key osmolyte

The importance of potassium (K+) lies in many inherently different contexts of plant functions since K^+ is listed among the 17 essential elements for plant growth and development according to Ajewole et al. (2018). K^+ is present in soil in four forms: ionic (0.1%–0.2%), exchangeable (1%–2%), nonexchangeable (1%–10%), and unavailable (90%–98%). According to Volf et al. (2021) and Grzebisz et al. (2013), K^+ cannot be mobilized easily but would sustain the supply over a longer time period; rather, the first two forms are the only ones that plants can utilize and are available for their immediate needs. To cope with variable soil concentrations, plants have evolved efficient K^+ transport mechanisms (Assaha et al., 2017). Once absorbed by roots, K^+ is now either stored in vacuoles or transported through xylem to shoots, while phloem will equally export K^+ from the shoots to different plant tissues (Sardans and Peñuelas, 2021). The timely release of K+ from exchangeable sources is lower than K+ uptake rates, making K+ concentration a limiting factor to plants. However, the K^+ availability from the nonlabile form will depend on the timing of deposition, intensity of weathering, as well as type and clay material proportion ($^+$). The availability of K^+ for plant use is known to be less than 1–2% of the total present, though it is the seventh most common element in nature. Important sources of K^+ include flue dust, wool waste, sugar beet industrial waste, and tobacco stems, which however are seldom utilized as fertilizers. Normally represented by the symbol K2O on fertilizer labels, K+ appears in the third position in the indicated nomenclature "N:P:K" Muriate of potash, also referred to as potassium chloride, constitutes 95% of all K+ fertilizers used almost everywhere (Zahoor et al., 2017). In 1868, Samuel William Jackson, a botanist from Connecticut, burned the plants and analyzed the ash to cross-check the K+ content in plants, and his finding was that it contained trace K^+ along with other minerals (Zhu et al., 2018;). In the plant system, K^+ acts as a critical regulator and is required for the activation of over 60 enzymes. K $^+$is really important in enabling plants to withstanda battery of biotic and abiotic environmental stressors (Borrelli et al., 2018). K^+ was observed by to be the most abundant cation in plants, accounting for about 10% of dry weight. Higher concentration lies within the cytosol and the chloroplast. It acts by temporary stabilizing the pH of K^+ between 7 and 8 through enzyme conformational modifications, thereby activating several enzymes (Naeem et al., 2019). It promotes cell elongation and osmoregulatory maintenance in plants (Zahoor et al., 2017). Moreover, K^+ absorption is a valuable tool that aids the detoxification of heavy metals in plants by synthesizing soluble protein, glucose, and compounds of soluble nitrogen in the plant cell sap. K^+ also confers tolerance to various diseases on plants (Ma et al., 2022). K^+ also regulates the opening and closing of the stomata, which control the plant's consumption of the water required for growth and development. K^+ helps with the formation of starch and regulates root development in plants. Remarkably, starchy plants have a high demand for K^+. K^+ further encourages stem strength, elongation, and shelf life in Gladiolus grandiflora, producing other positive qualitative and quantitative changes in the plants (Yasin et al., 2018).

Although K^+ does not appear in the chemical structure of plant proteins, it is involved in many kinds of regulations in plant growth. Under low availability of K^+, the stomatal response becomes sluggish, delaying closure during water stress (Nieves-Cordones et al., 2019). Additionally, K+ plays an important role in balancing electrical charges at the ATP production site (Zhang et al., 2019). Thus its deficiency reduces ATP production and thereby reduces photosynthesis rate in plants (Borrelli et al., 2018).

9.4.1 K+ and enzyme activation

9.4.1.1 K+-dependent enzymes involved in primary metabolism (e.g., starch synthesis)

K+ plays important roles in activating enzymes involved in primary metabolism, for instance, pyruvate and ATP are produced through the conversion of phosphoenolpyruvate (PEP) and adenosine diphosphate (ADP) in the glycolytic pathway by means of pyruvate kinase, which is activated by K+. Boyer (1953) studied the first enzyme activated by K^+, which was pyruvate kinase from rat muscle. Other studies later confirmed that univalent cations, when combined with magnesium (Mg^{2+}) or manganese (Mn^{2+}), are essential for the activity of this enzyme. Unlike the other two major nutrients, phosphorus (P) and nitrogen (N), which are integral components of biomolecules, potassium exists solely as a free ion, making it difficult to link directly to specific metabolic pathways (Sardans and Peñuelas, 2021). Rather, the role of K^+ is accentuated more in some of the physiological processes involving osmoregulation, cell growth, stomatal control, stimulation of enzyme activity, protein synthesis, photosynthesis, and phloem loading and transport (Tränkner et al., 2018). In particular, K^+ affects positively biosynthesis of ribulose-1,5-bisphosphate carboxylase/oxygenase (Rubisco), health of chloroplast, nutrition, and inflow of water and gases into chloroplasts (Cui and Tcherkez, 2021). These different roles tend to make a lack of K^+ injurious to plant growth as well as vulnerable to many abiotic stressors. Among the various challenges faced by crop plants is a constant decline in the yield levels owing to global climate change, which is rapidly becoming a major cause of concern in the environment these plants grow in. In addition, rising use of pesticides and fertilizers is becoming increasingly challenging for the current agriculture system (Anderson et al., 2020). Hence, it becomes very important to discover appropriate methods of practicing sustainable agriculture with fewer adverse implications on natural habitats while at the same time increasing crop productivity under stressful conditions. Given that K+ at high concentrations is not really harmful to the environment (Mahiwal and Pandey, 2022), optimal levels of K+ application may well be an emerging strategy in managing plant performance against abiotic stresses. Gradually increasing numbers of studies suggest that K^+ can amplify plant tolerance and survival in responses to a variety of abiotic stresses, including drought, high salinity, heat, cold, and metal toxicity (García-Martí et al., 2019; Naciri et al., 2021). K^+ induced improvement in abiotic stress tolerance is generally linked with enhanced physiological mechanisms of the plant. For example, K+ supplementation enhanced osmotic adjustments and WUE in a way that may benefit sugarcane (Saccharum officinarum) tolerance to water-shortage conditions (Silva et al., 2025).

9.4.1.2 *K+ flux in guard cells and its control over stomatal opening/closing*

Potassium homeostasis is central to stomatal movement and stress adaptation. The study conducted by Zhao et al. (2021) found that K+ treatment has reduced the effect of drought on cotton fiber strength in Gossypium hirsutum, regulating sucrose levels and carbohydrate-uptake rates during drought. Similarly, according to Almeida et al. (2017), K^+ and Na^+ should be maintained in a higher ratio to improve the plant efficacy against salinity for all major crops. It is interesting to note that even enzyme activity for the production of saturated fatty acids, vital for frost tolerance, relied on the correct cytoplasmic K^+ concentration. K^+ starvation significantly inhibits adaptations to external stimuli because of a large and sudden loss of a large number of stress-related proteins. According to Sardans and Peñuelas (2021) and Anschütz et al. (2014), stomatal closure is a defense strategy that curtails water loss when the internal K+ cellular values are low due to pressure applied by stressors such as salts and drought. For the purpose of coordinating stomatal closure to different abiotic challenges, K^+-channels and K^+ transporters are tightly integrated with a range of different phytohormone signaling elements in guard cells. For instance, drought gives rise to the endogenous ABA, which plays a pivotal role in stomatal control through K^+- transporters and K^+ channels. Under stress, Ca^{2+} oscillation induced by ABA in guard cells activates anion channels including rapid, R-type, and slow, S-type channels. This produces membrane depolarization that inactivates KAT1 and activates GORK, leading to turgor loss and closing of the stomata (Jalakas et al., 2021). Additionally, ABA could inactivate KAT1/AKT1 and facilitate stomatal closure through inhibition of H + ATPase activity. Inhibiting AKT1, an important K^+ transporter in guard cells, improved drought tolerance, and ABA-mediated stomatal closure are promoted in Arabidopsis (Nieves-Cordones et al., 2019). On the other hand, that KAT1 and KAT2 transcripts are downregulated in Arabidopsis due to drought stress indicates that ABA inactivates those channels directly under abiotic stresses. Conversely, K^+-selective channels TPK1 areinvolved in the release of vacuolar K^+ and completely dependent on ABA-mediated stomatal closure in Arabidopsis (Isner et al., 2018; Zhang et al., 2022). ABA signaling further involves the regulation of stress-responsive kinases like CPKs and Open Stomata 1 (OST1), which inhibit KAT1, KAT2, and GORK channels. Although OST1-mediated inhibitory effects on KAT1 were disputed by Zhang et al. (2016), it was found that ABA-induced inhibition of KAT1 was reduced in OST1 mutants, suggesting a suppressive role of OST1 on KAT1 activity. Thus the regulation of KAT1 activity by OST1 is still a mystery that needs to be investigated further. Additionally, stomatal closure mediated by KUP6 is regulated by ABA-sensitive sucrose nonfermenting (SNF)-1-related protein kinase 2 E (SRK2E). Transgenic plants overexpressing KUP6 exhibited superior efficiency in stomatal closure due to enhanced K^+ efflux than those in wild-type plants, indicating a likely role for this transporter in the stomatal closure induced by osmotic stresses. The role of HAK/KT/KUP transporters in K^+ efflux regulation across tonoplasts has been widely documented; however, their precise implication in guard cell tonoplast is yet unclear, particularly under abiotic stress conditions. Nonetheless, the specific stress-responsive function of this K^+ channel in H2S remains unknown. Hydrogen sulfide (H_2S) signaling interacts with ABA to regulate these transporters. The H_2S-MPK4 cascade is activated under drought to modulate stomatal movements, suggesting H_2S and ABA function synergistically (Du et al., 2019).

9.4.1.3 *Potassium in regulation of photosynthesis in plants*

Increased yield in crops, as well as postharvest quality of grains, is rooted in photo assimilation and the subsequent carbohydrate transport, and their metabolism in K +-dependent (Tränkner et al., 2018). Several research studies have demonstrated the importance of K^+ for keeping most of the gas exchange parameters in readiness for effective photosynthesis in abiotic stresses such as stomatal conductance (gs), transpiration rate (E), Chl a fluorescence, and photosynthetic rate (Pn) (Naciri et al., 2021; Verma et al., 2024; Sardans and Peñuelas, 2021).

Drought and other osmotic-type stresses usually induce a deficiency of endogenous K^+ in plants, which strongly represses the synthesis and activity of Rubisco, the universal carboxylation enzyme (Zahoor et al., 2017). This lower carboxylation efficiency of Rubisco in chloroplasts (Tränkner et al., 2018). However, higher foliar K+ application done during the vegetative and reproductive stages of the crop improved the photosynthetic rate, which eventually leads to increased seed yield and biological yield in canola (*Brassica napus*) under water deficit conditions (Waraich et al., 2020). Similarly, in maize (*Zea mays*), foliar K^+ enhanced photosynthesis and grain weight by optimizing the source-sink relationship (Luo et al., 2021).

When studied in soybeans under saline stress, K_2SO_4 application showed salt-tolerant soybeans (e.g., Pioneer) with significantly better attributes of photosynthetic as well as gas exchange features and thus enhanced sugar accumulation when compared with salt-sensitive cultivar (e.g., Jack) (Parveen et al., 2021). In stressed plants due to failure of photosynthetic CO_2 fixation, can lead to ROS formation, degrading photosynthetic pigments and membranes (Du et al., 2019; Zhang et al., 2020). K+ also maintains chloroplast structural integrity, and K+ facilitates the development of grana and stroma lamellae (Tränkner et al., 2018).

This structural role is critical for efficient photosynthesis. Specifically, the thylakoid K^+/H^+ antiporter KEA3 is essential for maintaining photosynthetic performance under fluctuating light conditions by promoting the relaxation of nonphotochemical quenching (NPQ). Overexpression of the Arabidopsis KEA3 isoform accelerates NPQ relaxation and maximizes Photosystem II (PSII) quantum efficiency, highlighting the transporter's key role in optimizing the photosynthetic apparatus. Conversely, triple mutants of atkea1/2/3 display abnormal chloroplast structure and reduced photosynthetic rates.

Under K+ deficiency, the PS electron transport chain becomes downregulated, leading to ROS production (Waraich et al., 2012). For instant when studied in low K^+ tolerant maize cultivar (e.g., D937) exhibited elevated levels of ROS, causing severe damage to chloroplasts and PSII reaction centers (Du et al., 2019). On the contrary, K^+ tolerant maize cultivars (e.g., 90-21-3) suffered less from oxidative damage to the photosynthetic apparatus because of increased stimulation of activities of the antioxidant enzymes, such as SOD, CAT, and APX, allowing them to maintain normal photosynthesis even under decreased K^+ conditions (Du et al., 2019). Under water-shortage conditions, addition of K+ not only increases the activities of nitrogen-metabolizing enzymes (e.g., ala and asp transaminases and nitrate reductase) but also the activities of SOD, GR, CAT, and APX, as well as the levels of AsA and GSH in sorghum (*Sorghum bicolor*). Therefore it is postulated by a number of scientists that a proper supply of K+ suppression under water stress enhances CO_2 fixation and exportation of photosynthates from source to sink (Raza et al., 2013) and inhibits photosynthetic electron transport to O_2. In tomato plants, direct links have been established between K^+ starvation and oxidative stress, whereby increases in ROS levels lead to massive alterations in the concentrations of anti-ROS enzymes under K+ starvation.

K+ also inhibits the generation of ROS by inhibiting NADPH oxidase, thus preventing lipid peroxidation under saline and drought conditions (Ahanger and Agarwal, 2017). Antioxidant enzymes such as DHAR, GPX, SOD, CAT, APX, POX, and reduced NADPH oxidase were able to ameliorate oxidative stress with higher accumulation of K+ in plant cells (García-Martí et al., 2019). The role of K+ in the alleviation of oxidative stress was substantiated by a reduction in the Na^+/K^+ ratio in wheat as described by Ahanger and Agarwal (2017). Also, with water deficit, well-timed supply of K+ decreased MDA content in sunflower (*Helianthus annuus* L.). Moreover, sodium toxicity, acting by the plants' Na^+/K^+ ratio, was responsible for ROS formation, which in turn obstructed stomatal closure, impaired photosynthesis of the plant, and escalated lipid peroxidation (AbdElgawad et al., 2016).

9.4.2 Key functions of K+ under heat stress

K+ serves pivotal functions in plants facing temperature stress by stimulating metabolic and physiological mechanisms such as alteration in photosystems, respiration, nutritional maintenance, and enhancement of tissue water potential to counter heat stress. Further, K+ plays a role as an osmotic regulator and helps in the retention of stomatal conductance to avoid injury to cells. The effect of moderately high temperatures on soil and plants also included the accumulation of K+ in plants and the influence of accumulation in the aerial parts and roots. (Elsayd et al., 2018) assessed with interest the mode of application of K+ fertilizer to palm trees: foliar application, soil application, and trunk injection into trees. In fact, trees injected by trunk showed the highest growth under 3% K_2SO_4, improved fruit quality, and yield under heat stress.

One of the important studies that Sarwar et al. (2022) conducted mentioned the damaging effects of heat stress with respect to plant defenses, water relations, physiology, and potential yield of cotton crops. Some of those minerals proved to be effective in alleviation of heat stress. Such minerals include potassium, zinc, and boron elevating heat stress during critical stages like squaring, flowering, and boll formation. Enhancements were observed in the physiological and biochemical systems, including the antioxidant systems of cotton (Sarwar et al., 2022). Application of K+ exogenously reduces negative heat stress effects in both terminal heat stress stages. Sustaining carbohydrate allocations may be explained by sufficient availability of K+ during heat stress for spike to develop for a longer time, prolonged time for grain fill, and enhanced grain yield (Shahid et al., 2019; Du et al., 2019). Additionally, K^+ has also been reported to increase the activity of SOD, which clears these ROS superoxide radicals and H_2O_2 (Shahid et al., 2019; Du et al., 2019). Similar findings from an experiment done by Shahid et al. (2020) on bread wheat showed reflectance of the biochemical and agronomic traits under heat stress when K^+ was applied in a foliar spray. Increased the synthesis of glycine betaine, a vital osmo-protectant that improves WUE and oxidative pressure.

9.4.3 Signaling of K+ under heat stress

The K+ content in the soil may not remain consistent throughout the crop-growing period, owing to the variations in the environment. These changes in K+ availability are sensed by the roots of the plant. In case of K+ deficiency, a series of molecular events and signaling components are activated in the plant at various levels, ranging from signal perception to adaptive responses. Under K^+ deficiency, AKT1 is activated via phosphorylation by CIPK23, a protein kinase activated by the calcium sensors CBL1 and CBL2. Subsequently, low K^+ concentration, through CBL1/9, induces distinct Ca^{2+} signals in A. thaliana, which in turn activate AKT1 by the formation of CBL1/9-CIPK23 complexes (Behera et al., 2017) (shown in Fig. 9.2). Overexpression of the type III peroxidase RCI3 resulted in the expression of HAK5 and an increase in ROS production. The K^+ channels GORK and NSCC and their ROS-mediated activation are significant to the decrease of K^+ levels in the cytosol under saline conditions (Anschütz et al., 2014; Demidchik, 2014).

9.5 Effects of calcium and potassium on agriculture crops in heat stress management

Overexpressing ZmCDPK7 in maize boosted heat tolerance by enhancing photosynthetic efficiency and CAT activity, which reduced Levels of ROS and MDA. the

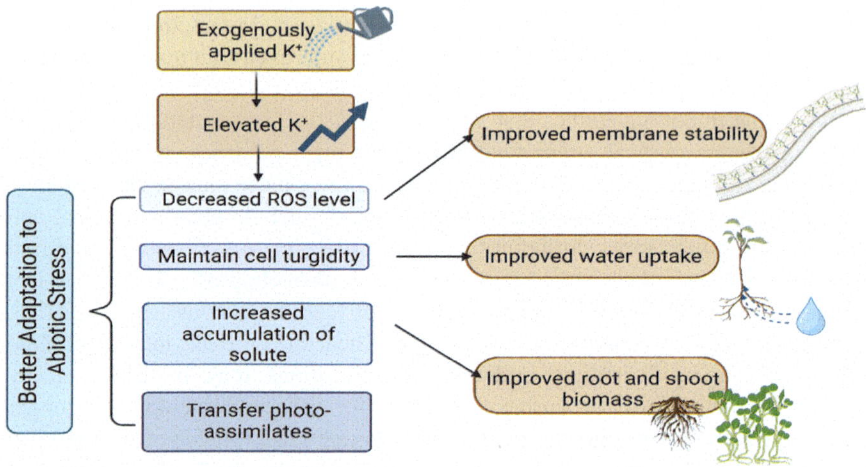

FIGURE 9.2

Role of exogenous potassium in abiotic stress regulation in plant system.

heat stress triggers ZmCDPK7 translocation and activation, leading to the phosphorylation of the chaperone sHSP17.4, further activating RBOHB for ROS signaling, integrating with ABA response. Conversely, knockdown mutants exhibited increased heat sensitivity, establishing ZmCDPK7 as a central hub for calcium signaling and a key target for engineering thermotolerance. Similarly, (Wei et al., 2023) reported in foxtail millet and Arabidopsis that overexpression of SiCDPK7 built up tolerance to both heat at 45°C and cold at -4°C. Transgenic lines survived better than wild types, supporting that CAT activity increased while MDA levels reduced. Exogenous application of Ca^{2+} and K+ proves to be effective in multiple crop species, as studied by Hussain et al. (2020) in Indian almond seedlings, treating with Ca^{2+} and K+ (250–500 mg per liter) significantly improved summer heat tolerance, improved growth, and enhanced cellular protection. Treated seedlings exhibited superior height with more no. of branches and leaf area. These treatments have maintained levels of Chlorophyll and carotenoid content. Suggesting osmotic and antioxidant balance while lowering levels of ABA. Foliar sprays of calcium, silicon, and potassium mitigated drought stress in cucumbers. By enhancing growth parameters like height, leaf area, and biomass. El-Beltagiet al. (2025) reported that combined treatments were most effective in protecting membranes and maintaining photosynthesis, upregulating antioxidant enzymes such as SOD, CAT, and POD, reducing ROS and lipid peroxidation. Combined treatments proved most effective.

Advances in nanotechnology (NT) have opened promising avenues toward improving plant stress tolerance through targeted nutrient delivery (Husen, 2025; Ayyaz et al., 2022) reported that foliar application of calcium nanoparticles significantly enhanced drought stress tolerance in *Brassica napus*, resulting in increased biomass production and improved water use efficiency hence counteracting the deletorious effect of moisture stress. Treated plants exhibited enhanced photosystem II efficiency, indicated by the increased Fv/Fm and nonphotochemical quenching values. The application of calcium nanoparticles further improved the uptake of essential nutrients, including potassium, magnesium, iron, and zinc, while antioxidant capacity and proline accumulation were markedly enhanced. These physiological improvements culminated in a significant reduction of reactive oxygen species, including hydrogen peroxide and malondialdehyde, under drought stress conditions. Correspondingly, Khushboo et al. (2018) confirmed that treatment of wheat plants with 10 mM calcium chloride was efficient in reducing the effects of drought stress caused by polyethylene glycol. Additionally, it helped in the rejuvenation of both drought-tolerant (HD 2733) and drought-sensitive (HD 3086) varieties, resulting in the promotion of plant development and growth. The drought stress was reduced by the augmentation of the activity levels of antioxidant enzymes such as superoxide dismutase, catalase, peroxidase, and ascorbate peroxidase, which helped in suppressing the levels of stress-related hormones and markers for oxidative damage by 40%–60%.

Zhang et al. (2019) reported that exogenous application of calcium mitigated salt-induced stress in maize by improving stomatal activity and photosynthesis. Consequently, Zrig et al. (2021) found that the simultaneous application of potassium and calcium enhanced salt tolerance in thyme. The combined effects of the two elements were effective in overcoming the negative effects of salinity stress. Reducing ion homeostasis raises K+ and Ca+2 levels while lowering Na. Antioxidants like SOD, CAT, and POD improved while H_2O_2 and superoxide reduced, leading to better growth and development when compared to the control plants. Foliar potassium at 2% kept flag leaf photosynthesis and grain filling steady in wheat under heat stress. Sarwar et al. (2022) noted that Potassium with zinc or boron helped cotton handle reproductive heat from 32°C to 45°C. The combination of K+ and Zn. restored relative water content and total soluble proteins. H_2O_2 and relative conductivity index decreased by 64% and 37%, while the yield of bolls, branches, and fiber length increased. Yadav et al. (2022) reviewed heat stress effects on wheat. Photosynthesis can drop 15% to 50%. Yields fall 10%–40% leading to Pollen sterility. Foliar potassium helps. Sowing time adjustments matter. Transgenics target the booting to anthesis stage. More recent work highlights how calcium and potassium work against various stresses. Elkelish et al. (2019) highlighted the synergistic effect of calcium (5–10 mM) and potassium in sorghum under combined heat and drought stress. The treatment stabilized membranes, increased proline and sugars, and activated APX and POD enzymes to limit lipid peroxidation. Zrig et al. (2021) also found that combined foliar K and Ca^{2+} improved salt tolerance in thyme by restoring ion homeostasis (higher K/Na and Ca/Na ratios) and boosting antioxidant defenses.

The integrated use of K^+ and Ca^{2+} was found to reduce the detrimental effects of soil salinity in sorghum, leading to observable enhancements in growth and morphological appearance (Jafari et al., 2009). Even casual additions of calcium have been shown to improve morpho-physiological function by counteracting the disruption in cell processes and optimal ion regulation in a saline environment. On model plants, the inhibition of cell function by salinity stress at the organ and plant level was found, by and large, to be the result of the competition of influx/efflux of Ca^{2+} and Na^{+} ion interactions, affecting membrane selectivity and ion regulation (). High Ca^{2+} to Na^{+} ratios were found to result in differential growth, which ranged from growth inhibition to significant anatomical and morphological changes, indicating an adaptation strategy to ion stress in maize (Evlagon et al., 1990; Cramer, 1992). Such observations highlight the important physiological function of calcium and potassium and their regulation of ion relations and plant growth.

In the context of the tomato, the efficacy of foliar treatment with Ca+2 and K +-based nano fertilizers at a concentration of 50–100 mg/L in reducing or mitigating heat stress, thereby resulting in increased fruit setting and yield by up to 25%–35% at temperatures of 40°C (Sharma et al., 2024). From the molecular perspective, the reduction of the heat stress effect was brought about by the increased expression of stress-responsive genes, such as LeCDPK2 and LeHSP70, among others. In spinach, foliar spray of calcium nitrate at 4 mM significantly mitigated the heat stress of 35°C by causing significant recovery in plant height and biomass accumulation (Uz Zaman et al., 2022). The photosynthetic performance improved while electrolyte leakage reduced, showing improved membrane stability. Calcium supplementation significantly enhanced the activities of superoxide dismutase, peroxidase, and catalase antioxidant enzymes by about twofold while stimulating the accumulation of osmolytes like proline and soluble sugar. Phytoconstituent levels were also higher under heat stress, especially phenolic compounds that increased between 25% and 40%. The 4 mM calcium was found to be most effective among different treatments for conferring physiological and biochemical protection against heat stress (Uz Zaman et al., 2022).

9.6 Conclusion

Heat stress is one of the major abiotic constraints that affect plant growth, physiological processes, and crop productivity under changing climatic conditions. Adequate mineral nutrition plays an important role in mitigating heat-induced damages and among the essential nutrients, potassium and calcium play an especially important role. Potassium provides osmotic regulation, stomatal functioning, activation of enzymes, and maintenance of cellular homeostasis to present high-temperature tolerance in plants. Ca+2 maintains membrane integrity, cell wall integrity, and stress signaling to maintain structural and metabolic stability under heat stress conditions. Coordinated action ofK+ andCa+2 enhances the efficiency of antioxidant defense systems, helps to maintain ion homeostasis, and promotes efficient stress sensing and signaling. Their cooperative action can boost the resistance of plants to various stresses by reducing oxidative damage, preserving the efficiency of photosynthesis, and improving the overall performance of plants under high temperatures. Physiological and molecular basis of heat stress tolerance by potassium and calcium in plants will help in the generation of nutritional approaches for its management. In conclusion, the optimization of potassium and calcium nutrition can be viewed as a sustainable and feasible method to upgrade the heat-resistant potential of plants. When proper nutrients are incorporated into breeding and agrotechnologies, this can make a considerable contribution to the increase of plant productivity and heat-resistant capacity in regions where high temperatures are prevalent.

References

Abdallah, M.M.S., El-Bassiouny, H.M.S., AbouSeeda, M.A., 2019. Potential role of kaolin or potassium sulfate as anti-transpirant on improving physiological, biochemical aspects and yield of wheat plants under different watering regimes. Bulletin of the National Research Centre 43 (1), 134.

AbdElgawad, H., Zinta, G., Hegab, M.M., Pandey, R., Asard, H., Abuelsoud, W., 2016. High salinity induces different oxidative stress and antioxidant responses in maize seedlings organs. Frontiers in Plant Science 7. Available at https://doi.org/10.3389/fpls.2016.00276.

Ahanger, M.A., Agarwal, R.M., 2017. Potassium up-regulates antioxidant metabolism and alleviates growth inhibition under water and osmotic stress in wheat (*Triticum aestivum* L). Protoplasma 254 (4), 1471–1486. Available at https://doi.org/10.1007/s00709-016-1037-0.

Ajewole, E., Santamaria-Kisiel, L., Pajak, A., Jaskolski, M., Marsolais, F., 2018. Structural basis of potassium activation in plant asparaginases. FEBS Journal 285 (8), 1528–1539. Available at https://doi.org/10.1111/febs.14428.

Aleynova, O.A., Dubrovina, A.S., Kiselev, K.V., 2017. Activation of stilbene synthesis in cell cultures of *Vitis amurensis* by calcium-dependent protein kinases VaCPK1 and VaCPK26. Plant Cell, Tissue and Organ Culture (PCTOC) 130 (1), 141–152. Available at https://doi.org/10.1007/s11240-017-1210 y.

Almeida, D.M., Oliveira, M.M., Saibo, N.J.M., 2017. Regulation of Na^+ and K^+ homeostasis in plants: towards improved salt stress tolerance in crop plants. Genetics and Molecular Biology 40 (1 suppl 1), 326–345. Available at https://doi.org/10.1590/1678-4685-gmb-2016-0106.

Anderson, R., Bayer, P.E., Edwards, D., 2020. Climate change and the need for agricultural adaptation. Current Opinion in Plant Biology 56, 197–202. Available at https://doi.org/10.1016/j.pbi.2019.12.006.

Anschütz, U., Becker, D., Shabala, S., 2014. Going beyond nutrition: regulation of potassium homoeostasis as a common denominator of plant adaptive responses to environment. Journal of Plant Physiology 171 (9), 670–687. Available at https://doi.org/10.1016/j.jplph.2014.01.009.

Apel, K., Hirt, H., 2004. Reactive oxygen species: metabolism, oxidative stress, and signal transduction. Annual Review of Plant Biology 55, 373–399. Available at https://doi.org/10.1146/annurev.arplant.55.031903.141701.

Arab, M., Najafi Zarrini, H., Nematzadeh, G., Heidari, P., Hashemipetroudi, S.H., Kuhlmann, M., 2023. Comprehensive analysis of calcium sensor families, CBL and CIPK, in *Aeluropus littoralis* and their expression profile in response to salinity. Genes 14 (3). Available at https://doi.org/10.3390/genes14030753.

Asai, S., Ichikawa, T., Nomura, H., Kobayashi, M., Kamiyoshihara, Y., Mori, H., et al., 2013. The variable domain of a plant calcium-dependent protein kinase (CDPK) confers subcellular localization and substrate recognition for NADPH oxidase. Journal of Biological Chemistry 288 (20), 14332–14340. Available at https://doi.org/10.1074/jbc.M112.448910.

Assaha, D.V.M., Ueda, A., Saneoka, H., Al-Yahyai, R., Yaish, M.W., 2017. The role of Na^+ and K^+ transporters in salt stress adaptation in glycophytes. Frontiers in Physiology 8. Available at https://doi.org/10.3389/fphys.2017.00509.

Atif, R.M., Shahid, L., Waqas, M., Ali, B., Rashid, M.A.R., Azeem, F., et al., 2019. Insights on calcium-dependent protein kinases (CPKs) signaling for abiotic stress tolerance in plants. International Journal of Molecular Sciences 20 (21). Available at https://doi.org/10.3390/ijms20215298.

Ayyaz, A., Fang, R., Ma, J., Hannan, F., Huang, Q., Athar, H.U.R., et al., 2022. Calcium nanoparticles (Ca-NPs) improve drought stress tolerance in Brassica napus by modulating the photosystem II, nutrient acquisition and antioxidant performance. NanoImpact 28. Available at https://doi.org/10.1016/j.impact.2022.100423.

Bachani, J., Mahanty, A., Aftab, T., Kumar, K., 2022. Insight into calcium signalling in salt stress response. South African Journal of Botany 151, 1–8. Available at https://doi.org/10.1016/j.sajb.2022.09.033.

Behera, S., Long, Y., Schmitz-Thom, I., Wang, X.P., Zhang, C., Li, H., et al., 2016. Two spatially and temporally distinct Ca^{2+} signals convey *Arabidopsis thaliana* responses to K^+ deficiency. New Phytologist 213, 739–750. Available at https://doi.org/10.1111/nph.14145.

Balfagón, D., Zandalinas, S.I., Mittler, R., Gómez-Cadenas, A., 2020. High temperatures modify plant responses to abiotic stress conditions. Physiologia Plantarum 170 (3), 335–344. Available at https://doi.org/10.1111/ppl.13151.

Borrelli, G.M., Fragasso, M., Nigro, F., Platani, C., Papa, R., Beleggia, R., et al., 2018. Analysis of metabolic and mineral changes in response to salt stress in durum wheat (*Triticum turgidum* ssp. durum) genotypes, which differ in salinity tolerance. Plant Physiology and Biochemistry 133, 57–70. Available at https://doi.org/10.1016/j.plaphy.2018.10.025.

Bouché, N., Yellin, A., Snedden, W.A., Fromm, H., 2005. Plant-specific calmodulin-binding proteins. Annual Review of Plant Biology 56, 435–466. Available at https://doi.org/10.1146/annurev.arplant.56.032604.144224.

Boyer, P.D., 1953. The activation by K^+ and occurrence of pyruvic phosphoferase in different species. Journal of Cellular and Comparative Physiology 42 (1), 71–77. Available at https://doi.org/10.1002/jcp.1030420105.

Cakmak, I., 2005. The role of potassium in alleviating detrimental effects of abiotic stresses in plants. Journal of Plant Nutrition and Soil Science 168 (4), 521–530.

Choudhury, F.K., Rivero, R.M., Blumwald, E., Mittler, R., 2017. Reactive oxygen species, abiotic stress and stress combination. The Plant Journal 90, 856–867.

Chaudhary, S., Jha, U.C., Paul, P.J., Prasad, P.V.V., Sharma, K.D., Kumar, S., et al., 2022. Assessing the heat sensitivity of Urdbean (*Vigna mungo* L. Hepper) genotypes involving physiological, reproductive and yield traits under field and controlled environment. Frontiers in Plant Science 13. Available at https://doi.org/10.3389/fpls.2022.1042999.

Cramer, G.R., 1992. Kinetics of maize leaf elongation: II. responses of a na-excluding cultivar and a na-including cultivar to varying na/ca salinities. Journal of Experimental Botany 43 (6), 857–864. Available at https://doi.org/10.1093/jxb/43.6.857.

Cui, J., Tcherkez, G., 2021. Potassium dependency of enzymes in plant primary metabolism. Plant Physiology and Biochemistry 166, 522–530. Available at https://doi.org/10.1016/j.plaphy.2021.06.017.

Dekomah, S.D., Bi, Z., Dormatey, R., Wang, Y., Haider, F.U., Sun, C., et al., 2022. The role of CDPKs in plant development, nutrient and stress signaling. Frontiers in Genetics 13. Available at https://doi.org/10.3389/fgene.2022.996203.

Del Río, L.A., Sandalio, L.M., Corpas, F.J., Palma, J.M., Barroso, J.B., 2006. Reactive oxygen species and reactive nitrogen species in peroxisomes. Production, scavenging, and role in cell signaling. American Society of Plant Biologists, Spain Plant Physiology 141 (2), 330–335. Available at https://doi.org/10.1104/pp.106.078204.

Demidchik, V., 2014. Mechanisms and physiological roles of K^+ efflux from root cells. Journal of Plant Physiology 171 (9), 696–707. Available at https://doi.org/10.1016/j.jplph.2014.01.015.

Demidchik, V., Shabala, S., Isayenkov, S., Cuin, T.A., Pottosin, I., 2018. Calcium transport across plant membranes: mechanisms and functions. New Phytologist 220 (1), 49–69. Available at https://doi.org/10.1111/nph.15266.

Du, Q., Zhao, X.H., Xia, L., Jiang, C.J., Wang, X.G., Han, Y., et al., 2019. Effects of potassium deficiency on photosynthesis, chloroplast ultrastructure, ROS, and antioxidant activities in maize (*Zea mays* L.). Journal of Integrative Agriculture 18 (2), 395–406. Available at https://doi.org/10.1016/S2095-3119(18)61953-7.

Elkelish, A.A., Alnusaire, T.S., Soliman, M.H., Gowayed, S., Senousy, H.H., Fahad, S., 2019. Calcium availability regulates antioxidant system, physio-biochemical activities and alleviates salinity stress mediated oxidative damage in soybean seedlings. Journal of Applied Botany and Food Quality 92, 258–266.

Elsayd, I.A.E.-R., El-Merghany, S., Zaen El-Dean, E.M.A., 2018. Influence of potassium fertilization on barhee date palms growth, yield and fruit quality under heat stress conditions. Journal of Plant Production 9 (1), 73–80. Available at https://doi.org/10.21608/jpp.2018.35258.

Evlagon, D., Ravina, I., Neumann, P., 1990. Interactive effects of salinity and calcium on hydraulic conductivity, osmotic adjustment, and growth in primary roots of maize seedlings. Israel Journal of Botany 39 (3), 239–247. Available at https://doi.org/10.1080/0021213X.1990.10677149.

El-Beltagi, H.S., Abdel-Haleem, M., El-Mogy, M.M., Khedr, E.H., 2025. Resilience of forests in a changing climate: Responses, mechanisms, and adaptations. Notulae Botanicae Horti Agrobotanici Cluj-Napoca 53 (1), 14360. Available at https://doi.org/10.15835/nbha53114360.

Fahad, S., Bajwa, A.A., Nazir, U., Anjum, S.A., Farooq, A., Zohaib, A., et al., 2017. Crop production under drought and heat stress: plant responses and management options. Frontiers in Plant Science 8. Available at https://doi.org/10.3389/fpls.2017.01147.

Feki, K., Tounsi, S., Masmoudi, K., Brini, F., 2017. The durum wheat plasma membrane Na^+/H^+ antiporter SOS1 is involved in oxidative stress response. Protoplasma 254 (4), 1725–1734. Available at https://doi.org/10.1007/s00709-016-1066-8.

Feng, D., Wang, X., Gao, J., Zhang, C., Liu, H., Liu, P., et al., 2023. Exogenous calcium: its mechanisms and research advances involved in plant stress tolerance. Frontiers in Plant Science 14. Available at https://doi.org/10.3389/fpls.2023.1143963.

García-Martí, M., Piñero, M.C., García-Sanchez, F., Mestre, T.C., López-Delacalle, M., Martínez, V., et al., 2019. Amelioration of the oxidative stress generated by simple or combined abiotic stress through the k^+ and ca^{2+} supplementation in tomato plants. Antioxidants 8 (4). Available at https://doi.org/10.3390/antiox8040081.

Goswami, S., Kumar, R.R., Sharma, S.K., Kala, Y.K., Singh, K., Gupta, R., et al., 2015. Calcium triggers protein kinases-induced signal transduction for augmenting the thermotolerance of developing wheat (*Triticum aestivum*) grain under heat stress. Journal of Plant Biochemistry and Biotechnology 24 (4), 441–452.

Grzebisz, W., Gransee, A., Szczepaniak, W., Diatta, J., 2013. The effects of potassium fertilization on water-use efficiency in crop plants. Journal of Plant Nutrition and Soil Science 176 (3), 355–374.

Harper, J.R., Breton, G., Harmon, A., 2004. Decoding Ca^{2+} signals through plant protein kinases. Annual Review of Plant Biology 55, 263–288. Available at https://doi.org/10.1146/annurev.arplant.55.031903.141627.

Hepler, P.K., 2005. Calcium: a central regulator of plant growth and development. The Plant Cell 17 (8), 2142–2155. Available at https://doi.org/10.1105/tpc.105.032508.

Hetherington, A.M., Trewavas, A., 1984. Activation of a pea membrane protein kinase by calcium ions. Planta 161 (5), 409–417. Available at https://doi.org/10.1007/BF00394571.

Husen, A., 2025. Nanobiotechnology for Abiotic Stress Adaptation and Mitigation in Agricultural Crops. Springer, Singapore. Available at https://doi.org/10.1007/978-981-96-9298-9.

Hussain, S., Maqsood, M., Ijaz, M., Ul-Allah, S., Sattar, A., Sher, A., et al., 2020. Combined application of potassium and zinc improves water relations, stay green, irrigation water use efficiency, and grain quality of maize under drought stress. Journal of Plant Nutrition 43 (15), 2214–2225.

Iqbal, Z., Shariq Iqbal, M., Singh, S.P., Buaboocha, T., 2020. Ca^{2+} /calmodulin complex triggers CAMTA transcriptional machinery under stress in plants: signaling cascade and molecular regulation. Frontiers in Plant Science 11, 598327 J.

Isner, J.C., Begum, A., Nuehse, T., Hetherington, A.M., Maathuis, F.J., 2018. KIN7 kinase regulates the vacuolar TPK1 K+ channel during stomatal closure. Current Biology 28 (3), 466–472. Available at https://doi.org/10.1016/j.cub.2017.12.046.

Jafari, M.H.S., Kafi, M., Astaraie, A., 2009. Interactive effects of nacl induced salinity, calcium and potassium on physiomorphological traits of Sorghum (*Sorghum bicolor* L.). Pakistan Journal of Botany 41 (6), 3053–3063.

Jalakas, P., Takahashi, Y., Waadt, R., Schroeder, J.I., Merilo, E., 2021. Molecular mechanisms of stomatal closure in response to rising vapour pressure deficit. New Phytologist 232 (2), 468–475. Available at https://doi.org/10.1111/nph.17592.

Jezek, M., Blatt, M.R., 2017. The membrane transport system of the guard cell and its integration for stomatal dynamics. Plant Physiology 174 (2), 487–519. Available at https://doi.org/10.1104/pp.16.01949.

Kang, X., Zhao, L., Liu, X., 2024. Calcium signaling and the response to heat shock in crop plants. International Journal of Molecular Sciences 25 (1). Available at https://doi.org/10.3390/ijms25010324.

Khan, N.A., Khan, S., Naz, N., Shah, M., Irfanullah, A.M.S., Sher, H., et al., 2017. Effect of heat stress on growth, physiological and biochemical activities of wheat (*Triticum aestivum* L.). International Journal of Biosciences 11 (4), 173–183.

Khushboo, K., Bhardwaj, P., Singh, M., Raina, V., Sharma, D., Kumar, 2018. Exogenous application of calcium chloride in wheat genotypes alleviates negative effect of drought stress by modulating antioxidant machinery and enhanced osmolyte accumulation. In Vitro Cellular and Developmental Biology - Plant 54 (5), 495–507. Available at https://doi.org/10.1007/s11627-018-9912-3.

Kim, J.S., Kidokoro, S., Yamaguchi-Shinozaki, K., Shinozaki, K., 2024. Regulatory networks in plant responses to drought and cold stress. Plant Physiology 195 (1), 170–189. Available at https://doi.org/10.1093/plphys/kiae105.

Klimecka, M., Muszyńska, G., 2007. Structure and functions of plant calcium-dependent protein kinases. Acta Biochimica Polonica, Poland Acta Biochimica Polonica 54 (2), 219–233. Available at https://doi.org/10.18388/abp.2007_3242.

Kolupaev, Y.E., Yastreb, T.O., Dmitriev, A.P., 2023. Signal mediators in the implementation of jasmonic acid's protective effect on plants under abiotic stresses. Plants 12 (14). Available at https://doi.org/10.3390/plants12142631.

La Verde, V., Dominici, P., Astegno, A., 2018. Towards understanding plant calcium signaling through calmodulin-like proteins: a biochemical and structural perspective. International Journal of Molecular Sciences 19 (5), 1331. Available at https://doi.org/10.3390/ijms19051331.

Li, Z., Howell, S.H., 2021. Heat stress responses and thermotolerance in maize. International Journal of Molecular Sciences 22 (2), 948.

Liu, Q., Kasuga, M., Sakuma, Y., Abe, H., Miura, S., Yamaguchi-Shinozaki, K., et al., 1998. Two transcription factors, DREB1 and DREB2, with an EREBP/AP2 DNA-binding domain, separate two cellular signal transduction pathways in drought- and low-temperature-responsive gene expression in *Arabidopsis*. The Plant Cell 10 (8), 1391–1406.

Liu, C., Tang, D., 2023. Comprehensive identification and expression analysis of CAMTA gene family in *Phyllostachys edulis* under abiotic stress. PeerJ 11, e15358.

Lu, S., Zhu, T., Luo, L., Ouyang, N., Hua, J., Zou, B., 2022. Divergent Roles of CNGC2 and CNGC4 in the Regulation of Disease Resistance, Plant Growth and Heat Tolerance in Arabidopsis. Agronomy 12, 2176. Available at 10.3390/agronomy12092176.

Luo, Y., Liu, X.Y., Fan, Y.Z., Fan, Y.H., Lv, Z.Y., Li, W.Q., et al., 2021. Exogenously supplied trehalose accelerates photosynthetic electron transport in heat-stressed maize. Russian Journal of Plant Physiology 68 (5), 857–866. Available at https://doi.org/10.1134/s1021443721050113.

Lynch, J., Polito, V.S., Läuchli, A., 1989. Salinity stress increases cytoplasmic ca activity in maize root protoplasts. Plant Physiology 90 (4), 1271–1274. Available at https://doi.org/10.1104/pp.90.4.1271.

Ma, C., Pei, Z.Q., Bai, X., Lu, S.H., Su, M., Kang, X., et al., 2023a. Exogenous melatonin and $CaCl_2$ alleviate cold-induced oxidative stress and photosynthetic inhibition in cucumber seedlings. Journal of Plant Growth Regulation 42 (6), 3441–3458. Available at https://doi.org/10.1007/s00344-022-10805-z.

Ma, L., Liu, X., Lv, W., Yang, Y., 2022. Molecular Mechanisms of Plant Responses to Salt Stress. Frontiers in Plant Science 13. Available at 10.3389/fpls.2022.934877.

Ma, Z., Lv, J., Wu, W., Fu, D., Lü, S., Ke, Y., et al., 2023b. Regulatory network of rice in response to heat stress and its potential application in breeding strategy. Molecular Breeding 43 (9), 68.

Marschner, H., 2011. Marschner's Mineral Nutrition of Higher Plants, third ed. Academic Press.

Meenakshi, Kumar, A., Kumar, V., Dubey, A.K., Narayan, S., Sawant, S.V., et al., 2022. CAMTA transcription factor enhances salinity and drought tolerance in chickpea (*Cicer arietinum* L). Plant Cell, Tissue and Organ Culture 148 (2), 319–330. Available at https://doi.org/10.1007/s11240-021-02191-3.

Mittler, R., 2006. Abiotic stress, the field environment and stress combination. Trends in Plant Science 11 (1), 15–19. Available at https://doi.org/10.1016/j.tplants.2005.11.002.

Mohanta, T.K., Bashir, T., Hashem, A., Abd Allah, E.F., Khan, A.L., Al-Harrasi, A.S., 2018. Early events in plant abiotic stress signaling: interplay between calcium, reactive oxygen species and phytohormones. Journal of Plant Growth Regulation 37 (4), 1033–1049. Available at https://doi.org/10.1007/s00344-018-9833-8.

Mori, K., Renhu, N., Naito, M., Nakamura, A., Shiba, H., Yamamoto, T., et al., 2018. Ca^{2+} permeable mechanosensitive channels MCA1 and MCA2 mediate cold-induced cytosolic Ca^{2+} increase and cold tolerance in *Arabidopsis*. Scientific Reports 8 (1), 550.

Mahiwal, S., Pandey, G.K., 2022. Potassium: a vital nutrient mediating stress tolerance in plants. Journal of Plant Biochemistry and Biotechnology 31, 705–719. Available at 10.1007/s13562-022-00775-4.

Mukherjee, S., 2022. Chemistry of soil nutrient. Current Topics in Soil Science. Springer, pp. 165–182.

Naciri, R., Lahrir, M., Benadis, C., Chtouki, M., Oukarroum, A., 2021. Interactive effect of potassium and cadmium on growth, root morphology and chlorophyll a fluorescence in tomato plant. Scientific Reports 11 (1). Available at https://doi.org/10.1038/s41598-021-84990-4.

Naeem, M., Amir, M., Manzoor, H., Rasul, S., Athar, H.U.R., 2019. Role of calcium in conferring abiotic stress tolerance. Plant Tolerance to Environmental Stress. CRC Press, pp. 205–218.

Navrot, N., Rouhier, N., Gelhaye, E., Jacquot, J.P., 2007. Reactive oxygen species generation and antioxidant systems in plant mitochondria. Physiologia Plantarum 129 (1), 185–195. Available at https://doi.org/10.1111/j.1399-3054.2006.00777.x.

Ndathe, R., Dale, R., Kato, N., 2022. Dynamic modeling of ABA-dependent expression of the *Arabidopsis* RD29A gene. Frontiers in Plant Science 13. Available at https://doi.org/10.3389/fpls.2022.928718.

Nieves-Cordones, M., Ródenas, R., Lara, A., Martínez, V., Rubio, F., 2018. The combination of K^+ deficiency with other environmental stresses: What is the outcome? Physiologia Plantarum 165, 264–276. Available at 10.1111/ppl.12827.

Nieves-Cordones, M., Ródenas, R., Lara, A., Martínez, V., Rubio, F., 2019. The combination of K^+ deficiency with other environmental stresses: what is the outcome? Physiologia Plantarum 165 (2), 264–276.

Nehra, A., Kundu, P., Ahlawat, K., Chhikara, A., Agarwala, N., Tuteja, N., et al., 2022. Comprehensive genomic insight deciphers significance of EF-hand gene family in foxtail millet [*Setaria italica* (L.) P. Beauv]. South African Journal of Botany 148, 652–665. Available at https://doi.org/10.1016/j.sajb.2022.01.008.

Parveen, Anwar-Ul-Haq, M., Aziz, T., Aziz, O., Maqsood, L., 2021. Potassium induces carbohydrates accumulation by enhancing morpho-physiological and biochemical attributes in soybean under salinity. Archives of Agronomy and Soil Science 67 (7), 946–959. Available at https://doi.org/10.1080/03650340.2020.1769075.

Poonia, A.K., Mishra, S.K., Sirohi, P., Chaudhary, R., Kanwar, M., Germain, H., et al., 2020. Overexpression of wheat transcription factor (TaHsfA6b) provides thermotolerance in barley. Planta 252 (4). Available at https://doi.org/10.1007/s00425-020-03457-4.

Poovaiah, B.W., Reddy, A.S.N., 1993. Calcium and signal transduction in plants. Critical Reviews in Plant Sciences 12 (3), 185–211. Available at https://doi.org/10.1080/07352689309701901.

Popescu, S.C., Popescu, G.V., Bachan, S., Zhang, Z., Seay, M., Gerstein, M., et al., 2007. Differential binding of calmodulin-related proteins to their targets revealed through high-density *Arabidopsis* protein microarrays. Proceedings of the National Academy of Sciences 104 (11), 4730–4735. Available at https://doi.org/10.1073/pnas.0611615104.

Prasad, K.V.S.K., Abdel-Hameed, A.A.E., Jiang, Q., Reddy, A.S.N., 2023. DNA-binding activity of CAMTA3 Is essential for its function: identification of critical amino acids for its transcriptional activity. Cells 12 (15), 1986. Available at https://doi.org/10.3390/cells12151986.

Raina, M., Kisku, A.V., Joon, S., Kumar, S., Kumar, D., 2021. Calmodulin and calmodulin-like Ca^{2+} binding proteins as molecular players of abiotic stress response in plants. Calcium Transport Elements in Plants. Elsevier, India, pp. 231–248. Available at http://doi.org/10.1016/B978-0-12-821792-4.00001-1.

Rajput, R.S., Singh, J., Singh, P., Vaishnav, A., Singh, H.B., 2021. Influence of seed biopriming and vermiwash treatment on tomato plant's immunity and nutritional quality upon sclerotium rolfsii challenge inoculation. Journal of Plant Growth Regulation 40 (4), 1493–1509. Available at https://doi.org/10.1007/s00344-020-10205-1.

Ranjan, A., Kalwan, G., Kaushik, S., Taneja, D., Yadav, P., Nehra, A., et al., 2025. Signaling cascades, transcription factors, and gene expression in ROS-mediated abiotic stress tolerance. Plant Signaling Molecules in Regulation of ROS-Scavenging System: Adapting to Climate Change. Springer Nature, Singapore, pp. 251–269.

Raza, S., Farrukh Saleem, M., Mustafa Shah, G., Jamil, M., Haider Khan, I., 2013. Potassium applied under drought improves physiological and nutrient uptake performances of wheat (*Triticum Aestivun* L.). Journal of Soil Science and Plant Nutrition 13 (1), 175–185.

Raza, H.Z., Shah, A.A., Noreen, Z., Usman, S., Zafar, S., Yasin, N.A., et al., 2024. Calcium oxide nanoparticles mitigate lead stress in *Abelmoschus esculentus* though improving the key antioxidative enzymes, nutritional content and modulation of stress markers. Plant Physiology and Biochemistry 206, 108171.

Reddy, V.S., Reddy, A.S.N., 2004. Proteomics of calcium-signaling components in plants. Phytochemistry 65 (12), 1745–1776. Available at https://doi.org/10.1016/j.phytochem.2004.04.033.

Ritonga, F.N., Chen, S., 2020. Physiological and molecular mechanism involved in cold stress tolerance in plants. Plants 9 (5), 560. Available at https://doi.org/10.3390/plants9050560.

Sanders, D., Pelloux, J., Brownlee, C., Harper, J.F., 2002. Calcium at the crossroads of signaling. The Plant Cell 14, S401. Available at https://doi.org/10.1105/tpc.002899.

Sardans, J., Peñuelas, J., 2021. Potassium control of plant functions: ecological and agricultural implications. Plants 10 (2), 419. Available at https://doi.org/10.3390/plants10020419.

Singh, T., Singh, C., C.V, N., Pansuriya, A.G., 2025. Influence of foliar applications of synthetic compounds on Wheat Grain Quality During Terminal Heat Stress. Journal of Cereal Research 17. Available at 10.25174/2582-2675/2025/159851.

Sarwar, M., Saleem, M.F., Ullah, N., Ali, S., Rizwan, M., Shahid, M.R., Alyemeni, M.N., Alamri, S.A., Ahmad, P., 2019. Role of mineral nutrition in alleviation of heat stress in cotton plants grown in glasshouse and field conditions. Scientific Reports 9. Available at 10.1038/s41598-019-49404-6.

Sarwar, M., Saleem, M.F., Maqsood, H., Ullah, N., Khan, A., Waqas, M., et al., 2022. Strengthening leaf physiological functioning and grain yield formation in heat-stressed wheat through potassium application. Frontiers in Plant Science 13. Available at https://doi.org/10.3389/fpls.2022.1005773.

Shabala, S., Pottosin, I.I., 2010. Potassium and potassium-permeable channels in plant salt tolerance. In: Demidchik, V., Maathuis, F.J. (Eds.), Ion channels and plant stress responses. Springer, pp. 87–110.

Shaffique, S., Injamum-Ul-Hoque, M., Husen, A., Kang, S.-M., Lee, I.-J., 2025. Revolutionizing heat stress tolerance in *Glycine max*: exploring the latest advances in microbial application. Plant Stress 15, 100725. Available at https://doi.org/10.1016/j.stress.2024.100725.

Shahid, M., Saleem, M.F., Saleem, A., Raza, M.A.S., Kashif, M., Shakoor, A., et al., 2019. Exogenous potassium–instigated biochemical regulations confer terminal heat tolerance in wheat. Journal of Soil Science and Plant Nutrition 19 (1), 137–147.

Shahid, M., Saleem, M.F., Saleem, A., Sarwar, M., Khan, H.Z., Shakoor, A., 2020. Foliar potassium-induced regulations in glycine betaine and malondialdehyde were associated with grain yield of heat-stressed bread wheat (*Triticum aestivum* L). Journal of Soil Science and Plant Nutrition 20 (4), 1785–1798. Available at https://doi.org/10.1007/s42729-020-00250-w.

Sharma, K., Jalota, K., Agarwal, C., Pal, P., Jindal, S., 2024. Biochemical responses of plants to individual and combined abiotic stresses. Plant Secondary Metabolites and Abiotic Stress 1–34. Available at https://doi.org/10.1002/9781394186457.ch1.

Shi, S., Li, S., Asim, M., Mao, J., Xu, D., Ullah, Z., et al., 2018. The *Arabidopsis* calcium-dependent protein kinases (CDPKs) and their roles in plant growth regulation and abiotic stress responses. International Journal of Molecular Sciences 19 (7). Available at https://doi.org/10.3390/ijms19071900.

Shkryl, Y.N., Veremeichik, G.N., Bulgakov, V.P., Zhuravlev, Y.N., 2011. Induction of anthraquinone biosynthesis in Rubia cordifolia cells by heterologous expression of a calcium-dependent protein kinase gene. Biotechnology and Bioengineering 108 (7), 1734–1738. Available at https://doi.org/10.1002/bit.23077.

Siddiqui, M.H., Alamri, S.A., Al-Khaishanya, M.Y.Y., Al-Qutami, M.A., Ali, H.M., Khan, M.N., 2017. Sodium nitroprusside and indole acetic acid improve the tolerance of tomato plants to heat stress by protecting against DNA damage. Journal of Plant Interactions 12 (1), 177–186. Available at https://doi.org/10.1080/17429145.2017.1310941.

Shen, H.F., Zhao, B., Xu, J.J., Liang, W., Huang, W.M. and Li, H., H., 2017. Effects of heat stress on changes in physiology and anatomy in two cultivars of Rhododendron. South African Journal of Botany 112, 338–345. Available at 10.1016/j.sajb.2017.06.018.

Silva, L.F., Barreto, K.F.M., Silva, H.C., de Souza, I.D., Meneses, C.H.S.G., Uchôa, A.F., et al., 2025. Insights of cellular and molecular changes in sugarcane response to oxidative signaling. BMC Plant Biology 25 (1). Available at https://doi.org/10.1186/s12870-024-06036-x.

Sakhonwasee, S., Phingkasan, W., 2017. Effects of the foliar application of calcium on photosynthesis, reactive oxygen species production, and changes in water relations in tomato seedlings under heat stress. Horticulture, Environment, and Biotechnology 58, 119–126. Available at 10.1007/s13580-017-0194-1.

Shen, J., Cheng, H., Li, X., Pan, X., Hu, Y., Jin, S., 2022. Beneficial Effect of Exogenously Applied Calcium Chloride on the Anatomy and Fast Chlorophyll Fluorescence in Rhododendron × pulchrum Leaves Following Short-Term Heat Stress Treatment. Agronomy 12, 3226. Available at 10.3390/agronomy12123226.

Singh, S.K., Reddy, V.R., 2017. Potassium starvation limits soybean growth more than the photosynthetic processes across CO_2 levels. Frontiers in Plant Science 8. Available at https://doi.org/10.3389/fpls.2017.00991.

Singh, R., 2020. Calcium in plant biology: nutrient and second messenger. International Journal of Biological Innovations 02 (01), 31–35. Available at https://doi.org/10.46505/ijbi.2020.2105.

Su, W., Huang, L., Ling, H., Mao, H., Huang, N., Su, Y., et al., 2020. Sugarcane calcineurin B-like (CBL) genes play important but versatile roles in regulation of responses to biotic and abiotic stresses. Scientific Reports 10 (1). Available at https://doi.org/10.1038/s41598-019-57058-7.

Suzuki, E., Masuda, S., 1961. Studies on muskmelon (*Cucumis melo* L. var. reticulates Naud.). V. The fluctuation of sugar contents in Earl's Favourite. Bulletin of the Faculty of Education, Sizuoka University 12, 205–213.

Tahtiharju, S., Sangwan, V., Monroy, A.F., Dhindsa, R.S., Borg, M., 1997. The induction of kin genes in cold-acclimating *Arabidopsis thaliana*: evidence of a role for calcium. Planta 203 (4), 442–447.

Tang, R.J., Zhao, F.G., Yang, Y., Wang, C., Li, K., Kleist, T.J., et al., 2020. A calcium signalling network activates vacuolar K^+ remobilization to enable plant adaptation to low-K environments. Nature Plants 6 (4), 384–393. Available at https://doi.org/10.1038/s41477-020-0621-7.

Tang, W., Arisha, M.H., Zhang, Z., Yan, H., Kou, M., Song, W., et al., 2023. Comparative transcriptomic and proteomic analysis reveals common molecular factors responsive to heat and drought stresses in sweetpotaoto (*Ipomoea batatas*). Frontiers in Plant Science 13. Available at https://doi.org/10.3389/fpls.2022.1081948.

Thoday-Kennedy, E.L., Jacobs, A.K., Roy, S.J., 2015. The role of the CBL–CIPK calcium signalling network in regulating ion transport in response to abiotic stress. Plant Growth Regulation 76 (1), 3–12. Available at https://doi.org/10.1007/s10725-015-0034-1.

Tiwari, A., Pandey Rai, S., Rai, K.K., Tiwari, A., Pandey, N., 2023. Molecular and epigenetic basis of heat stress responses and acclimatization in plants. The Nucleus 66 (1), 69–79. Available at https://doi.org/10.1007/s13237-022-00400-8.

Tong, T., Li, Q., Jiang, W., Chen, G., Xue, D., Deng, F., et al., 2021. Molecular evolution of calcium signaling and transport in plant adaptation to abiotic stress. International Journal of Molecular Sciences 22 (22). Available at https://doi.org/10.3390/ijms222212308.

Tovar, J.C., Berry, J.C., Quillatupa, C., Castillo, S.E., Acosta-Gamboa, L., Fahlgren, N., et al., 2022. Heat stress changes mineral nutrient concentrations in *Chenopodium quinoa* seed. Plant Direct 6 (2). Available at https://doi.org/10.1002/pld3.384.

Tränkner, M., Tavakol, E., Jákli, B., 2018. Functioning of potassium and magnesium in photosynthesis, photosynthate translocation and photoprotection. Physiologia Plantarum 163 (3), 414–431. Available at https://doi.org/10.1111/ppl.12747.

Tripathi, D.K., Singh, V.P., Chauhan, D.K., Kumar, A., Sharma, S., Rai, K., et al., 2014. Role of macronutrients in plant growth and acclimation: recent advances and future prospective. In: Ahmad, P., Wani, M.R., Azooz, M.M., Tran, L.-S.P. (Eds.), Improvement of Crops in the Era of Climatic Changes. Springer, pp. 197–216.

Tuteja, N., Mahajan, S., 2007. Calcium signaling network in plants: an overview. Plant Signaling and Behavior 2 (2), 79–85. Available at https://doi.org/10.4161/psb.2.2.4176.

Tuteja, N., Gill, S.S., Trivedi, P.K., Asif, M.H., Nath, P., 2010. Plant growth regulators and their role in stress tolerance. Plant Stress 4 (1), 1–18.

Tan, W., Meng, Q.W., Brestic, M., Olsovska, K., Yang, X., 2011. Photosynthesis is improved by exogenous calcium in heat-stressed tobacco plants. Journal of Plant Physiology 168, 2063–2071. Available at 10.1016/j.jplph.2011.06.009.

Uz Zaman, Q., Abbasi, A., Tabassum, S., Ashraf, K., Ahmad, Z., Siddiqui, M.H., et al., 2022. Calcium induced growth, physio-biochemical, antioxidants, osmolytes adjustments and phytoconstituents status in spinach under heat stress. South African Journal of Botany 149, 701–711. Available at https://doi.org/10.1016/j.sajb.2022.06.065.

Veremeichik, G.N., Grigorchuk, V.P., Silanteva, S.A., Shkryl, Y.N., Bulgakov, D.V., Brodovskaya, E.V., et al., 2019. Increase in isoflavonoid content in *Glycine max* cells transformed by the constitutively active Ca^{2+} independent form of the AtCPK1 gene. Phytochemistry 157, 111–120. Available at https://doi.org/10.1016/j.phytochem.2018.10.023.

Verma, L., Singh, A.K., Singh, S., Tiwari, D., Zaidi, S.T., Yadav, R.K., et al., 2024. Temperature stress its impact on yield of various wheat varieties at different growth stages. National Academy Science Letters 47 (3), 219–225. Available at https://doi.org/10.1007/s40009-023-01303-1.

Virdi, A.S., Singh, S., Singh, P., 2015. Abiotic stress responses in plants: roles of calmodulin-regulated proteins. Frontiers in Plant Science 6. Available at https://doi.org/10.3389/fpls.2015.00809.

Volf, M.R., Crusciol, C.A.C., de Azevedo, A.C., Thompson, M.L., Kovar, J.L., Rosolem, C.A., 2021. Potassium bioavailability in a tropical kaolinitic soil. Agronomy 11 (10). Available at https://doi.org/10.3390/agronomy11102016.

Wang, Y., Wu, W.-H., 2013. Potassium transport and signaling in higher plants. Annual Review of Plant Biology 64, 451–476.

Wang, Y., Tang, R.J., Yang, X., Zheng, X., Shao, Q., Tang, Q.L., et al., 2019. Golgi-localized cation/proton exchangers regulate ionic homeostasis and skotomorphogenesis in *Arabidopsis*. Plant Cell and Environment 42 (2), 673–687. Available at https://doi.org/10.1111/pce.13452.

Wang, H., Shen, Y., Wang, K., He, S., Kim, W.S., Shang, W., et al., 2022. Effect of exogenous calcium on the heat tolerance in rosa hybrida 'carolla' Horticulturae 8 (10). Available at https://doi.org/10.3390/horticulturae8100980.

Waraich, E.A., Ahmad, R., Halim, A., Aziz, T., 2012. Alleviation of temperature stress by nutrient management in crop plants: a review. Journal of Soil Science and Plant Nutrition 12 (2), 221–244. Available at https://doi.org/10.4067/S0718-95162012000200003.

Waraich, E.A., Rashid, F., Ahmad, Z., Ahmad, R., Ahmad, M., 2020. Foliar applied potassium stimulate drought tolerance in canola under water deficit conditions. Journal of Plant Nutrition 43 (13), 1923–1934. Available at https://doi.org/10.1080/01904167.2020.1758132.

Wei, J.T., Hou, Z.H., Wang, Y., Hao, J.M., Wang, J., Wang, W., et al., 2023. Foxtail millet SiCDPK7 gene enhances tolerance to extreme temperature stress in transgenic plants. Environmental and Experimental Botany 207. Available at https://doi.org/10.1016/j.envexpbot.2022.105197.

Xiao, X.H., Yang, M., Sui, J.L., Qi, J.Y., Fang, Y.J., Hu, S.N., et al., 2017. The calcium-dependent protein kinase (CDPK) and CDPK-related kinase gene families in Hevea brasiliensis—comparison with five other plant species in structure, evolution, and expression. FEBS Open Bio 7 (1), 4–24. Available at https://doi.org/10.1002/2211-5463.12163.

Xiao, C., Zhang, H., Xie, F., Pan, Z.Y., Qiu, W.M., Tong, Z., et al., 2022. Evolution, gene expression, and protein-protein interaction analyses identify candidate CBL-CIPK signalling networks implicated in stress responses to cold and bacterial infection in citrus. BMC Plant Biology 22 (1). Available at https://doi.org/10.1186/s12870-022-03809-0.

Xu, T., Niu, J., Jiang, Z., 2022. Sensing mechanisms: calcium signaling mediated abiotic stress in plants. Frontiers in Plant Science 13. Available at https://doi.org/10.3389/fpls.2022.925863.

Yang, H., Yang, Y., 2021. Effect of exogenous calcium and calcium inhibitors on the heat tolerance in *Phalaenopsis* seedlings. Chinese Journal of Tropical Crops 42 (2), 436–441.

Yang, Y., Wang, S., Leng, P., Wu, J., Hu, Z., 2022. Calcium and jasmonate signals mediate biosynthesis of the floral fragrance regulated by light quality in snapdragon. Plant Growth Regulation 97 (1), 91–100. Available at https://doi.org/10.1007/s10725-022-00807-y.

Yasin, N.A., Zaheer, M.M., Khan, W.U., Ahmad, S.R., Ahmad, A., Ali, A., et al., 2018. The beneficial role of potassium in Cd-induced stress alleviation and growth improvement in *Gladiolus grandiflora* L. International Journal of Phytoremediation 20 (3), 274–283. Available at https://doi.org/10.1080/15226514.2017.1374337.

Yadav, M.R., Choudhary, M., Singh, J., Lal, M.K., Jha, P.K., Udawat, P., et al., 2022. Impacts, tolerance, adaptation, and mitigation of heat stress on wheat under changing climates. International Journal of Molecular Sciences 23 (5), 2838. Available at https://doi.org/10.3390/ijms23052838.

Yip Delormel, T., Boudsocq, M., 2019. Properties and functions of calcium-dependent protein kinases and their relatives in *Arabidopsis thaliana*. New Phytologist 224 (2), 585–604. Available at https://doi.org/10.1111/nph.16088.

Yuan, L., Yuan, Y., Liu, S., Wang, J., Zhu, S., Chen, G., et al., 2017. Influence of high temperature on photosynthesis, antioxidative capacity of chloroplast, and carbon assimilation among heat-tolerant and heat-susceptible genotypes of nonheading Chinese cabbage. HortScience: a Publication of the American Society for Horticultural Science 52 (11), 1464–1470. Available at https://doi.org/10.21273/HORTSCI12259-17.

Zahoor, R., Dong, H., Abid, M., Zhao, W., Wang, Y., Zhou, Z., 2017. Potassium fertilizer improves drought stress alleviation potential in cotton by enhancing photosynthesis and carbohydrate metabolism. Environmental and Experimental Botany 137, 73–83. Available at https://doi.org/10.1016/j.envexpbot.2017.02.002.

Zhang, C.X., Fu, G.F., Yang, X.Q., Yang, Y.J., Zhao, X., Chen, T.T., et al., 2016. Heat stress effects are stronger on spikelets than on flag leaves in rice due to differences in dissipation capacity. Journal of Agronomy and Crop Science 202 (5), 394–408. Available at https://doi.org/10.1111/jac.12138.

Zhang, D.W., Vu, T.S., Huang, J., Chi, C.Y., Xing, Y., Fu, D.D., et al., 2019. Effects of calcium on germination and seedling growth in *Melilotus officinalis* L. (fabaceae) under salt stress. Pakistan Journal of Botany 51 (1), 1–9. Available at https://doi.org/10.30848/PJB2019-1(44).

Zhang, Z., Wu, P., Zhang, W., Yang, Z., Liu, H., Ahammed, G.J., et al., 2020. Calcium is involved in exogenous NO-induced enhancement of photosynthesis in cucumber

(*Cucumis sativus* L.) seedlings under low temperature. Scientia Horticulturae 261, 108953. Available at https://doi.org/10.1016/j.scienta.2019.108953.

Zhang, H., Zhu, J., Gong, Z., Zhu, J.K., 2022. Abiotic stress responses in plants. Nature Reviews. Genetics 23 (2), 104–119. Available at https://doi.org/10.1038/s41576-021-00413-0.

Zhao, Y., Du, H., Wang, Y., Wang, H., Yang, S., Li, C., et al., 2021. The calcium-dependent protein kinase ZmCDPK7 functions in heat-stress tolerance in maize. Journal of Integrative Plant Biology 63 (3), 510–527. Available at https://doi.org/10.1111/jipb.13056.

Zhu, X., Pan, T., Zhang, X., Fan, L., Quintero, F.J., Zhao, H., et al., 2018. K + efflux antiporters 4, 5, and 6 mediate pH and K^+ homeostasis in endomembrane compartments. Plant Physiology 178 (4), 1657–1678. Available at https://doi.org/10.1104/pp.18.01053.

Zrig, A., AbdElgawad, H., Touneckti, T., Ben Mohamed, H., Hamouda, F., Khemira, H., 2021. Potassium and calcium improve salt tolerance of Thymus vulgaris by activating the antioxidant systems. Scientia Horticulturae 277, 109812. Available at 10.1016/j.scienta.2020.109812.

Zhang, H, Zhu, J, Gong, Z, Zhu, JK, 2022. Abiotic stress responses in plants. Nat Rev Genet 23 (2), 104–119. Available at 10.1038/s41576-021-00413-0.

Further reading

Ali, L., Manzoor, N., Masood, H.A., Abbas, A., 2024. Nanotechnology-enabled approaches to mitigating abiotic stresses in agricultural crops. In: Shahid, M., Gaur, R. (Eds.), Molecular Dynamics of Plant Stress and its Management. Springer, Singapore, pp. 415–439.

Aslam, M., Greaves, J.G., Jakada, B.H., Fakher, B., Wang, X., Qin, Y., 2022. AcCIPK5, a pineapple CBL-interacting protein kinase, confers salt, osmotic and cold stress tolerance in transgenic *Arabidopsis*. Plant Science 320, 111284.

Boyer, J.S., 1982. Plant productivity and environment. Science (New York, N.Y.) 218 (4571), 443–448.

Cui, J., Nieves-Cordones, M., Rubio, F., Tcherkez, G., 2021. Involvement of salicylic acid in the response to potassium deficiency revealed by metabolomics. Plant Physiology and Biochemistry 163, 201–204.

Ho, L.H., Rode, R., Siegel, M., Reinhardt, F., Neuhaus, H.E., Yvin, J.C., et al., 2020. Potassium application boosts photosynthesis and sorbitol biosynthesis and accelerates cold acclimation of common plantain (*Plantago major* L). Plants 9 (10), 1259.

Liese, A., Romeis, T., 2013. Biochemical regulation of in vivo function of plant calcium-dependent protein kinases (CDPK). Biochimica et Biophysica Acta (BBA) - Proteins and Proteomics 1833 (7), 1582–1589.

Pathak, J., Ahmed, H., Kumari, N., Pandey, A., Sinha, R.P., 2020. Role of calcium and potassium in amelioration of environmental stress in plants. In: Roychoudhury, A., Tripathi, D.K. (Eds.), Protective Chemical Agents in the Amelioration of Plant Abiotic Stress: Biochemical and Molecular Perspectives. John Wiley and Sons Ltd, pp. 535–562.

Reddy, P.S., Chakradhar, T., Reddy, R.A., Nitnavare, R.B., Mahanty, S., Reddy, M.K., 2016. Role of heat shock proteins in improving heat stress tolerance in crop plants. Heat Shock Proteins and Plants. Springer International Publishing, pp. 283–307.
Römheld, V., Kirkby, E.A., 2010. Research on potassium in agriculture: needs and prospects. Plant and Soil 335 (1), 155–180.
Sima, N.A., Askari, H., Mirzaei, H.H., and colleagues, 2009. Genotype-dependent differential responses of three forage species to calcium supplement in saline conditions. Journal of Plant Nutrition 32 (4), 579–597.

CHAPTER

Iron nutrition and plant responses to abiotic stress

10

Syeda Irsa Mazhar Kazmi[1], **Iram Gul**[2], **Ahmad Mahmood**[1] **and Muhammad Arshad**[2]
[1]*School of Interdisciplinary Engineering and Sciences (SINES), National University of Sciences and Technology (NUST), Islamabad, Pakistan,*
[2]*Institute of Environmental Sciences and Engineering (IESE), School of Civil Engineering and Environmental Sciences (SCEE), National University of Sciences and Technology (NUST), Islamabad, Pakistan*

10.1 Introduction

The agricultural sector globally has suffered substantial setbacks due to the intensifying effects of climate change. Key environmental challenges such as drought, salinity, waterlogging, air pollutants, temperature limits, and heavy metals accumulation have made crops more vulnerable to secondary stresses like pest invasions and diseases. These abiotic stressors influence plants across their entire life cycle, particularly affecting their growth and developmental stages (Husen, 2021a,b,c, 2023, 2024, 2025). It is estimated that over 50% of global agricultural production losses are linked to various abiotic stresses, which impact approximately 91% of the planet's cultivable land (Younis et al., 2020). These stresses bring about notable changes in plant morphology, physiology, and biochemical functions as shown in Fig. 10.1, though the degree of impact largely depends on the specific crop, type of stressor, and duration of exposure (El-Saadony et al., 2022). Although climate change has been a progressive phenomenon, the rate of change has sharply accelerated in the last hundred years. The increase in greenhouse gas emissions from anthropogenic sources has caused a measurable rise in global temperatures by about 0.9°C since the 19th century (Fahad et al., 2017; Alotaibi, 2023). High insolation, predominantly in the daytime, enhances processes such as evapotranspiration, leading to further temperature spikes. Forecasts suggest that by the end of the 21st century, global temperatures may increase by as much as 4.5°C (Giordano et al., 2021), posing a significant risk to food production. By 2050, nearly 45% of the regions that produce maize may experience annual periods of extreme heat—exceeding 35°C for at least 5 consecutive days during the crop's reproductive stage. A mere 1°C increase in average seasonal temperature could reduce maize yields by 3%–13% (Waqas et al., 2021).

Mineral Nutrients to Mitigate Abiotic Stress in Crops. DOI: https://doi.org/10.1016/B978-0-443-44712-9.00003-7

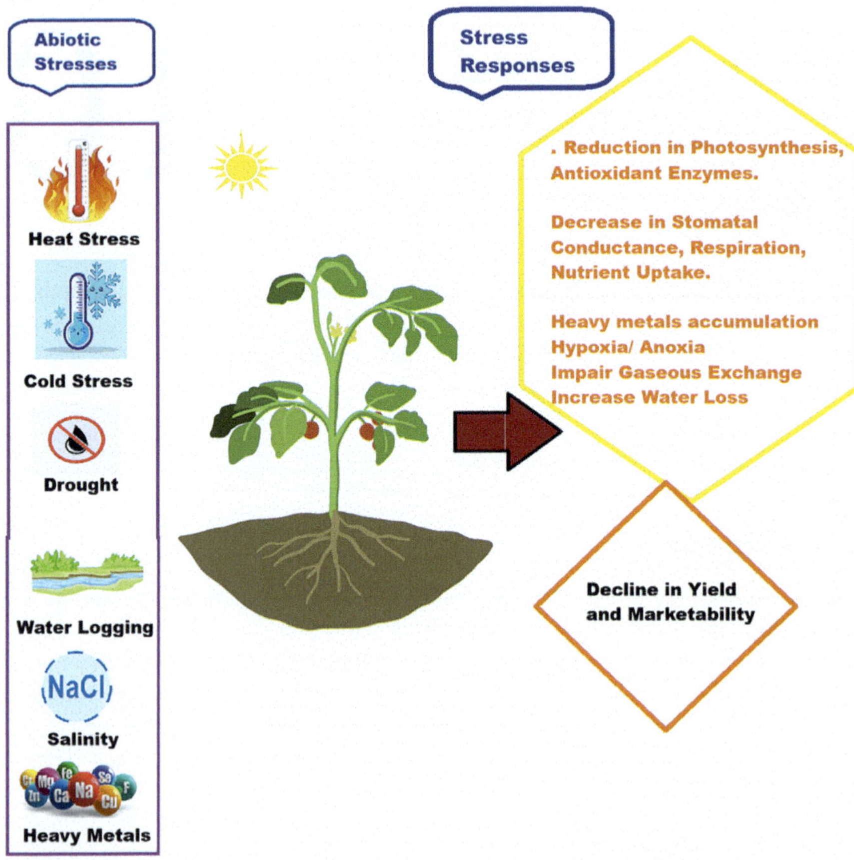

FIGURE 10.1

Detailed illustration of plants' response to abiotic stresses.

The unprecedented rise in global temperatures has been accompanied by various climate anomalies, including droughts, waterlogging, irregular rainfall, glacier retreat, stronger hurricanes, shifts in wildlife and plant habitats, and other extreme events (Alotaibi, 2023). With climate change intensifying, the frequency and severity of abiotic stress events are expected to grow, resulting in significant reductions in agricultural productivity and crop quality. These environmental stresses disrupt physiological, biochemical, and molecular processes in plants, ultimately affecting yield and market value. Studies on both model and nonmodel plants reveal common stress responses such as cell dehydration, stomatal closure, osmotic imbalance, impaired photosynthesis process, accumulation of reactive oxygen species (ROS), suppression of antioxidant enzymes, and altered gene expressions, all of which contribute to reduced plant growth and nutritional value (Waqas et al., 2021; dos Santos et al., 2022; Giordano et al., 2021). Plants exposed to saline or contaminated soil often accumulate excess sodium ions and heavy metals, which impair development and reduce yield. Due to their immobility, plants adapt to such stressors by altering their internal metabolic pathways to ensure survival, often at the cost of productivity (Rajput et al., 2021). Research indicates that high temperatures alone can reduce crop yield by 10%–17%, depending on the plant's growth stage (Faiz et al., 2020). Additionally, cold stress has resulted in drastic yield losses—up to 60%–70% in legumes such as chickpeas, soybeans, and mung beans. Overall, combined heat and cold stresses (Bhat et al., 2022) account for 10–40% of agricultural losses, while salinity can cause reductions ranging from 5% to 50% (Ahmad et al., 2023). The financial burden of climate-related damage has also been substantial, with global warming contributing to economic losses estimated at around US$125 billion (Kompas et al., 2018). Meeting the nutritional demands of an anticipated global population explosion of 9–10 billion near 2050 requires innovative strategies to enhance crop resilience under abiotic stress. Various approaches are being explored, including traditional and modern breeding techniques (Oladosu et al., 2019), improved agronomic practices, seed priming, microbial inoculation, grafting, and the application of plant growth regulators and osmoprotectants (Oyebamiji et al., 2023; Hassan et al., 2021). More recently, cost-effective and eco-friendly solutions such as the use of biochar, yeast extracts, and seaweed-based bio stimulants, as well as advancements in agricultural nanotechnology (NT), have gained attention for their role in enhancing stress tolerance. While there is growing evidence supporting the benefits of bio stimulants in improving yield and quality under stress, the precise morphological, physiological, and molecular mechanisms driving these benefits remain insufficiently understood. This chapter aims to provide a comprehensive and up-to-date overview of innovative strategies and technologies that can mitigate the impacts of climate change-induced abiotic stress (iron deficiency) on crop performance (Monteiro et al., 2022; Abideen et al., 2022).

10.2 Iron in plant physiology

Iron (Fe), though required in trace amounts, is vital for multiple plant metabolic processes. Its ability to shift between Fe^{2+} and Fe^{3+} states makes it a critical cofactor in enzymatic functions related to electron transport, redox balance, and energy metabolism. In photosynthesis, Fe contributes to photosystems I and II, cytochromes, and ferredoxins, while also supporting mitochondrial respiration through its presence in electron transport proteins and Fe–S clusters. A lack of iron commonly leads to chlorosis and reduced ATP production, negatively affecting nutrient uptake and stress responses (Oyebamiji et al., 2024; Briat et al., 2015). Iron also facilitates nitrogen metabolism by acting as a cofactor in nitrate and nitrite reductases and nitrogenase enzymes in legumes. Its role extends to DNA synthesis, amino acid metabolism, and antioxidant defense through enzymes like catalase and ascorbate peroxidase. When Fe is deficient, oxidative stress intensifies due to reduced enzyme activity (Tripathi et al., 2018). Plants absorb iron through two strategies: Strategy I (in nongrasses) involves acidification of the rhizosphere and Fe^{3+} reduction before uptake by IRT1, while Strategy II (in grasses) relies on phyto siderophores that chelate Fe^{3+}, taken up via YSL transporters. Internal Fe transport requires chelators like nicotiana amine, transport proteins (YSLs, OPT3), and storage in ferritin to avoid toxicity (Ishimaru et al., 2011; Kobayashi and Nishizawa, 2012; Dubeaux et al., 2015). Although a micronutrient, iron underpins many key physiological systems, and its scarcity can compromise plant health, particularly under environmental stress. Understanding iron dynamics is essential for enhancing nutrient efficiency and crop resilience (Tripathi et al., 2018).

10.3 Causes and symptoms of iron deficiency

Fe deficiency is one of the most widespread and agronomically significant micronutrient disorders affecting plants worldwide. Despite iron's abundance in the Earth's crust, its bioavailability to plants is often limited due to soil and environmental factors. The most common cause of iron deficiency is high soil pH (above 7.5), especially in calcareous and alkaline soils. In these soils, Fe^{3+} ions rapidly form insoluble ferric oxides and hydroxides, making them unavailable to plants. $Fe(OH)_3$ precipitation increases with rising pH. Plants growing in these soils often exhibit "induced" deficiency despite total soil Fe being sufficient (Shao et al., 2007). Poorly aerated soils or waterlogged conditions lead to oxygen deficiency, which impairs the function of Fe(III) reductase enzymes and disrupts iron reduction and uptake mechanisms in plant roots. Waterlogging can cause Fe oxidation or microbial transformations that reduce Fe availability. Iron deficiency can paradoxically co-occur with Fe toxicity in such environments. Bicarbonate (HCO_3^-) ions interfere with Fe acquisition by inhibiting root proton extrusion, suppressing the activity of ferric chelate reductase enzymes, and reducing expression of Fe uptake genes such as FRO2 and IRT1. These effects exacerbate Fe deficiency in calcareous soils. Excessive levels of other elements can interfere with Fe uptake: High phosphorus (P) binds with Fe, forming insoluble Fe–P complexes, Zinc (Zn), manganese (Mn), and copper (Cu) compete with Fe for transporter proteins, Calcium (Ca^{2+}) and magnesium (Mg^{2+}) in excess reduce Fe solubility and transport (Anjum et al., 2015). While organic matter generally improves micronutrient availability, immature or undecomposed organic matter can immobilize Fe or stimulate microbial populations that oxidize Fe^{2+}, reducing its plant availability. Conversely, insufficient natural chelators like nicotiana amine or low root exudation of organic acids in some species also limit iron solubilization. Some plant species and cultivars are inherently inefficient at acquiring iron, particularly under stress or high pH conditions. Symptoms of iron deficiency in plants include visual symptoms. The classical and most diagnostic indicator of Fe deficiency is interveinal chlorosis, yellowing, while the veins themselves remain green. First appears in young leaves due to Fe's immobility in the phloem. As deficiency progresses, leaves may become entirely yellow or white. In severe cases, leaf necrosis, stunted growth, and abscission may occur. Iron deficiency leads to:

- Reduced photosynthetic capacity due to lower chlorophyll content.
- Shortened internodes and reduced leaf expansion.
- Poor root development and reduced root hair formation.
- Delayed flowering and fruit set.

Iron-deficient plants show: Poor pollen viability and seed set. Smaller and less viable fruits in crops like tomato, grapevine, and soybean. This adversely affects both yield quantity and quality. Biochemical and molecular responses include: Decreased activities of chlorophyll biosynthesis enzymes (such as ALA synthase). Reduced expression of Fe-dependent proteins such as ferredoxins and cytochromes. Overproduction of ROS due to impaired antioxidant defenses. Upregulation of Fe-deficiency markers like IRT1, FRO_2, and bHLH38/39 in roots (Sharma et al., 2012).

10.3.1 Diagnostic tools for iron deficiency

- **Soil testing:** Measures extractable Fe and pH.
- **Tissue analysis:** Leaf Fe concentration below 50 mg/kg dry weight indicates deficiency.
- **SPAD meter readings:** Nondestructive chlorophyll estimation; lower SPAD values correlate with Fe deficiency.
- **Molecular markers:** Expression levels of Fe-regulated genes (e.g., FIT, IRT1, FRO2).

Iron deficiency arises not due to scarcity in the environment, but due to factors that limit its bioavailability. Its impacts span from leaf chlorosis to metabolic dysfunction and yield loss. Early detection and understanding of the specific causal factors—whether environmental, soil-based, or genetic—are essential for effective diagnosis and targeted intervention (Tripathi et al., 2018).

10.4 Abiotic stress tolerance and iron in plants

10.4.1 Salinity stress and iron

Salinity is among the most challenging issues of the recent era that puts great control on worldwide agricultural productivity (Ashraf and Harris, 2004). Arid and semi-arid territories of the world are experiencing more pronounced salinity effects.

Estimated annually 20% of the total cultivated land area is affected by salt stress. Equally, nearly half of the world's irrigated land is being affected by waterlogging and salinity settings (Talei et al., 2012; Parihar et al., 2015). Furthermore, extensive salinity problems in crop lands are mainly due to inappropriate water management practices in the field (Tripathi et al., 2018).

When plants are exposed to high salinity, they experience many harmful effects at the physiological, biochemical, and molecular levels. Salt stress mainly causes ion toxicity, nutrient deficiency, and both oxidative and osmotic stress in plants (FAO, 2008). It also disrupts plant growth stages (Khan et al., 2009). High salt levels in soil lower the osmotic potential, making it harder for roots to take up water. As a result, plants struggle to absorb both minerals and water from the soil (Jithesh et al., 2006). Hormonal signals in root cells lead to quick changes in growth, productivity, and metabolism. Salinity also lowers the rate of photosynthesis, slows development, and causes both ionic and/or nutrient imbalances in plant cells (Talei et al., 2011). Reduced photosynthesis may be due to too many sodium ions and chlorine ions in leaves, stomatal and vice versa limitations, blocked biochemical processes, and oxidative damage from reactive oxygen species (ROS). Too much ROS can damage membranes, DNA, proteins, and chloroplasts, and disrupt biochemical processes. In severe cases, high salinity can kill plant cells. Plants defend themselves against salinity through osmotic adjustments, salt separation in cells, and other changes in form. Research shows that salinity also reduces nutrient uptake and affects how nutrients are distributed in plants. Limited iron supply under salinity can cause chlorosis. Many studies have found that salinity harms iron uptake in plants (Yousfi et al., 2007; Sharma et al., 2012; Heidari et al., 2011).

Okçu et al. (2005) found that salinity reduces iron content in the shoots of Pea, barley, and corn. Heidari and Sarani (2012) also observed that salinity causes stunted growth and lowers biochemical components and ion content in chamomile. Research now shows that the interaction between salinity and Fe can reduce salt noxiousness in plants. Over the past few decades, many studies have shown that Fe helps lessen the effects of salinity. Fe helps reduce salt stress by producing antioxidative enzymes. In addition, Ghasemia et al. (2014) highlighted the protective role of iron amino acid chelates in tomato plants under high salt content. Overall, these studies show that Fe is important in reducing the negative effects of salinity.

10.4.2 Iron and drought stress

Drought is acknowledged as one of the most serious threats to global food security due to the scarcity of water resources (Farooq et al., 2009, 2012; Monjezi et al., 2013; Pourgholam et al., 2013) (Table 10.1). Additionally, legumes have been shown to respond favorably to Fe nutrition (Rotaru, 2011). Additionally, Elanz et al. (2011) suggested applying Fe foliar to sunflowers under drought stress,

Table 10.1 Iron nutrition and plant response to drought stress and heavy metal stress tolerance.

Plant species	Mode of iron application	Heavy metals	Effects on plants	References
Soyabean	Foliar	-	Productivity increase	Kobraee et al. (2011), Afshar et al. (2013)
Wheat crop	Foliar	-	Increase in seed weight	Afshar et al. (2013)
Cumin	Foliar with Zn	-	Diminishes oxidative stress and lessens lipid peroxidation	Akbari et al. (2013)
Calendula	Iron with Zn spray	-	Improves leaf attributes, resulting in an enhancement in the effects triggered by drought stress	Pirzad and Shokrani (2012)
Sesame	Iron with sulphur	-	Improves growth, nutrients, and yield	Mostafa et al. (2011)
Cowpea	Nano iron	-	Protein quality improvement also helps in drought stress resistance	Afshar et al. (2012)
Creeping bent grass	Iron spray	-	Modifies drought resistance by enhancing root growth	Tripathi et al. (2018)
Turf grasses	Iron	-	color enrichment, growth improvement in iron-deficient conditions	Tripathi et al. (2018)
Legumes	Iron	-	Positive response toward iron	Rotaru (2011)
Rapeseed	Iron with Zn	-	Influence on proline, protein, and nitrogen enhances leaf metabolism	Pourgholam et al. (2013)
Sunflower	Iron foliar fertilization	-	Improves oil yield, growth, and seed development	Elanz et al. (2011)
Rice	Iron fertilizer	Cd	Soil/foliar application of iron fertilizer reduces the adverse effect of Cadmium on plant roots, shoots, and grains. Cadmium uptake and accumulation inside the plant could be reduced by modifying the iron status of the soil	Yadav (2010), Singh et al. (2016)
Spirodela polyrhiza L.	— —	As	Arsenate uptake occurred through the phosphate uptake pathways in *S. polyrhiza* by physico-chemical adsorption on the plant surface	Rahman et al. (2008)
Pilea cadierei	— —	Pb	Plants in wetland conditions remove the phosphorus from Fe-rich soil, suitable for the construction of an artificial wetland	Huang et al. (2009), Yang et al. (2011)

demonstrating the significant roles that Fe foliar application plays in seed and oil production under both normal and stress conditions (Table 10.1). Foliar application of Fe increased the yield of soybeans and increased the intensity of seed set in wheat crops (Kobraee et al., 2011; Afshar et al., 2013). Akbari et al. (2013) proposed that under drought stress, foliar application of iron with zinc reduces oxidative stress by lowering H_2O_2 content and decreasing lipid peroxidation by boosting antioxidant enzymes (Table 10.1). Additionally, it was suggested that overexpression of iron-SOD improves drought tolerance and lessens the symptoms of secondary injuries (Pirzad and Shokrani, 2012).

10.4.3 Heavy metal stress and iron

Rapid advancements (technical) in altered spheres of agriculture, industries, and anthropogenic activities have resulted in a profuse release of heavy metals (Oliveira, 2012; Kumar et al., 2013). Heavy metals move in the soil ecosystem as pollutants, which pose a fundamental concern to the environment. These heavy metals contaminate soil and vegetation health. Agricultural soils are mostly contaminated with heavy metals/ metal ions like Lead, Chromium, Cadmium, and arsenic due to extensive use of phosphate-based fertilizers, industrial wastes, and sewage sludge applications (Tripathi et al., 2018; Yadav et al., 2016). Consequently, the fate of these heavy metals is plants, animals, and humans through the food chain by physiosorption and accumulation (Tchounwou et al., 2012). For instance, the mayhem of "Itai-Itai disease" in Japan in rice and soybean fields was due to the presence of Cd contamination (Yadav, 2010; Singh et al., 2016). Those plants that are mostly affected by heavy metals show defects like morphological, physiological, and biochemical (Emamverdian et al., 2015). Key detrimental effects of heavy metal stress in plants include lipid peroxidation, enzymatic and transport endeavors. Loss of potassium ions in the plant cell membrane is also due to the penetration of several metal ions, like cadmium and copper. Sulfhydryl reactions in the plant cells are also because of plant cells' affinity toward sulfhydryl groups and heavy metals. Nonetheless, heavy metals in plants cause oxidative stress due to increased production of reactive oxygen species, which further causes membrane damage due to the disturbed antioxidant defense system in plants. Heavy metal stress not only fades essential metabolic activities but also affects the electron transport chain in plant cells (Table 10.1) (Garnier et al., 2010). To cope with the deleterious effects of heavy metals, plants adopt special mechanisms at various levels, like mineral nutrient management in the plant cell body (Table 10.1). Among all mineral nutrients, Iron is being recognized as the most efficient and potentially essential micronutrient that plays a paramount role in mitigating the damaging effects of metal stress (Table 10.1) (Rahman et al., 2008). In plants, one of the important biologically active components is iron, which serves as an important micronutrient in cell redox systems. Garnier et al. (2010) described the role of iron plaque with groundwater-irrigated paddy fields and suggested weakening uptake and accumulation of Arsenic in rice plants (Table 10.1). Rahman et al. (2008) utilized iron with phosphate for the organization of water fern Arsenic toxicity (Table 10.1).

Huang et al. (2009) proved that in rice plants, three heavy metals, namely cadmium, lead, and copper, contamination is controlled by iron. Yang et al. (2011) showed iron significance in *Pilea cadierei* against phosphorus (P) (Table 10.1).

10.4.4 Physiological impacts of iron imbalance and redox homeostasis in plants

Fe toxicity can occur in plants under two circumstances: either an excess of Fe or a deficiency of Fe. Iron is essential for many aspects of plant metabolism, including respiration, photosynthesis, nitrogen fixation, and electron transport systems. Additionally, roughly one-third of the cultivated land may be affected by the Fe deficiency.

Fe deficiency also results in the decline of photosynthetic components, particularly the chloroplast's Fad (Fe–S protein ferredoxin). Fe deficiency in plants results in growth retardation, poor root development, interveinal chlorosis, and ultimately plant death (Tognetti et al., 2007). Furthermore, many photosynthetic pigment levels are down-regulated, and the expression of chlorophyll-binding protein CAT2 is altered as a result of Fe deficiency (Rout and Sahoo, 2015). Within the rural soils, Fe insufficiency may moreover happen either at amazingly high pH or at amazingly low pH levels. illustrated that the expanded levels of chlorosis in youthful Avocado clears out was correlated with a decay within the level of iron in leaf parts, chloroplasts, and the movement of catalase. The Fe insufficiency is additionally characterized by interveinal chlorosis in youthful clears out instead of primary veins, a sort of indication known as chlorosis (Rout and Sahoo, 2015). Essentially, intemperate sums of Fe additionally cause harm in plants. It appears that within the nearness of an abundance of Fe, plant roots increase their capability and translocate the Fe in more youthful takes off. High levels of Fe irritate the basal level of magnesium and potassium in plants and cause wholesome unsettling influences (Tripathi et al., 2018). In spite of the fact that high concentrations of Fe can lead to harmful results in plants. Overabundance of free Fe particles like Fe^{2+} and Fe^{3+} can cause ROS formation by taking part in the Fenton reaction and leading to oxidative stress (Mai and Bauer, 2016). In addition, Fe deficiency in the roots of *M. truncatula*, *P. dulcis*, and *P. persica* (Rodríguez-Celma et al., 2013) was characterized by the superoxide dismutase. Furthermore, nonenzymatic ROS generation also occurs due to Fe deficiency. In this process, two enzymes reported in the roots and shoots of *Arabidopsis thaliana* were GST1 and MDAR1 (Mai et al., 2015).

10.5 Strategies to alleviate iron deficiency under abiotic stress

In the recent technological era, various management strategies have been implemented to reduce the effects of abiotic stresses on plants and improve plant productivity. These procedures range from commencing breeding styles, for

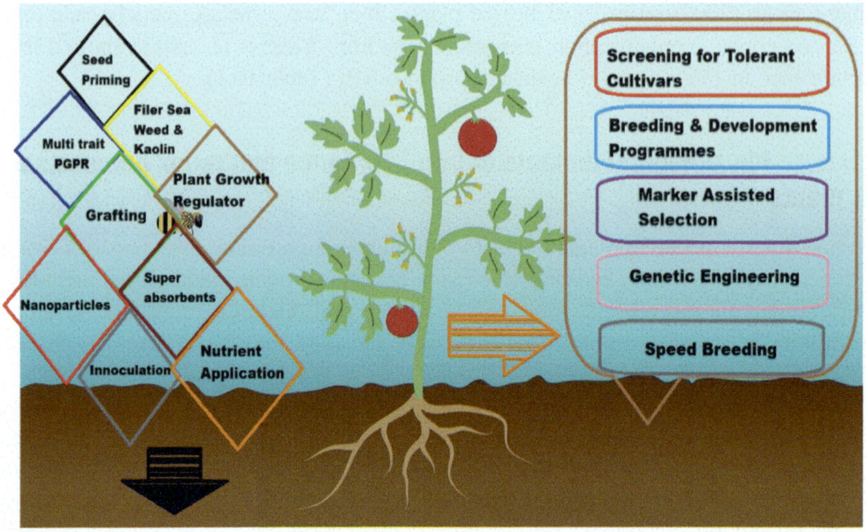

FIGURE 10.2

Detailed modern and nano management strategies of abiotic stresses of plants toward iron.

instance, genetic engineering (conventional/molecular), speed breeding techniques, to agronomical approaches such as nutrient management (Fig. 10.2) (Oyebamiji et al., 2024).

10.6 Conclusions and future prospects

Abiotic stressors significantly disrupt a plant's morphological, biochemical, and physiological functions, negatively affecting germination, photosynthetic efficiency, hormonal regulation, enzyme activities, and ultimately crop yields. Plants exhibit diverse adaptive responses at physiological, molecular, and cellular levels to counteract stress-induced damage. A deep understanding of these mechanisms is vital for developing climate-resilient cultivars. Although significant progress has been made in elucidating the underlying pathways, field-based studies are still needed to investigate how nutrient acquisition and transport are affected under variable stress conditions.

Multiple breeding strategies have been utilized to strengthen plant tolerance to abiotic pressures. However, for these programs to be more effective, it is imperative to continue screening genetic resources under natural environmental conditions to identify novel stress-tolerant genotypes. These accessions serve as valuable input for future breeding initiatives. To safeguard these resources from genetic erosion and the threats posed by climate variability, conservation strategies should be implemented using both in situ (e.g., seed banks, pollen storage, botanical gardens) and ex situ (e.g., on-farm conservation, biodiversity reserves) methods.

Beyond traditional breeding, advanced approaches such as QTL mapping, marker-assisted selection, genetic transformation, and speed breeding have considerably improved the capacity to enhance crop performance under stressful environments. These tools not only expedite breeding timelines but also help uncover the intricate molecular basis of stress tolerance. Nevertheless, further evaluations are necessary to confirm the adaptability and stability of newly developed genotypes, particularly those containing specific QTLs across diverse environmental conditions. Such assessments will shed light on genotype-by-environment interactions and help pinpoint location-specific varietal performance. Moreover, transgenic cultivars must undergo rigorous testing under open-field settings to ensure their viability and safety. There is also considerable promise in combining speed breeding with other molecular and conventional techniques to quicken the development of elite, stress-resilient lines.

Innovative agronomic interventions such as the application of nanoparticles, plant growth regulators, grafting, osmoprotectants, and optimized nutrient regimes have shown potential in enhancing tolerance. Additionally, emerging methods like biochar incorporation, kaolin spraying, super absorbents, chitosan treatments, and seaweed extract application have demonstrated positive outcomes in mitigating stress impacts, though the mechanistic insights behind their effects remain under-explored and demand deeper investigation.

In moving forward, collaborative, interdisciplinary research is essential. This includes fostering global partnerships, participating in scientific forums, sharing datasets, and publishing findings across disciplines. Such integrative efforts will not only enrich the scientific understanding of abiotic stress responses but also support the development of comprehensive, practical solutions to promote sustainable agriculture in a changing climate.

References

Abideen, Z., Waqif, H., Munir, N., El-Keblawy, A., Hasnain, M., Radicetti, E., et al., 2022. Algal mediated nanoparticles, phycochar, and biofertilizers for mitigating abiotic stresses in plants: a review. Agronomy 12, 1788.

Afshar, R.M., Hadi, H., Pirzad, A., 2012. Effect of nano-iron foliar application on qualitative and quantitative characteristics of cowpea, under end season drought stress. International Research Journal of Applied and Basic Sciences. 3, 1709–1717.

Afshar, R.M., Hadi, H., Pirzad, A., 2013. Effect of nano-iron on the yield and yield component of cowpea (Vigna unguiculata) under end season water deficit. International Journal of Agriculture Research and Review. 3, 27–34.

Ahmad, N., Naeem, M., Ali, H., Alabbosh, K.F., Hussain, H., Khan, I., et al., 2023. From challenges to solutions: the impact of melatonin on abiotic stress synergies in horticultural plants via redox regulation and epigenetic signaling. Scientia Horticulturae 321. Available at https://doi.org/10.1016/j.scienta.2023.112369.

Akbari, G.A., Amirinejad, M., Baghizadeh, A., Allahdadi, I., Shahbazi, M., 2013. Effect of Zn and Fe foliar application on yield, yield components and some physiological traits of cumin (*Cuminum cyminum*) in dry farming. International Journal of Agronomy and Plant Production 4, 3231–3237.

Alotaibi, M., 2023. Climate change, its impact on crop production, challenges, and possible solutions. Notulae Botanicae Horti Agrobotanici Cluj-Napoca 51 (1), 13020. Available at https://doi.org/10.15835/nbha51113020.

Anjum, N.A., Singh, H.P., Khan, M.I.R., Masood, A., Per, T.S., Negi, A., et al., 2015. Too much is bad—an appraisal of phytotoxicity of elevated plant-beneficial heavy metal ions. Environmental Science and Pollution Research 22 (5), 3361–3382. Available at https://doi.org/10.1007/s11356-014-3849-9.

Ashraf, M., Harris, P.J.C., 2004. Potential biochemical indicators of salinity tolerance in plants. Plant Science 166 (1), 3–16. Available at https://doi.org/10.1016/j.plantsci.2003.10.024.

Bhat, K.A., Mahajan, R., Pakhtoon, M.M., Urwat, U., Bashir, Z., Shah, A.A., et al., 2022. Low Temperature stress tolerance: an insight into the omics approaches for legume crops. Frontiers in Plant Science 13. Available at https://doi.org/10.3389/fpls.2022.888710.

Briat, J.F., Dubos, C., Gaymard, F., 2015. Iron nutrition, biomass production, and plant product quality. Trends in Plant Science 20 (1), 33–40. Available at https://doi.org/10.1016/j.tplants.2014.07.005.

dos Santos, T.B., Ribas, A.F., de Souza, S.G.H., Budzinski, I.G.F., Domingues, D.S., 2022. Physiological responses to drought, salinity, and heat stress in plants: a review. Stresses 2 (1), 113–135. Available at https://doi.org/10.3390/stresses2010009.

Dubeaux, G., Zelazny, E., Vert, G., 2015. Getting to the root of plant iron uptake and cell-cell transport: polarity matters!. Communicative & Integrative Biology 8 (3), e1038441. Available at https://doi.org/10.1080/19420889.2015.1038441.

Elanz, E., Mohsen, R., Ahmed, B., 2011. Influence of salt stress on cations concentration quality and quantity of sunflower cultivars. Journal of Food Agriculture Environment. 2, 469–476.

El-Saadony, M.T., Saad, A.M., Soliman, S.M., Salem, H.M., Desoky, E.S.M., Babalghith, A.O., et al., 2022. Role of nanoparticles in enhancing crop tolerance to abiotic stress: a comprehensive review. Frontiers in Plant Science 13. Available at https://doi.org/10.3389/fpls.2022.946717.

Emamverdian, A., Ding, Y., Mokhberdoran, F., Xie, Y., Sebastiani, L., 2015. Heavy metal stress and some mechanisms of plant defense response. The Scientific World Journal 2015 (1). Available at https://doi.org/10.1155/2015/756120.

Fahad, S., Bajwa, A.A., Nazir, U., Anjum, S.A., Farooq, A., Zohaib, A., et al., 2017. Crop production under drought and heat stress: plant responses and management options. Frontiers in Plant Science 8. Available at https://doi.org/10.3389/fpls.2017.01147.

Faiz, H., Ayyub, C.M., Khan, R.W., Ahmad, R., 2020. Morphological, physiological and biochemical responses of eggplant (*Solanum melongena* L.) seedling to heat stress. Pakistan Journal of Agricultural Sciences 57 (2), 371–380. Available at https://doi.org/10.21162/PAKJAS/20.9433.

FAO (2008). Land Resources, Management, Planning and Use. Available at http://www.fao.org/ag/agl/agll/spush/.

Farooq, M., Hussain, M., Wahid, A., Siddique, K.H.M., 2012. Drought stress in plants: an overview. In: Aroca, R. (Ed.), Plant Responses to Drought Stress. Springer, Berlin; Heidelberg, pp. 1–33. Available at https://doi.org/10.1007/978-3-642-32653-0_1.

Farooq, M., Basra, S.M.A., Wahid, A., Ahmad, N., Saleem, B.A., 2009. Improving the drought tolerance in rice (*Oryza sativa* L.) by exogenous application of salicylic acid. Journal of Agronomy and Crop Science 195 (4), 237–246. Available at https://doi.org/10.1111/j.1439-037x.2009.00365.x.

Garnier, J.-M., Travassac, F., Lenoble, V., Rose, J., Zheng, Y., Hossain, M.S., et al., 2010. Temporal variations in arsenic uptake by rice plants in Bangladesh: the role of iron plaque in paddy fields irrigated with groundwater. Science of The Total Environment 408 (19), 4185–4193. Available at https://doi.org/10.1016/j.scitotenv.2010.05.019.

Ghasemia, S., Khoshgoftarmanesha, A.H., Afyunia, M., Hadadzadehb, H., 2014. Iron (II)–amino acid chelates alleviate salt-stress induced oxidative damages on tomato grown in nutrient solution culture. Scientia Horticulturae 165, 91–98.

Giordano, M., Petropoulos, S.A., Rouphael, Y., 2021. Response and defence mechanisms of vegetable crops against drought, heat and salinity stress. Agriculture 11 (5), 463. Available at https://doi.org/10.3390/agriculture11050463.

Hassan, M.U., Chattha, M.U., Khan, I., Chattha, M.B., Barbanti, L., Aamer, M., et al., 2021. Heat stress in cultivated plants: nature, impact, mechanisms, and mitigation strategies—a review. Plant Biosystems 155 (2), 211–234. Available at https://doi.org/10.1080/11263504.2020.1727987.

Heidari, M., Sarani, S., 2012. Growth, biochemical components and ion content of Chamomile (*Matricaria chamomilla* L.) under salinity stress and iron deficiency. Journal of the Saudi Society of Agricultural Sciences 11 (1), 37–42. Available at https://doi.org/10.1016/j.jssas.2011.05.002.

Heidari, M., Galavi, M., Hassani, M., 2011. Effect of sulfur and iron fertilizers on yield, yield components and nutrient uptake in sesame (*Sesamum indicum* L.) under water stress. African Journal of Biotechnology 10 (44), 8816–8822. Available at https://doi.org/10.5897/ajb11.854.

Huang, Y., Hu, Y., Liu, Y., 2009. Heavy metal accumulation in iron plaque and growth of rice plants upon exposure to single and combined contamination by copper, cadmium and lead. Acta Ecologica Sinica 29 (6), 320–326. Available at https://doi.org/10.1016/j.chnaes.2009.09.011.

Husen, A., 2021. The harsh environment and resilient plants: an overview. In: Husen, A. (Ed.), Harsh Environment and Plant Resilience. Springer International Publishing AG, Cham, pp. 1–23. Available at https://doi.org/10.1007/978-3-030-65912-7_1.

Husen, A., 2021. Harsh Environment and Plant Resilience (Molecular and Functional Aspects). Springer Nature Switzerland AG, Cham. Available at https://doi.org/10.1007/978-3-030-65912-7.

Husen, A., 2021. Plant Performance Under Environmental Stress (Hormones, Biostimulants and Sustainable Plant Growth Management). Springer Nature Switzerland AG, Cham. Available at https://doi.org/10.1007/978-3-030-78521-5.

Husen, A., 2023. Plants and their Interaction to Environmental Pollution: Damage detection, adaptation, Tolerance, Physiological and Molecular Responses. Elsevier Inc., Cambridge, MA. Available at https://doi.org/10.1016/C2020-0-03319-0.

Husen, A., 2024. Essential Minerals in Plant-Soil System (Coordination, Signalling, and Interaction under Adverse Situations). Elsevier Inc., Cambridge, MA. Available at https://doi.org/10.1016/C2022-0-02437-5.

Husen, A., 2025. Role of Antioxidants in Mitigating Plant Stress. Elsevier Inc., Cambridge, MA. Available at https://doi.org/10.1016/C2023-0-51268-6.

Ishimaru, Y., Kakei, Y., Shimo, H., Bashir, K., Sato, Y., Sato, Y., et al., 2011. A rice phenolic efflux transporter is essential for solubilizing precipitated apoplasmic iron in the plant stele. Journal of Biological Chemistry 286 (28), 24649–24655. Available at https://doi.org/10.1074/jbc.m111.221168.

Jithesh, M.N., Prashanth, S.R., Sivaprakash, K.R., Parida, A.K., 2006. Antioxidative response mechanisms in halophytes: their role in stress defence. Journal of Genetics 85 (3), 237–254. Available at https://doi.org/10.1007/BF02935340.

Khan, M.A., Ansari, R., Ali, H., Gul, B., Nielsen, B.L., 2009. Panicum turgidum: a sustainable feed alternative for cattle in saline areas. Agriculture, Ecosystems & Environment 129, 542–546.

Kobayashi, T., Nishizawa, N.K., 2012. Iron uptake, translocation, and regulation in higher plants. Annual Review of Plant Biology 63, 131–152. Available at https://doi.org/10.1146/annurev-arplant-042811-105522.

Kobraee, S., Shamsi, K., Rasekhi, B., 2011. Effect of micronutrients application on yield and yield components of soybean. Annual Biology Research 2, 476–482.

Kompas, T., Pham, V.H., Che, T.N., 2018. The effects of climate change on gdp by country and the global economic gains from complying with the paris climate accord. Earth's Future 6 (8), 1153–1173. Available at https://doi.org/10.1029/2018EF000922.

Kumar, S., Asif, M.H., Chakrabarty, D., Tripathi, R.D., Dubey, R.S., Trivedi, P.K., 2013. Expression of a rice Lambda class of glutathione S-transferase, OsGSTL2, in *Arabidopsis* provides tolerance to heavy metal and other abiotic stresses. Journal of Hazardous Materials 248-249 (1), 228–237. Available at https://doi.org/10.1016/j.jhazmat.2013.01.004.

Mai, H.J., Bauer, P., 2016. From the proteomic point of view: integration of adaptive changes to iron deficiency in plants. Current Plant Biology 5, 45–56. Available at https://doi.org/10.1016/j.cpb.2016.02.001.

Mai, H.J., Lindermayr, C., von Toerne, C., Fink-Straube, C., Durner, J., Bauer, P., 2015. Iron and fer-like iron deficiency-induced transcription factor-dependent regulation of proteins and genes in *Arabidopsis thaliana* roots. Proteomics 15 (17), 3030–3047. Available at https://doi.org/10.1002/pmic.201400351.

Monjezi, F., Vazin, F., Hassanzadehdelouei, M., 2013. Effects of iron and zinc spray on yield and yield components of wheat (*Triticum aestivum* L.) in drought stress. Cercetari agronomice in Moldova 46 (1), 23–32. Available at https://doi.org/10.2478/v10298-012-0072-z.

Monteiro, E., Gonçalves, B., Cortez, I., Castro, I., 2022. The role of biostimulants as alleviators of biotic and abiotic stresses in grapevine: a review. Plants 11 (3), 396. Available at https://doi.org/10.3390/plants11030396.

Mostafa, M.G., Chen, Y.H., Jean, J.S., Liu, C.C., Lee, Y.C., 2011. Kinetics and mechanism of arsenate removal by nanosized iron oxide-coated perlite. Journal of Hazardous Materials 187 (1-3), 89–95. Available at https://doi.org/10.1016/j.jhazmat.2010.12.117.

Okçu, G., Kaya, M.D., Atak, M., 2005. Effects of salt and drought stresses on germination and seedling growth of pea (*Pisum sativum* L.). Turkish Journal of Agriculture and Forestry 29 (4), 237–242. Available at https://doi.org/10.3906/tar-0407-2.

Oladosu, Y., Rafii, M.Y., Samuel, C., Fatai, A., Magaji, U., Kareem, I., et al., 2019. Drought resistance in rice from conventional to molecular breeding: a review. International

Journal of Molecular Sciences 20 (14). Available at https://doi.org/10.3390/ijms20143519.

Oliveira, H., 2012. Chromium as an environmental pollutant: insights on induced plant toxicity. Journal of Botany 2012, 1–8. Available at https://doi.org/10.1155/2012/375843.

Oyebamiji, Y.O., Shamsudin, N.A.A., Ikmal, A.M., Yusop, M.R., 2023. Heat stress in vegetables: impacts and management strategies - a review. Malaysia Sains Malaysiana 52 (7), 1925–1938. Available at https://doi.org/10.17576/jsm-2023-5207-03.

Oyebamiji, Y.O., Adigun, B.A., Shamsudin, N.A.A., Ikmal, A.M., Salisu, M.A., Malike, F.A., et al., 2024. Recent advancements in mitigating abiotic stresses in crops. Horticulturae 10 (2). Available at https://doi.org/10.3390/horticulturae10020156.

Parihar, P., Singh, S., Singh, R., Singh, V.P., Prasad, S.M., 2015. Effect of salinity stress on plants and its tolerance strategies: a review. Environmental Science and Pollution Research 22 (6), 4056–4075. Available at https://doi.org/10.1007/s11356-014-3739-1.

Pirzad, A., Shokrani, F., 2012. Effects of iron application on growth characters and flower yield of calendula officinalis L. under water stress. World Applied Sciences Journal 18 (9), 1203–1208. Available at https://doi.org/10.5829/idosi.wasj.2012.18.09.538.

Pourgholam, M., Nemati, N., Oveysi, M., 2013. Effect of zinc and iron under the influence of drought on yield and yield components of rapeseed (*Brassica napus*). Annual Biology Research 4, 186–189.

Rahman, M.A., Hasegawa, H., Ueda, K., Maki, T., Rahman, M.M., 2008. Arsenic uptake by aquatic macrophyte *Spirodela polyrhiza* L.: interactions with phosphate and iron. Journal of Hazardous Materials 160 (2-3), 356–361. Available at https://doi.org/10.1016/j.jhazmat.2008.03.022.

Rajput, V.D., Minkina, T., Kumari, A., Harish, Singh, V.K., Verma, K.K., et al., 2021. Coping with the challenges of abiotic stress in plants: new dimensions in the field application of nanoparticles. Plants 10 (6). Available at https://doi.org/10.3390/plants10061221.

Rodríguez-Celma, J., Lattanzio, G., Jiménez, S., Briat, J.F., Abadía, J., Abadía, A., et al., 2013. Changes induced by fe deficiency and fe resupply in the root protein profile of a peach-almond hybrid rootstock. Journal of Proteome Research 12 (3), 1162–1172. Available at https://doi.org/10.1021/pr300763c.

Rotaru, V., 2011. The effect of phosphorus and iron on plant growth and Nutrient status of two soybean (*Glycine max* L.) cultivars under suboptimal water regime of soil. Lucrări Ştiinţifice, Supliment, Seria Agronomie. 54, 11–16.

Rout, G.R., Sahoo, S., 2015. Role of iron in plant growth and metabolism. Reviews in Agricultural Science 3 (0), 1–24. Available at https://doi.org/10.7831/ras.3.1.

Shao, G., Chen, M., Wang, W., Mou, R., Zhang, G., 2007. Iron nutrition affects cadmium accumulation and toxicity in rice plants. Plant Growth Regulation 53 (1), 33–42. Available at https://doi.org/10.1007/s10725-007-9201-3.

Sharma, P., Jha, A.B., Dubey, R.S., Pessarakli, M., 2012. Reactive oxygen species, oxidative damage, and antioxidative defense mechanism in plants under stressful conditions. Journal of Botany 2012, 1–26. Available at https://doi.org/10.1155/2012/217037.

Singh, S., Parihar, P., Singh, R., Singh, V.P., Prasad, S.M., 2016. Heavy metal tolerance in plants: role of transcriptomics, proteomics, metabolomics, and ionomics. Frontiers in Plant Science 6. Available at https://doi.org/10.3389/fpls.2015.01143.

Talei, D., Mihdzar, A.K., Khanif, M.Y., Saad, M.S., Valdiani, A., 2011. Effects of different surface sterilizers on seed germination and contamination of king of bitters

(*Andrographis paniculata Nees.*). American-Eurasian Journal of Agricultural & Environmental Sciences. 10, 639–643.

Talei, D., Kadir, M.A., Yusop, M.K., Abdullah, M.P., Valdiani, A., 2012. Salinity effects on macro and micronutrients uptake in medicinal plant King of Bitters (*Andrographis paniculata* Nees.). Plant OMICS 5 (3), 271–278.

Tchounwou, P.B., Yedjou, C.G., Patlolla, A.K., Sutton, D.J., 2012. Heavy metal toxicity and the environment. In: Luch, A. (Ed.), Molecular, Clinical and Environmental Toxicology. Experientia Supplementum, vol 101. Springer, Basel, pp. 133–164. Available at https://doi.org/10.1007/978-3-7643-8340-4_6.

Tognetti, V.B., Zurbriggen, M.D., Morandi, E.N., Fillat, M.F., Valle, E.M., Hajirezaei, M.R., et al., 2007. Enhanced plant tolerance to iron starvation by functional substitution of chloroplast ferredoxin with a bacterial flavodoxin. Proceedings of the National Academy of Sciences of the United States of America 104 (27), 11495–11500. Available at https://doi.org/10.1073/pnas.0704553104.

Tripathi, D.K., Singh, S., Gaur, S., Singh, S., Yadav, V., Liu, S., et al., 2018. Acquisition and homeostasis of Iron in higher plants and their probable role in abiotic stress tolerance. Frontiers in Environmental Science 5. Available at https://doi.org/10.3389/fenvs.2017.00086.

Waqas, M.A., Wang, X., Zafar, S.A., Noor, M.A., Hussain, H.A., Azher Nawaz, M., et al., 2021. Thermal stresses in maize: effects and management strategies. Plants 10 (2), 1–23. Available at https://doi.org/10.3390/plants10020293.

Yadav, S.K., 2010. Heavy metals toxicity in plants: an overview on the role of glutathione and phytochelatins in heavy metal stress tolerance of plants. South African Journal of Botany 76 (2), 167–179. Available at https://doi.org/10.1016/j.sajb.2009.10.007.

Yadav, V., Arif, N., Singh, S., Srivastava, P.K., Sharma, S., Tripathi, D.K., et al., 2016. Exogenous mineral regulation under heavy metal stress: advances and prospects. Biochemistry & Pharmacology: Open Access 5 (5). Available at https://doi.org/10.4172/2167-0501.1000220.

Yang, L., Li, Y., Yang, X., Xiao, H., Peng, H., Deng, S., 2011. Effects of iron plaque on phosphorus uptake by *Pilea cadierei* cultured in constructed wetland. Procedia Environmental Sciences 11 (C), 1508–1512. Available at https://doi.org/10.1016/j.proenv.2011.12.227.

Younis, A., Ramzan, F., Ramzan, Y., Zulfiqar, F., Ahsan, M., Lim, K.B., 2020. Molecular markers improve abiotic stress tolerance in crops: a review. Plants 9 (10), 1374. Available at https://doi.org/10.3390/plants9101374.

Yousfi, S., Wissal, M., Mahmoudi, H., Abdelly, C., Gharsalli, M., 2007. Effect of salt on physiological responses of barley to iron deficiency. Plant Physiology and Biochemistry 45 (5), 309–314. Available at https://doi.org/10.1016/j.plaphy.2007.03.013.

CHAPTER

Micronutrients in plant abiotic stress tolerance

11

Adalberto Benavides-Mendoza[1,2], Paola Fincheira[3], Gonzalo Tortella[3], Rebeca Casique Valdés[1], Fernando Borrego Escalante[4] and Laura Olivia Fuentes Lara[5]

[1]*Department of Horticulture, Antonio Narro Autonomous Agrarian University, Saltillo, Coahuila, Mexico,*

[2]*Conahcyt Laboratory of Plant Ecophysiology and Food Safety (LANCEVSA), Antonio Narro Autonomous Agrarian University, Saltillo, Coahuila, Mexico,*

[3]*Facultad de Ingeniería y Ciencias, Centro de Excelencia en Investigación Biotecnológica Aplicada al Medio Ambiente (CIBAMA), Universidad de La Frontera, Temuco, Región de la Araucanía, Chile,*

[4]*Department of Plant Breeding, Antonio Narro Autonomous Agrarian University, Saltillo, Coahuila, Mexico,*

[5]*Antonio Narro Autonomous Agrarian University, Department of Animal Nutrition, Saltillo, Coahuila, Mexico*

Abbreviations

APX	ascorbate peroxidase
CAT	catalase
EC	electric conductivity
GR	glutathione reductase
MDA	malondialdehyde
NUE	nitrogen use efficiency
NPs	nanoparticles
PEG	polyethylene glycol
POD	peroxidase
PSI	photosystem I
PPO	polyphenol Oxidase
PSII	photosystem II
RDA	recommended daily allowance values
RG-II	rhamnogalacturonan II
ROS	reactive oxygen species
RWC	relative water content
SOD	superoxide dismutase
WUE	water-use efficiency

Mineral Nutrients to Mitigate Abiotic Stress in Crops. DOI: https://doi.org/10.1016/B978-0-443-44712-9.00014-1

11.1 Introduction

Sustainable agricultural and food production is increasingly challenged by the combined pressures of climate change, land degradation, and the growing global demand for safe and nutritious food. The rising average global temperatures, more frequent heat waves and cold episodes, irregular patterns of precipitation, and salinization of agricultural soils contribute to yield instability and reduced crop quality. Abiotic stresses have an unfavorable impact on global crop yields, and the incidence and intensity of abiotic stresses are projected to increase in the coming decades, particularly in rainfed systems and marginal environments. Regarding the above, several works emphasize that mineral nutrient disorders often co-occur with drought, salinity, or temperature extremes and can aggravate their negative effects on plant growth and productivity (Kumari et al., 2022; Waraich et al., 2011; Husen, 2024; Hossain et al., 2024).

Crop resilience cannot rely solely on plant genetics. Adequate agronomic management of crops, soils, and water resources is essential to resist environmental variability and to maintain yield stability under stress. Measures such as optimizing sowing date and plant density, improving soil conservation and management, and irrigation scheduling, as well as conserving soil organic matter and biological activity, must be integrated with complete and balanced fertilization (Srivastav et al., 2021). Among the abovementioned agronomic actions, plant mineral nutrition stands out as one of the most effective and rapidly adjustable tools to mitigate the impacts of abiotic stress. Both experimental and field studies show that adequate nutrient supply enhances root development, improves water-use efficiency, strengthens antioxidant defense systems, and supports the repair of stress-induced cellular damage, leading to better performance under adverse conditions (Assunção et al., 2022; Waraich et al., 2012).

Plant mineral nutrition includes the supply and internal homeostasis of macro, secondary, and micronutrients. While deficiencies of macronutrients such as nitrogen, phosphorus, and potassium clearly limit yield in many agroecosystems, there is mounting recognition that micronutrients exert a significant influence on plant health and stress tolerance, even at their typically low tissue concentrations. The above causes, especially in extensive agriculture, require great attention to macronutrient nutrition compared to micronutrient nutrition (Jeevanraj et al., 2025).

Essential micronutrients for higher plants include Fe, Zn, Cu, Mn, B, Mo, Cl, and Ni. These micronutrients function as structural components or cofactors of key enzymes involved in photosynthesis, respiration, nitrogen and sulfur metabolism, cell wall formation, and hormone biosynthesis (Table 11.1). However, under stress conditions, they are critically involved in redox regulation, detoxification of reactive oxygen species (ROS), osmolyte accumulation, and maintenance of membrane integrity. Hence, micronutrients significantly influence crop stress tolerance (Jeevanraj et al., 2025; Kumari et al., 2022). At the cellular level, the various metabolic functions promoted by micronutrients result in a specific distribution as shown in Fig. 11.1.

Table 11.1 Metalloproteins with biological importance and their roles.

Metal	Metalloenzymes
Co	Dioldehydrase, glycerol dehydratase, isobutyryl-CoA mutase, methionyl aminopeptidase, methylmalonyl-CoA mutase, nitrile hydratase, ribonucleotide reductase.
Cu	Cytochrome oxidase, dopamine beta hydroxylase, laccase, nitrite reductase, plastocyanin, superoxide dismutase.
Fe	Aconitase, catalase, chitochrome P-450, ferritin, hydrogenase, NADH dehydrogenase, nitrogenase, peroxidase, succinate dehydrogenase, transferrin, xanthine oxidase.
Mn	Arginase, histidine ammonia lyase, pyruvate carboxylase, superoxide dismutase.
Mo	Nitrate reductase, nitrogenase, sulfite oxidase, xanthine oxidase.
Ni	M reductase, hydrogenase, urease.
Zn	Alcohol dehydrogenase, alkaline phosphatase, carbonic anhydrase, carboxypeptidase, DNA polymerase, RNA polymerase, superoxide dismutase, thermolysin.

Table 1 from Juárez-Maldonado, A., González-Morales, S., Cabrera-De la Fuente, M., Medrano-Macías, J., Benavides-Mendoza, A., 2018. Nanometals as promoters of nutraceutical quality in crop plants. In: Impact of Nanoscience in the Food Industry. Elsevier Inc., Mexico, pp. 277–310.

Globally, large fractions of arable land are affected by insufficient availability of at least one micronutrient -frequently Zn, Fe, or B- due to the inherent composition of parent material, soil pH, organic matter depletion, salinity, or antagonistic interactions with other elements. The inadequate nutrition of crops with micronutrients, particularly metal micronutrients, due to environmental factors or inadequate agronomic management, limits crop growth in a substantial proportion of the world's cultivated soils and is particularly widespread in dryland and calcareous regions (Ofori et al., 2022; Pérez-Labrada et al., 2020).

The above-cited micronutrient imbalances not only reduce yields but also alter sink–source relationships, reproductive success, grain filling, or fruit set, making plants more vulnerable to episodes of drought, heat, or salinity. Furthermore, abiotic stresses can impair micronutrient uptake and translocation by reducing root growth, altering rhizosphere chemistry, or inhibiting the synthesis of transporter proteins, thereby establishing an unfavorable feedback loop between stress and nutritional imbalance (Kumari et al., 2022; Hossain et al., 2024).

Beyond strictly essential elements, several mineral elements are considered "beneficial" because they consistently improve growth or stress tolerance in many species, even though a classical deficiency symptom has not been described (Brown et al., 2022). Se, Co, and I are prominent examples. This manuscript will include these nutrients as part of its content.

The importance of micronutrient nutrition extends beyond crop productivity and plant stress tolerance to human and animal health. Insufficient micronutrient concentrations in edible plant organs contribute to "hidden hunger" in human populations, a form

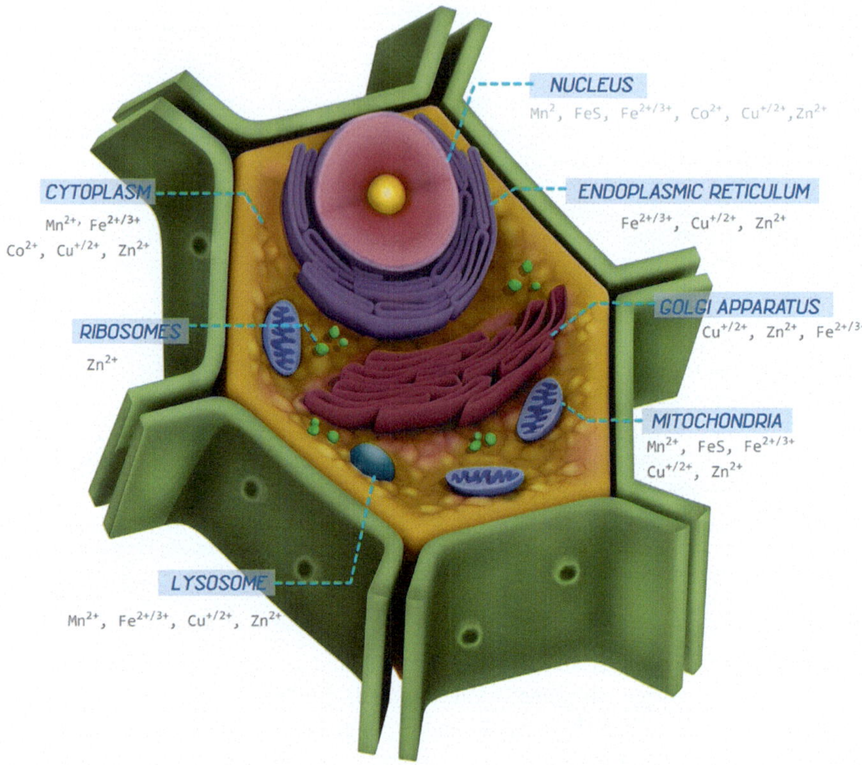

FIGURE 11.1

Distribution of metal micronutrients at the plant cell level. The differential distribution among the various organelles reflects their specific metabolic functions.

Figure from Juárez-Maldonado, A., González-Morales, S., Cabrera-De la Fuente, M., Medrano-Macías, J., Benavides-Mendoza, A., 2018. Nanometals as promoters of nutraceutical quality in crop plants. In: Impact of Nanoscience in the Food Industry. Elsevier Inc., Mexico, pp. 277–310.

of malnutrition characterized by deficiencies of Zn, Fe, Se, and other trace elements despite adequate caloric or protein intake. Biofortification—through conventional breeding, genetic engineering, or agronomic strategies—aims to increase the micronutrient density in crops. The adequate metal micronutrient contents found in plants, their minimum levels, and toxicity thresholds are shown in Fig. 11.2.

Recent reviews highlight agronomic biofortification with micronutrients such as Zn, Fe, and Se as a cost-effective and rapidly deployable strategy that can simultaneously improve plant stress resilience and the nutritional quality of harvested products (Consentino et al., 2023; Szerement et al., 2022). In many cases, the same micronutrient-related physiological mechanisms that enhance crop stress tolerance—improved root architecture, stronger antioxidant systems, and

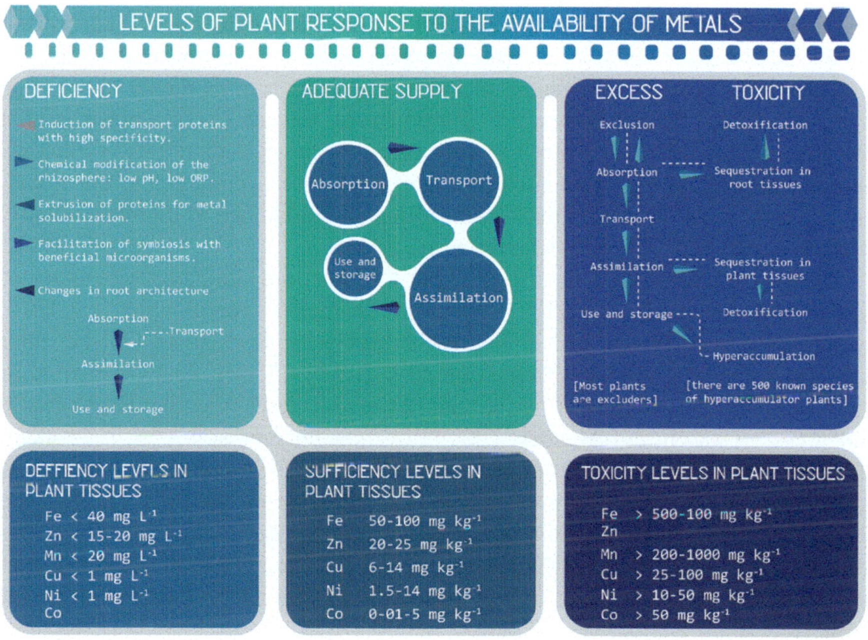

FIGURE 11.2

The concentrations of micronutrient metals in plant tissues. The responses to deficiency or excess are shown in the right and left sections above.

Figure from Juárez-Maldonado, A., Gonzalez-Morales, S., Cabrera-De la Fuente, M., Medrano-Macías, J., Benavides Mendoza, A., 2018. Nanometals as promoters of nutraceutical quality in crop plants. In: Impact of Nanoscience in the Food Industry. Elsevier Inc., Mexico, pp. 277–310.

more efficient transport and storage of micronutrients—also underpin successful biofortification outcomes (Nasim et al., 2025).

The objective of this chapter is to present an updated view of the current knowledge on the roles of micronutrients—encompassing both essential elements Fe, Zn, Cu, Mn, B, Mo, Cl, and Ni, and the beneficial elements Se, Co, and I—in promoting tolerance to common abiotic stresses such as drought, salinity, and temperature extremes and in enabling crops to maintain productivity and nutritional quality under nonideal growing conditions.

11.2 Metabolic functions of metal and semimetal micronutrients in plants and their impact on stress tolerance

The micronutrients Fe, Zn, Cu, Mn, B, Mo, Cl, and Ni, together with beneficial elements such as Se, Co, and I, participate in hundreds of enzymatic and structural functions in plants. Many of these functions are related to redox metabolism, photosynthetic and respiratory electron transport, osmotic adjustment, and hormone signaling, so that small changes in micronutrient supply can markedly redirect the way plants perceive and respond to abiotic stress. Recent works emphasize that adequate micronutrient nutrition not only prevents classical deficiency symptoms but also actively primes antioxidant systems, osmolyte accumulation, ion homeostasis, and metabolic flexibility, thereby enhancing tolerance to drought, salinity, heat, cold, and heavy metals (Kumari et al., 2022).

Redox-active transition metals (Fe, Zn, Cu, Mn) are obligatory cofactors of stress-responsive enzymes such as superoxide dismutases, peroxidases, catalases, cytochromes, ferredoxins, and alternative oxidase (Pandey, 2018). On the other hand, the so-called "nonredox" micronutrients (B, Mo, Cl, Ni) occupy structural or catalytic positions in cell walls, nitrate and sulfite reductases, vacuolar H^+-ATPases, and urease (Hänsch and Mendel, 2009; Roychoudhury et al., 2025). Beneficial elements such as Se, Co, and I further modulate glutathione-dependent peroxidases, legume symbioses, phytohormone balance, and phenolic metabolism. Through the abovementioned roles, micronutrients influence both the generation and scavenging of reactive oxygen species (ROS), the maintenance of membrane integrity, stomatal behavior, and the reprogramming of primary and secondary metabolism under stress (Rai et al., 2023).

Below are some examples of the role of micronutrients during stress events in plants.

11.2.1 Iron

Iron functions in many enzymes either as a redox-active metal or as a bridge between enzyme and substrate, so Fe deficiency lowers their activity and disrupts metabolism. Key heme proteins include cytochromes, which carry electrons in the chloroplast and mitochondrial electron transport chains and in the redox chain of nitrate reductase. Fe is also essential in leghemoglobin and nitrogenase. In thylakoid membranes, Fe is directly involved in electron transport. The chloroplasts contain up to 80% of the total leaf Fe. Photosystem I is a major Fe sink (approximately 12 Fe atoms per complex) compared with photosystem II (approximately 3) and the cytochrome b_6f complex (approximately 5). Together with the Fe required for ferredoxin and chlorophyll biosynthesis, the high demand explains the particular sensitivity of chloroplasts, especially thylakoids, to Fe deficiency (Cakmak et al., 2023).

Otherwise, nonheme Fe–S proteins coordinate Fe with cysteine thiols and/or inorganic sulfur. The best-known is ferredoxin, an electron carrier in multiple metabolic pathways. Another example is Fe–SOD, a superoxide dismutase isoenzyme that uses Fe in its prosthetic group (Cakmak et al., 2023).

Related to stress events, Fe is the core of heme-containing peroxidases and catalases and of Fe–S proteins in photosystems and mitochondria. Under salinity and drought, sufficient Fe limits photooxidative damage by stabilizing photosystem II and reinforcing ROS-scavenging enzymes. Integrated analyses show that Fe fertilization or Fe-enriched seed priming in cereals under water or salt stress increases the activities of CAT, POD, and Fe–SOD, lowers MDA accumulation, and maintains chlorophyll and quantum efficiency, thereby sustaining carbon gain under stress (Kumari et al., 2022; Tripathi et al., 2018).

In another example, Mannan et al. (2022) carried out a pot experiment to evaluate whether soil or foliar applications of zinc sulfate ($ZnSO_4 \cdot 7H_2O$) and/or iron sulfate ($FeSO_4 \cdot 7H_2O$) mitigate drought stress in winter wheat. Plants were grown under drought (40% field capacity) or well-watered conditions (80% field capacity). Drought unfavorably affected the plants. However, in contrast, soil and foliar applications of $ZnSO_4$ and/or $FeSO_4$ significantly improved relative water content, leaf chlorophyll, leaf number, and leaf area, while increasing proline accumulation, water saturation deficit, water uptake capacity, physiological and morphological traits, and, consequently, increased yield under drought. Foliar treatments were the most effective, particularly the combined Zn and Fe application.

11.2.2 Zinc

Zn is the second most abundant transition metal in plants after Fe and is taken up mainly as the divalent cation Zn^{2+}. Within plant tissues, Zn occurs exclusively as Zn^{2+} and does not participate directly in redox reactions. Bioinformatic analyses of the *Arabidopsis* genome indicate that Zn can bind to approximately 13% of all proteins, corresponding to approximately 2500 Zn-binding proteins. These proteins bind Zn with very high affinity to maintain their structure, activity, and interactions. Representative Zn-containing or Zn-activated enzymes include carbonic anhydrase, Cu/Zn-SOD, RNA polymerase, phospholipases, and pyrophosphatases. In addition, Zn plays an important role in the biosynthesis and metabolism of auxins. Likewise, Zn is a structural element in Zn-finger transcription factors, alcohol dehydrogenases, and many dehydrogenases (Cakmak et al., 2023).

In wheat seedlings subjected to salt stress, Zn supplementation enhanced the activities of SOD, catalase, and ascorbate peroxidase, reduced Na^+ accumulation, and improved membrane stability, resulting in higher relative growth and photosynthetic performance (Xu et al., 2014). On the other hand, under salinity, Zn application enhances seed germination, seedling growth, water uptake, plant water status, and nutrient uptake and homeostasis, thereby improving overall performance under saline conditions. It also protects the photosynthetic apparatus from oxidative damage and promotes stomatal function, chlorophyll synthesis, carbon fixation, and the accumulation of osmolytes and hormones. In addition, Zn increases secondary metabolite production, upregulates stress-responsive genes, and boosts antioxidant activity to counteract salt-induced toxicity (Shao et al., 2023).

In another study designed to elucidate how exogenous Zn enhances drought tolerance in sunflower, a hydroponic experiment was conducted using PEG-6000–induced drought (0, 7, 14 days) and Zn treatments (0, 1, 3 μM ZnSO4) in a completely randomized factorial design. Under PEG-6000-induced drought, Zn application markedly increased chlorophyll a (31.2%), chlorophyll b (35.9%), photosynthetic rate (41.7%), and shoot and root dry matter (44.0% and 38.7%) compared with plants without Zn. At 3 μM $ZnSO_4$, the activities of APX, SOD, CAT, and ascorbic acid content increased by 57.1%, 26.6%, 62.3%, and 38.3%, respectively, while MDA decreased by 83.5%. The same Zn treatment enhanced soluble proteins (79.5%) and free proline (70.5%) (Jan et al., 2022).

An additional example of Zn impact on plant stress response is reported by Nadeem et al. (2020). In a pot experiment, rice was grown in saline soil (EC 7.5 dS m^{-1}, and the control) or the same soil supplemented with 3.5 mg kg^{-1} Zn or Mn. Zn application reduced leaf Na^+ and Cl^- and increased K^+ compared with the control. Increases in SOD, GR, and CAT activities decreased salinity-induced oxidative damage, leading to a higher membrane stability index and RW. A fivefold increase in soil-solution Zn^{2+} ensured adequate availability, raising SPAD values, aboveground biomass, and grain yield. In contrast, in this experiment, Mn had a weaker ameliorative effect than Zn.

11.2.3 Copper

This micronutrient acts primarily as a redox-active cofactor cycling between Cu^+ and Cu^{2+}, enabling one-electron transfer reactions in numerous metalloproteins associated with photosynthesis, respiration, oxidative stress control, cell wall metabolism, and hormone and lignin biosynthesis. Because of its strong redox activity, plants tightly regulate Cu uptake, complexation, and compartmentation to balance its metabolic roles against the oxidative damage potential.

In photosynthesis, Cu is central to electron transport in chloroplasts. Plastocyanin, a soluble Cu protein in the thylakoid lumen, transfers electrons from the cytochrome b_6f complex to photosystem I, thereby sustaining linear electron flow and photochemical energy conversion. In mitochondria, Cu-containing cytochrome c oxidase (complex IV) catalyzes the terminal step of the respiratory chain, contributing to H^+ gradient generation and ATP synthesis. In the apoplast, several Cu-dependent oxidases, especially laccases and ascorbate oxidase, are involved in cell wall modification, lignin polymerization, and control of apoplastic redox status (Cakmak et al., 2023).

Regarding the impact of copper on stress tolerance, Cu is crucial in oxidative stress management. For example, Cu/Zn-SOD is a key enzyme in ROS scavenging in chloroplasts, cytosol, and peroxisomes. Additionally, Cu oxidases participate in redox signaling, modulating pathways involved in acclimation to abiotic and biotic stresses (Cakmak et al., 2023).

By catalyzing the oxidative crosslinking of phenolic compounds and structural proteins, Cu enzymes influence mechanical strength, xylem development, and defense barrier formation. Copper amine oxidases contribute to the catabolism of polyamines, linking Cu metabolism to nitrogen turnover, H_2O_2 production, and signaling. Copper also plays roles in hormone biosynthesis and signaling, for example, via Cu-dependent enzymes contributing to ethylene perception and phenolic metabolism, thereby affecting growth, morphogenesis, and stress adaptation. Overall, Cu is integrated into a complex network of redox and signaling pathways crucial for plant development and environmental acclimation (Chen et al., 2022).

In maize exposed to polyethylene glycol-induced drought, combined Cu and water deficit treatments caused a strong upregulation of SOD, peroxidase, and catalase and a reduction in lipid peroxidation compared with drought alone, indicating that Cu nutrition can condition the antioxidant system to better face dehydration stress (Cetinkaya et al., 2014). In another experiment, exogenous CuSO4 (0.5 and 1.0 mM) application in eggplant under salinity improved plant water status, chlorophyll content, and net photosynthesis while enhancing antioxidant enzyme activities and maintaining a more favorable K^+/Na^+ ratio, jointly contributing to higher biomass under salt stress (Imran et al., 2024).

11.2.4 Manganese

This micronutrient functions primarily as a redox-active cofactor in numerous enzymes and as a structural component of key protein complexes. Its most prominent role is in photosynthesis. Mn forms part of the Mn_4CaO_5 cluster of the oxygen-evolving complex in photosystem II, where Mn catalyzes the oxidation of H_2O to O_2 and sustains linear electron transport. Additionally, Mn participates as a cofactor or activator for enzymes of the tricarboxylic acid cycle and glycolysis, thereby influencing respiration, carbon partitioning, and the supply of organic acids. Mn-dependent enzymes are also involved in the shikimate and phenylpropanoid pathways, which provide precursors for lignin, flavonoids, and other phenolics important for structural integrity and plant defense (Cakmak et al., 2023).

In the same way, Mn plays a central role in oxidative stress responses. Mn-SOD of mitochondria and peroxisomes catalyzes the dismutation of O_2^- to H_2O_2. Other Mn-dependent oxidases contribute to apoplastic H_2O_2 generation, which is required for cell wall crosslinking and functions as a signaling molecule in response to abiotic stresses. An additional Mn-dependent process is the deposition of cuticular waxes in leaves, a key activity for plant WUE (Alejandro et al., 2020).

In rice seedlings, 0.5 mM $MnSO_4$ supplementation under NaCl stress (150 mM) increased SOD, catalase, and the antioxidant system, improved K^+/Na^+ homeostasis, and raised proline and soluble sugar contents, translating into higher biomass and less oxidative damage (Rahman et al., 2016). On the other hand, a broad survey of Mn salts and Mn-nanoparticles under diverse stresses revealed improved PSII photochemistry, reduced hydrogen peroxide and lipid peroxidation, and upregulation of antioxidant and stress-responsive genes, supporting a role for Mn in fine-tuning redox homeostasis and photoprotection under abiotic constraints (Ye et al., 2019).

11.2.5 Boron

It is a micronutrient that displays metallic and nonmetallic properties. It is taken up by plant roots predominantly in the form of small uncharged boric acid molecules $B(OH)_3$, which easily enter the cell by passing through the wall's phospholipid bilayer (Brdar-Jokanović, 2020). At sufficient or supra-optimal boron levels in the external solution (>20 μM), boron influx increases in direct proportion to the surrounding soluble B concentration. This largely reflects the high permeability of plant membranes to undissociated boric acid, together with the presence of numerous intracellular boron-binding ligands that sequester boric acid, lowering its free concentration in the cytosol and thereby sustaining a favorable concentration gradient for passive entry. In contrast, when boron availability is low (<1 μM), its acquisition is no longer purely diffusion-driven but becomes tightly controlled by specific transport systems. Under these conditions, B uptake is mediated primarily by two types of membrane proteins: boric acid channels belonging to the major intrinsic protein family (e.g., NIP5;1) and borate transporters of the BOR family, which together ensure efficient uptake and distribution of boron under scarcity (Cakmak et al., 2023).

The primary, unequivocally demonstrated metabolic role of B in higher plants is structural. At the molecular level, borate forms diester bonds with apiosyl residues of RG-II in the primary cell wall, generating RG-II dimers that are crucial for wall architecture, porosity, mechanical strength, and cell adhesion. Crosslinking allows the normal growth of meristems, vascular tissue, and pollen tubes, and contributes to the ability of tissues to withstand mechanical and osmotic challenges (Funakawa and Miwa, 2015).

Beyond the above structural role, B interacts with cis-diol–containing molecules in membranes, glycoproteins, glycolipids, and soluble metabolites, thereby influencing membrane integrity, ion transport, sugar loading into the phloem, cytoskeletal organization, and hormone metabolism. Adequate B supply sustains long-distance transport of sucrose and polyols, supports reproductive development, and maintains the functionality of vascular tissues (Brdar-Jokanović, 2020).

Boron is closely linked to plant responses to abiotic stress. By stabilizing the cell wall–plasma membrane continuum, B helps limit membrane leakage and structural damage under drought, freezing, salinity, and high irradiance (Tenhaken, 2015). B deficiency or excess frequently enhances ROS accumulation. In turn, B status modulates SOD, POD, CAT, and APX as well as nonenzymatic antioxidants and phenylpropanoid metabolism, which generates phenolics and lignin that reinforce cell walls and act as ROS scavengers. Under salinity or drought, an appropriate B supply has been shown to improve Na^+/K^+ homeostasis, maintain photosynthesis, and reduce oxidative damage, thereby enhancing tolerance (Shah et al., 2017).

In soybean, exogenous B under salinity enhanced ascorbate and glutathione pools and the activities of ascorbate peroxidase, monodehydroascorbate and dehydroascorbate reductases, as well as glyoxalase enzymes, while lowering methylglyoxal, H_2O_2 and malondialdehyde; these changes were associated with improved growth and chlorophyll content under salt stress (Alharby et al., 2021).

In cotton, B supply under NaCl stress promoted proline, soluble protein, and soluble sugar accumulation, reduced Na^+ and Cl^-, and increased K^+ and Ca^{2+} in roots, and remodeled cell-wall pectin and cellulose, jointly enhancing osmotic adjustment and ion homeostasis (Lu et al., 2023). Similar B-mediated reinforcement of drought tolerance in coffee (in synergy with Ca) involves improved water status, antioxidant capacity, and long-term yield stability (Ramirez-Builes et al., 2024).

11.2.6 Molybdenum

The plant demand for Mo is lower than that for any other essential nutrient. Its biological role is linked to its ability to shift between the +4, +5, and +6 oxidation states as a metal component of enzymes, thereby catalyzing two-electron transfer reactions. In higher plants, only a few enzymes use Mo as a cofactor, notably nitrate reductase, xanthine dehydrogenase, aldehyde oxidase, and sulfite reductase, which link N and S metabolism with ABA biosynthesis and ROS turnover. Mo is also a cofactor of nitrogenase in N_2-fixing bacteria. Consequently, Mo function is tightly connected to N metabolism, and plant Mo requirements depend strongly on the form and mode of nitrogen supply (Cakmak et al., 2023).

Additionally, Mo is associated with tolerance to certain abiotic stresses in plants. For example, cold stress studies in wheat show that Mo application enhances antioxidant enzymes and induces COR15a protein accumulation, thereby improving freezing tolerance (Al-Issawi et al., 2016). On the other hand, (Wu et al., 2014) reported in winter wheat that, under limited water supply induced with 7.5% PEG-6000, Mo 1 μM [as $(NH_4)_6Mo_7O_{24}{\cdot}4H_2O$] fertilization significantly improved WUE by increasing net photosynthetic rate, chlorophyll content, dry matter, grain yield, total biomass, relative water content and water-use efficiency, while reducing transpiration rate, stomatal conductance and water loss. In addition, Mo increased osmotic adjustment solutes such as soluble proteins, proline, and soluble sugars. In the same way, Mo treatment enhanced the activities of the antioxidant enzymes SOD, CAT, POD, and APX and the levels of the nonenzymatic antioxidants ascorbic acid, reduced glutathione, and carotenoids while decreasing MDA, suggesting a link between adequate Mo nutrition and the ability to scavenge reactive species

11.2.7 Chloride

Although in most plant species the minimum Cl- requirement for plant growth is in the micronutrient range of 200–400 mg kg^{-1} dw, Cl- is readily taken up by plants, reaching concentrations similar to those of macronutrients and becoming, together with NO_3^-, the most abundant anion (Cakmak et al., 2023). Chloride is the predominant inorganic anion in the vacuole, where it contributes to osmotic adjustment, cell expansion, and turgor-driven growth, and it participates in charge balance for cation uptake and phloem loading (Raven, 2017).

At the chloroplast level, Cl^- is required for the activity and structural stability of the oxygen-evolving complex of photosystem II PSII, where it, together with Ca^{2+}, optimizes water oxidation kinetics and supports efficient electron transport (Imaizumi and Ifuku, 2022). Within chloroplasts, Cl^- is the predominant anion in the stroma, with concentrations typically in the range of 50–90 mM. Therefore both the chloroplast envelope and the thylakoid membranes exhibit a marked permeability to Cl^-. The roles played by Cl^- inside the chloroplast cannot be adequately substituted by other anions, such as nitrate (NO_3^-), most likely because these membranes contain transport systems that are specifically selective for Cl^- (Cakmak et al., 2023). Chloride also modulates the activity of tonoplast proton pumps (V-ATPase and V-PPase), thereby helping to establish proton gradients that energize secondary transport of nutrients and metabolites, and the sequestering of ions into the vacuole during salinity stress (Mansour, 2023).

When exposed to moderate external Cl^- levels (1–5 mM), glycophytic species, including many traditionally classified as chloride-sensitive, often exhibit maintenance of photosynthetic performance under low water availability, reflected in greater biomass production, better plant water status and cell turgence, enhanced cell elongation, and increased leaf expansion. Current evidence indicates that, at concentrations comparable to those of macronutrients, Cl^- fulfills substantial and specific physiological functions that cannot be completely substituted by other inorganic or organic anions.

In addition, an appropriate Cl^- supply can improve NUE, partly by reducing excessive NO_3^- accumulation and rebalancing anion/cation relations, which indirectly improves stress tolerance. However, at high concentrations, especially under salinity, Cl^- becomes toxic, disrupting photosynthesis and causing ion imbalances, so its beneficial versus harmful effects are strongly concentration-dependent (Cakmak et al., 2023). As an example, in the study of (Lucas et al., 2024), the authors found that supplying 5 mM Cl^- in the nutrient solution allows the NO_3^- concentration to be reduced by as much as 50% without impairing growth or yield in tomato, lettuce, red chard, and spinach plants. Furthermore, the addition of Cl^- alleviates stress symptoms and promotes growth under low-NO_3^- conditions. This Cl^--mediated tolerance to N limitation was associated with more efficient utilization of the available NO_3^-, improved osmotic adjustment and plant water status, enhanced stomatal conductance and CO_2 assimilation, and more effective antioxidant activity.

11.2.8 Nickel

Nickel is chemically close to Fe and Co, and in plants, it occurs predominantly in the divalent state Ni^{2+}. Ni forms stable coordination complexes with ligands such as histidine, cysteine, and citrate, and in Ni-dependent enzymes, it is bound within diverse ligand environments. Functionally, Ni acts as a redox-active metal cofactor, comparable to Fe and Cu, participating in electron transfer and other redox-related catalytic processes. Along with Zn, Ni is among the most competitive metal ions after Cu in terms of binding to cellular metal-binding sites. Consequently, the concentration of freely available Ni^{2+} in the cytosol must be kept extremely low to prevent toxicity arising from mismetallation, that is, the inappropriate occupation of metal-binding sites in proteins by Ni instead of their correct physiological metal. To ensure correct metal insertion while minimizing such deleterious interactions, the delivery of Ni to its target enzymes is tightly controlled (Cakmak et al., 2023).

In plants, the abovementioned targeted delivery occurs by at least two main mechanisms: (1) specific metallochaperone proteins, which bind Ni and guide it directly to the newly assembled Ni-dependent apoproteins, and (2) the prior assembly of complex metalloprotein cofactors that already contain Ni in the correct coordination environment before being inserted into the final holoenzyme. These mechanisms collectively ensure that Ni is efficiently utilized in metabolism while limiting the risk of Ni-induced protein malfunction and cellular damage (Cakmak et al., 2023).

Nickel participates as an essential metal cofactor in at least nine plant and bacterial enzymes. These include urease and methyleneurease, which are central to urea and related nitrogen compound metabolism; methyl-coenzyme M reductase (in transgenic plants) (Su et al., 2019), a key enzyme in methanogenesis; and NiFe-hydrogenase (Russell et al., 2020), which catalyzes reversible hydrogen oxidation. Ni is also required for bacterial carbon monoxide dehydrogenase, possibly present in plants (Bidwell and Bebee, 1974), and the bacterial acetyl-CoA decarbonylase synthase complex (Gencic et al., 2010), both of which are involved in C_1 and acetyl-CoA metabolism. In addition, Ni is associated with Ni-SOD, Ni-dependent glyoxalase, and aci-reductone dioxygenase (Cakmak et al., 2023).

The link between Ni and abiotic stress responses is dual and concentration-dependent. At adequate but subtoxic levels, Ni can modulate antioxidant defenses, often increasing the activities of SOD, CAT, POD, and components of the ascorbate–glutathione cycle. This response helps limit oxidative damage associated with drought, salinity, and temperature extremes, especially when these stresses perturb N metabolism and promote urea accumulation. Proper Ni nutrition therefore contributes to redox homeostasis and to the maintenance of photosynthesis and membrane integrity under stress (Guedes et al., 2025; Kováčik and Vydra, 2024).

In contrast, excess Ni is strongly phytotoxic: it can displace other essential metals from their binding sites, inhibit key metalloenzymes, and promote over-production of reactive species (Kumar et al., 2025).

11.2.9 Selenium

This micronutrient is not considered essential for most higher plants; however, at low doses, it exerts beneficial effects on growth and stress tolerance. Because of its close chemical similarity to S, Se is absorbed mainly as selenate via sulfate transporters and is metabolized through the S-assimilation pathway to selenite, selenide, and then to organic forms such as selenocysteine, selenomethionine, and Se-methylselenocysteine (Reis et al., 2025). In hyperaccumulator species, these organoselenium compounds can reach very high levels and function as a defense against herbivores and pathogens through toxicity and deterrence (Trippe and Pilon-Smits, 2021).

In no accumulator crops, Se at micromolar levels often stimulates photosynthesis, nitrogen and sulfur metabolism, and the synthesis of selected secondary metabolites, while excessive Se becomes phytotoxic by misincorporation into proteins and by disturbing redox balance. At the genomic level, higher plants appear to have lost canonical selenoprotein genes; instead, Se is incorporated unspecifically into S-containing amino acids or is metabolized into nonprotein forms such as Se-methylselenocysteine (Schiavon and Pilon-Smits, 2017).

A topic of current interest is the role of Se in abiotic stress tolerance. Numerous studies have shown that foliar or root Se supply mitigates the effects of drought, salinity, heat, and heavy-metal stress by enhancing antioxidant capacity, maintaining membrane integrity, and sustaining photosynthesis. Selenium often upregulates the activities of SOD, CAT, POD, and glutathione-based enzymes, increases pools of nonenzymatic antioxidants (ascorbate, glutathione, phenolics), and promotes the accumulation of osmoprotectants such as proline. These changes reduce reactive species formation, lipid peroxidation, and electrolyte leakage, thereby improving water status, carbon fixation, and ultimately yield under stress (Medrano-Macías and Narvaéz-Ortiz, 2022; Reis et al., 2025).

In sunflowers with 25% and 50% irrigation restriction, foliar Se application (30, 60, 90 mg L^{-1}) significantly lowered H_2O_2 accumulation and improved glycine betaine and total soluble protein, improved the accumulation of ascorbic acid, total phenolics, flavonoids, and anthocyanin, and improved the activity of SOD, POD, and CAT. The application improved sunflower growth by 36.65% during drought stress. Additionally, supplemental Se significantly increased shoot Se content (93.86%) and enhanced Ca, K, and Na in roots (Ameen et al., 2024).

On the other hand, a pot experiment (Fatahiyan et al., 2025) evaluated whether soil-applied Se alleviates salt stress in *Carthamus tinctorius*. The plants received three Na_2SeO_4 doses (0, 0.01, 0.02 g kg^{-1}) combined with four NaCl levels (0, 0.5, 1.5, 2.5 g kg^{-1}). Increasing NaCl concentrations depressed seed germination indices, vegetative growth, K^+ content, relative water content, and photosynthetic pigments while promoting Na^+ accumulation, soluble carbohydrates, H_2O_2, and MDA, indicating oxidative and osmotic stress. Se supply significantly counteracted the above NaCl-induced effects. Additionally, anatomical observations showed that Se-treated plants under salinity developed thicker stems, enlarged cortical parenchyma, and wider pith, whereas changes in xylem and phloem thickness were not statistically significant. At 0.02 g kg^{-1} Se under 2.5 g kg^{-1} NaCl, plants exhibited improved ion balance, water status, pigments, photosynthetic capacity, and soluble carbohydrates. Concomitantly, Se reduced oxidative damage in plants

11.2.10 Cobalt

This element is a transition metal closely related to Fe and Ni and is now regarded as a potentially essential micronutrient for many plants, particularly legumes. Its essentiality in higher plants is mainly indirect, because Co is the central atom of cobalamin (vitamin B_{12}), which is required by symbiotic and associative diazotrophs for efficient biological N_2 fixation. In legumes, Co enhances nodulation, leghemoglobin formation, and nitrogenase activity, thereby improving N status, biomass accumulation, and grain yield (Hu et al., 2023).

Beyond its role in N_2-fixing symbioses, Co influences several metabolic processes in nonlegume species. At low concentrations, it can stimulate chlorophyll content, photosynthetic CO_2 assimilation, and carbohydrate metabolism, partly through improved N acquisition and assimilation (Hu et al., 2021). Co also interacts strongly with phytohormone metabolism: Co^{2+} is a well-known inhibitor of aminocyclopropane-1-carboxylate (ACC) oxidase, the terminal enzyme in ethylene biosynthesis, and low doses of $CoCl_2$ have been shown to reduce ethylene overproduction and promote root growth and nutrient acquisition under P deficiency. This modulation of ethylene signaling can delay senescence, alter root system architecture, and improve source–sink relations under stress (Shukla, 2025).

Cobalt is increasingly recognized for its contribution to abiotic stress tolerance. At optimal, subtoxic levels, Co enhances the activity of the antioxidant enzymes SOD, CAT, POD, and PPO, thereby improving the detoxification of reactive species generated under drought, salinity, metal toxicity, or nutrient stress. Co can also promote the accumulation of compatible solutes and improve relative water content, contributing to osmotic adjustment and maintenance of turgor under water deficit. Studies in plants show that moderate Co supply increases antioxidant capacity and membrane stability, whereas higher doses become clearly phytotoxic (Inayat et al., 2024; Reis et al., 2025).

11.2.11 Iodine

This micronutrient is traditionally not regarded as an essential element for higher plants, yet accumulating evidence shows that it behaves as a nutrient with specific metabolic and regulatory functions. Plants take up iodine mainly as iodide (I^-) or iodate (IO_3^-); after uptake, it can be incorporated into organoiodine compounds or remain as inorganic I^- in apoplastic and vacuolar compartments (Medrano-Macías et al., 2016; Riyazuddin et al., 2023).

At low concentrations, iodine frequently stimulates photosynthesis, nutrient acquisition, and carbohydrate metabolism, and modifies the abundance of chloroplast and defense-related proteins. In many species, I supply upregulated genes for antioxidant enzymes and pathogenesis-related proteins, suggesting a tight integration of iodine metabolism with plant defense pathways. In aquatic macrophytes and brown algae, iodine reaches very high concentrations and functions as a redox buffer and halogen reservoir, strongly influencing local biogeochemical cycles (Gonzali et al., 2017).

The contribution of iodine to abiotic stress tolerance is increasingly documented. Foliar or root application of micromolar I^-/IO_3^- mitigates the effects of drought, salinity, and metal toxicity by enhancing the activities of SOD, CAT, POD, and components of the ascorbate–glutathione cycle, increasing nonenzymatic antioxidants and osmolytes, and lowering lipid peroxidation and electrolyte leakage (Riyazuddin et al., 2023; Medrano Macías et al., 2021). Iodide can also directly detoxify H_2O_2 via peroxidase-catalyzed formation of hypoiodous acid, acting as a diffusible antioxidant system in the apoplast (Arnhold and Malle, 2022). These effects can maintain membrane integrity, photosynthetic performance, and water status under stress, thereby improving plant performance and, in crop systems, yield stability.

Strict iodine-containing cofactors have not been demonstrated in higher plants; however, several enzymes in plants and plant-associated photosynthetic organisms are functionally iodine dependent, or they require iodide as a substrate (Kiferle et al., 2021).

The above explanation and examples of micronutrients show that metal and semimetal micronutrients induce abiotic stress tolerance through convergent mechanisms: activation of enzymatic and nonenzymatic antioxidant systems, modulation of methylglyoxal and other toxic metabolites, osmotic adjustment via proline, soluble carbohydrates, and other osmolytes, maintenance of photosynthetic integrity, and fine regulation of ion transport and cell-wall architecture. These processes support not only survival but also the maintenance of yield and quality under stress, linking micronutrient management to both the agronomic resilience and nutritional value of crops.

11.3 Factors associated with micronutrient deficiency in crops and its mitigation

Except for beneficial micronutrients such as Se and Co, the traditional view of micronutrient application in crops focuses strictly on nutrition—that is, meeting a crop's needs for metabolic activities, growth, and yield. However, it is now known that both essential and beneficial micronutrients play additional roles in improving crop resilience. Therefore their use should be promoted, and alternatives should be developed to ensure the affordable availability of fertilizers or methods to make micronutrients bioavailable to crops.

An important part, although not the only one, of providing that crops have access to micronutrients involves fertilization, whether applied to the soil, foliage, or directly to the seed or tuber through priming. Equally important are agronomic practices that prevent micronutrient leaching or volatilization and ensure that micronutrients remain bioavailable to crops. This section addresses this latter topic and discusses environmental factors related to nutrient bioavailability, as well as those that induce leaching or volatilization.

The bioavailability of micronutrients in agroecosystems is controlled by interacting climatic, edaphic, and irrigation-water factors that regulate their speciation, solubility, mobility, and losses. In many regions, rising temperatures and more intense rainfall events accelerate the weathering and leaching of mobile species such as B, Mo, Cl, Se, and I from coarse-textured soils, while episodic drought concentrates salts and bicarbonates that raise pH and depress the availability of Fe, Zn, Cu, and Mn. Soil pH is a key variable: in calcareous or overlimed soils (pH > 7.5), Fe, Mn, Zn, and Cu precipitate as carbonates, oxides, or phosphates, sharply reducing their concentration in soil solution; in strongly acidic soils, the opposite problem, Mn and Cu toxicity, is more frequent (Dhaliwal et al., 2019).

Redox conditions, which are strongly dependent on organic matter and soil aeration, further modulate availability. The redox microenvironments created by the organic matter and the soil microbiome promote the bioavailability of micronutrients (Dhaliwal et al., 2019; Medrano-Macías et al., 2016). On the other hand, poorly aerated or waterlogged soils reduce Fe and Mn oxides, mobilizing these elements but simultaneously increasing the sorption of Zn and Cu onto freshly formed oxides and carbonates, or promoting their leaching from sandy soils (Iu et al., 1981). Organic matter and carbonates bind Cu, Zn, and Mn, sometimes buffering availability but also immobilizing them when reactive functional groups are saturated. An important aspect, particularly in relation to soil organic matter, is that the addition of new organic matter can markedly modify the availability of mineral nutrients. The general affinity sequence for the formation of organic complexes in soils (in both solid and dissolved phases) is: $Fe^{3+} > Cu^{2+} > Co^{2+} > Zn^{2+} > Fe^{2+} > Mn^{2+}$. As organic amendments decompose, they release dissolved organic matter that can chelate or complex these cations, effectively sequestering them and thereby decreasing their bioavailability for plant uptake (Galić et al., 2025).

Soil organic matter has a favorable effect on micronutrient availability, provided that the biological activity of microorganisms and plant roots is adequate. If biological activity is insufficient, the organic matter, or part of it, enters a recalcitrant state. This metastable state allows carbon to be stored in the soil for extended periods. When soil organic matter is in a recalcitrant form, micronutrients are strongly complexed and unavailable to plants. For the above reason, it is advisable to avoid fallow periods whenever possible, maintain plant cover, and review pesticide management practices that could impact the soil microbiome (Bashir et al., 2021).

For anions such as $B(OH)_4^-$, MoO_4^{2-}, SeO_4^{2-}, and IO_3^-, high rainfall or overirrigation in low-cation exchange capacity sandy soils favors leaching, whereas irrigation with bicarbonate-rich or saline water elevates pH and ionic strength, impairing root uptake through antagonistic effects (Assunção et al., 2022). Additionally, plants and the soil microbiome can volatilize Se and I as methylated compounds, especially in flooded or organic-rich soils, representing a true elemental loss from the crop-soil system (Consentino et al., 2023).

Beyond direct micronutrient fertilization (soil, foliar, or seed priming), several agronomic practices can markedly improve plant micronutrient acquisition, indirectly promoting crop stress tolerance. Adjusting soil pH toward acid-neutral values (6.0–7.0) through careful liming of acidic soils or application of elemental S (an acidifying fertilizer) in calcareous soils strongly affects Fe, Zn, Cu, and Mn solubility. Incorporation of organic amendments (compost, manures, green manures, biochar) can enhance cation exchange capacity and provide natural complexes that stabilize Fe and Zn in the rhizosphere while reducing leaching of B, Mo, and Se (Galić et al., 2025).

For crops whose micronutrient needs are entirely met by soil reserves or by fertilizer applied to the soil, two key factors must be considered: pH and redox potential. These are influenced, on the one hand, by the soil parent material, climatic conditions, and water availability. However, they also depend on soil organic matter and microbial activity. Organic matter acts as a reservoir of protons and electrons, which it can contribute or receive in soil reactions. Therefore it acts as a pH and redox potential buffer, regulating the bioavailability of micronutrients.

For this reason, many farmers prefer foliar applications of micronutrients, as this ensures a direct supply to meet plant needs without the complexities of the soil. However, this technique does not account for the fact that micronutrients also play an important role in soil metabolism and are essential for preventing soil degradation (Lal, 2009).

Cropping system design is another powerful agronomic tool. Intercropping cereals with legumes or broadleaf species that exude organic acids and phytosiderophores increases Fe and Zn mobilization in calcareous soils. Using rotations, including deep-rooted or mycotrophic species, maintain mycorrhizal networks that support the uptake of Zn, Cu, Mn, and B. Inoculation with arbuscular mycorrhizal fungi and rhizosphere-competent bacteria can improve the acquisition of poorly mobile micronutrients (Fe, Zn, Cu, Mn) and beneficial elements (Se, Co, I) while also modulating root architecture. Adequate irrigation management, avoiding waterlogging, limiting excessive leaching, and reducing bicarbonate and salinity loads, helps maintain stable soil redox conditions and pH, which is particularly relevant for Fe, Mn, and Cl. In addition, cultivar and rootstock selection for micronutrient-efficient genotypes, with enhanced transporter expression, root exudation, and xylem loading capacity, represents a long-term and cost-effective strategy to promote micronutrient absorption and assimilation (Assunção et al., 2022; Dhaliwal et al., 2019).

11.4 Micronutrient sources for crops: fertilizers, nanofertilizers, and crop stress tolerance

In modern agriculture, micronutrient management is widespread in horticultural crops, both leafy and fruit. However, micronutrient management is less common in staple crops; in these cases, NPK fertilizers are typically used, and calcium, magnesium, and sulfur are used less commonly. For this reason, although staple crops are much more important in terms of the amount of nutrients consumed by the global population, they generally have limitations in terms of micronutrient supply (Aziz et al., 2019).

Globally, a small set of staple crops, including maize, rice, wheat, sorghum, potatoes, and cassava, provides most of the dietary energy and a large share of protein and minerals for humankind. The FAO reports that just three cereals, rice, maize, and wheat, supply approximately 60% of the world's food energy intake, while roots and tubers such as potato and cassava dominate caloric intake in many low-income regions (FAO, 2025). Diets in which these staples account for most calories, but access to animal-source foods, pulses, fruits, and vegetables is limited, often provide sufficient energy yet remain poor in essential micronutrients (Biro and Menon, 2014).

Because maize, wheat, sorghum, potato, and cassava constitute the bulk of daily intake for hundreds of millions of people, even small increases in their grain or tuber concentrations of Fe, Zn, and other micronutrients can deliver large, low-cost public health gains. This is the central rationale of biofortification programmes, which deliberately enrich staple crops through breeding and agronomic practices and have been shown to be technically feasible and highly cost-effective for improving micronutrient intake in vulnerable populations (Nestel et al., 2006). Directing efforts to increase micronutrient density toward these dominant staples, therefore maximizes the impact on human nutrition, as they are the primary vehicle for calories, protein, minerals, vitamins, and phytochemicals in most diets (Harvest Plus, 2025).

Prioritizing the above crops is equally crucial from an agronomic and climate-adaptation perspective. As explained earlier, a growing body of physiological and field evidence shows that adequate micronutrient supply enhances plant tolerance to drought, heat, salinity, and nutrient-poor or degraded soils by reinforcing antioxidant defense, maintaining membrane integrity and photosynthetic capacity, and supporting osmolyte accumulation and hormone-mediated stress signaling. These reasons indicate that research and technology transfer on the use of micronutrient fertilizers can provide greater benefits to human quality of life if focused on staple crops.

Adequate fertilization is one of the central pillars of modern agriculture and global food security. Mineral fertilizers account for a substantial fraction of yield gains achieved during the Green Revolution, and projections indicate that closing current yield gaps in cereals will remain impossible without continued but smarter nutrient inputs. However, the use efficiency of conventional fertilizers is often below 50%, and for several micronutrients it can be markedly lower due to strong soil fixation, leaching, or volatilization, especially in marginal environments with degraded or highly weathered soils (Dimkpa and Bindraban, 2018). The above inefficiencies drive both environmental damage (eutrophication, contamination) and "hidden hunger" in human populations because low micronutrient density in cereal-based diets is tightly linked to soil and plant deficiencies of Fe, Zn, Se, and other trace elements (Benavides-Mendoza et al., 2024).

11.4.1 Conventional fertilizers versus nanofertilizers and crop stress tolerance

In cereals such as maize, wheat, sorghum, potato, and cassava grown on calcareous, saline, or nutrient-depleted soils, deficiencies in Fe, Zn, B, Mn, and other micronutrients are common and exacerbate the effects of drought, heat, and low NPK inputs. Conventional strategies such as soil application of metal salts or chelates, foliar sprays, and seed coating have improved yields but are constrained by large application doses, rapid precipitation or sorption in soil, and short residence times on leaf surfaces (Benavides-Mendoza et al., 2024).

Traditional micronutrient fertilizers supply elements mainly as soluble salts, oxides/hydroxides, or synthetic chelates (e.g., Fe-EDDHA, Zn-EDTA). These products are applied to soil, foliage, or seeds, and their agronomic performance depends strongly on soil pH, redox conditions, organic matter, and competing ions. In alkaline and calcareous soils, Fe and Zn rapidly precipitate as poorly soluble carbonates or oxides; in highly weathered tropical soils, Mn and B can be leached or strongly sorbed to Fe/Al oxides (Boyd, 2020).

To circumvent these constraints, agronomists have adopted band placement, fertigation, foliar sprays, or seed priming to temporarily increase micronutrient concentrations in the rhizosphere or directly on the leaf surface. Such practices can improve grain yield and micronutrient density, but often require repeated applications and do not fully resolve the mismatch between nutrient supply and crop demand, particularly under stress conditions that restrict root growth or transpiration (Singh et al., 2025).

Nanofertilizers are fertilizer products in which the nutrient itself is present at the nanoscale (e.g., ZnO NPs, Fe_2O_3 NPs) or where nutrients are loaded into/onto nanoscale carriers (e.g., layered double hydroxides, polymeric nanoparticles, nanobiochar). Compared with bulk formulations, nanofertilizers differ in several key physicochemical and functional aspects (Ojeda-Barrios et al., 2019) (Fig. 11.3):

- High specific surface area and reactivity. A particle size <100 nm dramatically increases the surface area per unit mass and enhances the free surface energy, thus enhancing the dissolution rates of sparingly soluble oxides and enabling stronger interactions with root or leaf surfaces and penetration through cell walls.
- Controlled release and targeted delivery. Nutrients can be encapsulated or adsorbed onto nanocarriers that release ions slowly in response to pH, moisture, or root exudates, better synchronizing supply with crop demand and reducing leaching or fixation.
- Improved foliar penetration and uptake. Nanoparticles can pass through cuticular pores or stomatal apertures more readily than bulk particles, facilitating translocation within the plant and enhancing biofortification of grains or tubers.

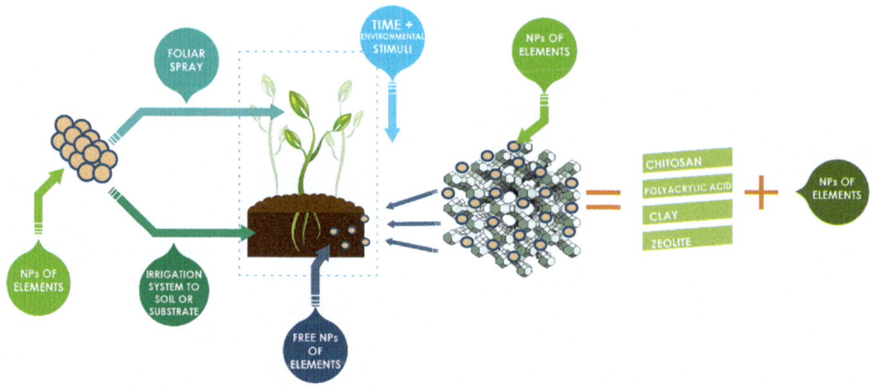

FIGURE 11.3

Mechanisms of application and absorption of micronutrient nanoparticles (Morales-Díaz et al., 2017).

Figure from Morales-Díaz, A.B., Ortega-Ortíz, H., Juárez-Maldonado, A., Cadenas-Pliego, G., González-Morales, S., Benavides-Mendoza, A., 2017. Application of nanoelements in plant nutrition and its impact in ecosystems. Advances in Natural Sciences: Nanoscience and Nanotechnology. 8, 013001. https://doi.org/10.1088/2043-6254/8/1/013001.

Crucially, many nanofertilizers exhibit a dual function: they act not only as nutrient sources but also as biostimulants that modulate plant adaptive responses at doses sometimes below classical nutritional requirements (Naz and Benavides-Mendoza, 2021). Reviews of nanostimulants emphasize that metallic and semi-metallic nanoparticles (e.g., ZnO, Fe_2O_3, Se, Si) can alter hormonal balance, enhance antioxidant enzyme systems, upregulate stress-responsive genes, and remodel root architecture, thereby improving stress tolerance beyond what is expected from simple nutrient addition (Juárez-Maldonado et al., 2018). Nanofertilizers occupy an intermediate conceptual space between fertilizers and classical biostimulants, simultaneously addressing nutrient limitation and stress tolerance signaling.

For example, studies with Fe NPs in maize under salt stress similarly report improved shoot growth, enhanced antioxidant capacity, and better ionic balance compared with conventional Fe sources, highlighting both metabolic and signaling dimensions of Fe-mediated tolerance (Alsamadany et al., 2024). The study reported by Gupta et al. (2024) describes the use of ZnO NPs for alleviating salt stress in common beans. Salt-stressed plants were treated with ZnO NPs at 25, 50, 100, and 200 mg/L via foliar spray, nanopriming, or soil application. Salinity markedly reduced fresh weight (24%), relative water content (27%), plant height (33%), and chlorophyll content (37%) and increased proline (over 100%), sodium accumulation, and antioxidant enzyme activity. ZnO NP treatments improved physiological growth traits and reduced salt damage by regulating nutrient homeostasis, limiting ROS generation, and enhancing chlorophyll fluorescence. All application methods mitigated salt stress, but nanopriming was most effective, producing values close to those of nonstresses control plants.

For Se, multiple studies have shown that Se NPs function as both micronutrient sources and potent biostimulants (Medrano-Macías and Narvaéz-Ortiz, 2022). For example, foliar delivery of Se NPs enhanced wheat growth, morpho-physiological performance, biochemical status, and the expression of stress-responsive genes under both heat and drought conditions. Drought caused more severe reductions in growth than heat, whereas heat stress exerted a stronger influence on the transcriptional regulation of stress-related genes. Exogenous spraying with 10 mg L-1 Se NPs significantly improved photosynthetic pigments, net photosynthetic rate, gas-exchange parameters, and transpiration rate in stressed plants. These improvements were accompanied by enhanced activity of key antioxidant enzymes, including CAT, APX, and SOD, as well as upregulation of multiple stress-responsive genes (Omar et al., 2023).

Cobalt-containing nanoparticles have been less studied, but Novikova et al. (2025) showed that Co and B nanoparticles applied to durum and bread wheat improved germination and early growth under flooding, likely by modulating respiratory metabolism and antioxidant defenses.

Iodine has been increasingly recognized as a beneficial element and biostimulant for plants (Nascimento et al., 2022). Foliar iodine fertilization with iodide or iodate improves antioxidant status, but nanoenabled formulations are only now emerging (Marcial-Salvador et al., 2024).

For example, (Huliaieva et al., 2025) evaluated iodine nanocitrates in juvenile wheat exposed to short-term cold stress and phytopathogenic infection, reporting improved maintenance of pigment content, altered H_2O_2 dynamics, and modest stimulation of growth in stressed plants, suggesting that nano-I functions as an inorganic antioxidant and stress modulator. In lettuce, Treviño-Ruiz et al. (2024) applied nanochitosan–iodine complexes and found 30.8% increases in head yield alongside enhanced I biofortification, confirming that nanometric I can act simultaneously as a nutrient carrier and growth promoter.

The arguments and examples above suggest that, for micronutrients, further exploration of nanofertilizers, especially for staple crops, for agricultural production would be advisable.

11.5 Impact of the synergy between micronutrients on plant stress tolerance

As with biostimulant mixes, micronutrient mixtures are expected to work synergistically. Synergy refers to the fact that the combined impact of two or more components is greater than the sum of their individual impacts. In plants, the synergistic effects of micronutrients are expected, given that each micronutrient has specific metabolic or regulatory functions and each micronutrient specifically influences multiple responses related to stress tolerance. The combination of the above specific impacts generates emergent metabolic and gene-regulatory responses that cannot be predicted from the study of each individual micronutrient (Jeevanraj et al., 2025).

In metabolism, signaling, and stress physiology, synergy arises because living systems are built as interconnected metabolic and gene networks rather than as independent processes. Applied to micronutrients, synergy means that plant responses under stress cannot be fully predicted by managing the elements in isolation. Instead, the benefits of one nutrient on growth and defense often depend on the sufficiency, chemical form, and timing of application of others. This system's logic becomes even more important when beneficial elements such as Se, Co, and I are included, because they can shift redox homeostasis and nutrient partitioning in ways that amplify essential micronutrient effects (Kumar et al., 2021).

A further consideration is regulatory crosstalk among nutrients. The P-Zn-Fe network is a well-studied example: deficiencies or excesses in one element can reprogram transporter expression, root exudation, and allocation of others (Gao et al., 2023). Thus balanced mixes of nutrients can be more resilient than single-nutrient corrections because they reduce the risk of inducing secondary deficiencies and because they support entire pathways (e.g., photosynthetic electron flow plus ROS detox plus osmoprotection) rather than a single plant response. The central idea is that plant stress tolerance is frequently limited by several micronutrients that jointly enable defense rather than by any single micronutrient.

Considering the above and because plant stressors rarely act individually in the field, probably the most robust approach is to manage micronutrients as crop species-condition-designed combinations (crop x chemical form × dose × timing) matched to soil constraints and climatic stress probability (Mikkelsen, 2011). Obviously, this approach requires significant support for farmers, including extension technicians and real-time databases accessible through networks or apps to support decisions during crop management.

As an example, two illustrative target environments are as follows:

High irradiance and calcareous soils. Fe and Zn deficiencies frequently co-occur because solubility and root uptake are strongly limited at high pH. In such settings, combined foliar Zn and Fe can simultaneously raise grain micronutrient concentrations and sustain yield under water limitation, effectively treating nutrient deficiency as an abiotic stress syndrome rather than as a single-element problem. In wheat grown under contrasting water regimes, combined foliar Zn-Fe increased yield and reduced yield loss under water stress while also increasing grain Zn and Fe contents (Shoormij et al., 2022).

High temperature and drought in acidic or marginal soils. Heat and moisture stress disrupt photosynthesis and reproduction and intensify oxidative damage. In rainfed lentil exposed to late-season heat and moisture stress, foliar sprays of B, Zn, and Fe

Table 11.2 A sample of literature between 2015 and 2025 on micronutrient synergy and its impact on plant stress responses.

Plant micronutrients	Summary of results	References
B, Se, Co, and Ti	In saline alluvial soil, salinity severely impaired potato growth, nutrition, yield, and tuber quality while increasing oxidative damage. Combining banana or soybean compost with foliar B, Se, Co, and Ti markedly mitigated these effects. The most effective treatment was banana compost plus B and Se, which maximized productivity and salt tolerance by reducing oxidative stress.	El-Ghamry et al. (2024)
Co, Mo, and Ni	High-vigor and low-vigor eed treatment with Co, Mo, and Ni (2–4 mL kg^{-1}) enhanced soybean N metabolism and growth, especially in high-vigor seeds. Urease rose 191% (high vigor) and 65% (low vigor), nitrogenase was greater in high-vigor plants, and low-vigor plants showed 42% higher nodule biomass. Shoot biomass increased 30% at 2 mL kg^{-1}. Excess doses induced phytotoxic effects.	Carmo-Filho et al. (2025)
I and Se	Carrot foliar fertilization (400 g I + 20 g Se ha^{-1}) with iodine and selenium (alone or combined) effectively biofortified roots, safely providing 4.16–4.37% of the human RDA for each element per 100 g fresh weight, without reducing yield or quality. Iodide was incorporated into diverse organoiodine metabolites (e.g., iodosalicylates or iodobenzoates), while no synergistic or antagonistic I-Se interaction occurred; root I:Se ratios remained close to control.	Rakoczy-Lelek et al. (2021)
Se and B	In soybean, increasing NaCl (150–450 mM) impaired growth, lowered RWC, and elevated proline, H_2O_2, and MDA, while differentially altering antioxidant and glyoxalase activities. Foliar Se, B, and especially the combination Se +B, consistently enhanced enzymatic antioxidant activities, reduced H_2O_2 and MDA accumulation, and improved growth.	Rahman et al. (2021)
Se and I	A valuable review on joint Se-I biofortification, a topic that remains until now not explored enough; mechanisms and controlling factors require further study, making it an emerging area in plant physiology and human nutrition. Nevertheless, existing evidence demonstrates feasible large-scale Se-I enrichment of several crops, and consumer acceptance plus nutritionally meaningful Se and I levels in produce underpin strong potential for developing Se-I biofortified functional foods.	Golubkina et al. (2021)

(Continued)

Table 11.2 A sample of literature between 2015 and 2025 on micronutrient synergy and its impact on plant stress responses. (*Continued*)

Plant micronutrients	Summary of results	References
Se and Zn	Combined $ZnSO_4.H_2O$ (10 mM) and Na_2SeO_4 (0.075 mM) seed priming (12 and 24 h) in maize greatly improved germination indices (up to 52–54% increases), seedling vigor, and antioxidant enzyme activities under 60% field capacity in pots with calcareous soil (pH=7.4). It also raised shoot Se and Zn by 154% and 114%, respectively, enhancing drought tolerance while simultaneously contributing to Se and Zn biofortification.	Nawaz et al. (2021)
Se and Zn	Over two seasons in semiarid conditions in Spain, combined foliar Zn (8 kg ha^{-1}) and Se (10 g ha^{-1}) fertilization in bread wheat most strongly enhanced biofortification, raising straw Zn and Se 12.3- and 2.7-fold and grain 1.3- and 4.3-fold, while increasing grain yield by almost 7%, confirming wheat's suitability for Zn-Se biofortification of grain for human food and straw for livestock feed.	Reynolds-Marzal et al. (2021)
Zn, B, and Cu NPs	In greenhouse-grown soybean at 50% field capacity, drought reduced shoot growth by 27% and grain yield by 54%. Soil or foliar application of composite Zn, B, and Cu nanoparticles or their salts mitigated these losses, increasing shoot growth by 33% and yield by 36%, and markedly enhancing shoot and grain N, K, and micronutrient contents. To mitigate drought stress, the soil-applied salts were most effective than foliar application. For foliar spraying, the effects of nanoparticles and salts were similar.	Dimkpa et al. (2017)
Zn and Mn	Maize at 8°C–14°C showed 133% greater oxidative leaf injury and 25% lower shoot biomass under NO_3 nutrition, linked to reduced P, Zn, and Mn uptake. Providing NH4 and a Zn/Mn-fortified microbial consortium (*Trichoderma harzianum* and *Bacillus* spp.) improved Zn/Mn nutrition, hormone (ABA, IAA, JA, SA) signaling, ROS detoxification, root colonization, and growth; extra Zn/Mn further maximized root length, antioxidants, biomass, and minimized cold damage.	Moradtalab et al. (2020)
Zn, NPK, Mg, and B	In tomatoes, a pot trial showed that combining Zn (5 mg kg^{-1}) with recommended NPK, but not consistently with Mg or B, most effectively suppressed early blight (25% incidence, 13% severity), while enhancing growth,	Awan et al. (2019)

(*Continued*)

Table 11.2 A sample of literature between 2015 and 2025 on micronutrient synergy and its impact on plant stress responses. (*Continued*)

Plant micronutrients	Summary of results	References
	photosynthetic pigments, phenolics, proteins, and defense enzymes (SOD, CAT, POD, PPO, PAL), indicating Zn+NPK-driven metabolic crosstalk and induced resistance to *Alternaria*.	
Zn and Se	In *Billbergia zebrina*, three Zn levels (0–300 μM) with or without 4 μM Se were tested for 75 days. High Zn reduced total chlorophyll and altered Chl a fluorescence. Se increased total chlorophyll, lowered the Chl a/b ratio, and alleviated Zn toxicity.	Martins et al. (2020)

applied individually or in combination improved photosynthetic function, water relations, and yield, with the combined strategy explicitly framed as stress alleviation via multielement support of plant defense and reproduction (Venugopalan et al., 2022).

Although the majority of information on the impact of micronutrients on crops focuses on a purely nutritional approach, some literature reports their effects on stress tolerance in plants; in this case, however, a substantial part refers to nanomicronutrients. Table 11.2includes literature from 2015 to 2025 regarding micronutrient synergy and its impact on stress responses.

The synergy points to a shift from single-deficiency correction to scenario-based micronutrient management with at least three components: (1) diagnose co-occurring limitations and stress probabilities, (2) select compatible chemical forms (e.g., chelates or nanofertilizers for high pH or marginal soils; seed priming for establishment stress), and (3) time multielement delivery may be combined with biostimulants to stress-sensitive stages, for example, reproduction under heat and early growth under drought.

Agronomic practice increasingly uses multielement foliar blends because deficiency patterns commonly overlap in the field. On the other hand, the scientific literature shows that paired or bundled micronutrients can simultaneously increase productivity and strengthen stress tolerance (Aziz et al., 2019).

11.6 Conclusion and future consideration

Micronutrient deficiency in crops, associated with a detrimental impact on stress tolerance, is rarely a simple lack of fertilizer input; instead, it emerges from complex soil–plant–climate interactions that must be addressed through integrated management of pH, organic matter, cropping systems, water, biostimulants, and genotype choice, alongside appropriately designed micronutrient fertilization strategies. In practice, this implies integrated management moving beyond single-element correction toward site-specific packages that coordinate fertilizer source selection (soil, foliar, seed/tuber priming), timing at sensitive phenological stages, and agronomic practices that sustain rhizosphere function and reduce nutrient losses.

Future progress depends on closing knowledge gaps and translating mechanisms into decision-ready recommendations. A current frontier is the implementation of environment-based multinutrient management. Because plant metabolic responses and gene expression regulation are networked, it is assumed that the most robust approach is to design crop–condition combinations (crop x chemical form × dose × timing) matched to soil constraints and the probability of co-occurring stresses, supported by extension and real-time decision tools. However, evidence remains uneven across environments and nutrient bundles, and a stronger multiseason field validation is needed.

For the above, the authors propose the following priority research topics:

1. Predictive diagnostics: integrated soil–plant indicators that jointly capture soil pH/redox status, organic matter, irrigation-water chemistry, and crop demand dynamics.
2. Mechanisms of interaction: rules for when mixes are synergistic, additive, or antagonistic (including transporter regulation and crosstalk among elements) to prevent unintended secondary deficiencies.
3. Nanoenabled delivery: standardized characterization and dose–response relationships linking particle properties to dissolution, uptake routes, and translocation, alongside environmental fate, soil-health impacts, and food-safety assessment under realistic field use.
4. Rhizosphere biology: how organic amendments, microbial consortia, and mycorrhizae reshape element chemistry and acquisition under differing irrigation, soil, and climatic contexts.
5. Genotype x management integration: plant breeding for uptake and translocation efficiency combined with tailored fertilization regimes.

Finally, the agronomic agenda should be explicitly aligned with human nutrition. Low trace-element density in edible organs contributes to hidden hunger, and biofortification offers a practical bridge between resilient production and public health gains. Given their dietary dominance and current underuse of micronutrient management, more support and coordinated initiatives should prioritize staple crops, particularly maize, rice, wheat, sorghum, potato, and cassava, because modest increases in edible-organ density can generate outsized benefits for both nutrition and climate adaptation.

References

Alejandro, S., Höller, S., Meier, B., Peiter, E., 2020. Manganese in plants: from acquisition to subcellular allocation. Frontiers in Plant Science 11. Available at https://doi.org/10.3389/fpls.2020.00300.

Alharby, H.F., Nahar, K., Al-Zahrani, H.S., Hakeem, K.R., Hasanuzzaman, M., 2021. Enhancing salt tolerance in soybean by exogenous boron: Intrinsic study of the ascorbate-glutathione and glyoxalase pathways. Plants 10 (10). Available at https://doi.org/10.3390/plants10102085.

Al-Issawi, M., Rihan, H.Z., Al-Shmgani, H., Fuller, M.P., 2016. Molybdenum application enhances antioxidant enzyme activity and cor15a protein expression under cold stress in wheat. Journal of Plant Interactions 11 (1), 5–10. Available at https://doi.org/10.1080/17429145.2015.1129074.

Alsamadany, H., Anayatullah, S., Zia-ur-Rehman, M., Usman, M., Ameen, T., Alharby, H.F., et al., 2024. Residual efficiency of iron-nanoparticles and different iron sources on growth, and antioxidants in maize plants under salts stress: life cycle study. Heliyon 10 (7). Available at https://doi.org/10.1016/j.heliyon.2024.e28973.

Ameen, M., Zia, M.A., Najeeb Alawadi, H.F., Naqve, M., Mahmood, A., Shahzad, A.N., et al., 2024. Exogenous application of selenium on sunflower (*Helianthus annuus* L.) to enhance drought stress tolerance by morpho-physiological and biochemical adaptations. Frontiers in Plant Science 15. Available at https://doi.org/10.3389/fpls.2024.1427420.

Arnhold, J., Malle, E., 2022. Halogenation activity of mammalian heme peroxidases. Antioxidants 11 (5), 890. Available at https://doi.org/10.3390/antiox11050890.

Assunção, A.G.L., Cakmak, I., Clemens, S., González-Guerrero, M., Nawrocki, A., Thomine, S., 2022. Micronutrient homeostasis in plants for more sustainable agriculture and healthier human nutrition. Journal of Experimental Botany 73 (6), 1789–1799. Available at https://doi.org/10.1093/jxb/erac014.

Awan, Z.A., Shoaib, A., Khan, K.A., 2019. Crosstalk of zn in combination with other fertilizers underpins interactive effects and induces resistance in tomato plant against early blight disease. Plant Pathology Journal 35 (4), 330–340. Available at https://doi.org/10.5423/PPJ.OA.01.2019.0002.

Aziz, M.Z., Yaseen, M., Abbas, T., Naveed, M., Mustafa, A., Hamid, Y., et al., 2019. Foliar application of micronutrients enhances crop stand, yield and the biofortification essential for human health of different wheat cultivars. Journal of Integrative Agriculture 18 (6), 1369–1378. Available at https://doi.org/10.1016/S2095-3119(18)62095-7.

Bashir, O., Ali, T., Baba, Z.A., Rather, G.H., Bangroo, S.A., Mukhtar, S.D., et al., 2021. Soil organic matter and its impact on soil properties and nutrient status. Microbiota and Biofertilizers, Vol 2: Ecofriendly Tools for Reclamation of Degraded Soil Environs. Springer International Publishing, India, pp. 129–159.

Benavides-Mendoza, A., Juárez-Maldonado, A., Tortella, G., Ojeda-Barrios, D.L., Fuentes-Lara, L.O., 2024. Differences in crop responses to fertilizers and nanofertilizers. Nanofertilizer Delivery, Effects and Application Methods. Elsevier, Mexico, pp. 299–329.

Bidwell, R.G.S., Bebee, G.P., 1974. Carbon monoxide fixation by plants. Canadian Journal of Botany 52 (8), 1841–1847. Available at https://doi.org/10.1139/b74-236.

Biro, E., Menon, P., 2014. Addressing the challenge of hidden hunger. Global Hunger Index (GHI) [online]. Available from: https://www.globalhungerindex.org/issues-in-focus/2014.html (accessed 12.11.25).

Boyd, C.E., 2020. Micronutrients and other trace elements. In: Boyd, C.E. (Ed.), Water Quality: An Introduction. Springer International Publishing, Cham, pp. 335–378. Available at https://doi.org/10.1007/978-3-030-23335-8_17.

Brdar-Jokanović, M., 2020. Boron toxicity and deficiency in agricultural plants. International Journal of Molecular Sciences 21 (4), 1424. Available at https://doi.org/10.3390/ijms21041424.

Brown, P.H., Zhao, F.J., Dobermann, A., 2022. What is a plant nutrient? Changing definitions to advance science and innovation in plant nutrition. Plant and Soil 476 (1–2), 11–23. Available at https://doi.org/10.1007/s11104-021-05171-w.

Cakmak, I., Brown, P., Colmenero-Flores, J.M., Husted, S., Kutman, B.Y., Nikolic, M. et al., 2023. Chapter 7 - Micronutrients☆☆This chapter is a revision of the third edition chapter by M. Broadley, P. Brown, I. Cakmak, Z. Rengel, and F.J. Zhao, pp. 191–248. DOI: https://doi.org/10.1016/B978-0-12-384905-2.00007-8. © Elsevier Ltd.**The individual micronutrient sections were written as follows: Fe (M.N.), Mn (S.H. and S.B.S.), Cu (Z. R.), Zn (I.C.), Ni (P.B. and B.Y.K.), Mo (F-J.Z.), B (P.B. and I.C.), and Cl (J.M.C.F.)., in: Rengel, Z., Cakmak, I., White, P.J. (Eds.), Marschner's Mineral Nutrition of Plants (Fourth Edition). Academic Press, San Diego, pp. 283–385. https://doi.org/10.1016/B978-0-12-819773-8.00017-4.

Carmo-Filho, Ad.S., Rego, C.H.Q., Ribeiro, Gd.F.R., Alves, R.M., Almeida, L.Ad, Rodak, B.W., et al., 2025. Seed-applied cobalt, molybdenum, and nickel improve nitrogen metabolism in soybean plants across seed vigor levels. Plants 14 (21). Available at https://doi.org/10.3390/plants14213368.

Cetinkaya, H., Dinler, B.S., Tasci, E., 2014. Investigation of comparative regulation on antioxidant enzyme system under copper treatment and drought stress in maize (*Zea mays* L.). Notulae Botanicae Horti Agrobotanici Cluj-Napoca 42 (2), 363–371. Available at https://doi.org/10.1583/nbha4229632.

Chen, G., Li, J., Han, H., Du, R., Wang, X., 2022. Physiological and molecular mechanisms of plant responses to copper stress. International Journal of Molecular Sciences 23 (21), 12950. Available at https://doi.org/10.3390/ijms232112950.

Consentino, B.B., Ciriello, M., Sabatino, L., Vultaggio, L., Baldassano, S., Vasto, S., et al., 2023. Current acquaintance on agronomic biofortification to modulate the yield and functional value of vegetable crops: a review. Horticulturae 9 (2), 219. Available at https://doi.org/10.3390/horticulturae9020219.

Dhaliwal, S.S., Naresh, R.K., Mandal, A., Singh, R., Dhaliwal, M.K., 2019. Dynamics and transformations of micronutrients in agricultural soils as influenced by organic matter build-up: a review. Environmental and Sustainability Indicators 1–2, 100007. Available at https://doi.org/10.1016/j.indic.2019.100007.

Dimkpa, C.O., Bindraban, P.S., 2018. Nanofertilizers: new products for the industry? Journal of Agricultural and Food Chemistry 66 (26), 6462–6473. Available at https://doi.org/10.1021/acs.jafc.7b02150.

Dimkpa, C.O., Bindraban, P.S., Fugice, J., Agyin-Birikorang, S., Singh, U., Hellums, D., 2017. Composite micronutrient nanoparticles and salts decrease drought stress in soybean. Agronomy for Sustainable Development 37 (1). Available at https://doi.org/10.1007/s13593-016-0412-8.

El-Ghamry, A.M., El-Sherpiny, M.A., Alkharpotly, A.E.A., Ghazi, D.A., Helmy, A.A., Siddiqui, M.H., et al., 2024. The synergistic effects of organic composts and microelements co-application in enhancing potato productivity in saline soils. Heliyon 10 (12). Available at https://doi.org/10.1016/j.heliyon.2024.e32694.

FAO, 2025. Dimensions of need - Staple foods: What do people eat? [WWW Document]. Available from: https://www.fao.org/4/u8480e/u8480e07.htm (accessed 12.11.25).

Fatahiyan, F., Najafi, F., Shirkhani, Z., 2025. Enhancing salt stress tolerance in *Carthamus tinctorius* L. through selenium soil treatment: anatomical, biochemical, and physiological insights. BMC Plant Biology 25 (1). Available at https://doi.org/10.1186/s12870-025-06078-9.

Funakawa, H., Miwa, K., 2015. Synthesis of borate cross-linked rhamnogalacturonan II. Frontiers in Plant Science 6. Available at https://doi.org/10.3389/fpls.2015.00223.

Galić, L., Vukadinović, V., Nikolin, I., Lončarić, Z., 2025. Soil properties and microelement availability in crops for human health: an overview. Crops 5 (4), 40. Available at https://doi.org/10.3390/crops5040040.

Gao, D., Ran, C., Dang, K., Wang, X., Zhang, Y., Geng, Y., et al., 2023. Effect of phosphorus, iron, zinc, and their combined deficiencies on photosynthetic characteristics of rice (*Oryza sativa* L.) seedlings. Agronomy 13 (6), 1657. Available at https://doi.org/10.3390/agronomy13061657.

Gencic, S., Duin, E.C., Grahame, D.A., 2010. Tight coupling of partial reactions in the acetyl-CoA decarbonylase/synthase (ACDS) multienzyme complex from *Methanosarcina thermophila*: acetyl C-C bond fragmentation at the A cluster promoted by protein conformational changes. Journal of Biological Chemistry 285 (20), 15450–15463. Available at https://doi.org/10.1074/jbc.M109.080994.

Golubkina, N., Moldovan, A., Kekina, H., Kharchenko, V., Sekara, A., Vasileva, V., et al., 2021. Joint biofortification of plants with selenium and iodine: new field of discoveries. Plants 10 (7), 1352. Available at https://doi.org/10.3390/plants10071352.

Gonzali, S., Kiferle, C., Perata, P., 2017. Iodine biofortification of crops: agronomic biofortification, metabolic engineering and iodine bioavailability. Current Opinion in Biotechnology 44, 16–26. Available at https://doi.org/10.1016/j.copbio.2016.10.004.

Guedes, F.R.C.M., Pereira Junior, E.M., Batista, B.L., Guedes Lobato, E.M.S., Lobato, A.K.da S., 2025. Nickel boots antioxidant defense and nutrient uptake, reducing deleterious effects on biomass in soybean plants copper-stressed. Discover Plants 2 (1). Available at https://doi.org/10.1007/s44372-025-00354-6.

Gupta, A., Bharati, R., Kubes, J., Popelkova, D., Praus, L., Yang, X., et al., 2024. Zinc oxide nanoparticles application alleviates salinity stress by modulating plant growth, biochemical attributes and nutrient homeostasis in *Phaseolus vulgaris* L. Frontiers in Plant Science 15. Available at https://doi.org/10.3389/fpls.2024.1432258.

Hänsch, R., Mendel, R.R., 2009. Physiological functions of mineral micronutrients (Cu, Zn, Mn, Fe, Ni, Mo, B, Cl). Current Opinion in Plant Biology 12 (3), 259–266. Available at https://doi.org/10.1016/j.pbi.2009.05.006.

Harvest Plus, 2025. Enriching 100 Million Lives [WWW Document]. HarvestPlus. Availeble from: https://www.harvestplus.org/ (accessed 12.11.25).

Hossain, M.S., Garcia Caparros, P., Mühling, K.H., 2024. Editorial: mineral nutrition and plant stress tolerance. Frontiers in Plant Science 15. Available at https://doi.org/10.3389/fpls.2024.1461651.

Hu, X., Wei, X., Ling, J., Chen, J., 2021. Cobalt: an essential micronutrient for plant growth? Frontiers in Plant Science 12. Available at https://doi.org/10.3389/fpls.2021.768523.

Hu, X., Wei, X., Ling, J., Chen, J., 2023. Cobalt in plant life. Beneficial Chemical Elements of Plants. John Wiley & Sons, Ltd, pp. 182–204. Available at https://doi.org/10.1002/9781119691419.ch8.

Huliaieva, H., Tokovenko, I., Kharchuk, M., Bohdan, M., Pasichnyk, L., 2025. The influence of iodine nano citrates on juvenile wheat plants at phytopathogenic infection and cooling stress. Plant Nano Biology 13, 100175. Available at https://doi.org/10.1016/j.plana.2025.100175.

Husen, A., 2024. Essential Minerals in Plant-Soil System (Coordination, Signalling, and Interaction under Adverse Situations). Elsevier Inc., Cambridge, MA. Available at https://doi.org/10.1016/C2022-0-02437-5.

Imaizumi, K., Ifuku, K., 2022. Binding and functions of the two chloride ions in the oxygen-evolving center of photosystem II. Photosynthesis Research 153 (3), 135–156. Available at https://doi.org/10.1007/s11120-022-00921-y.

Imran, S., Sarker, P., Mahamud, M.A., Paul, N.C., Chakrobortty, J., Harine, I.J., et al., 2024. Copper mitigates salinity stress by regulating water status, photosynthetic pigments and ion homeostasis and increases the yield of Eggplant (*Solanum melongena*). BMC Plant Biology 24 (1). Available at https://doi.org/10.1186/s12870-024-05625-0.

Inayat, H., Mehmood, H., Danish, S., Alharbi, S.A., Ansari, M.J., Datta, R., 2024. Impact of cobalt and proline foliar application for alleviation of salinity stress in radish. BMC Plant Biology 24 (1). Available at https://doi.org/10.1186/s12870-024-04998-6.

Iu, K.L., Pulford, I.D., Duncan, H.J., 1981. Influence of waterlogging and lime or organic matter additions on the distribution of trace metals in an acid soil. Plant and Soil 59 (2), 327–333. Available at https://doi.org/10.1007/bf02184204.

Jan, A.U., Hadi, F., Ditta, A., Suleman, M., Ullah, M., 2022. Zinc-induced anti-oxidative defense and osmotic adjustments to enhance drought stress tolerance in sunflower (*Helianthus annuus* L.). Environmental and Experimental Botany 193, 104682. Available at https://doi.org/10.1016/j.envexpbot.2021.104682.

Jeevanraj, R., Sivakumar, R., Boominathan, P., Kavitha, P.S., Sendhilvel, V., 2025. The Essential role of macronutrients and micronutrients in improving crop resilience to biotic stress. Russian Journal of Plant Physiology 72 (4). Available at https://doi.org/10.1134/s1021443724608838.

Juárez-Maldonado, A., González-Morales, S., Cabrera-De la Fuente, M., Medrano-Macías, J., Benavides-Mendoza, A., 2018. Nanometals as promoters of nutraceutical quality in crop plants. Impact of Nanoscience in the Food Industry. Elsevier Inc., Mexico, pp. 277–310.

Kiferle, C., Martinelli, M., Salzano, A.M., Gonzali, S., Beltrami, S., Salvadori, P.A., et al., 2021. Evidences for a nutritional role of iodine in plants. Frontiers in Plant Science 12. Available at https://doi.org/10.3389/fpls.2021.616868.

Kováčik, J., Vydra, M., 2024. The impact of nickel on plant growth and oxidative balance. Physiologia Plantarum 176 (6). Available at https://doi.org/10.1111/ppl.14595.

Kumar, S., Kumar, S., Mohapatra, T., 2021. Interaction between macro- and micro-nutrients in plants. Frontiers in Plant Science 12. Available at https://doi.org/10.3389/fpls.2021.665583.

Kumar, U., Kumar, I., Singh, P.K., Dwivedi, A., Singh, P., Mishra, S., et al., 2025. Nickel Contamination in terrestrial ecosystems: insights into impacts, phytotoxicity mechanisms, and remediation technologies. Reviews of Environmental Contamination and Toxicology 263 (1). Available at https://doi.org/10.1007/s44169-024-00075-z.

Kumari, V.V., Banerjee, P., Verma, V.C., Sukumaran, S., Chandran, M.A.S., Gopinath, K.A., et al., 2022. Plant nutrition: an effective way to alleviate abiotic stress in agricultural crops. International Journal of Molecular Sciences 23 (15). Available at https://doi.org/10.3390/ijms23158519.

Lal, R., 2009. Soil degradation as a reason for inadequate human nutrition. Food Security 1 (1), 45–57. Available at https://doi.org/10.1007/s12571-009-0009-z.

Lu, K., Yan, L., Riaz, M., Babar, S., Hou, J., Zhang, Y., et al., 2023. Exogenous boron alleviates salt stress in cotton by maintaining cell wall structure and ion homeostasis. Plant Physiology and Biochemistry 201, 107858. Available at https://doi.org/10.1016/j.plaphy.2023.107858.

Lucas, M., Diaz-Espejo, A., Romero-Jimenez, D., Peinado-Torrubia, P., Delgado-Vaquero, A., Álvarez, R., et al., 2024. Chloride reduces plant nitrate requirement and alleviates low nitrogen stress symptoms. Plant Physiology and Biochemistry 212. Available at https://doi.org/10.1016/j.plaphy.2024.108717.

Mannan, M.A., Tithi, M.A., Islam, M.R., Al Mamun, M.A., Mia, S., Rahman, M.Z., et al., 2022. Soil and foliar applications of zinc sulfate and iron sulfate alleviate the destructive impacts of drought stress in wheat. Cereal Research Communications 50 (4), 1279–1289. Available at https://doi.org/10.1007/s42976-022-00262-5.

Mansour, M.M.F., 2023. Role of vacuolar membrane transport systems in plant salinity tolerance. Journal of Plant Growth Regulation 42, 1364–1401. Available at https://doi.org/10.1007/s00344-022-10655-9.

Marcial-Salvador, L., Medrano-Macías, J., Enríquez-Medrano, F.J., Benavides-Mendoza, A., 2024. Synthesis and characterization of submicron iodinated polymer particles with biostimulant effect on the germination and vigor of corn seedlings. Revista Chapingo, Serie Horticultura 30 (3), 47–60. Available at https://doi.org/10.5154/r.rchsh.2024.04.004.

MARTINS, J.P.R., SOUZA, A.F.C., RODRIGUES, L.C.A., BRAGA, P.C.S., GONTIJO, A.B.P.L., FALQUETO, A.R., 2020. Zinc and selenium as modulating factors of the anatomy and physiology of Billbergia zebrina (Bromeliaceae) during in vitro culture. Photosynthetica 58 (5), 1068–1077. Available at https://doi.org/10.32615/ps.2020.058.

Medrano-Macías, J., Narvaéz-Ortiz, W.A., 2022. Selenium and nano-selenium as a new frontier of plant biostimulant. In: Hossain, M.A., Ahammed, G.J., Kolbert, Z., El-Ramady, H., Islam, T., Schiavon, M. (Eds.), Selenium and Nano-Selenium in Environmental Stress Management and Crop Quality Improvement, Sustainable Plant Nutrition in a Changing World. Springer International Publishing, Cham, pp. 41–54. Available at https://doi.org/10.1007/978-3-031–07063-1_3.

Medrano-Macías, J., Leija-Martínez, P., González-Morales, S., Juárez-Maldonado, A., Benavides-Mendoza, A., 2016. Use of iodine to biofortify and promote growth and stress tolerance in crops. Frontiers in Plant Science 7. Available at https://doi.org/10.3389/fpls.2016.01146 (2016).

Medrano Macías, J., López Caltzontzit, M.G., Rivas Martínez, E.N., Narváez Ortiz, W.A., Benavides Mendoza, A., Martínez Lagunes, P., 2021. Enhancement to salt stress tolerance in strawberry plants by iodine products application. Agronomy 11 (3). Available at https://doi.org/10.3390/agronomy11030602.

Mikkelsen, R.L., 2011. The "4R" nutrient stewardship framework for horticulture. Horttechnology 21 (6), 658–662. Available at https://doi.org/10.21273/horttech.21.6.658.

Moradtalab, N., Ahmed, A., Geistlinger, J., Walker, F., Höglinger, B., Ludewig, U., et al., 2020. Synergisms of microbial consortia, N forms, and micronutrients alleviate oxidative damage and stimulate hormonal cold stress adaptations in maize. Frontiers in Plant Science 11. Available at https://doi.org/10.3389/fpls.2020.00396.

Morales-Díaz, A.B., Ortega-Ortíz, H., Juárez-Maldonado, A., Cadenas-Pliego, G., González-Morales, S., Benavides-Mendoza, A., 2017. Application of nanoelements in plant nutrition and its impact in ecosystems. Advances in Natural Sciences: Nanoscience and Nanotechnology 8, 013001. Available at https://doi.org/10.1088/2043–6254/8/1/013001.

Nadeem, F., Azhar, M., Anwar-ul-Haq, M., Sabir, M., Samreen, T., Tufail, A., et al., 2020. Comparative response of two rice (*Oryza sativa* L.) cultivars to applied zinc and manganese for mitigation of salt stress. Journal of Soil Science and Plant Nutrition 20 (4), 2059–2072. Available at https://doi.org/10.1007/s42729-020-00275-1.

Nascimento, V.L., Souza, B.C.O.Q., Lopes, G., Guilherme, L.R.G., 2022. On the role of iodine in plants: a commentary on benefits of this element. Frontiers in Plant Science 13. Available at https://doi.org/10.3389/fpls.2022.836835.

Nasim, A., Hao, J., Tawab, F., Jin, C., Zhu, J., Luo, S., et al., 2025. Micronutrient biofortification in wheat: QTLs, candidate genes and molecular mechanism. International Journal of Molecular Sciences 26 (5). Available at https://doi.org/10.3390/ijms26052178.

Nawaz, F., Zulfiqar, B., Ahmad, K.S., Majeed, S., Shehzad, M.A., Javeed, H.M.R., et al., 2021. Pretreatment with selenium and zinc modulates physiological indices and antioxidant machinery to improve drought tolerance in maize (*Zea mays* L.). South African Journal of Botany 138, 209–216. Available at https://doi.org/10.1016/j.sajb.2020.12.016.

Naz, M., Benavides-Mendoza, A., 2021. Nanofertilizers as tools for plant nutrition and plant biostimulation under adverse environment. Plant Performance Under Environmental Stress: Hormones, Biostimulants and Sustainable Plant Growth Management. Springer International Publishing, China, pp. 387–415.

Nestel, P., Bouis, H.E., Meenakshi, J.V., Pfeiffer, W., 2006. Biofortification of staple food crops. Journal of Nutrition 136 (4), 1064–1067. Available at https://doi.org/10.1093/jn/136.4.1064.

Novikova, A.A., Podlasova, E.Y., Lebedev, S.V., Latushkin, V.V., Glushchenko, N.N., Sudarikov, K.A., et al., 2025. Can boron and cobalt nanoparticles be beneficial effectors to prevent flooding-induced damage in durum and bread wheat at germination and tillering stage? Plants 14 (7). Available at https://doi.org/10.3390/plants14071044.

Ofori, K.F., Antoniello, S., English, M.M., Aryee, A.N.A., 2022. Improving nutrition through biofortification–a systematic review. Frontiers in Nutrition 9. Available at https://doi.org/10.3389/fnut.2022.1043655.

Ojeda-Barrios, D.L., Morales, I., Juárez-Maldonado, A., Sandoval-Rangel, A., Fuentes-Lara, L.O., Benavides-Mendoza, A., 2019. Importance of nanofertilizers in fruit nutrition. Fruit Crops: Diagnosis and Management of Nutrient Constraints. Elsevier, Mexico, pp. 497–508. Available at http://doi.org/10.1016/B978-0-12–818732-6.00035-6

Omar, A.A., Heikal, Y.M., Zayed, E.M., Shamseldin, S.A.M., Salama, Y.E., Amer, K.E., et al., 2023. Conferring of drought and heat stress tolerance in wheat (*Triticum aestivum* L.) genotypes and their response to selenium nanoparticles application. Nanomaterials 13 (6). Available at https://doi.org/10.3390/nano13060998

Pandey, S.N., 2018. Biomolecular functions of micronutrients toward abiotic stress tolerance in plants. In: Hasanuzzaman, M., Fujita, M., Oku, H., Nahar, K., Hawrylak-Nowak, B. (Eds.), Plant Nutrients and Abiotic Stress Tolerance. Springer, Singapore, pp. 153–170. Available at https://doi.org/10.1007/978-981-10-9044-8_6.

Pérez-Labrada, F., Benavides-Mendoza, A., Juárez-Maldonado, A., Solís-Gaona, S., González-Morales, S., 2020. Organic acids combined with Fe-chelate improves ferric nutrition in tomato grown in calcisol soil. Journal of Soil Science and Plant Nutrition 20 (2), 673–683. Available at https://doi.org/10.1007/s42729-019-00155-3.

Rahman, A., Hossain, M.S., Mahmud, J.A., Nahar, K., Hasanuzzaman, M., Fujita, M., 2016. Manganese-induced salt stress tolerance in rice seedlings: regulation of ion homeostasis, antioxidant defense and glyoxalase systems. Physiology and Molecular Biology of Plants 22 (3), 291–306. Available at https://doi.org/10.1007/s12298-016-0371-1.

Rahman, M., Rahman, K., Sathi, K.S., Alam, M.M., Nahar, K., Fujita, M., et al., 2021. Supplemental selenium and boron mitigate salt-induced oxidative damages in glycine max l. Plants 10 (10). Available at https://doi.org/10.3390/plants10102224.

Rai, G.K., Kumar, P., Choudhary, S.M., Singh, H., Adab, K., Kosser, R., et al., 2023. Antioxidant potential of glutathione and crosstalk with phytohormones in enhancing abiotic stress tolerance in crop plants. Plants 12 (5). Available at https://doi.org/10.3390/plants12051133.

Rakoczy-Lelek, R., Smoleń, S., Grzanka, M., Ambroziak, K., Pitala, J., Skoczylas, Ł., et al., 2021. Effectiveness of foliar biofortification of carrot with iodine and selenium in a field condition. Frontiers in Plant Science 12. Available at https://doi.org/10.3389/fpls.2021.656283.

Ramirez-Builes, V.H., Küsters, J., Thiele, E., Leal-Varon, L.A., 2024. Boron nutrition in coffee improves drought stress resistance and, together with calcium, improves long-term productivity and seed composition. Agronomy 14 (3). Available at https://doi.org/10.3390/agronomy14030474.

Raven, J.A., 2017. Chloride: essential micronutrient and multifunctional beneficial ion. Journal of Experimental Botany 68 (3), 359–367. Available at https://doi.org/10.1093/jxb/erw421.

Reis, A.Rd, Gonzalez-Porras, C.V., Ferreira, P.M., Silva, P.Gd, Freire, F.B.S., Morais, E.G., 2025. Unlocking resilience: physiological roles of benefit elements silicon, selenium and cobalt to enhance tolerance of crop plants against abiotic stresses. Plant Physiology and Biochemistry 229. Available at https://doi.org/10.1016/j.plaphy.2025.110515.

Reynolds-Marzal, D., Rivera-Martin, A., Santamaria, O., Poblaciones, M.J., 2021. Combined selenium and zinc biofortification of bread-making wheat under mediterranean conditions. Plants 10 (6), 1209. Available at https://doi.org/10.3390/plants10061209.

Riyazuddin, R., Singh, K., Iqbal, N., Nisha, N., Rani, A., Kumar, M., et al., 2023. Iodine: an emerging biostimulant of growth and stress responses in plants. Plant and Soil 486 (1–2), 119–133. Available at https://doi.org/10.1007/s11104-022-05750-5.

Roychoudhury, A., Mukherjee, S., Ghosh, A., 2025. Molecular and physiological studies of the role of micronutrients in sustainable growth and development of crop plants. Agricultural Crop Improvement: Plant and Soil Relationships. CRC Press, India, pp. 135–147.

Russell, G., Zulfiqar, F., Hancock, J.T., 2020. Hydrogenases and the role of molecular hydrogen in plants. Plants 9 (9), 1136. Available at https://doi.org/10.3390/plants9091136.

Schiavon, M., Pilon-Smits, E.A.H., 2017. The fascinating facets of plant selenium accumulation – biochemistry, physiology, evolution and ecology. New Phytologist 213 (4), 1582–1596. Available at https://doi.org/10.1111/nph.14378.

Shah, A., Wu, X., Ullah, A., Fahad, S., Muhammad, R., Yan, L., et al., 2017. Deficiency and toxicity of boron: alterations in growth, oxidative damage and uptake by citrange orange plants. Ecotoxicology and Environmental Safety 145, 575–582. Available at https://doi.org/10.1016/j.ecoenv.2017.08.003.

Shao, J., Tang, W., Huang, K., Ding, C., Wang, H., Zhang, W., et al., 2023. How does zinc improve salinity tolerance? Mechanisms and future prospects. Plants 12 (18). Available at https://doi.org/10.3390/plants12183207.

Shoormij, F., Mirlohi, A., Saeidi, G., Shirvani, M., 2022. Combined foliar application of Zn and Fe increases grain micronutrient concentrations and alleviates water stress across diverse wheat species and ploidal levels. Scientific Reports 12 (1). Available at https://doi.org/10.1038/s41598-022-24868-1.

Shukla, D., 2025. Role of ethylene in plant tissue culture. In: Shukla, D. (Ed.), Ethylene Sensing and Signaling. Springer US, New York, NY, pp. 245–260. Available at https://doi.org/10.1007/978-1-0716-4650-2_18.

Singh, K., Singh, R., Nwosu, N.J., Omara, P., Sharma, L., Dunn, B.L., et al., 2025. Optimizing nutrient delivery in agronomic crops: a review of enhanced efficiency fertilizers. Journal of Plant Nutrition and Soil Science 188 (4), 570–584. Available at https://doi.org/10.1002/jpln.12010.

Srivastav, A.L., Dhyani, R., Ranjan, M., Madhav, S., Sillanpää, M., 2021. Climate-resilient strategies for sustainable management of water resources and agriculture. Environmental Science and Pollution Research 28 (31), 41576–41595. Available at https://doi.org/10.1007/s11356-021-14332-4.

Su, J., Yang, X., He, J., Zhang, Y., Duan, X., Wang, R., et al., 2019. Methyl-coenzyme M reductase-dependent endogenous methane enhances plant tolerance against abiotic stress and alters ABA sensitivity in Arabidopsis thaliana. Plant Molecular Biology 101 (4–5), 439–454. Available at https://doi.org/10.1007/s11103-019-00914-x.

Szerement, J., Szatanik-Kloc, A., Mokrzycki, J., Mierzwa-Hersztek, M., 2022. Agronomic biofortification with Se, Zn, and Fe: an effective strategy to enhance crop nutritional quality and stress defense—a review. Journal of Soil Science and Plant Nutrition 22 (1), 1129–1159. Available at https://doi.org/10.1007/s42729-021-00719-2.

Tenhaken, R., 2015. Cell wall remodeling under abiotic stress. Frontiers in Plant Science 5. Available at https://doi.org/10.3389/fpls.2014.00771.

Treviño-Ruiz, K.S., Ortega-Ortiz, H., Benavides-Mendoza, A., González-Morales, S., 2021. Aplicación de nanoquitosán-yodo en lechuga y su efecto en la biofortificación, crecimiento y rendimiento. Ecosistemas y Recursos Agropecuarios 11 (2). Available at https://doi.org/10.19136/era.a11n2.3615.

Tripathi, D.K., Singh, S., Gaur, S., Singh, S., Yadav, V., Liu, S., et al., 2018. Acquisition and homeostasis of Iron in higher plants and their probable role in abiotic stress tolerance. Frontiers in Environmental Science 5. Available at https://doi.org/10.3389/fenvs.2017.00086.

Trippe, R.C., Pilon-Smits, E.A.H., 2021. Selenium transport and metabolism in plants: phytoremediation and biofortification implications. Journal of Hazardous Materials 404. Available at https://doi.org/10.1016/j.jhazmat.2020.124178.

Venugopalan, V.K., Nath, R., Sengupta, K., Pal, A.K., Banerjee, S., Banerjee, P., et al., 2022. Foliar spray of micronutrients alleviates heat and moisture stress in lentil (*Lens culinaris* Medik) grown under rainfed field conditions. Frontiers in Plant Science 13. Available at https://doi.org/10.3389/fpls.2022.847743.

Waraich, E.A., Ahmad, R., Saifullah, Ashraf, M.Y., Ehsanullah, 2011. Role of mineral nutrition in alleviation of drought stress in plants. Australian Journal of Crop Science 5 (6), 764–777.

Waraich, E.A., Ahmad, R., Halim, A., Aziz, T., 2012. Alleviation of temperature stress by nutrient management in crop plants: a review. Journal of Soil Science and Plant Nutrition 12 (2), 221–244. Available at https://doi.org/10.4067/s0718-95162012000200003.

Wu, S., Hu, C., Tan, Q., Nie, Z., Sun, X., 2014. Effects of molybdenum on water utilization, antioxidative defense system and osmotic-adjustment ability in winter wheat (*Triticum aestivum*) under drought stress. Plant Physiology and Biochemistry 83, 365–374. Available at https://doi.org/10.1016/j.plaphy.2014.08.022.

Xu, L.H., Wang, W.Y., Guo, J.J., Qin, J., Shi, D.Q., Li, Y.L., et al., 2014. Zinc improves salt tolerance by increasing reactive oxygen species scavenging and reducing Na+ accumulation in wheat seedlings. Biologia Plantarum 58 (4), 751–757. Available at https://doi.org/10.1007/s10535-014-0442-5.

Ye, Y., Medina-Velo, I.A., Cota-Ruiz, K., Moreno-Olivas, F., Gardea-Torresdey, J.L., 2019. Can abiotic stresses in plants be alleviated by manganese nanoparticles or compounds? Ecotoxicology and Environmental Safety 184. Available at https://doi.org/10.1016/j.ecoenv.2019.109671.

SECTION

III

Mechanisms, interactions, and physiological pathways

CHAPTER

Soil fertility management for stress resilience

12

Hina Alim[1], Sandeep K. Behera[1], Gayatri A. Talekar[1], Ahmad Ali[1] and Nimisha Patel[2]

[1]*Department of Life Science, University of Mumbai, Vidyanagari, Santacruz (East), Mumbai, Maharashtra, India,*

[2]*Department of Life Science, J C Bose University of Science and Technology, YMCA, Faridabad, Haryana, India*

Abbreviations

ATP	adenosine triphosphate
DSTs	decision-support tools
EIA	Environmental Impact Assessment
ISFM	Integrated Soil Fertility Management
IPCC	Intergovernmental Panel on Climate Change
NUE	nutrient-use efficiency
PGPR	plant growth promoting rhizobacteria
ROS	reactive oxygen species
SOC	soil organic carbon
SOM	soil organic matter
SFM	soil fertility management

12.1 Introduction

Soil fertility is the natural ability of the soil to provide essential nutrients in sufficient amounts and proper ratios for the growth of plants over time. It goes beyond the availability of nutrients and takes into account the physical, chemical, and biological properties of the soil that together provide for healthy plant growth. Healthy soils have good structure, enough organic matter, nutrient stores in equilibrium, lively microbial communities, and suitable pH, conditions which are all very important for the continuous production of crops.

Soil fertility is the main factor that determines yield stability, resource-use efficiency, and environmental quality for sustainable agriculture. The goal of sustainable soil fertility management is to have a balance between nutrient inputs and outputs while at the same time keeping soil organic carbon, soil biodiversity,

Mineral Nutrients to Mitigate Abiotic Stress in Crops. DOI: https://doi.org/10.1016/B978-0-443-44712-9.00013-X

and soil ecosystem functions intact, thus making it possible to have soil that is productive both now and in the future (Brady and Weil, 2017;).

Stress resilience in agro-ecosystems is the capacity of crops and the related soil systems to endure, adjust to, and get back from a situation of trouble without a significant drop in their production. The aforementioned stresses can be abiotic ones like drought, salinity, temperature extremes, flooding, or nutrient imbalances, or biotic ones, such as pests, pathogens, and weeds. A resilient agro-ecosystem is one that can still function stably under pressure through various adaptive strategies such as selective nutrient uptake, water-use efficiency, a robust root system, or effective plant–microbe interactions. Soil thus becomes a major factor in such resilience when it works as a stabilizer against environmental variations and at the same time, provides a medium for the growth of plants even when the weather is not favorable (FAO, 2017).

Soil fertility management plays a major role in the plant tolerance capacity to biotic and abiotic stresses. When nutrients are available in an adequate and balanced way, physiological processes such as photosynthesis, enzyme activity, osmotic regulation, and antioxidant defense systems are enhanced. To give an example, potassium is one of the elements that improve drought and salinity tolerance by regulating stomatal function and osmotic balance, whereas phosphorus is a promoter of root growth, thus the plants will get water from deeper soil during drought (Marschner, 2012). In the same way, organic matter changes soil structure, makes the soil more water-retentive, and accommodates a large number of microbial populations that protect from the pathogens. Good soil fertility practices lead to the formation of a favorable rhizosphere environment, which results in the activation of the plant defense mechanisms and, consequently, the plant gets stress resistance at a higher level (Smith and Smith, 2011).

12.2 Types of stresses affecting crop productivity

Crop productivity is highly dependent on the occurrence of various environmental and biological stresses. Such factors interfere with the growth, metabolism, and reproduction of the plants that are necessary to yield and quality. Basically, these can be grouped as abiotic and biotic stresses, and the interaction of these modifies and deepens their effect in the context of global climate change (Table 12.1).

12.2.1 Abiotic stresses

Abiotic stresses are the consequences of harsh environmental factors that are physical or chemical in nature. These have been identified as the primary factor behind the loss of yield worldwide, especially in rainfed and marginal agricultural areas. Drought, salinity, pollution, high or low temperatures, and nutrient deficiency in plants are examples of abiotic stress and can adversely influence plant growth.

Table 12.1 Types of stresses affecting crop productivity.

Category	Type of stress	Major causes	Effects on soil	Effects on crops
Abiotic	Drought	Low rainfall, high evapotranspiration	Reduced soil moisture, limited nutrient mobility	Wilting, reduced photosynthesis, poor yield
	Salinity	Irrigation with saline water, poor drainage	High salt accumulation, osmotic stress	Ion toxicity, reduced nutrient uptake
	Heat stress	Rising temperatures, heat waves	Accelerated organic matter decomposition	Enzyme denaturation, reduced grain filling
	Cold stress	Low temperature, frost	Reduced microbial activity	Slow growth, membrane damage
	Flooding/ Waterlogging	Heavy rainfall, poor drainage	Anaerobic conditions, nutrient losses	Root suffocation, reduced nutrient absorption
	Nutrient deficiency	Imbalanced fertilization, poor soils	Depleted nutrient reserves	Stunted growth, chlorosis
	Nutrient toxicity	Excess fertilization, acidic soils	Elemental imbalance	Leaf burn, reduced productivity
Biotic	Soil-borne pathogens	Fungi, bacteria, viruses	Disruption of soil biological balance	Root rot, wilting, disease incidence
	Insect pests	Climate-driven pest outbreaks	Indirect soil degradation	Tissue damage, yield loss
	Nematodes	Monocropping, poor soil health	Root damage	Reduced nutrient and water uptake
	Weeds	Poor management practices	Competition for nutrients and water	Reduced crop growth and yield

Abiotic stress leads to a reduction in plant biomass and yield as a result of disorders in physiological processes such as photosynthesis, water, and nutrient uptake (Husen, 2021a; Husen, 2021b; Husen, 2021c; Husen, 2023; Husen, 2024; Husen, 2025a; Husen, 2025b).

12.2.1.1 Drought stress

Drought stress is a condition when water in the soil is not enough to satisfy the water needs of the crops. The impact of the drought extends to the biological, biochemical, and molecular aspects of plants. When the soil is dry, the plant closes its stomata, photosynthesis is reduced, nutrient uptake is impaired, and cell expansion is weakened. If the drought persists, reactive oxygen species (ROS) would build up and harm enzymes and cellular membranes, which could result in oxidative stress. Plant developmental stages that are particularly vulnerable to water stress, such as flowering and grain filling, are particularly dangerous during drought. As a result, there will be significant output losses. Drought effects can be mitigated by making the soil more fertile, particularly by increasing the amount of organic matter and potassium available. This will improve both the soil's ability to hold water and the efficiency with which plants use it (Farooq et al., 2020; IPCC, 2022).

12.2.1.2 Salinity stress

Salinity stress is the cause of the condition to be the presence of soluble salts mainly sodium chloride, in the soil or irrigation water. High salinity significantly lowers the osmotic potential of soil water, making it hard for plants to absorb moisture even in the presence of water. Additionally, ion toxicity and nutrient imbalances, primarily a decrease in potassium and calcium uptake, may result from the overabundance of sodium and chloride ions. At the same time, salinity negatively impacts seed germination, root development, photosynthesis, and enzyme activity (Husen, 2021a; Husen, 2021b; Husen, 2021c). Plants grown in saline soils are likely to show the symptoms of stunted growth, leaf chlorosis, and low yield. The proper management of soil fertility, such as among others delivering enough calcium, potassium, and organic amendments, is the factor that leads to improving plant tolerance to saline conditions (Shrivastava and Kumar, 2015; FAO, 2021).

12.2.1.3 Heat and cold stress

One of the significant limiting factors of crop productivity is temperature extremes (Shaffique et al., 2025). When the temperature rises above the optimal level for plant development, heat stress takes place. Among others, heat stress is accompanied by protein denaturation, membrane instability, pollen sterility, and reduced grain set; as a consequence, lowered yields will be obtained. Moreover, heat stress quickens crop phenology, thus making the growth period shorter and limitation of the biomass accumulation possible.

The adverse effects of cold stress include, among others, changes in membrane fluidity, enzyme activity, and photosynthetic efficiency. This happens due to the fact that a cold-stressed plant is not only able to delay germination, but it will also have reduced nutrient uptake, and its tissues will undergo the damage process as well. Good soil nutrition, especially phosphorus and microelements, is the support system for energy metabolism and membrane stability, and thus the plants are able to cope with temperature extremes (Hatfield and Prueger, 2015; Wahid et al., 2007).

12.2.1.4 *Flooding and waterlogging*

The environmental factors that lead to flooding and waterlogging are, among others, heavy rains, snowmelt, and insufficient drainage or poor irrigation practices. These EMPs come together to cause the over-saturation of the soil with water to the point that the oxygen supply to the root zone is cut off. Due to the lack of oxygen, the roots' breathing is limited and so is the nutrient uptake; at the same time, however, the generation of toxic compounds such as ethanol and iron (II) may occur.

Waterlogging leads to growth inhibition of roots, reduction of nodulation in legumes, and adversely affects microbial nutrient transformations. Denitrification is the major way nitrogen is lost in waterlogged soils. By improving soil structure, drainage, and organic matter content, one not only enhances aeration but also reduces the intensity of waterlogging stress (Parent et al., 2016).

12.2.1.5 *Nutrient deficiencies and toxicities*

Nutrient stress refers to a state in which the essential nutrients are either lacking or present in toxic amounts. Deficiencies of nitrogen, phosphorus, and potassium, or even of microelements such as zinc and iron, lead to the limitation of photosynthesis, protein synthesis, and enzyme activities. On the other hand, the toxicities of such elements as aluminum, manganese, or excess boron may result in the inhibition of root growth and metabolic processes. Soil pH is an important determinant of nutrient availability and toxicity. Good soil fertility management, which comprises balanced fertilization and pH correction, is the only way to keep nutrient-related stresses at bay and guarantee the best performance of crops (Marschner, 2012; Alloway, 2013).

12.2.2 Biotic stresses

Biotic stresses are those that are brought about by living creatures, that feed on the nutrients of crops or that damage crop tissues directly. Such troubles always become more severe when plants are stressed by an abiotic factor.

12.2.2.1 *Soil-borne pathogens*

Diseases like root rot, wilt, and damping-off are caused by soil-borne pathogens that include fungi, bacteria, and oomycetes. They go after roots and vascular tissues, so the whole water and nutrient supply becomes disrupted. The soil fertility, low organic matter, and nutrient imbalances are some of the factors that often increase the severeness of diseases. The soil that is good and rich in diverse microbes can suppress soil pathogens through competition, antagonism, and induced systemic resistance in plants. Besides, organic amendments and balanced nutrition not only activate the plant defense mechanisms but also lower the occurrence of diseases (Berendsen, Pieterse and Bakker, 2012).

12.2.2.2 *Insect pests*

Insect pests distort plants by eating their leaves, branches, roots, or the parts of plants used for reproduction, thus causing direct yield loss and indirect damage through disease transmission. Nutrient imbalances, and in particular excessive nitrogen fertilization, can increase pest vulnerability by making plants grow lusher and tender, thus more attractive for pests. By changing soil fertility management for the better, the plant's vigor and tolerance are enhanced so that it can brave a pest attack. Moreover, soils in good condition are habitats for the natural enemies of insect pests, who thus biologically control the latter and make a contribution toward their reduction (Altieri et al., 2015).

12.2.2.3 *Nematodes and weeds*

The plant-parasitic nematodes are those that attack roots, thereby causing the appearance of lesions, the formation of galls, and, at the same time, a reduction of nutrient uptake. Weeds compete with crops for water, nutrients, light, and space, and as these causes of growth have been taken away, an increase in yield reduction, that is, decrease in crop production would occur, especially in their early growth stages (time of seedling/germination). The integration of soil fertility and crop management practices via the use of crop rotation, application of organic resources, and balanced fertilization will result in both the suppression of nematode populations as well as weed competition by virtue of crop establishment and growth (Nicol et al., 2011; Liebman et al., 2018).

(Table 12.1)

12.3 Role of soil fertility in stress resilience

Soil fertility is a major factor that determines the way crops will feel, react, and get back to normal from environmental stresses. Properly fertile soils not only offer the necessary nutrients but also a physically and biologically supportive environment that the plant can draw strength from in both abiotic and biotic stress situations.

12.3.1 Influence of soil nutrients on plant physiological and biochemical responses

Soil nutrients are the main regulators of the key physiological and biochemical processes that decide plant performance under stress. In particular, sufficient nitrogen is involved in the chlorophyll synthesis, protein production, and photosynthetic capacity, thus giving the plants the ability to keep growing even during a period of stress of moderate intensity. Phosphorus is an essential element for energy transfer reactions (ATP formation), root development, and membrane integrity, and these factors are extremely significant when the plant faces drought and temperature stress. Potassium is considered to be the most reliable stress-protective nutrient; it controls stomatal movement, osmotic balance, and enzyme activation; therefore it facilitates water-use efficiency and invulnerability to drought, salinity, and heat stress (Marschner, 2012; Zörb, Senbayram and Peiter, 2014).

Moreover, at the biochemical level, a well-balanced nutrient supply elevates the plant antioxidant defense system through the promotion of the synthesis of enzymes like superoxide dismutase, catalase, and peroxidases. These enzymes remove reactive oxygen species produced by the stress conditions, lowering the damage to cells. Also, micronutrients, namely zinc, iron, and manganese, become cofactors in enzymatic reactions and stabilize protein structures; thus they contribute to the reinforcement of stress tolerance mechanisms (Raza et al., 2022).

12.3.2 Soil organic matter and its role in moisture retention and nutrient buffering

Soil organic matter (SOM) is the major contributor to soil fertility that greatly determines the ability to resist stress. Elevated SOM content results in better soil structure and soil aggregates, which leads to more porosity and water-holding capacity. This, in turn, enables soils to hold water for longer during drought and to get rid of the excess water more effectively during flooding or waterlogging. Therefore plants are less exposed to drastic changes in soil moisture availability (Lal, 2020). Besides moisture regulation, SOM serves as a nutrient buffer that, through mineralization processes, temporarily stores and then releases nutrients that are indispensable for plants. This ability to buffer keeps a low level of nutrients lost by leaching or volatilization, especially under situations of heavy rainfall or heat. Apart from that, organic matter improves cation exchange capacity, which ensures that there is better holding of nutrients such as potassium, calcium, and magnesium that, combined, are the mainstay of plant metabolic stability under stress (FAO, 2021).

12.3.3 Nutrient balance and stress tolerance mechanisms

One of the conditions necessary for a plant to be able to resist stress is the balanced supply of nutrients rather than just their availability. An unbalanced diet of nutrients may lead a plant to be more sensitive to stress through the disruption of metabolic pathways and lowering of the defense responses. To illustrate, a huge amount of nitrogen without a sufficient supply of potassium could cause a plant to grow rapidly in the vegetative phase, but, at the same time, its drought tolerance will be lowered, and it will be more susceptible to pests and diseases. In the same manner, deficiencies as well as excessive amounts of certain micronutrients may cause the impeding of enzyme function and cellular signaling under stress conditions. Properly balanced nutrient management aids in the process of osmotic adjustment, hormonal regulation, and the use of carbon for energy, all of which result in higher stress resistance. For instance, calcium helps in cell walls and membranes and, thus the plant gets more resistance to heat and salinity stress while sulfur is a contributor to the synthesis of stress-related proteins and antioxidants. So, the maintenance of nutrient balance is a prerequisite for the activation of several physiological pathways that, in general, lead to stress resilience (Hawkesford et al., 2023).

12.3.4 Interactions among soil health, root growth, and plant stress adaptation

Soil fertility has a strong impact on root growth as well as the development of root system, which are essential for stress adaptation. Fertile soils characterized with good structure, sufficient organic matter, and lively microbial populations encourage roots to grow deeper and more extensively. Deep-rooted plants gain more water and nutrients not only from the upper but also from deeper soil layers, thus they become more drought and nutrient-stress resistant. Biological health of the soil is equally important for the enhancement of root-mediated stress adaptation. For instance, mycorrhizal fungi and plant growth–promoting rhizobacteria are some of the beneficial microorganisms that can make nutrient uptake more efficient, especially for phosphorus and micronutrients, and at the same time, they can improve the hormonal balance of the plant under stress. These processes increase the connection between the root and the soil as well as the capability of the plant to receive and implement the environmental stress signals (Trivedi et al., 2020; Berendsen, Pieterse and Bakker, 2012). Therefore healthy, fertile soils are able to form an active root–soil system, which is not only a source of stress resilience still but also a foundation for sustainable crop productivity in the long run.

12.4 Integrated soil fertility management

Integrated soil fertility management (ISFM) is the integration of various soil nutrient management methods in a comprehensive and environmentally friendly manner that aims at not only increasing crop yields but also at the crops being more resistant to stress. The methods introduced are focused on the efficient use of the existing resources in harmony with nature.

12.4.1 Concept and principles of integrated soil fertility management

ISFM is using site-specific soil fertility management practices that combine the use of improved crop varieties, organic input, the application of fertilizers, and good agronomic practices in a given area or farm. In essence, ISFM aims at nutrient management at maximum efficiency while also increasing the common goal of crop productivity and soil quality. ISFM, as a less radical method of fertilization, also brings up the point that soil fertility is a mutable characteristic that is influenced by not only the physical structure but also the chemical balance and biological activity as well. Some of the major points in ISFM are keeping or even upgrading the soil organic matter, giving the nutrient in balanced portions, matching the nutrient supply to the demand of the crop, and tailoring management to the local soil and weather conditions as shown in (Fig. 12.1) Furthermore, farmer knowledge and adaptability to decision-making are prolonged by ISFM, thus implying that soil fertility problems are very different from one agro-ecological region to another (FAO, 2022; Vanlauwe et al., 2015).

12.4.2 Integration of organic, inorganic, and biological nutrient sources

One of the most prominent features of ISFM is the joint utilization of organic, inorganic, and biological nutrient sources in order to take advantage of the complementary aspects of these sources. Organic inputs such as farmyard manure, compost, green manures, crop residues, and biochar not only improve soil structure but also increase water retention and stimulate microbial activity. Besides, these materials also deliver nutrients in slow and continuous forms, thus reducing the losses and making the soil more adaptable to stress conditions.

On the other hand, inorganic fertilizers provide crops with nutrients that are easily and quickly available; thus the fertilizer should be used on the most nutrient-starved soils first, where nutrient demand is high. In the case of chemical fertilizers applied to the soil along with organic material, it is a safe bet that they will perform better in terms of efficiency and bring less harm to the environment. To complete the trilogy, biofertilizers, mycorrhizal fungi, and plant growth–promoting rhizobacteria are some of the biological sources of nutrients that can, among other things, fix atmospheric nitrogen symbiotically, solubilize phosphorus, and stimulate root growth. By using these nutrient sources, the synergies created have great potential not only to improve soil fertility status, but also to increase food crop tolerance to stresses (Trivedi et al., 2020; Chivenge, Vanlauwe and Six, 2011). (Fig. 12.1).

12.4.3 Role of integrated soil fertility management in improving stress resilience

ISFM is the solution to the global problem of the feed-demanded population increasing faster than the supply of food. ISFM develops farming systems that can continuously produce food over a long period of time, which means soil fertility is not sacrificed to achieve higher yields in the short term. ISFM employs organic matter in soil health management, which is the key to combating desertification caused by drought and soil erosion. In this way, the future of food production in drought-prone areas has been secured thanks to ISFM. ISFM involves and

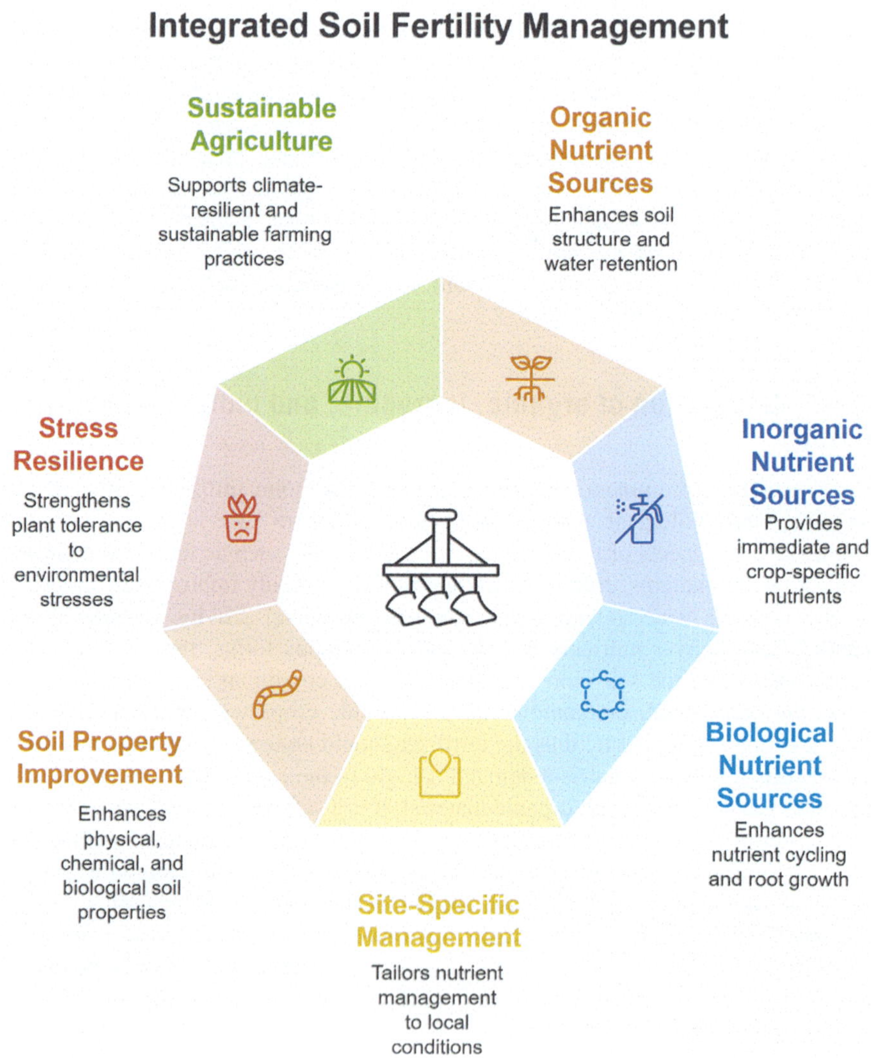

FIGURE 12.1

Applications of integrated soil fertility management (ISFM).

incorporates the entire soil food web, which is a biological bank of all-natural matter that feeds on one another and provides each other with nutrients. Healthy soils dominated by natural symbiotic relationships within soil food webs are often less vulnerable to pests and diseases. ISFM thereby contributes indirectly to the protection of the environment by minimizing the use of pesticides and fertilizers, hence cutting emissions of harmful gases.

ISFM is very specific and uniquely designed to local conditions therein managing to combine technologies, practices, knowledge, and social skills of farmers. ISFM remains a very notable and significant base of knowledge for a gradual and cautious increase of agriculture productions in agro-ecological regions that experience food insecurity and malnutrition simultaneously (Vanlauwe et al., 2010; Lal, 2020).

12.5 Soil biological management

Soil biological management is about improving nutrient availability, plant health, and resistance to environmental stress by increasing the activity and diversity of soil organisms. Beneficial soil microorganisms are the biological base of soil fertility and are the main agents that keep crop productivity at a good level under both normal and stressed situations.

12.5.1 Role of beneficial microorganisms

Microorganisms beneficial to plants, like plant growth–promoting rhizobacteria (PGPR), mycorrhizal fungi, and nitrogen-fixing bacteria, are the main contributors to plant nutrition and stress tolerance. PGPR energize the growth of plants by the production of phytohormones, phosphorus solubilization, nitrogen atmospheric fixation, and suppression of soil-borne pathogens. In addition, these microorganisms facilitate the induction of systemic tolerance in plants, thus when these plants are exposed to drought, salinity, or temperature stress, they activate their defense mechanisms more rapidly (Backer et al., 2018; Kapoor et al., 2024).

Mycorrhizal fungi associate symbiotically with plant roots and thus can access a larger area of soil through their hyphal networks. The association not only allows for the uptake of phosphorus, zinc, and copper, which are relatively immobile nutrients but also helps in the absorption of water under drought conditions. Apart from that, nitrogen-fixing bacteria such as symbiotic forms (e.g., Rhizobium) and free-living diazotrophs are capable of converting nitrogen in the atmosphere into nitrogen available to plants, thus the use of chemical fertilizers is lessened, and plant vigor improves under nutrient-limited environments as well as stress conditions (Smith and Read, 2008; Trivedi et al., 2020).

12.5.2 Microbial-mediated nutrient cycling under stress conditions

Microorganisms in the soil are the main characters in nutrient cycling through processes such as mineralization, immobilization, nitrification, and denitrification. When stress like drought or salinity is present, the microbial community will struggle to maintain nutrient levels by stabilizing nutrient pools and increasing nutrient-use efficiency. Nutrient release is facilitated by microbial exudates and extracellular enzymes during the breakdown of organic matter, and the release is in sync with plant demand. The microbial populations that have adapted to the stress can endure the harsh conditions and continue nutrient transformations, thus they are able to buffer crops against nutrient shortages. On top of that, the microbial interactions with plant roots determine root structure and exudation patterns, which in turn nutrient acquisition and stress adaptation are improved. These processes are very important, especially in soils that have high organic inputs and diverse microbial communities (Schimel et al., 2022).

12.5.3 Use of biofertilizers and biostimulants

Biofertilizers and biostimulants are seen more and more as tools for soil biological management in the field of sustainable agriculture. Biofertilizers are sources of living microorganisms that contribute to nutrient availability by means of one or more of the following mechanisms: biological nitrogen fixation, phosphorus solubilization, or potassium mobilization. Their correct application results in the improvement of nutrient-use efficiency and decrease in the environmental impact of chemical fertilizers. Biostimulants that contain microbial-based products as well as natural compounds like humic substances and seaweed extracts, promote plant physiological processes and help the plant resist various types of abiotic stress. These products make the root grow better, improve nutrient uptake, and promote antioxidant activity, thereby the plants become more resistant to drought, salinity, and temperature extremes. They get the maximum effectiveness if they are used together with good soil fertility and organic matter management (du Jardin, 2015; Rouphael and Colla, 2020; Gautam et al., 2024).

12.5.4 Soil biodiversity and ecosystem resilience

Soil biodiversity refers to the community of bacteria, fungi, protozoa, nematodes, arthropods, and earthworms that live in soil, and it is a major factor for soil health as well as ecosystem resilience. Diverse soil food webs are the mainstays for nutrient cycling, organic matter decomposition, and biological regulation of pests and pathogens. One of the benefits of high biodiversity is the presence of functional redundancy, thus essential soil functions can still be performed even if some soil organisms are affected by stress. Soils that have a biologically rich community are more resistant to different kinds of disturbances like drought, flooding, and intensive land use. These soils bounce back quickly after a stress event and support stable crop yields over time. Therefore the provision of soil biodiversity through less tillage, use of organic amendments, crop diversification, and biological inputs is the cornerstone of resilient agro-ecosystems (Bardgett and Van Der Putten, 2014; FAO, 2020, Lehmann et al., 2020)

12.6 Soil physical properties and stress resilience

The physical properties of soil are the structural elements of soil health and thus have a great influence on the extent to which crops can resist and recover from environmental stresses. Good soil physical conditions contribute to water availability, root growth, gas exchange, and nutrient uptake, thus raising the level of stress resilience in agro-ecosystems.

12.6.1 Soil structure, aggregation, and porosity

Soil structure is the way soil particles are combined into aggregates, and aggregation and porosity indicate the manner in which these particles interact and the pore spaces of different sizes that result. Soil particles bound together into stable aggregates become less susceptible to erosion and pressure and can keep their functional integrity even when exposed to different stress situations like heavy rainfall or long-lasting drought. Proper aggregation ensures that there is a balance between macro-pores (used for air movement and drainage) and micropores (used for water retention), which in turn play a pivotal role in plant growth under a variable environment (Bronick and Lal, 2005; Vogel et al., 2025).

Well-aggregated soils are hardly ever subjected to surface crusting, and thus runoff is also avoided, which in turn leads to the infiltration becoming better and the stress caused by water in excess or deficit alleviated. Soil organic matter and soil life are two indispensable factors that impel aggregate stability since the microbial by-products and root exudates are the bonding agents that strengthen soil structure. Therefore soils that have good structural integrity turn out to be more buffered environments for plant roots during stress events (Six et al., 2014).

12.6.2 Water-holding capacity and infiltration rate

The water-holding capacity of the soil is the volume of water that the soil can hold and at the same time make it available to plants. On the other hand, the infiltration rate is the one that determines the efficiency of water getting into the soil profile. Their importance cannot be overstated when it comes to drought and flood resilience. A soil with a great water-holding capacity will be able to keep the crops alive during the dry periods as it will let the stored water out slowly, thus the occurrence of drought will be less severe (Lal, 2020). Simultaneously, enough infiltration will stop surface runoff and water stagnation even when it is raining heavily, thus the occurrence of flooding and waterlogging will also be less. Soils that are compacted or in bad condition are usually characterized by a low infiltration rate, and thus they are likely to experience rapid runoff and erosion. Soil physical properties restoration through the use of organic matter and lessening the disturbance will not only improve infiltration but also moisture retention, thus the yield will be more stable even under irregular rainfall patterns (FAO, 2021).

12.6.3 Root penetration and aeration

Root penetration and soil aeration are two factors that are very much dependent on the state of the soil physically. Soils that are compacted and have high bulk density will not only stop roots from lengthening but will also limit the access to moisture and nutrients in the deeper soil, and at the same time, the crops will become more sensitive to drought and nutrient deficiency. On the other hand, those soils that are well aerated will make the roots grow longer and be able to spread even more, thus the water and nutrients that the plant will need during the time of stress will be more readily available (Hamza and Anderson, 2005).

At the same time, the importance of aeration cannot be underestimated. This is because the availability of oxygen in the root zone is what keeps the root respiration and other microbial activities going. Soils that are poorly aerated, which is mostly as a result of compaction or waterlogging, will lead to the development of anaerobic conditions that, in turn, will weaken root function and nutrient transformations. Therefore if one maintains physical conditions at an optimum level, it will be very difficult for roots to lose their functions, and they will still be able to respond even when environmental conditions are not favorable (Parent et al., 2016).

12.6.4 Management practices to improve soil physical health

Improving soil physical properties and intensifying stress resilience through management practices has been possible by adopting methods that are geared toward these goals. Adding organic amendments like farmyard manure, compost, and crop residues not only increases the organic matter in the soil but also improves aggregation, porosity, and moisture retention. Conservation agriculture, which is characterized by practices such as less ploughing and retention of residues, helps in reducing the disturbance of the soil, thereby keeping the structure intact for a long time. Rotating the crops, using cover crops, and introducing deep-rooted species are some of the ways that can be used to get rid of the problem of soil compaction as well as improve soil porosity, since through them, natural channels for water and roots are formed. Controlled traffic farming and the nonuse of heavy machinery at times when the ground is wet are some of the ways through which compaction can be avoided. Individually, these methods act as a thermostat for the physical condition of the soil; in the long run, their combined operation will give the soil the capability to become a buffer for the crops against drought, flooding, and temperature extremes (Hobbs, Sayre and Gupta, 2008; Lal, 2020).

12.7 Climate-smart soil fertility

Climate-smart soil fertility management involves the integration of nutrient stewardship with climate adaptation and mitigation objectives to continue to provide crop productivity under climate change-induced weather variability. The main idea is to increase nutrient-use efficiency, promote soil health, decrease greenhouse gas emissions, and ensure that farming systems are resilient to extreme weather events.

12.7.1 Adaptive nutrient management under climate variability

Climate variability, by changing the precipitation pattern, temperature, and the number of extreme events, affects nutrient dynamics in soils. Adaptive nutrient management adjusts nutrient inputs in terms of sources, amounts, timing, and location according to crop needs in variable conditions. For instance, it also helps reduce nitrogen losses by precipitation through divided application during heavy rain, whereas an appropriate supply of nutrients enables the plant to use the short period of development affected by heat or drought stress. Moreover, adaptation measures also focus on adjusting fertilizer recommendations depending on weather forecasts, soil moisture, and growth stage. Besides that, organic fertilizers combined with mineral ones can serve as a reservoir of nutrients, coming from improving soil moisture and decreasing nutrient leaching during periods of stress. These adaptive measures not only make the yields more stable but also lower the environmental losses when the climate is unpredictable (Lipper et al., 2014; Smith et al., 2020).

12.7.2 Carbon sequestration and soil fertility

One of the main points that connects soil fertility with climate change mitigation is soil organic carbon (SOC). By increasing SOC through the retention of residues, application of organics, cover cropping, and minimum tillage, the soil not only becomes structurally sound but also nutrient retention and water-holding ability are improved. These water and nutrient-related changes lead to the strengthening of the plants against drought, heat, and nutrient stress. From a climate point of view, carbon sequestration in soil lowers the carbon dioxide concentration in the atmosphere, and thus the soil acts as a solution provider to the climate change problem. However, carbon-rich soils are not only good for plants, but also for microbial life, which can be more efficient in nutrient cycling. Therefore efforts that lead to an increase in SOC have a mutual benefit of improving soil fertility as well as climate resilience, thus carbon management being a crucial part of climate-smart agriculture (Lal, 2020; FAO, 2021).

12.7.3 Use of decision-support tools and precision nutrient management

The developments in digital agriculture have given rise to a variety of decision-support tools (DSTs) and precision nutrient management techniques that can be used to optimize fertilizer use under climate uncertainty. Tools such as soil testing kits, remote sensing, crop growth models, and nutrient decision algorithms enable farmers to put nutrients at the most effective rate, time, and location. Through these technologies, farmers are able to cut nutrient losses, reduce their production costs, and alleviate their contribution to climate change. In the case of precision nutrient management, the greatest need arises under climate stress situations, whereby real-time decisions can be made based on changing weather, soil, and crop nutrient status. The implementation of site-specific nutrient application enhances nutrient-use efficiency, and thus there is a potential for lowering nitrous oxide emissions as a result of less nitrogen usage. Due to that, as climate risks become more severe, DSTs will be more and more important for soil fertility management that is not only adaptive but also resilient (Dobermann et al., 2002; Zhang, Chen and Vitousek, 2013).

12.7.4 Integration with climate-resilient cropping systems

One condition under which climate-smart soil fertility management can yield the greatest benefits is its integration with climate-resilient cropping systems. Crop diversification, intercropping, crop rotation, and stress-tolerant cultivars are some of the activities that can be implemented to improve nutrient cycling and decrease the risk of pests, diseases, and climatic extremes. In addition, legume-based rotations enhance biological nitrogen fixation, and thus synthetic fertilizers become less used, whereas plants with long roots not only improve soil structure but also allow it to be replenished with nutrients from deep layers. There is a synergistic effect whereby combining resilient cropping systems with better soil fertility management leads to improvements in both. Resilient soils are able to support resilient crops and diversified cropping systems, which in turn facilitate soil biological activity and nutrient balance. Such a holistic system approach bolsters the agro-ecosystem's capacity to withstand stress, increases the stability of yields, and makes possible sustainable agricultural intensification under the climate change scenario (Altieri et al., 2015; FAO, 2022).

12.8 Future perspective and research needs

12.8.1 Innovations in nutrient-use efficiency

Improving nutrient-use efficiency (NUE) is a main focus of research aimed at increasing plant tolerance to stress in combination with the reduction of negative impacts on the environment. As innovations, such as slow- and controlled-release fertilizers, enhanced-efficiency fertilizers, and nanofertilizers, seem to be promising in the aspect of synchronizing nutrient supply with crop demand under climate change. Besides that, breeding crops with better nutrient uptake and higher tolerance to stress is a good complement to the soil-based fertility strategies. There is still a necessity for continued investigation to evaluate the agronomic and environmental impacts of these novel solutions in different agro-ecosystems in the long run (Hawkesford et al., 2023).

12.8.2 Role of digital tools and soil sensors

Digital agriculture tools and soil sensors are changing the way soil fertility is managed by giving the possibility of immediate and exact local decisions. A set of technologies involving remote sensing, geographic information systems, rapid soil testing via smartphones, and nutrient decision-support models affects optimal fertilizer utilization under fluctuating weather conditions for farmers. Implementation of the instruments leads to the reduction of nutrient losses, increasing of yields, and strengthening of climate stress resistance. Upcoming inquiry should be engaged in making the digital tools affordable, accessible, and user-friendly, primarily for smallholder farmers (Basso and Antle, 2020).

12.8.3 Long-term soil fertility monitoring

Permanent monitoring of soil fertility indicators, among which are soil organic carbon, nutrient stocks, pH, and biological activity, is necessary to grasp the concept of soil health and stress resistance. Nonstop observation grants the possibility of the first recognition of soil degradation and is instrumental in the implementation of the adaptive management strategy under climatic variability. It is going to be of utmost importance to create long-term experimental sites and well-structured national soil health databases for laying the foundation of soil management policies based on the collected evidence (Lal, 2020; FAO, 2021).

12.8.4 Policy support and extension strategies

The support of good policy structures and extensional services is the key to the implementation of the climate-smart soil fertility management on a grand scale. Thus promotional policies of the balanced use of fertilizers, the giving of financial incentives for organic amendments, the support of soil testing, and the encouragement of carbon sequestration are the most effective ways to accelerate the process. By conversing and exchanging information with the agents of the agricultural extension service, farmers and the agricultural extension workers get the opportunity to strengthen their knowledge, and through it, the adoption of farming practices that are site-specific and integrated is facilitated. The interaction of researchers, policy-makers, extension workers, and farmers allows for better comprehension and application of scientific findings, leading to advanced solutions at the ground level (Lipper et al., 2014; FAO, 2022).

12.9 Conclusion

Management of soil fertility is the main factor that can lead to the increase of crop resistance to gene and environmental stress as well as to improve nutrient supply, soil physical structure, biological activity, and water-holding capacity. The implementation of integrated nutrient management, organic amendments, soil biological management, pH correction, and climate-smart practices, to mention a few, overall, they energize the agro-ecosystems' capacity to resist and recover from the environmental stresses. Integrated, site-specific soil fertility management is indispensable to the proper handling of soil heterogeneity and climate variability. These measures enhance nutrient-use efficiency, diminish environmental losses, and increase yield stability in conditions prone to stress. As for the future, the combination of technological innovations, long-term soil monitoring, supportive policies, and efficient extension services will be indispensable for the continuation of soil fertility and food security maintenance.

In brief, the promotion of soil fertility management is the basis for the realization of sustainable and climate-resilient agriculture goals. When healthy soils are funded, they become a source of agricultural systems' resilience to climate change, while at the same time, they can keep productivity, environmental protection, and long-term ecosystem stability.

References

Alloway, B.J., 2013. Heavy Metals in soils: trace metals and metalloids in soils and their bioavailability. Springer, Dordrecht. Available at https://doi.org/10.1007/978-94-007-4470-7.

Altieri, M.A., Nicholls, C.I., Henao, A., Lana, M.A., 2015. Agroecology and the design of climate change-resilient farming systems. Agronomy for Sustainable Development 35 (3), 869–890. Available at https://doi.org/10.1007/s13593-015-0285-2.

Backer, R., Rokem, J.S., Ilangumaran, G., et al., 2018. Plant growth-promoting rhizobacteria: mechanisms and applications. Frontiers in Plant Science 9, 1473.

Bardgett, R.D., Van Der Putten, W.H., 2014. Belowground biodiversity and ecosystem functioning. Nature 515 (7528), 505–511. Available at https://doi.org/10.1038/nature13855.

Basso, B., Antle, J., 2020. Digital agriculture to design sustainable agricultural systems. Nature Sustainability, 3(4), 254–256.

Berendsen, R.L., Pieterse, C.M.J., Bakker, P.A.H.M., 2012. The rhizosphere microbiome and plant health. Trends in Plant Science 17 (8), 478–486. Available at https://doi.org/10.1016/j.tplants.2012.04.001.

Brady, N.C., Weil, R.R., 2017. The Nature and Properties of Soils, fifteenth ed. Pearson Education.

Bronick, C.J., Lal, R., 2005. Soil structure and management: a review. Geoderma 124 (1-2), 3–22. Available at https://doi.org/10.1016/j.geoderma.2004.03.005.

Chivenge, P., Vanlauwe, B., Six, J., 2011. Does the combined application of organic and mineral nutrient sources influence maize productivity? A meta-analysis. Plant and Soil 342 (1-2), 1–30. Available at https://doi.org/10.1007/s11104-010-0626-5.

Dobermann, A., Witt, C., Dawe, D., Abdulrachman, S., Gines, H.C., Nagarajan, R., et al., 2002. Site-specific nutrient management for intensive rice cropping systems in Asia. Field Crops Research 74 (1), 37–66. Available at https://doi.org/10.1016/s0378-4290(01)00197-6.

FAO, 2017. Soil Organic Carbon: The Hidden Potential. Food and Agriculture Organization of the United Nations.

FAO, 2020. State of Knowledge of Soil Biodiversity: Status, Challenges and Potentialities. Food and Agriculture Organization of the United Nations.

FAO, 2021. Soil Organic Carbon: The Hidden Potential for Climate-Smart Agriculture. Food and Agriculture Organization of the United Nations.

FAO, 2022. Climate-Smart Agriculture Sourcebook. Food and Agriculture Organization of the United Nations.

Farooq, M., Wahid, A., Siddique, K., H., M., 2020. Drought stress in plants: an overview. In: Farooq, M., Siddique, K.H.M. (Eds.), Plant Responses to Drought Stress. Springer, pp. 1–33.

Gautam, A., Chauhan, A., Singh, A., Mundepi, S., Pant, M., Husen, A., 2024. Chapter 9 - Use of seaweed extract-based biostimulants in plant growth, biochemical constituents, and productions. In: Husen, A. (Ed.), Biostimulants in Plant Protection and Performance. Elsevier, pp. 129–148. Available at https://doi.org/10.1016/B978-0-443-15884-1.00022-1.

Hamza, M.A., Anderson, W.K., 2005. Soil compaction in cropping systems: a review of the nature, causes and possible solutions. Soil and Tillage Research 82 (2), 121–145. Available at https://doi.org/10.1016/j.still.2004.08.009.

Hatfield, J.L., Prueger, J.H., 2015. Temperature extremes: effect on plant growth and development. Weather and Climate Extremes 10, 4–10. Available at https://doi.org/10.1016/j.wace.2015.08.001.

Hawkesford, M.J., Kopriva, S., De Kok, L.J., McGrath, S.P., 2023. Nutrient use efficiency and resilience to abiotic stress. Journal of Experimental Botany 74, 589–602.

Hobbs, P.R., Sayre, K., Gupta, R., 2008. The role of conservation agriculture in sustainable agriculture. Philosophical Transactions of the Royal Society B: Biological Sciences 363 (1491), 543–555. Available at https://doi.org/10.1098/rstb.2007.2169.

Husen, A., 2021. Harsh Environment and Plant Resilience: Molecular and Functional Aspects. Springer, Cham. Available at https://doi.org/10.1007/978-3-030-65912-7.

Husen, A., 2021. The harsh environment and resilient plants: an overview. In: Husen, A. (Ed.), Harsh Environment and Plant Resilience. Springer, Cham, pp. 1–23. Available at https://doi.org/10.1007/978-3-030-65912-7_1.

Husen, A., 2021. Plant Performance Under Environmental Stress (Hormones, Biostimulants and Sustainable Plant Growth Management). Springer Nature, Cham. Available at https://doi.org/10.1007/978-3-030-78521-5.

Husen, A., 2023. Plants and Their Interaction to Environmental Pollution: Damage Detection, Adaptation, Tolerance, Physiological and Molecular Responses. Elsevier Inc., Cambridge, MA. Available at https://doi.org/10.1016/C2020-0-03319-0.

Husen, A., 2024. Essential Minerals in Plant-Soil System: Coordination, Signalling, and Interaction under Adverse Situations. Elsevier Inc., Cambridge, MA. Available at https://doi.org/10.1016/C2022-0-02437-5.

Husen, A., 2025a. Roles of Osmolytes in Changing Environment. Elsevier Inc., Cambridge, MA. Available at https://doi.org/10.1016/C2023-0-51382-5.

Husen, A., 2025b. Role of Antioxidants in Mitigating Plant Stress. Elsevier Inc., Cambridge, MA. Available at https://doi.org/10.1016/C2023-0-51268-6.

IPCC, 2022. Climate Change 2022: Impacts, Adaptation and Vulnerability. Cambridge University Press.

du Jardin, P., 2015. Plant biostimulants: definition, concept, main categories and regulation. Scientia Horticulturae 196, 3–14. Available at https://doi.org/10.1016/j.scienta.2015.09.021.

Kapoor, D., Sharma, P., Sharma, M.M.M., Yadav, S., Husen, A., 2024. Exploring soil microbiota and their role in plant growth, stress tolerance, disease control and nutrient immobilizer. Biocatalysis and Agricultural Biotechnology 61, 103358. Available at https://doi.org/10.1016/j.bcab.2024.103358.

Lal, R., 2020. Managing soils for resolving the conflict between agriculture and nature. European Journal of Soil Science 71 1–9.

Lehmann, J., Bossio, D.A., Kögel-Knabner, I., Rillig, M.C., 2020. The concept and future prospects of soil health. Nature Reviews Earth & Environment. 1, 544–553. Available at https://doi.org/10.1038/s43017-020-0080-8.

Liebman, M., Baraibar, B., Buckley, Y.M., et al., 2018. Ecologically sustainable weed management: how do we get from proof-of-concept to adoption? Ecological Applications 28, 1354–1369.

Lipper, L., Thornton, P., Campbell, B.M., Baedeker, T., Braimoh, A., Bwalya, M., et al., 2014. Climate-smart agriculture for food security. Nature Climate Change 4 (12), 1068–1072. Available at https://doi.org/10.1038/nclimate2437.

Marschner, P., 2012. Marschner's Mineral Nutrition of Higher Plants, third ed. Academic Press.

Nicol, J.M., Turner, S.J., Coyne, D.L., den Nijs, L., Hockland, S., Tahna Maafi, Z., 2011. Current nematode threats to world agriculture. In: Jones, J. (Ed.), Genomics and Molecular Genetics of Plant–nematode Interactions. Springer, pp. 21–43.

Parent, C., Capelli, N., Berger, A., Crèvecoeur, M., Dat, J.F., 2016. An overview of plant responses to soil waterlogging. Plant Stress 10, 1–14.

Raza, A., Tabassum, J., Zahid, Z., et al., 2022. Advances in "omics" approaches for improving stress tolerance in plants. Plant Physiology and Biochemistry 172, 197–214.

Rouphael, Y., Colla, G., 2020. Biostimulants in agriculture. Frontiers in Plant Science 11, 40.

Schimel, J.P., Schaeffer, S.M., Drake, J.E., 2022. Microbial control over carbon and nutrient cycling in soils under climate stress. Nature Reviews. Microbiology 20, 401–414.

Shaffique, S., Injamum-Ul-Hoque, M., Husen, A., Kang, S.-M., Lee, I.-J., 2025. Revolutionizing heat stress tolerance in Glycine max: exploring the latest advances in microbial application. Plant Stress 15, 100725. Available at https://doi.org/10.1016/j.stress.2024.100725.

Shrivastava, P., Kumar, R., 2015. Soil salinity: a serious environmental issue and plant growth promoting bacteria as one of the tools for its alleviation. Saudi Journal of Biological Sciences 22 (2), 123–131. Available at https://doi.org/10.1016/j.sjbs.2014.12.001.

Six, J., Bossuyt, H., Degryze, S., Denef, K., 2014. A history of research on the link between soil biota, aggregation, and soil organic matter dynamics. Soil & Tillage Research 79, 7–31.

Smith, S.E., Read, D.J., 2008. Mycorrhizal Symbiosis, 3rd ed. Academic Press.

Smith, S.E., Smith, F.A., 2011. Roles of arbuscular mycorrhizas in plant nutrition and growth: new paradigms from cellular to ecosystem scales. Annual Review of Plant Biology 62, 227–250. Available at https://doi.org/10.1146/annurev-arplant-042110-103846.

Smith, P., Soussana, J.F., Angers, D., Schipper, L., Chenu, C., Rasse, D.P., et al., 2020. How to measure, report and verify soil carbon change to realize the potential of soil carbon sequestration for atmospheric greenhouse gas removal. Global Change Biology 26 (1), 219–241. Available at https://doi.org/10.1111/gcb.14815.

Trivedi, P., Leach, J.E., Tringe, S.G., Sa, T., Singh, B.K., 2020. Plant–microbiome interactions: from community assembly to plant health. Nature reviews. Microbiology 18 (11), 607–621. Available at https://doi.org/10.1038/s41579-020-0412-1.

Vanlauwe, B., Bationo, A., Chianu, J., Giller, K.E., Merckx, R., Mokwunye, U., et al., 2010. Integrated soil fertility management: operational definition and consequences for implementation and dissemination. Outlook on Agriculture 39 (1), 17–24. Available at https://doi.org/10.5367/000000010791169998.

Vanlauwe, B., Wendt, J., Diels, J., 2015. Combined application of organic matter and fertilizer. In: Tian, G., Ishida, F., Keatinge, D., Carsky, R., Wendt, J. (Eds.), Sustaining Soil Fertility in West Africa. SSSA Special Publications, pp. 247–279. Available at https://doi.org/10.2136/sssaspecpub58.ch12.

Vogel, H.J., Eynard, A., Jarvis, N., 2025. Soil structure dynamics and their role in water flow, aeration, and root growth under climate extremes. Soil & Tillage Research 244, 105829.

Wahid, A., Gelani, S., Ashraf, M., Foolad, M.R., 2007. Heat tolerance in plants: an overview. Environmental and Experimental Botany 61 (3), 199–223. Available at https://doi.org/10.1016/j.envexpbot.2007.05.011.

Zhang, F., Chen, X., Vitousek, P., 2013. An experiment for the world. Nature 497 (7447), 33–35. Available at https://doi.org/10.1038/497033a.

Zörb, C., Senbayram, M., Peiter, E., 2014. Potassium in agriculture – status and perspectives. Journal of Plant Physiology 171 (9), 656–669. Available at https://doi.org/10.1016/j.jplph.2013.08.008.

CHAPTER

13 Nutrient imbalance and plant stress responses

Shifa Shaffique[1,*], Atia Iqbal[2,*], Muhammad Aaqil Khan[3], Muhammad Imran[1], Sang Mo Kang[1], Javier Trujillo Rogmas[1] and In-Jung Lee[1]

[1]*Department of Applied Biosciences, College of Agriculture and Life Sciences, Kyungpook National University, Daegu, South Korea,*
[2]*Department of Microbiology and Molecular Genetics, The Women University Multan, Multan, Pakistan,*
[3]*International Joint Research Laboratory for Global Change Ecology, School of Life Sciences, Henan University, Kaifeng, Henan, P.R. China*

Abbreviations

N	nitrogen
P	phosphorus
K Potassium	
Fe	iron
Zn	zinc
Mn	manganese
B	boron
ROS	reactive oxygen species

13.1 Introduction

For optimal plant growth and regulation, it requires a balanced supply of the essential nutrients to sustain growth, cellular homeostasis, and stimulate the antioxidant defense system under physiological or adverse environmental conditions (Pearson and Rengel, 2022; Reddy, 2006). Seventeen essential elements are broadly categorized into macro- and micronutrients based on their requirement to plant cells. Macronutrients are nitrogen (N), phosphorus (P), and potassium (K), and micronutrients such as iron (Fe), zinc (Zn), manganese (Mn), and boron (B). All these macro- and micronutrients play a crucial role in maintaining structural integrity and act as signaling molecules (Hasanuzzaman et al., 2018; Kaur et al.,

[*] Contributed equally.

Mineral Nutrients to Mitigate Abiotic Stress in Crops. DOI: https://doi.org/10.1016/B978-0-443-44712-9.00012-8

2023; Singhal et al., 2023). Nutrient imbalance arises when the concentration, bioavailability, uptake, and transport of these nutrients exceed or fall below the threshold level from the physiological range. This imbalance can manifest as a nutrient deficiency, where supply is insufficient, or nutrient toxicity nutrient deficiency and toxicity are lethal for plant physiology and biochemistry. An excess amount of nutrients causes toxicity, while an insufficient supply can lead to nutrient deficiency (Abbas et al., 2021; Hajiboland, 2018; Unit, I., 1976; Vitousek et al., 2009; Kumari et al., 2021).

Nutrient imbalance is a kind of abiotic stress that is getting increasingly prevalent all over agricultural land. The increasing prevalence of climatic change conditions, along with global warming, significantly affects agricultural land, as well as posing a nutrient imbalance in plants (Manzoor Alam and Alam, 1999; Moshood et al., 2025; Noordwijk and Cadisch, 2002). Moreover, modern agricultural techniques, such as rhizospheric engineering (Singh et al., 2022; Vega, 2007), nanotechnology (Amjad and Serajuddin, 2021; Chugh et al., 2021), and other fertilizer applications (Lu and Tian, 2017; Noulas et al., 2023), also contribute to the nutrient imbalance in plants. For example, drought stress is a significant abiotic stress that, with a reduction in relative water content it also contributes toward the lesser nutrient mobility (Ahanger et al., 2016; Da Silva et al., 2011). Similarly, salinity stress significantly affects ionic uptake and their balance, such as nitrogen and phosphorus (Awais et al., 2023; Chakraborty et al., 2018; Miura, 2013), and heavy metals stress affects nutrient uptake. For instance, in common bean (*Phaseolus vulgaris* L.) seeds were exposed to a heavy metal solution that contained the copper sulfate, magnesium sulfate, lead nitrate, nickel silicate, cadmium nitrate, and zinc sulfate with two different concentrations, 150 and 350 mg/L, for 7 days. The DNA fingerprint was analyzed, and results suggested that mineral nutrient imbalance causes stress in plants at the DNA level; the highest number of dispersion bands was observed in plants exposed to zinc and nickel (Gjorgieva et al., 2013). Collectively, these studies suggested that nutrient imbalance rarely occurs alone; it is a consequence or side effect of the abiotic stress, and it is lethal for the optimal growth of the plant.

In response to nutrient imbalance, plants respond through highly coordinated pathways that involve physiological, biochemical, antioxidant, phytohormone, and molecular responses (Hajiboland, 2012; Pandey, 2018; Romera et al., 2021; Kumari et al., 2021). Plants alter the root architecture, activate the nutrient transporter, modulate the phytohormone pathway, and rapidly accumulate the reactive oxygen species. The ROS acts as a signaling molecule that further tends to stimulate the enzymatic and nonenzymatic antioxidant defense system. The nutrient imbalance is sensed by plants by specific transporters, regulatory complexes, and signaling proteins. These all act as a nutrient detector in plants (Hajiboland, 2012; Tarkowski et al., 2023). Upon sensing nutrient imbalance, they tend to downregulate the transcription factors such as *WRKY*, *NAC* (NAM, ATAF, and CUC), dehydration-responsive element binding protein (*DREB*), *bZIP*, etc. These transcription factors reorganize the molecular expression and tend to regulate the nutrient stress (Iswanto et al., 2025; Xiong et al., 2025). For example, under Fe deficiency, the translocation of Fe is regulated by the *WRKY* transcription factor (Yan et al., 2016), while in sulfate deficiency, the integrated regulatory network of transcription factors involving *DREB* is responsible for the regulation and stress tolerance in the experimental plant tomato (Landaeta-Sepúlveda et al., 2025).

After sensing the nutrient imbalance, it is regulated by the NRT1.1 (nitrate) (Fang et al., 2021; Maghiaoui et al., 2020), PHR1-SPX (phosphate), (Motte and Beeckman, 2017; Thakur et al., 2025), HAK/KUP/KT (potassium) (Wang and Wu, 2010; Yang et al., 2020), IRT1/FRO2(iron) (Fourcroy et al., 2016; Yamaguchi and Satoh, 2014).

13.2 Photosynthesis and nutrient imbalance

Nutrient imbalance, either toxicity or deficiency both affects every cellular organelle, such as chloroplast, mitochondria, plasma membrane, vacuole, and cytosol, because the nutrient ions are required for physiological, biochemical, enzymatic, and electro transport functionality (Akhtar et al., 2023; Gupta et al., 2024).

Photosynthesis is one of the important nutrient-sensitive physiological processes that takes place in chloroplasts. For a smooth and regular photosynthetic process, chloroplasts require N, Mg, Fe, Mn, copper (Cu), and Zinc (Zn) for structure and function. Deficiencies of these nutrients impair chlorophyll synthesis at several points (Cakmak and Engels, 2024; Lu et al., 2025). For example, nitrogen plays an important role in plant productivity via facilitating the process of photosynthesis. Nitrogen imbalance leads to investing more nitrogen supply in bioenergetics to keep a rapid electron transport rate (Mu and Chen, 2021). Similarly, Mg deficiency also impairs the process of photosynthesis. Mg forms the central atoms of the chlorophyll porphyrin ring that is important for Rubisco activation. When there is an insufficient level of Mg in the plant, the chlorophyll molecule degrades rapidly, and the thylakoid membranes lose their structural integrity and energy transfer within photosystem II, and photosystem I becomes insufficient (Peng et al., 2019; Samborska et al., 2018). In a study on the experimental plant tomato, Mg deficiency reduced the electron transfer capacity and weakened the electron flow during the process of photosynthesis in the tomato plant (Zhu et al., 2025).

On the other hand, the iron (Fe) deficiency also has an adverse effect on the process of photosynthesis of plants, as it negatively affects the cytochrome, ferredoxin, and components of the electron transport chain. In the absence of iron, the cofactor that is required for electron flow from photosystem II to photosystem I becomes blocked and ultimately leads to the accumulation of excess excitation energy within photosystem. Following this, the chloroplast can't dissipate excess energy, it converted to oxygen and generates the reactive oxygen species. ROS causes damage to the D1 protein of PSII, causes lipid peroxidation, and leads to chronic photoinhibition (Roncel et al., 2016; Abadía et al., 1999). Recently, Wu et al. (2025) reported that long-term deficiency of iron in grapevine impairs the chlorophyll contents and photosynthetic performance. The mechanism of impairing the photosynthetic process revealed that iron deficiency may lead to enzymatic and metabolic dysfunction and ultimately lead to growth-negative effects.

Nutrient toxicity adversely affects photosynthesis. An excess amount of iron, Mn, and B accumulates within the chloroplast and induces ionic stress in plant cells. This ionic stress further destabilizes the pigment protein complex and impairs the enzymatic function. For example, higher levels of Fe catalyze the Fenton reaction and tend to produce the reactive hydroxyl radicals, ultimately causing the degradation of the chlorophyll molecule and affecting membrane integrity. Similarly, the excess amount of Mn also disturbs chlorophyll biosynthesis. An excess amount of B causes adverse effects on chloroplast osmoregulation, etc. (Cakmak and Engels, 2024; Han et al., 2008; Jezek and Geilfus, 2015; Kering et al., 2009).

13.3 Effects of nutrient imbalance on biochemical reactions

Nutrient imbalance effect on the functionality of photosystem PSI and PSII. Particularly, the PSI is more vulnerable (because of the existing D1 protein) to oxidative stress. Nutrient deficiencies limit the repair cycle to replace DNA. Following the compromised activity of photosystem II, the electron flow to PSI becomes interrupted, ultimately causing photoinhibition. Sometimes it may cause a worse condition that is irreversible. This further reduces the formation of ATP, NADPH, and potential energy carriers for the Calvin cycle. Moreover, the enzymatic reaction of the Calvin cycle also slows down, most importantly, catalyzed by Rubisco (Kono et al., 2022; Ohnishi et al., 2021). For example, in a study, it was reported that a deficiency of nitrogen causes a decrease in the process of repair and enhances the photoinhibition of PSII (Li et al., 2021).

13.4 Anatomical changes under nutrient imbalance

Nutrient stress also affects the structure of the chloroplast. In cases of deficiencies, it may lead to poorly developed grana stacks, reduced thickness of thylakoid membrane density, and increased plastoglobuli. On the other hand, the toxic concentration causes the swelling of the thylakoid membrane, loss of luminal space, and disorganization of lamellae. The structural changes further reduce the efficiency of electron transport. This further tends to allow the leakage of the ions and

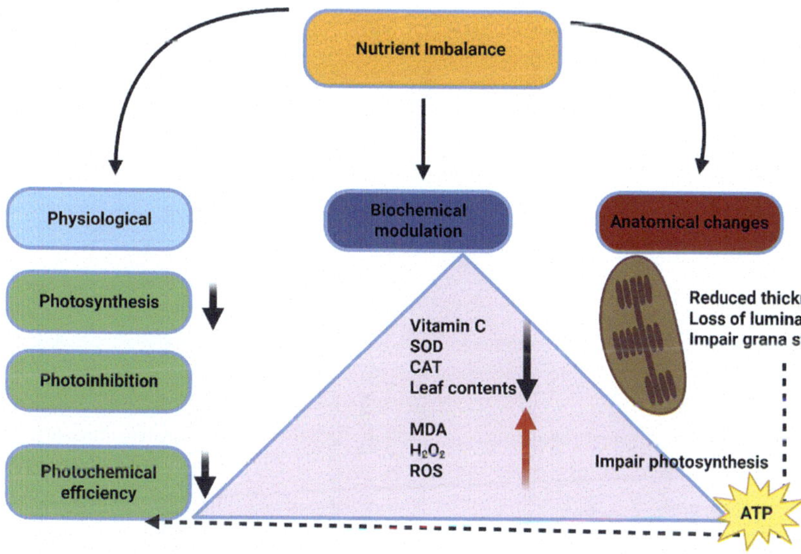

FIGURE 13.1

Effects of nutrient imbalance on physiological, biochemical, and anatomical levels. The symbol ↓ represents a decrease, and ↑ represents an increase in the specific parameter.

metabolites and affects the proton gradient needed for ATP synthesis. All these changes lead to impaired photochemical efficiency. (Fig. 13.1).

13.5 Case studies

Nutrients are important for plant production. Iron is an essential element for plant growth. In a study, iron deficiency was induced to see the response of Chinese cabbage at different levels of iron. The results were surprising; upon sensing the iron deficiency, it severely slowed plant growth and lowered the level of vitamin C, soluble sugar, and protein contents in plants. Meanwhile, the level of cellulose and nitrate was increased. The deficient level of iron also causes a lower photosynthesis rate. Upon exposure to iron deficiency stress, there is rapid production of reactive oxygen species ROS, and they tend to activate the antioxidant defense system. As a result, the MDA and LPO levels were significantly higher (Wang et al., 2022). In another study, it was observed that sorghum, upon exposure to nitrogen deficiency, showed a reduction in leaf area and reduced biomass. However, the physiological traits, such as reduced photosynthesis (Zhao et al., 2005). Some case studies are given in Table 13.1.

Table 13.1 Nutrient imbalance in different plant species and their outcome. The symbol ↓ represents a decrease, and ↑ represents an increase in the specific parameter.

Nutrient imbalance	Plant	Effects	Outcome	Key references
Iron deficiency	Chinese cabbage	Vitamin ↓C ↓Soluble protein↓ Soluble sugar↓ Cellulose and nitrate	Photosynthesis↓ Nitrate ↓ Reductase↓ SOD ↓ CAT↓ POD↑ MDA↑	Wang et al. (2022)
Nitrogen deficiency	Sorghum	↓Leaf area ↓Leaf contents ↓Net photosynthesis Lower stomatal conductance	Reduced biomass	Zhao et al. (2005)
Nitrogen deficiency	Maize	Photosynthetic capacity ↓	Decrease in grain yield Plant height ↓ Dry matter production reduces	Ding et al. (2005)
Nitrogen deficiency	Tomato	↓Light-saturated photosynthetic rate ↓ ↓Stomatal conductance	Growth reduction	Guidi et al. (1997)
Phosphorus deficiency	Rice	Reduction in net photosynthesis Decrease CO_2 assimilation	Photoinhibition	Xu et al. (2007)
Calcium deficiency	Coffee plant	Shoot growth decreases Leaf protein concentration decreases Increase the concentration of the potassium, magnesium, and sodium	Decrease photochemical efficiency	Ramalho et al. (1995)

13.6 Conclusion

Nutrient imbalance, either excess or a deficiency, causes impaired nutrient homeostasis and poses an abiotic stress that affects plant biology. Nutrients serve structural, metabolic, and regulatory roles in plant physiological, molecular, and biochemical functions. Deficiencies in these nutrients, such as N, Mg, P, K, and Fe, restrict the process of chlorophyll synthesis, electron transport, and energy metabolism, ultimately leading to restricted plant growth. Similarly, the excess amount of these nutrients induces oxidative stress in plants and is lethal for plant metabolic reprogramming. Overall, understanding anatomical, physiological, and biochemical responses provides insight into nutrient imbalance and critical insight for improving nutrient use efficiency for stress tolerance. Future prospective and continued research will open new approaches to optimize the nutrient use efficiency in plants.

References

Abadía, J., Morales, F., Abadía, A., 1999. Photosystem II efficiency in low chlorophyll, iron-deficient leaves. Plant and Soil 215 (2), 183–192. Available at https://doi.org/10.1023/a:1004451728237.

Abbas, S., Amna, Javed, M.T., Ali, Q., Azeem, M., Ali, S., 2021. Nutrient deficiency stress and relation with plant growth and development. Engineering Tolerance in Crop Plants Against Abiotic Stress. CRC Press, Boca Raton, pp. 239–262. Available at 10.1201/9781003160717-12.

Ahanger, M.A., Morad-Talab, N., Abd-Allah, E.F., Ahmad, P., Hajiboland, R., 2016. Plant growth under drought stress: significance of mineral nutrients. In: Ahmad, P. (Ed.), Water Stress and Crop Plants: A Sustainable Approach. Wiley, pp. 649–668. Available at https://doi.org/10.1002/9781119054450.ch37.

Akhtar, I., Javad, S., Jabeen, K., Tariq, A., Nawaz, K., Shah, A.A., et al., 2023. Adverse impact of ROS on nutrient accumulation and distribution in plants. In: Faizan, M., Hayat, S., Ahmed, S.M. (Eds.), Reactive Oxygen Species: Prospects in Plant Metabolism. Springer, Singapore, pp. 263–292. Available at https://doi.org/10.1007/978-981-19-9794-5_15.

Amjad, S., Serajuddin, M., 2021. Applications of nanobiotechnology to mitigate mineral nutrients deficiency stress in crop plants. In: Al-Khayri, J.M., Ansari, M.I., Singh, A.K. (Eds.), Nanobiotechnology: Mitigation of Abiotic Stress in Plants. Springer, Cham, pp. 437–452. Available at https://doi.org/10.1007/978-3-030-73606-4_19.

Awais, M., Rasheed, Z., Sadiq, M.T., 2023. Salinity Stress Effects on Nutrient Uptake in Plants and its Influence on Plant Growth Efficiency. Trends in Animal and Plant Sciences 1, 64–72. Available at https://doi.org/10.62324/taps/2023.009.

Cakmak, I., Engels, C., 2024. Role of mineral nutrients in photosynthesis and yield formation. In: Rengel, Z. (Ed.), Mineral Nutrition of Crops: Fundamental Mechanisms and Implications. CRC Press, Boca Raton, pp. 141–168. Available at 10.1201/9781003578468-6.

Chakraborty, K., Basak, N., Bhaduri, D., Ray, S., Vijayan, J., Chattopadhyay, K., et al., 2018. Ionic basis of salt tolerance in plants: nutrient homeostasis and oxidative stress tolerance. In: Hasanuzzaman, M., Fujita, M., Oku, H., Nahar, K., Hawrylak-Nowak, B. (Eds.), Plant Nutrients and Abiotic Stress Tolerance. Springer, Singapore, pp. 325–362. Available at https://doi.org/10.1007/978-981-10-9044-8_14.

Chugh, G., Siddique, K.H.M., Solaiman, Z.M., 2021. Nanobiotechnology for agriculture: smart technology for combating nutrient deficiencies with nanotoxicity challenges. Sustainability 13 (4), 1781. Available at https://doi.org/10.3390/su13041781.

Da Silva, E.C., Nogueira, R., Da Silva, M.A., De Albuquerque, M.B., 2011. Drought stress and plant nutrition. Plant Stress 5 (1), 32–41.

Ding, L., Wang, K.J., Jiang, G.M., Biswas, D.K., Xu, H., Li, L.F., et al., 2005. Effects of nitrogen deficiency on photosynthetic traits of maize hybrids released in different years. Annals of Botany 96 (5), 925–930. Available at https://doi.org/10.1093/aob/mci244.

Fang, X.Z., Fang, S.Q., Ye, Z.Q., Liu, D., Zhao, K.L., Jin, C.W., 2021. NRT1. 1 dual-affinity nitrate transport/signalling and its roles in plant abiotic stress resistance. Frontiers in Plant Science 12, 715694.

Fourcroy, P., Tissot, N., Gaymard, F., Briat, J.F., Dubos, C., 2016. Facilitated Fe nutrition by phenolic compounds excreted by the *Arabidopsis* ABCG37/PDR9 transporter requires the IRT1/FRO2 high-affinity root Fe^{2+} transport system. Molecular Plant 9 (3), 485–488. Available at https://doi.org/10.1016/j.molp.2015.09.010.

Gjorgieva, D., Kadifkova Panovska, T., Ruskovska, T., Bačeva, K., Stafilov, T., 2013. Mineral nutrient imbalance, total antioxidants level and DNA damage in common bean (*Phaseolus vulgaris* L.) exposed to heavy metals. Physiology and Molecular Biology of Plants 19 (4), 499–507. Available at https://doi.org/10.1007/s12298-013-0196-0.

Guidi, L., Lorefice, G., Pardossi, A., Malorgio, F., Tognoni, F., Soldatini, G.F., 1997. Growth and photosynthesis of *Lycopersicon esculentum* (L.) plants as affected by nitrogen deficiency. Academy of Sciences of the Czech Republic, Italy Biologia Plantarum 40 (2), 235–244. Available at https://doi.org/10.1023/A:1001068603778.

Gupta, R., Verma, N., Tewari, R.K., 2024. Micronutrient deficiency-induced oxidative stress in plants. Plant Cell Reports 43 (9). Available at https://doi.org/10.1007/s00299-024-03297-6.

Hajiboland, R., 2012. Effect of micronutrient deficiencies on plants stress responses. Abiotic Stress Responses in Plants: Metabolism, Productivity and Sustainability. Springer, New York, pp. 283–329. Available at 10.1007/9781461406341_16.

Hajiboland, R., 2018. Nutrient Deficiency and Abundance in Tea Plants: Metabolism to Productivity. In: Han, W.Y., Li, X., Ahammed, G. (Eds.), Stress Physiology of Tea in the Face of Climate Change. Springer, Singapore, pp. 173–215. Available at https://doi.org/10.1007/978-981-13-2140-5_9.

Han, S., Chen, L.S., Jiang, H.X., Smith, B.R., Yang, L.T., Xie, C.Y., 2008. Boron deficiency decreases growth and photosynthesis, and increases starch and hexoses in leaves of citrus seedlings. Journal of Plant Physiology 165 (13), 1331–1341. Available at https://doi.org/10.1016/j.jplph.2007.11.002.

Hasanuzzaman, M., Fujita, M., Oku, H., Nahar, K., Hawrylak-Nowak, B., 2018. Plant Nutrients and Abiotic Stress Tolerance. Springer, Singapore, pp. 1–590. Available at https://doi.org/10.1007/978-981-10-9044-8.

Iswanto, A.B.B., Kang, H., Park, S., Son, G.H., Pike, S.M., Kim, S.H., 2025. Recent advances in improving yield and immunity through transcription factor engineering.

Journal of Integrative Plant Biology 67 (8), 2005–2027. Available at https://doi.org/10.1111/jipb.13932.

Jezek, M., Geilfus, C.M., Bayer, A., Muhling, K.H., 2015. Photosynthetic capacity, nutrient status, and growth of maize (*Zea mays* L.) upon $MgSO_4$ leaf-application. Frontiers in Plant Science 5. Available at https://doi.org/10.3389/fpls.2014.00781.

Kaur, H., Kaur, H., Kaur, H., Srivastava, S., 2023. The beneficial roles of trace and ultratrace elements in plants. Plant Growth Regulation 100 (2), 219–236. Available at https://doi.org/10.1007/s10725-022-00837-6.

Kering, M.K., Lukaszewska, K., Blevins, D.G., 2009. Manganese requirement for optimum photosynthesis and growth in NAD-malic enzyme C-4 species. Plant and Soil 316 (1-2), 217–226. Available at https://doi.org/10.1007/s11104-008-9772-4.

Kono, M., Oguchi, R., Terashima, I., 2022. Photoinhibition of PSI and PSII in nature and in the laboratory: Ecological approaches. In: Lüttge, U., Cánovas, F.M., Risueño, M.-C., Leuschner, C., Pretzsch, H. (Eds.), Progress in Botany (Vol. 84). Springer, Cham, pp. 241–292. Available at https://doi.org/10.1007/124_2022_67.

Kumari, A., Sharma, B., Singh, B.N., Hidangmayum, A., Jatav, H.S., Chandra, K., et al., 2021. Chapter 10 - Physiological mechanisms and adaptation strategies of plants under nutrient deficiency and toxicity conditions. In: Aftab, T., Roychoudhury, A. (Eds.), Plant Perspectives to Global Climate Changes: Developing Climate-Resilient Plants. Academic Press, pp. 173–194. Available at https://doi.org/10.1016/B978-0-323-85665-2.00012-1.

Landaeta-Sepúlveda, D., Johnson, N.R., Morales-Espinoza, J., Tobar, M., Sánchez, E., Fernández, J.D., et al., 2025. Sulfate deficiency-responsive MicroRNAs in tomato uncover an expanded and functionally integrated regulatory network. International Journal of Molecular Sciences 26 (17), 8392. Available at https://doi.org/10.3390/ijms26178392.

Li, Z., Lan, T., Zhang, J., Gao, K., Beardall, J., Wu, Y., 2021. Nitrogen limitation decreases the repair capacity and enhances photoinhibition of photosystem ii in a diatom. Photochemistry and Photobiology 97 (4), 745–752. Available at https://doi.org/10.1111/php.13386.

Lu, C., Tian, H., 2017. Global nitrogen and phosphorus fertilizer use for agriculture production in the past half century: shifted hot spots and nutrient imbalance. Earth System Science Data 9 (1), 181–192. Available at https://doi.org/10.5194/essd-9-181-2017.

Lu, Z., Ren, T., Li, Y., Cakmak, I., Lu, J., 2025. Nutrient limitations on photosynthesis: from individual to combinational stresses. Trends in Plant Science 30 (8), 872–885. Available at https://doi.org/10.1016/j.tplants.2025.03.006.

Maghiaoui, A., Gojon, A., Bach, L., Gifford, M., 2020. NRT1.1-centered nitrate signaling in plants. Journal of Experimental Botany 71 (20), 6226–6237. Available at https://doi.org/10.1093/jxb/eraa361.

Manzoor Alam, S., Alam, S.M., 1999. Nutrient uptake by plants under stress conditions. Books in Soils. Plants, and the Environment 285–313.

Yamaguchi S., Satoh S. (2014). Gibberellin-induced expression of Fe uptake-related genes in Arabidopsis. Plant Cell Physiol., 55(1):87-98. Available at https://doi.org/10.1093/pcp/pct160. Epub 2013 Nov 4. PMID: 24192296.

Miura, K., 2013. Nitrogen and phosphorus nutrition under salinity stress. In: Ahmad, P., Azooz, M., Prasad, M. (Eds.), Ecophysiology and Responses of Plants under Salt Stress. Springer, New York, NY, pp. 425–441. Available at https://doi.org/10.1007/978-1-4614-4747-4_16.

Moshood, A.Y., Abdulraheem, M.I., Li, L., Zhang, Y., Raghavan, V., Hu, J., 2025. Deciphering nutrient stress in plants: integrative insight from metabolomics and proteomics. Functional and Integrative Genomics 25 (1). Available at https://doi.org/10.1007/s10142-025-01551-y.

Motte, H., Beeckman, T., 2017. PHR1 balances between nutrition and immunity in plants. Developmental Cell 41 (1), 5–7. Available at https://doi.org/10.1016/j.devcel.2017.03.019.

Mu, X., Chen, Y., 2021. The physiological response of photosynthesis to nitrogen deficiency. Plant Physiology and Biochemistry 158, 76–82. Available at https://doi.org/10.1016/j.plaphy.2020.11.019.

Noordwijk, M.van, Cadisch, G., 2002. Access and excess problems in plant nutrition. Plant and Soil 247 (1), 25–39. Available at https://doi.org/10.1023/a:1021194628358.

Noulas, C., Torabian, S., Qin, R., 2023. Crop nutrient requirements and advanced fertilizer management strategies. Agronomy 13 (8), 2017. Available at https://doi.org/10.3390/agronomy13082017.

Ohnishi, M., Furutani, R., Sohtome, T., Suzuki, T., Wada, S., Tanaka, S., et al., 2021. Photosynthetic parameters show specific responses to essential mineral deficiencies. Antioxidants 10 (7), 996. Available at https://doi.org/10.3390/antiox10070996.

Pandey, N., Hasanuzzaman, M., Fujita, M., Oku, H., Nahar, K., Hawrylak-Nowak, B., 2018. Role of plant nutrients in plant growth and physiology. In: Hasanuzzaman, M., Fujita, M., Oku, H., Nahar, K., Hawrylak-Nowak, B. (Eds.), Plant Nutrients and Abiotic Stress Tolerance. Springer, Singapore, pp. 51–93. Available at https://doi.org/10.1007/978-981-10-9044-8_2.

Pearson, J.N., Rengel, Z., 2022.). Mechanisms of plant resistance to nutrient deficiency stress. Mechanisms of Environmental Stress Resistance in Plants. Routledge, pp. 213–240.

Peng, Y.Y., Liao, L.L., Liu, S., Nie, M.M., Li, J., Zhang, L.D., et al., 2019. Magnesium deficiency triggers sgr-mediated chlorophyll degradation for magnesium remobilization. Plant Physiology 181 (1), 262–275. Available at https://doi.org/10.1104/pp.19.00610.

Ramalho, J.C., Rebelo, M.C., Santos, M.E., Antunes, M.L., Nunes, M.A., 1995. Effects of calcium deficiency on *Coffea arabica*. Nutrient changes and correlation of calcium levels with some photosynthetic parameters. Plant and Soil 172 (1), 87–96. Available at https://doi.org/10.1007/BF00020862.

Reddy, K.J., 2006. Nutrient stress. In: Madhava Rao, K.V., Raghavendra, A.S., Janardhan Reddy, K. (Eds.), Physiology and Molecular Biology of Stress Tolerance in Plants. Springer, Netherlands, pp. 187–217. Available at https://doi.org/10.1007/1-4020-4225-6_7.

Romera, F.J., Lucena, C., García, M.J., Alcántara, E., Angulo, M., Aparicio, M.Á., et al., 2021. Plant hormones and nutrient deficiency responses. Hormones and Plant Response. Springer, pp. 29–65.

Roncel, M., González-Rodríguez, A.A., Naranjo, B., Bernal-Bayard, P., Lindahl, M., Hervás, M., et al., 2016. Iron deficiency induces a partial inhibition of the photosynthetic electron transport and a high sensitivity to light in the diatom *Phaeodactylum tricornutum*. Frontiers in Plant Science 7 (2016). Available at https://doi.org/10.3389/fpls.2016.01050.

Samborska, I.A., Kalaji, H.M., Sieczko, L., Goltsev, V., Borucki, W., Jajoo, A., 2018. Structural and functional disorder in the photosynthetic apparatus of radish plants under magnesium deficiency. Functional Plant Biology 45 (6), 668–679. Available at https://doi.org/10.1071/FP17241.

Singh, S.K., Wu, X., Shao, C., Zhang, H., 2022. Microbial enhancement of plant nutrient acquisition. Stress Biology 2 (1). Available at https://doi.org/10.1007/s44154-021-00027-w.

Singhal, R.K., Fahad, S., Kumar, P., Choyal, P., Javed, T., Jinger, D., et al., 2023. Beneficial elements: new players in improving nutrient use efficiency and abiotic stress tolerance.

Plant Growth Regulation 100 (2), 237–265. Available at https://doi.org/10.1007/s10725-022-00843-8.

Tarkowski, Ł.P., Signorelli, S., Considine, M.J., Montrichard, F., 2023. Integration of reactive oxygen species and nutrient signalling to shape root system architecture. Plant, Cell and Environment 46 (2), 379–390. Available at https://doi.org/10.1111/pce.14504.

Thakur, P.P., Jadhav, P.V., Dasgupta, D., Mote, G., Bhagat, R., Kale, P.B., et al., 2025. Mineral nutrient acquisition under stress: sensing, signaling, and transportation. Plant Science 360. Available at https://doi.org/10.1016/j.plantsci.2025.112702.

Unit, I., 1976. Mineral nutrition.

Vega, N.W.O., 2007. A review on beneficial effects of rhizosphere bacteria on soil nutrient availability and plant nutrient uptake. Revista Facultad Nacional de Agronomía-Medellín 60 (1), 3621–3643.

Vitousek, P.M., Naylor, R., Crews, T., David, M.B., Drinkwater, L.E., Holland, E., et al., 2009. Nutrient imbalances in agricultural development. Science 324 (5934), 1519–1520. Available at https://doi.org/10.1126/science.1170261.

Wang, Y., Wu, W.H., 2010. Plant sensing and signaling in response to K^+-deficiency. Molecular Plant 3 (2), 280–287. Available at https://doi.org/10.1093/mp/ssq006.

Wang, Y., Kang, Y., Zhong, M., Zhang, L., Chai, X., Jiang, X., et al., 2022. Effects of iron deficiency stress on plant growth and quality in flowering chinese cabbage and its adaptive response. Agronomy 12 (4). Available at https://doi.org/10.3390/agronomy12040875.

Wu, Y., Zhao, T., Zhang, R., Wang, Y., 2025. Iron deficiency stress impairs iron absorption and accumulation, compromising photosynthetic performance and antioxidant defense to limit grapevine growth. Plant Physiology and Biochemistry 229 (Pt C), 110557. Available at https://doi.org/10.1016/j.plaphy.2025.110557.

Xiong, H., He, H., Chang, Y., Miao, B., Liu, Z., Wang, Q., et al., 2025. Multiple roles of NAC transcription factors in plant development and stress responses. Journal of Integrative Plant Biology 67 (3), 510–538. Available at https://doi.org/10.1111/jipb.13854.

Xu, H.X., Weng, X.Y., Yang, Y., 2007. Effect of phosphorus deficiency on the photosynthetic characteristics of rice plants. Russian Journal of Plant Physiology 54 (6), 741–748. Available at https://doi.org/10.1134/S1021443707060040.

Yan, J.Y., Li, C.X., Sun, L., Ren, J.Y., Li, G.X., Ding, Z.J., et al., 2016. A WRKY transcription factor regulates Fe translocation under Fe deficiency. Plant Physiology 171 (3), 2017–2027. Available at https://doi.org/10.1104/pp.16.00252.

Yang, T., Lu, X., Wang, Y., Xie, Y., Ma, J., Cheng, X., et al., 2020. HAK/KUP/KT family potassium transporter genes are involved in potassium deficiency and stress responses in tea plants (*Camellia sinensis* L.): expression and functional analysis. BMC Genomics 21 (1). Available at https://doi.org/10.1186/s12864-020-06948-6.

Zhao, D., Reddy, K.R., Kakani, V.G., Reddy, V.R., 2005. Nitrogen deficiency effects on plant growth, leaf photosynthesis, and hyperspectral reflectance properties of sorghum. European Journal of Agronomy 22 (4), 391–403. Available at https://doi.org/10.1016/j.eja.2004.06.005.

Zhu, Z., Zhang, H., Tian, H., Chai, G., Muhammad, R., Wang, Q., et al., 2025. Comprehensive analysis of the effects on photosynthesis and energy balance in tomato leaves under magnesium deficiency. Plant Physiology and Biochemistry 222, 109671. Available at https://doi.org/10.1016/j.plaphy.2025.109671.

CHAPTER

Mineral nutrients in plant antioxidant defense

14

Ana Cristina Ramírez-Anguiano[1], Sandra Fabiola Velasco-Ramírez[1], Ana Paulina Velasco-Ramírez[2], Alejandro Velasco-Ramírez[2], Rosalba Mireya Hernández-Herrera[3] and Milagros Melissa Flores Fonseca[4]

[1]Department of Chemistry, University Center for Exact Sciences and Engineering (CUCEI), University of Guadalajara, Guadalajara, Jalisco, Mexico, [2]Department of Agricultural Production, University Center for Biological and Agricultural Sciences (CUCBA), University of Guadalajara, Zapopan, Jalisco, Mexico, [3]Department of Botany and Zoology, University Center for Biological and Agricultural Sciences (CUCBA), University of Guadalajara, Zapopan, Jalisco, Mexico, [4]Medical Science, University of Colima, Colima, Jalisco, Mexico

Abbreviations

ROS	reactive oxygen species;
SOD	superoxide dismutase
CAT	catalase
APX	ascorbate peroxidase
GPX	glutathione peroxidases
GR	glutathione reductase
AsA–GSH	ascorbate–glutathione
N	nitrogen
P	phosphorus
K	potassium
Ca	calcium
Mg	magnesium
S	sulfur
Fe	iron
Zn	zinc
Cu	copper
Mn	manganese
Ni	nickel
Mo	molybdenum
B	boron
Cl^-	chloride
Se	selenium
$O_2^{\bullet -}$	superoxide anion
$OH^\bullet$	hydroxyl radical

Mineral Nutrients to Mitigate Abiotic Stress in Crops. DOI: https://doi.org/10.1016/B978-0-443-44712-9.00018-9

H_2O_2	hydrogen peroxide
$^1O_2$	singlet oxygen
DNA	deoxyribonucleic acid
NADPH	nicotinamide adenine dinucleotide phosphate (reduced form)
Cu/ZnSOD	copper/zinc superoxide dismutase
MnSOD	manganese superoxide dismutase
FeSOD	iron superoxide dismutase
AtGPX6	*Arabidopsis thaliana* glutathione peroxidase 6
GSSG	oxidized glutathione
GR2	glutathione reductase 2
ABA	abscisic acid
NO	nitric oxide
Na^+	sodium ion
Ca^{2+}	calcium ion
CAMTAs	calmodulin-binding transcription activators
CDPKs	calcium-dependent protein kinases
CO_2	carbon dioxide
RSS	reactive sulphur species
H_2S	*hydrogen sulfide*
MDA	malondialdehyde
SOS pathway	salt overly sensitive pathway
CBF/DREB	*CBF/DREB signaling pathways(CBF: C-repeat binding factor*
DREB	*dehydration-responsive element-binding)*
COR genes	cold-responsive genes
Si	silicon
POD	*peroxidase*
ATP	adenosine triphosphate
PSII	photosystem II
UV	ultraviolet radiation
Cd	cadmium
Pb	lead
Cr	chromium
As	arsenic
miRNAs	*microRNAs* (*micro ribonucleic acids*)
AMF	arbuscular mycorrhizal fungi
PGPR	plant growth-promoting rhizobacteria
ISR	induced systemic resistance
RNS	reactive nitrogen species
DW	dry weight

14.1 Introduction

Agricultural crops are exposed to adverse environmental conditions such as drought, salinity, air pollutant, cold temperatures, and heat. These conditions limit their growth and production quality (You and Chan, 2015; Husen, 2021a, 2021b, 2021c, 2022, 2023a). These abiotic stresses are characterized by excessive accumulation of reactive oxygen species (ROS), such as superoxide anion ($O_2{\bullet}^-$), hydroxyl radical (OH•), hydrogen peroxide (H_2O_2), and singlet oxygen ($^1O_2$). These ROS are mainly found in chloroplasts, mitochondria, and peroxisomes (Panda et al., 2024; Gill and Tuteja, 2010). An imbalance in ROS levels leads to lipid peroxidation, protein carbonylation, and DNA damage, compromising cellular integrity and resulting in programmed cell death. However, at controlled concentrations, ROS act as key signaling molecules in the activation of adaptive response pathways and the induction of antioxidant defenses (Nadarajah, 2020). Redox homeostasis is maintained by a highly coordinated network of both enzymatic and nonenzymatic antioxidants including superoxide dismutase (SOD), catalase (CAT), ascorbate peroxidase (APX), glutathione reductase (GR), and nonenzymatic antioxidants such as ascorbate, glutathione, tocopherols, as well as carotenoids (Hasanuzzaman et al., 2020). Different types of stress induce specific ROS "signatures," which act as molecular fingerprints in acclimation processes.

Within this defensive framework, mineral nutrition plays a central role. Under conditions of drought, salinity, or extreme temperatures, macronutrients such as nitrogen, potassium, calcium, and magnesium strengthen antioxidant capacity by modulating gene expression and the activity of key enzymes (SOD, peroxidases, CAT), promoting the reduction of intracellular ROS (Ashokan et al., 2024). Meanwhile, micronutrients such as iron, zinc, copper, and manganese act as cofactors of antioxidant enzyme systems and redox proteins, including cytochromes, plastocyanin, metalloenzymes, and "zinc finger proteins" (Gupta et al., 2024a). Their deficiency destabilizes redox homeostasis, increasing susceptibility to oxidative damage, whereas an adequate supply increases resilience to stress. Furthermore, the interaction between potassium and zinc alongside NADPH oxidases limits the excessive generation of ROS, adding an additional regulatory layer (Cakmak, 2006). Together, the interdependence between mineral nutrition and antioxidant defense forms a strategic axis for strengthening crop tolerance in conditions of climate change and increasing environmental stress.

Understanding how each nutrient contributes physiologically to maintaining redox homeostasis allows for the precise design of strategies for fertilization, biofortification, and soil management aimed at strengthening plant endogenous antioxidant systems. From an agronomic perspective, optimizing nutritional status is a cost-effective and environmentally viable method of mitigating yield losses, reducing reliance on agrochemicals, and enhancing the resilience of agricultural systems (Dutta et al., 2024). Therefore strengthening the connection between mineral nutrition and antioxidant defense integrates plant physiology with innovative management practices that ensure the productivity and sustainability of strategic crops, forming the basis for the material in this chapter.

14.2 Antioxidant systems in plants: an overview

Plant antioxidant systems are complex defense mechanisms that limit damage caused by ROS (Husen, 2025a). These systems contain both enzymatic components, such as SOD, catalase and ascorbate peroxidase, and nonenzymatic components, including glutathione, ascorbate, flavonoids, and proline (Dumanović et al., 2021; Rao et al., 2025). As well as detoxifying ROS, these networks perform a signaling function that aims to activate stress responses, enabling plants to tolerate hydrogen peroxide concentrations that would be lethal to animal cells (Dumanović et al., 2021; Cannea and Padiglia, 2025). These systems play a particularly important role under conditions of drought, salinity, and extreme temperatures, supporting plant molecular resilience (Ansari et al., 2024). Emerging technologies such as gene editing and artificial intelligence are improving our understanding of, and ability to manipulate, these antioxidant networks.

14.2.1 Enzymatic antioxidants (SOD, CAT, APX, GPX, GR) as the first line of defense against oxidative stress

The enzymatic antioxidants SOD, CAT, APX, GPX, and GR constitute the major line of defense against oxidative stress, as they neutralize ROS in a coordinated manner. SOD converts superoxide into hydrogen peroxide; CAT and GPX detoxify H_2O_2 into water (the latter using reduced glutathione as a cofactor); APX uses ascorbate as an electron donor to eliminate H_2O_2, and GR regenerates reduced glutathione, thus sustaining the antioxidant cycle (Rajput et al., 2021; Sahoo and Tiwari, 2022). Multiple studies agree that these enzyme systems are essential for preventing cell damage in plants and other organisms, emphasizing their significance in redox homeostasis (Gałecka et al., 2008).

14.2.1.1 Superoxide dismutases in plants

SOD are essential metalloenzymes that constitute the front-line of defense against oxidative stress by dismuting the superoxide anion ($O_2{\bullet}^-$) into O_2 and H_2O_2. In plants, there are three main types (Cu/ZnSOD, MnSOD, and FeSOD), distributed in chloroplasts, mitochondria, peroxisomes, and cytosol, allowing fine spatial control of ROS generated both during basal metabolism and under biotic and abiotic stresses. This subcellular organization and high catalytic efficiency place SODs at the center of maintaining redox homeostasis and protecting the structural and functional integrity of plant cells (Stephenie et al., 2020).

14.2.1.1.1 Variation in SOD expression among plant species under oxidative stress

SOD expression varies widely between species, subfamilies (Cu/ZnSOD, FeSOD, MnSOD), and cellular compartments, with specific regulatory patterns in response to different types of stress. Comparisons between monocotyledons and dicotyledons reveal distinctive strategies under conditions of drought or heavy metal exposure, while extremophile plants and tolerant cultivars (e.g., frost-resistant potatoes) show higher SOD activities than sensitive genotypes. In many cases, increases of two- to threefold in expression and activity under stress compared to control conditions have been reported. This reflects a highly context-dependent response and the key role of SODs in tolerance and adaptation to challenging environments (Zameer et al., 2022).

14.2.1.2 *Catalase as a key antioxidant enzyme in plants*

Catalase (CAT) is a central antioxidant enzyme that breaks down H_2O_2 into water and oxygen, protecting cells from oxidative damage and participating in development and stress response processes. Genomic analyses in 29 species of the "green line" show that CAT genes are grouped into three phylogenetic clades and have evolved through horizontal transfer (Pan et al., 2022). Furthermore, their function exceeds peroxisomal metabolism, interacting with multiple proteins and cellular pathways (Baker et al., 2023). In most species, between 6 and 10 CAT genes per genome have been identified, whose modulated activity improves tolerance to salinity, cold, heat, and heavy metals (Zhang et al., 2022; Ghorbel et al., 2023).

14.2.1.2.1 Variation in catalase expression under oxidative stress

CAT expression is highly species- and gene-specific. In *Arabidopsis thaliana*, CAT1, CAT2, and CAT3 show different responses: CAT1 acts in the elimination of H_2O_2 under various stresses, while CAT2 and CAT3 function as scavengers under specific conditions (Du et al., 2008). In tobacco, seven CAT genes show contrasting patterns of overexpression or repression depending on the type of stress (cold, drought) (Liu et al., 2023), and in potato and tomato, constitutive and inducible isoforms coexist (Jbir Koubaa et al., 2023). These data reveal a complex and finely tuned regulation of CAT, adapted to the physiology and ecology of each species.

14.2.1.3 *Ascorbate peroxidase as the center piece of the plant antioxidant system*

Ascorbate peroxidase (APX) detoxifies H_2O_2 by converting it into water using ascorbate as an electron donor. It is located in chloroplasts, mitochondria, peroxisomes, and cytosol, providing distributed redox surveillance. A large body of evidence documents its central role in stress response and redox state regulation. APX isoforms are differentially and flexibly regulated in response to environmental changes and also participate in signaling and development, integrating antioxidant defense and growth regulation. Its expression is dynamically modulated by water, salt, and heat stresses, excess light, and pollutants, as well as by developmental processes (differentiation, maturation, senescence), highlighting its great functional versatility (Hoque et al., 2024).

14.2.1.3.1 Variation in APX expression between species under oxidative stress

APX expression differs between species and between subcellular isoforms. In spinach, strong induction of cytosolic APX is observed under high luminous intensity and treatment with methyl viologen, while in *Vigna unguiculata* different cultivars show contrasting expression patterns under drought, reflecting genotypic differences in antioxidant capacity. In model species such as Arabidopsis, rice, and maize, APXs share conserved domains but exhibit species-specific responses to stress, determined by their location (chloroplasts, mitochondria, peroxisomes, cytosol) and by particular signals (intense light, desiccation, salinity, xenobiotics) and stages of development. This reveals a highly sophisticated defense system in which APX acts as a central node integrating environmental and developmental signals (Li, 2023).

14.2.1.4 Plant glutathione peroxidases and their role in redox regulation and stress response

Plant glutathione peroxidases (GPX) are essential enzymes in redox regulation and stress response. They catalyze the reduction of H_2O_2 and hydroperoxides using thioredoxin or glutathione as electron donors (Bela et al., 2022) and, unlike their animal counterparts, are non-selenoprotein monomeric proteins. In "*Arabidopsis thaliana*", seven isoforms have been described. These are located in chloroplasts, mitochondria, and cytosol (Rodriguez Milla et al., 2003) and are involved in stress tolerance (saline, oxidative, heavy metal), developmental processes, and redox signaling, where they act as nodes connecting ROS metabolism and signaling pathways.

14.2.1.4.1 Variation in GPX expression between species under oxidative stress

GPX genes exhibit species-specific and context-dependent expression patterns. Ozyigit et al. (2016) identified GPX in 18 species, which have diverse regulatory profiles (Ozyigit et al., 2016). In *Phoenix dactylifera*, five GPX genes respond differentially to salinity and oxidative stress (Jana and Yaish, 2020), whereas in *Thellungiella salsuginea*, the eight GPX genes are coordinately regulated in response to specific stresses (Gao et al., 2014). Certain isoforms, such as AtGPX6, show particularly intense responses in Arabidopsis, integrating into multiple signaling pathways (Rodriguez Milla et al., 2003). These findings suggest a finely tuned system that varies depending on the species, tissue, and type of stress.

14.2.1.5 Glutathione reductase as a key antioxidant enzyme in the response to oxidative stress

Glutathione reductase (GR) catalyzes the reduction of oxidized glutathione (GSSG) to its reduced form (GSH), maintaining cellular redox potential and sustaining the ascorbate–glutathione (AsA–GSH) pathway, which is crucial for ROS detoxification (Dorion et al., 2021). Studies in rice and *Arabidopsis* show high conservation of GR genes, and more recent work confirms their involvement in the response to drought, heat, salinity, and arsenic. The increased (GR) activity under stress is interpreted as an adaptive mechanism that preserves cellular integrity in adverse environments (Harshavardhan et al., 2017).

14.2.1.5.1 Variation in glutathione reductase expression among species under oxidative stress

GR expression is species- and cultivar-specific. In *Vigna unguiculata*, drought induces diverse expression patterns in tolerant and sensitive cultivars, evidencing its contribution to the molecular basis of water stress tolerance (Contour-Ansel et al., 2006). In tobacco, overexpression of Arabidopsis GR2 increases GR activity approximately sevenfold and improves tolerance to oxidative stress (Wang et al., 2019). In wheat, seven GR genes have been identified with differential expression under drought, heat, salinity, and arsenic stress (Madhu et al., 2022), and at least two main isoforms (cytoplasmic and chloroplastic) with different sensitivities to stress, with cytosolic isoforms being the most inducible (Tahmasebi et al., 2012). These variations reflect complex molecular adaptation mechanisms in which GR plays a central role in the antioxidant capacity and stress tolerance of plants.

14.2.2 Nonenzymatic antioxidants (ascorbate, glutathione, tocopherols, carotenoids)

Nonenzymatic antioxidants are an essential defense against ROS, protecting macromolecules and contributing to stress tolerance. The most relevant are ascorbate, glutathione, flavonoids, α-tocopherol, and carotenoids. They are all involved in free radical scavenging, membrane stabilization, and stress response modulation (Abdulfatah, 2022). Several studies agree that these compounds interact with key physicochemical processes, vital for plant survival under temperature, salinity, and water availability variation (Ramadoss et al., 2021).

14.2.2.1 Ascorbate as a multifunctional metabolite in plants

Ascorbate is a key multifunctional metabolite whose primary function is as an antioxidant, taking part in metabolism, photosynthesis, and stress response (Tóth, 2023). It accumulates at high levels (1–5 mM in leaves and up to 25 mM in chloroplasts), reflecting its adaptive importance. In addition to eliminating ROS, ascorbate acts as an enzyme cofactor, a redox modulator, and a regulator of processes such as growth, photosynthesis, and stress resistance (Zechmann, 2018).

14.2.2.1.1 Variation in ascorbate expression and metabolism under oxidative stress

The content and metabolism of ascorbate vary between species and tissues, with differences in the number of ascorbate oxidase (AAO) genes and their regulation under stress 3 genes in "*Arabidopsis*" and up to 7 in "Glycine max" (Batth et al., 2017). In rice, for example, some AAO genes are specifically expressed in roots or sprouts and respond dynamically to drought and salinity. This variability, together with their role as antioxidants, cofactors, and redox regulators (Xiao et al., 2021), results in species- and stress-dependent changes in ascorbate levels.

14.2.2.2 ***Glutathione: an essential tripeptide for defense and metabolic regulation***

Glutathione is an essential tripeptide for plant survival, considered a central node of defense and metabolic regulation. Thanks to its thiol group, it detoxifies ROS and toxic compounds, facilitating adaptation to stresses caused by extreme temperatures, heavy metals, and pathogens. It is synthesized by the enzymes glutamate-cysteine ligase and glutathione synthase (Galant et al., 2011).

14.2.2.2.1 Variation in glutathione response under oxidative stress

The glutathione response to oxidative stress is dynamic and species-specific. It involves changes in content, GSH accumulation rate, synthesis/recycling enzyme activity, and subcellular distribution (cytoplasm, chloroplasts, mitochondria) (Gasperl et al., 2022). Although systematic comparisons between species are lacking, evidence coincides in indicating that glutathione is involved in the chelation of toxins, antioxidant defense and cellular signaling (Koh et al., 2021).

14.2.2.3 ***Tocopherols: fat-soluble antioxidants in membrane protection***

Tocopherols, synthesized exclusively by photosynthetic organisms, are fat-soluble antioxidants that play a key role in membrane protection and cell signaling (Fritsche et al., 2017). Studies of vte mutants and transgenic plants reveal their protective role in photosynthetic membranes, preventing lipid peroxidation, and possible involvement in signaling pathways (Falk and Munné-Bosch, 2010). α-Tocopherol, synthesized in plastids from homogentisic acid, is the predominant form. It contributes to the elimination of ROS, the maintenance of membrane integrity, and the regulation of growth and stress responses.

14.2.2.3.1 Variation in tocopherol levels under oxidative stress

Total levels and the profile of tocopherol isoforms vary between species and depending on the type and intensity of stress. Studies have shown that α-tocopherol response varies according to species sensitivity and that α- and γ-tocopherol exhibit different response patterns in tobacco (Abbasi et al., 2007). Transgenic *Brassica juncea* plants with higher α-tocopherol show superior stress tolerance (Kumar et al., 2013), while mutations in tocopherol biosynthesis can induce oxidative stress in Arabidopsis (Semchuk et al., 2009), thereby highlighting its essential role.

14.2.2.4 Carotenoids: multifunctional isoprenoid pigments

Carotenoids are C_{40} isoprenoid pigments essential for photosynthesis, photoprotection, photomorphogenesis, hormone synthesis (ABA, strigolactones), and visual attraction for pollination and seed dispersal. Their biosynthetic pathway is widely characterized, and their production is regulated by developmental and environmental signals, reflecting sophisticated metabolic control (Cazzonelli, 2011).

14.2.2.4.1 Variation in carotenoid accumulation and expression under oxidative stress

The accumulation and expression of carotenogenic genes vary widely between species and stress conditions, driven by evolutionary changes in enzymes, mutations, and transcriptional regulation (Luan et al., 2020). In *Nannochloropsis oculata*, oxidative stress can induce up to a sixfold increase in the expression of carotenoid biosynthesis genes (Zainal Abidin et al., 2021). Furthermore, carotenoid oxidation generates bioactive apocarotenoids that act as acclimation signals, and factors such as cultivar, location, and stress type modulate their accumulation (Othman et al., 2014).

14.2.2.5 The ascorbate–glutathione cycle as a central antioxidant defense mechanism

This cycle plays a pivotal role in antioxidant plant defense by detoxifying ROS in different compartments, particularly in chloroplasts and mitochondria. Four enzymes coordinate the use of ascorbate and glutathione to neutralize ROS. These enzymes also perform "moonlighting" functions in plant signaling (Foyer and Kunert, 2024), and their activation is associated with enhanced stress tolerance.

14.2.2.5.1 Variation in the AsA–GSH cycle between species under oxidative stress

The AsA–GSH network exhibits context- and stress-dependent variations in gene number, subcellular localization, and regulation. In soybean (*Glycine max*), 24 genes related to this cycle were identified, which is more than in other legumes, due to gene duplications and possible functional specialization (Costa et al., 2021). *Euphorbiaceae* species exhibit differential expression patterns during drought (Jardim-Messeder et al., 2023). These differences include cytosolic, chloroplastic, and mitochondrial isoforms, stage-specific expression, and differential activation depending on the type of stress. This confirms that the AsA–GSH cycle is a conserved yet highly modifiable core component of the plant antioxidant response (Fig. 14.1

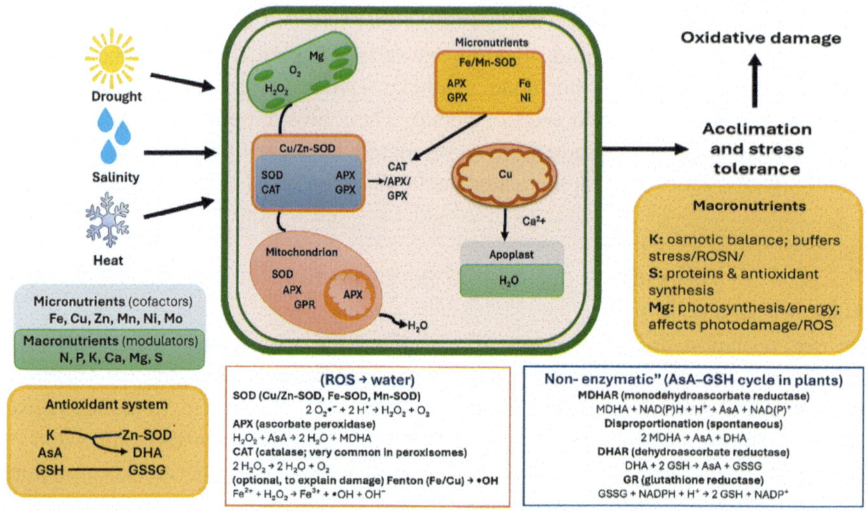

FIGURE 14.1

Integrative schematic of the role of macro and micronutrients in antioxidant defense and stress tolerance in plant cells. Abiotic stresses (drought, salinity, and temperature) increase ROS generation (e.g., $O_2^{\bullet -}$ and H_2O_2) in compartments such as the chloroplast and mitochondrion, as well as in the apoplast. Micronutrients (Fe, Cu, Zn, Mn, Ni, Mo) act as cofactors of antioxidant enzymes (Cu/ZnSOD; Fe/MnSOD; APX; GPX), facilitating the conversion of ROS into less reactive products, ultimately to water (ROS → H_2O) via SOD, APX/GPX, and CAT. In parallel, detoxification is sustained by the AsA–GSH redox cycle (AsA↔DHA; GSH↔GSSG), which regenerates non-enzymatic antioxidants and maintains cellular reducing capacity. Macronutrients (N, P, K, Ca, Mg, S) modulate these processes: K contributes to osmotic balance and buffering of stress/ROS, N/S support the synthesis of proteins and antioxidant compounds, and Mg is linked to photosynthesis and energy metabolism, influencing photodamage and ROS production. The balance between these pathways determines the physiological outcome: acclimation and stress tolerance versus oxidative damage.

14.3 Macronutrients and micronutrients in plant antioxidant systems

Macronutrients and micronutrients are essential for activating components of plant antioxidant systems, as they activate enzymes that regulate reactive oxygen species (ROS) and reduce oxidative stress. Micronutrients such as Fe, Zn, Cu, Mn, and Ni work as vital cofactors for antioxidant enzymes, including SODs (Tavanti et al., 2021; Gupta et al., 2024a), whereas macronutrients such as N, P, and K provide structural and metabolic elements necessary for immune signaling and the synthesis of defensive compounds (Dutta et al., 2024). Since the antioxidant system functions as a coordinated network deficiencies in micronutrients compromise ROS management and cell protection. However, an adequate fertilization can improve plants' ability to mitigate oxidative stress and increase their tolerance to adverse conditions.

14.3.1 Role of macronutrients in antioxidant defense

Macronutrients such as N, P, K, Ca, Mg, and S contribute to antioxidant defense by modulating the activity and expression of antioxidant enzymes, sustaining energy metabolism and the biosynthesis of defensive metabolites, and participating in redox homeostasis (Dutta et al., 2024). N regulates levels of enzymes associated with systemic resistance, K increases polyphenolic compounds involved in plant immunity, and P plays a more ambiguous role, even increasing susceptibility to pathogens depending on the context (Tripathi et al., 2022). While the functional link is evident, the specific molecular mechanisms connecting each macronutrient to particular antioxidant pathways remain unclear.

14.3.1.1 Macronutrients and antioxidant defense in organisms (comparative view)

Macronutrients play a decisive role in plant antioxidant defense by modulating the generation and control of reactive oxygen species (ROS) and sustaining the underlying metabolic machinery. Nitrogen, phosphorus, and potassium significantly influence the synthesis of defensive compounds and immune signaling by providing essential structural and energetic building blocks for these pathways (Dutta et al., 2024). When macronutrient supply is deficient, ROS balance is disrupted, and both cellular integrity and stress responsiveness are compromised, increasing plant vulnerability under adverse conditions (Hasanuzzaman et al., 2020).

14.3.1.1.1 Nitrogen

Nitrogen (N) shows responses that vary greatly depending on the species, the source (nitrate vs. ammonium), and the type of stress. Different forms of inorganic N modulate nitric oxide (NO) levels and ROS homeostasis and associated signaling networks (da Silva et al., 2024; Allagulova et al., 2023). In *Arabidopsis thaliana*, N deficiency triggers metabolic adaptations that prioritize defense over growth (Cadena-Zamudio et al., 2023). In lettuce, salt stress reduces N, P, and K and increases Na^+, altering nutrition and antioxidant response (Sardar et al., 2023).

14.3.1.1.2 Phosphorus

Phosphorus (P) deficiency consistently induces oxidative stress, characterized by increased ROS, lipid peroxidation, and activation of antioxidant enzymes (Hernández and Munné-Bosch, 2015; Oubenali et al., 2022). In mulberry, SOD activity and H_2O_2 levels increase, while in chickpea, a reduction in photosynthesis, an increase in malondialdehyde, and greater antioxidant activity are observed

(Oubenali et al., 2022). The magnitude of these responses varies between species and genotypes, reflecting a strong influence of genetic diversity (Zeng et al., 2025).

14.3.1.1.3 Potassium

Potassium (K) status decisively modulates the response to oxidative stress and interspecific tolerance. Adequate K supply reduces ROS production in barley (Tavakol et al., 2022) and improves the mitigation of oxidative damage in salt-tolerant tomato genotypes (Amjad et al., 2016). In the halophyte *Cakile maritima*, K deficiency strongly activates the antioxidant system, reflecting specific adaptations to saline and nutrient-poor environments. Although its importance is well established, the fine molecular mechanisms of K in the control of oxidative stress are not yet fully defined (Houmani et al., 2022).

14.3.1.1.4 Calcium

Ca^{2+} primarily acts as an oxidative stress signal rather than a direct antioxidant, and its response patterns are species- and context-dependent. In tobacco, H_2O_2 induces transient Ca^{2+} signals with characteristic refractory periods, whereas a biphasic elevation of Ca^{2+} was observed in different tissues of *Arabidopsis* under oxidative stress. Sensor proteins such as calmodulins/CAMTAs and CDPKs translate these "calcium signatures" into changes in gene expression associated with stress response (Gupta et al., 2023).

14.3.1.1.5 Magnesium

Magnesium (Mg) deficiency systematically causes oxidative stress, resulting in an increase in ROS and antioxidant machinery alterations. In barley, an increase of up to 55% in H_2O_2 concentrations has been observed (Tränkner et al., 2016), while in cotton, changes in ROS-related isoenzyme patterns were reported (Özgür Uzilday et al., 2017). Given its central role in photosynthesis, Mg deficiency mainly induces oxidative photostress caused by an imbalance between energy capture and CO_2 assimilation (Jamali Jaghdani et al., 2021). This response is widely preserved among species (wheat, barley, cotton, grapevine), with particular physiological nuances, though.

14.3.1.1.6 Sulfur

Sulfur (S) is essential for forming of sulfur-containing defensive compounds and in the generation of reactive sulfur species (RSS), such as thiols, H_2S, and glutathione, which participate in ROS detoxification and redox signaling (Ul Haq and Ijaz, 2019). S deprivation in *Chlorella sorokiniana* causes immediate increases in H_2O_2 and changes in antioxidant enzyme activity (Salbitani et al., 2015). Secondary S metabolism produces sulfated compounds that modulate responses ranging from chloroplastic signaling to stomatal opening (Chan et al., 2019). Their defensive efficacy is closely dependent on S supply, absorption, and assimilation (Capaldi et al., 2015), highlighting their central role in plant adaptation to adverse environments.

14.3.2 Micronutrients in antioxidant defense: key cofactors and modulators of oxidative stress

Micronutrients are essential components of the antioxidant defense system, acting both as cofactors of key enzymes (SOD, CAT, APX) and for their ability to directly modulate the generation and detoxification of ROS. Iron (Fe), zinc (Zn), copper (Cu), manganese (Mn), and nickel (Ni), among others, activate metalloenzymes that protect membranes, DNA, and proteins from oxidative damage (Gupta et al., 2024a). As antioxidant defenses operate as a coordinated system, a deficiency in a single micronutrient compromises the entire protective response. However, high doses of some of these elements could have pro-oxidant effects, highlighting the need for a careful balance in their availability and supplementation strategies.

14.3.2.1 *Interspecific variability in micronutrient availability and its role in response to oxidative stress*

Accumulation, availability, and function of micronutrients under oxidative stress vary significantly between plant species. Fe, Zn, Cu, Mn, and Ni are essential cofactors of enzymes such as SOD and CAT, and fertilization with micronutrients can activate these enzymes and improve stress tolerance (Tavanti et al., 2021; Gupta et al., 2024a). Some elements play particularly nuanced roles: Mn is essential for photosynthetic processes while Se contributes to protection against free radicals (Kanth et al., 2020). Furthermore, trace metal concentrations range from deficient to toxic, resulting in responses heavily dependent on physiological, environmental, and genetic context (Andresen et al., 2018).

14.3.2.1.1 Iron

Iron (Fe) plays a dual role in oxidative metabolism: it is an essential cofactor; however, both deficiency and excess can induce oxidative stress. In *Nicotiana plumbaginifolia*, excess Fe reduced photosynthesis by ~40% within a few hours and doubled catalase activity, indicating redox overload (Kampfenkel et al., 1995). Plants have evolved distinct acquisition strategies (Strategies I and II) based on rhizosphere acidification, chelator secretion, and specific transport systems (Zheng, 2010). In *Arabidopsis thaliana*, species-specific homeostasis mechanisms have been described, where genetic variations determine tolerance to Fe deficiency or toxicity (Briat et al., 2010). However, comprehensive comparative studies between species are still lacking.

14.3.2.1.2 Copper

Although essential, excess Copper (Cu) can become highly toxic, with responses that are strongly species- and dose-dependent. Depending on the concentration and context, plants can exhibit anything from antioxidant adaptations to severe cellular damage (Xu et al., 2024). Specific responses to Cu in *Brassica rapa* have been observed compared to other metals (Moreira et al., 2015), whereas concentrations of 10 μM are toxic (García et al., 1999). In *Arabidopsis*, low concentrations of Cu (approximately 5 μM) can stimulate growth, but levels of 20–60 μM induce marked oxidative stress (Wang et al., 2023). Excess Cu inhibits growth and increases ROS formation in rice (Thounaojam et al., 2012).

14.3.2.1.3 Zinc

Zinc (Zn) is an essential micronutrient whose toxicity and tolerance vary widely between species. On this basis, plants were classified as sensitive, resistant, tolerant, and hypertolerant to metals (Lin and Aarts, 2012). *Brassica juncea* shows greater tolerance to Zn stress than *B. napus*, with clear differences in ROS and RNS metabolism (Feigl et al., 2015). In aquatic macrophytes such as *Lemna gibba*, *L. minor,* and *Spirodela polyrrhiza*, different strategies have been described for enzyme activation, cell damage prevention, Zn accumulation/compartmentalization, and protection of the photosynthetic machinery (Uruç Parlak and Demirezen Yilmaz, 2012).

14.3.2.1.4 Manganese

Manganese (Mn) is essential for photosystem II and stress response, but its excessive accumulation induces toxicity and oxidative damage. Different species show contrasting strategies: *Alhagi camelthorn* accumulates Mn in leaf vacuoles, while *Chenopodium ambrosioides* and *Datura stramonium* respond mainly through increased antioxidant activity (Boojar and Goodarzi, 2008). At the population level, *Populus cathayana* from humid climates was more susceptible to excess Mn than populations from dry climates, accumulating more of the metal and exhibiting greater oxidative damage (Lei et al., 2007). Plants use a combination of absorption regulation, subcellular compartmentalization, and antioxidant system activation to mitigate Mn toxicity (Li et al., 2019).

14.3.2.1.5 Molybdenum

Molybdenum (Mo), a cofactor of several oxidoreductases, can reduce oxidative stress in a species-dependent manner. In rice, its application improves cadmium stress tolerance (≈58%–73%) and stimulates the activity of antioxidant enzymes (Imran et al., 2020). In wheat, Mo increases antioxidant activity and resistance to drought and cold (Wu et al., 2017). In soybeans, differential accumulation patterns between roots and leaves under excess Mo have been described (Xu et al., 2018). In *Arabidopsis*, wide natural differences in Mo content controlled by a mitochondrial transporter have been identified (Baxter et al., 2008).

14.3.2.1.6 Boron

Boron (B) is essential for cell wall structure and sugar transport, but both its deficiency and excess can induce oxidative stress. In tomatoes, B toxicity increases ascorbate levels, reflecting a compensatory antioxidant response (Cervilla et al., 2007). In pistachios, moderate amounts of B reduce markers of oxidative stress, while high doses increase them (Tavallali et al., 2018). In mulberry, it increases ROS and antioxidant enzyme activity (Tewari et al., 2010), while in olive, different B compounds trigger species- and form-dependent antioxidant expression patterns (Çelik and Kaya, 2024).

14.3.2.1.7 Chloride

Traditionally considered a "beneficial" micronutrient, chloride (Cl^-) plays contrasting roles in the redox context. In rice, it can contribute to the generation of ROS under saline stress, resulting in a net pro oxidant effect (Khare et al., 2015). In canola, however, it displays a positive function, enhancing antioxidant defense (Sun et al., 2024). Chloride channels are fundamental to ionic homeostasis and exhibit species-specific sequestration strategies (Liu et al., 2023). Glycophytes and halophytes widely differ in their management of Cl^-: the former tend to exclude it, while the latter can use Na^+ and Cl^- as osmoprotectants (Chakraborty et al., 2021).

14.3.2.1.8 Nickel

At very low concentrations, nickel (Ni) is essential (e.g., for urease), but it becomes toxic in the short term. Furthermore, its accumulation varies by more than three orders of magnitude between species. This variation enables plants to be categorized as nonaccumulators, hemiaccumulators, hyperaccumulators, or hypernickelophores (Boyd and Jaffré, 2009). Overall tolerance is higher in dicotyledons than in monocotyledons, with increases in H_2O_2 and other markers of oxidative stress, being proportional to dose and exposure time (Kováčik and Vydra, 2024). Different species depend to a greater extent on enzymatic or nonenzymatic systems. In order to cope with Ni-induced stress, revealing the highly specific nature of this trait in each species (Amari et al., 2020).

14.3.2.1.9 Selenium

Depending on the species and concentration, selenium (Se) can act as a beneficial element or a stress factor. While most plants accumulate less than 100 mg of Se per kg of dry matter, a few hyperaccumulators can reach concentrations of 1000–15,000 mg per kg (White, 2015). Comparisons between *Arabidopsis thaliana* (sensitive) and *Brassica juncea* (tolerant) revealed differences in stomatal density, oxidative responses, and nitrosative modifications (Molnár et al., 2014). Within the genus *Astragalus*, sensitive species such as *A. membranaceus* contrast with hyperaccumulators such as *A. bisulcatus*, which exhibit advanced tolerance mechanisms (Kolbert et al., 2018). In summary, Se modulates responses to oxidative stress remarkably in a species- and dose-dependent manner (Hasanuzzaman et al., 2020) (Table 14.1

Table 14.1 Integrated role of macro- and micronutrients in plant antioxidant systems and reactive oxygen species (ROS) homeostasis.

Nutrients	Type	Principal role in antioxidant/redox defense	Deficit (trend)	Excess/ Ambivalence	Comparative evidence (examples)	Key references
N	Macro	Modulates ROS–RNS homeostasis (NO) and reprograms growth vs defense; strong dependence on N source (NO_3^- vs NH_4^+)	ROS/NO imbalance; metabolic shifts toward defense (species-/stress-dependent)	Highly context-dependent (species, N form, stress)	Arabidopsis: deficiency prioritizes defense over growth; lettuce under salinity shows ↓N (and P, K) and altered antioxidant responses	Dutta et al. (2024), Allagulova et al. (2023), da Silva et al. (2024), Cadena-Zamudio et al. (2023), Sardar et al. (2023)
P	Macro	Supports energy metabolism/ photosynthesis (capacity to buffer ROS); deficiency activates antioxidant responses	↑ROS, ↑lipid peroxidation; compensatory activation of antioxidant enzymes	May increase pathogen susceptibility depending on context	Mulberry: ↑SOD and ↑H_2O_2; chickpea: ↓ photosynthesis, ↑ MDA, ↑antioxidant activity; magnitude varies among genotypes	Hernández and Munné-Bosch (2015), Oubenali et al. (2022), Tripathi et al. (2022), Zeng et al. (2025)
K	Macro	Decreases ROS production; promotes polyphenols and immunity; shapes interspecific tolerance	In adapted species, deficiency can strongly activate antioxidant systems	Fine molecular mechanisms not fully resolved	Barley: adequate K ↓ ROS; salt-tolerant tomato genotypes show improved oxidative damage mitigation; Cakile maritima: deficiency strongly activates antioxidants (adaptation)	Dutta et al. (2024), Tavakol et al. (2022), Amjad et al. (2016), Houmani et al. (2022)

Ca^{2+}	Macro	Primarily an oxidative-stress signal (Ca^{2+} "signatures" → gene regulation) rather than a direct antioxidant	Less coordinated stress responses if Ca^{2+} signaling is impaired	Strong tissue/species dependence; temporal patterns (transient peaks)	Tobacco: H_2O_2 elicits transient Ca^{2+} signals; Arabidopsis: biphasic Ca^{2+} elevation; CAMTAs/CDPKs decode signatures	Gupta et al. (2023)
Mg	Macro	Central to photosynthesis, deficiency drives oxidative photostress (energy capture vs CO_2 assimilation imbalance)	↑ROS (e.g., ↑ H_2O_2), altered "antioxidant machinery"	Conserved across species with physiological nuances	Barley: H_2O_2 increases up to ~55%; cotton: changes in ROS-related isoenzyme patterns; similar trends across species	Tränkner et al. (2016), Özgür Uzilday et al. (2017), Jamali Jaghdani et al. (2021)
S	Macro	Basis for GSH and RSS (thiols, H_2S): ROS detoxification + redox signaling; sulfated secondary metabolites	Deprivation: rapid ↑H_2O_2; changes in antioxidant enzyme activity	Defensive efficacy depends on supply/uptake/assimilation	Chlorella: immediate ↑H_2O_2 under S deprivation; sulfated metabolism modulates chloroplastic signals and stomatal regulation	Ul Haq and Ijaz (2019), Salbitani et al. (2015), Chan et al. (2019), Capaldi et al. (2015)
Fe	Micro	Key cofactor; dual role (essential yet can induce oxidative stress when unbalanced)	Limits metalloenzymes → weaker ROS control	Excess: redox overload (↓ photosynthesis, ↑CAT)	Nicotiana: excess Fe ↓photosynthesis ~40% within hours and doubles CAT; Strategy I/II acquisition; genetic variation in Arabidopsis	Gupta et al. (2024b), Kampfenkel et al. (1995), Zheng (2010), Briat et al. (2010)

(Continued)

Table 14.1 Integrated role of macro- and micronutrients in plant antioxidant systems and reactive oxygen species (ROS) homeostasis. (*Continued*)

Nutrients	Type	Principal role in antioxidant/redox defense	Deficit (trend)	Excess/ Ambivalence	Comparative evidence (examples)	Key references
Cu	Micro	Essential; narrow dose–response window	Compromised metalloenzymes	Excess: highly toxic; ↑ROS, growth inhibition	Arabidopsis: ~5 μM may stimulate growth, 20–60 μM induces oxidative stress; rice: excess increases ROS; thresholds vary by species	Xu et al. (2024), A. García et al. (1999), Wang et al. (2023), Thounaojam et al. (2012), Moreira et al. (2015)
Zn	Micro	Essential; tolerance/ toxicity highly variable; distinct accumulation/ compartmentation strategies	Weakens coordinated defense	Excess: oxidative stress; tolerance classes (sensitive→ hypertolerant)	*B. juncea* more tolerant than *B. napus*; aquatic macrophytes show divergent strategies (enzyme activation, damage prevention, compartmentation)	Lin and Aarts (2012), Feigl et al. (2015), Uruç Parlak and Demirezen Yilmaz (2012)
Mn	Micro	Essential for PSII and stress responses; excess induces oxidative damage	Compromised photosynthesis/ defenses	Excess: toxicity; mitigation via compartmentation + antioxidants	Contrasting strategies: Alhagi vacuolar sequestration vs Chenopodium/Datura antioxidant upregulation; Populus populations differ in susceptibility	Kanth Verma and Gour (2020), Boojar and Goodarzi (2008), Lei et al. (2007), Li et al. (2019)

Mo	Micro	Cofactor of oxidoreductases; can reduce oxidative stress (species-dependent)	Limits Mo-dependent enzymes	Excess: tissue partitioning differs (root vs leaf)	Rice: improves Cd tolerance (~58%–73%) and boosts antioxidant enzymes; wheat: increases drought/cold resistance; Arabidopsis: natural variation linked to a mitochondrial transporter	Imran et al. (2020), Wu et al. (2017); Xu et al. (2018), Baxter et al. (2008)
B	Micro	Cell wall structure and sugar transport; redox-sensitive under deficiency/excess	Deficiency induces oxidative stress	Excess also induces oxidative stress; depends on dose/form	Tomato: toxicity increases ascorbate; pistachio: moderate B reduces oxidative markers, high B increases them; mulberry: ↑ROS/enzymes; olive: responses depend on B compound	Cervilla et al. (2007), Tavallali et al. (2018), Tewari et al. (2010), Çelik and Kaya (2024)
Cl^-	Micro (beneficial)	Ionic homeostasis; contrasting redox roles (pro- vs antioxidant depending on species/context)	(not detailed)	Under salinity can promote ROS (rice) or enhance defense (canola)	Rice: net pro-oxidant effect under salinity; canola: positive effect on antioxidant defense; glycophytes exclude vs halophytes use Na^+/Cl^- as osmoprotectants	Khare et al. (2015), Sun et al. (2024), Liu et al. (2023), Chakraborty et al. (2021)
Ni	Micro	Essential at ultra-low levels (urease) but rapidly toxic; very large interspecific accumulation range	Affects Ni-dependent functions	Excess: ↑H_2O_2 and oxidative damage (dose/time); dicots often show higher overall tolerance	>3 orders of magnitude variation; categories (non-/hemi-/hyperaccumulators); enzymatic vs non-enzymatic defense reliance differs by species	Boyd and Jaffré (2009), Kováčik and Vydra (2024), Amari et al. (2020)

(Continued)

Table 14.1 Integrated role of macro- and micronutrients in plant antioxidant systems and reactive oxygen species (ROS) homeostasis. (*Continued*)

Nutrients	Type	Principal role in antioxidant/redox defense	Deficit (trend)	Excess/ Ambivalence	Comparative evidence (examples)	Key references
Se	Micro	Beneficial or stress-inducing depending on species/dose; modulates oxidative/nitrosative responses; hyperaccumulation in few taxa	(not detailed)	Excess may stress; hyperaccumulators tolerate very high levels	Most plants <100 mg/kg DW; hyperaccumulators 1000–15,000 mg/kg; Arabidopsis (sensitive) vs B. juncea (tolerant) differ in stomata and redox; Astragalus sensitive vs hyperaccumulator	White (2015), Molnár et al. (2014); Kolbert et al. (2018), Hasanuzzaman et al. (2020)
Integrative micronutrient framework	Micro	Micronutrients act as cofactors of SOD/CAT/APX; antioxidant defense is a coordinated network—one deficiency can compromise system-wide protection; adequate fertilization improves tolerance	Single-element deficiency → impaired ROS control and cell protection	High doses → potential pro-oxidant effects (critical balance)	Enzyme activation and stress tolerance increase with micronutrient supply; outcomes depend on physiology, environment, and genetics	Gupta et al. (2024a), Tavanti et al. (2021), Andresen et al. (2018)

14.4 Minerals and plant response to abiotic stress

Mineral nutrients play a crucial role in enabling plants to cope with adverse environmental conditions. They are also fundamental metabolic pathways, contributing to the modulation of stress signaling, sustaining antioxidant systems, and maintaining the structural integrity of membranes and organelles. The availability and mobility of minerals such as K, Ca, Mg, Fe, Zn, Mn, Si, and B directly influence the efficiency by which plants can limit oxidative damage, also helping to preserve cellular homeostasis, as well as adjusting their physiology in response to different types of abiotic stress. The main mechanisms through which minerals contribute to enhancing plant tolerance under different environmental stress scenarios are described below.

14.4.1 Drought and heat stress

Drought drastically reduces the absorption and distribution of mineral nutrients, destabilizing metabolic processes and promoting the accumulation of reactive oxygen species (ROS). Heat exacerbates these effects by promoting protein denaturation and photosynthetic overexcitement.

- Potassium (K): Maintains osmotic adjustment and stomatal opening, preventing excessive water loss and reducing ROS formation.
- Calcium (Ca): Acts as a key second messenger in water and heat stress signaling, activating antioxidant pathways regulated by SOD, CAT, and APX.
- Magnesium (Mg): Protects chlorophyll and maintains photochemical efficiency, minimizing photooxidative damage.
- Zinc (Zn) and manganese (Mn): Essential cofactors of SOD isoforms, fundamental for eliminating superoxides generated during heat-induced electron leakage.
- Iron (Fe): Contributes to the functioning of redox enzymes and electron transport, the alteration of which is critical under drought conditions

Together, these minerals maintain membrane stability and strengthen stress signaling by limiting the accumulation of ROS (D'Oria et al., 2022; Kopecká et al., 2023).

14.4.2 Salinity and sodicity

Saline stress causes osmotic imbalance and toxic accumulation of Na^+ and Cl^-, affecting the absorption of essential minerals and causing oxidative damage.

- Potassium (K): Competes with Na^+ for transportation and maintains cytosolic K^+/Na^+ balance, crucial for turgor and enzyme activity.
- Calcium (Ca): Stabilizes membranes by activating components of the SOS pathway (SOS3-SOS2-SOS1), reducing the influx of Na^+.

- Magnesium (Mg): Supports chloroplast function and prevents excessive ROS generation under salinity.
- Boron (B): Contributes to cell wall integrity and membrane transport, processes compromised by excess salts.
- Micronutrients (Zn, Cu, Fe, Mn): Maintain the activity of antioxidant enzymes affected by oxidative stress secondary to salinity.

Adequate provision of Ca and K is one of the most effective strategies for enhancing salt tolerance (Kopecká et al., 2023).

14.4.3 Cold and frost stress

Low temperatures decrease membrane fluidity, alter electron transport, and increase ROS generation.

- Calcium (Ca): Participates in the activation of CBF/DREB pathways through Ca^{2+} signals, essential for inducing cold response of COR genes.
- Magnesium (Mg): Preserves the efficiency of the photosynthetic apparatus, preventing photoinhibition at low temperatures.
- Iron (Fe), manganese (Mn), and copper (Cu): Promote electron transfer reactions and sustain the activity of antioxidant enzymes.
- Silicon (Si): Reduces lipid peroxidation, promotes the synthesis of osmoprotectants, and stimulates the activities of SOD, CAT, and POD.

In synergy, these minerals strengthen structural stability, regulate redox homeostasis mitigating the effects of frost (Zhang et al., 2023).

14.4.4 Hypoxia and flooding

Flooding limits oxygen availability, decreases root respiration, and alters soil redox cycles.

- Iron (Fe) and manganese (Mn): Their availability increases in reducing soils, potentially reaching toxic levels that promote ROS-generating reactions.
- Calcium (Ca): Maintains membrane integrity, facilitating signaling during hypoxic stress.
- Potassium (K): Helps stabilize membrane potentials and osmotic processes impaired by anoxia.
- Magnesium (Mg): Sustains ATP production under partially functional mitochondria.
- Silicon (Si): Mitigates oxidative damage by activating antioxidant enzymes under flooding conditions.

Mineral nutritional balance contributes to maintaining redox homeostasis and modulating adaptive responses under hypoxia (Mareri et al., 2022).

14.4.5 UV radiation and excessive light

UV-B radiation and excessive light generate photosynthetic overexcitation and high levels of ROS in chloroplasts.

- Magnesium (Mg): Optimizes photochemical efficiency reducing overexcitation in PSII.
- Manganese (Mn), iron (Fe), and copper (Cu): Critical components of SOD, CAT, and peroxidases responsible for detoxifying ROS generated by intense light.
- Zinc (Zn): Stabilizes proteins and promotes DNA repair mechanisms.
- Silicon (Si): Generates deposits in the cuticle that attenuate UV penetration and reduce ROS formation.

Minerals strengthen photoprotection mechanisms and preserve the structural integrity of tissues exposed to high irradiation (Kopecká et al., 2023).

14.4.6 Toxicity caused by heavy metals and metalloids

Excessive levels of metals such as Cd, Pb, Ni, Cr, and metalloids such as As cause severe oxidative stress, alteration of nutritional homeostasis, and structural damage.

- Calcium (Ca) and magnesium (Mg): Reduce membrane permeability and limit the absorption of toxic metals.
- Zinc (Zn): Competes with toxic metals for transporters and promotes the synthesis of phytochelatins and metallothioneins
- Iron (Fe) and manganese (Mn): Maintain electron transport pathways, reducing Fenton-type reactions.
- Silicon (Si): Decreases metal absorption, strengthens cell walls, increasing antioxidant activity.
- Regulation by miRNAs: miR398, miR397, and miR408 modulate Cu- and Zn-dependent antioxidant enzymes, affecting tolerance.

As a whole, minerals reduce the availability of toxic metals, protect cell structures, and strengthen the antioxidant response (Yang et al., 2022; Angulo-Bejarano et al., 2021) (Fig. 14.2).

14.5 Nutritional management strategies to enhance antioxidant defense

The antioxidant capacity of plants is closely dependent on their nutritional status, as enzymes such as SOD, CAT, APX, POD, and GPX require mineral cofactors (Zn, Cu, Mn, Fe, Ni, Mo, Se) to function efficiently. Balanced mineral nutrition helps support enzyme activity modulating redox signaling pathways linked to stress

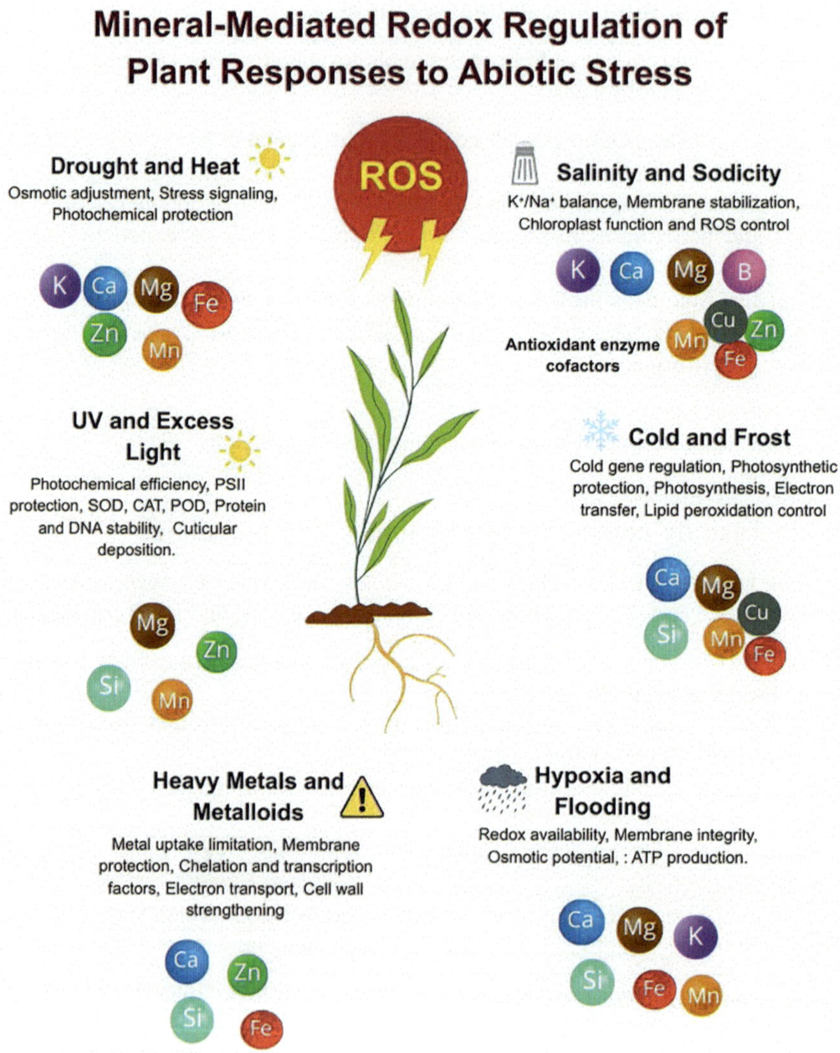

FIGURE 14.2

Mineral-mediated redox regulation of plant responses to abiotic stress. This figure summarizes the role of mineral nutrients in regulating reactive oxygen species (ROS) homeostasis and enhancing plant tolerance to abiotic stresses. Under drought, heat, salinity, sodicity, cold, flooding, UV radiation, and metal toxicity, minerals such as K, Ca, Mg, Fe, Zn, Mn, Si, and B support osmotic adjustment, membrane stability, photosynthetic efficiency, and antioxidant enzyme activity. These nutrients modulate redox signaling, electron transport, and gene regulation, limiting oxidative damage. Overall, mineral-dependent mechanisms converge to maintain cellular homeostasis and strengthen plant resilience under adverse environmental conditions.

perception. In this context, mineral fertilization, biofortification, bioinputs, nano-fertilizers, and the rhizosphere microbiome are essential tools for enhancing antioxidant protection and improving resilience to abiotic stresses.

14.5.1 Mineral fertilization (soil, foliar, and fertigation)

Mineral fertilization is the most direct way to ensure the provision of essential nutrients and maintain the functionality of metal-dependent antioxidant systems.

14.5.1.1 Soil fertilization

Provides macronutrients such as K, Ca, Mg, and S, necessary for membrane stability, osmotic adjustment, and nonenzymatic antioxidants synthesis. Likewise, adequate availability of Zn, Cu, Fe, Mn, Mo, and Ni in soil supports the activity of SOD, CAT, POD, and AsA–GSH cycle enzymes, reducing lipid peroxidation and excessive ROS accumulation under drought, salinity, or extreme temperatures (Tavanti et al., 2021).

14.5.1.2 Foliar fertilization

Allows rapid correction of micronutrient deficiencies when root absorption is limited (hypoxia, salinity, cold). The application of chelated formulations of Cu, Fe, Mn, Zn, and Mo increases the activity of SOD, POD, and catalase, decreases the formation of malondialdehyde, protects membrane integrity, and improves photosynthetic efficiency under stress (Pakhomova and Daminova, 2020).

14.5.1.3 Fertigation

It integrates nutrient supply with water management, providing a continual supply of minerals and reducing fluctuations in nutritional status. This stabilizes electron transport in chloroplasts and mitochondria, restricting the overproduction of ROS, especially under controlled drought conditions (Tavanti et al., 2021)

Overall, balanced fertilization maintains mineral homeostasis, strengthening antioxidant metabolism and increasing crop tolerance to several abiotic stresses.

14.5.2 Biofortification and bioinputs

Biofortification, mainly agronomic, and the use of microbial bioinputs represent complementary strategies for increasing micronutrient content and enhancing antioxidant defense.

14.5.2.1 Agronomic biofortification

Targeted supplementation with Zn, Fe, and Se increases the tissue content of these elements, which are essential for SOD (Zn, Mn, Fe), catalase (Fe), GPX (Se), and AsA–GSH cycle enzymes. Improved Zn and Fe status results in greater photosynthetic efficiency, membrane stability, and greater ROS detoxification capacity. Biofortified Se protects peroxidases and glutathione, reducing oxidative damage (Szerement et al., 2022; Consentino et al., 2023).

14.5.2.2 *Microbial bioinputs*

Arbuscular mycorrhizal fungi (AMF) and plant growth-promoting rhizobacteria (PGPR) increase the solubility and absorption of P, Fe, Zn, and Mn, reinforcing antioxidant pathways dependent on these minerals. In addition, they produce siderophores, ROS-degrading enzymes, and phenolic compounds which complement plant defense and modulate hormonal and redox signals that activate AsA–GSH metabolism and delay stress-induced senescence (Tripathi et al., 2022).

This way, biofortification and bio-inputs act synergistically to increase the availability of mineral cofactors and strengthen the antioxidant machinery.

14.5.3 Emerging use of nanofertilizers

Nanofertilizers optimize the release, absorption, and efficiency of essential nutrients, making them a promising tool for sustaining antioxidant mechanisms under stress. They gradually release micronutrients such as Fe, Zn, Cu, and Mn, ensuring a more constant supply of cofactors for antioxidant enzymes. They improve absorption and reduce losses due to leaching, which is critical in degraded soils or soils with limited water availability. They can induce antioxidant response, activating SOD, APX, and CAT in early stages (Husen, 2023b, 2025b). Nanoforms of phosphates and metal oxides promote electron transport and stabilize membranes, reducing ROS generation (Basavegowda and Baek, 2021; Verma et al., 2022). In short, nanofertilisers facilitate more efficient nutrient management and help strengthen the antioxidant system of crops.

14.5.4 Synergies with the rhizosphere microbiome

The rhizosphere microbiome decisively modulates mineral nutrition and antioxidant defense by interacting with the root and the edaphic environment.

14.5.4.1 *Mineral solubilization and mobilization*

Rhizospheric microorganisms increase the availability of P, Fe, Zn, and Mn by releasing siderophores, organic acids, and phosphatase enzymes, which promote the activity of metal-dependent antioxidant enzymes.

14.5.4.2 *Activation of antioxidant defense and ISR*

Various bacteria and fungi induce the accumulation of SOD, CAT, APX, and phenolic compounds, reducing lipid peroxidation and increasing stress tolerance. At the same time, induced systemic resistance (ISR) activates ROS- and RNS-based signaling pathways that contribute to maintaining oxidative balance.

14.5.4.3 *Optimization of root architecture*

Microorganism-mediated root expansion and branching facilitate access to soil areas with greater availability of Ca, Mg, and K, which are essential for membrane stability and osmotic homeostasis (Balakrishnan and Subramanian, 2017; Szerement et al., 2022; Tripathi et al., 2022; Consentino et al., 2023; Thepbandit and Athinuwat, 2024). In a whole, the rhizosphere microbiome acts as a dynamic modulator of nutritional status and redox balance, strengthening antioxidant defenses and increasing plant resilience to multiple abiotic stresses.

14.6 Conclusion

Plant antioxidant defense operates as an integrated network of enzymatic and nonenzymatic components with dual roles in ROS detoxification and stress signaling, and its performance is strongly shaped by environmental context. Nutritional status is a primary determinant of ROS control efficiency because many core antioxidant enzymes require mineral cofactors; therefore nutrient imbalance (deficiency or excess) destabilizes redox homeostasis and reduces stress tolerance. From an applied perspective, effective mitigation should combine balanced fertilization, timely foliar correction, biofortification, bioinputs, nanofertilizers, and rhizosphere microbiome management to sustain cofactor supply and activate protective pathways. Looking forward, progress will increasingly rely on molecular improvement (transporters, redox sensors, transcriptional regulators) and gene editing/omics approaches to jointly optimize mineral nutrition, redox regulation, and multistress resilience.

14.7 Future considerations

Future research should (1) clarify the molecular mechanisms linking specific nutrients to ROS signaling and antioxidant pathways across tissues and organelles; (2) define optimal dose windows (deficiency/sufficiency/toxicity) under combined, field-relevant stresses using integrated ionomics and redox biomarkers; (3) identify and validate key genetic targets (transporters, chelators, redox sensors, transcriptional regulators) through multiomics and high-throughput phenotyping; and (4) translate findings into scalable solutions by testing microbiome-based strategies, biofortification, and fertilizer/nanofertilizer approaches with robust efficacy and environmental-safety assessments, ultimately supporting breeding and gene editing for nutrient use efficiency and multistress redox resilience.

References

Abbasi, A.R., Hajirezaei, M., Hofius, D., Sonnewald, U., Voll, L.M., 2007. Specific roles of α- and γ-tocopherol in abiotic stress responses of transgenic tobacco. Plant Physiology 143 (4), 1720–1738. Available at https://doi.org/10.1104/pp.106.094771.

Abdulfatah, H., 2022. Non-enzymatic antioxidants in stressed plants: a review. Journal of University of Anbar for Pure Science 16 (2), 25–37. Available at https://doi.org/10.37652/juaps.2022.176435.

Allagulova, C.R., Lubyanova, A.R., Avalbaev, A.M., 2023. Multiple ways of nitric oxide production in plants and its functional activity under abiotic stress conditions. International Journal of Molecular Sciences 24 (14). Available at https://doi.org/10.3390/ijms241411637.

Amari, T., Souid, A., Ghabriche, R., Porrini, M., Lutts, S., Sacchi, G.A., et al., 2020. Why does the halophyte mesembryanthemum crystallinum better tolerate ni toxicity than *Brassica juncea*: implication of antioxidant defense systems. Plants 9 (3), 312. Available at https://doi.org/10.3390/plants9030312.

Amjad, M., Akhtar, J., Anwar-ul-Haq, M., Riaz, M.A., Saqib, Z.A., Murtaza, B., et al., 2016. Effectiveness of potassium in mitigating the salt-induced oxidative stress in contrasting tomato genotypes. Journal of Plant Nutrition 39 (13), 1926–1935. Available at https://doi.org/10.1080/01904167.2016.1201107.

Andresen, E., Peiter, E., Küpper, H., 2018. Trace metal metabolism in plants. Journal of Experimental Botany 69 (5), 909–954. Available at https://doi.org/10.1093/jxb/erx465.

Angulo-Bejarano, P.I., Puente-Rivera, J., Cruz-Ortega, R., 2021. Metal and metalloid toxicity in plants: an overview on molecular aspects. Plants 10 (4), 635. Available at https://doi.org/10.3390/plants10040635.

Ansari, M.K.A., Iqbal, M., Ahmad, M., Munir, M., Gaffar, S.A., Chaachouay, N., 2024. Heavy metal stress and cellular antioxidant systems of plants: a review. Agricultural Reviews. (Of). Available at https://doi.org/10.18805/ag.rf-321.

Ashokan, A., Mini, V., Rani, B., Anand, S., 2024. Mitigation of abiotic stresses in plants through nutrient management. International Journal of Environment and Climate Change 14 (7), 254–267. Available at https://doi.org/10.9734/ijecc/2024/v14i74267.

Baker, A., Lin, C.C., Lett, C., Karpinska, B., Wright, M.H., Foyer, C.H., 2023. Catalase: a critical node in the regulation of cell fate. Free Radical Biology and Medicine 199, 56–66. Available at https://doi.org/10.1016/j.freeradbiomed.2023.02.009.

Balakrishnan, N., Subramanian, K.S., 2017. Arbuscular mycorrhizal fungus inoculation on antioxidant enzyme activities in maize plants at different levels of Fe and Zn fertilization. International Journal of Current Microbiology and Applied Sciences 6 (6), 1754–1768. Available at https://doi.org/10.20546/ijcmas.2017.606.204.

Basavegowda, N., Baek, K.H., 2021. Current and future perspectives on the use of nanofertilizers for sustainable agriculture: the case of phosphorus nanofertilizer. 3 Biotech 11 (7). Available at https://doi.org/10.1007/s13205-021-02907-4.

Batth, R., Singh, K., Kumari, S., Mustafiz, A., 2017. Transcript profiling reveals the presence of abiotic stress and developmental stage specific ascorbate oxidase genes in plants. Frontiers in Plant Science 8. Available at https://doi.org/10.3389/fpls.2017.00198.

Baxter, I., Muthukumar, B., Hyeong, C.P., Buchner, P., Lahner, B., Danku, J., et al., 2008. Variation in molybdenum content across broadly distributed populations of *Arabidopsis thaliana* is controlled by a mitochondrial molybdenum transporter (MOT1). PLoS Genetics 4 (2). Available at https://doi.org/10.1371/journal.pgen.1000004.

Bela, K., Riyazuddin, R., Csiszár, J., 2022. Plant glutathione peroxidases: non-heme peroxidases with large functional flexibility as a core component of ROS-processing mechanisms and signalling. Antioxidants 11 (8). Available at https://doi.org/10.3390/antiox11081624.

Boojar, M.M.A., Goodarzi, F., 2008. Comparative evaluation of oxidative stress status and manganese availability in plants growing on manganese mine. Ecotoxicology and Environmental Safety 71 (3), 692–699. Available at https://doi.org/10.1016/j.ecoenv.2007.10.011.

Boyd, R.S., Jaffré, T., 2009. Elemental concentrations of eleven new caledonian plant species from serpentine soils: Elemental correlations and leaf-age effects. Northeastern Naturalist 16 (5), 93–110. Available at https://doi.org/10.1656/045.016.0508.

Briat, J.F., Ravet, K., Arnaud, N., Duc, C., Boucherez, J., Touraine, B., et al., 2010. New insights into ferritin synthesis and function highlight a link between iron homeostasis and oxidative stress in plants. Annals of Botany 105 (5), 811–822. Available at https://doi.org/10.1093/aob/mcp128.

Cadena-Zamudio, J.D., Monribot-Villanueva, J.L., Pérez Torres, C.A., Alatorre-Cobos, F., Guerrero-Analco, J.A., Ibarra-Laclette, E., 2023. Non-targeted metabolomic analysis of *Arabidopsis thaliana* (L.) Heynh: metabolic adaptive responses to stress caused by N starvation. Metabolites 13 (9). Available at https://doi.org/10.3390/metabo13091021.

Cakmak, I., 2006. Role of mineral nutrients in tolerance of crop plants to environmental stress factors. Plant and Cell Physiology 38 (Special Issue), 975–982.

Cannea, F.B., Padiglia, A., 2025. Antioxidant defense systems in plants: mechanisms, regulation, and biotechnological strategies for enhanced oxidative stress tolerance. Life (Chicago, Ill.: 1978) 15 (8). Available at https://doi.org/10.3390/life15081293.

Capaldi, F.R., Gratão, P.L., Reis, A.R., Lima, L.W., Azevedo, R.A., 2015. Sulfur metabolism and stress defense responses in plants. Tropical Plant Biology 8 (3–4), 60–73. Available at https://doi.org/10.1007/s12042-015-9152-1.

Cazzonelli, C.I., 2011. Carotenoids in nature: insights from plants and beyond. Functional Plant Biology 38 (11), 833–847. Available at https://doi.org/10.1071/FP11192.

Çelik, O., Kaya, E., 2024. mRNA transcription analyses of ROS genes of *Olea europaea* L. in vitro cultures treated with different boron salts. ANADOLU Ege Tarımsal Araştırma Enstitüsü Dergisi 34 (1), 24–32. Available at https://doi.org/10.18615/anadolu.1457244.

Cervilla, L.M., Blasco, B., Rios, J.J., Romero, L., Ruiz, J.M., 2007. Oxidative stress and antioxidants in tomato (*Solanum lycopersicum*) plants subjected to boron toxicity. Annals of Botany 100 (4), 747–756. Available at https://doi.org/10.1093/aob/mcm156.

Chakraborty, K., Mondal, S., Bhaduri, D., Mohanty, A., Paul, A., 2021. Interplay between sodium and chloride decides the plant's fate under salt and drought stress conditions. Plant Nutrition and Food Security in the Era of Climate Change. Elsevier, India, pp. 271–314 10.1016/B978-0-12-822916-3.00020-2.

Chan, K.X., Phua, S.Y., Breusegem, F.V., 2019. Secondary sulfur metabolism in cellular signalling and oxidative stress responses. Journal of Experimental Botany 70 (16), 4237–4250. Available at https://doi.org/10.1093/jxb/erz119.

Consentino, B.B., Ciriello, M., Sabatino, L., Vultaggio, L., Baldassano, S., Vasto, S., et al., 2023. Current acquaintance on agronomic biofortification to modulate the yield and functional value of vegetable crops: a review. Horticulturae 9 (2). Available at https://doi.org/10.3390/horticulturae9020219.

Contour-Ansel, D., Torres-Franklin, M.L., Cruz De Carvalho, M.H., D'arcy-Lameta, A., Zuily-Fodil, Y., 2006. Glutathione reductase in leaves of cowpea: cloning of two cDNAs, expression and enzymatic activity under progressive drought stress, desiccation and abscisic acid treatment. Annals of Botany 98 (6), 1279–1287. Available at https://doi.org/10.1093/aob/mcl217.

Costa, J.H., Roque, A.L.M., Aziz, S., dos Santos, C.P., Germano, T.A., Batista, M.C., et al., 2021. Genome-wide identification of ascorbate-glutathione cycle gene families in soybean (*Glycine max*) reveals gene duplication events and specificity of gene members linked to development and stress conditions. International Journal of Biological Macromolecules 187, 528–543. Available at https://doi.org/10.1016/j.ijbiomac.2021.07.103.

D'Oria, A., Courbet, G., Billiot, B., Jing, L., Pluchon, S., Arkoun, M., et al., 2022. Drought specifically downregulates mineral nutrition: plant ionomic content and associated gene expression. Plant Direct 6 (8). Available at https://doi.org/10.1002/pld3.402.

da Silva, R.C., Oliveira, H.C., Igamberdiev, A.U., Stasolla, C., Gaspar, M., 2024. Interplay between nitric oxide and inorganic nitrogen sources in root development and abiotic stress responses. Journal of Plant Physiology 297. Available at https://doi.org/10.1016/j.jplph.2024.154241.

Dorion, S., Ouellet, J.C., Rivoal, J., 2021. Glutathione metabolism in plants under stress: beyond reactive oxygen species detoxification. Metabolites 11 (9). Available at https://doi.org/10.3390/metabo11090641.

Du, Y.-Y., Wang, P.-C., Chen, J., Song, C.-P., 2008. Comprehensive functional analysis of the catalase gene family in *Arabidopsis thaliana*. Journal of Integrative Plant Biology 50 (10), 1318–1326. Available at https://doi.org/10.1111/j.1744-7909.2008.00741.x.

Dumanović, J., Nepovimova, E., Natić, M., Kuča, K., Jaćević, V., 2021. The significance of reactive oxygen species and antioxidant defense system in plants: a concise overview. Frontiers in Plant Science 11. Available at https://doi.org/10.3389/fpls.2020.552969.

Dutta, A., Dracatos, P.M., Khan, G.A., 2024. Balancing act: the dynamic relationship between nutrient availability and plant defence. Plant Journal 120 (5), 1724–1734. Available at https://doi.org/10.1111/tpj.17098.

Falk, J., Munné-Bosch, S., 2010. Tocochromanol functions in plants: antioxidation and beyond. Journal of Experimental Botany 61 (6), 1549–1566. Available at https://doi.org/10.1093/jxb/erq030.

Feigl, G., Lehotai, N., Molnár, Á., Ördög, A., Rodríguez-Ruiz, M., Palma, J.M., et al., 2015. Zinc induces distinct changes in the metabolism of reactive oxygen and nitrogen species (ROS and RNS) in the roots of two *Brassica* species with different sensitivity to zinc stress. Annals of Botany 116 (4), 613–625. Available at https://doi.org/10.1093/aob/mcu246.

Foyer, C.H., Kunert, K., 2024. The ascorbate-glutathione cycle coming of age. Journal of Experimental Botany 75 (9), 2682–2699. Available at https://doi.org/10.1093/jxb/erae023.

Fritsche, S., Wang, X., Jung, C., 2017. Recent advances in our understanding of tocopherol biosynthesis in plants: an overview of key genes, functions, and breeding of vitamin E improved crops. Antioxidants 6 (4), 99. Available at https://doi.org/10.3390/antiox6040099.

Galant, A., Preuss, M.L., Cameron, J.C., Jez, J.M., 2011. Plant glutathione biosynthesis: diversity in biochemical regulation and reaction products. Frontiers in Plant Science 2. Available at https://doi.org/10.3389/fpls.2011.00045.

Gałecka, Jacewicz, Mrowicka, Florkowski, Gałecki, 2008. [Antioxidative enzymes—structure, properties, functions]. Polski Merkuriusz Lekarski: Organ Polskiego Towarzystwa Lekarskiego 25 (147), 266–268.

Gao, F., Chen, J., Ma, T., Li, H., Wang, N., Li, Z., et al., 2014. The glutathione peroxidase gene family in *Thellungiella salsuginea*: genome-wide identification, classification, and gene and protein expression analysis under stress conditions. International Journal of Molecular Sciences 15 (2), 3319–3335. Available at https://doi.org/10.3390/ijms15023319.

Gasperl, A., Zellnig, G., Kocsy, G., Müller, M., 2022. Organelle-specific localization of glutathione in plants grown under different light intensities and spectra. Histochemistry and Cell Biology 158 (3), 213–227. Available at https://doi.org/10.1007/s00418-022-02103-2.

Ghorbel, M., Zribi, I., Haddaji, N., Siddiqui, A.J., Bouali, N., Brini, F., 2023. Genome-wide identification and expression analysis of catalase gene families in triticeae. Plants 13 (1), 11.

Gupta, O., Shailaja, D., Dashrathrao, K.V., 2024b. Bioenergy and biofuels potential in India: current and future prospects. International Journal of Research in Agronomy 7 (10), 233–237.

Gupta, R., Verma, N., Tewari, R.K., 2024a. Micronutrient deficiency-induced oxidative stress in plants. Plant Cell Reports 43 (9). Available at https://doi.org/10.1007/s00299-024-03297-6.

Gupta, S., Kaur, N., Kant, K., Jindal, P., Ali, A., Naeem, M., 2023. Calcium: a master regulator of stress tolerance in plants. South African Journal of Botany 163, 580–594. Available at https://doi.org/10.1016/j.sajb.2023.10.047.

Harshavardhan, V.T., Wu, T.-M., Hong, C.-Y., 2017. Glutathione reductase and abiotic stress tolerance in plants. In: En, M.A., Hossain, M.G., Mostofa, P., Diaz-Vivancos, D.J., Burritt, M., Fujita, L.-S.P., Tran (Eds.), Glutathione in Plant Growth, Development, and Stress Tolerance. Springer International Publishing, pp. 265–286.

Hasanuzzaman, M., Bhuyan, M.H.M.B., Raza, A., Hawrylak-Nowak, B., Matraszek-Gawron, R., Mahmud, J.A., et al., 2020. Selenium in plants: boon or bane? Environmental and Experimental Botany 178. Available at https://doi.org/10.1016/j.envexpbot.2020.104170.

Hernández, I., Munné-Bosch, S., 2015. Linking phosphorus availability with photo-oxidative stress in plants. Journal of Experimental Botany 66 (10), 2889–2900. Available at https://doi.org/10.1093/jxb/erv056.

Hoque, M., Jony, M., Das, A., Shaffique, S., Rahman, M., Sharif, I., et al., 2024. Coordination of elicitors and ascorbate-glutathione cycle: a vital nexus for mitigating post-harvest injury and losses of cultivated fruits. Plant Trends 2 (2), 24. Available at https://doi.org/10.5455/pt.2024.03.

Houmani, H., Debez, A., de Freitas-Silva, L., Abdelly, C., Palma, J.M., Corpas, F.J., 2022. Potassium (K+) starvation-induced oxidative stress triggers a general boost of antioxidant and NADPH-generating systems in the halophyte *Cakile maritima*. Antioxidants 11 (2), 401. Available at https://doi.org/10.3390/antiox11020401.

Husen, A., 2021a. Harsh Environment and Plant Resilience (Molecular and Functional Aspects). Springer Nature Switzerland AG, Cham https://doi.org/10.1007/978-3-030-65912-7.

Husen, A., 2021b. The harsh environment and resilient plants: an overview. In: Husen, A. (Ed.), Harsh Environment and Plant Resilience. Springer International Publishing AG, Cham, pp. 1–23 https://doi.org/10.1007/978-3-030-65912-7_1.

Husen, A., 2021c. Plant Performance Under Environmental Stress (Hormones, Biostimulants and Sustainable Plant Growth Management). Springer Nature Switzerland AG, Cham https://doi.org/10.1007/978-3-030-78521-5.

Husen, A., 2022. Environmental Pollution and Medicinal Plants. CRC Press, Boca Raton, FL, pp. 33487. 2742 https://doi.org/10.1201/9781003178866.

Husen, A., 2023b. Nanomaterials and Nanocomposites Exposures to Plants (Response, Interaction, Phytotoxicity and Defense Mechanisms). Springer Nature, Singapore https://doi.org/10.1007/978-981-99-2419-6.

Husen, A., 2023a. Plants and Their Interaction to Environmental Pollution (Damage Detection, Adaptation, Tolerance, Physiological and Molecular Responses). Elsevier Inc, Cambridge, MA https://doi.org/10.1016/C2020-0-03319-0.

Husen, A., 2025b. Role of Antioxidants in Mitigating Plant Stress. Elsevier Inc., Cambridge, MA https://doi.org/10.1016/C2023-0-51268-6.

Husen, A., 2025a. Nanobiotechnology for Abiotic Stress Adaptation and Mitigation in Agricultural Crops. Springer Nature, Singapore https://doi.org/10.1007/978-981-96-9298-9.

Imran, M., Hussain, S., El-Esawi, M.A., Rana, M.S., Saleem, M.H., Riaz, M., et al., 2020. Molybdenum supply alleviates the cadmium toxicity in fragrant rice by modulating oxidative stress and antioxidant gene expression. Biomolecules 10 (11), 1–17. Available at https://doi.org/10.3390/biom10111582.

Jamali Jaghdani, S., Jahns, P., Tränkner, M., 2021. Mg deficiency induces photo-oxidative stress primarily by limiting CO_2 assimilation and not by limiting photosynthetic light utilization. Plant Science 302, 110751. Available at https://doi.org/10.1016/j.plantsci.2020.110751.

Jana, G.A., Yaish, M.W., 2020. Genome-wide identification and functional characterization of glutathione peroxidase genes in date palm (*Phoenix dactylifera* L.) under stress conditions. Plant Gene 23. Available at https://doi.org/10.1016/j.plgene.2020.100237.

Jardim-Messeder, D., de Souza-Vieira, Y., Lavaquial, L.C., Cassol, D., Galhego, V., Bastos, G.A., et al., 2023. Ascorbate-glutathione cycle genes families in euphorbiaceae: characterization and evolutionary analysis. Biology 12 (1). Available at https://doi.org/10.3390/biology12010019.

Jbir Koubaa, R., Ayadi, M., Saidi, M.N., Charfeddine, S., Gargouri-Bouzid, R., Nouri-Ellouz, O., 2023. Comprehensive genome-wide analysis of the catalase enzyme toolbox in potato (*Solanum tuberosum* L.). Potato Research 66 (1), 23–49. Available at https://doi.org/10.1007/s11540-022-09554-z.

Kampfenkel, K., Van Montagu, M., Inze, D., 1995. Effects of iron excess on *Nicotiana plumbaginifolia* plants (implications to oxidative stress). Plant Physiology 107 (3), 725–735. Available at https://doi.org/10.1104/pp.107.3.725.

Kanth, P.C., Verma, S.K., Gour, N., 2020. Functionalized nanomaterials for biomedical and agriculture industries. Handbook of Functionalized Nanomaterials for Industrial Applications. Elsevier, India, pp. 231–265. Available at http://doi.org/10.1016/B978-0-12-816787-8.00010-7.

Khare, T., Kumar, V., Kishor, P.B.K., 2015. Na+ and Cl- ions show additive effects under NaCl stress on induction of oxidative stress and the responsive antioxidative defense in rice. Protoplasma 252 (4), 1149–1165. Available at https://doi.org/10.1007/s00709-014-0749-2.

Koh, Y.S., Wong, S.K., Ismail, N.H., Zengin, G., Duangjai, A., Saokaew, S., et al., 2021. Mitigation of environmental stress-impacts in plants: role of sole and combinatory exogenous application of glutathione. Frontiers in Plant Science 12. Available at https://doi.org/10.3389/fpls.2021.791205.

Kolbert, Z., Molnï R, Ï.R.D., Szőllősi, R.K., Feigl, G.B., Erdei, L.S., Rdï G, A.Ï., 2018. Nitro-oxidative stress correlates with Se tolerance of astragalus species. Plant and Cell Physiology 59 (9), 1827–1843. Available at https://doi.org/10.1093/pcp/pcy099.

Kopecká, R., Kameniarová, M., Černý, M., Brzobohatý, B., Novák, J., 2023. Abiotic stress in crop production. International Journal of Molecular Sciences 24 (7), 6603. Available at https://doi.org/10.3390/ijms24076603.

Kováčik, J., Vydra, M., 2024. The impact of nickel on plant growth and oxidative balance. Physiologia Plantarum 176 (6). Available at https://doi.org/10.1111/ppl.14595.

Kumar, D., Yusuf, M.A., Singh, P., Sardar, M., Sarin, N.B., 2013. Modulation of antioxidant machinery in α-tocopherol-enriched transgenic Brassica juncea plants tolerant to abiotic stress conditions. Protoplasma 250 (5), 1079–1089. Available at https://doi.org/10.1007/s00709-013-0484-0.

Lei, Y., Korpelainen, H., Li, C., 2007. Physiological and biochemical responses to high Mn concentrations in two contrasting *Populus cathayana* populations. Chemosphere 68 (4), 686–694. Available at https://doi.org/10.1016/j.chemosphere.2007.01.066.

Li, J., Jia, Y., Dong, R., Huang, R., Liu, P., Li, X., et al., 2019. Advances in the mechanisms of plant tolerance to manganese toxicity. International Journal of Molecular Sciences 20 (20), 5096. Available at https://doi.org/10.3390/ijms20205096.

Li, S., 2023. Novel insight into functions of ascorbate peroxidase in higher plants: more than a simple antioxidant enzyme. Redox Biology 64, 102789. Available at https://doi.org/10.1016/j.redox.2023.102789.

Lin, Y.F., Aarts, M.G.M., 2012. The molecular mechanism of zinc and cadmium stress response in plants. Cellular and Molecular Life Sciences 69 (19), 3187–3206. Available at https://doi.org/10.1007/s00018-012-1089-z.

Liu, Z., Wang, D., Tang, H., Li, H., Zhang, X., Dong, S., et al., 2023. Identification and analysis of the catalase gene family response to abiotic stress in *Nicotiana tabacum* L. Agronomy 13 (3), 936. Available at https://doi.org/10.3390/agronomy13030936.

Luan, Y., Fu, X., Lu, P., Grierson, D., Xu, C., 2020. Molecular mechanisms determining the differential accumulation of carotenoids in plant species and varieties. Critical Reviews in Plant Sciences 39 (2), 125–139. Available at https://doi.org/10.1080/07352689.2020.1768350.

Madhu, Kaur, A., Tyagi, S., Shumayla, K., Singh, S.K., Upadhyay, 2022. Exploration of glutathione reductase for abiotic stress response in bread wheat (*Triticum aestivum* L.). Plant Cell Reports 41 (3), 639–654. Available at https://doi.org/10.1007/s00299-021-02717-1.

Mareri, L., Parrotta, L., Cai, G., 2022. Environmental Stress and Plants. International Journal of Molecular Sciences 23 (10), 5416. Available at https://doi.org/10.3390/ijms23105416.

Molnár, G., Salmon, L., Nicolazzi, W., Terki, F., Bousseksou, A., 2014. Emerging properties and applications of spin crossover nanomaterials. Journal of Materials Chemistry C 2 (8), 1360–1366. Available at https://doi.org/10.1039/c3tc31750a.

Moreira, I.N., Mourato, M.P., Reis, R., Martins, L.L., 2015. Oxidative stress induced by cadmium and copper in *Brassica rapa* leaves: indicators of stress, oxidative damage, and antioxidant mechanisms. Communications in Soil Science and Plant Analysis 46 (19), 2475–2489. Available at https://doi.org/10.1080/00103624.2015.1085554.

Nadarajah, K.K., 2020. ROS homeostasis in abiotic stress tolerance in plants. International Journal of Molecular Sciences 21 (15), 5208. Available at https://doi.org/10.3390/ijms21155208.

Othman, R., Mohd Zaifuddin, F.A., Hassan, N.M., 2014. Carotenoid biosynthesis regulatory mechanisms in plants. Journal of Oleo Science 63 (8), 753–760. Available at https://doi.org/10.5650/jos.ess13183.

Oubenali, A., Moukhtari, A.E., Lamsaadi, N., Farssi, O., Mouradi, M., Lazali, M., et al., 2022. Availability disturbs growth and photosynthetic-related parameters and induced oxidative stress in chickpea (*Cicer Arietinum* L.). Applied Ecology and Environmental Research 20 (6), 4785–4805. Available at https://doi.org/10.15666/aeer/2006_47854805.

Özgür Uzilday, R., Uzilday, B., Yalçinkaya, T., Türkan, İ., 2017. Mg deficiency changes the isoenzyme pattern of reactive oxygen species-related enzymes and regulates NADPH-oxidase-mediated ROS signaling in cotton. Turkish Journal of Biology 41 (6), 868–880. Available at https://doi.org/10.3906/biy-1704-39.

Ozyigit, I.I., Filiz, E., Vatansever, R., Kurtoglu, K.Y., Koc, I., Öztürk, M.X., et al., 2016. Identification and comparative analysis of H_2O_2-scavenging enzymes (ascorbate peroxidase and glutathione peroxidase) in selected plants employing bioinformatics approaches. Frontiers in Plant Science 7. Available at https://doi.org/10.3389/fpls.2016.00301 2016.

Pakhomova, V.M., Daminova, A.I., 2020. EDP sciences Russian federation chelated micronutrient fertilizers as effective antioxidants applied for foliar plant treatment. BIO Web of Conferences 17. Available at https://doi.org/10.1051/bioconf/20201700057.

Pan, L., Luo, Y., Wang, J., Li, X., Tang, B., Yang, H., et al., 2022. Evolution and functional diversification of catalase genes in the green lineage. BMC Genomics 23 (1). Available at https://doi.org/10.1186/s12864-022-08621-6.

Panda, S.K., Gupta, D., Patel, M., Vyver, C.V.D., Koyama, H., 2024. Functionality of reactive oxygen species (ROS) in plants: toxicity and control in poaceae crops exposed to abiotic stress. Plants 13 (15). Available at https://doi.org/10.3390/plants13152071.

Rajput, V.D., Harish, Singh, R.K., Verma, K.K., Sharma, L., Quiroz-Figueroa, F.R., et al., 2021. Recent developments in enzymatic antioxidant defence mechanism in plants with special reference to abiotic stress. Biology 10 (4). Available at https://doi.org/10.3390/biology10040267.

Ramadoss, B.R., Subramanian, U., Alagarsamy, M., Gangola, M.P., 2021. Non-enzymatic antioxidants' significant role in abiotic stress tolerance in crop plants. In: Hamed Abdel Latef, A.A. (Ed.), Organic Solutes, Oxidative Stress, and Antioxidant Enzymes Under Abiotic Stressors, 1st ed. CRC Press, pp. 365–392.

Rao, M.J., Duan, M., Zhou, C., Jiao, J., Cheng, P., Yang, L., et al., 2025. Antioxidant defense system in plants: reactive oxygen species production, signaling, and scavenging during abiotic stress-induced oxidative damage. Horticulturae 11 (5). Available at https://doi.org/10.3390/horticulturae11050477.

Rodriguez Milla, M.A., Maurer, A., Huete, A.R., Gustafson, J.P., 2003. Glutathione peroxidase genes in *Arabidopsis* are ubiquitous and regulated by abiotic stresses through diverse signaling pathways. Plant Journal 36 (5), 602–615. Available at https://doi.org/10.1046/j.1365-313X.2003.01901.x.

Sahoo, A., Tiwari, S., 2022. Antioxidants and antioxidative enzymes as potential biomarkers for assessing stress in plants. International Journal of Plant and Environment 8 (02), 95–105. Available at https://doi.org/10.18811/ijpen.v8i02.01.

Salbitani, G., Vona, V., Bottone, C., Petriccione, M., Carfagna, S., 2015. Sulfur deprivation results in oxidative perturbation in *Chlorella sorokiniana* (211/8k. Plant and Cell Physiology 56 (5), 897–905. Available at https://doi.org/10.1093/pcp/pcv015.

Sardar, H., Khalid, Z., Ahsan, M., Naz, S., Nawaz, A., Ahmad, R., et al., 2023. Enhancement of salinity stress tolerance in lettuce (*Lactuca sativa* L.) via foliar application of nitric oxide. Plants 12 (5), 1115. Available at https://doi.org/10.3390/plants12051115.

Semchuk, N.M., Lushchak, O.V., Falk, J., Krupinska, K., Lushchak, V.I., 2009. Inactivation of genes, encoding tocopherol biosynthetic pathway enzymes, results in oxidative stress in outdoor grown *Arabidopsis thaliana*. Plant Physiology and Biochemistry 47 (5), 384–390. Available at https://doi.org/10.1016/j.plaphy.2009.01.009.

Stephenie, S., Chang, Y.P., Gnanasekaran, A., Esa, N.M., Gnanaraj, C., 2020. An insight on superoxide dismutase (SOD) from plants for mammalian health enhancement. Journal of Functional Foods 68. Available at https://doi.org/10.1016/j.jff.2020.103917.

Sun, L., Cao, X., Du, J., Wang, Y., Zhang, F., Turkan, I., 2024. Canola (*Brassica napus*) enhances sodium chloride and sodium ion tolerance by maintaining ion homeostasis, higher antioxidant enzyme activity and photosynthetic capacity fluorescence parameters. Functional Plant Biology 51 (8). Available at https://doi.org/10.1071/fp23089.

Szerement, J., Szatanik-Kloc, A., Mokrzycki, J., Mierzwa-Hersztek, M., 2022. Agronomic Biofortification with Se, Zn, and Fe: an effective strategy to enhance crop nutritional quality and stress defense—a review. Journal of Soil Science and Plant Nutrition 22 (1), 1129–1159. Available at https://doi.org/10.1007/s42729-021-00719-2.

Tahmasebi, A., Aram, F., Ebrahimi, M., Mohammadi-Dehcheshmeh, M., Ebrahimie, E., 2012. Genome-wide analysis of cytosolic and chloroplastic isoforms of glutathione reductase in plant cells. Plant OMICS 5 (2), 94–102.

Tavakol, E., Jákli, B., Cakmak, I., Dittert, K., Senbayram, M., 2022. Optimization of potassium supply under osmotic stress mitigates oxidative damage in barley. Plants 11 (1), 55. Available at https://doi.org/10.3390/plants11010055.

Tavallali, V., Karimi, S., Espargham, O., 2018. Boron enhances antioxidative defense in the leaves of salt-affected *Pistacia vera* seedlings. The Horticulture Journal 87 (1), 55–62. Available at https://doi.org/10.2503/hortj.okd 062.

Tavanti, T.R., Melo, A.A.R.D., Moreira, L.D.K., Sanchez, D.E.J., Silva, R.D.S., Silva, R.M.D., et al., 2021. Micronutrient fertilization enhances ROS scavenging system for alleviation of abiotic stresses in plants. Plant Physiology and Biochemistry 160, 386–396. Available at https://doi.org/10.1016/j.plaphy.2021.01.040.

Tewari, R.K., Kumar, P., Sharma, P.N., 2010. Morphology and oxidative physiology of boron-deficient mulberry plants. Tree Physiology 30 (1), 68–77. Available at https://doi.org/10.1093/treephys/tpp093.

Thepbandit, W., Athinuwat, D., 2024. Rhizosphere microorganisms supply availability of soil nutrients and induce plant defense. Microorganisms 12 (3), 558. Available at https://doi.org/10.3390/microorganisms12030558.

Thounaojam, T.C., Panda, P., Mazumdar, P., Kumar, D., Sharma, G.D., Sahoo, L., et al., 2012. Excess copper induced oxidative stress and response of antioxidants in rice. Plant Physiology and Biochemistry 53, 33–39. Available at https://doi.org/10.1016/j.plaphy.2012.01.006.

Tóth, S.Z., 2023. The functions of chloroplastic ascorbate in vascular plants and algae. International Journal of Molecular Sciences 24 (3), 2537. Available at https://doi.org/10.3390/ijms24032537.

Tränkner, M., Jákli, B., Tavakol, E., Geilfus, C.M., Cakmak, I., Dittert, K., et al., 2016. Magnesium deficiency decreases biomass water-use efficiency and increases leaf water-use efficiency and oxidative stress in barley plants. Plant and Soil 406 (1-2), 409–423. Available at https://doi.org/10.1007/s11104-016-2886-1.

Tripathi, R., Tewari, R., Singh, K.P., Keswani, C., Minkina, T., Srivastava, A.K., et al., 2022. Plant mineral nutrition and disease resistance: a significant linkage for sustainable crop protection. Frontiers in Plant Science 13. Available at https://doi.org/10.3389/fpls.2022.883970.

Ul Haq, I., Ijaz, S., 2019. Use of metallic nanoparticles and nanoformulations as nanofungicides for sustainable disease management in plants. In: Prasad, E.R., Kumar, V., Kumar, M., Choudhary, D. (Eds.), Nanobiotechnology in Bioformulations. Springer International Publishing, pp. 289–316.

Uruç Parlak, K., Demirezen Yilmaz, D., 2012. Response of antioxidant defences to Zn stress in three duckweed species. Ecotoxicology and Environmental Safety 85, 52–58. Available at https://doi.org/10.1016/j.ecoenv.2012.08.023.

Verma, K.K., Song, X.P., Joshi, A., Tian, D.D., Rajput, V.D., Singh, M., et al., 2022. Recent trends in nano-fertilizers for sustainable agriculture under climate change for global food security. Nanomaterials 12 (1). Available at https://doi.org/10.3390/nano12010173.

Wang, B., Ding, H.Y., Chen, Q., Ouyang, L., Li, S., Zhang, J., 2019. Enhanced tolerance to methyl viologen-mediated oxidative stress via AtGR2 expression from chloroplast genome. Frontiers in Plant Science 10. Available at https://doi.org/10.3389/fpls.2019.01178.

Wang, J., Moeen-ud-din, M., Yin, R., Yang, S., 2023. ROS homeostasis involved in dose-dependent responses of *Arabidopsis* seedlings to copper toxicity. Genes 14 (1), 11. Available at https://doi.org/10.3390/genes14010011.

White, P.J., 2015. Selenium accumulation by plants. Annals of Botany 117 (2). Available at https://doi.org/10.1093/aob/mcv180 mcv180.

Wu, S., Hu, C., Tan, Q., Xu, S., Sun, X., 2017. Nitric oxide mediates molybdenum-induced antioxidant defense in wheat under drought stress. Frontiers in Plant Science 8. Available at https://doi.org/10.3389/fpls.2017.01085.

Xiao, M., Li, Z., Zhu, L., Wang, J., Zhang, B., Zheng, F., et al., 2021. The multiple roles of ascorbate in the abiotic stress response of plants: antioxidant, cofactor, and regulator. Frontiers in Plant Science 12. Available at https://doi.org/10.3389/fpls.2021.598173.

Xu, E., Liu, Y., Gu, D., Zhan, X., Li, J., Zhou, K., et al., 2024. Molecular mechanisms of plant responses to copper: from deficiency to excess. International Journal of Molecular Sciences 25 (13), 6993. Available at https://doi.org/10.3390/ijms25136993.

Xu, S., Hu, C., Tan, Q., Qin, S., Sun, X., 2018. Subcellular distribution of molybdenum, ultrastructural and antioxidative responses in soybean seedlings under excess molybdenum stress. Plant Physiology and Biochemistry 123, 75–80. Available at https://doi.org/10.1016/j.plaphy.2017.11.023.

Yang, Y., Huang, J., Sun, Q., Wang, J., Huang, L., Fu, S., et al., 2022. microRNAs: key players in plant response to metal toxicity. International Journal of Molecular Sciences 23 (15), 8642. Available at https://doi.org/10.3390/ijms23158642.

You, J., Chan, Z., 2015. ROS regulation during abiotic stress responses in crop plants. Frontiers in Plant Science 6. Available at https://doi.org/10.3389/fpls.2015.01092.

Zainal Abidin, A.A., Yokthongwattana, C., Balia, Z.N., 2021. Carotenogenesis in *Nannochloropsis oculata* under oxidative and salinity stress. Sains Malaysiana 50 (2), 327–337. Available at https://doi.org/10.17576/jsm-2021-5002-05.

Zameer, R., Fatima, K., Azeem, F., ALgwaiz, H.I.M., Sadaqat, M., Rasheed, A., et al., 2022. Genome-wide characterization of superoxide dismutase (SOD) genes in *Daucus carota*: novel insights into structure, expression, and binding interaction with hydrogen peroxide

(H_2O_2) under abiotic stress condition. Frontiers in Plant Science 13. Available at https://doi.org/10.3389/fpls.2022.870241.

Zechmann, B., 2018. Compartment-specific importance of ascorbate during environmental stress in plants. Antioxidants & Redox Signaling 29 (15), 1488–1501. Available at https://doi.org/10.1089/ars.2017.7232.

Zeng, H., Chen, F., Zhu, Q., Ali, S., Du, J., Zhu, Y., et al., 2025. The interplay between phosphorus nutrition and abiotic stresses in plants. Journal of Genetics and Genomics 52 (12), 1507–1523. Available at https://doi.org/10.1016/j.jgg.2025.08.008.

Zhang, Y., Zheng, L., Yun, L., Ji, L., Li, G., Ji, M., et al., 2022. Catalase (CAT) gene family in wheat (*Triticum aestivum* L.): evolution, expression pattern and function analysis. International Journal of Molecular Sciences 23 (1), 542. Available at https://doi.org/10.3390/ijms23010542.

Zhang, Y., Xu, J., Li, R., Ge, Y., Li, Y., Li, R., 2023. Plants' response to abiotic stress: mechanisms and strategies. International Journal of Molecular Sciences 24 (13), 10915. Available at https://doi.org/10.3390/ijms241310915.

Zheng, S.J., 2010. Iron homeostasis and iron acquisition in plants: maintenance, functions and consequences. Annals of Botany 105 (5), 799–800. Available at https://doi.org/10.1093/aob/mcq082.

CHAPTER

15 Mineral nutrition and ethylene in plant stress tolerance

Shahnoush Nayeri[1] and Maryam Zakavi[2]

[1]*Muller Lab, Department of Biology, Pennsylvania State University, University Park, PA, United States,*
[2]*Department of Cellular and Molecular Biology, Faculty of Life Sciences and Biotechnology, Shahid Beheshti University, Tehran, Iran*

Abbreviations

ATP	adenosine triphosphate
ACC	1-aminocyclopropane-1-carboxylic acid
ACS	ACC synthase
ACO	ACC oxidase
AB	abscisic acid
SAM	S-adenosyl-L-methionine
ROS	reactive oxygen species
SOD	superoxide dismutase
ER	endoplasmic reticulum
ERF	ethylene response factor
EIN	ethylene insensitive
CTR1	constitutive triple response 1
CaM	calmodulin
CDPK	Ca^{2+}-dependent protein kinase
HSP	heat shock protein
NUE	nutrient use efficiency
SOS	salt overly sensitive
P	phosphorus
N	nitrogen
K	potassium
Ca	alcium
Mg	magnesium
Zn	zinc
Cu	copper
Fe	iron
Mn	manganese
B	boron
Se	selenium

Mineral Nutrients to Mitigate Abiotic Stress in Crops. DOI: https://doi.org/10.1016/B978-0-443-44712-9.00015-3

15.1 Introduction

Plants are doomed to a life of environmental conflict due to their sessile nature. Since they are unable to escape, they have developed an advanced set of physiological, biochemical, and molecular defenses to deal with the ongoing challenge of abiotic stresses. Two essential and inextricably linked key mechanisms of plant defense, mineral nutrition and the phytohormone ethylene, are at the core of this resilience. For many years, physiologists deciphered the intricate hormone signaling pathways, while agronomists concentrated on the quantitative advantages of fertilizers. However, a paradigm shift is occurring, demonstrating that these systems are not merely next to one another but are intimately interconnected in a dynamic interaction. To overcome these obstacles, this chapter examines the critical intersection where specific mineral nutrients directly influence ethylene synthesis and function, ultimately controlling the plant's viability under abiotic stress conditions.

15.1.1 Abiotic stresses in contemporary agriculture: an unavoidable challenge

The increasing effects of climate change in the 21st century make it more difficult to feed a growing population. Drought, salinity, high temperatures, heavy metal pollution, air pollution, and nutrient imbalances are examples of abiotic stresses that have spread widely and are becoming more severe barriers to agricultural productivity worldwide (Husen, 2022, 2023, 2024). Abiotic stresses are unavoidable environmental realities, in contrast to biotic stresses brought on by pests and diseases, which are frequently under control. They cause severe yield losses that jeopardize global food security because they disturb cellular homeostasis, harm macromolecules like proteins and lipids, prevent photosynthesis, and stunt growth (Husen, 2021b, 2021a, 2021c, 2025).

Although effective, the conventional breeding method of creating stress-tolerant crops is a slow process that cannot keep up with the speed of environmental change. As a result, a more detailed and mechanistic understanding of how plants detect stress and activate their natural defense systems has become necessary (Bailey-Serres et al., 2019). Plant responses are now considered to be sophisticated, integrated signaling cascades that coordinate, ranging from gene expression to plant development, instead of simple reactions. The availability of mineral nutrients and the management of hormonal pathways, particularly ethylene, are critical survival determinants in those complex pathways (Zhu, 2016).

15.1.2 Mineral nutrients: not just components

As previously reported, the function of mineral nutrients—potassium (K) for osmotic regulation, phosphorus (P) for adenosine triphosphate (ATP) and nucleic acids, and nitrogen (N) for proteins and chlorophyll—has been investigated primarily in terms of their essential roles. Although there is no denying in their pivotal roles in plant growth and development, this perspective is insufficiently comprehensive. We now know that mineral nutrients are directly involved in the stress response mechanism as dynamic, active signaling molecules (Maathuis, 2009). For instance, potassium (K) is an important micronutrient that functions beyond being an osmoticum; it is also essential for the regulation of turgor pressure, stomatal aperture, and the activation of key enzymes that detoxify reactive oxygen species (ROS) (Wang et al., 2013). The availability of potassium has a direct impact on a plant's capacity to retain water during drought and salt stress conditions (Assaha et al., 2017). Likewise, calcium (Ca^{2+}) is a universal second messenger that plays key role in the first cellular reactions to almost any abiotic stresses by a transient influx into the cytoplasm. A variety of sensors, including calmodulins (CaMs) and CDPKs (Ca^{2+}-dependent protein kinases), initiate this "calcium signature" to induce subsequent defense mechanisms (Kudla et al., 2018; Dodd et al., 2010). Additionally, sulfur (S) plays a crucial role in the antioxidant and heavy metal detoxification systems as a component of glutathione and phytochelatins (Capaldi et al., 2015). According to (Hänsch and Mendel, 2009), micronutrients like copper (Cu) and zinc (Zn) are essential as cofactors for vital enzymes like superoxide dismutase (SOD), which scavenges the harmful superoxide radical (Hänsch and Mendel, 2009). Consequently, nutrient deficiencies disrupt the signaling and defense systems necessary for stress tolerance, in addition to precipitating nutritional deficiencies (Huber et al., 2012).

15.1.3 Ethylene: the dual roles of the "stress hormone"

Ethylene has a special and somewhat contradictory status among plant hormones as the "stress hormone." Numerous abiotic stressors, such as flooding, drought, wounding, heat, cold, and high salinity, can quickly induce its production (Wang et al., 2002). However, its physiological effects are well known to be context-dependent and show a "dual personality" that can make the difference between survival and mortality in plants (Kazan, 2015).

Ethylene has the potential to be a potent acclimation material. Ethylene accumulation triggers the "low-oxygen escape syndrome" during flooding or submersion, which causes petiole elongation and hyponastic growth, which enables leaves to rise above the water's surface (Voesenek and Bailey-Serres, 2015). It can assist in overcoming mechanical impedance during root growth (Khan et al., 2022; Pandey, 2020). Ethylene coordinates adaptive morphological changes in these situations to improve the survival rate of plants under different abiotic stress conditions.

On the other hand, ethylene hormone is probably involved in cell maturity and aging, which induces the programmed cell death mechanism in response to extreme or prolonged stress conditions. A strategy that potentially benefits the normal and healthy growth of plants by reallocating resources from older tissues to preserve reproductive structures, speed up leaf yellowing, chlorophyll degradation, and the remobilization of nutrients from tissues under stress conditions (Brodersen et al., 2019). The "survival versus mortality" model, which indicates that an unregulated, excessive ethylene response causes senescence, while a modulated and controlled response fosters tolerance, succinctly elucidates this dual function (Dubois et al., 2018). Therefore the crucial query is: What influences the plant's choice of path? There is mounting evidence that the plant's nutritional status acts as a key regulatory mechanism (Iqbal et al., 2011).

This chapter discusses our knowledge associated with plant responses and tolerance to significant abiotic stresses by investigating the effects of mineral nutrients on the regulation of ethylene synthesis and signaling. The mineral nutrient profile of plant serves as a basic modulator of the ethylene pathway, supplying the contextual cues that ultimately dictate the function of the hormone. We will examine the precise mechanistic connections, showing that nutrients are active modulators of hormonal fate rather than passive resources. The conversation will be organized around important nutrients and their distinct functions:

- Nitrogen and sulfur: We will investigate how the availability of N affects the synthesis of 1-aminocyclopropane-1-carboxylic acid (ACC), a precursor to ethylene, and how S-assimilation combines with ethylene to strengthen antioxidant defense (Jahan et al., 2021; Khan et al., 2022).
- Potassium and calcium: We will investigate how the activity of important enzymes in the ethylene biosynthesis pathway is directly impacted by K^+ status, and how Ca^{2+} signaling interacts with ethylene perception to either enhance or reduce stress signals (Dodd et al., 2010; Jung et al., 2009; Kudla et al., 2018; Wang et al., 2013; Yu et al., 2023).
- Micronutrients (Cu, Zn): As cofactors for enzymes such as ACC oxidase and superoxide dismutase (SOD), we will examine the crucial functions of micronutrients (Cu, Zn). This will establish a direct connection between micronutrient nutrition and the production of ethylene and the control of oxidative stress (Hänsch and Mendel, 2009; Maksymiec, 2007; Mikula et al., 2020; Yang and Hoffman, 1984).
- A comprehensive framework for understanding how mineral nutrition shapes the ethylene-mediated stress response (Srivastava et al., 2018).
- Developing innovative agronomic techniques, like precision nutrient management, that can strategically control this nutrient-hormone interaction to improve crop resilience in a challenging environment (Forni et al., 2017).

15.2 Ethylene signaling and biosynthesis

A comprehensive understanding of the hormone's synthesis and perception pathways is necessary to investigate how mineral nutrients affect stress tolerance through ethylene biosynthesis. While ethylene signaling is a linear cascade that converts the hormonal message into particular physiological outcomes, ethylene biosynthesis is a

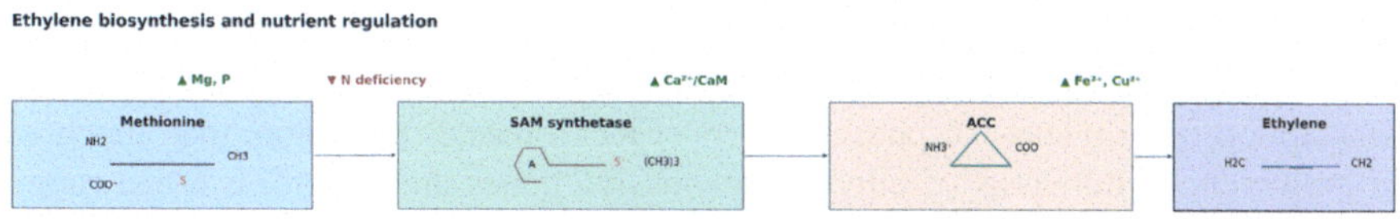

FIGURE 15.1

Ethylene biosynthesis with nutrient control points and regulation direction. Methionine → S-adenosyl-L-methionine (SAM) → 1-aminocyclopropane-1-carboxylic acid (ACC) → Ethylene. Green ▲ indicates upregulation (e.g., Mg^{2+}/P supporting ATP & SAM synthetase; Ca^{2+}/CaM stabilizing ACC synthase; Fe^{2+}/ascorbate/Cu^{2+} enabling ACC oxidase), red ▼ indicates down-regulation/limitation (e.g., N deficiency/NH_4^+ dominance).

strictly controlled, three-step process as follows: (1) Methionine is an essential amino acid that is converted into S-adenosyl-L-methionine (SAM) by ATP-dependent SAM synthetase enzyme. SAM is the universal precursor for ethylene and polyamines, a class of compounds frequently related to stress tolerance, and is an essential methyl group donor in many cellular reactions (Yang and Hoffman, 1984). SAM is changed into 1-aminocyclopropane-1-carboxylic acid (ACC), the direct precursor to ethylene, by ACC Synthase (ACS), which is responsible for committing ethylene synthesis. This is the pathway's main bottleneck and is tightly controlled by auxin, cytokinin, and different stressors are some of the factors that regulate ACS activity, along with gene transcription and protein stability (Wang et al., 2002). ACC oxidase (ACO), which catalyzes the last step, transforms ACC into ethylene through a reaction that needs ascorbate, oxygen, and Fe^{2+} (Fig. 15.1). Although ACO is typically more constitutively expressed, stress can also increase its activity, and its reliance on cofactors like iron establishes a direct connection between the production of ethylene and the availability of micronutrients (Lin et al., 2009).

15.2.1 From perception to reaction: the ethylene signaling pathway

A family of membrane-bound receptors detects ethylene presence, initiating a signaling cascade that eventually changes the expression of certain genes. Receptors in the endoplasmic reticulum (ER) membrane such as ethylene response 1 (ETR1) and Ethylene Insensitive 4 (EIN4) are active when ethylene is not present. Constitutive Triple Response 1 (CTR1), a Raf-like serine/threonine kinase, is the subsequent negative regulator in the pathway that they physically interact with and activate (Bleecker and Kende, 2000). Ethylene binds to the receptors, which leads to the inactivation of the receptors and, consequently, the inactivation of CTR1. To stop the ethylene response, active CTR1 phosphorylates and negatively regulates the downstream protein Ethylene Insensitive 2 (EIN2). Consequently, CTR1 is rendered inactive, alleviating the inhibition of EIN2. Transcriptional reprogramming by EIN2/EIN3 cascade is another regulatory system in the ethylene biosynthesis and signaling in plants. EIN2's C-terminal end can cleave and move to the nucleus when it is de-repressed. It stabilizes important transcription factors in the nucleus, mainly Ethylene Insensitive 3 (EIN3) and its analogs. These EIN3-like factors then trigger the transcription of Ethylene Response Factors (ERFs), master regulators that attach to the promoters of hundreds of ethylene-responsive genes and control a broad range of physiological reactions, including defense activation and growth inhibition (Chang et al., 1993; Merchante et al., 2013) (Fig. 15.1).

15.2.2 The ethylene dilemma: acclimation versus growth inhibition

Ethylene signaling has a highly context-dependent and unpredictable outcome, which leaves the plant with a choice between starting acclimatory processes or stopping growth to save resources. During growth inhibition and to overcome mechanical impedance in soil, seedlings employ the well-known "triple response": thickened hypocotyl, increased apical hook, and decreased hypocotyl and root elongation. Ethylene can accelerate fruit ripening, abscission, and leaf and flower senescence in mature plants. This encourages a "programmed retreat," whereby older tissues are sacrificed to preserve resources for seeds or younger tissues under extreme, irreversible stress (Brodersen et al., 2019). On the other hand, ethylene plays a crucial role in mediating adaptive reactions. It is necessary for the development of air-filled tissue, or aerenchyma, in roots during flooding, which promotes the diffusion of oxygen. In order to explore less compacted soil layers, it modifies root architecture, increases antioxidant production, and controls stomatal closure during drought (Pandey, 2020; Voesenek and Bailey-Serres, 2015). The severity and duration of stress, the presence of other hormones (e.g., high cytokinin can promote acclimation, high ABA can promote senescence), and most importantly, the plant's nutritional status all have an impact on the transition between these fates (Kazan, 2015). The schematic overview of the ethylene signaling pathway in plant cells is illustrated in Fig. 15.2.

15.3 Macronutrients as ethylene pathway regulators

The ethylene biosynthesis and signaling pathway are significantly influenced by the availability of macronutrients, which frequently determines whether the hormone's action results in beneficial senescence or successful acclimation. These substances directly interact with the essential elements of the ethylene network as metabolic substrates, enzymatic cofactors, and important signaling molecules.

15.3.1 Nitrogen: the metabolism master regulator

The availability of nitrogen, a basic component of amino acids and nucleic acids, controls the total metabolic flux and has a direct effect on the substrates needed to produce ethylene.

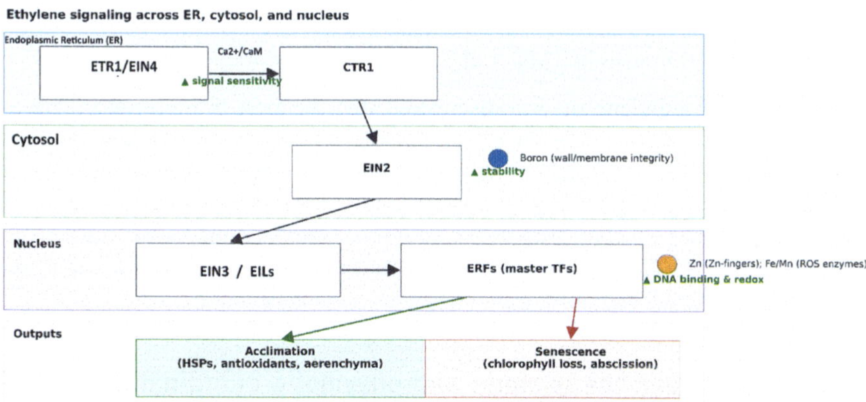

FIGURE 15.2

Ethylene signaling in a plant cell with nutrient modulation. Ethylene (C_2H_4) diffuses to the ER and binds ETR1/EIN4, inactivating CTR1 and de-repressing EIN2. The EIN2 C-end translocates to the nucleus, stabilizing EIN3/EILs and activating ERFs, which reprogram gene expression. Nutrient checkpoints tune signal strength: Ca^{2+}/CaM (ER sensitivity), Boron (cell wall/membrane integrity), and Zn/Fe/Mn (ERF DNA binding and redox). Moderated signaling drives acclimation (HSPs, antioxidants, aerenchyma), whereas an excessive response promotes senescence (chlorophyll loss, abscission).

15.3.1.1 Impact on the biosynthesis of polyamines and S-adenosyl-L-methionine

SAM is a key precursor for ethylene and polyamines such as spermidine and spermine, which are frequently linked to growth promotion and stress tolerance. High metabolic activity under ideal N conditions ensures that both pathways produce enough SAM. However, SAM becomes a limited resource when there is a lack of N (Fig. 15.1). The growth inhibition and early senescence that characterize N-starved plants are caused by the plant's frequent shifting of this balance, favoring the production of ethylene over the synthesis of polyamines (Alcázar et al., 2010).

15.3.1.2 The different ways that ethylene production impacted by N-form (NO_3^- vs. NH_4^+)

The evolution of ethylene can be directly influenced by the type of nitrogen that the plant has access to. It has been demonstrated that whereas ammonium (NH_4^+) nutrition can increase ethylene production, nitrate (NO_3^-) nutrition suppresses it. This is partially due to the fact that NO_3^- assimilation has the opposite effect and can also function as a signal to repress ethylene-related genes, while NH_4^+ assimilation acidifies the rhizosphere and cytosol, producing a stress-like condition that stimulates ACC synthase (ACS) activity (Ma et al., 2023).

- By regulating ethylene, it was discovered that the best N nutrition improves wheat tolerance during drought. A more moderate and controlled ethylene signal was made possible by adequate N, which suppressed a runaway ethylene burst (Fig. 15.1). By sensitizing stomatal guard cells to closure signals and encouraging deeper root growth—a traditional ethylene-mediated response to reach deeper soil layers for water—this moderated signal increased water use efficiency without hastening leaf senescence (Wang et al., 2024).

15.3.2 Potassium: the osmotic and enzymatic coordinator

Potassium plays a key role in stress reactions and the hormonal changes that accompany them because it is essential for stomatal function, osmotic regulation, and enzyme activation. K^+ is necessary for the activity of more than 50 enzymes, including those in primary metabolism that produce the ATP and reductants needed for the energetically expensive process of ethylene biosynthesis, even though it is not a direct cofactor for ACS or ACO. This metabolic basis is disturbed by K^+ deficiency (Wang et al., 2013; Yu et al., 2023; Jung et al., 2009). When the plant recognizes a K^+ deficiency as a strong abiotic stressor, the transcription and activity of ACS enzymes rapidly increase, which results in a remarkable production of ethylene and ACC. This ethylene burst is a crucial signal in Arabidopsis that rewires root architecture, preventing the growth of primary roots and encouraging the formation of lateral roots to improve foraging in the soil for small K^+ patches (Jung et al., 2009). By regulating the "ethylene burst," K supplementation reduces salt stress. Plants under salt stress suffer from K^+ deficiency and Na^+ toxicity. By avoiding the severe K^+ depletion that would otherwise cause a massive stress ethylene burst, research in cotton has shown that K^+ supplementation mitigates salt damage. The plant preserves photosynthetic tissue and increases survival rate by avoiding excessive ethylene production, which causes oxidative damage and cell death, by maintaining K^+ homeostasis (Yu et al., 2023).

15.3.3 Calcium: the second messenger link

The signaling pathways of calcium ions, which are ubiquitous intracellular second messengers, are intricately linked to the biosynthesis of ethylene. The cytosolic Ca^{2+} concentration briefly increases in response to almost all abiotic stressors. Different sensor proteins decode this "calcium signature" and have the ability to directly affect the ethylene pathway (Fig. 15.1). For example, posttranslational regulation of ACS proteins involves Ca^{2+} fluxes (Dodd et al., 2010). There is a significant relationship between ACS enzymes and calmodulin

(CaM). Calmodulin (CaM), a major Ca^{2+} sensor, controls a number of ACC synthase isoforms. Certain ACS isoforms, such as ACS6 in Arabidopsis, can be physically interacted with by Ca^{2+}-bound CaM, stabilizing the protein and increasing its activity. This offers a clear mechanistic connection, whereby an increase in ethylene production is immediately transduced from a stress-induced Ca^{2+} signal (Snedden and Fromm, 2001). Through a coordinated Ca^{2+}-ethylene cascade, calcium priming improves resistance to heat stress. Pretreating tall fescue with calcium chloride ($CaCl_2$) prior to applying heat stress greatly increased their thermotolerance. A controlled activation of ACS and a moderate increase in ethylene were the results of the primed plants' quick and temporary increase in cytosolic Ca^{2+}. In turn, this ethylene signal triggered the production of antioxidant enzymes and heat-shock proteins (HSPs), which offered defense against oxidative damage and protein denaturation. Ethylene inhibitors prevented this process (Virdi et al., 2015).

15.3.4 Magnesium: the primary atom of adenosine triphosphate and chlorophyll

Magnesium is essential to energy metabolism because it is a key component of chlorophyll and a cofactor for ATP-dependent enzymes. It also has a significant but indirect impact on the biosynthesis of ethylene. SAM synthetase uses ATP to convert methionine to SAM. Stabilizing the structure of ATP (creating Mg-ATP, the actual enzyme substrate) requires Mg^{2+}. Consequently, the rate of SAM synthesis is directly limited by magnesium deficiency, which also limits the amount of substrate available to produce ethylene and polyamines (Cakmak and Kirkby, 2008) (Fig. 15.1). A magnesium deficiency significantly hinders photosynthesis, which causes sugars and reactive oxygen species (ROS) to build up. The internal environment becomes extremely stressful as a result. The expression and activity of ACS and ACO can be significantly upregulated by stress signals (ROS, photo-inhibition), even though the substrate (SAM) may be limited. According to Farhat et al. (2016), the outcome is frequently a complex interplay in which magnesium deficiency both causes stress ethylene and may limit its maximum production capacity, resulting in the distinctive interveinal chlorosis and early senescence.

15.4 Micronutrients: inconspicuous but important factors

Micronutrients are needed in smaller amounts, but because of their unique and frequently nonredundant functions as structural elements and cofactors for enzymes, they have a disproportionate impact on the ethylene pathway. Their absence can either directly hinder the biosynthesis of ethylene or cause internal oxidative stress, which in turn causes the production of ethylene.

15.4.1 Zinc: a component of structure and regulation

Zinc's status is directly related to the metabolic and signaling pathways of the ethylene response because it is an essential component of many enzymes and transcription factors.

- The role of zinc as a cofactor for superoxide dismutase (SOD) in maintaining the homeostasis of reactive oxygen species (ROS), which affects the synthesis of ethylene. A crucial structural element of Cu/Zn-SOD, a significant enzyme that scavenges the superoxide radical (O_2^-) in the cytosol and chloroplasts, is zinc (Zhou et al., 2022). O_2^- and other ROS build up because of Cu/Zn-SOD activity being compromised by a zinc deficiency. As part of the stress alarm response, this oxidative stress causes a strong increase in ethylene production by inducing the expression of the ACS and ACO genes (Bastakoti, 2023). Therefore by upsetting cellular redox homeostasis, zinc deficiency indirectly encourages the evolution of ethylene.
- Zn-finger proteins play a direct or indirect part in ethylene signaling. Zinc finger domains are present in a wide variety of transcription factors, including many Ethylene Response Factors (ERFs), and are crucial to their ability to bind DNA. Zn ions are necessary for the correct structural folding of these domains. The function of these transcription factors may be compromised by zinc deficiency, which would hinder the plant's capacity to mount a suitable acclimation response and interfere with the downstream transcriptional reprogramming required by the ethylene signal (Kiełbowicz-Matuk, 2012).

15.4.2 Copper: direct contribution to the synthesis of ethylene

Of all the micronutrients, copper has the strongest and clearest connection to the biosynthesis of ethylene (Fig. 15.1); it is a necessary cofactor for the pathway's last stage.

- Cu is a necessary cofactor for the activity of the ACO enzyme. The ferrous iron and ascorbate-dependent dioxygenase enzyme ACC oxidase (ACO) catalyzes the transformation of ACC into ethylene. Importantly, to function fully, it also needs Cu^{2+} as a cofactor. The structural integrity and catalytic efficiency of the ACO enzyme are thought to be influenced by copper (Ruduś et al., 2013).
- Even if the ACC substrate has accumulated because of upstream stress signals, a lack of Cu directly impairs ACO activity, resulting in a decreased ability to produce ethylene. Stress signaling may be interfered with. Cu deficiency, for instance, has been demonstrated to decrease ethylene emission in tomatoes, which may change processes such as fruit ripening and stress-induced senescence (Czékus et al., 2022). On the other hand, the plant's enzymatic ability to produce the ethylene required for defense signaling is guaranteed by optimal Cu nutrition.

15.4.3 Boron: preserving signaling and membrane integrity

The primary function of boron is to cross-link pectin in the cell wall, which has significant secondary effects on membrane function and, in turn, stress perception.

- How stress ethylene and membrane damage are brought on by a B deficiency. Cell wall and membrane integrity quickly deteriorate when there is a lack of boron. Ions and metabolites leak out as a result, and membrane-bound receptors and transporters stop working properly (Kohli et al., 2023). The production of stress ethylene and ROS rises sharply because of this physical damage, which is interpreted as a severe stress (Fatma et al., 2022). The normal necrosis of meristematic tissues and the inhibition of root elongation are frequently caused by the ethylene generated during B deficiency (Camacho-Cristóbal et al., 2015).
- The function of B in cell wall structure and how it indirectly affects how stress is perceived. The first line of defense against abiotic stressors is a strong cell wall. Boron indirectly affects how stress is perceived by fortifying the cell wall. Under mild stress, a B-sufficient plant might sustain less physical membrane damage, which would lessen the stress signal that would otherwise cause a massive ethylene to burst. This demonstrates how B nutrition can change the plant's initial perception of the stress, thereby modulating the intensity of the ethylene response (Lynch and Brown, 1997).

15.4.4 Manganese and iron: cofactors for enzymes and redox players

Ethylene is largely affected by the disruption of redox balance caused by the deficiencies of both Fe and Mn, which are essential components in electron transfer reactions.

- Correlates with oxidative stress and ROS metabolism, which are key ethylene triggers.
- **Iron (Fe):** Catalase and other peroxidases that break down H_2O_2 require Fe as a cofactor. H_2O_2 builds up when there is a deficiency of Fe. Additionally, through the Fenton reaction, Fe itself can catalyze the production of extremely hazardous hydroxyl radicals (-OH). According to Briat et al. (2015), the oxidative stress that results is a potent inducer of the ethylene biosynthesis pathway.
- **Manganese (Mn):** Mn-SOD, which scavenges O_2^- in the mitochondria and peroxisomes, requires manganese (Mn) as a cofactor. This vital antioxidant defense is compromised by Mn deficiency, which causes O_2^- to build up and oxidative damage to follow. As seen in Mn-deficient plants, this oxidative burst increases ethylene production by activating the expression of the ACS gene (Alejandro et al., 2020).

In both situations, even in the absence of an external stressor, the micronutrient deficit induces an internal oxidative stress state that the plant perceives as an alarm signal and causes an upregulation of ethylene synthesis as part of a generalized defense activation.

15.5 Combined strategies to address particular abiotic stressors

The interaction of ethylene and mineral nutrients determines a plant's resistance to environmental stresses. The plant's nutritional state functions as a rheostat, adjusting the ethylene signal to either have a beneficial or negative effect. Fig. 15.3 shows the integrated abiotic stress–nutrient–ethylene framework.

15.5.1 Stress from drought

A careful balance between water conservation and ongoing resource acquisition is necessary for drought tolerance; this balance is regulated by phytohormones and impacted by plant nutrition.

- Root growth (deepening) mediated by K and N through ethylene regulation. A regulated, low-level ethylene signal encourages root elongation to reach deeper soil water reserves during moderate drought stress. Maintaining root turgor and promoting this growth depends on optimal potassium (K) status. Similar to how severe N deficiency or ammonium dominance can suppress root growth, sufficient nitrogen (N), especially in the nitrate form, can encourage a root architecture that is more responsive to this ethylene-mediated deepening signal (Pandey et al., 2023).

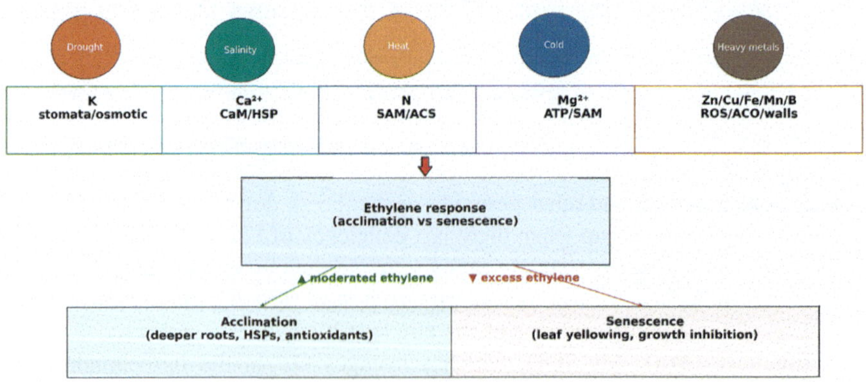

FIGURE 15.3

Integrated abiotic stress–nutrient–ethylene framework. Stress inputs (drought, salinity, heat, cold, heavy metals) feed nutrient nodes—K (stomatal/osmotic), Ca^{2+} (CaM/HSP priming), N (SAM/ACS flux), Mg^{2+} (ATP/SAM supply), Zn/Cu/Fe/Mn/B (ROS homeostasis, ACO cofactor, wall integrity). The enlarged ethylene-response panel summarizes outcomes: moderated ethylene (▲) → acclimation (deeper roots, aerenchyma, antioxidants/HSPs); excess ethylene (▼) → senescence (chlorophyll loss, abscission, growth inhibition).

- In response to nutritional conditions, ABA and ethylene crosstalk coordinate stomatal closure. Abscisic Acid (ABA) is the main hormone responsible for stomatal closure during drought. A complex crosstalk exists between ethylene and ABA; moderate ethylene can increase ABA sensitivity and encourage closure. But in this case, K is crucial. The primary osmolyte for stomatal opening is K^+. Because of the dysfunctional stomatal regulation caused by K-deficiency, the plant is unable to effectively execute the ABA/ethylene-mediated closure signal, which results in unnecessary water loss. Sufficient K guarantees that the stomata can react to the hormonal signal to close (Beguerisse-Dıaz et al., 2012; Müller, 2021).

15.5.2 Stress from salinity

The nutrient-ethylene interaction is essential for controlling the osmotic stress and ionic stress caused by salinity (Na^+ toxicity).

- Ethylene's function is in triggering the SOS pathway and antioxidant systems, as well as Na/K homeostasis. Elevated cytosolic Na+ is harmful. Salt stress causes ethylene to be produced quickly and contributes to the activation of the SOS (Salt Overly Sensitive) pathway, which causes Na+ to be effluxed from the cell. To counteract related oxidative stress, ethylene signaling also increases the synthesis of antioxidants. Sodium (Na^+) and potassium (K) directly compete for uptake. Maintaining the cytosolic K^+/Na^+ ratio, a crucial factor in determining salinity tolerance, requires high K status, which reduces Na^+ influx (Assaha et al., 2017). The systems that K^+ nutrition enables are activated in part by the ethylene signal.
- How dietary K and Ca^{2+} can encourage acclimatory signals and inhibit harmful ethylene peaks. Under extreme salt stress, senescence may be triggered by a large, sharp "ethylene burst." K and calcium (Ca^{2+}) are important in reducing this. It has been demonstrated that applying Ca^{2+} suppresses the transcript levels of particular ACS genes, reducing the initial ethylene surge and enabling a more regulated, acclimatory signal. Likewise, as was previously mentioned, K supplementation stops the ethylene burst caused by K-deficiency in salinity, which changes the ethylene response from one that causes senescence to one that promotes survival (Li et al., 2013; Zhu et al., 2022).

15.5.3 Heat and cold stress

The nutrient-ethylene nexus plays a role in stabilizing both protein function and membrane integrity, which are both at risk from temperature extremes. Damage to the membrane from either heat or cold conditions results in significant changes in fluidity and leakage. When it comes to temperature stress perception, calcium (Ca^{2+}) serves as the main messenger, stabilizing membranes and starting defense signaling. To prevent leakiness that would otherwise cause a strong stress ethylene response, boron (B) cross-links pectins in the cell wall, providing structural support that aids in maintaining membrane integrity under stress (Camacho-Cristóbal et al., 2015; Dodd et al., 2010). Under ideal nutrient conditions, ethylene acts as a modulator of heat-shock protein (HSP) expression. Chaperones known as HSPs stop other proteins from becoming denaturated at high temperatures. The upregulation of HSP genes has been linked to ethylene signaling. A healthy diet optimizes this process. One obvious mechanism by which a nutrient (Ca^{2+}) primes a beneficial ethylene response (HSP induction) to confer thermotolerance is the Ca^{2+}-ethylene cascade previously described (Larkindale and Huang, 2005). Magnesium (Mg) is also indirectly essential for this response since it supplies the ATP needed for the synthesis of proteins, including HSP.

15.5.4 Toxicity of heavy metals

Heavy metals that cause extreme oxidative stress, such as arsenic (As), lead (Pb), and cadmium (Cd), need to be detoxified. By competing with Cd for protein binding sites and uptake transporters, zinc (Zn) directly lowers the toxicity of Cd. Iron (Fe) homeostasis is essential because Cd interferes with Fe metabolism, resulting in chlorosis that resembles a deficiency. This can be lessened with proper Fe nutrition. Although not necessary for every plant, selenium (Se) can help plants cope with heavy metal stress by lowering metal bioavailability, strengthening the antioxidant system, and forming nontoxic Se-metal complexes (Nayeri et al., 2023; Wu et al., 2016).

The function of ethylene in initiating antioxidant defense and phytochelatin synthesis. Ethylene synthesis is strongly stimulated by heavy metals (Nayeri et al., 2023). This ethylene burst is an active signaling element of the defense, not just a sign of damage. It has been demonstrated that ethylene signaling increases the genes that produce phytochelatins, which are cysteine-rich peptides that detoxify heavy metals by chelating them and enclosing them in vacuoles (Biswas and Pal, 2025; Vitelli et al., 2024). Additionally, ethylene forms an integrated defense network against metal-induced oxidative stress by collaborating with antioxidant systems that rely on micronutrients such as Zn (SOD) and Fe (peroxidases) (Sana et al., 2025).

15.6 The "Yin and Yang": stress-inducing nutrient deficiency ethylene

One of the basic regulatory nexuses in plant stress physiology is the connection between ethylene synthesis and mineral nutrition. The lack of vital mineral nutrients themselves acts as a strong endogenous stress signal that sets off intricate ethylene-mediated reactions, even though external abiotic stressors are usually the ones that cause ethylene production (Iqbal et al., 2013). This section investigates the dual nature of these responses in plant adaptation and survival, as well as how ethylene biosynthesis pathways are activated by nutrient deficiencies.

15.6.1 Abiotic stressor: nutrient deficiency

One important type of abiotic stress that upsets cellular homeostasis via a variety of mechanisms is mineral nutrient deficiencies (Maathuis, 2009). Depending on the particular nutrient, the physiological effects can vary, but typically include:

Metabolic disruption: While magnesium deficiency impairs chlorophyll synthesis and photosynthetic efficiency, nitrogen deficiency hinders amino acid synthesis and protein production (Cakmak and Kirkby, 2008).
Energy constraints: Almost all energy-dependent cellular functions are impacted by phosphorus deficiency, which restricts ATP synthesis and energy transfer (Carstensen et al., 2018).
Oxidative damage: By reducing the activity of important enzymes like superoxide dismutase, micronutrient deficiencies, especially those of zinc, copper, and manganese, jeopardize antioxidant defense systems (Roychoudhury and Tripathi, 2020).
Membrane integrity loss: Loss of membrane integrity: A lack of boron quickly impairs the structure of cell walls and the function of membranes, resulting in cellular leakage (Camacho-Cristóbal et al., 2015).

According to the previous reports, the plant interprets these disturbances as stress signals that set off complex reaction mechanisms, with ethylene emerging as a crucial signaling element in nutrient deficiency responses (Table 15.1) (Czékus et al., 2022; Fatma et al., 2022; Iqbal et al., 2013; Iqbal et al., 2011; Jahan et al., 2021).

15.6.2 The ethylene biosynthesis pathway's quick induction

One of the first physiological reactions to restricted nutrient availability is frequently the remarkably quick changes in ethylene biosynthesis brought on by nutrient deficiencies:

Regulation of transcription: Deficits in nitrogen, potassium, phosphorus, magnesium, and other micronutrients have been shown in numerous studies to trigger the expression of genes encoding the essential enzymes for ethylene biosynthesis, ACC synthase (ACS) and ACC oxidase (ACO) (Wang et al., 2002). In Arabidopsis, for example, potassium deficiency causes the upregulation of particular ACS isoforms (ACS7 and ACS11) within hours of K^+ deprivation (Jung et al., 2009).
Enzyme activation: In addition to transcriptional regulation, dietary deficiencies can have a direct impact on the stability and activity of the enzymes involved in ethylene biosynthesis. Through phosphorylation/dephosphorylation processes, calcium signaling pathways are impacted by nutritional status, controlling the stability of ACS proteins (Li et al., 2012). In a similar vein, ethylene production capacity is directly

Table 15.1 Ethylene signaling cascade and nutrient modulation points.

Component	Location	Core role	Nutrient modulation	Outcome when perturbed	Examples and study cases	Key references
ETR1/EIN4 (receptors)	ER membrane	Perceive ethylene; repress pathway in absence	B maintains wall/membrane integrity; ionic homeostasis affects stability	B deficiency exaggerates stress signal → ethylene surge	Arabidopsis seedlings: B deficiency inhibits root elongation via ethylene/auxin/ROS	Bleecker and Kende (2000), Camacho-Cristóbal et al. (2015)
CTR1 (Raf-like kinase)	ER-associated	Negative regulator; phosphorylates EIN2	Ion stress (Na^+/K^+) and Ca^{2+} signaling shift responsiveness	CTR1 inactivation allows EIN2 cleavage → ERF activation	Ethylene-triggered CTR1 trafficking enhances response	Ju et al. (2012), Park et al. (2023)
EIN2 (C-end)	Cytosol → nucleus	Transduces signal; stabilizes EIN3	Ca^{2+} fluxes and micronutrient-dependent ROS affect stability	Robust ethylene response if antioxidant capacity adequate	Mustard with exogenous ethylene under varied N shows improved photosynthesis and conductance	Merchante, Alonso and Stepanova (2013), Dodd, Kudla and Sanders (2010), Iqbal et al. (2011)
EIN3/EILs	Nucleus	Activate ERFs	Zn stabilizes TF domains; N form modulates network	Poor Zn impairs ERF binding → maladaptation	Zn deficiency increases ROS and induces ethylene	Kiełbowicz-Matuk (2012), Bastakoti (2023)
ERFs	Nucleus	Activate stress genes	Zn-SOD, Fe-peroxidases, Mn-SOD status calibrate ROS	Balanced ERF activity drives acclimation; excess ROS favors senescence	Poplar ERF72 overexpression coordinates stress and metabolism	Zhou et al. (2022), Briat Dubos and Gaymard (2015), Dong et al. (2025)

impacted by the availability of cofactors like copper for ACO and magnesium for SAM synthetase (Cakmak and Kirkby, 2008; Yang and Hoffman, 1984).
Systemic signaling: Whole-plant responses are coordinated by the ethylene burst caused by a localized nutrient deficit. This has been especially well-documented for deficiencies in phosphorus and nitrogen, where ethylene signals from roots inform shoot tissues of nutrient status (Iqbal et al., 2013).

Ethylene is a primary integrator of nutrient status information due to the rapidity and magnitude of its production in response to nutrient deficiency (Sun et al., 2025).

15.6.3 Distinguishing between beneficial and harmful effects

Ethylene's dual nature in plant stress adaptation is demonstrated by the functional outcome of deficiency-induced ethylene production, which shows remarkable context dependency (Table 15.2).

- **Adaptive reactions (The "Yang"):** Ethylene signaling triggers complex acclimation mechanisms under moderate deficiency conditions:
  Remodeling the root architecture: When phosphorus or potassium levels are low, ethylene and auxin work together to prevent primary root elongation and encourage the growth of lateral roots and root hairs, thereby expanding the root surface area available for nutrient uptake (López-Vicente and Navas, 2009; Jung et al., 2009).
  Remobilization of nutrients: Deficits in magnesium and nitrogen cause ethylene-mediated senescence in older leaves, which makes it easier for nutrients to move to developing tissues and maximizes the plant's use of its resources (Schildhauer et al., 2008).
  Symbiotic associations: Under nutrient-limited circumstances, ethylene signaling regulates the formation of mycorrhizal associations and root nodulation, improving the plant's ability to absorb nutrients (Guinel, 2015).
- **Negative effects (The "Yin"):** The ethylene response turns maladaptive in cases of severe or protracted deficiency:
  Accelerated senescence: Excessive ethylene production in the presence of a severe nitrogen or magnesium deficiency causes premature and uncontrollable senescence, which seriously impairs photosynthetic capacity and growth (Jibran et al., 2013).
  Growth inhibition: A chronic nutrient deficiency causes persistently high ethylene levels, which inhibit cell division and expansion over time and cause severe growth stunting (Millward, 2017).
  Oxidative damage: When antioxidant systems are already weakened, the ethylene burst linked to micronutrient deficiencies can worsen oxidative stress, resulting in a damaging feedback loop (Gupta et al., 2024).

Deficit severity, duration, plant developmental stage, and interactions with other hormonal pathways are some of the variables that influence the shift from adaptive to detrimental responses (Kazan, 2015). Gaining an understanding of this delicate

Table 15.2 Ethylene biosynthesis: steps, enzyme requirements, nutrient dependencies, and stress modulation.

Step	Substrate → product	Enzyme (cofactors)	Nutrient dependency and control points	Stress/defense modulation	Examples and study cases	Key references
1	Methionine → SAM	SAM synthetase (ATP)	Mg^{2+} forms Mg-ATP; P controls ATP supply; N controls methionine pool	Low Mg^{2+}/P reduce SAM; N deficiency shifts SAM toward ethylene vs polyamines	Mg-deficiency limits SAM and increases ROS leading to chlorosis	Yang and Hoffman (1984), Cakmak and Kirkby (2008), Carstensen et al. (2018), Alcázar et al. (2010)
2	SAM → ACC	ACC Synthase (ACS)	Ca^{2+}/CaM stabilize ACS; transcription induced by K, N, P, Mg deficiencies	K-deficiency triggers "ethylene burst" via ACS7/11 upregulation	Arabidopsis under K-deficiency shows rapid ACS induction and root architectural change	Wang et al. (2002); Snedden and Fromm (2001), Jung et al. (2009); Ma et al. (2023)
3	ACC → Ethylene	ACC Oxidase (ACO) (O_2, Fe^{2+}, ascorbate, Cu^{2+})	Cu^{2+} critical for ACO; Fe and ascorbate required	Cu deficiency lowers ethylene emission; adequate micronutrients enable acclimatory signaling	Tomato shows altered ER stress and unfolded protein response with ethylene modulation	Lin et al. (2009); Ruduś et al. (2013); Czékus et al. (2022)

Abbreviations: *SAM*, S-adenosyl-L-methionine; *ACC*, 1-aminocyclopropane-1-carboxylic acid; *ACS*, ACC synthase; *ACO*, ACC oxidase.

balance is essential for creating precision nutrient management plans that maximize ethylene-mediated reactions for increased crop resilience.

15.7 Implications for agronomy and biotechnology

The complex interrelationship between ethylene biology and mineral nutrition offers revolutionary prospects for agricultural innovation. We can create cutting-edge strategies to improve crop resilience and resource efficiency in harsh environments by carefully controlling this nutrient-hormone interface (Iqbal et al., 2013).

15.7.1 Precision nutrient management: adapting fertilizer schedules to maximize hormonal equilibrium in response to anticipated stressors

By actively modifying the plant's hormonal landscape for improved stress readiness, precision nutrient management signifies a paradigm shift away from merely satisfying crop nutrient demands (Sun et al., 2025). Predictive nutrient applications are made possible by sophisticated soil sensing and weather forecasting (Raza et al., 2023). Predrought potassium application, for example, has shown impressive effectiveness in preserving osmotic balance and averting the harmful ethylene burst linked to water stress (Sun et al., 2025). According to Wang et al. (2017), field research on maize revealed that optimal K status prior to the onset of drought led to a 25%–30% increase in water use efficiency through moderated ethylene signaling that encouraged deeper root proliferation and optimized stomatal regulation.

Formulations containing supplemental calcium and optimized K:Na ratios efficiently suppress salinity-induced ethylene peaks in salinity-prone areas while preserving ionic homeostasis. Studies on tomatoes showed that under saline conditions, Ca-supplemented nutrition improved fruit yield while reducing stress ethylene production by 40% (Saleh et al., 2019). Aligning nutrient applications with crucial developmental windows and predictable stress periods is known as "temporal application synchronization." Particularly in legumes and cereals, nitrogen deficiency-induced ethylene surges that result in flower abortion can be avoided by applying controlled-release nitrogen fertilizers prior to reproductive stages (Pandey et al., 2023).

15.7.2 Nutrient priming: applying particular nutrients to seeds or leaves to "precondition" plants for increased resistance to stress

Through physiological and epigenetic preconditioning, nutrient priming creates a "metabolic memory" that increases plant resilience:

Seed priming mechanisms: By creating a memory for Ca^{2+}-ethylene-HSP, osmotic priming with calcium chloride improves heat tolerance. Ca-primed plants showed 50% higher survival rates under heat stress due to increased ethylene-mediated HSP expression, according to studies conducted on creeping bent grass (Larkindale and Huang, 2005). Likewise, the activation of antioxidant systems through zinc and selenium priming results in a primed state that regulates ethylene-mediated oxidative stress responses (El-Badri et al., 2021).

Foliar priming applications: Magnesium and boron prestress foliar treatments improve the integrity of the cell wall and photosynthetic apparatus, respectively. In comparison to nonprimed plants, studies on soybeans revealed that magnesium priming prior to heat stress preserved photosynthetic efficiency and decreased stress ethylene evolution by 35% (Farhat et al., 2016). Inoculation of salt-tolerant *Bacillus sp.* is a promising approach to increase maize yield under salt stress condition (Zakavi et al., 2022a, 2024, 2022b).

Combined priming methods: Synergistic effects are produced by sequential or combined nutrient priming. By coordinating the regulation of ethylene biosynthesis and antioxidant defense pathways, a zinc–manganese copriming strategy in rice improved submergence tolerance (Nadeem et al., 2020; Qiao et al., 2019). Our research identified a promising strategy to enhance crop growth by optimizing photosynthesis through the use of an Ag/ZnO nanoparticle mixture. The integration of Ag/ZnO nanoparticles into accelerated breeding programs holds significant potential to revolutionize crop production by lowering costs and augmenting agricultural productivity (Nayeri et al., 2023).

15.7.3 Synergy in genetic engineering

Crop improvement through strategic genetic manipulation at the nutrient-ethylene interface has revolutionary potential:

- **Modified ethylene sensitivity with nutrient use efficiency:** Resilient genotypes are produced by pyramid NUE traits with ethylene insensitivity. Under low nitrogen conditions, when conventional varieties exhibited significant ethylene-mediated senescence, the ethylene-insensitive *etr1-1* mutant-maintained yield during introgression of high-NUE rice lines (Sharma and Kumari, 2025). According to Park et al. (2023), altering EIN2 expression in genotypes that are phosphorus-efficient enhanced root foraging in the presence of P deficiency without causing growth penalties.
- **Targeted genetic interventions:**

Transporter–hormone fusions: Feedback-regulated uptake systems are produced by combining ethylene-responsive promoters with nutrient transporters. In Arabidopsis, the ethylene-responsive promoter-controlled NRT1.1 nitrate transporter demonstrated enhanced N uptake coordination with stress status (Yao et al., 2025).

Signaling network nodes: Changing important integrators that coordinate ethylene and nutrient signaling, such as ERF transcription factors and EIN3. Overexpression of

ERF72 in poplar enhanced both drought tolerance and nitrogen utilization efficiency through coordinated regulation of stress and metabolic genes (Dong et al., 2025).
Nutrient-sensing circuits: Developing synthetic genetic circuits that connect ethylene response modulation and nutrient sensing. In response to environmental stimuli, the tomato's calcium-CaMK-ACS regulatory module dynamically controlled the production of ethylene (Lee and Kang, 2024).

By strategically manipulating the nutrient-ethylene nexus, these integrated approaches show that crop performance can be significantly improved. This opens the door to climate-resilient agriculture by optimizing the body rather than just increasing input intensity

15.8 Conclusion

An important paradigm shift in our understanding of plant stress physiology is epitomized by the work presented in this chapter: mineral nutrition and hormonal signaling are recognized not as separate systems, but rather as highly complex components of a unitary stress response system. It is exemplified that in order for precision agriculture under climate change scenarios to become a practical possibility, methods for plant diagnoses must shift to account for these considerations.

Mineral nutrients are pivotal regulators and modulators of the stress tolerance response mediated by ethylene based on the synthesis of research evidence supporting the major argument. This regulatory action occurs through four closely interrelated processes. (1) Metabolic basis: through the methionine-SAM-ACC pathway, macronutrients like nitrogen and sulfur govern substrate availability for ethylene biosynthesis, and magnesium plays a pivotal role in facilitating ATP availability in energy-demanding reactions. (2) Enzymatic regulation: through calmodulin binding, calcium signaling modulates ACS protein stability and/or activity, and potassium acts as an essential cofactor in enzymes. (3) Cofactor requirements: ethylene biosynthesis (ACO activity) and its integrated antioxidant systems delineating the redox environment of ethylene action are all reliant on micronutrient copper, zinc, and iron. (4) Signal integration: sensing or signaling of acclimation and/or senescence by ethylene activity and responses are conditioning actions of plant nutritional status that governs not just acclimation responses but also defines specificity or thresholds of ethylene sensing. The tools of diagnostics have to quantify or fail to detect the dynamically integrated effect of plant nutrition and ethylene signaling that makes plants have an "advanced" stress response modification mechanism based on plant metabolism.

15.9 Future considerations

Notwithstanding notable progress, there are still important knowledge gaps that present obstacles as well as chances for further study. Although nutrients and ethylene have different roles in different tissues, most diagnostic tools available today offer evaluations at the whole plant or organ level. How does ethylene-mediated potassium deficiency specific to roots change the physiology of shoots? To resolve these spatial dynamics, sophisticated methods such as tissue-specific biosensors and single-cell RNA sequencing are required. Although copper is known to play a part in ACO activity, little is known about how other micronutrients might affect ethylene receptor function. Does the ethylene receptor-CTR1 complex assemble or remain stable in response to the presence of zinc, manganese, or other metals? Advanced metal imaging methods in conjunction with structural biology approaches may uncover new regulatory layers. Although the interaction between ethylene and ABA has been studied, little is known about how it interacts with strigolactones, jasmonic acid, and brassinosteroids in nutritional contexts. These intricate relationships could be deciphered, and novel diagnostic biomarkers could be found through multihormone profiling using sophisticated mass spectrometry under specific nutritional conditions.

Integrated methods that consider crops' physiological conditions holistically are the way of the future for agricultural diagnostics and management. The next-generation diagnostic platforms creating field-deployable instruments that evaluate hormonal markers and nutritional status at the same time. These include machine learning algorithms that combine soil, tissue, and remote sensing data for predictive stress diagnostics, hyperspectral imaging linked to ethylene precursor levels, and nanosensors for real-time hormone monitoring. Using dynamic management systems, the decision-support tools are developed that use integrated nutrient-hormone indices to enable adaptive management as opposed to static fertilizer recommendations. To produce precise recommendations that maximize nutrition and stress resilience, these systems would integrate real-time environmental data, crop models, and hormonal status. Pyramiding ideal nutrient-uptake traits with tuned hormone sensitivity should be the focus of crop improvement in the future. This entails creating high-throughput phenotyping techniques for nutrihormonal traits and identifying naturally occurring variations of ethylene signaling components that preserve adaptive responses while reducing yield penalties under stress. To develop context-specific management protocols, adoption calls for long-term monitoring networks that track hormonal and nutritional dynamics across various agroecological zones.

With regards to the consequences, a new area of plant stress physiology with significant ramifications for agricultural sustainability is the combination of ethylene biology and mineral nutrition. We can move closer to climate-resilient agriculture that maximizes plant performance through physiological understanding rather than input intensification by creating diagnostic tools and management strategies that take this integration into account.

References

Alcázar, R., Altabella, T., Marco, F., Bortolotti, C., Reymond, M., Koncz, C., et al., 2010. Polyamines: molecules with regulatory functions in plant abiotic stress tolerance. Planta 231 (6), 1237–1249. Available at https://doi.org/10.1007/s00425-010-1130-0.

Alejandro, S., Höller, S., Meier, B., Peiter, E., 2020. Manganese in plants: from acquisition to subcellular allocation. Frontiers in Plant Science 11. Available at https://doi.org/10.3389/fpls.2020.00300.

Assaha, D.V.M., Ueda, A., Saneoka, H., Al-Yahyai, R., Yaish, M.W., 2017. The role of Na+ and K+ transporters in salt stress adaptation in glycophytes. Frontiers in Physiology 8. Available at https://doi.org/10.3389/fphys.2017.00509.

Bailey-Serres, J., Parker, J.E., Ainsworth, E.A., Oldroyd, G.E.D., Schroeder, J.I., 2019. Genetic strategies for improving crop yields. Nature 575 (7781), 109–118. Available at https://doi.org/10.1038/s41586-019-1679-0.

Bastakoti, S., 2023. Role of zinc in management of plant diseases: a review. Cogent Food & Agriculture 9 (1). Available at https://doi.org/10.1080/23311932.2023.2194483.

Beguerisse-Díaz, M., Hernández-Gómez, M.C., Lizzul, A.M., Barahona, M., Desikan, R., 2012. Compound stress response in stomatal closure: a mathematical model of ABA and ethylene interaction in guard cells. BMC Systems Biology 6 (1). Available at https://doi.org/10.1186/1752-0509-6-146.

Biswas, A., Pal, S., 2025. Role of phytohormones in developing heavy metal stress resilience in rice: physiological–molecular adjustments, crosstalk, and prospects. Journal of Plant Growth Regulation 44 (9), 4960–4980. Available at https://doi.org/10.1007/s00344-025-11757-w.

Bleecker, A.B., Kende, H., 2000. Ethylene: a gaseous signal molecule in plant. Annual Review of Cell and Developmental Biology 16, 1–18. Available at https://doi.org/10.1146/annurev.cellbio.16.1.1.

Briat, J.F., Dubos, C., Gaymard, F., 2015. Iron nutrition, biomass production, and plant product quality. Trends in Plant Science 20 (1), 33–40. Available at https://doi.org/10.1016/j.tplants.2014.07.005.

Brodersen, C.R., Roddy, A.B., Wason, J.W., McElrone, A.J., 2019. Functional status of xylem through time. Annual review of plant biology 70, 407–433. Available at https://doi.org/10.1146/annurev-arplant-050718-100455.

Cakmak, I., Kirkby, E.A., 2008. Role of magnesium in carbon partitioning and alleviating photooxidative damage. Physiologia Plantarum 133 (6), 692–704. Available at https://doi.org/10.1111/j.1399-3054.2007.01042.x.

Camacho Cristóbal, J.J., Martín-Rejano, E.M., Herrera-Rodríguez, M.B., Navarro-Gochicoa, M.T., Rexach, J., González-Fontes, A., 2015. Boron deficiency inhibits root cell elongation via an ethylene/auxin/ROS-dependent pathway in *Arabidopsis* seedlings. Journal of Experimental Botany 66 (13), 3831–3840. Available at https://doi.org/10.1093/jxb/erv186.

Capaldi, F.R., Gratão, P.L., Reis, A.R., Lima, L.W., Azevedo, R.A., 2015. Sulfur metabolism and stress defense responses in plants. Tropical Plant Biology 8 (3-4), 60–73. Available at https://doi.org/10.1007/s12042-015-9152-1.

Carstensen, A., Herdean, A., Schmidt, S.B., Sharma, A., Spetea, C., Pribil, M., et al., 2018. The impacts of phosphorus deficiency on the photosynthetic electron transport chain1[OPEN]. American Society of Plant Biologists, Denmark Plant Physiology 177 (1), 271–284. Available at https://doi.org/10.1104/PP.17.01624.

Chang, C., Kwok, S.F., Bleecker, A.B., Meyerowitz, E.M., 1993. Arabidopsis ethylene-response gene ETR1: similarity of product to two-component regulators. Science 262 (5133), 539–544. Available at https://doi.org/10.1126/science.8211181.

Czékus, Z., Szalai, G., Tari, I., Khan, M.I.R., Poór, P., 2022. Role of ethylene in ER stress and the unfolded protein response in tomato (*Solanum lycopersicum* L.) plants. Plant Physiology and Biochemistry 181, 1–11. Available at https://doi.org/10.1016/j.plaphy.2022.03.031.

Dodd, A.N., Kudla, J., Sanders, D., 2010. The language of calcium signaling. Annual Review of Plant Biology 61, 593–620. Available at https://doi.org/10.1146/annurev-arplant-070109-104628.

Dong, W., Wang, X., Yang, K., Zhang, H., Ding, Y., Gao, W., et al., 2025. The dynamic regulation response to salt stress in poplar. Trees 39 (3). Available at https://doi.org/10.1007/s00468-025-02619-9.

Dubois, M., Van den Broeck, L., Inzé, D., 2018. The pivotal role of ethylene in plant growth. Trends in Plant Science 23 (4), 311–323. Available at https://doi.org/10.1016/j.tplants.2018.01.003.

El-Badri, A.M., Batool, M., Wang, C., Hashem, A.M., Tabl, K.M., Nishawy, E., et al., 2021. Selenium and zinc oxide nanoparticles modulate the molecular and morpho-physiological processes during seed germination of *Brassica napus* under salt stress. Academic Press, China Ecotoxicology and Environmental Safety 225. Available at https://doi.org/10.1016/j.ecoenv.2021.112695.

Farhat, N., Elkhouni, A., Zorrig, W., Smaoui, A., Abdelly, C., Rabhi, M., 2016. Effects of magnesium deficiency on photosynthesis and carbohydrate partitioning. Acta Physiologiae Plantarum 38 (6). Available at https://doi.org/10.1007/s11738-016-2165-z.

Fatma, M., Asgher, M., Iqbal, N., Rasheed, F., Sehar, Z., Sofo, A., et al., 2022. Ethylene signaling under stressful environments: analyzing collaborative knowledge. Plants 11 (17), 2211. Available at https://doi.org/10.3390/plants11172211.

Forni, C., Duca, D., Glick, B.R., 2017. Mechanisms of plant response to salt and drought stress and their alteration by rhizobacteria. Plant and Soil 410 (1-2), 335–356. Available at https://doi.org/10.1007/s11104-016-3007-x.

Guinel, F.C., 2015. Ethylene, a hormone at the center-stage of nodulation. Frontiers in Plant Science 6. Available at https://doi.org/10.3389/fpls.2015.01121.

Gupta, R., Verma, N., Tewari, R.K., 2024. Micronutrient deficiency-induced oxidative stress in plants. Plant Cell Reports 43 (9). Available at https://doi.org/10.1007/s00299-024-03297-6.

Hänsch, R., Mendel, R.R., 2009. Physiological functions of mineral micronutrients (Cu, Zn, Mn, Fe, Ni, Mo, B, Cl). Current Opinion in Plant Biology 12 (3), 259–266. Available at https://doi.org/10.1016/j.pbi.2009.05.006.

Huber, D., Römheld, V., Weinmann, M., 2012. Relationship between nutrition, plant diseases and pests. Marschner's Mineral Nutrition of Higher Plants. Elsevier Ltd., United States, pp. 283–298. Available from. Available at 10.1016/B978-0-12-384905-2.00010-8.

Husen, A., 2021a. Harsh Environment and Plant Resilience: Molecular and Functional Aspects. Springer Nature, Cham. Available at https://doi.org/10.1007/978-3-030-65912-7.

Husen, A., 2021. The harsh environment and resilient plants: an overview. In: Husen, A. (Ed.), Harsh Environment and Plant Resilience. Springer, Cham, pp. 1–23. Available at https://doi.org/10.1007/978-3-030-65912-7_1.

Husen, A., 2021c. Plant Performance Under Environmental Stress: Hormones, Biostimulants and Sustainable Plant Growth Management. Springer Nature, Cham. Available at https://doi.org/10.1007/978-3-030-78521-5.

Husen, A., 2022. Environmental Pollution and Medicinal Plants. CRC Press, Boca Raton, FL. Available at https://doi.org/10.1201/9781003178866.

Husen, A., 2023. Plants and Their Interaction to Environmental Pollution: Damage Detection, Adaptation, Tolerance, Physiological and Molecular Responses. Elsevier Inc., Cambridge, MA. Available at https://doi.org/10.1016/C2020-0-03319-0.

Husen, A., 2024. Essential Minerals in Plant-Soil System (Coordination, Signalling, and Interaction under Adverse Situations. Elsevier Inc., Cambridge, MA. Available at https://doi.org/10.1016/C2022-0-02437-5.

Husen, A., 2025a. Roles of Osmolytes in Changing Environment. Elsevier Inc., Cambridge, MA. Available at https://doi.org/10.1016/C2023-0-51382-5.

Iqbal, N., Trivellini, A., Masood, A., Ferrante, A., Khan, N.A., 2013. Current understanding on ethylene signaling in plants: the influence of nutrient availability. Plant Physiology and Biochemistry 73, 128–138. Available at https://doi.org/10.1016/j.plaphy.2013.09.011.

Iqbal, N., Nazar, R., Syeed, S., Masood, A., Khan, N.A., 2011. Exogenously-sourced ethylene increases stomatal conductance, photosynthesis, and growth under optimal and deficient nitrogen fertilization in mustard. Journal of Experimental Botany 62 (14), 4955–4963. Available at https://doi.org/10.1093/jxb/err204.

Jahan, B., Iqbal, N., Fatma, M., Sehar, Z., Masood, A., Sofo, A., et al., 2021. Ethylene supplementation combined with split application of nitrogen and sulfur protects salt-inhibited photosynthesis through optimization of proline metabolism and antioxidant system in mustard (*Brassica juncea* L). Plants 10 (7). Available at https://doi.org/10.3390/plants10071303.

Jibran, R., A. Hunter, D., P. Dijkwel, P., 2013. Hormonal regulation of leaf senescence through integration of developmental and stress signals. Plant Molecular Biology 82 (6), 547–561. Available at https://doi.org/10.1007/s11103-013-0043-2.

Jung, J.Y., Shin, R., Schachtman, D.P., 2009. Ethylene mediates response and tolerance to potassium deprivation in arabidopsis. The Plant Cell 21 (2), 607–621. Available at https://doi.org/10.1105/tpc.108.063099.

Kazan, K., 2015. Diverse roles of jasmonates and ethylene in abiotic stress tolerance. Trends in Plant Science 20 (4), 219–229. Available at https://doi.org/10.1016/j.tplants.2015.02.001.

Khan, F., Pandey, P., Upadhyay, T.K., 2022. Applications of nanotechnology-based agrochemicals in food security and sustainable agriculture: an overview. Agriculture 12 (10), 1672. Available at https://doi.org/10.3390/agriculture12101672.

Kiełbowicz-Matuk, A., 2012. Involvement of plant C_2H_2-type zinc finger transcription factors in stress responses. Plant Science 185-186, 78–85. Available at https://doi.org/10.1016/j.plantsci.2011.11.015.

Kohli, S.K., Kaur, H., Khanna, K., Handa, N., Bhardwaj, R., Rinklebe, J., et al., 2023. Boron in plants: uptake, deficiency and biological potential. Plant Growth Regulation 100 (2), 267–282. Available at https://doi.org/10.1007/s10725-022-00844-7.

Kudla, J., Becker, D., Grill, E., Hedrich, R., Hippler, M., Kummer, U., et al., 2018. Advances and current challenges in calcium signaling. New Phytologist 218 (2), 414–431. Available at https://doi.org/10.1111/nph.14966.

Larkindale, J., Huang, B., 2005. Effects of abscisic acid, salicylic acid, ethylene and hydrogen peroxide in thermotolerance and recovery for creeping bentgrass. Plant Growth Regulation 47 (1), 17–28. Available at https://doi.org/10.1007/s10725-005-1536-z.

Lee, K., Kang, H., 2024. Recent insights into the physio-biochemical and molecular mechanisms of low temperature stress in tomato. Plants 13 (19), 2715. Available at https://doi.org/10.3390/plants13192715.

Li, C., Wei, Z., Liang, D., Zhou, S., Li, Y., Liu, C., et al., 2013. Enhanced salt resistance in apple plants overexpressing a *Malus* vacuolar Na+/H+ antiporter gene is associated with differences in stomatal behavior and photosynthesis. Plant Physiology and Biochemistry 70, 164–173. Available at https://doi.org/10.1016/j.plaphy.2013.05.005.

Li, Z., Yue, H., Xing, D., 2012. MAP Kinase 6-mediated activation of vacuolar processing enzyme modulates heat shock-induced programmed cell death in *Arabidopsis*. New Phytologist 195 (1), 85–96. Available at https://doi.org/10.1111/j.1469-8137.2012.04131.x.

Lin, Z., Zhong, S., Grierson, D., 2009. Recent advances in ethylene research. Journal of Experimental Botany 60 (12), 3311–3336. Available at https://doi.org/10.1093/jxb/erp204.

López-Vicente, M., Navas, A., 2009. Predicting soil erosion with RUSLE in mediterranean agricultural systems at catchment scale. Soil Science 174 (5), 272–282. Available at https://doi.org/10.1097/ss.0b013e3181a4bf50.

Lynch, J., Brown, K.M., 1997. Ethylene and plant responses to nutritional stress. Physiologia Plantarum 100 (3), 613–619. Available at https://doi.org/10.1034/j.1399-3054.1997.1000324.x.

Ma, B., Ma, T., Xian, W., Hu, B., Chu, C., 2023. Interplay between ethylene and nitrogen nutrition: How ethylene orchestrates nitrogen responses in plants. Journal of Integrative Plant Biology 65 (2), 399–407. Available at https://doi.org/10.1111/jipb.13355.

Maathuis, F.J., 2009. Physiological functions of mineral macronutrients. Current Opinion in Plant Biology 12 (3), 250–258. Available at https://doi.org/10.1016/j.pbi.2009.04.003.

Maksymiec, W., 2007. Signaling responses in plants to heavy metal stress. Acta Physiologiae Plantarum 29 (3), 177–187. Available at https://doi.org/10.1007/s11738-007-0036-3.

Merchante, C., Alonso, J.M., Stepanova, A.N., 2013. Ethylene signaling: simple ligand, complex regulation. Current Opinion in Plant Biology 16 (5), 554–560. Available at https://doi.org/10.1016/j.pbi.2013.08.001.

Mikula, K., Izydorczyk, G., Skrzypczak, D., Mironiuk, M., Moustakas, K., Witek-Krowiak, A., et al., 2020. Controlled release micronutrient fertilizers for precision agriculture – a review. Science of The Total Environment 712, 136365. Available at https://doi.org/10.1016/j.scitotenv.2019.136365.

Millward, D.J., 2017. Nutrition, infection and stunting: the roles of deficiencies of individual nutrients and foods, and of inflammation, as determinants of reduced linear growth of children. Nutrition Research Reviews 30 (1), 50–72. Available at https://doi.org/10.1017/S0954422416000238.

Müller, M., 2021. Foes or friends: ABA and ethylene interaction under abiotic stress. Plants 10 (3), 448. Available at https://doi.org/10.3390/plants10030448.

Nadeem, F., Azhar, M., Anwar-ul-Haq, M., Sabir, M., Samreen, T., Tufail, A., et al., 2020. Comparative response of two rice (*Oryza sativa* L.) cultivars to applied zinc and manganese for mitigation of salt stress. Journal of Soil Science and Plant Nutrition 20 (4), 2059–2072. Available at https://doi.org/10.1007/s42729-020-00275-1.

Nayeri, S., Dolatyari, M., Mouladoost, N., Nayeri, S., Zarghami, A., Mirtagioglu, H., et al., 2023. Ag/ZnO core–shell NPs boost photosynthesis and growth rate in wheat seedlings under simulated full sun spectrum. Scientific Reports 13 (1). Available at https://doi.org/10.1038/s41598-023-41575-7.

Nayeri, S., Dehghanian, Z., Asgari Lajayer, B., Thomson, A., Astatkie, T., Price, G.W., 2023. CRISPR/Cas9-mediated genetically edited ornamental and aromatic plants: a promising technology in phytoremediation of heavy metals. Journal of Cleaner Production 428, 139512. Available at https://doi.org/10.1016/j.jclepro.2023.139512.

Pandey, A., Khan, M.K., Hamurcu, M., Athar, T., Yerlikaya, B.A., Yerlikaya, S., et al., 2023. Role of exogenous nitric oxide in protecting plants against abiotic stresses. Agronomy 13 (5). Available at https://doi.org/10.3390/agronomy13051201.

Pandey, G., 2020. Agri-nanotechnology for sustainable agriculture. In: Bauddh, K., Kumar, S., Singh, R., Korstad, J. (Eds.), Ecological and Practical Applications for Sustainable Agriculture. Springer, Singapore, pp. 229–249. Available at https://doi.org/10.1007/978-981-15-3372-3_11.

Park, H.L., Seo, D.H., Lee, H.Y., Bakshi, A., Park, C., Chien, Y.C., et al., 2023. Ethylene-triggered subcellular trafficking of CTR1 enhances the response to ethylene gas. Nature Communications 14 (1). Available at https://doi.org/10.1038/s41467-023-35975-6.

Qiao, K., Wang, F., Liang, S., Wang, H., Hu, Z., Chai, T., 2019. New biofortification tool: wheat TaCNR5 enhances zinc and manganese tolerance and increases zinc and manganese accumulation in rice grains. Journal of Agricultural and Food Chemistry 67 (35), 9877–9884. Available at https://doi.org/10.1021/acs.jafc.9b04210.

Raza, I., Zubair, M., Zaib, M., Khalil, M.H., Haidar, A., Sikandar, A., et al., 2023. Precision nutrient application techniques to improve soil fertility and crop yield: a review with future prospect. International Research Journal of Educational and Tecnology 5 (8), 109–123.

Roychoudhury, A., Tripathi, D.K., 2020. Role of selenium and manganese in mitigating oxidative damages. In: Roychoudhury, A., Tripathi, D.K. (Eds.), Protective Chemical Agents in the Amelioration of Plant Abiotic Stress: Biochemical and Molecular Perspectives. Wiley, pp. 597–621. Available at https://doi.org/10.1002/9781119552154.ch30.

Rudus, I., Sasiak, M., Kępczyński, J., 2013. Regulation of ethylene biosynthesis at the level of 1-aminocyclopropane-1-carboxylate oxidase (ACO) gene. Acta Physiologiae Plantarum 35 (2), 295–307. Available at https://doi.org/10.1007/s11738-012-1096-6.

Saleh, S., Liu, G., Liu, M., Liu, W., Gruda, N., He, H., 2019. Reducing the salinity impact on soilless culture of tomatoes using supplemental ca and foliar micronutrients. Acta Scientiarum Polonorum, Hortorum Cultus 18 (3), 187–200. Available at https://doi.org/10.24326/asphc.2019.3.18.

Sana, U., Aqeel, T., Aftab, R., Gill, S.S., Gill, M., Naeem, 2025. Harnessing gasotransmitters for enhanced plant resilience: strategies for managing metalloid (s) stress. Journal of Plant Growth Regulation 44 (8), 4203–4226. Available at https://doi.org/10.1007/s00344-025-11676-w.

Schildhauer, J., Wiedemuth, K., Humbeck, K., 2008. Supply of nitrogen can reverse senescence processes and affect expression of genes coding for plastidic glutamine synthetase and lysine-ketoglutarate reductase/saccharopine dehydrogenase. Plant Biology 10 (s1), 76–84. Available at https://doi.org/10.1111/j.1438-8677.2008.00075.x.

Sharma, P., Kumari, A., 2025. Approaches to enhance abiotic and biotic stress tolerance in leguminous crops and microgreens. In: Mathur, P., Gupta, A. (Eds.), Recent Trends and Applications of Leguminous Microgreens as Functional Foods. Springer Nature, Cham, pp. 179–215. Available at https://doi.org/10.1007/978-3-031-75678-8_9.

Snedden, W.A., Fromm, H., 2001. Calmodulin as a versatile calcium signal transducer in plants. New Phytologist 151 (1), 35–66. Available at https://doi.org/10.1046/j.1469-8137.2001.00154.x.

Srivastava, A.K., Dev, A., Karmakar, S., 2018. Nanosensors and nanobiosensors in food and agriculture. Environmental Chemistry Letters 16 (1), 161–182. Available at https://doi.org/10.1007/s10311-017-0674-7.

Sun, X., Miao, Q., Gu, Y., Yang, L., Wang, P., 2025. Research on the physiological mechanisms of nitrogen in alleviating plant drought tolerance. Plants 14 (18), 2928. Available at https://doi.org/10.3390/plants14182928.

Virdi, A.S., Singh, S., Singh, P., 2015. Abiotic stress responses in plants: roles of calmodulin-regulated proteins. Frontiers Research Foundation, India Frontiers in Plant Science 6. Available at https://doi.org/10.3389/fpls.2015.00809.

Vitelli, V., Giamborino, A., Bertolini, A., Saba, A., Andreucci, A., 2024. Cadmium stress signaling pathways in plants: molecular responses and mechanisms. Current Issues in Molecular Biology 46 (6), 6052–6068. Available at https://doi.org/10.3390/cimb46060361.

Voesenek, L.A.C.J., Bailey-Serres, J., 2015. Flood adaptive traits and processes: an overview. New Phytologist 206 (1), 57–73. Available at https://doi.org/10.1111/nph.13209.

Wang, J., Renninger, H.J., Ma, Q., Jin, S., 2024. Measuring stomatal and guard cell metrics for plant physiology and growth using StoManager1. Plant Physiology 195 (1), 378–394. Available at https://doi.org/10.1093/plphys/kiae049.

Wang, K.L.C., Li, H., Ecker, J.R., 2002. Ethylene biosynthesis and signaling networks. The Plant Cell 14, S131. Available at https://doi.org/10.1105/tpc.001768.

Wang, L., Ran, L., Hou, Y., Tian, Q., Li, C., Liu, R., et al., 2017. The transcription factor MYB115 contributes to the regulation of proanthocyanidin biosynthesis and enhances fungal resistance in poplar. New Phytologist 215 (1), 351–367. Available at https://doi.org/10.1111/nph.14569.

Wang, M., Zheng, Q., Shen, Q., Guo, S., 2013. The critical role of potassium in plant stress response. International Journal of Molecular Sciences 14 (4), 7370–7390. Available at https://doi.org/10.3390/ijms14047370.

Wu, Z., Yin, X., Bañuelos, G.S., Lin, Z.Q., Liu, Y., Li, M., et al., 2016. Indications of selenium protection against cadmium and lead toxicity in oilseed rape (*Brassica napus* L.). Frontiers in Plant Science 7, 2016. Available at https://doi.org/10.3389/fpls.2016.01875.

Yang, S.F., Hoffman, N.E., 1984. Ethylene biosynthesis and its regulation in higher plants. Annual Review of Plant Physiology 35 (1), 155–189. Available at https://doi.org/10.1146/annurev.pp.35.060184.001103.

Yao, G., Mu, C., Yan, Z., Ma, S., Liu, X., Sun, Y., et al., 2025. The AP2/ERF transcription factor ERF56 negatively regulating nitrate-dependent plant growth in *Arabidopsis*. International Journal of Molecular Sciences 26 (2), 613. Available at https://doi.org/10.3390/ijms26020613.

Yu, K., Ju, F., Wang, Z., Sun, L., Huo, Y., Zhu, J., et al., 2023. Potassium ameliorates cotton (*Gossypium hirsutum* L.) fiber length by regulating osmotic and K + /Na + homeostasis under salt stress. Physiologia Plantarum 175 (1). Available at https://doi.org/10.1111/ppl.13842.

Zakavi, M., Askari, H., Shahrooei, M., 2022a. Characterization of bacterial diversity between two coastal regions with heterogeneous soil texture. Scientific Reports 12 (1). Available at https://doi.org/10.1038/s41598-022-23487-0.

Zakavi, M., Askari, H., Shahrooei, M., 2024. Isolation and characterization of a resistance *Bacillus subtilis* for soil stabilization and dust alleviation purposes. Scientific Reports 14 (1). Available at https://doi.org/10.1038/s41598-024-77613-1.

Zakavi, M., Askari, H., Shahrooei, M., 2022b. Maize growth response to different *Bacillus* strains isolated from a salt-marshland area under salinity stress. BMC Plant Biology 22 (1). Available at https://doi.org/10.1186/s12870-022-03702-w.

Zhou, G., Liu, C., Cheng, Y., Ruan, M., Ye, Q., Wang, R., et al., 2022. Molecular evolution and functional divergence of stress-responsive Cu/Zn superoxide dismutases in plants. International Journal of Molecular Sciences 23 (13), 7082. Available at https://doi.org/10.3390/ijms23137082.

Zhu, J.K., 2016. Abiotic stress signaling and responses in plants. Cell 167 (2), 313–324. Available at https://doi.org/10.1016/j.cell.2016.08.029.

Zhu, J., Sun, L., Ju, F., Wang, Z., Xiong, C., Yu, H., et al., 2022. Potassium application increases cotton (*Gossypium hirsutum* L.) fiber length by improving K+/Na+ homeostasis and potassium transport capacity in the boll-leaf system under moderate salinity. Agronomy 12 (12), 2962. Available at https://doi.org/10.3390/agronomy12122962.

CHAPTER

Mineral nutrition and phytohormones in plant stress tolerance

16

Seyyed Amirreza Mousavi[1], Ghorban Badali[2], Salar Farhangi-Abriz[3], Morteza Alami-Milani[1] and Parisa Aghaei-Gharachorlou[1]

[1]*Department of Plant Eco-physiology, Faculty of Agriculture, University of Tabriz, Tabriz, Iran,*
[2]*Department of Horticulture and Medicinal Plants, Miyaneh Branch, Islamic Azad University, Miyaneh, Iran,*
[3]*Cotton Research Institute of Iran, Agricultural Research, Education and Extension Organization (AREEO), Gorgan, Iran*

Abbreviations

ABA	abscisic acid
ACC	1-aminocyclopropane-1-carboxylic acid
ACO	1-aminocyclopropane-1-carboxylic acid oxidase
ACS	1-aminocyclopropane-1-carboxylic acid synthase
AMF	arbuscular mycorrhizal fungi
Ca	calcium
CBLs	calcineurin B-like proteins
CDPKs	calcium-dependent protein kinases
CIPKs	CBL-interacting protein kinases
Fe	iron
GAs	gibberellins
GGPP	geranylgeranyl diphosphate
HAK5	high-affinity K+ transporter 5
IAA	indole-3-acetic acid
IPA	indole-3-pyruvic acid
IPTs	isopentenyltransferases
JA	jasmonic acid
JA-Ile	jasmonoyl-isoleucine
K	potassium
LEA	late embryogenesis abundant
Mg	magnesium
Mn	manganese
N	nitrogen
NCED	9-cis-epoxycarotenoid dioxygenase
NPR1	nonexpressor of pathogenesis-related genes 1.
NRT1.1	nitrate transporter 1.1

Mineral Nutrients to Mitigate Abiotic Stress in Crops. DOI: https://doi.org/10.1016/B978-0-443-44712-9.00017-7

OPDA	12-oxophytodienoic acid
P	phosphorus
PAL	phenylalanine ammonia-lyase
PGPR	plant growth-promoting rhizobacteria
Pi	inorganic phosphate
ROS	reactive oxygen species
SA	salicylic acid
SAM	S-adenosylmethionine
SAR	systemic acquired resistance
SOD	superoxide dismutase
TAA	tryptophan aminotransferase of arabidopsis
TAR	tryptophan aminotransferase related
YUCCA	flavin monooxygenases
Zn	zinc

16.1 Introduction

As sessile organisms, plants, which are fixed in place, deal with diverse biotic and abiotic stresses that affect their survival, growth, and ability to reproduce. By changing water potential, ion balance, and metabolic pathways, abiotic stresses like drought, salinity, heavy metal toxicity, nutrient imbalances, extreme temperatures, and flooding disrupt cellular homeostasis, which frequently results in oxidative damage due to the buildup of reactive oxygen species (ROS) (Mittler et al., 2022). Biotic stresses, including attacks from pathogens like fungi, bacteria, and viruses, as well as herbivores and parasitic plants, trigger defense mechanisms that demand significant energy reallocation from growth to immunity (Du et al., 2024; Farhangi-Abriz and Torabian, 2018). Through signaling networks that merge environmental cues with internal developmental programs, these responses are interconnected, not isolated. As essential elements, including nitrogen (N), phosphorus (P), potassium (K), calcium (Ca), magnesium (Mg), and trace metals such as iron (Fe), zinc (Zn), and manganese (Mn), are critical for enzyme function, membrane stability, and osmotic regulation, mineral nutrition is foundational (Fageria, 2016). Deficiencies or excesses in these minerals can exacerbate stress effects; for instance, potassium deficiency impairs drought tolerance by weakening stomatal control, while excess salinity induces sodium toxicity that competes with potassium uptake (Fageria, 2016). Furthermore, hormonal biosynthesis emerges as a key orchestrator, with hormones like abscisic acid (ABA) promoting drought acclimation through gene expression modulation, ethylene accelerating senescence under flooding, and jasmonic acid (JA) and salicylic acid (SA) coordinating defenses against biotic threats (Fahad et al., 2014). Understanding these dynamics is crucial for agriculture, especially as climate change intensifies stress frequencies, threatening global food security and necessitating resilient crop varieties.

Mineral nutrition forms the bedrock of plant resilience, supplying the raw materials and cofactors necessary for metabolic processes that underpin stress tolerance. Macronutrients like nitrogen are integral to amino acid and protein synthesis, enabling the production of stress-protective compounds such as proline and glycine betaine, which act as osmoprotectants during osmotic stress (Hernandez-Leon and Valenzuela-Soto, 2023). Phosphorus, vital for ATP and nucleic acid formation, supports energy-intensive responses like root proliferation in nutrient-scarce soils or the activation of phosphatases in signal transduction pathways (Fageria, 2016). Potassium, often dubbed the master regulator of cellular functions, maintains turgor pressure, facilitates enzyme activation, and counters ion imbalances under saline conditions by regulating Na^+/K^+ ratios via transporters like HKT1 (Fageria, 2016; Asins et al., 2022; Ma et al., 2025).

As a dynamic regulatory hub in plant stress responses, hormonal biosynthesis functions where environmental signals initiate the de novo synthesis or modulation of critical phytohormones to precisely control physiological outcomes (Kurepa and Smalle, 2025). Due to the fact that ABA is primarily synthesized from carotenoids by the enzyme NCED, it rapidly accumulates during drought and salinity, triggering stomatal closure, modifications in root hydraulic conductivity, and the expression of LEA proteins for cellular protection (Gupta et al., 2022; Li et al., 2025). Ethylene, produced from methionine through the Yang cycle involving ACS and ACO, mediates responses to hypoxia and wounding by accelerating adventitious root formation and programmed cell death in aerenchyma tissues (Chen et al., 2025). By inducing protease inhibitors and volatile organic compounds that attract predators, JA and its derivatives, produced from linolenic acid through lipoxygenases and allene oxide synthases, bolster defenses against necrotrophic pathogens and herbivores (Farhangi-Abriz and Ghassemi-Golezani, 2019). Mineral nutrition profoundly influences hormonal pathways; calcium acts as a second messenger in ABA signaling via calcineurin CBLs and CIPKs, while magnesium is crucial for chlorophyll synthesis, indirectly affecting gibberellins (GAs) and cytokinin balances that regulate growth under nutrient stress (Fageria, 2016).

Plant stress resilience is amplified by a complex feedback loop, wherein hormones regulate nutrient acquisition and allocation while nutrients function as substrates, cofactors, or signals that influence hormone levels, thus establishing a relationship between mineral nutrition and hormonal biosynthesis (Romera et al., 2021). During a drought, for example, ABA signaling boosts potassium uptake via KUP transporters, preserving osmotic balance, while nitrogen availability influences ABA catabolism via nitrate reductases, thus preventing an overabundance of the hormone (Yang et al., 2020). Similarly, in biotic stress, SA induced defenses require adequate sulfur for glutathione synthesis, a ROS scavenger, and zinc for NPR1 protein function in SAR activation (Fang et al., 2025). Fig. 16.1 illustrates how essential mineral nutrients contribute to regulating major plant hormones under stress conditions. Understanding these mechanisms allows for precision farming, which can include tailored nutrient applications coordinated with hormonal priming by using bio-stimulants or microbial inoculants, within sustainable agriculture.

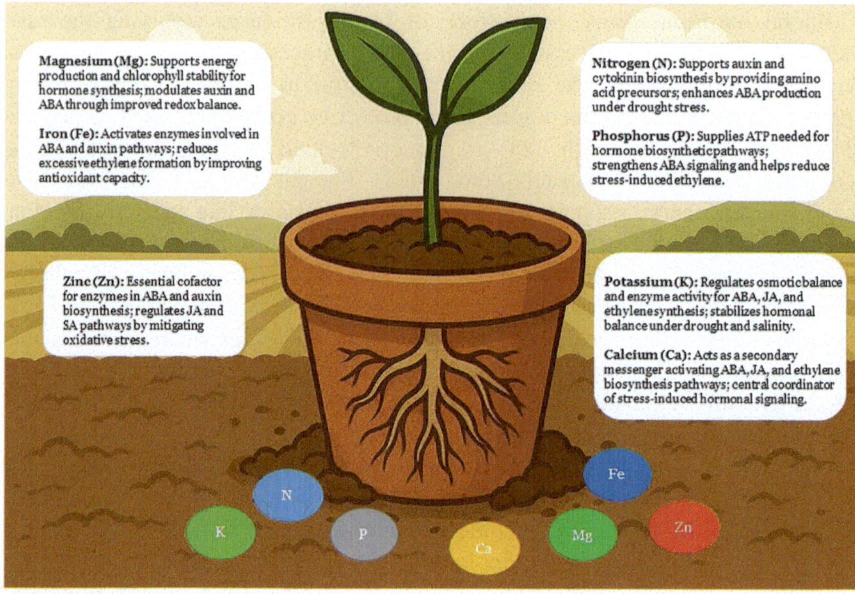

FIGURE 16.1

Roles of essential mineral nutrients in regulating plant hormone biosynthesis under stress conditions.

There are still challenges, like deciphering tissue-specific dynamics and environmental variability; however, interdisciplinary research is expected to bring breakthroughs for climate-resilient crops. Ultimately, harnessing this synergy could mitigate yield losses in stress-prone regions, supporting food production for a growing global population amid escalating environmental pressures.

16.2 The interplay of mineral nutrition and hormonal biosynthesis in plant stress responses

Because the complex interaction between mineral nutrition and hormonal biosynthesis is crucial, integrated plant stress responses are coordinated, allowing plants to adjust to various challenges (Bittner et al., 2022; Wei et al., 2021; Zhao et al., 2021). Mineral nutrients, obtained from the soil, are incorporated into vital compounds and metabolic pathways, while phytohormones, acting as signaling molecules, regulate virtually every aspect of plant life, including stress acclimation (Furman et al., 2021).

16.3 Mineral nutrition: essentiality and physiological roles

The processes of absorption, translocation, and assimilation of inorganic ions from the soil are all part of plant mineral nutrition, which is essential for plant growth and development. Essential mineral nutrients serve as structural components, enzyme cofactors, and osmoregulators, participating in critical biochemical pathways (Arif et al., 2017). While micronutrients such as iron, zinc, and boron are essential but in smaller amounts, macronutrients like nitrogen, phosphorus, and potassium are required in greater quantities. Vital for plant growth and defense, proper mineral nutrition is essential (Khan et al., 2023a).

16.4 Phytohormones: regulators of growth and stress adaptation

At very low concentrations, phytohormones, which are organic compounds, control plant growth, development, and reactions to environmental stimuli. Key classes are comprised of auxins, cytokinins, GAs, ABA, ethylene, JAs, and SA (Farhangi-Abriz and Ghassemi-Golezani, 2019; Poonam et al., 2015). These hormones form complex signaling networks that mutually regulate metabolic systems essential for both plant development and responses to biotic and abiotic stresses.

16.4.1 Plant stress: abiotic and biotic challenges

Plant stress factors are external environmental challenges that negatively impact plant growth and productivity (Husen, 2021; Husen, 2022; Husen, 2023; Husen, 2024; Husen, 2025a, 2025b). Drought, salinity, extreme temperatures, heavy metal toxicity, pollution, and nutrient deficiencies are among the nonliving factors that constitute abiotic stresses. Interactions with living organisms, like pathogens, insects, and pests, cause biotic stresses to arise (Kumari et al., 2024). Due to the stresses, plants begin adaptive molecular, physiological, and biochemical responses, which are frequently influenced by mineral nutrient status and facilitated by phytohormones (EL Sabagh et al., 2022).

16.5 Biosynthesis pathways of key phytohormones and their regulation under stress

Plants employ distinct biosynthetic pathways for each major phytohormone, and these pathways are dynamically regulated in response to stress conditions.

16.5.1 Abscisic acid biosynthesis and stress responses

ABA, a crucial phytohormone, plays a key role in tolerating abiotic stress, especially drought and salinity (Tiwari et al., 2017). Its biosynthesis occurs via the cleavage of carotenoids in plastids, where 9-cis-epoxycarotenoids are oxidatively cleaved to xanthoxin, which is subsequently converted to ABA. The NCED enzyme is a key regulator in this pathway (Wang et al., 2011). Osmotic stress, such as drought, significantly induces ABA biosynthesis; this then causes stomata to close, stress-responsive genes to activate, and water to be conserved. In addition to its other functions, the plant cuticle has a surprising role in osmotic stress-induced ABA biosynthesis, where it mediates both signaling and tolerance (Wang et al., 2011).

16.5.2 Ethylene biosynthesis and stress regulation

Involving itself in various physiological processes and stress responses, such as mechanical stress, hypoxia, chilling, and nutritional disorders, ethylene is a gaseous plant hormone. Starting with SAM, the biosynthesis process converts it to ACC, which then yields ethylene (Romera et al., 2016). Environmental stresses, including water deficit, nutrient deficiencies (Fe, P, K, S), heavy metal stress, and salinity, typically increase ethylene production, often due to the induction of ACC synthase, a rate-limiting enzyme (Naing et al., 2022). Increased ethylene production is typically associated with an elevation of transcript abundance for genes involved in ethylene biosynthesis and signaling. Protein kinases and phosphatases also regulate ethylene biosynthesis and signaling, highlighting a complex control network. Because of excessive ethylene, plant growth can be inhibited, and the impact depends on the context (Romera et al., 2016).

16.5.3 Jasmonic acid biosynthesis and stress responses

Jasmonates are a family of fatty acid-derived compounds that act as potent physiological and developmental regulators, particularly in response to wounding and pathogen attack. Via lipoxygenase, allene oxide synthase, and cyclase enzymes, α-linolenic acid is converted to OPDA, which initiates the biosynthesis of Jas (Shi et al., 2023). These initial steps occur in plastids, with further metabolism in peroxisomes leading to JA and its active conjugates like jasmonoyl-isoleucine (JA-Ile). Rapid activation of JA biosynthesis, caused by various stresses, leads to crucial roles in defense mechanisms and developmental processes (Farhangi-Abriz and Ghassemi-Golezani, 2018, 2019).

16.5.4 Salicylic acid biosynthesis and defense signaling

SA is a vital plant hormone primarily known for mediating host responses to pathogen infection and activating plant defense. Originating from chorismate, two pathways, the isochorismate pathway and the PAL pathway, mainly facilitate SA biosynthesis (Mishra and Baek, 2021). The importance of each pathway varies among plant species. Environmental stresses significantly elevate SA levels, contributing to local immune responses and systemic acquired resistance (Farhangi-Abriz et al., 2020). SA also plays roles in other biological processes beyond defense (Li et al., 2022).

16.5.5 Auxin biosynthesis and stress adaptations

Plant growth and development are greatly influenced by auxins, especially indole-3-acetic acid (IAA). The IPA pathway, using TAA/TAR aminotransferases and YUCCA flavin monooxygenases, is primarily responsible for the local biosynthesis of auxin, thus playing a critical role in several developmental processes (Luo and Di, 2023). Integrating abiotic stress signals and controlling downstream stress responses, like those caused by temperature, salt, and drought, are also pivotal roles of auxin (Jing et al., 2023). At transcriptional, post-transcriptional, and protein levels, its biosynthesis is regulated, which in turn influences root morphology and overall plant architecture, allowing adaptation to environmental challenges (Salehin, 2024).

16.5.6 Cytokinin biosynthesis and stress responses

Cytokinins are hormones involved in cell division and differentiation, and their biosynthesis occurs via a de novo pathway involving IPTs acting on adenine nucleotides, as well as through tRNA degradation (Chen, 1997). Cytokinin levels are maintained in hormonal balance through intricate regulation involving biosynthesis, interconversion, and degradation. Under stress conditions, cytokinin biosynthesis and signaling are often modulated, affecting growth and stress tolerance (Li et al., 2021).

16.5.7 Gibberellinss biosynthesis and stress modulation

Modulating plant growth and development is a key function of gibberellins, which are isoprenoid plant hormones (Castro-Camba et al., 2022). GA biosynthesis initiates from GGPP and involves a series of enzymatic steps to produce bioactive gas (Hernández-García et al., 2021). Environmental conditions, including salinity and drought, influence GA biosynthesis, thereby affecting plant growth and adaptation. For instance, high salinity can inhibit seed germination by modulating ABA signaling and GA biosynthesis, demonstrating a delicate balance (Muhammad et al., 2024). Table 16.1 provides a comprehensive summary of the biosynthesis pathways of major phytohormones and also their regulatory responses to environmental stresses, which emphasize the dynamic mechanisms that help plants adapt and maintain homeostasis in harsh conditions.

Table 16.1 Biosynthesis pathways and stress-induced regulation of key phytohormones.

Phytohormone	Biosynthesis pathway overview	Key enzymes/Regulators	Stress-induced regulation
Abscisic acid	Cleavage of carotenoids in plastids; 9-cis-epoxycarotenoids oxidized to xanthoxin, then converted to ABA.	9-cis-epoxycarotenoid dioxygenase (NCED).	Induced by osmotic stresses (e.g., drought, salinity); leads to stomatal closure, activation of stress-responsive genes, and water conservation; plant cuticle mediates signaling and tolerance.
Ethylene	From S-adenosylmethionine (SAM) to 1-aminocyclopropane-1-carboxylic acid (ACC), then to ethylene.	ACC synthase (rate-limiting); regulated by protein kinases and phosphatases.	Increased production under water deficit, nutrient deficiencies (Fe, P, K, S), heavy metals, salinity, mechanical stress, hypoxia, chilling; elevates transcript abundance for biosynthesis and signaling genes; effects context-dependent, excessive levels may inhibit growth.
Jasmonic acid	From α-linolenic acid to 12-oxophytodienoic acid (OPDA) in plastids, then metabolized in peroxisomes to JA and active conjugates like JA-Ile.	Lipoxygenase, allene oxide synthase, allene oxide cyclase.	Rapidly activated by wounding, pathogen attack, and various stresses; crucial for defense mechanisms and developmental processes.
Salicylic acid	From chorismate via isochorismate pathway or phenylalanine ammonia-lyase (PAL) pathway (pathway importance varies by species).	Isochorismate synthase, PAL.	Elevated by pathogen infection and environmental stresses; mediates local immune responses, systemic acquired resistance, and other biological processes.
Auxin (indole-3-acetic acid, IAA)	Primarily tryptophan-dependent indole-3-pyruvic acid (IPA) pathway.	TAA/TAR aminotransferases, YUCCA flavin monooxygenases; regulated at transcriptional, post-transcriptional, and protein levels.	Integrates signals from abiotic stresses (temperature, salt, drought); controls downstream responses, influences root morphology and plant architecture for adaptation.
Cytokinin	De novo pathway via isopentenyltransferases (IPTs) on adenine nucleotides; also from tRNA degradation.	IPTs; regulated through biosynthesis, interconversion, and degradation for homeostasis.	Modulated under stress conditions; affects growth, cell division/differentiation, and stress tolerance.
Gibberellin	From geranylgeranyl diphosphate (GGPP) through a series of enzymatic steps to bioactive GAs.	Multiple enzymes in the isoprenoid pathway.	Influenced by salinity and drought; affects plant growth and adaptation (e.g., high salinity inhibits seed germination via modulation of ABA signaling and GA biosynthesis).

16.6 Influence of mineral nutrition on hormonal biosynthesis and signaling under stress

The availability and balance of mineral nutrients profoundly influence the biosynthesis and signaling of plant hormones, shaping stress responses.

16.6.1 Macronutrient impact on hormonal regulation

Nitrogen (N): Nitrogen availability is closely linked with hormonal signaling, particularly involving auxin, ABA, and cytokinins (Xing et al., 2023). These hormones integrate nitrogen signals to adjust plant morphology and physiology, crucial for sustained plant growth under stressful environments (Khan et al., 2023b). With the effect of increasing ABA, SA, and JA, and decreasing auxin, nitrogen supplementation enhances salt tolerance in grape seedlings and activates stress-related physiological responses (Kaya et al., 2019). Supporting amino acid metabolism, maintaining energy balance, and influencing key enzymatic pathways, nitrogen supplementation is critical for regulating plant hormone biosynthesis under stress (Wang et al., 2024a; Chen et al., 2021a). Because nitrogen is a major part of proteins, nucleic acids, and cofactors, the availability of it directly influences the synthesis of hormone precursors, including tryptophan for auxin and methionine for ethylene (Zheng et al., 2013). Under stress conditions like drought or salinity, adequate nitrogen enhances ABA production, which improves stomatal regulation and stress signaling (Kamada-Nobusada et al., 2013; Ayub et al., 2025; Chen et al., 2021a). Furthermore, nitrogen influences JA and SA pathways by sustaining metabolic activity and strengthening defense responses (Zhou et al., 2024; Chen et al., 2021b).

Phosphorus (P): As a result of phosphorus deficiency, ethylene production rises, and this is associated with activating morphological and physiological responses, like increased acid phosphatase activity and the development of cluster roots (Romera et al., 2016). Under diverse abiotic stress situations, like drought, salinity, and nutrient deficiency, phosphorus, which is mainly Pi, regulates plant hormone biosynthesis through metabolic pathways, gene expression, and signaling cascades, which helps the plant adapt and become more resilient, playing a key part (Khan et al., 2023b). Under Pi starvation, which often exacerbates other stresses like drought, plants activate phosphate starvation responses that precede and interact with hormone signaling, leading to upregulated biosynthesis of stress-responsive hormones including ABA, ethylene, JA, and SA, while downregulating growth-promoting hormones like IAA to prioritize survival over growth (Isidra-Arellano et al., 2021; Puga et al., 2024). For instance, low Pi levels trigger ABA biosynthesis through enzymes like NCED, promoting stomatal closure and osmotic adjustment to conserve water during drought, with ABA in turn regulating phosphate starvation response genes via cis-regulatory elements in Pi transporter promoters, creating a feedback loop that sustains Pi homeostasis and stress tolerance (Jia et al., 2022). Similarly, ethylene production is induced under Pi deficiency via increased activity of 1-aminocyclopropane-1-carboxylic acid synthase and oxidase, facilitating root architectural changes like enhanced lateral root formation to improve Pi scavenging, while JA accumulation, driven by upregulation of lipoxygenase and allene oxide synthase genes, bolsters antioxidant defenses and mitigates oxidative stress from ROS buildup (Khan et al., 2024; Li et al., 2025). SA biosynthesis, activated through phenylalanine ammonia-lyase, enhances systemic resistance and enzyme activities under combined Pi limitation and biotic or abiotic pressures (Mohammadi et al., 2021). Adequate Pi supply alleviates these stresses by downregulating ABA and IAA synthesis, reducing ROS accumulation, and supporting energy-dependent processes like ATP production essential for hormone metabolic pathways, as seen in genes such as HvPIP aquaporins and PHO1 that facilitate water movement and ABA signaling under salinity (Abou Seeda et al., 2022; Olukayode, 2023; Lv et al., 2025). In mutants deficient in PSRs, such as phr1 phl1 in Arabidopsis, growth halts under mild drought, highlighting Pi's indispensable role in maintaining hormone balance for rapid recovery upon stress relief (Puga et al., 2024; Wang et al., 2024b).

Potassium (K): Adequate potassium supply helps maintain cellular osmotic balance and enzyme activation, which supports the synthesis of stress-related hormones such as ABA, ethylene, and JA (Che et al., 2022; Kumar et al., 2025). Under drought or salinity stress, potassium enhances ABA biosynthesis by stabilizing precursor enzymes and improving the transport of ABA from roots to shoots, thereby strengthening stomatal regulation and stress tolerance (Hussain et al., 2021; Che et al., 2022). Maintaining cellular redox homeostasis and reducing oxidative damage also influences ethylene and jasmonate production, which in turn helps these hormones to coordinate defense responses more effectively (Kazan, 2015). Overall, potassium availability helps plants fine-tune their hormonal signaling networks, leading to improved resilience against abiotic stresses. Potassium deficiency also promotes ethylene synthesis, contributing to responses like the upregulation of the HAK5 potassium transporter and root hair development (Romera et al., 2016).

Calcium (Ca): Calcium plays a crucial regulatory role in plant hormone biosynthesis under stress, acting primarily as a secondary messenger that activates key enzymes and signaling pathways (Fageria, 2016). When faced with drought, salinity, cold, or heat stress, fluctuations in cytosolic Ca levels activate calcium-binding proteins, like calmodulins and CDPKs, to regulate the biosynthesis of major stress-related hormones (Fageria, 2016). Calcium enhances ABA production by activating enzymes involved in carotenoid cleavage, strengthening stomatal closure and drought tolerance (Singh et al., 2017). It also influences auxin biosynthesis and transport by stabilizing cell walls and membrane integrity, helping the plant maintain controlled growth under stress (Klimecka and Muszyńska, 2007). Moreover, calcium regulates ethylene and JA pathways by modulating stress-responsive gene expression and reducing oxidative damage (Farooq et al., 2024;Zhao et al., 2023).

Magnesium (Mg): As the central atom in chlorophyll and a key activator of many enzymes, magnesium ensures sufficient ATP production, which is required for the synthesis of hormone precursors and signaling molecules (Fageria, 2016). By supporting carotenoid metabolism and preserving chloroplast function, magnesium aids in maintaining ABA biosynthesis and thereby enhancing stomatal control and stress signaling, especially under stress conditions like drought, salinity, or nutrient deficiency (Mao et al., 2022; Li et al., 2025). It also influences auxin production and transport by stabilizing cell membranes and maintaining proton gradients essential for auxin mobility (Zhang et al., 2020). Additionally, magnesium contributes to cytokinin and ethylene balance by reducing oxidative stress and supporting the activity of antioxidant enzymes. Overall, adequate magnesium nutrition helps stabilize hormonal pathways and strengthens plant resilience against environmental stress (Farhangi-Abriz and Ghassemi-Golezani, 2022; Ghassemi-Golezani et al., 2021).

16.6.2 Micronutrient impact on hormonal regulation

Iron (Fe): As an essential micronutrient, iron is crucial for tightly regulating plant processes and significantly contributing to stress adaptation (Ghassemi-Golezani et al., 2024). Iron deficiency influences auxin transport and production, which is vital for root morphological adaptations under stress (Bacaicoa et al., 2011; Singh and Siddiqui, 2023). Furthermore, dynamic iron signaling regulates iron uptake and homeostasis, with transcription factors like FER-LIKE IRON DEFICIENCY-INDUCED TRANSCRIPTION FACTOR (FIT) acting as regulatory hubs that cross-connect iron nutrition with ABA mediated osmotic stress responses (Kanwar et al., 2021; Dutta et al., 2024).

Zinc (Zn): Under stress, zinc's role in plant hormone biosynthesis regulation is crucial, which it performs by acting as an essential cofactor for many enzymes and transcription factors involved in hormonal pathways (Umair Hassan et al., 2020). During abiotic stresses such as drought, salinity, or heat, zinc supports the synthesis of ABA by stabilizing dehydrogenases and other key enzymes responsible for converting carotenoid precursors into ABA, thereby enhancing stomatal control and stress signaling (Umair Hassan et al., 2020). Zinc also influences auxin biosynthesis and transport by maintaining tryptophan metabolism and supporting the function of zinc-finger proteins that regulate auxin-related genes (Casanova-Sáez et al., 2021). Furthermore, zinc contributes to the modulation of ethylene and JA pathways by reducing oxidative stress, since it is a structural component of antioxidant enzymes like superoxide SOD (Marreiro et al., 2017). Through these combined roles, zinc helps maintain hormonal balance and strengthens plant adaptive responses during stressful environmental conditions.

Boron (B): Boron is crucial for cell wall integrity and lignin biosynthesis (Zhou et al., 2016). Boron deficiency significantly disturbs phytohormone homeostasis, leading to reduced IAA concentrations in shoots and roots, alongside altered expression of auxin biosynthesis and efflux genes (Kohli et al., 2023). Conversely, it increases JAs and ABA concentrations, inducing the expression of ABA biosynthesis genes like BnNCED3. These hormonal changes modulate root growth and overall stress adaptation (Zhou et al., 2016).

Selenium (Se): Cellular functions are enhanced by selenium supplementation, such as membrane stability, mineral nutrition homeostasis, and antioxidant responses. By stimulating hormone biosynthesis, root architecture is remodeled, thereby reducing metal uptake and enhancing plant growth when under metal/metalloid stress (Hasanuzzaman et al., 2022).

Selenium's low concentrations can mitigate metal and metalloid toxicity because of how they restrict metal uptake, improve mineral homeostasis, enhance antioxidant responses, and boost photosynthesis (Hasanuzzaman et al., 2022). Selenium also stimulates hormone biosynthesis to remodel root architecture, further decreasing metal uptake and improving plant growth under stress (Hasanuzzaman et al., 2022). Selenium supplementation plays an important regulatory role in plant hormone biosynthesis under stress by enhancing antioxidant protection and modulating key metabolic pathways (Zhou et al., 2025). Under abiotic stresses such as drought, salinity, or heavy metal toxicity, selenium improves redox homeostasis and reduces oxidative damage, which helps stabilize enzymes involved in hormone production. Selenium has been shown to promote ABA biosynthesis, strengthening stomatal regulation and improving plant drought tolerance (Nazir et al., 2023). It can also influence auxin and cytokinin pathways by supporting better root growth, nutrient uptake, and cell division under stress (Jiang et al., 2019; Nazir et al., 2023). Additionally, selenium modulates ethylene and JA levels by mitigating oxidative stress, preventing excessive hormone accumulation that could otherwise lead to growth inhibition (Sehar et al., 2022; Jia et al., 2024). (Table 16.2).

16.7 Crosstalk and integrated responses

The effectiveness of plant stress responses relies on the intricate crosstalk among different phytohormones and their interaction with mineral nutrient signals (Deb et al., 2016).

16.7.1 Interhormonal crosstalk

Because hormones build a signaling network, they mutually regulate several signaling and metabolic systems (Munné-Bosch and Müller, 2013). For example, ABA and brassinosteroids can regulate cytokinin conjugation, while brassinosteroid biosynthesis is influenced by both ABA and cytokinin. Jasmonic acid and ethylene biosynthesis may be modulated by ABA through DREB transcription factors (Deb et al., 2016).

Table 16.2 Influence of mineral nutrients on phytohormone biosynthesis and signaling under stress.

Nutrient	Category	Key hormones influenced	Stress conditions	Regulatory mechanisms and effects
Nitrogen (N)	Macronutrient	Auxin, ABA, cytokinins, SA, JA, ethylene	Drought, salinity, nutrient deficiency	Enhances ABA, SA, JA biosynthesis; supports amino acid metabolism (e.g., tryptophan for auxin, methionine for ethylene); promotes cytokinin production in roots; maintains energy balance and enzymatic pathways; activates stress-related physiological responses and defense mechanisms.
Phosphorus (P)	Macronutrient	ABA, ethylene, JA, SA, IAA	Drought, salinity, Pi starvation, nutrient deficiency	Increases ABA, JA, SA under deficiency; activates PSRs interacting with hormone signaling; triggers ABA via NCED for stomatal closure; induces ethylene via ACC synthase/oxidase for root changes; upregulates JA via LOX/AOS for antioxidant defenses; enhances SA via PAL for resistance; reduces ROS, supports ATP production.
Potassium (K)	Macronutrient	ABA, Ethylene, JA	Drought, salinity, K deficiency	Enhances ABA biosynthesis and transport for stomatal regulation, maintains osmotic balance and enzyme activation; influences ethylene/JA by reducing oxidative damage; promotes ethylene under deficiency for uptake responses (e.g., HAK5 transporter, root hair development); fine-tunes hormonal signaling for resilience.
Calcium (Ca)	Macronutrient	ABA, auxin, ethylene, JA	Drought, salinity, cold, heat	Acts as secondary messenger via calmodulins/CDPKs; enhances ABA via carotenoid cleavage enzymes for stomatal closure; stabilizes cell walls/membranes for auxin transport; modulates ethylene/JA pathways via gene expression and oxidative damage reduction; integrates stress signals with hormonal biosynthesis.
Magnesium (Mg)	Macronutrient	ABA, auxin, cytokinins, ethylene	Drought, aalinity, nutrient deficiency	Supports ATP production and enzyme activity; maintains ABA via carotenoid metabolism and chloroplast function; stabilizes membranes for auxin mobility; balances cytokinins/ethylene by reducing oxidative stress and supporting antioxidants; stabilizes hormonal pathways for resilience.

(Continued)

Table 16.2 Influence of mineral nutrients on phytohormone biosynthesis and signaling under stress. (*Continued*)

Nutrient	Category	Key hormones influenced	Stress conditions	Regulatory mechanisms and effects
Iron (Fe)	Micronutrient	Auxin, ABA	Nutrient deficiency, osmotic stress	Influences auxin transport/production for root adaptations; regulates uptake/homeostasis via FIT transcription factor; cross-connects with ABA-mediated stress responses.
Zinc (Zn)	Micronutrient	ABA, auxin, ethylene, JA	Drought, salinity, heat	Cofactor for enzymes/transcription factors; supports ABA via dehydrogenases; maintains tryptophan metabolism and zinc-finger proteins for auxin; modulates ethylene/JA by reducing oxidative stress via SOD; maintains hormonal balance for adaptive responses.
Boron (B)	Micronutrient	Auxin (IAA), JA, ABA	B deficiency	Disturbs homeostasis: reduces IAA in shoots/roots, alters biosynthesis/efflux genes; increases JA/ABA, induces ABA genes (e.g., BnNCED3); modulates root growth and stress adaptation.
Selenium (Se)	Micronutrient (beneficial)	ABA, auxin, cytokinins, ethylene, JA	Drought, salinity, heavy metal toxicity	Enhances antioxidant protection and redox homeostasis; promotes ABA for stomatal regulation; supports auxin/cytokinin for root growth and cell division; modulates ethylene/JA to prevent excessive accumulation; remodels root architecture to reduce metal uptake; improves resilience and adaptive responses.

Ethylene interacts with auxin and phloem signals to regulate responses to iron and phosphorus deficiency. To regulate salinity responses, it interacts with ABA, and it interacts with ROS to regulate potassium deficiency and heavy metal responses (Romera et al., 2016). These complex regulatory networks ensure an integrated response to various stressors.

16.7.2 Role of root–microbe interactions

Under stress, the interplay between roots and microbes, which includes helpful microorganisms such as AMF and PGPR, greatly affects mineral nutrition and hormone production (Nikpour Rashidabad et al., 2025). AMF enhance mineral nutrient uptake through symbiosis with plant roots, which is especially helpful during drought, salinity, and heavy metal stress. Regulating hormone synthesis, this symbiosis also contributes to improved plant growth and stress tolerance (Ahmed et al., 2025). PGPR promote plant growth because they regulate nutritional and hormonal balance, produce plant growth regulators, and induce resistance against pathogens (Aloo et al., 2023). These microbes can maintain hormonal homeostasis under stress by influencing biosynthesis, degradation, and signaling pathways of plant hormones (Timofeeva et al., 2024).

16.8 Mechanisms linking mineral nutrition and hormonal biosynthesis in biotic stress

Mineral nutrition critically modulates plant responses to biotic stresses like pathogen and pest attacks by influencing hormonal biosynthesis and signaling pathways (Dutta et al., 2024).

16.8.1 Ethylene and nutrient deficiency in pathogen responses

Ethylene's direct role in regulating responses to nutrient deficiencies indirectly influences an organism's susceptibility to pathogens. High ethylene production under stress generally plays positive roles in activating responses to nutrient deficiencies, salinity, and heavy metal stress (Romera et al., 2016).

16.8.2 Role of salicylic acid and abscisic acid in immune responses

Plant immunity hinges on the balance between SA and ABA signaling. SA is key for activating innate immune responses against biotrophic pathogens, while ABA can suppress these defenses. Pathogens often exploit this antagonism to promote disease (Dutta et al., 2024). Mineral nutrient availability affects these hormonal dynamics, thereby shaping the efficiency of immune signaling cascades. For instance, nitrogen, phosphorus, and iron significantly influence defense mechanisms by providing building blocks for defense compounds and immune signaling (Fageria, 2016; Dutta et al., 2024).

16.8.3 Micronutrients in balancing reactive oxygen species and immune signaling

Micronutrients such as zinc, copper, and iron balance ROS, which is crucial for plant immune responses. Acting as signaling molecules and playing a dual role in stress responses, ROS's balance is crucial for effective defense (Dutta et al., 2024;Suzuki, 2016). Micronutrient homeostasis thus directly links to signal transduction in stress responses, enhancing plant resilience against pathogens.

16.9 Future directions and knowledge gaps

Despite significant progress, several knowledge gaps remain in fully elucidating the intricate relationship between mineral nutrition and hormonal biosynthesis in plant stress biology. Future research trends emphasize an integrated molecular understanding and biotechnological interventions. A more comprehensive understanding of hormone crosstalk and the integration of mineral nutrition with hormonal pathways at the molecular and systems biology levels is crucial. This includes detailed studies on how different hormone signaling pathways interact and influence each other under specific nutrient statuses and stress conditions. For example, the exact molecular mechanisms by which melatonin alters endogenous levels in response to nutrient imbalances and its full role in optimizing nutrient acquisition remain to be thoroughly investigated.

Leveraging advanced biotechnological tools such as genome editing and transgenic technologies holds immense promise for modulating hormone biosynthesis genes and signaling components. This precision engineering can fine-tune stress responses, potentially balancing the trade-off between growth and defense without compromising crop yield. Further research is needed to identify specific genes and regulatory elements that can be targeted for enhanced stress resilience. Future efforts must bridge fundamental research with applied strategies like biofortification to address both crop productivity and human nutritional security. Integrative studies combining physiological, genetic, and biochemical approaches are necessary to advance understanding and develop crops with tailored mineral and hormonal profiles optimized for adaptation to changing climates and nutrient-poor soils. This includes exploring how the root microbiome can be manipulated to enhance both nutrient uptake and hormonal regulation under stress conditions.

16.10 Conclusions

The interplay between mineral nutrition and hormonal biosynthesis is central to plant adaptation to abiotic and biotic stresses. Essential macro- and micronutrients support metabolic resilience and modulate phytohormone pathways, including ABA, ethylene, JA, SA, auxins, cytokinins, and GAs. Adequate mineral availability enhances stress responses such as osmotic adjustment, root remodeling, ROS detoxification, and immune activation, while deficiencies increase vulnerability. This bidirectional crosstalk highlights the importance of integrated regulatory networks for effective stress acclimation. With climate change intensifying stresses and threatening global food security, optimizing mineral nutrition through precision fertilization, biofortification, and microbial symbioses can strengthen hormonal responses, improve crop resilience, and support sustainable agriculture. Future research should focus on omics-driven approaches to uncover novel nutrient-hormone interactions and develop predictive models. Bridging fundamental insights with applied strategies will enable the development of nutrient-efficient, stress-tolerant crops, advancing plant biology and promoting eco-friendly farming aligned with global sustainability goals.

References

Abou Seeda, M., Abou El-Nour, E.Z., Abdallah, M., El-Bassiouny, H., 2022. Impacts of metal, metalloid and their effects in plant physiology: a review. Middle East Journal of Agricultural Research 11, 838–931.

Ahmed, N., Li, J., Li, Y., Deng, L., Deng, L., Chachar, M., et al., 2025. Symbiotic synergy: how arbuscular mycorrhizal fungi enhance nutrient uptake, stress tolerance, and soil health through molecular mechanisms and hormonal regulation. IMA Fungus 16. Available at https://doi.org/10.3897/imafungus.16.144989.

Aloo, B.N., Dessureault-Rompré, J., Tripathi, V., Nyongesa, B.O., Were, B.A., 2023. Signaling and crosstalk of rhizobacterial and plant hormones that mediate abiotic stress tolerance in plants. Frontiers in Microbiology 14. Available at https://doi.org/10.3389/fmicb.2023.1171104.

Arif, M., Kumar, A., Kadam, S.S., 2017. Mineral nutrition for plant health. Rashtriya Krishi 12, 61–64.

Asins, M.J., Romero-Aranda, M.R., Espinosa, J., González-Fernández, P., Jaime-Fernández, E., Traverso, J.A., et al., 2022. HKT1;1 and HKT1;2 Na+ transporters from solanum galapagense play different roles in the plant Na+ distribution under salinity. International Journal of Molecular Sciences 23 (9). Available at https://doi.org/10.3390/ijms23095130.

Ayub, A., Nayab, A., Yunyou, N., Yuyu, X., Derong, S., Hussan, M.U., et al., 2025. Integration of transcription factors, photosynthesis, and nitrogen metabolic genes modulates nitrogen stress with abscisic acid in rapeseed. Physiologia Plantarum 177 (5). Available at https://doi.org/10.1111/ppl.70486.

Bacaicoa, E., Mora, V., Zamarreño, A.M., Fuentes, M., Casanova, E., García-Mina, J.M., 2011. Auxin: a major player in the shoot-to-root regulation of root Fe-stress physiological responses to Fe deficiency in cucumber plants. Plant Physiology and Biochemistry 49 (5), 545–556. Available at https://doi.org/10.1016/j.plaphy.2011.02.018.

Bittner, A., Cieśla, A., Gruden, K., Lukan, T., Mahmud, S., Teige, M., et al., 2022. Organelles and phytohormones: a network of interactions in plant stress responses. Journal of Experimental Botany 73 (21), 7165–7181. Available at https://doi.org/10.1093/jxb/erac384.

Casanova-Sáez, R., Mateo-Bonmatí, E., Ljung, K., 2021. Auxin metabolism in plants. Cold Spring Harbor Perspectives in Biology 13 (3), a039867. Available at https://doi.org/10.1101/cshperspect.a039867.

Castro-Camba, R., Sánchez, C., Vidal, N., Vielba, J.M.[a], 2022. Plant development and crop yield: the role of gibberellins. Plants 11 (19), 2650. Available at https://doi.org/10.3390/plants11192650.

Che, Y., Yao, T., Wang, H., Wang, Z., Zhang, H., Sun, G., et al., 2022. Potassium ion regulates hormone, Ca^{2+} and H_2O_2 signal transduction and antioxidant activities to improve salt stress resistance in tobacco. Plant Physiology and Biochemistry 186, 40–51. Available at https://doi.org/10.1016/j.plaphy.2022.06.027.

Chen, C.-maw, 1997. Cytokinin biosynthesis and interconversion. Physiologia Plantarum 101 (4), 665–673. Available at https://doi.org/10.1034/j.1399-3054.1997.1010402.x.

Chen, H., Zhao, Z., Chen, J., Mertens, J., Van de Poel, B., Li, D., et al., 2025. The Yang cycle in plants: a journey of methionine recycling with fascinating metabolites and enzymes. Plant Hormones 1 (1), e007. Available at https://doi.org/10.48130/ph-0025-0007.

Chen, J., Liu, S., Zhang, S., Ge, C., Shen, Q., Ma, H., et al., 2021a. Nitrogen modulates cotton root morphology by affecting abscisic acid (ABA) and salicylic acid (SA) content. Archives of Agronomy and Soil Science 67 (12), 1722–1738. Available at https://doi.org/10.1080/03650340.2020.1807518.

Chen, J., Wang, Z., Liu, S., Zhang, S., Ge, C., Shen, Q., Ma, H., Zhang, X., Dong, H., Zhao, X., Liu, R., Pang, C., 2021b. Nitrogen stress inhibits root growth by regulating cell wall and hormone changes in cotton (*Gossypium hirsutum* L.). Journal of Agronomy and Crop Science 207 (6), 1006–1023. Available at https://doi.org/10.1111/jac.12538.

Deb, A., Grewal, R.K., Kundu, S., 2016. Regulatory cross-talks and cascades in rice hormone biosynthesis pathways contribute to stress signaling. Frontiers in Plant Science 7 (2016). Available at https://doi.org/10.3389/fpls.2016.01303.

Du, B., Haensch, R., Alfarraj, S., Rennenberg, H., 2024. Strategies of plants to overcome abiotic and biotic stresses. Biological Reviews 99 (4), 1524–1536. Available at https://doi.org/10.1111/brv.13079.

Dutta, A., Dracatos, P.M., Khan, G.A., 2024. Balancing act: the dynamic relationship between nutrient availability and plant defence. Plant Journal 120 (5), 1724–1734. Available at https://doi.org/10.1111/tpj.17098.

Fageria, N.K., 2016. The Use of Nutrients in Crop Plants. CRC Press.

Fahad, S., Nie, L., Chen, Y., Wu, C., Xiong, D., Saud, S., Hongyan, L., Cui, K., Huang, J., 2014. Crop plant hormones and environmental stress. In: Lichtfouse, E. (Ed.), Sustainable Agriculture Reviews. Sustainable Agriculture Reviews, vol 15. Springer, Cham, pp. 371–400. Available at https://doi.org/10.1007/978-3-319-09132-7_10.

Fang, X., Xie, Y., Yuan, Y., Long, Q., Zhang, L., Abid, G., et al., 2025. The role of salicylic acid in plant defense responses against biotic stresses. Plant Hormones 1 (1), 0. Available at https://doi.org/10.48130/ph-0025-0003 –0.

Farhangi-Abriz, S., Tavasolee, A., Ghassemi-Golezani, K., Torabian, S., Monirifar, H., Rahmani, H.A., 2020. Growth-promoting bacteria and natural regulators mitigate salt toxicity and improve rapeseed plant performance. Protoplasma 257 (4), 1035–1047. Available at https://doi.org/10.1007/s00709-020-01493-1.

Farhangi-Abriz, S., Ghassemi-Golezani, K., 2018. How can salicylic acid and jasmonic acid mitigate salt toxicity in soybean plants? Ecotoxicology and Environmental Safety 147, 1010–1016. Available at https://doi.org/10.1016/j.ecoenv.2017.09.070.

Farhangi-Abriz, S., Ghassemi-Golezani, K., 2019. Jasmonates: mechanisms and functions in abiotic stress tolerance of plants. Biocatalysis and Agricultural Biotechnology 20, 101210. Available at https://doi.org/10.1016/j.bcab.2019.101210.

Farhangi-Abriz, S., Ghassemi-Golezani, K., 2022. The modified biochars influence nutrient and osmotic statuses and hormonal signaling of mint plants under fluoride and cadmium toxicities. Frontiers in Plant Science 13. Available at https://doi.org/10.3389/fpls.2022.1064409.

Farhangi-Abriz, S., Torabian, S., 2018. Biochar increased plant growth-promoting hormones and helped to alleviates salt stress in common bean seedlings. Journal of Plant Growth Regulation 37 (2), 591–601. Available at https://doi.org/10.1007/s00344-017-9756-9.

Farooq, M.A., Ayyaz, A., Zou, H.X., Zhou, W., Hannan, F., Yan, X., 2024. Jasmonic acid mediates Ca^{2+} dependent signal transduction and plant immunity. Plant Science 348. Available at https://doi.org/10.1016/j.plantsci.2024.112239.

Furman, B., Noorani, A., Mba, C., 2021. On-farm crop diversity for advancing food security and nutrition. Landraces - Traditional Variety and Natural Breed. IntechOpen. Available at https://doi.org/10.5772/intechopen.96067.

Ghassemi-Golezani, K., Farhangi-Abriz, S., Abdoli, S., Ghassemi-Golezani, K., Farhangi-Abriz, S., 2021. Biochar-based metal oxide nanocomposites of magnesium and manganese improved root development and productivity of safflower under salt stress. Rhizosphere 19, 100416. Environmental Geochemistry and Health 43 (5), 2007–2023. Available at https://doi.org/10.1007/s10653-020-00780-3.

Ghassemi-Golezani, K., Farhangi-Abriz, S., Rahimzadeh, S., 2024. Iron signaling in oxidative stress of plants under saline conditions: a meta-analysis. In: Aftab, T. (Ed.), Metals and Metalloids in Plant Signaling. Signaling and Communication in Plants. Springer, Cham, pp. 63–72. Available at https://doi.org/10.1007/978-3-031-59024-5_4.

Gupta, P., Mann, V., Hirschberg, J., 2022. Two roads diverge in the yellow pigments metabolic pathway for ABA synthesis. Molecular Plant 15 (1), 3–5. Available at https://doi.org/10.1016/j.molp.2021.11.002.

Hasanuzzaman, M., Nahar, K., García-Caparrós, P., Parvin, K., Zulfiqar, F., Ahmed, N., et al., 2022. Selenium supplementation and crop plant tolerance to metal/metalloid toxicity. Frontiers in Plant Science 12. Available at https://doi.org/10.3389/fpls.2021.792770.

Hernandez-Leon, S.G., Valenzuela-Soto, E.M., 2023. Glycine betaine is a phytohormone-like plant growth and development regulator under stress conditions. Journal of Plant Growth Regulation 42 (8), 5029–5040. Available at https://doi.org/10.1007/s00344-022-10855-3.

Hernández-García, J., Briones-Moreno, A., Blázquez, M.A., 2021. Origin and evolution of gibberellin signaling and metabolism in plants. Seminars in Cell and Developmental Biology 109, 46–54. Available at https://doi.org/10.1016/j.semcdb.2020.04.009.

Husen, A., 2022. Environmental Pollution and Medicinal Plants. CRC Press, Boca Raton, FL. Available at https://doi.org/10.1201/9781003178866.

Husen, A., 2023. Plants and Their Interaction to Environmental Pollution: Damage Detection, Adaptation, Tolerance, Physiological and Molecular Responses. Elsevier. Available at https://doi.org/10.1016/C2020-0-03319-0.

Husen, A., 2024. Essential Minerals in Plant-Soil System: Coordination, Signalling, and Interaction under Adverse Situations. Elsevier. Available at https://doi.org/10.1016/C2022-0-02437-5.

Husen, A., 2025. Role of Antioxidants in Mitigating Plant Stress. Academic Press. Available at https://doi.org/10.1016/C2023-0-51268-6.

Husen, A., 2025. Roles of Osmolytes in Changing Environment. Academic Press. Available at https://doi.org/10.1016/C2023-0-51382-5.

Husen, A., 2021. The harsh environment and resilient plants: an overview. In: Husen, A. (Ed.), Harsh Environment and Plant Resilience. Springer International Publishing AG, Cham, pp. 1–23. Available at https://doi.org/10.1007/978-3-030-65912-7_1.

Hussain, Q., Asim, M., Zhang, R., Khan, R., Farooq, S., Wu, J., 2021. Transcription factors interact with ABA through gene expression and signaling pathways to mitigate drought and salinity stress. Biomolecules 11 (8), 1159. Available at https://doi.org/10.3390/biom11081159.

Isidra-Arellano, M.C., Delaux, P.-M., Valdés-López, O., 2021. The phosphate starvation response system: its role in the regulation of plant–microbe interactions. Plant and Cell Physiology 62 (3), 392–400. Available at https://doi.org/10.1093/pcp/pcab016.

Jia, K.P., Mi, J., Ali, S., Ohyanagi, H., Moreno, J.C., Ablazov, A., et al., 2022. An alternative, zeaxanthin epoxidase-independent abscisic acid biosynthetic pathway in plants. Molecular Plant 15 (1), 151–166. Available at https://doi.org/10.1016/j.molp.2021.09.008.

Jia, Y., Kang, L., Wu, Y., Zhou, C., Cai, R., Zhang, H., et al., 2024. Nano-selenium foliar intervention-induced resistance of cucumber to *Botrytis cinerea* by activating jasmonic acid biosynthesis and regulating phenolic acid and cucurbitacin. Pest Management Science 80 (2), 554–568. Available at https://doi.org/10.1002/ps.7784.

Jiang, L., Liu, C., Cao, H., Chen, Z., Yang, J., Cao, S., et al., 2019. The role of cytokinin in selenium stress response in *Arabidopsis*. Plant Science 281, 122–132. Available at https://doi.org/10.1016/j.plantsci.2019.01.028.

Jing, H., Wilkinson, E.G., Sageman-Furnas, K., Strader, L.C., 2023. Auxin and abiotic stress responses. Journal of Experimental Botany 74 (22), 7000–7014. Available at https://doi.org/10.1093/jxb/erad325.

Kamada-Nobusada, T., Makita, N., Kojima, M., Sakakibara, H., 2013. Nitrogen-dependent regulation of de novo cytokinin biosynthesis in rice: the role of glutamine metabolism as an additional signal. Plant and Cell Physiology 54 (11), 1881–1893. Available at https://doi.org/10.1093/pcp/pct127.

Kanwar, P., Baby, D., Bauer, P., Sperotto, R., 2021. Interconnection of iron and osmotic stress signalling in plants: is FIT a regulatory hub to cross-connect abscisic acid responses? Plant Biology 23 (S1), 31–38. Available at https://doi.org/10.1111/plb.13261.

Kaya, C., Ashraf, M., Sonmez, O., Polat, T., Tuna, A.L., 2019. RETRACTED ARTICLE: the combined effects of nitric oxide and thiourea on plant growth and mineral nutrition of salt-stressed plants of two maize cultivars with differential salt tolerance. Journal of Plant Nutrition 42, 901–917.

Kazan, K., 2015. Diverse roles of jasmonates and ethylene in abiotic stress tolerance. Trends in Plant Science 20 (4), 219–229. Available at https://doi.org/10.1016/j.tplants.2015.02.001.

Khan, F., Siddique, A.B., Shabala, S., Zhou, M., Zhao, C., 2023a. Phosphorus plays key roles in regulating plants' physiological responses to abiotic stresses. Plants 12 (15), 2861. Available at https://doi.org/10.3390/plants12152861.

Khan, M.I.R., Nazir, F., Maheshwari, C., Chopra, P., Chhillar, H., Sreenivasulu, N., 2023b. Mineral nutrients in plants under changing environments: a road to future food and nutrition security. Plant Genome 16 (4). Available at https://doi.org/10.1002/tpg2.20362.

Khan, S., Alvi, A.F., Saify, S., Iqbal, N., Khan, N.A., 2024. The ethylene biosynthetic enzymes, 1-aminocyclopropane-1-carboxylate (ACC) synthase (ACS) and ACC oxidase (ACO): the less explored players in abiotic stress tolerance. Biomolecules 14 (1). Available at https://doi.org/10.3390/biom14010090.

Klimecka, M., Muszyńska, G., 2007. Structure and functions of plant calcium-dependent protein kinases. Acta Biochimica Polonica 54 (2), 219–233. Available at https://doi.org/10.18388/abp.2007_3242.

Kohli, S.K., Kaur, H., Khanna, K., Handa, N., Bhardwaj, R., Rinklebe, J., et al., 2023. Boron in plants: uptake, deficiency and biological potential. Plant Growth Regulation 100 (2), 267–282. Available at https://doi.org/10.1007/s10725-022-00844-7.

Kumar, A., Singh, J., Samota, M.K., Behera, P.R., 2025. Nutritional and hormonal strategies for plant stress tolerance: mechanistic insights into mineral uptake and signal transduction. Journal of Plant Growth Regulation 44, 1–29.

Kumari, S., Nazir, F., Maheshwari, C., Kaur, H., Gupta, R., Siddique, K.H.M., et al., 2024. Plant hormones and secondary metabolites under environmental stresses: enlightening defense molecules. Plant Physiology and Biochemistry 206. Available at https://doi.org/10.1016/j.plaphy.2023.108238.

Kurepa, J., Smalle, J., 2025. The evolution of plant hormones: from metabolic byproducts to regulatory hubs. International Journal of Molecular Sciences 26 (15), 7190. Available at https://doi.org/10.3390/ijms26157190.

Li, A., Sun, X., Liu, L., 2022. Action of salicylic acid on plant growth. Frontiers in Plant Science 13. Available at https://doi.org/10.3389/fpls.2022.878076.

Li, L., Rodríguez-Concepción, M., Al-Babili, S., 2025. Advances in carotenoid and apocarotenoid metabolisms and functions in plants. Plant Physiology 198 (4). Available at https://doi.org/10.1093/plphys/kiaf304.

Li, S.M., Zheng, H.X., Zhang, X.S., Sui, N., 2021. Cytokinins as central regulators during plant growth and stress response. Plant Cell Reports 40 (2), 271–282. Available at https://doi.org/10.1007/s00299-020-02612-1.

Luo, P., Di, D.-W., 2023. Precise regulation of the TAA1/TAR-YUCCA auxin biosynthesis pathway in plants. International Journal of Molecular Sciences 24 (10), 8514. Available at https://doi.org/10.3390/ijms24108514.

Lv, W., Bai, Y., Zhu, D., He, C., Bu, F., Luo, Y., et al., 2025. Innovative application of nanomaterials in vegetable cultivation: recent advances in growth promotion and stress tolerance. Nanomaterials 15 (21), 1659. Available at https://doi.org/10.3390/nano15211659.

Ma, X., Wang, W., Zhang, J., Jiang, Z., Xu, C., Zhu, W., et al., 2025. NRT1.1B acts as an abscisic acid receptor in integrating compound environmental cues for plants. Cell 188 (19), 5231. Available at https://doi.org/10.1016/j.cell.2025.07.027.

Mao, Y., Chai, X., Zhong, M., Zhang, L., Zhao, P., Kang, Y., et al., 2022. Effects of nitrogen and magnesium nutrient on growth, quality, photosynthesis, antioxidant metabolism, and endogenous hormones of Chinese kale (*Brassica albograbra*). Scientia Horticulturae 303, 111243.

Marreiro, D.D.N., Cruz, K.J.C., Morais, J.B.S., Beserra, J.B., Severo, J.S., Soares de Oliveira, A.R., 2017. Zinc and oxidative stress: current mechanisms. Antioxidants 6 (2). Available at https://doi.org/10.3390/antiox6020024.

Mishra, A.K., Baek, K.H., 2021. Salicylic acid biosynthesis and metabolism: a divergent pathway for plants and bacteria. Biomolecules 11 (5). Available at https://doi.org/10.3390/biom11050705.

Mittler, R., Zandalinas, S.I., Fichman, Y., Van Breusegem, F., 2022. Reactive oxygen species signalling in plant stress responses. Nature Reviews. Molecular Cell Biology 23 (10), 663–679. Available at https://doi.org/10.1038/s41580-022-00499-2.

Mohammadi, M.A., Cheng, Y., Aslam, M., Jakada, B.H., Wai, M.H., Ye, K., et al., 2021. ROS and oxidative response systems in plants under biotic and abiotic stresses: revisiting the crucial role of phosphite triggered plants defense response. Frontiers in Microbiology 12. Available at https://doi.org/10.3389/fmicb.2021.631318.

Muhammad, M., Basit, A., Arooj, A., Dixit, G., Majeed, M., Sinha, D., et al., 2024. Roles of gibberellins in plant defense against biotic and abiotic stress. In: Abd-Elsalam, K.A., Mohamed, H.I. (Eds.), Plant Growth Regulators to Manage Biotic and Abiotic Stress in Agroecosystems. CRC Press, Boca Raton, pp. 183–210. Available at 10.1201/9781003389507-9.

Munné-Bosch, S., Müller, M., 2013. Hormonal cross-talk in plant development and stress responses. Frontiers in Plant Science 4. Available at https://doi.org/10.3389/fpls.2013.00529.

Naing, A.H., Campol, J.R., Kang, H., Xu, J., Chung, M.Y., Kim, C.K., 2022. Role of ethylene biosynthesis genes in salt and drought stress tolerance in *Petunia*. Frontiers in Plant Science 13, 844449.

Nazir, F., Kumari, S., Mahajan, M., Khan, M.I.R., 2023. An explicit story of plant abiotic stress resilience: overtone of selenium, plant hormones and other signaling molecules. Plant and Soil 486 (1-2), 135–163. Available at https://doi.org/10.1007/s11104-022-05826-2.

Nikpour Rashidabad, N., Farhangi-Abriz, S., Naughton, H., Hashemi, M., 2025. Deciphering PGPR mechanisms in saline soils: enhancing plant productivity. Advances in Agronomy 192, 293–340. Available at https://doi.org/10.1016/bs.agron.2025.03.006.

Olukayode, T.O., 2023. Phosphate deficiency responsive transcription factors in cucumber and long-distance mobility of MYB44 mRNA in Arabidopsis. PhD thesis, University of Saskatchewan.

Poonam, Bhardwaj, R., Kaur, R., Bali, S., Kaur, P., Sirhindi, G., et al., 2015. Role of various hormones in photosynthetic responses of green plants under environmental stresses. Current Protein and Peptide Science 16 (5), 435–449. Available at https://doi.org/10.2174/1389203716666150330125215.

Puga, M.I., Poza-Carrión, C., Martinez-Hevia, I., Perez-Liens, L., Paz-Ares, J., 2024. Recent advances in research on phosphate starvation signaling in plants. Journal of Plant Research 137 (3), 315–330. Available at https://doi.org/10.1007/s10265-024-01545-0.

Romera, F.J., Lucena, C., García, M.J., Alcántara, E., Angulo, M.Á., Aparicio, M.Á., et al., 2021. Plant hormones and nutrient deficiency responses. In: Gupta, D.K., Corpas, F.J. (Eds.), Hormones and Plant Response. Plant in Challenging Environments, vol 2. Springer International Publishing, Cham, pp. 29–65. Available at https://doi.org/10.1007/978-3-030-77477-6_2.

Romera, F.J., Smith, A.P., Pérez-Vicente, R., 2016. Editorial: ethylene's role in plant mineral nutrition. Frontiers in Plant Science 7. Available at https://doi.org/10.3389/fpls.2016.00911.

EL Sabagh, A., Islam, M.S., Hossain, A., Iqbal, M.A., Mubeen, M., Waleed, M., et al., 2022. Phytohormones as growth regulators during abiotic stress tolerance in plants. Frontiers in Agronomy 4. Available at https://doi.org/10.3389/fagro.2022.765068.

Salehin, M., 2024. Emerging roles of auxin in plant abiotic stress tolerance. Physiologia Plantarum 176 (6). Available at https://doi.org/10.1111/ppl.14601.

Sehar, Z., Iqbal, N., Fatma, M., Rather, B.A., Albaqami, M., Khan, N.A., 2022. Ethylene suppresses abscisic acid, modulates antioxidant system to counteract arsenic-inhibited photosynthesis in the presence of selenium in mustard. Frontiers in Plant Science 13, 878704.

Shi, R., Yu, J., Chang, X., Qiao, L., Liu, X., Lu, L., 2023. Recent advances in research into jasmonate biosynthesis and signaling pathways in agricultural crops and products. Processes 11 (3), 736. Available at https://doi.org/10.3390/pr11030736.

Singh, A., Sagar, S., Biswas, D.K., 2017. Calcium dependent protein kinase, a versatile player in plant stress management and development. Critical Reviews in Plant Sciences 36 (5-6), 336–352. Available at https://doi.org/10.1080/07352689.2018.1428438.

Singh, V.P., Siddiqui, M.H., 2023. Nutrient availability regulates root system behavior. Plant Ionomics: Sensing, Signaling, and Regulation. Wiley, pp. 96–119. Available at https://doi.org/10.1002/9781119803041.ch6.

Suzuki, N., 2016. Hormone signaling pathways under stress combinations. Plant Signaling & Behavior 11 (11), e1247139. Available at https://doi.org/10.1080/15592324.2016.1247139.

Timofeeva, A.M., Galyamova, M.R., Sedykh, S.E., 2024. How do plant growth-promoting bacteria use plant hormones to regulate stress reactions? Plants 13 (17), 2371. Available at https://doi.org/10.3390/plants13172371.

Tiwari, S., Lata, C., Chauhan, P.S., Prasad, V., Prasad, M., 2017. A functional genomic perspective on drought signalling and its crosstalk with phytohormone-mediated signalling pathways in plants. Current Genomics 18 (6), 469–482. Available at https://doi.org/10.2174/1389202918666170605083319.

Umair Hassan, M., Aamer, M., Umer Chattha, M., Haiying, T., Shahzad, B., Barbanti, L., et al., 2020. The critical role of zinc in plants facing the drought stress. Agriculture 10 (9), 396. Available at https://doi.org/10.3390/agriculture10090396.

Wang, Q., Li, S., Li, J., Huang, D., 2024a. The utilization and roles of nitrogen in plants. Forests 15 (7), 1191. Available at https://doi.org/10.3390/f15071191.

Wang, Z., Zheng, Z., Liu, D., 2024b. Comparative functional analyses of PHR1, PHL1, and PHL4 transcription factors in regulating *Arabidopsis* responses to phosphate starvation. Frontiers in Plant Science 15. Available at https://doi.org/10.3389/fpls.2024.1379562.

Wang, Z.Y., Xiong, L., Li, W., Zhu, J.K., Zhu, J., 2011. The plant cuticle is required for osmotic stress regulation of abscisic acid biosynthesis and osmotic stress tolerance in *Arabidopsis*. The Plant Cell 23 (5), 1971–1984. Available at https://doi.org/10.1105/tpc.110.081943.

Wei, H., Jing, Y., Zhang, L., Kong, D., Lawson, T., 2021. Phytohormones and their crosstalk in regulating stomatal development and patterning. Journal of Experimental Botany 72 (7), 2356–2370. Available at https://doi.org/10.1093/jxb/erab034.

Xing, J., Cao, X., Zhang, M., Wei, X., Zhang, J., Wan, X., 2023. Plant nitrogen availability and crosstalk with phytohormones signallings and their biotechnology breeding application in crops. Plant Biotechnology Journal 21 (7), 1320–1342. Available at https://doi.org/10.1111/pbi.13971.

Yang, T., Lu, X., Wang, Y., Xie, Y., Ma, J., Cheng, X., et al., 2020. HAK/KUP/KT family potassium transporter genes are involved in potassium deficiency and stress responses in tea plants (*Camellia sinensis* L.). BMC Genomics 21, 556.

Zhang, Z., Liu, D., Meng, H., Li, S., Wang, S., Xiao, Z., et al., 2020. Magnesium alleviates aluminium toxicity by promoting polar auxin transport and root alkalization in *Populus*. Plant and Soil 448, 565–585.

Zhao, B., Liu, Q., Wang, B., Yuan, F., 2021. Roles of phytohormones and their signaling pathways in leaf development and stress responses. Journal of Agricultural and Food Chemistry 69 (12), 3566–3584. Available at https://doi.org/10.1021/acs.jafc.0c07908.

Zhao, Q., Liu, R., Zhou, Q., Ye, J., Meng, F., Liu, J., et al., 2023. Calcium-binding protein OsANN1 regulates rice blast disease resistance by inactivating jasmonic acid signaling. Plant Physiology 192 (2), 1621–1637. Available at https://doi.org/10.1093/plphys/kiad174.

Zheng, Z., Guo, Y., Novák, O., Dai, X., Zhao, Y., Ljung, K., et al., 2013. Coordination of auxin and ethylene biosynthesis by the aminotransferase VAS1. Nature Chemical Biology 9 (4), 244–246. Available at https://doi.org/10.1038/nchembio.1178.

Zhou, S., Zi, X., Rao, D., Liu, K., Yang, L., Shen, P., et al., 2024. Jasmonic acid mediates maize (*Zea mays* L.) roots response to soil nitrogen heterogeneity. Journal of Plant Biology 67 (3), 241–255. Available at https://doi.org/10.1007/s12374-024-09428-6.

Zhou, T., Hua, Y., Huang, Y., Ding, G., Shi, L., Xu, F., 2016. Physiological and transcriptional analyses reveal differential phytohormone responses to boron deficiency in *Brassica napus* genotypes. Frontiers in Plant Science 7, 2016. Available at https://doi.org/10.3389/fpls.2016.00221.

Zhou, Y., Nie, K., Geng, L., Wang, Y., Li, L., Cheng, H., 2025. Selenium's role in plant secondary metabolism: regulation and mechanistic insights. Agronomy 15 (1). Available at https://doi.org/10.3390/agronomy15010054.

CHAPTER

Osmotic regulation and nutrient role

17

Prasann Kumar[1], Amit Raj[1] and Sudhir Kumar Upadhyay[2]

[1]*Department of Agronomy, School of Agriculture, Lovely Professional University, Phagwara, Punjab, India,*

[2]*School of Chemical Engineering and Physical Sciences, Lovely Professional University, Phagwara, Punjab, India*

17.1 Introduction

Plant osmotic regulation is a vital physiological process that allows cells and tissues to keep hydrated, maintain membrane stability, and support growth under changing environmental conditions. Water potential and osmotic gradients govern the passive and active movement of water across cell membranes, whereas solutes, aquaporin channels, and membrane transporters regulate cellular osmotic balance. The components of water potential, including osmotic, pressure, matric, and gravitational potentials, interact dynamically to ensure that water flows in the appropriate direction, enabling cells to expand and maintain hydraulic conductivity. Cellular osmotic buffering relies on compatible solutes such as proline, glycine betaine, trehalose, raffinose family oligosaccharides, and polyols, which stabilize macromolecules, regulate cytoplasmic viscosity, and prevent protein unfolding during dehydration (Giri et al., 2024). The restructuring of lipid microdomains, increased sterol content, phospholipid remodeling, and activation of desaturases help adjust membrane fluidity in response to osmotic pressure gradients, aiding membrane stability. Vacuolar ion flow, tonoplast H^+-ATPases, Na^+/H^+ antiporters, and mechanosensitive ion channels work together to maintain dynamic turgor in response to osmotic compression and expansion (Joshi et al., 2022). Early signal transduction begins with osmosensors, including OSCA channels, two-component signal transduction via histidine kinases, and plasma membrane stretch receptors. These sensors trigger downstream calcium waves, MAPK cascades, ROS waves, and ABA-mediated transcriptional reprogramming. Osmotic gradients in guard cells, which change with environmental conditions, help regulate stomatal opening and closing, integrating blue-light signaling, phototropins, and proton pump activity to conserve water. The link between osmotic sensing and metabolic activation is evident in the way primary carbon metabolism connects to osmolyte production, with precursors generated through glycolysis, the TCA cycle, and the pentose phosphate pathway, thereby maintaining cellular redox balance (Fujii, 2024).

Mineral Nutrients to Mitigate Abiotic Stress in Crops. DOI: https://doi.org/10.1016/B978-0-443-44712-9.00016-5

Metabolic reorganization involves increased sucrose mobilization, changes in amino acid turnover, enhanced glutathione cycling, and activation of an alternative pathway that supplies energy and reducing power during osmotic stress. Hormonal interactions among ABA, ethylene, salicylic acid, jasmonates, brassinosteroids, strigolactones, and cytokinins coordinate osmotic adjustment by affecting stress-responsive gene expression, root development patterns, stomatal behavior, and detoxification processes (Thakur et al., 2025). Root plasticity traits such as hydrotropism, xerobranching, and water transport via aquaporins contribute to whole-plant osmotic homeostasis (Kang et al., 2022). The interaction between osmotic signal amplification and transcription factor activation occurs at the cellular signaling level, involving calcium signatures, phospholipase D activation, inositol phosphate turnover, and NO-mediated posttranslational modifications. The regulation of osmotic stress genes, including dehydrins, late embryogenesis abundant proteins, heat shock proteins, ROS-scavenging enzymes, and ion transporters, is governed by key transcription factors from families such as DREB, NAC, bZIP, MYB, and WRKY (Kumar et al., 2023). Mitochondrial and chloroplast responses to osmotic stress include changes in electrolyte balance related to electron transport, regulation of ATP synthase, stromal pH buffering, and the production of alternative oxidase pathways to prevent ROS formation.

Mechanical toughness in response to osmotic shrinkage or expansion is achieved through cell wall remodeling, catalyzed by expansins, pectin methylesterases, cellulose synthases, and xyloglucan endotransglucosylases. Apoplastic and symplastic communication, driven by changes in plasmodesmatal conductance and the regulation of pH, coordinate the tissues' osmotic responses (Cui et al., 2023). New pathways include circadian rhythm regulation, DNA methylation, histone acetylation, and small RNA-mediated processes that fine-tune osmotic stress memory and adaptive resilience. These multilayered processes highlight the complexity and integrative nature of osmotic regulation in plants, forming a highly coordinated network that supports homeostasis, growth, and stress adaptation.

17.2 Compatible solutes and osmoprotective metabolites

17.2.1 Biosynthesis of proline, glycine betaine, and polyols

A key osmoprotective pathway that helps plants resist dehydration, salinity, and osmotic stresses is the biosynthesis of compatible solutes like proline, glycine betaine, and polyols. The glutamate and ornithine pathways are the main routes for proline production, guided by the activity of pyrroline-5-carboxylate synthetase (P5CS), pyrroline-5-carboxylate reductase (P5CR), and the negative regulation of proline dehydrogenase (ProDH) (Yan et al., 2025). Redox-sensitive enzymes become more active under osmotic stress, thereby affecting NADPH-dependent transformations and linking proline accumulation to oxidative buffering. Choline monooxygenase (CMO) and betaine aldehyde dehydrogenase (BADH) are involved in producing glycine betaine and in importing choline into plastids; methylation cycles with S-adenosylmethionine-dependent and phosphoethanolamine methyltransferases increase substrate flow (Chen et al., 2025a). The polyols mannitol, sorbitol, pinitol, and galactinol are produced by dehydrogenases of mannitol-1-phosphate and sorbitol-6-phosphate, as well as by galactinol synthase, and they aid osmoprotection by preventing protein denaturation and maintaining enzyme activity under hyperosmotic conditions. These pathways are interconnected through ABA-responsive element-mediated transcriptional activation, ROS-triggered promoter binding, metabolic channeling, and plastid-to-cytosol shuttling, illustrating the functional flexibility of these mechanisms in osmotic resilience.

17.2.2 Molecular regulation of osmoprotective pathways

Multilayered networks mediating the molecular regulation of osmoprotective metabolite biosynthesis include transcriptional, posttranscriptional, and posttranslational regulation. DREB2A, AREB/ABF, MYB60, bZIP17, NAC29, and WRKY33 are transcription factors that regulate osmolyte biosynthetic gene expression in response to dehydration (Dong et al., 2023). Stress-induced promoter accessibility and osmoprotectant gene upregulation occur through chromatin remodeling involving histone acetylation, H3K4 trimethylation, and DNA demethylation. The phosphorylation of metabolic enzymes and regulatory scaffolds is controlled by Ca^{2+}-dependent protein kinases (CDPKs), MAPK cascades, SnRK2s, and TOR signaling (Li et al., 2025). Noncoding RNAs (miR169, miR398) and long intergenic RNAs modulate osmoprotectant synthesis by cleaving mRNAs and suppressing translation. Thioredoxin-sensitive redox interactions help enzymes change their shape, while ubiquitin-proteasome processes regulate enzyme turnover, linking metabolic activity to cellular energy levels (Kuczyńska et al., 2022). Second messengers generated in membranes via lipid-derived pathways such as phosphatidic acid, inositol triphosphate, and sphingolipid intermediates are essential for activating osmosensory pathways. Crosstalk among hormonal networks, including ABA, jasmonates, salicylic acid, ethylene, and brassinosteroids, further integrates metabolic activation with whole-plant homeostasis, enabling rapid adaptation to drought, salinity, and osmotic stress (Fig. 17.1).

17.2.3 Role of solute compartmentalization in stress tolerance

Compartmentalization of solutes is essential for maintaining the osmotic balance of cells during stress and ensuring proper metabolism. Proline, betaine, and polyols are differentially partitioned into the apoplast, cytosol, vacuole, and chloroplasts, depending on osmotic gradients, the proton motive force, and transporter specificity. Tonoplast transporters, such as NHX antiporters, TPC1 channels, and sugar transporters, regulate the accumulation of osmolytes in vacuoles, thereby maintaining functional cytosolic enzymes and regulating glycosylated turgor pressure

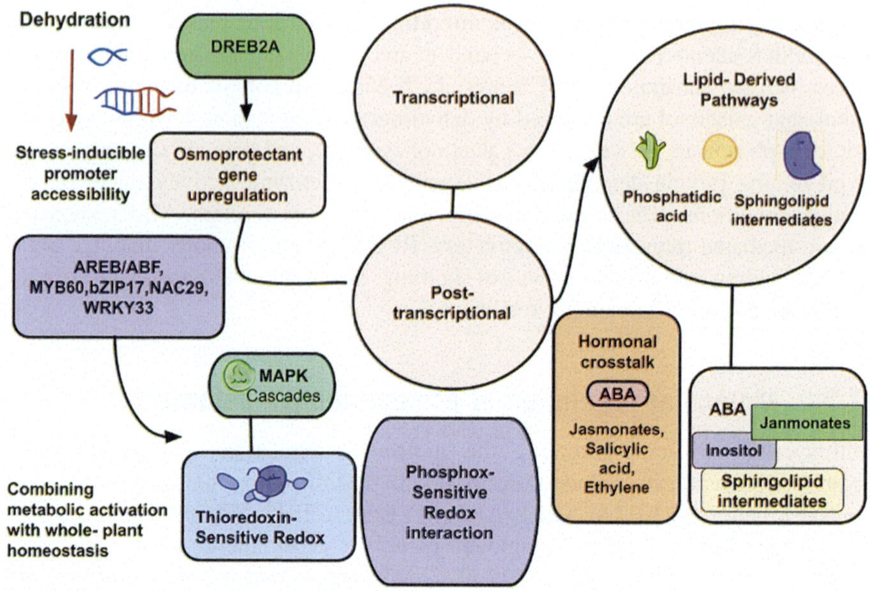

FIGURE 17.1

Integrated molecular networks regulating osmoprotective metabolite biosynthesis under stress.

(Pottosin and Dobrovinskaya, 2022). Osmolyte uptake in chloroplasts stabilizes thylakoid membrane stacking, maintains stromal pH, and prevents photoinhibition of the photosystems. The compatibility of solutes in the cytosol prevents protein misfolding, supports ribosome assembly, and buffers redox changes during stress induced by ROS formation. In the apoplast, osmolyte distribution influences cell wall elasticity, extensin cross-linking, and pectin reconfiguration, enabling controlled mechanical adjustments (Obomighie et al., 2025). Recent evidence highlights the roles of vesicle trafficking, SNARE complexes, autophagic recycling, and ER-vacuolar contact sites in the spatial routing of osmolytes. This multicompartmental coordination equips different tissues with distinct osmotic buffers, enhancing localized stress tolerance.

17.2.4 Metabolomic signatures of osmolyte accumulation

Metabolomic profiling reveals signatures of osmotic stress tolerance through unique osmolyte profiles, indicating changes in pools of amino acids, sugar alcohols, quaternary ammonium compounds, and sulfur-containing metabolites. Mass spectrometry and NMR-based metabolomics show rapid increases in proline, betaine, mannitol, and galactinol in response to water deprivation, along with secondary metabolic reprogramming of flavonoids, alkaloids, glucosinolates, and phenylpropanoid derivatives (Gupta, 2025). The glutamate cycle, methyl-donor pathway, and

PPP exhibit higher fluxes, as indicated by metabolite correlations, demonstrating a tightly regulated network for osmoprotectant synthesis. Chronic stress leads to more extensive metabolic rewiring, with the emergence of biomarkers such as isosmotic peptides, organic acid conjugates, and oxidative lipid derivatives (Acierno et al., 2025). Spatial metabolomics also identifies tissue-specific hotspots of osmolytes in roots, guard cells, vascular bundles, and mesophyll cells. These dynamic metabolic signatures can offer predictive information for developing drought-resilient genotypes.

17.3 Ion transport machinery and electrochemical balance

17.3.1 Potassium transporters and membrane depolarization control

Maintaining potassium (K^+) homeostasis is essential for regulating membrane potential, osmotic gradients, and enzyme activation under changing osmotic conditions. Cytosolic K^+ fluxes are controlled by high-affinity K^+ transport proteins (known here as HKT1-like), Shaker-type channels (including AKT1, SKOR, and GORK), and cyclic nucleotide-gated channels (CNGCs), which respond to electrochemical differences (Chatterjee et al., 2022). Membrane depolarization induced by hyperosmotic stress triggers K^+ efflux via outward-rectifying channels. In response, plants enhance H^+-ATPase activity, electrogenic pumping, and phosphorylation-dependent gating of K^+ channels. Calcium spiking patterns, the interaction between CBLs and CIPKs, and channel phosphorylation are coordinated by CBLs, CBL-interacting protein kinases (CIPKs), and calcium oscillations, enabling rapid K^+ adjustments (Bihani et al., 2025). Stress-induced reactive oxygen species (ROS) sensors, such as HPCA1, modulate K^+ permeability via redox-mediated changes. Recent mechanistic studies highlight the roles of lipid raft microdomains, phosphoinositide-mediated transporter recruitment, and mechano-sensitive gating by MSL and MCA proteins (Bobkov and Semenova, 2022). These regulatory mechanisms sustain membrane excitability and stabilize metabolic functions during osmotic stress.

17.3.2 Sodium exclusion, vacuolar sequestration, and antiporters

Salt and osmotic stress cause harmful buildup of sodium (Na^+), which requires the coordinated actions of exclusion, intracellular sequestration, and regulation of xylem loading. The SOS (salt overly sensitive) pathway, including SOS1 (a plasma-membrane Na^+/H^+ antiporter), SOS2 (CIPK24 kinase), and SOS3 (CBL4 Ca^{2+} sensor), continues to focus on limiting Na^+ buildup (Steinhorst et al., 2022). SOS-triggered phosphorylation boosts Na^+ efflux from the apoplast, while H^+ -PPases and V-PPases strengthen proton gradients to fuel vacuolar Na^+ sequestration via NHX antiporters (NHX1-NHX4) (Albloushi, 2023). Trapping Na^+ in the vacuole maintains cytosolic ionic balance, thereby preventing metabolic inhibition of Na^+-sensitive enzymes. At the same time, HKT-type transporters control Na^+ retrieval in the xylem, protecting photosynthetic tissues. Recent mechanistic models highlight the roles of endosomal–vacuolar sorting, clathrin-coated vesicle sorting, SNARE-dependent vesicle sorting, and retromer-mediated sorting as key to keeping antiporter levels stable during stress (Gopaldass et al., 2024). Additionally, the effectiveness of Na^+ compartmentalization is improved by sphingolipid-induced membrane rigidification and ABA-driven phosphorylation cascades. These processes help stabilize ion selectivity, prevent cytotoxicity, and maintain osmotic balance during high salinity or drought.

17.3.3 Chloride, sulfate, and phosphate ion dynamics

Anionic species such as chloride (Cl^-), sulfate $(SO_4)_{2-}$, and phosphate (Pi) are essential for maintaining a stable electrochemical state and metabolic balance under osmotic stress. Chloride channels (CLC family) and SLAC/SLAH channels, along with anion exchangers, facilitate Cl- transport across membranes and are involved in maintaining turgor pressure (Bian et al., 2025), including stomatal closure and charge compensation. ABA-dependent protein kinases activated by dehydration phosphorylate SLAC1, causing quick guard-cell Cl- efflux in stomatal control. SULTR1;1 and SULTR1;2 facilitate sulfate uptake, providing substrates needed for glutathione and cysteine synthesis, which enhances redox buffering during osmotic and oxidative stresses (Alvi et al., 2023). SULTR3 subgroup transporters regulate sulfate distribution within plastids and vacuoles, thereby maintaining sulfur metabolic fluxes. Phosphate regulation involves the PHT1 transporter, SPX-domain regulators, and PHO1 export proteins, coordinating Pi availability, ATP synthesis, membrane remodeling, and polyphosphate metabolism (Ding et al., 2025). Recent findings indicate that contact sites between organelles and the plasma membrane, like ER-vacuolar tethers, facilitate direct anion exchange. Additionally, anion homeostasis is controlled by phospholipid turnover, inositol pyrophosphate signaling, and redox-sulfur interactions, which support adaptive responses to osmotic pressure.

17.3.4 Coordination of transport proteins under osmotic stress

Osmotic stress triggers systemic reprogramming of the ion transport system by linking hormonal, transcriptional, and metabolic networks. Transporter coordination involves ABA, jasmonate, ethylene, and Ca^{2+} signals that cause successive changes in channel gating, transporter location, and turnover via SNRK2s, MAPKs, CPKs, and CIPKs (Song et al., 2025). Transporter endocytosis, recycling, and delivery to specific regions along a stress gradient are regulated by membrane-trafficking complexes comprising exocyst subunits, Rab GTPases, ARF-GEFs, and ESCRT machinery (Qi et al., 2024). ROS bursts affect redox-sensitive cysteine residues in transporters, thereby modifying channel conductivity and stress selectivity. Mechanosensitive responses allow sensing of membrane tension and rapid shifts in ion flow. Metabolic feedback mechanisms involving ATP supply, proton-motive force, and carbon-nitrogen regulation of transporter activity optimize energy use. Systems biology studies show that transportosomes, dynamic multi-protein complexes of K^+, Na^+, Cl^-, and H^+ transporters, function as units of electrochemical control (Rioux et al., 2024). This stabilization of the membrane potential through orchestrated ion transport ensures cell survival under extreme osmotic conditions and facilitates osmotic buffering (Table 17.1).

17.4 Nutrient elements as drivers of osmotic stability

17.4.1 Potassium and calcium in osmotic buffering

Potassium (K^+) and calcium (Ca^{2+}) are the main ionic determinants of osmotic stability that regulate membrane potential, turgor responsiveness, and fidelity to cytosolic signaling (Joshi et al., 2022). K^+ also controls osmotic potential by acting as the dominant inorganic osmolyte and by regulating guard cell activity, metabolic enzyme activity, and vacuolar osmotic equilibria. K^+ fluxes in changing osmotic gradients are mediated by high-affinity transporters (HKT-like), Shaker-family channels (AKT1, SKOR), tandem-pore channels, and CNGCs. Ca^{2+} functions as a second messenger that creates decoding modules through Ca^{2+} signatures, calmodulin (CaM) interactions, calcineurin B-like (CBL) proteins, and Ca^{2+}-dependent protein kinases (CPK) (Negi et al., 2023). MSL, MCA, ER-plasma membrane tethered channels, and stretch-activated conductances mediate cytosolic Ca^{2+} spikes induced by osmotic shock: Transporter phosphorylation states are reconfigured by osmotic shock through stimulation of these channels. K^+/Ca^{2+} synergy maintains plasma membrane selectivity by regulating phospholipid order, ROS-activated ion channels, and membrane nanodomain assembly (Perez-Rivera et al., 2025). Such structural changes, such as Ca^{2+}-pectate cross-linking in the apoplast, increase cell wall rigidity, thereby reducing water efflux during osmotic contraction. The combined action of these factors creates hydrodynamic buffering loops that are essential for osmotic acclimation.

17.4.2 Micronutrient cofactors in redox–osmotic integration

Redox dynamics are integrated into micronutrients such as zinc (Zn), manganese (Mn), iron (Fe), copper (Cu), molybdenum (Mo), and boron (B), which also help in osmotic stabilization. These nutrients act as catalytic or structural cofactors in antioxidant enzymes, transporter complexes, and metabolic nodes. Zn^{2+} stabilizes membranes via zinc-finger transcription factors that regulate osmotic-stress genes and serves as a cofactor for SOD isoforms that reduce ROS accumulation during dehydration-induced oxidative bursts (Moulick et al., 2023). Mn^{2+} supports the

Table 17.1 Ion transport machinery and electrochemical balance under osmotic stress.

Transport module	Key components	Core mechanisms	Stress	Functional outcomes	Key references
K^+ homeostasis	HKT1-like, AKT1, SKOR, GORK	High/low-affinity K^+ flux control	Hyperosmotic stress	Membrane potential stability	Chatterjee et al. (2022)
K^+ channel gating	CNGCs, H^+-ATPase	Depolarization-driven K^+ efflux	Osmotic shock	Electrogenic pumping and charge balance	Bihani et al. (2025)
CBL–CIPK regulation	CBLs, CIPKs	Ca^{2+}-coupled phosphorylation of channels	Rapid osmotic fluctuations	Fast K^+ oscillation tuning	Bobkov and Semenova (2022)
Redox control of K^+ permeability	HPCA1, ROS sensors	Redox-triggered gating adjustments	ROS-associated osmotic stress	Protection of excitability	Albloushi (2023)
Mechanical and lipid control of K^+ channels	MSL, MCA, phosphoinositides	Mechanosensitive gating & lipid raft recruitment	Membrane tension changes	Stabilized K^+ dynamics	Gopaldass et al. (2024)
Na^+ exclusion	SOS1, SOS2, SOS3	Na^+/H^+ antiport via SOS phosphorylation	High salinity	Cytosolic Na^+ detoxification	Bian et al. (2025)
Vacuolar Na^+ sequestration	NHX1-NHX4, H^+-PPase	Proton-gradient–driven Na^+ compartmentalization	Salinity, drought	Enzyme protection and ionic balance	Alvi et al. (2023)
Xylem Na^+ retrieval	HKT transporters	Na^+ removal from xylem sap	Salt-induced leaf toxicity	Protection of photosynthetic tissue	Ding et al. (2025)
Vesicular regulation of antiporters	SNAREs, retromer, clathrin, endosomal sorting	Trafficking & abundance control of antiporters	Salt and osmotic stress	Sustained transporter availability	Qi et al. (2024)
Cl^- transport	CLC channels, SLAC/SLAH	Anion flux for charge compensation	Dehydration, turgor shifts	Stomatal regulation via Cl^- efflux	Rioux et al. (2024)

Sulfate uptake and distribution	SULTR1;1, SULTR1;2, SULTR3 group	Uptake and plastid/vacuole partitioning	Oxidative and osmotic stress	Boosted glutathione/cysteine biosynthesis	Chatterjee et al. (2022)
Phosphate dynamics	PHT1 SPX PHO1	Pi uptake, export, and signaling	Pi limitation and osmotic imbalance	ATP stability and lipid remodeling	Bihani et al. (2025)
Membrane contact–mediated anion exchange	ER–vacuole tethers	Direct interorganelle anion transfer	High osmotic demand	Rapid anion balancing	Bobkov and Semenova (2022)
Transporter coordination networks	SnRK2, MAPK, CPK, CIPK	Hormonal and Ca^{2+}-dependent gating control	Systemic osmotic stress	Transporter relocation and turnover	Albloushi (2023)
Transportosomes in electrochemical control	$K^+/Na^+/Cl^-/H^+$ multiprotein assemblies	Integrated channel/antiporter synchronization	Severe osmotic gradients	Membrane potential stabilization	Gopaldass et al. (2024)

activity of the oxygen-evolving complex (OEC) in chloroplasts, preventing thylakoid membrane damage caused by osmotic swelling. Fe-S cluster proteins facilitate electron transport and avoid redox leakage in mitochondria during osmotic collapse (Yamasaki et al., 2025). Copper-dependent oxidases help regulate lignification, strengthen cell walls, and enhance osmotic resistance. Molybdenum enzymes also contribute to nitrate reduction, thereby influencing osmolyte pathways under nitrogen-limiting conditions. Boron maintains membrane architecture by cross-linking rhamnogalacturonan-II with borate and preserves vesicular fidelity under osmotic tension. Recently discovered pathways highlight micronutrient-dependent regulation of peroxisomal β-oxidation, redox nanodomains, and metal-reactive transcription factors that coordinate osmotic and redox homeostasis.

17.4.3 Nitrogen forms influencing osmolyte biosynthesis

The availability of nitrogen (N), including its chemical forms nitrate (NO_{3-}), ammonium (NH_{4+}), and organic nitrogen, plays a crucial role in regulating the biosynthesis of osmolytes and maintaining metabolic resilience under osmotic stress. NIT1/NIT2-mediated uptake of NO_{3-} activates nitrate reductase cascades, which generate signaling intermediates that promote the synthesis of proline, glycine betaine, and polyamines (Das et al., 2025). The GS-GOGAT pathway of NH_{4+} assimilation increases glutamine and glutamate levels, serving as primary carbon-nitrogen donors for osmoprotectants such as proline, γ-aminobutyric acid (GABA), and tertiary amines (Lal and Bhatla, 2023). Differential nitrogen levels reprogram transcriptional networks via NLP transcription factors, TOR-based metabolic checkpoints, and C/N balance sensors that coordinate the osmoprotective response. Nitrate-responsive kinases also regulate aquaporin gating and membrane hydraulic conductivity under osmotic stress conditions. Organic N sources, such as amino acids, induce a redistribution of metabolites toward TCA cycle intermediates, thereby enhancing osmotic adjustment capabilities (Mandal et al., 2022). Additionally, new pathways involving nitric oxide (NO)-mediated redox signaling, amino-acid-derived phytohormones, and N-mediated activation of stressosome complexes integrating osmotic signaling with metabolic remodeling have recently been discovered (e.g., polyamine-ethylene crosstalk) (Fig. 17.2).

17.4.4 Phosphorus-mediated energy support for osmotic repair

Phosphorus (P) is the source of energy for osmotic repair, which relies on ATP regeneration, membrane-lipid remodeling, and phosphorylation-mediated signaling cascades, primarily involving inorganic phosphate (Pi). The cytosolic Pi used in ATP synthesis, fueled by Pi uptake through PHT1 transporters and intracellular distribution via SPX-domain proteins, is necessary to regulate osmotic rebalancing by H^+ ATPases, V-ATPases, and vesicle-trafficking motors (Ding et al., 2025). During osmotic stress, Pi promotes the exchange of structural phospholipids for

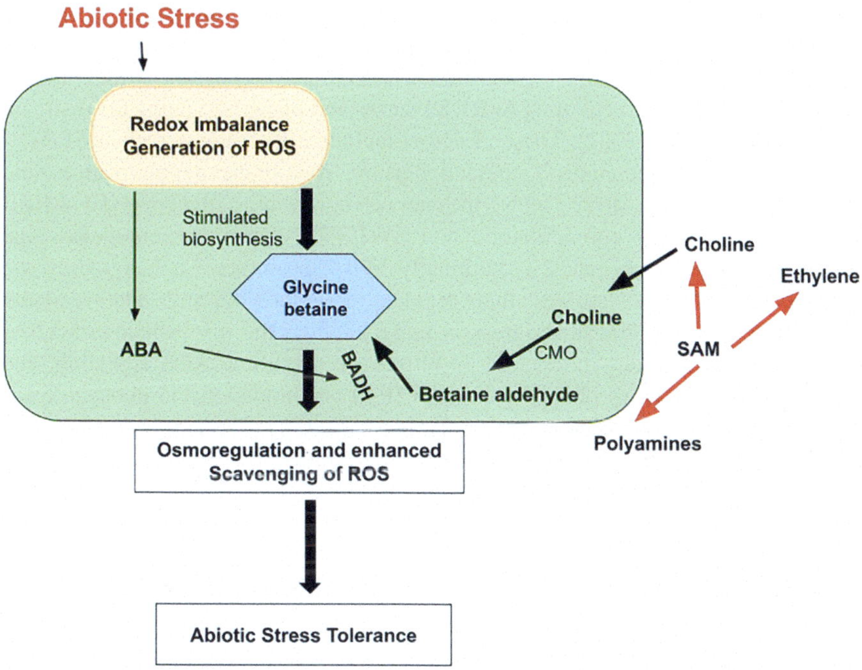

FIGURE 17.2

Nitrogen forms modulate osmolyte biosynthesis through metabolic, transcriptional, and redox-regulatory pathways.

galactolipids, thereby reducing bilayer lecithin content and maintaining bilayer stability. Pi export, controlled by PHO1, provides metabolic protection to growing tissues with phloem Pi. SnRK2 kinases, MAPK modules, and CDPK networks form a phosphorylation cascade that alters transporter activity or detoxification pathways in response to osmotic stress. A deficiency in Pi triggers alternative glycolytic bypasses, increasing resource utilization under stress conditions. Pyrophosphate-driven vacuolar pumps, phosphoinositide signaling (PIP2, IP3, IP7), and Pi-mediated mitochondrial biogenesis are involved in osmotic repair mechanisms (Zhao and Sheng, 2025). The interconnectedness of these processes indicates that phosphorus supplies the energy needed for ion transport, osmolyte synthesis, and membrane restoration.

17.5 Hormonal crosstalk in osmotic and nutrient regulation

17.5.1 Abscisic acid-mediated osmotic perception

Abscisic acid (ABA) is the primary regulator of osmotic perception, integrating hydrodynamic stress responses with metabolic and transcriptional activation. ABA is triggered by osmotic stress, which is mediated by NCED-catalyzed cleavage of carotenoids, and it rapidly accumulates in the vascular parenchyma and guard cells. ABA detection occurs through PYR/PYL/RCAR receptors, which inhibit clade A PP2Cs, thereby activating SnRK2 kinases and phosphorylating downstream effectors (Fidler et al., 2022). These effectors include ion channels such as SLAC1, KAT1, and TPKs that control stomatal closure, membrane depolarization, and turgor regulation. ABA-driven transcriptional reprogramming involves ABF/AREB factors, DREB–AREB comodulators, and SWI/SNF chromatin remodelers that regulate osmoprotective genes. Additionally, ABA promotes Ca^{2+} signaling via CPKs, generating Ca^{2+} spikes that regulate aquaporin gating and vesicular trafficking (Lin et al., 2025). Progress in understanding the mechanisms of ABA-induced priming includes mitochondrial retrograde signaling, ROS signaling, and lipid remodeling, such as phosphatidic acid (PA) accumulation via phospholipase D, which enhances membrane stress tolerance. These multiple layers of control enable ABA to establish a dynamic osmotic surveillance system that manages ion balance, carbon partitioning, and transcriptional memory.

17.5.2 Cytokinin and auxin redistribution under nutrient shifts

Cytokinin (CK) and auxin are nutrient-sensitive regulators whose redistribution patterns control root structure, nutrient uptake, and osmotic plasticity. CK synthesis via IPT genes is increased by sufficient nitrate availability, and the sixth signal transduces growth in response to nutrients; however, osmotic stress reduces CK levels through type-B ARRs, which activate the CKX pathway. CK deficiency promotes root growth through increased branching, transporter gene expression (NRTs, PHTs), and osmotic foraging. PIN, AUX1/LAX, and ABCB transporters are dynamically required to regulate auxin distribution and generate polar auxin fluxes that govern cell expansion and root gravitropic responses to nutrient changes (Vanneste et al., 2025). Nutrient deprivation triggers auxin maxima at root tips, encouraging elongation and lateral root outgrowth to maximize uptake of N, P, and K. MAPK cascades and changes in vesicle trafficking of PIN proteins, along with membrane domain reorganization, stabilize the effects of auxin-CK antagonism. Nutrient-dependent regulation of TIR1/AFB auxin receptor turnover, histidine phosphotransfer interactions between CK and aquaporin expression, and axonometric loops of NO, sugars, and auxin transport have been identified through newly discovered mechanisms (Vanneste et al., 2025).

17.5.3 Ethylene and jasmonate signaling in osmo-adaptation

Regulation of cellular plasticity, detoxification systems, and ion homeostasis under stress, mediated by ethylene and jasmonates (JA, JA-Ile), controls osmo-adaptive behavior. ACS/ACO ethylene biosynthesis increases during osmotic compression, leading to ethylene bursts that activate the EIN2/EIN3 transcription modules

(Khan et al., 2024). These regulators are involved in root hair elongation, membrane fluidity, and ROS detoxification. Ethylene also contributes to the interaction between Ca^{2+} influx channels and G-protein complexes in osmotic sensing. Octadecanoid biosynthesis of JA remains elevated under osmotic and nutrient fluctuations, producing JA-Ile, which binds to the COI1-JAZ coreceptor complex to release MYC transcription factors, thereby promoting defense-related metabolic genes (Kaur et al., 2025). JA crosstalk affects Na^+ exclusion, K^+ retention, and polyamine synthesis and influences lipid turnover enzymes that stabilize membrane bilayers. The interaction between ethylene and JA enhances osmotic hardening through WRKY, ERF, and NAC transcriptional networks, thereby activating this process (Li et al., 2024a, 2024b). Recent findings reveal JA's role in regulating mitochondrial quality, ER-Golgi trafficking loops, and autophagy during osmotic stress; ethylene also modulates mechanosensitive channels and cytoskeletal reorganization, supporting adaptive remodeling.

17.5.4 Hormone–nutrient–ion interaction networks

There are networks of hormone–nutrient–ion interactions that are multilayered regulation circuits coordinating metabolic allocation with osmotic regulation. ABA-K^+ signaling controls stomatal hydraulics, whereas CK-NO_3 signaling modules optimize root transport capacity. Auxin enhances phosphate scavenging through PHR1 and Pi-responsive miRNAs, thereby linking Pi homeostasis and root system architecture (Zulfiqar et al., 2025). The interaction between JA-Ca^{2+} and Ca^{2+} oscillation signatures affects ion channel stability, and ethylene influences Fe acquisition via FIT-controlled transcriptional modules. TOR and SnRK1 energy sensors are complex regulatory hubs integrating carbohydrate status with nutrient-regulated hormone signaling (Feng et al., 2025). Crosstalk is transmitted to ROS, where hormone-mediated redox nodes regulate transporter phosphorylation, autophagy induction, and vesicle trafficking. Recent models include nutrient-driven epigenetic marking, microRNA-controlled hormonal regulation, and ion-hormone coregulated stressosomes that manage rapid osmo-adaptive responses. This connects nutrient acquisition, ionic balance, and hormonal dynamics within a comprehensive osmotic-resilience system through these networks (Table 17.2).

17.6 Aquaporins and membrane water relations

17.6.1 Diversity and function of plasma membrane intrinsic proteins and tonoplast intrinsic proteins

Plant aquaporins are multidimensional water channels that include plasma membrane intrinsic proteins (PIPs), tonoplast intrinsic proteins (TIPs), nodulin-26-like intrinsic proteins (NIPs), small basic intrinsic proteins (SIPs), and uncharacterized

Table 17.2 Hormonal crosstalk in osmotic and nutrient regulation.

Regulatory module	Key molecular components	Primary mechanisms	Stress	Functional outcomes	Key references
ABA-mediated osmotic perception	NCED, PYR/PYL/RCAR, PP2Cs, SnRK2s	Carotenoid cleavage, ABA sensing, kinase activation	Osmotic depression, dehydration	Stomatal closure, turgor maintenance	Fidler et al. (2022)
ABA signaling to ion channels	SLAC1, KAT1, TPKs	Channel phosphorylation, membrane depolarization	Drought, osmotic imbalance	Reduced transpiration, ion homeostasis	Lin et al. (2025)
ABA transcriptional control	ABF/AREB, DREB–AREB, SWI/SNF	Chromatin remodeling, osmoprotective gene induction	Water deficit, salt stress	Enhanced stress gene expression	Vanneste et al. (2025)
ABA–calcium coupling	CPKs, Ca^{2+} spikes, aquaporins	Ca^{2+}-dependent signaling, aquaporin gating	Osmotic shifts, vesicle trafficking stress	Adjusted hydraulic conductivity	(Kaur et al., 2025)
ABA lipid–ROS integration	PLD–PA pathway, ROS nodes	Retrograde signaling and membrane remodeling	Mitochondrial stress, ROS bursts	Improved membrane stability	Khan et al. (2024)
Cytokinin redistribution	IPT, CKX, type-B ARRs	Nutrient-responsive CK synthesis/degradation	Nitrate availability, osmotic stress	Root branching, nutrient uptake shifts	Li et al. (2024a, 2024b)
Auxin transport modulation	PIN, AUX1/LAX, ABCB	Polar auxin flux, vesicle trafficking	Nutrient deprivation, osmotic gradients	Elongation, lateral root formation	Fidler et al. (2022)
CK–auxin antagonism	TIR1/AFB, histidine phosphotransfer	Receptor turnover, transporter regulation	N, P, and K limitation	Balanced root system architecture	Lin et al. (2025)
Nutrient–auxin integration	NO–sugar–auxin loops	Metabolic cues shaping auxin maxima	Multinutrient shifts	Enhanced nutrient foraging capacity	Zulfiqar et al. (2025)
Ethylene stress signaling	ACS/ACO, EIN2/EIN3	Ethylene bursts, transcriptional activation	Osmotic compression	Root hair elongation, ROS detox support	Feng et al. (2025)

Ethylene–calcium–G-protein nodes	Ca^{2+} channels, G-protein complexes	Mechanosensing and osmotic perception	Mechanical & osmotic strain	Improved adaptive plasticity	Fidler et al. (2022)
JA signaling under stress	COI1-JAZ, MYC TFs	JA-Ile perception, defense-metabolic reprogramming	Osmotic and nutrient fluctuations	Na^{+} exclusion, K^{+} retention	Lin et al. (2025)
JA–lipid remodeling	Polyamine and lipid turnover enzymes	Membrane bilayer stabilization	Salinity, osmotic stress	Enhanced membrane fluidity control	Vanneste et al. (2025)
Ethylene–JA crosstalk	WRKY, ERF, NAC	Joint transcriptional hardening programs	Combined abiotic stresses	Stronger osmotic resilience	Kaur et al. (2025)
Hormone–nutrient–ion networks	TOR, SnRK1, PHR1, FIT modules	Energy sensing, ion-hormone coordination	Nutrient shifts, redox imbalance	Integrated ion balance and osmotic resilience	Khan et al. (2024)

XIPs (Hewawitharanage and Sarvananda, 2024). Among these, PIPs and TIPs mainly control water movement through highly selective channels with an hourglass-shaped pore, characterized by two NPA motifs: the aromatic/arginine (ar/R) selectivity filter and the froger site. PIP1 and PIP2 isoforms form heterotetramers, with PIP2 mediating cooperative gating that enhances PIP1 permeability, creating conductance patterns essential during osmotic loading (Li et al., 2024a, 2024b). TIPs, such as TIP1;1, TIP2;1, and TIP4;1, regulate vacuolar water flow, osmotic buffering, and lumenal hydrostatic responses to dehydration and rehydration. PIPs combine hydrostatic gating and substrate promiscuity with water–solute copermeation, and their rapid phosphorylation-dephosphorylation allows plants to match external osmotic signals with internal hydraulic balance, as shown by recent proteomic mapping.

17.6.2 pH and nutrient-dependent aquaporin modulation

Apoplastic and cytosolic pH signatures, nutrient fluxes, and ion-driven conformational microstates significantly influence aquaporin gating. External acidification protonates key histidines, causing pore constriction, while an open conformation can induce alkaline shifts through electrostatic repulsion at the cytosolic vestibule (Rodnin et al., 2023). Nutrients like nitrate, ammonium, potassium, and calcium regulate aquaporin activity by modulating ion-coupled allostery; for example, nitrate increases PIP2 levels via NRT1.1-mediated phosphorylation cascades. Conversely, ammonium inhibits water flux by inducing cytosolic acidification. Calcium acts as a dual regulator: when Ca^{2+} is low, it promotes channel opening through calmodulin-binding activation loops; when high, it causes channels to close via calcium-dependent protein kinase (CDPK) phosphorylation (Negi et al., 2023). CLC antiporters, H^+-ATPases, V-ATPases, and nutrient-sensing TOR-SnRK1 circuits participate in integrated nutrient-pH sensing to reorganize membrane water permeability, ensuring ionic and osmotic balance (Fig. 17.3).

17.6.3 Membrane localization and gating mechanisms

Aquaporins are trafficked to their subcellular destinations by ER export motifs, COPII vesicle docking, SNARE-mediated docking, and clathrin-dependent endocytosis, generating a high turnover rate required for osmotic homeostasis. Ubiquitin-tagged internalization is induced by environmental stress factors, which cause the translocation of aquaporins between the plasma and endosomal compartments (Berger et al., 2022). Mechanisms Gating is seen to require loop D conformational bending and helix a2 protonation dynamics, oxidation-dependent cysteine cross-linking, and mechanosensitive pore deformation by membrane tension. H_2O_2-mediated sulfenylation of aquaporins regulates water permeability via reactive oxygen species, a redox checkpoint in aquaporins (Mukherjee et al., 2024). Moreover, phosphorylation of the C-terminal domain, mediated by MAPKs, CDPKs, and OST1, regulates the open probability of PIP2 isoforms and coordinates

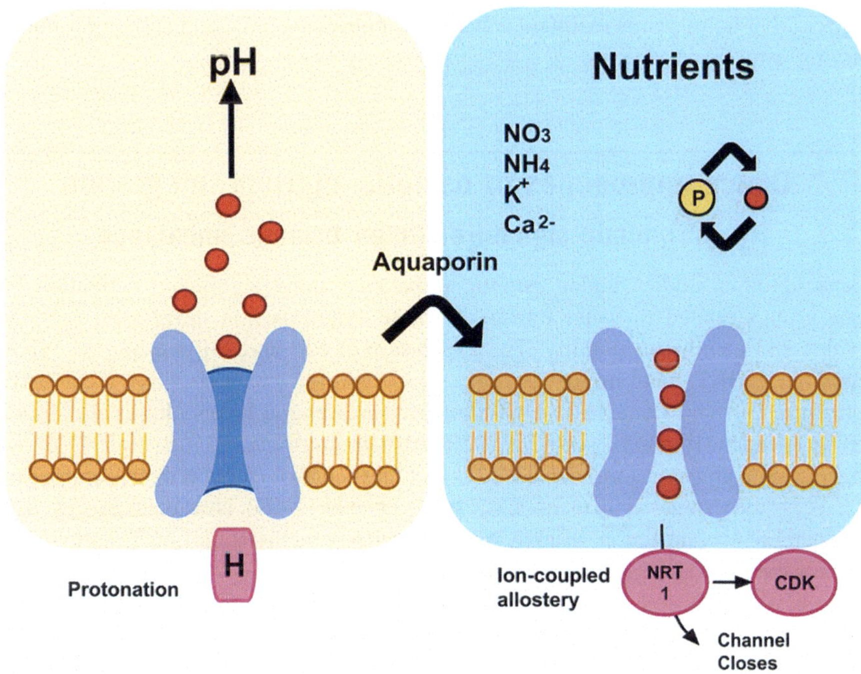

FIGURE 17.3

pH and nutrient cues regulate aquaporin gating through ion-dependent allosteric modulation.

water transport with cellular metabolic conditions. TIP proteins have lumen-sensing gating and respond to the steepness of the osmotic gradient, tonoplast stretch, and metabolite accumulation by modulating vacuolar permeability.

17.6.4 Regulatory interplay between ions and water channels

Ionic dynamics are closely linked to water transport through hydro-ionic coupling, where aquaporins operate alongside H^+ ATPases, K^+ channels, SOS1-NHX antiporters, and anion transporters. Shaker channel potassium flux causes membrane hyperpolarization, which enhances PIP2 opening, while Na^+ toxicity activates SOS2-SOS3-NHX1 signaling that redirects water flow to vacuoles to offset osmotic stress (Akhter et al., 2022). Aquaporin activity is influenced by chloride and sulfate, which engage in anion-induced charge screening to modify the electrostatics of the pore vestibule. Additionally, boron, urea, ammonia, and silicic acid are transported across membranes via aquaporins, linking solute homeostasis with water regulation. Plants subjected to osmotic stress integrate ABA-Ca^+ ROS-ion signaling, coordinating ion flux changes, aquaporin gating, and membrane potential stabilization (Liu et al., 2022). These interactions form a layered, robust hydro-ionic regulatory network that helps plants maintain water flow, nutrient supply, and cell turgor under varying osmotic conditions.

17.7 Omics approaches to osmotic–nutrient integration

17.7.1 Transcriptomic signatures under osmotic imbalance

Transcriptomic studies under osmotic imbalance show significant activation of gene groups related to water transport, solute synthesis, ion movement, nutrient uptake, and membrane stability. The expression of PIP and TIP aquaporins, NHX antiporters, HKT sodium transporters, CLC chloride channels, and KUP/HAK potassium carriers was assessed by comparing expression levels (Jin et al., 2024). ABA-dependent regulons, such as AREB/ABF transcription factors, SnRK2 phosphorylation pathways, and DREB2A/2C modules, control osmoprotective transcriptional cascades triggered by osmotic stress (Ayub et al., 2025). New mechanisms, like chromatin nanocompaction, histone lactylation, RNA methylation (m6A marks), and spliceosome remodeling, also influence osmotic gene expression. Single-cell RNA sequencing reveals cell-type-specific responses, including ion flux networks enriched in the stele, hydraulic adjustment modules in the epidermis, and guard-cell-specific transcriptional nodes that involve ROS bursts, turgor-regulation genes, and membrane-cytoskeletal interaction factors (Liu et al., 2022). These transcriptomic signatures form the molecular framework that reconfigures osmotic sensing in plants through nutrient-controlled metabolic activation.

17.7.2 Ionomic profiling for nutrient transport efficiency

The ionomics framework provides a quantitative method for analyzing the pattern of elemental distribution and the efficiency of nutrient transfer under osmotic pressure. High-throughput ICP-MS and XRF ion profiling detect changes in K^+/Na^+, Ca^{2+}, miRNA, Mg^{2+}, stabilization pools, and micronutrient redox cofactors, including Fe, Mn, Zn, and Cu (Joshi et al., 2024). Osmotic disturbances create ionomic fingerprints characterized by selective nutrient loss in leaves, compensatory increase in roots, and enhanced vacuolar sequestration of toxic ions. These changes are supported by mechanisms such as ion channel gating, fluxomics, nutrient-lipid raft recruitment, charge redistribution hotspots, and electrodynamic membrane conduction states. In ionomic interactions with transporter expression matrices, functional coupling occurs between ABC transporters, K^+ inward rectifiers, nitrate shuttle transporter NRT1.5/1.8, phosphate carriers (PHT1 family), and hydrostatic changes mediated by aquaporins (Sena Falero, 2022). Multielement interaction networks highlight synergistic interactions such as Ca^{2+} stabilizing PIP aquaporins, K^+ facilitating membrane repolarization to enhance nutrient mobilization, and boron-sucrose cross-linking supporting cell wall osmotic elasticity. These ionomic data provide mechanistic insights into how plants maintain electrochemical precision under osmotic variation.

17.7.3 Metabolomic mapping of stress-induced solute reprogramming

Metabolomics reveals a broad range of osmoprotective metabolites and nutrient-specific solutes that respond to osmotic pressure. LC-MS, GC-MS, and NMR analyses, both targeted and untargeted, show increased levels of proline, glycine betaine, raffinose family oligosaccharides, polyols, and organic acids, which act as osmotic stabilizers (Głuchowska et al., 2025). Stress-induced metabolic reprogramming involves rerouting the TCA cycle, accelerating the pentose phosphate pathway, amino acid transamination loops, glycolytic bypasses, and lipidomic remodeling, all of which help maintain membrane fluidity. Interactions between nutrient availability and solute biosynthesis include nitrate supplying carbon skeletons, sulfur helping redox buffering of thiols, and phosphorus influencing biosynthetic flux. New metabolomic markers include osmolyte nanocomplexes, ROS-quenching metabolites, phospholipid headgroup transitions, and sugar-ion costabilization clusters (Silva et al., 2023). Integrated space metabolomics compares solute gradients across tissues to show nitrogen-containing osmolytes accumulating for osmotic balance in roots and sugar alcohol-ion synergy promoting turgor in shoots. These data illustrate how plants respond to osmotic stress by linking nutrient metabolism through metabolomic mechanisms.

17.7.4 Systems modeling of nutrient–water flux coordination

Systems modeling integrates multiomics data to reconstruct nutrient–water coordination of fluxes across tissues, membranes, and developmental stages. The computational models rely on hydraulic flux equations, transporter kinetics, metabolic flux balance analysis (FBA), ion lattice models, and ODE-based signaling network models of whole-plant osmotic behavior (Gao and Zhao, 2024). These systems-level predictions have identified key regulatory centers, such as the ABA-Ca^{2+}-ROS three-way, TOR-SnRK1 metabolic switches, SOS ion homeostasis nodes, and PIP/NHX interaction clusters, which act as control points in osmoregulation. Advanced modeling also considers the propagation of nanodomain signaling, mechano-hydraulic feedbacks, modules of phloem–xylem nutrient–water cotransport, and root–soil interface conductivity fields. Combining ionomics with hydraulic modeling shows that the optimal K^+ membrane potential, calcium-gating threshold levels, and Na^+ exclusion efficiency influence water flow pathways (Contreras et al., 2024). Machine learning systems trained on multiomics data have been used to predict osmotic responses under various nutrient conditions, revealing emergent properties like hydro-ionic modularity, transporter synchronicity, solute-membrane resonance, and energy-efficient osmotic repair cycles. These models provide a mechanistic framework for understanding osmotic and nutrient coregulation in complex plant systems.

17.8 Agronomic and biotechnological strategies

17.8.1 Optimized fertilizer regimes for osmotic protection

Improved fertilizer programs are also used as supporting interventions to strengthen plant osmotic protection by promoting ion homeostasis, turgor stability, and metabolic resilience. Electrochemical gradients, enhanced osmotic buffering ability, and membrane polarization under osmotic pressure are preserved using potassium, calcium, magnesium, and sulfur (Saleem et al., 2022). Precise timing of fertilizer application requires slow-release carriers, nanofertilizer formulations, polymer-coated nutrients, and rhizosphere-activated nutrient-dispersal systems to prevent ionic surges that disrupt hydraulic conductivity. New mechanisms include ion-aquaporin coregulation, nutrient-dependent plasmodesmata adjustments, microdomain-based membrane lipid remodeling, and osmotic-sensorimimetic priming of cells by nutrient-active kinases, all of which boost cellular buffering capacity (Li et al., 2024a, 2024b). By applying fertigation alongside soil moisture sensors and real-time electrical conductivity mapping, it is possible to align nutrient availability with osmotic demand, which, under certain heterogeneous field conditions, can be more accurate and thus enhance hydro-ionic stability.

17.8.2 Nutrient formulations enhancing membrane stability

Expansive nutrient preparations are also being developed to strengthen membrane stability during osmotic pressure, considering structural and biochemical defenses. Formulations enriched with calcium and boron reinforce pectin cross-linking in the cell wall, limit membrane porosity, and stabilize phospholipid bilayers. The activity of membrane-bound enzymes is regulated by magnesium and zinc to enhance redox reactions and reduce lipid peroxidation caused by ROS. Improved formulations, including amino-acid-chelated nutrients, nano-emulsion micronutrients, mineral-biopolymer complexes, and ionic buffer-enhanced blends, promote more consistent nutrient absorption under varying osmotic conditions (Madhupriyaa et al., 2024). Emerging processes include nutrient-induced membrane microviscosity correlations, harmonization of sterol-phospholipid domains, vesicle trafficking anchoring, and SNARE protein-mediated membrane repair cascades (Chen et al., 2025b). These nutrient designs remain hydraulically intact through adaptation of H^+-ATPase proton gradients, membrane electrostatic shielding, and solute-membrane permeability compatibility.

17.8.3 Genetic improvement of nutrient-efficient osmotic regulators

Biotechnological approaches aim to optimize plant nutrient efficiency and osmotic regulation, with a focus on ion transport and solute biosynthesis pathways. Editing gene clusters such as HKT, NHX, AKT, SOS1, and V-ATPase through CRISPR allows precise control of ion exclusion, sequestration, and vacuolar loading. Simultaneously, modifications in genes involved in osmolyte biosynthesis, including the P5CS, BADH, MIPS, and RS gene families, improve osmotic tolerance by increasing the accumulation of proline, glycine betaine, and polyols (Kumar et al., 2024). New techniques lead to regulatory changes in ABA receptor components (PYL/PYR), calcium signaling centers (CAX channels), TOR-SnRK1 metabolic switches, and membrane-tension-regulated kinases, which coordinate nutrient and osmotic responses (Fidler et al., 2022). Stress-responsive nutrient uptake synchronization achieved through multigene stacking and a synthetic promoter system, as well as alleles linked to ion partitioning efficiency, hydraulic plasticity, osmotic memory formation, and nutrient mobilization kinetics, are identified through genome-wide association studies. These genetic advances support the development of nutrient-efficient, osmotic-tolerant crop varieties.

17.8.4 Integration of soil, plant, and nutrient diagnostics

Integrated diagnostics involves soil, plant, and nutrient assessments to guide precise osmotic management. Tools such as ion-specific sensors, electrical resistivity imaging, hyperspectral nutrient mapping, and microbial functional assays based on soil samples identify nutrient imbalances underlying osmotic responses. Plant-level diagnostic sensors, such as chlorophyll fluorescence kinetics, thermal imaging, sap ionometry, ionomic scanning, and root-zone hydraulic profiling, enable dynamic measurement of osmotic stress and nutrient-use efficiency (Bulacio Fischer et al., 2025). Mechanistic insights are gained through nutrient–water flux modeling, real-time osmotic stress telemetry, multisensor data fusion, and AI-driven nutrient forecasting, grounded in hydro-ionic feedback loops. The diagnostic approach also includes soil-plant interface conductivity modeling, rhizosphere nutrient cloud imaging, and root exudate-ion interaction mapping, which improve the accuracy of osmotic intervention strategies (Zeng et al., 2025). This comprehensive diagnostic platform enhances resilience by aligning nutrient delivery with actual osmotic and metabolic needs.

17.9 Conclusion

Osmotic and nutrient dynamics constitute a closely coupled system that regulates plant resilience in challenging environments. This chapter emphasizes the importance of cellular osmotic homeostasis, which involves multiple physiological responses, including ionic flows, solute transport, membrane fluidity, nutrient signaling, and molecular reorganization. New evidence highlights that nutrients, particularly potassium, calcium, nitrogen, phosphorus, sulfur, and micronutrient cofactors, serve as both structural or metabolic inputs and as key factors influencing osmotic perception, ion conductance, aquaporin activity, and cellular redox balance. These elements act as biochemical anchors that enhance the plant's ability to maintain turgor pressure, cellular hydration, and stability of macromolecular complexes during drought, salinity, dehydration, and mineral imbalances. The sophisticated nature of nutrient-osmotic interactions has also been demonstrated through omics research, revealing control at levels such as gene expression, transporter activation, metabolite buildup, and signaling pathways. Integrated approaches show that nutrient-regulated pathways drive osmotic adjustment, which in turn manages energy supply, stress detection, and metabolic adaptation. Additionally, the interplay between hormonal regulation and nutrient availability indicates that complex crosstalk, rather than single-factor pathways, sustains plant responses. Moving forward, crop improvement will rely on balancing nutrient management with osmotic defense strategies. Developing more advanced nutrient formulations, ion-efficient crops, genome-guided transporter engineering, and effective agronomic practices will be crucial to boosting crop resilience. A systems-level approach that integrates ionomics, metabolomics, and nutrient-responsive transcriptomics will provide valuable insights for the development of climate-smart agriculture. Ultimately, strengthening osmotic regulation through nutrient innovation will be vital for maintaining productivity amid increasing environmental uncertainties, ensuring crops remain vigorous, adaptable, and high-yielding across diverse agricultural systems.

References

Acierno, C., Barletta, F., Nevola, R., Rinaldi, L., Sasso, F.C., Adinolfi, L.E., et al., 2025. Metabolic rewiring of bacterial pathogens in response to antibiotic pressure—a molecular perspective. International Journal of Molecular Sciences 26 (12). Available at https://doi.org/10.3390/ijms26125574.

Akhter, A., Bibi, G., Rasti, N., Rasheed, H., Noor, Z., Hussain, J., 2022. Na+ sensing, transport, and plant salt tolerance. Sustainable Plant Nutrition Under Contaminated Environments. Springer International Publishing, Cham, pp. 257–285. Available at https://doi.org/10.1007/978-3-030-91499-8_13.

Albloushi, M.A., 2023. Using bioengineering approaches to express halophyte genes that enhance salt tolerance gene expression in tobacco (Doctoral dissertation, University of Bristol).

Alvi, A.F., Iqbal, N., Albaqami, M., Khan, N.A., 2023. The emerging role of reactive sulfur species in plant abiotic stress tolerance. Physiologia Plantarum 175 (3), e13945. Available at https://doi.org/10.1111/ppl.13945.

Ayub, A., Javed, T., Nayab, A., Nan, Y., Xie, Y., Hussain, S., et al., 2025. AREB/ABF/ABI5 transcription factors in plant defense: regulatory cascades and functional diversity. Critical Reviews in Biotechnology 45 (7), 1414–1434. Available at https://doi.org/10.1080/07388551.2025.2475127.

Berger, N., Demolombe, V., Hem, S., Rofidal, V., Steinmann, L., Krouk, G., et al., 2022. Root membrane ubiquitinome under short-term osmotic stress. International Journal of Molecular Sciences 23 (4), 1956. Available at https://doi.org/10.3390/ijms23041956.

Bian, X., Xie, T., Chen, J., Li, C., Yin, D., Zhang, W., 2025. Genome-wide identification of key genes related to chloride ion (Cl–) channels and transporters in response to salt stress in birch. BMC Genomics 26 (1). Available at https://doi.org/10.1186/s12864-025-11795-4.

Bihani, S.C., Tarushi, Srivastava, A.K., 2025. Decoding the calcium signal: structural insights into CBL-CIPK pathway in plants. Biochimica et Biophysica Acta - General Subjects 1869 (8). Available at https://doi.org/10.1016/j.bbagen.2025.130819.

Bobkov, D., Semenova, S., 2022. Impact of lipid rafts on transient receptor potential channel activities. Journal of Cellular Physiology 237 (4), 2034–2044. Available at https://doi.org/10.1002/jcp.30679.

Bulacio Fischer, P.T., Carella, A., Massenti, R., Fadhilah, R., Lo Bianco, R., 2025. Advances in monitoring crop and soil nutrient status: proximal and remote sensing techniques. Horticulturae 11 (2), 182. Available at https://doi.org/10.3390/horticulturae11020182.

Chatterjee, M., Ghosh, P., Sen, S., Sinha, D., Ganguly, S., 2022. The Function of HAK as K+ Transporter and AKT as Inward-Rectifying Agent in the K+ Channel. In: Kumar, K., Srivastava, S. (Eds.), Plant Metal and Metalloid Transporters. Springer, Singapore, pp. 227–243. Available at https://doi.org/10.1007/978-981-19-6103-8_11.

Chen, J., Zhang, J., Liu, Y., Zhang, K., Zhu, F., Xie, Y., 2025a. Advances in the biosynthetic regulation and functional mechanisms of glycine betaine for enhancing plant stress resilience. International Journal of Molecular Sciences 26 (16), 7971. Available at https://doi.org/10.3390/ijms26167971.

Chen, S., Liu, Y., Yu, H., 2025b. Uncovering the mechanisms of intracellular membrane trafficking by reconstituted membrane systems. Membranes 15 (5), 154. Available at https://doi.org/10.3390/membranes15050154.

Contreras, R.G., Torres-Carrillo, A., Flores-Maldonado, C., Shoshani, L., Ponce, A., 2024. Na+/K+-ATPase: more than an electrogenic pump. International Journal of Molecular Sciences 25 (11), 6122. Available at https://doi.org/10.3390/ijms25116122.

Cui, Y., He, M., Liu, D., Liu, J., Liu, J., Yan, D., 2023. Intercellular communication during stomatal development with a focus on the role of symplastic connection. International Journal of Molecular Sciences 24 (3), 2593. Available at https://doi.org/10.3390/ijms24032593.

Das, D., Kashtoh, H., Panda, J., Rustagi, S., Mohanta, Y.K., Singh, N., et al., 2025. From hormones to harvests: a pathway to strengthening plant resilience for achieving sustainable development goals. Plants 14 (15). Available at https://doi.org/10.3390/plants14152322.

Ding, Y., She, B., Hu, X., Xing, Y., Wang, Z., Ge, C., et al., 2025. Study of the regulatory mechanisms of phosphorus homeostasis and phosphate (Pi)-starvation rescue in rice. Russian Journal of Plant Physiology 72 (2). Available at https://doi.org/10.1134/s1021443724610061.

Dong, S., Ling, J., Song, L., Zhao, L., Wang, Y., Zhao, T., 2023. Transcriptomic profiling of tomato leaves identifies novel transcription factors responding to dehydration stress. International Journal of Molecular Sciences 24 (11), 9725. Available at https://doi.org/10.3390/ijms24119725.

Feng, L., Li, X., Zheng, X.A., Zheng, Z., Liu, Q.R., Liu, C., et al., 2025. SnRK1 and TOR: central regulators of autophagy in plant energy stress responses. aBIOTECH 6 (4), 663–679. Available at https://doi.org/10.1007/s42994-025-00218-3.

Fidler, J., Graska, J., Gietler, M., Nykiel, M., Prabucka, B., Rybarczyk-Płońska, A., et al., 2022. PYR/PYL/RCAR receptors play a vital role in the abscisic-acid-dependent responses of plants to external or internal stimuli. Cells 11 (8), 1352. Available at https://doi.org/10.3390/cells11081352.

Fujii, J., 2024. Redox remodeling of central metabolism as a driving force for cellular protection, proliferation, differentiation, and dysfunction. Free Radical Research 58 (10), 606–629. Available at https://doi.org/10.1080/10715762.2024.2407147.

Gao, Y., Zhao, C., 2024. Development and applications of metabolic models in plant multi-omics research. Frontiers in Plant Science 15. Available at https://doi.org/10.3389/fpls.2024.1361183.

Giri, S., Singh, P., Biswas, M., Mishra, R., Poddar, N.K., 2024. Harnessing the power of osmolytes for industrial and pharmaceutical applications. Cellular Osmolytes: From Chaperoning Protein Folding to Clinical Perspectives. Springer, Singapore, pp. 129–160. Available at https://doi.org/10.1007/978-981-97-6001-5_8.

Gopaldass, N., Chen, K.E., Collins, B., Mayer, A., 2024. Assembly and fission of tubular carriers mediating protein sorting in endosomes. Nature Reviews. Molecular Cell Biology 25 (10), 765–783. Available at https://doi.org/10.1038/s41580-024-00746-8.

Gupta, V., 2025. Metabolomic studies of halophytes under salinity stress. Physiology of halophytes. Apple Academic Press, pp. 1–37. Available at https://doi.org/10.1201/9781003504085.

Głuchowska, A., Zieniuk, B., Pawełkowicz, M., 2025. Unlocking plant resilience: metabolomic insights into abiotic stress tolerance in crops. Metabolites 15 (6), 384. Available at https://doi.org/10.3390/metabo15060384.

Hewawitharanage, H., Sarvananda, L., 2024. Aquaporins: multifunctional players in plant growth, development and stress responses. Silva Balcanica 25 (1), 73–86. Available at https://doi.org/10.3897/silvabalcanica.25.e117347.

Jin, L., Yin, X., Wen, M., Huang, B., Liu, F., Li, X., et al., 2024. Genome-wide identification, characterization, and expression of the HAK/KUP/KT potassium transporter gene family in *Poncirus trifoliata* and functional analysis of PtKUP10 under salt stress. Horticulturae 10 (6), 628. Available at https://doi.org/10.3390/horticulturae10060628.

Joshi, S., Nath, J., Singh, A.K., Pareek, A., Joshi, R., 2022. Ion transporters and their regulatory signal transduction mechanisms for salinity tolerance in plants. Physiologia Plantarum 174 (3). Available at https://doi.org/10.1111/ppl.13702.

Joshi, A., Kaur, S., Taneja, S.K., Mandal, R., 2024. Review article on molecular mechanism of regulation of hypertension by macro-elements (Na, K, Ca and Mg), micro-elements/trace metals (Zn and Cu) and toxic elements (Pb and As). Biological Trace Element Research 202 (4), 1477–1502. Available at https://doi.org/10.1007/s12011-023-03784-z.

Kang, J., Peng, Y., Xu, W., 2022. Crop root responses to drought stress: molecular mechanisms, nutrient regulations, and interactions with microorganisms in the rhizosphere. International Journal of Molecular Sciences 23 (16), 9310. Available at https://doi.org/10.3390/ijms23169310.

Kaur, J., Viswanathan P, A., Bari, V.K., 2025. CRISPR/Cas9-mediated editing of jasmonic acid pathways to enhance biotic & abiotic stress tolerance: an overview & prospects. Functional and Integrative Genomics 25 (1). Available at https://doi.org/10.1007/s10142-025-01623-z.

Khan, S., Alvi, A.F., Saify, S., Iqbal, N., Khan, N.A., 2024. The ethylene biosynthetic enzymes, 1-aminocyclopropane-1-carboxylate (ACC) synthase (ACS) and ACC Oxidase (ACO): the less explored players in abiotic stress tolerance. Biomolecules 14 (1). Available at https://doi.org/10.3390/biom14010090.

Kuczyńska, M., Jakubek, P., Bartoszek, A., 2022. More than just antioxidants: redox-active components and mechanisms shaping redox signalling network. Antioxidants 11 (12), 2403. Available at https://doi.org/10.3390/antiox11122403.

Kumar, P., Kumar, P., Suniti, Kumar, U., Avni, Mann, A., 2023. Transcriptional Regulatory Network Involved in Drought and Salt Stress Response in Rice. In: Kumar, A., Dhansu, P., Mann, A. (Eds.), Salinity and Drought Tolerance in Plants: Physiological

Perspectives. Springer, Singapore, pp. 237–274. Available at https://doi.org/10.1007/978-981-99-4669-3_13.

Kumar, P., Kumar, A., Agrawal, A., Bhusan, R., Kumar, S., Rehman, S., et al., 2024. Osmolytes as stress sensors in plants: acclimatizing plants under stress conditions. Cellular Osmolytes: From Chaperoning Protein Folding to Clinical Perspectives. Springer, Singapore, pp. 199–210. Available at https://doi.org/10.1007/978-981-97-6001-5_11.

Lal, M.A., Bhatla, S.C., 2023. Nitrogen metabolism. Plant Physiology, Development and Metabolism. Springer Nature Singapore, Singapore, pp. 295–334. Available at https://doi.org/10.1007/978-981-99-5736-1_11.

Li, S., Khoso, M.A., Xu, H., Zhang, C., Liu, Z., Wagan, S., et al., 2024a. WRKY transcription factors (TFs) as key regulators of plant resilience to environmental stresses: current perspective. Agronomy 14 (10), 2421. Available at https://doi.org/10.3390/agronomy14102421.

Li, X., Guo, Y., Ling, Q., Guo, Z., Lei, Y., Feng, X., et al., 2024b. Advances in the structure, function, and regulatory mechanism of plant plasma membrane intrinsic proteins. Genes 16 (1), 10. Available at https://doi.org/10.3390/genes16010010.

Li, Q., Hu, T., Lu, T., Yu, B., Zhao, Y., 2025. Calcium-dependent protein kinases CPK3/4/6/11 and 27 respond to osmotic stress and activate SnRK2s in *Arabidopsis*. Developmental Cell 60 (10), 1423. Available at https://doi.org/10.1016/j.devcel.2024.12.036.

Lin, Z., Guo, Y., Zhang, R., Li, Y., Wu, Y., Sheen, J., et al., 2025. ABA-activated low-nanomolar Ca^{2+}–CPK signalling controls root cap cycle plasticity and stress adaptation. Nature Research, China Nature Plants 11 (1), 90–104. Available at https://doi.org/10.1038/s41477-024-01865-y.

Liu, H., Song, S., Zhang, H., Li, Y., Niu, L., Zhang, J., et al., 2022. Signaling transduction of ABA, ROS, and Ca^{2+} in plant stomatal closure in response to drought. International Journal of Molecular Sciences 23 (23), 14824. Available at https://doi.org/10.3390/ijms232314824.

Madhupriyaa, D., Baskar, M., Sherene Jenita Rajammal, T., Kuppusamy, S., Rathika, S., Umamaheswari, T., et al., 2024. Efficacy of chelated micronutrients in plant nutrition. Communications in Soil Science and Plant Analysis 55 (22), 3609–3637. Available at https://doi.org/10.1080/00103624.2024.2397019.

Mandal, S.K., Mukherjee, D., Roy, S., Banerjee, S., Bose, S., 2022. Modulation of TCA cycle and enzymes during different environmental stresses. Photosynthesis and Respiratory Cycles during Environmental Stress Response in Plants. Apple Academic Press, pp. 309–336. Available at https://doi.org/10.1201/9781003315162.

Moulick, D., Bhutia, K.L., Sarkar, S., Roy, A., Mishra, U.N., Pramanick, B., et al., 2023. The intertwining of Zn-finger motifs and abiotic stress tolerance in plants: current status and future prospects. Frontiers in Plant Science 13. Available at https://doi.org/10.3389/fpls.2022.1083960.

Mukherjee, S., Roy, S., Corpas, F.J., 2024. Aquaporins: a vital nexus in H_2O_2-gasotransmitter signaling. Trends in Plant Science 29 (6), 681–693. Available at https://doi.org/10.1016/j.tplants.2023.11.021.

Negi, N.P., Prakash, G., Narwal, P., Panwar, R., Kumar, D., Chaudhry, B., et al., 2023. The calcium connection: exploring the intricacies of calcium signaling in plant-microbe interactions. Frontiers in Plant Science 14. Available at https://doi.org/10.3389/fpls.2023.1248648.

Obomighie, I., Prentice, I.J., Lewin-Jones, P., Bachtiger, F., Ramsay, N., Kishi-Itakura, C., et al., 2025. Understanding pectin cross-linking in plant cell walls. Communications Biology 8 (1). Available at https://doi.org/10.1038/s42003-025-07495-0.

Perez-Rivera, E., Plasencia, C., Ros, U., 2025. The multifaceted role of calcium signaling in regulated necrosis. Biomolecules 15 (6), 854. Available at https://doi.org/10.3390/biom15060854.

Pottosin, I., Dobrovinskaya, O., 2022. Major vacuolar TPC1 channel in stress signaling: what matters, K+, Ca2+ conductance or an ion-flux independent mechanism? Stress Biology 2 (1). Available at https://doi.org/10.1007/s44154-022-00055-0.

Qi, W., Zhang, Y., Li, M., Zhang, P., Xing, J., Chen, Y., et al., 2024. Endocytic recycling in plants: pathways and regulation. Journal of Experimental Botany 75 (16), 4712–4728. Available at https://doi.org/10.1093/jxb/erae188.

Rioux, A.V., Nsimba-Batomene, T.R., Slimani, S., Bergeron, N.A.D., Gravel, M.A.M., Schreiber, S.V., et al., 2024. Navigating the multifaceted intricacies of the Na + -Cl – cotransporter, a highly regulated key effector in the control of hydromineral homeostasis. Physiological Reviews 104 (3), 1147–1204. Available at https://doi.org/10.1152/physrev.00027.2023.

Rodnin, M.V., Vasques-Montes, V., Kyrychenko, A., Oliveira, N.F.B., Kashipathy, M.M., Battaile, K.P., et al., 2023. Histidine protonation and conformational switching in diphtheria toxin translocation domain. Toxins 15 (7). Available at https://doi.org/10.3390/toxins15070410.

Saleem, S., Mushtaq, N.U., Rasool, A., Shah, W.H., Tahir, I., Rehman, R.U., 2022. Chapter two - Plant nutrition and soil fertility: physiological and molecular avenues for crop improvement. In: Aftab, T., Hakeem, K.R. (Eds.), Sustainable Plant Nutrition: Molecular Interventions and Advancements for Crop Improvement. Academic Press, pp. 23–49. Available at https://doi.org/10.1016/B978-0-443-18675-2.00009-2.

Sena Falero, M.F., 2022. Role of the nitrate/potassium transporter, NRT1. 5/NPF7. 3, in the regulatory network of ion homeostasis in *Arabidopsis thaliana* (Doctoral dissertation). https://doi.org/10.17169/refubium-31049.

Silva, S., Dias, M.C., Pinto, D.C.G.A., Silva, A.M.S., 2023. Metabolomics as a tool to understand nano-plant interactions: the case study of metal-based nanoparticles. Plants 12 (3), 491. Available at https://doi.org/10.3390/plants12030491.

Song, Y., Li, F., Ali, M., Li, X., Zhang, X., Ahmed, Z.F.R., 2025. Advances in protein kinase regulation of stress responses in fruits and vegetables. International Journal of Molecular Sciences 26 (2), 768. Available at https://doi.org/10.3390/ijms26020768.

Steinhorst, L., He, G., Moore, L.K., Schültke, S., Schmitz-Thom, I., Cao, Y., et al., 2022. A Ca^{2+}-sensor switch for tolerance to elevated salt stress in *Arabidopsis*. Developmental Cell 57 (17), 2081. Available at https://doi.org/10.1016/j.devcel.2022.08.001.

Thakur, G., Singh, P., Sharma, V., Sharma, A., Singh, J., Kumar, S., 2025. Phytohormonal crosstalk with flowering genes regulating drought stress response in citrus: a systematic review. Journal of Plant Growth Regulation 44 (11), 6313–6332. Available at https://doi.org/10.1007/s00344-025-11838-w.

Vanneste, S., Pei, Y., Friml, J., 2025. Mechanisms of auxin action in plant growth and development. Nature Reviews. Molecular Cell Biology 26 (9), 648–666. Available at https://doi.org/10.1038/s41580-025-00851-2.

Yamasaki, H., Itoh, R.D., Mizumoto, K.B., Yoshida, Y.S., Otaki, J.M., Cohen, M.F., 2025. Spatiotemporal characteristics determining the multifaceted nature of reactive oxygen, nitrogen, and sulfur species in relation to proton homeostasis. Antioxidants and Redox Signaling 42 (7-9), 421–441. Available at https://doi.org/10.1089/ars.2023.0544.

Yan, L., Lu, M., Riaz, M., Tong, K., Yu, H., Gao, G., et al., 2025. Exogenous proline enhances salt acclimation in soybean seedlings: modifying physicochemical properties and controlling proline metabolism through the ornithine-glutamate dual pathway. Ecotoxicology and Environmental Safety 294, 118012. Available at https://doi.org/10.1016/j.ecoenv.2025.118012.

Zeng, Y., Verhoef, A., Vereecken, H., Ben-Dor, E., Veldkamp, T., Shaw, L., et al., 2025. Monitoring and modeling the soil-plant system toward understanding soil health. Reviews of Geophysics 63 (1). Available at https://doi.org/10.1029/2024rg000836.

Zhao, W.B., Sheng, R., 2025. The correlation between mitochondria-associated endoplasmic reticulum membranes (MAMs) and Ca^{2+} transport in the pathogenesis of diseases. Acta Pharmacologica Sinica 46 (2), 271–291. Available at https://doi.org/10.1038/s41401-024-01359-9.

Zulfiqar, S., Gu, R., Liu, Y., Zhang, Y., 2025. From genes to traits: maximizing phosphorus utilization efficiency in crop plants. Frontiers in Plant Science 16. Available at https://doi.org/10.3389/fpls.2025.1527547.

SECTION

IV

Stress-specific nutrient management strategies

CHAPTER

Nutrient management in drought-prone regions

18

Mohamed Zouari[1,2], Lina Trabelsi[1], Ameni Ben Hassena[1], Imed Mezghani[3], Nada Elloumi[3], Nacim Zouari[4], Bechir Ben Rouina[1], Pascal Labrousse[2] and Chedliya Ben Ahmed[5]

[1]*Laboratory of Improvement of Olive Productivity and Fruit Trees, Olive Tree Institute of Sfax, University of Sfax, Sfax, Tunisia,*

[2]*Laboratory of Botany and Cryptogamy, Faculty of Pharmacy, University of Limoges, Limoges, France,*

[3]*Laboratory of Water, Energy and Environment, Higher Institute of Biotechnology of Sfax, University of Sfax, Sfax, Tunisia,*

[4]*Laboratory of Biodiversity, Molecules, and Applications (LR22ES02), Higher Institute of Applied Biology of Medenine, University of Gabes, Medenine, Tunisia,*

[5]*Laboratory of Ecosystems and Biodiversity in Arid Areas in Tunisia, Faculty of Sciences of Sfax, University of Sfax, Sfax, Tunisia*

Abbreviations

AMF	arbuscular mycorrhizal fungi
BNF	biological nitrogen fixation
BSCU	bioactive sulfur-coated urea
CA	conservation agriculture
CEC	cation exchange capacity
CRF	controlled-release fertilizers
C/N ratio	carbon-to-nitrogen ratio
DAP	diammonium phosphate
EC	electrical conductivity
FYM	Farmyard manure
NCU	neem-coated urea
NT	no-till
PGPR	plant growth-promoting rhizobacteria
PSB	phosphate-solubilizing bacteria
PSMs	phosphate-solubilizing microorganisms
RT	reduced tillage
RWEQ	revised wind erosion equation
SCU	sulfur-coated urea
SOC	soil organic carbon
SRF	slow-release fertilizers
TN	total nitrogen

Mineral Nutrients to Mitigate Abiotic Stress in Crops. DOI: https://doi.org/10.1016/B978-0-443-44712-9.00020-7

TT	traditional tillage
4R	right-source, right-rate, right-time, right-place

18.1 Introduction

Global drylands encompass approximately 41% of the terrestrial land surface and support over 2 billion people, many of whom depend directly on rain-fed agriculture for food security and livelihoods. These agroecosystems are characterized by chronic water scarcity, high evapotranspiration rates, erratic precipitation patterns, and extreme temperature fluctuations that collectively constrain crop productivity and soil fertility. Climate projections indicate that drought frequency, intensity, and duration will increase substantially across many dryland regions during the coming decades, further exacerbating agricultural vulnerabilities and threatening sustainable development goals (Oueriemmi et al., 2021; Carneiro et al., 2025; Zhang et al., 2024a; Kugedera and Kokerai, 2024).

Nutrient management in drought-prone regions presents unique challenges that distinguish these systems from more humid agricultural environments. Water limitation fundamentally alters nutrient cycling processes by restricting microbial mineralization of organic matter, reducing nutrient diffusion through the soil matrix toward plant roots, increasing nutrient fixation into unavailable chemical forms, and promoting nutrient losses through wind erosion, surface runoff during sporadic high-intensity rainfall, and volatilization under elevated temperatures. Consequently, even when soil contains substantial total nutrient reserves, plant-available fractions remain critically low, and fertilizer use efficiency is markedly reduced compared to well-watered conditions (Shirinbayan et al., 2019; Sileshi, 2022).

Conventional nutrient management approaches developed for temperate or irrigated systems frequently fail when transferred to drylands without substantial adaptation. Blanket fertilizer recommendations that disregard pronounced spatial and temporal variability in soil properties and moisture availability often lead to nutrient imbalances, accelerated depletion of secondary and micronutrients, economic losses from poor fertilizer recovery, and environmental contamination through excess nutrient application. Furthermore, the widespread practice of continuous monocropping without adequate crop diversification or organic matter inputs has progressively degraded soil structure, depleted soil biological communities essential for nutrient transformations, and increased vulnerability to both nutrient losses and climate stress (Sileshi, 2022; Mahmood et al., 2025; Yang et al., 2023).

Sustainable intensification of dryland agriculture requires a paradigm shift toward integrated nutrient management strategies explicitly designed to address the water–nutrient nexus. These approaches must simultaneously enhance nutrient availability and use efficiency while improving soil water retention capacity, restoring soil organic matter and biological functionality, minimizing nutrient losses through multiple pathways, and building overall system resilience to drought stress (Xie et al., 2025; Lopez et al., 2025; Zhou et al., 2025). This chapter synthesizes current scientific knowledge on constraints to nutrient management in drought-prone regions and evaluates evidence-based agronomic interventions that show promise for improving nutrient stewardship, crop productivity, and environmental sustainability under water-limited conditions.

18.2 Constraints to nutrient management in drought-prone regions

Dryland cropping systems are confronted with multiple, interrelated constraints that severely limit the effectiveness of nutrient management strategies. These constraints include chronic nutrient imbalances, restricted crop diversification, accelerated soil erosion, and declined soil organic matter, which could result in a progressive deterioration of soil biological activity and nutrient cycling processes. Together, these factors reduce fertilizer use efficiency, exacerbate nutrient depletion, and increase the vulnerability of crops to water stress (Nawaz and Farooq, 2017; Bogati and Walczak, 2022; Perera et al., 2025). Consequently, ensuring sustainable nutrient management in dryland agroecosystems remains a major challenge. In the following section, the principal limitations affecting nutrient management in drought-prone regions are discussed.

18.2.1 Nutrient imbalance and inadequate fertilization practices

Nutrient imbalance and inadequately adapted fertilization practices represent major constraints to effective nutrient management in drought-prone regions. In fact, they interact synergistically with water scarcity to markedly reduce both fertilizer use efficiency and crop productivity. In many dryland agroecosystems, fertilization strategies remain largely dominated by blanket nitrogen–phosphorus–potassium (N–P–K) recommendations that disregard the pronounced spatial variability of soil physicochemical properties and the critical role of micronutrients (Moreno-Jiménez et al., 2019; Rehman et al., 2019). In sub-Saharan Africa, fertilizer application rates have been historically low while nutrient depletion rates are high, leading to nutrient mining that has gradually exhausted soil nutrient reserves and caused widespread deficiencies of macronutrients and micronutrients (Sileshi, 2022). Dargie et al. (2022) showed that blanket fertilizer recommendations promote nutrient imbalances in Ethiopian wheat systems. The nationwide application of uniform N and P rates (64 kg N and 20 kg P ha^{-1}), without K, S, or micronutrients, disregards soil variability and accelerates the depletion of essential secondary and micronutrients such as K, S, Zn, and B, leading to yield decline and soil fertility degradation. Although derived from submoist highland systems, this state is fully transferable to drylands, where exclusive reliance on N and P is even more detrimental due to water-limited nutrient uptake and heightened sensitivity to nutrient imbalances.

18.2.2 Monocropping and lack of crop diversification

In most dryland agroecosystems, monocropping persists as the dominant cultivation practice because precipitation scarcity and temporal variability restrict the optimal sowing period and discourage agricultural producers from implementing diversified crop rotations. As a result, fields are often planted with the same crop for several continuous years, leaving soil unprotected and highly vulnerable to nutrient losses *via* wind and water erosion, as well as leaching or volatilization during episodic rainfall events (Nawaz and Farooq, 2017). Perera et al. (2025) and Choudhary et al. (2025) reported that the widespread practice of monocropping in dryland agricultural systems poses major challenges for nutrient management and soil health. According to these authors, continuous cultivation of a single crop under limited water availability leads to rapid depletion of essential soil nutrients, a reduction in soil organic matter, and deterioration of soil structure. In addition, monocropping systems also disrupt soil microbial diversity and nutrient cycling, thereby diminishing soil fertility and reducing resilience to climatic stress. In contrast, diversified cropping systems through crop rotation or intercropping with legumes and cover crops have been shown to improve nutrient retention, enhance biological nitrogen fixation, and stimulate soil microbial activity, ultimately contributing to greater soil productivity and ecological sustainability in dryland regions (Nawaz and Farooq, 2017; Perera et al., 2025; Choudhary et al., 2021).

18.2.3 Soil erosion

Soil erosion in dryland regions is a major driver of nutrient depletion that threatens agricultural sustainability and ecosystem functioning. By preferentially removing fine particles rich in organic matter and adsorbed nutrients, erosion causes nutrient losses, leading to rapid fertility decline even under moderate erosion (Zhang et al., 2024a; Sun et al., 2024; Yang et al., 2023).

Drought-prone regions are particularly susceptible to water-driven erosion due to the lack of continuous soil cover. High-intensity rainstorms, even for a short period, further intensify this vulnerability by removing and transporting soil particles at rapid rates. Ramos (2024) examined the influence of rainfall intensity on water erosion and nutrient losses in Mediterranean rainfed vineyards under bare-soil conditions, comparing compost-amended and nonamended plots. Compost application reduced runoff rates, increased soil water content, and lowered sediment concentrations in runoff, particularly during intermediate erosion events. Similarly, (Yang et al., 2023) investigated nutrient transport from water and wind erosion across New South Wales, Australia, over a 20-year period (2001–21). The authors report that average annual nutrient transport rates were 2.2 kg N ha^{-1}, 0.3 kg P ha^{-1}, and 41.6 kg SOC ha^{-1}, with substantial spatial variability ranging from near zero to 395 kg ha^{-1} yr^{-1} depending on the region. They found that water erosion dominated the nutrient transport process, accounting for approximately 98% of the total and being about 90 times higher than wind erosion, though wind erosion contributed up to 12% in spring months in drier western regions (Sun et al., 2024) conducted a global assessment of wind-driven soil erosion using the Revised Wind Erosion Equation (RWEQ) model for 1982–2019, reporting an average global rate of 312.5 Pg yr^{-1}. They observed that drylands dominated erosion patterns, accounting for 82% of total wind erosion (256.4 Pg yr^{-1}), with the highest rates in arid desert climates (93.68 Mg ha^{-1} yr^{-1}), including the Sahara, Arabian Peninsula, and Australian deserts. Likewise, (Zhang et al., 2024a) evaluated the spatial and temporal dynamics of wind erosion across Africa's drylands from 2001 to 2020 using the RWEQ model combined with machine learning. Their analysis revealed an average annual wind erosion modulus of 16,672 t km^{-2} yr^{-1}, with hyper-arid and arid regions accounting for more than 90% of total erosion. Wind speed appeared as the dominant driver, contributing over 80% of erosion variability, while vegetation cover played a significant mitigating role, particularly in arid and semi-arid zones. The study concluded that both increasing vegetation cover and reducing near-surface wind speed are essential strategies for mitigating land degradation and desertification in drought-affected regions of Africa.

18.2.4 Decline in soil biological function and microbial activity

In dryland agricultural systems, repeated mechanical disturbance, inadequate organic matter inputs, and extreme moisture–temperature fluctuations progressively diminish microbial diversity and disrupt key nutrient-cycling communities, including nitrifiers, phosphate-solubilizing bacteria, and arbuscular mycorrhizal fungi. Working on the impact of drought on soil microbial communities and nutrient cycling processes, De Silva et al. (2025) reported that severe drought conditions significantly compromise nutrient availability in soil by reducing microbial metabolic activity. The research indicates that the decreased metabolic activity of the soil microbial community under water limitations leads to a reduction in the mineralization of carbon, nitrogen, and phosphorus, thereby affecting their biological cycles and pathways. Consequently, even when soil contains substantial reserves of organic N or P, the reduced capacity for biological transformation limits crop's nutrient availability and constrains fertilizer use efficiency. In another study conducted by Bogati and Walczak (2022), these authors reported that prolonged drought significantly suppresses microbial abundance and activity in dryland soils, leading to significant decreases in nutrient mineralization and soil respiration. Further, activities of major soil enzymes, including β-glucosidase, phosphatase, chitinase, and cellulases, decline substantially under low soil moisture, disrupting organic matter decomposition and nutrient cycling. Drought also impairs arbuscular mycorrhizal fungi and phosphate-solubilizing bacteria, thereby reducing phosphorus mobilization and limiting plant nutrient acquisition. However, organic amendments such as compost, sewage sludge, and municipal solid waste can restore enzyme activity and protect microbial communities by improving substrate availability, safeguarding soil moisture, and providing refugia for drought-sensitive populations. These findings demonstrate that soils suffering from drought present disruption of soil biological function and require nutrient management practices that explicitly support microbial communities responsible for nutrient transformations.

18.2.5 Soil-specific nutrient constraints

In a study conducted by Donovan et al. (2025), soil nitrogen (N) dynamics in water-limited environments were shown to be strongly constrained by the interaction between drought and soil biogeochemical processes. These findings demonstrate that limited soil moisture strongly reduces organic matter mineralization and nitrification, thereby restricting the formation of plant-available nitrogen forms. Under water deficit, microbial activity declines, which in turn suppresses net nitrogen mineralization. Indeed, adequate water in the root zone is essential for efficient nutrient uptake, since low soil water content simultaneously limits nitrogen mineralization and plant nitrogen acquisition. Furthermore, nitrogen availability in dryland soils is governed not only by fertilizer inputs but also by water-mediated controls on microbial processes and root functioning.

Beyond nitrogen, drought-induced constraints extend to phosphorus cycling, where fixation processes further limit nutrient bioavailability. Recently, Mahmood et al. (2025) demonstrated that drought significantly alters soil phosphorus (P) dynamics in tropical ecosystems, primarily through increased P fixation with calcium (apatite P), which becomes the dominant P form under water-limited conditions. The same study revealed that labile inorganic P available for plant uptake decreased markedly with intensifying drought due to phosphatase enzyme suppression, while organic P fractions increased substantially due to reduced microbial mineralization. Moreover, Xu et al. (2021) reported that water limitation sharply constraints K nutrition because low moisture reduces the pool of available K and its movement toward roots.

Extending beyond macronutrients, soil aridity imposes severe constraints on micronutrient availability across global drylands. Moreno-Jiménez et al. (2019) reported that in a global survey of 143 dryland ecosystems, the obtained results revealed that both total and available concentrations of essential micronutrients (Fe, Zn, Cu, Mn) were remarkably low when compared to global averages for natural and agricultural soils. The same researchers reported that aridity increase is associated with soil organic matter declines, reducing, as a consequence, its ability to enhance metal solubility through complexation. Concurrently, soil pH rises due to the accumulation of soluble salts and carbonates, leading to the metal precipitation into insoluble phases. This geochemical constraint was particularly severe for iron and zinc, with the available Fe/Zn ratio decreasing exponentially along aridity gradients, indicating a decoupling of nutrient cycles that may exacerbate nutritional imbalances for plants and higher trophic levels in dryland ecosystems.

18.2.6 Secondary salinization and poor-quality irrigation water

Secondary salinization and poor-quality irrigation water are major contrasting environmental factors in water-limited drylands, where continuous use of saline water results in soil structure degradation, impaired plant water uptake, and reduced availability and uptake efficiency of key nutrients such as K, P, and several micronutrients. According to Tarolli et al. (2024), secondary salinization in drought-prone regions arises from the combined effects of natural factors, including low precipitation, high evaporation rates, salt-rich parent materials, and poor natural drainage, in association with inappropriate irrigation practices, excessive groundwater extraction, and insufficient drainage infrastructure. The study by Zhang et al. (2022) indicated that in the arid and semi-arid climate zones of northwest China, the use of poor-quality irrigation water combined with inadequate drainage resulted in continuously accumulated salts in the upper soil profile. Soil electrical conductivity frequently exceeded the tolerance thresholds of most crops, while alkaline pH values above 8.0 further reduced nutrient availability and impaired soil biological functioning.

18.3 Strategies for nutrient management in drought-prone regions

To overcome the multifaceted constraints described previously, particularly the synergistic negative effects of water scarcity and nutrient depletion, the adoption of tailored nutrient management strategies is imperative in drought-prone regions. Sustainable agricultural intensification in these fragile agroecosystems requires moving beyond blanket fertilization recommendations toward adaptive practices that enhance both nutrient use efficiency and soil water retention. The primary objective is to synchronize nutrient supply with crop demand while simultaneously restoring soil organic matter and biological functionality. Accordingly, this section evaluates key agronomic practices to optimize nutrient uptake *via* the application of organic amendments, biofertilizers, and precise fertilization technologies (Table 18.1).

Table 18.1 Illustrative summary of findings from selected studies on the impact of nutrient management strategies on soil and plant parameters under water-limited cropping systems.

Plant species	Practice	Impact on soil	Impact on plant	Location/Country	References
Maize (*Zea mays*)	Green manuring (*Centrosema pubescens*)	- Increased nutrient content in the soil	- Increased leaf area, stem diameter, cob length, nutrient uptake, grain yield	Dryland Indonesia	Idham et al. (2021)
Maize (*Zea mays*)	Legume Green Manure (Vetch)	- Increased soil water storage - Decline soil evaporation	- Reduced maize evapotranspiration - Increased maize aboveground and root biomass, grain yield and crop water productivity	Hexi Desert-Oasis, China	Wang et al. (2023)
Barley (*Hordeum vulgare*)	Farmyard manure	- Increased organic matter, cation exchange capacity and available P, Ca, Mg and K	- Increased the grain yield (up to 51%), and the barley plants' nutrient contents	Arid Regions Institute, Medenine, Tunisia	Oueriemmi et al. (2021)
Wheat (*Triticum aestivum*)	Compost	- Alleviated negative environmental indicators	- Increased wheat yield with reduced N fertilizer	Hamedan state of Iran. Semi-arid climate	Amirahmadi et al. (2024)
Cauliflower-Maize	Biochar	- Increased soil organic C and nutrients (N, P, K) - Enhanced C-, N-, and P-acquiring enzyme activities and C:N enzyme ratio	-	Station for dryland soil, Shanghai, China	Wang et al. (2025)

Table 18.1 Illustrative summary of findings from selected studies on the impact of nutrient management strategies on soil and plant parameters under water-limited cropping systems. (*Continued*)

Plant species	Practice	Impact on soil	Impact on plant	Location/Country	References
Faba bean (*Vicia faba*)	Reduced/No tillage	- No effect on soil water storage	- Increased leaf photosynthesis, micronutrients (Fe, Mn, Zn) - Increased arbuscular mycorrhizal colonization and Rhizobium nodulation in roots	The experimental site has a Mediterranean climate, with soil water supplied solely by natural rainfall, partially reduced using rainout shelters to simulate drought.	Madejón et al. (2023)
Maize–soybean intercropping: 2 cereals (*Zea mays*, *Sorghum bicolor*) and 3 legumes (*Glycine max*, *Medicago sativa*, *Vigna radiata*).	Crop rotation	-	- Increased biomass, N transfer to cereals	Huailai Experiment Station of the State Key Northwestern Hebei, China. This area is characterized by a semi-arid continental monsoon climate.	Zhao et al. (2022)
Rice (*Oryza sativa*) and wheat (*Triticum aestivum*)	Crop residue retention	- Improved soil structure (aggregation, bulk density, infiltration) - Increased soil organic carbon, nutrients (N, P, K), and microbial activity	- Increased effective tillers, spike length, grains spike, grain weight, straw yield, biological yield, grain yield	Indo-Gangetic Plains, India	Gupta et al. (2024)

(*Continued*)

Table 18.1 Illustrative summary of findings from selected studies on the impact of nutrient management strategies on soil and plant parameters under water-limited cropping systems. (*Continued*)

Plant species	Practice	Impact on soil	Impact on plant	Location/Country	References
Maize-Wheat	4R nutrient stewardship (subsurface drip)	- Increased mineral N	- Increased grain yields	Borlaug Institute for South Asia, Ladhowal, Punjab, India, which has a semi-arid subtropical climate	Gupta et al. (2023)
Vegetables	Slow-release fertilizer (polymer-coated urea)	- Reduced cumulative N_2O and NH_3 emissions	- Yield maintained with 10–20% less N	Liuyang Research Station, Changsha, China	Xie et al. (2025)
Spring wheat	Nanofertilizer (nano-DAP foliar)	- Improved P availability, reduced soil P requirement	- Yield maintained with 25% less soil P; increased P use efficiency, plant height	New Delhi, India	Reddy et al. (2025)
Olive (*olea europaea*)	AMF inoculation (*Rhizophagus intraradices*)	-	Increased photosynthesis, improve the maximum quantum efficiency of photosystem II (Fv/Fm), and enhance leaf water potential	A pot experiment was conducted at the China West Normal University	Zhou et al. (2025)
-	Phosphate-solubilizing bacteria (*Pantoea agglomerans*)	Increased soluble P microbial activity	-	Tunisia	Saadouli et al. (2021)

Source: Adhikari, R., 2024. Microbial phosphorus solubilization under drought conditions. Journal of Soil and Future Research. 5, 45–47.

18.3.1 Organic soil amendment

Organic soil amendments constitute a suitable strategy for enhancing agricultural productivity and sustainability in resource-limited agroecosystems. Unlike conventional inorganic fertilizers, which are highly soluble and can be rapidly lost from the soil profile, organic materials are decomposed gradually. This slow-release characteristic synchronizes nutrient availability with crop demand over the growing season, substantially reducing the risk of nutrient leaching and gaseous losses into the environment (Santoso et al., 2021; Ren et al., 2024; Kugedera and Kokerai, 2024). In addition to nutrient supply, the application of organic amendments confers broad improvements to soil health and functionality. Their primary contribution is the buildup of soil organic matter, which enhances soil structure and increases cation exchange capacity (CEC). Crucially for drought-prone regions, these amendments significantly improve soil moisture retention. Furthermore, organic materials support beneficial soil microbial communities and help micronutrients chelate to make them more available for plant uptake. Common and effective organic sources, such as farmyard manure, compost, green manuring, and biochar, contribute efficiently to the long-term soil fertility and resilience (Fig. 18.1).

18.3.1.1 *Green manuring*

Fertilizer use strongly affects crop yields, soil nutrient status, and environmental quality. The recent increase in mineral fertilizer prices, combined with progressive soil fertility decline, has promoted the use of legume-based green manures as a cost-effective organic strategy to restore soil fertility in the long term. In this context,

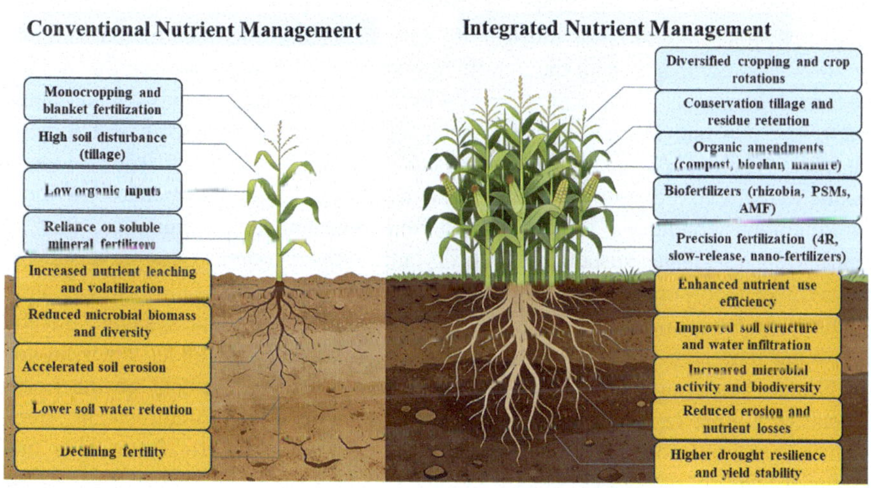

FIGURE 18.1

Key functional contributions of organic soil amendment to sustainable crop productivity.

green manures, regarded as a safe and environmentally friendly source of organic fertilizer in conventional agriculture, are primarily incorporated as soil amendments to supply nutrients to subsequent crops, particularly in drought-prone regions. Idham et al. (2021) demonstrated that green manure application significantly improved soil fertility and maize performance in dryland conditions. Among the species tested, *Centrosema pubescens* at 10 t ha^{-1} produced the greatest benefits in maize plants, including the highest leaf area, stem diameter, cob length, nutrient uptake (N, P, K), and overall grain yield. These results highlighted that green manure, particularly *C. pubescens*, effectively boosts nutrient availability, enhances maize nutrient uptake, and increases productivity in nutrient-poor dryland soils. In a recent study, Wang et al. (2023) reported that the use of legume-based green manure in the Hexi desert–oasis region of northwestern China significantly improved soil physical and hydrological properties under a maize-based system. Specifically, a vetch-type leguminous green manure decreased soil bulk density and increased field capacity, reflecting an improvement in soil structure and water-holding capacity in the root zone. Green manure treatments also enhanced soil water storage within the 0–100 cm profile throughout the maize growing season, thereby improving water availability for crops under arid irrigated conditions and contributing to a more efficient use of limited irrigation resources. In another study, Ren et al. (2024) investigated the long-term effects of leguminous green manure crops on soil fertility and mineral nutrition in the Loess Plateau of China, a semi-arid dryland region characterized by severe soil erosion, low organic matter content (below 1.0% in two-thirds of the area), and degraded soil structure resulting from intensive cultivation. These authors reported that, after 13 years of leguminous green manure cultivation, soil fertility indicators increased markedly in the 0–20 cm layer. Nitrate nitrogen (NO_3^--N) varied by 96%–138%, total nitrogen (TN) by 10%–21%, soil organic carbon (SOC) by 14%–20%, microbial biomass nitrogen by 45%–62%, and microbial biomass carbon by 26%–57% compared with the bare fallow system. The study showed that among the three leguminous green manure species evaluated (soybean, Huai bean, and mung bean), soybean and Huai bean were the most effective in enhancing soil mineral nutrition. Their superior performance was attributed to higher root biomass production, improved nutrient cycling and availability, and the creation of more stable abiotic conditions that supported soil mesofaunal communities, thereby contributing to greater sustainability in dryland agroecosystems. Xue et al. (2025) showed that incorporating green manure crops in place of bare fallow significantly enhances soil organic carbon sequestration in semi-arid dryland systems, thereby improving long-term soil fertility and nutrient-holding capacity under water-limited conditions. These authors reported that in a 9-year field experiment on the Loess Plateau of China, replacing fallow periods with soybean or sorghum-sudan, green manure increased soil organic carbon (SOC) content in the 0–20 cm layer and produced SOC sequestration rates of 189 and 328 kg C ha^{-1} yr^{-1}, respectively, in the 0–40 cm profile. Sorghum-sudan provided substantially higher total carbon inputs (approximately 5.1–6.2 Mg C ha^{-1} yr^{-1}) than fallow (3.4 Mg C ha^{-1} yr^{-1}), while soybean supplied lower carbon but 2.3 times more nitrogen than sorghum-sudan, illustrating the complementary roles of leguminous versus nonleguminous green manures in dryland nutrient management. The research indicates that selecting green manure species in semi-arid calcareous soils: sorghum-sudan is recommended where maximizing carbon sequestration and water-use efficiency are priorities, whereas leguminous green manures such as soybean are preferred where biological nitrogen fixation and protein-rich residues are needed to address nitrogen deficits and reduce dependence on synthetic N fertilizers in drought-prone cropping systems.

18.3.1.2 Farm yard manure

Farm yard manure (FYM) consists of a well-decomposed mixture of livestock urine, dung, bedding materials, and residual roughage or fodder fed to cattle. This organic amendment substantially enhances soil chemical, physical, and biological properties, including nutrient content, structure, and microbial activity, thereby supporting long-term fertility and productivity in drought-prone systems. Oueriemmi et al. (2021) reported that applying FYM at rates of 20–60 t ha^{-1} markedly improved soil fertility in arid sandy soils, with the applied FYM containing 28.9% C, 1.2% N, and 577.7 mg kg^{-1} available P. According to these authors, FYM application induced dose-dependent increases in soil organic carbon (up to a 3.3-fold rise), total nitrogen (by 14%–48%), available phosphorus (up to an 11-fold increase), and cation exchange capacity (by 32%), with beneficial effects persisting over two consecutive cropping cycles. In parallel, FYM significantly enhanced barley performance, increasing grain yield by 51% at the first harvest and 77% at the second harvest, while improving plant nutrient uptake, with reported increases in N (up to 30% in grains), K (61%), Ca (24%), and Mg concentrations in plant tissues. In another study, Santoso et al. (2021) showed that the application of livestock manures (chicken, goat, and cow) significantly improved soil chemical properties and nutrient availability in acidic dryland soils. Chicken manure, in particular, increased soil organic carbon by about 38% and supplied higher amounts of phosphorus and potassium than goat and cow manures, while all manure types contributed with substantial nitrogen inputs. Manure application also enhanced cation exchange capacity and base saturation, reduced aluminum toxicity, and improved nitrogen availability, thereby better matching crop nutrient demand and strengthening soil fertility and nutrient retention in drought-prone acidic environments.

In semi-arid regions of Africa, sorghum productivity is critically constrained by the dual challenges of soil nutrient depletion and severe water stress, often resulting in grain yields below 500–650 kg ha^{-1}, notably in countries such as Mozambique and Zimbabwe. Beyond simply supplying nutrients, FYM plays a vital role in ameliorating soil physicochemical properties and enhancing moisture retention. However, the application of only fibrous cattle manure, often rich in recalcitrant plant residues, poses a challenge due to potential nitrogen immobilization by decomposers. Consequently, the integrated application of FYM with mineral fertilizers emerges as the optimal strategy; this synergistic approach can potentially increase yields from 500 to 5000 kg ha^{-1} by balancing immediate nutrient release with long-term soil health improvement.

18.3.1.3 **Compost**

Harvested plant residues, converted into a compost, represent a valuable organic resource for enhancing soil fertility levels and plant productivity. This process biologically converts residues into a stable, humus-like product easily mineralized, pathogen-free, and with low risk for weeds, while reducing bulk volume, odors, and transport costs (Adeleke et al., 2021). Compost lowers the C/N ratio to produce nutrient-rich material with slow mineralization, ensuring sustained availability throughout the growing season, while improving soil structure *via* cation bridges and stimulating microbial activity and root growth.

In a study on organic dryland wheat in northern Utah (Adeleke et al., 2021), a one-time compost application at 12.5, 25, and 50 Mg ha^{-1} dry weight significantly improved soil health and mineral levels. Mineralizable carbon, microbial biomass, and phosphatase enzyme activity increased significantly at all compost treatments, while available soil phosphorus increased at 25 and 50 Mg ha^{-1} rates and total soil nitrogen at only 50 Mg ha^{-1} rate. Wheat yield was significantly greater three years after application at the 50 Mg ha^{-1} compost rate only, with no response to conventional fertilizer (50 kg N ha^{-1} anhydrous ammonia) (Adeleke et al., 2021). In a recent study, Amirahmadi et al. (2024) found that compost application at 3000 kg ha^{-1} in dryland winter wheat farming significantly improved grain yield by 27.4% (from 6.7 to 8.6 ton ha^{-1}); while reducing chemical nitrogen fertilizer use from 69.42 to 58.64 kg ha^{-1}, global warming potential by 33.8%, water consumption by 34.4%, fossil resource scarcity by 28.2%, and mineral resource scarcity by 31.1% compared to dryland farming without compost. Their findings indicate that the compost provided essential macronutrients and micronutrients through enhanced microbial activity and organic matter decomposition.

18.3.1.4 **Biochar**

Biochar is a carbon-rich solid material formed through the pyrolysis of crop residues, animal manure, biological wastes, and other organic feedstocks. Pyrolysis refers to the thermochemical decomposition of organic matter at elevated temperatures in the absence of oxygen. The resulting biochar is highly heterogeneous in composition, containing both labile fractions that decompose readily and stable aromatic structures persisting in the soil (Santos et al., 2025; Bekchanova et al., 2024). The composition of biochar depends strongly on the type of feedstock and the pyrolysis conditions. Biochar derived from animal manure is typically richer in nutrients and more readily decomposed by soil microbes, whereas biochar produced from crop residues such as straw or grain husks tends to contain higher ash content (Qi et al., 2024; Wang et al., 2025; Zhang et al., 2024b). Temperature plays a key role because higher pyrolysis temperatures can cause volatilization of nutrients originally present in the feedstock, altering the final chemical profile of the biochar. Previous studies have shown that biochar application significantly increases soil moisture as well as the availability of N, P, and K (Santos et al., 2025; Bekchanova et al., 2024). Among the properties of biochar that contribute to improved soil moisture and nutrient retention is its highly porous structure, with micro- and mesopores that store water and nutrients. In addition, biochar has a high specific surface area, which facilitates greater interaction with soil water and nutrients (Qi et al., 2024; Zhang et al., 2024b). It also exhibits a high cation exchange capacity, as it develops negative charges that help retain cationic nutrients such as Ca^{2+}, Mg^{2+}, K^{+}, and NH^{4+}, thereby reducing leaching. Biochar produced at low temperatures (<500 °C) is generally more hydrophilic, enhancing water retention in dry soils (Wang et al., 2025; Zhang et al., 2024b) found that after seven years of biochar-based fertilizer substitution at 30% (LBF) and 60% (HBF) a significant improvement of soil organic carbon by 8.36% and 13.20%, respectively, was recorded if compared to chemical fertilizer alone, while enhancing total and available nitrogen, phosphorus, and potassium in dryland cauliflower-maize rotation systems. Similarly, in Brazilian semi-arid region, (Santos et al., 2025) found that biochar application derived from sewage sludge-cashew pruning (SPB) and cashew bagasse (CBB) at rates of 5 and 40 Mg ha^{-1} significantly increased total organic carbon by 59%–70% and soil carbon stocks by 59%–72%, in comparison to control in degraded dryland soils. Further, biochar incorporation enhanced soil nutrient levels including nitrogen, phosphorus, and potassium, while improving soil structure, water retention capacity, and fostering beneficial microbial activity in sandy degraded soils. Moreira Cevallos et al. (2024) found that biochar application at 1 t ha^{-1} (equivalent to 50 g per linear meter of planting row) combined with liquid or conventional fertilization significantly improved yield components, nutrient use efficiency, and economic profitability in rainfed peanut production in Ecuador's drylands. The same authors reported that biochar derived from carob wood (*Prosopis pallida*) through pyrolysis at 350°C–550°C contained 38.8% carbon, 1.0% nitrogen, 0.22% potassium, 1.01% calcium, and substantial micronutrients including 194 mg kg^{-1} iron, with a pH of 7.5 and high cation exchange capacity.

18.3.2 Conservation agriculture

Conservation agriculture (CA) represents a systems-level approach to sustainable crop production that is particularly relevant to nutrient management in drought-prone regions because it simultaneously addresses the three constraints identified earlier: water conservation, soil fertility maintenance and erosion control. CA is defined by three interlinked core principles that must be applied together with locally adapted complementary practices: (1) Minimum mechanical soil disturbance; (2) permanent soil organic cover with crop residues and/or cover crops; and (3) species diversification through varied crop rotations and associations. Reduction in soil losses and production cost and improvement in soil organic matter and plant available soil water have been reported in various dryland regions, due to implementation of CA (Madejón et al., 2023).

18.3.2.1 Conservation tillage

Conservation tillage systems, including no-till and reduced tillage, are widely recognized for improving nutrient management in drought-prone regions by enhancing soil structure, limiting erosion, and increasing water-use efficiency. Minimizing soil disturbance helps preserve organic matter and maintain essential nutrients such as nitrogen and phosphorus within the upper soil horizons. However, several studies have reported that long-term no-till may gradually increase soil bulk density and promote surface compaction. Such physical constraints can restrict root penetration and reduce nutrient uptake efficiency (Madejón et al., 2023; Lv et al., 2023; Deng et al., 2023). These findings suggest that sustainable conservation tillage in dryland environments may require the periodic integration of shallow tillage or subsoiling to alleviate compaction while maintaining the benefits of soil conservation and nutrient availability. Working on a global meta-analysis of 203 studies, Lv et al. (2023) reported that reduced tillage significantly improved soil nutrient status compared with conventional tillage by minimizing mechanical disturbance and promoting nutrient retention. Specifically, reduced tillage increased topsoil organic matter, organic carbon, total nitrogen, and available potassium. These improvements resulted from reduced soil compaction, lower nutrient leaching losses (particularly nitrate-nitrogen), enhanced aggregate stability, and improved water conductivity, all of which contributed to better nutrient cycling and accumulation in the soil profile. Madejón et al. (2023) reported that in a Mediterranean rainfed faba bean system in southwestern Spain, conservation tillage no-till NT and reduced tillage RT mitigated drought effects compared to traditional tillage. NT and RT conditions showed lower biomass yield losses under 30% rainfall exclusion (30% NT, 50% RT vs. 20% TT), higher leaf photosynthesis and stomatal conductance, and increased mycorrhizal root colonization, especially under drought, linked to elevated micronutrients (Fe, Mn, Zn) in NT leaves. Working in the arid drylands of the northern Loess Plateau of China, Deng et al. (2023) reported that rotary tillage significantly improved soil aggregate stability compared with conventional tillage. This improvement was reflected in higher mean weight diameter, geometric mean diameter, and a greater proportion of water-stable macroaggregates within the 0–40 cm soil layer. In addition, it contributed to more stable maize yields over the 5-year experimental period by reducing yield fluctuations under drought conditions. These results indicate that rotary tillage improves soil water retention, strengthens soil hydraulic functioning, and promotes more efficient nutrient cycling in arid agricultural systems.

18.3.2.2 Crop residue management

Several studies have reported that retaining crop residues on the soil surface or incorporating them into the topsoil constitutes a central component of conservation agriculture that helps address key nutrient management constraints in drought-prone regions. Residue retention increases soil organic matter and microbial biomass, promotes nitrogen mineralization from organic pools, and enhances nutrient availability by sustaining microbial activity while reducing erosion and volatilization losses. Leguminous residues are particularly beneficial because they supply biologically fixed nitrogen, and this nitrogen is incorporated into the soil organic pool more efficiently than nitrogen derived from mineral fertilizers (Dinesh et al., 2024; Chen et al., 2025). Working on a long-term rice-wheat system in the semi-arid Indo-Gangetic Plains of India (Gupta et al., 2024) demonstrated that soil organic carbon, available nutrient content (N, P, and K), microbial count, and enzyme activities were also significantly higher under conservation tillage and residue-treated plots than in soil deprived of residue. The residue derived from both crops, when either retained or incorporated, improved the soil organic carbon 0.80%) and resulted in a significant increase in SOC (73.9%) in the topsoil layer as compared to the conventional practice. Therefore, results suggested that crop residue retention following the conservation tillage approach improved the yield of wheat cultivated in the rice–wheat cropping system. In semi-arid rainfed potato systems Chen et al. (2022, 2025) reported that straw strip mulching significantly increased yields by 14.7% and water use efficiency by 9.2% while reducing early-stage water consumption by 8%–10%, thereby conserving moisture for critical tuber expansion phases. This residue management enhances nutrient retention by minimizing erosion losses, improving soil aggregation, and promoting microbial decomposition that synchronizes nutrient release with crop demand, while also suppressing evaporation to maintain nutrient availability in water-limited profiles. These dual benefits optimize both water and nutrient cycling for sustainable dryland productivity.

However, several studies have shown that the effectiveness of residue retention for soil carbon sequestration depends strongly on sufficient mineral nutrient availability, particularly nitrogen, phosphorus, and sulfur, to maintain microbial stoichiometric requirements. Long-term experiments indicate that residue incorporation alone does not ensure carbon sequestration, and soil organic carbon may remain unchanged or decline when accelerated mineralization occurs. Climate and soil texture further modulate these outcomes, with evidence showing that in hot, dry semi-arid regions with sandy-loam soils, the benefits of residue retention diminish in the absence of sufficient rainfall or irrigation to sustain decomposition and nutrient cycling (Cheng et al., 2024; Hu et al., 2018).

18.3.2.3 *Crop rotation and intercropping*

Crop rotation, the practice of alternating crops in the same field each season to avoid monocropping, has recently been shown to improve belowground benefits such as nutrient cycling, soil organic carbon, microbial biomass, and activity, and soil structure through better aggregation in water-limited dryland systems. According to a comprehensive meta-analysis by Mudare et al. (2025) examining 3663 paired observations from 738 experiments worldwide, crop rotations increased total system yields by 23%, dietary energy by 24%, protein by 14%, and micronutrients (iron, magnesium, zinc) by 17%–27% compared to continuous monoculture. Moreover, (Zhao et al., 2022), working in a semi-arid continental monsoon climate (370 mm annual rainfall), reported that intercropping legumes with cereals proved the most promising approach to reduce nitrogen fertilizer application, with fixed N transfer to cereals negatively correlated with legume biomass and positively with cereal biomass. These authors indicated that Glycine max-maize intercropping was optimal, as G. max fixed more atmospheric N_2 than *Medicago sativa* or *Vigna radiata*, while *Zea mays* obtained greater N benefits than Sorghum bicolor through enhanced biological N fixation and rhizosphere facilitation.

In a five-year study in semi-arid Kansas, Crews et al. (2022) examined perennial intermediate wheatgrass (*Thinopyrum intermedium*)-alfalfa (*Medicago sativa*) intercropping on Cozad silt loam soils. They indicated that Alfalfa fixed 54–162 kg N/ha/year. Intercrop grain yields remained stable at 250 kg/ha/year while unfertilized monocultures declined by 78%–87%. The system achieved positive nitrogen balance (188 kg N/ha) with 77%–99% less nitrate leaching and reduced greenhouse gas emissions, demonstrating effective ecological nutrient management for drylands without synthetic fertilizers. Yang et al. (2024) have shown that diversifying the conventional winter wheat–summer maize double-cropping system with cash crops (sweet potato) and legumes (peanut, soybean) markedly enhanced multiple agroecosystem functions. In fact, crop rotations such as sweet potato–wheat–maize increased wheat-equivalent yields by up to about 38% when compared to the wheat–maize control.

18.3.3 Integrated nutrient management and fertilizer efficiency

In drought-prone regions, sustainable crop production is limited by low soil moisture, erratic rainfall, and poor nutrient use efficiency. Continuous use of conventional mineral fertilizers results in low nutrient recovery and significant losses caused by high temperatures accompanied by irregular rainfall. Sustainable intensification through integrated nutrient management is essential to synchronize nutrient supply with crop demand, improve nutrient retention in the root zone, and reduce environmental losses under water scarcity. Fertilizer efficiency depends on applying the 4R nutrient stewardship principles, soil-test-based balanced fertilization, and advanced fertilizer technologies such as slow- and controlled-release fertilizers, microbe-coated fertilizers, nanofertilizers, foliar feeding, and fertigation. These strategies collectively enhance nutrient use efficiency, crop resilience, and environmental sustainability in dryland agriculture (Samijan et al., 2024; Ahmad et al., 2025).

18.3.3.1 Fertilizer application at right source, right rate, right time, right place (4R nutrient)

Recent studies show that optimal fertilizer management in dryland crops requires applying fertilizer at the right source, rate, time, and place to maximize nutrient use efficiency under water-limited conditions, following the 4R nutrient stewardship framework (Gupta et al., 2023), working on maize and wheat in drought-prone northwest India, have demonstrated that implementing 4R nutrient stewardship through subsurface drip irrigation significantly enhanced nitrogen management compared to conventional flood irrigation. Urea delivered *via* fertigation at 20 cm depth directly to root zones reduced volatilization losses and improved nitrogen availability versus surface broadcasting. Split-timing of nitrogen in five equal applications at 10-day intervals starting 20 days after sowing synchronized nutrient supply with crop demand, increasing agronomic efficiency in maize and wheat compared to conventional two-split applications. Optimizing nitrogen rates at 150 kg N/ha for maize and 120 kg N/ha for wheat under subsurface drip irrigation achieved 25% nitrogen savings in maize while maintaining yields 15.8% higher than conventional systems. Moreover, Ayankojo and Morgan (2021) reported that the interaction between nitrogen application timing and irrigation regime had a strong influence on tomato performance, with the treatment combining 25% pre-plant N and 75% N *via* fertigation producing significantly greater early-season growth and higher marketable yields (mean 56.00 Mg ha^{-1}) than the treatment with 100% N supplied through fertigation and no pre-plant application (mean 40.23 Mg ha^{-1}). Ahmad et al. (2025) showed that nitrogen is the key nutrient controlling the growth and development of cotton in semi-arid southern Punjab, Pakistan. A wide range of cotton traits were measured, including canopy height, leaf area index, leaf and stem nitrogen status, seed cotton yield, and time-series biomass data. Higher nitrogen rates of 240 and 280 kg N/ha resulted in better performance for most traits; however, the highest leaf nitrogen content was observed at 210 kg N/ha. Based on these findings, it is suggested that under semi-arid conditions, applying slightly higher than the recommended nitrogen rates may be a viable strategy for cotton growers to achieve increased seed cotton yield.

Several authors have emphasized that foliar fertilization in water-limited systems can be particularly advantageous, because it allows precise, temporally targeted nutrient delivery to the leaf canopy at stages of high demand, precisely when soil-applied basal fertilizers remain inaccessible due to insufficient moisture at sowing and restricted mass flow or diffusion in the dry soil matrix (Selvakumar et al., 2022; Shakoor et al., 2024). Recent research demonstrates that foliar application of concentrated urea solution substantially enhances crop yields. In wheat production, foliar spray of 4% urea solution at the heading stage increased grain yield by 30% compared to control treatments, proving most effective when applied at critical growth stages (Shakoor et al., 2024). Similarly, foliar application of 2% urea at flowering and boll development stages in hybrid cotton significantly enhanced growth traits, yield attributes, and final seed cotton yield (Selvakumar et al., 2022).

18.3.3.2 Nutrient application based on soil analysis and balanced fertilization

Accurate soil analysis remains essential in dryland regions to assess nutrient deficiencies and adjust fertilizer applications according to crop needs. Indeed, soil nutrient concentrations in these areas may decline over time due to continuous nutrient uptake by successive crops and the low organic matter content of the soil.

Meena et al. (2023) reported that soil analysis-based nutrient management in drought zones involves comprehensive precrop characterization (pH, organic carbon, N, P, K, S, micronutrients) and multi-depth sampling (0–15, 15–30, 30–45, 45–60 cm) to monitor fertility changes. These authors reported that science-based soil testing combined with balanced fertilization using both organic and inorganic sources represents a long-term strategy for sustaining soil fertility and ensuring food security in drought zones. Working on dryland Inceptisols, Samijan et al. (2024) reported that balanced fertilization based on soil tests (organic fertilizer 1600 kg ha^{-1} + NPK 300 kg ha^{-1} + urea 200 kg ha^{-1}) substantially improved sorghum performance compared with existing farmer practice (organic 1600 kg ha^{-1} + carbontilizer 300 kg ha^{-1}). This balanced fertilization provided adequate soil nutrients, enhanced N, P, and K uptake, and was associated with yield increases of approximately 21.4%–36.1%. The same authors also reported economic returns of about USD 930–1958 per hectare, with revenue–cost ratios greater than 1.0.

Most farmers in dryland regions still rely on a single nutrient, typically nitrogen, which limits nutrient use efficiency and crop productivity. Jiang et al. (2024) found that combined application of organic and inorganic fertilizers in dryland maize not only boosted yield but also enhanced water-use efficiency by improving soil moisture retention during the fallow period. A meta-analysis by Li et al. (2023) confirmed that NPK fertilization in dryland spring maize reduced soil bulk density, increased porosity, and raised water-use efficiency compared to N alone. Additionally, Shi et al. (2023) demonstrated that combined organic–inorganic fertilization (organic fertilizer + NPK) in dryland spring maize increased water-use efficiency by 116%, nearly twice that of NPK alone (46%) or organic fertilizer alone (56%) in the dryland of northern China. These results emphasize that balanced fertilization strategies are essential for improving nutrient-water synergies, raising yields, and enhancing resource-use efficiency in dryland cropping systems.

18.3.3.3 Use of slow/controlled release fertilizers

Conventional granular fertilizers present significant challenges in drought-prone agricultural systems. Upon application, these standard formulations release nutrients rapidly, making them highly susceptible to losses through multiple pathways, including leaching during sporadic rainfall events, ammonia volatilization under high temperature conditions, and surface runoff when intense precipitation occurs after extended dry periods. In water-limited regions where rainfall patterns are unpredictable and irrigation resources are constrained, these nutrient losses not only reduce fertilizer use efficiency but also increase production costs and environmental contamination risks (Kaur et al., 2023; Song et al., 2024; Xie et al., 2025). Modern fertilizer technologies aimed to synchronize nutrient release with crop demand can be systematically classified into two principal categories based on their release control mechanisms:

Slow-release fertilizers (SRF) employ materials that inherently delay nutrient availability through: (1) Chemical or biological decomposition pathways: Formulations such as isobutylidenediurea (IBDU), urea-formaldehyde compounds, and nitrogen-lignin complexes require enzymatic or microbial degradation to convert nutrients into plant-available forms, with release rates governed by soil temperature, moisture, and microbial activity. (2) Physical size and form modifications: Large-particle formulations including urea super granules and deep-placement briquettes reduce dissolution surface area, thereby extending the nutrient release period through slower diffusion rates into the soil solution.

Controlled-release fertilizers (CRF) are based on physical barriers that regulate nutrient diffusion through: (1) Semipermeable coating technologies: polymer-encapsulated fertilizers feature synthetic resin membranes that control water penetration and nutrient release *via* osmotic gradients and temperature-dependent diffusion rates. (2) Degradable coating systems: sulfur-coated urea, shellac-coated formulations, and biodegradable polymer coatings progressively break down under field conditions, releasing nutrients as the protective layer degrades.

While the terms "slow-release" and "controlled-release" are frequently used interchangeably in agricultural practice, critical functional distinctions exist between these fertilizer classes. Slow-release fertilizers exhibit nutrient release patterns that vary substantially depending on unpredictable soil environmental factors, including microbial population dynamics, temperature fluctuations, and moisture variability. Conversely, controlled-release fertilizers are engineered to maintain relatively predictable nutrient release kinetics even under fluctuating biotic and abiotic conditions, with release rates predominantly determined by coating properties rather than soil biological or chemical processes.

Xie et al. (2025) showed that polymer-coated urea with 10%–20% nitrogen reduction maintained comparable vegetable yields in subtropical hilly drylands of southern China while significantly improving nitrogen use efficiency relative to conventional fertilization. The controlled nitrogen release timing from polymer-coated formulations minimized rainfall-induced losses, with cumulative N_2O and NH_2 emissions reduced by 32%–47% and 20%–31%, respectively, compared to conventional urea application. On the other hand, Ghafoor et al. (2022) reported that slow-release nitrogenous fertilizers including bioactive sulfur-coated urea (BSCU), neem-coated urea (NCU), and sulfur-coated urea (SCU) significantly improved wheat production under arid climatic conditions in Pakistan. Bioactive sulfur-coated urea at 130 kg N/ha demonstrated the highest irrigation water use efficiency (20.92 kg ha^{-1} mm^{-1}) and crop index (25.52%), if compared to conventional urea at 94 kg N/ha, and thus superior performance in water-scarce environments. The slow-release mechanism maintained nutrient supply to plants at critical growth stages, enhancing dry matter accumulation, grain nutritional quality, and overall water productivity. According to Song et al. (2024), starch-based superabsorbent polymers as carriers for controlled-release fertilizers provided dual functionality by controlling nutrient release while simultaneously improving soil moisture retention in drought and desertification-affected regions. These eco-friendly hydrogel formulations enhanced soil aeration, mitigated soil degradation, reduced carbon emissions, and improved nutrient retention capacity in soils under water-limited conditions. Rafique et al. (2025) reported that the incorporation of biochar and semi-interpenetrating chitosan biopolymer to synthesize novel slow-release phosphorus fertilizers significantly enhanced both soil moisture retention and nutrient availability in drought-stressed agricultural systems. The biochar-based slow-release formulations increased soil water-holding capacity while providing controlled phosphorus release over extended periods, addressing the dual challenge of water scarcity and phosphorus fixation characteristic of alkaline dryland soils with pH values exceeding 7.5, where conventional phosphate fertilizers rapidly precipitate into unavailable calcium-phosphate forms.

18.3.3.4 *Polymer-entrapped microbe-coated fertilizer*

Polymer-entrapped microbe-coated fertilizers represent an advanced strategy for nutrient management in drought-prone regions by integrating controlled nutrient release with biological activation of the rhizosphere. In this system, beneficial microorganisms such as nitrogen-fixing bacteria, phosphate-solubilizing bacteria, and plant growth-promoting rhizobacteria (PGPR) are immobilized within a polymer matrix that also encapsulates mineral nutrients. The polymer coating protects microbial viability under severe drought and high-temperature stress while regulating nutrient diffusion according to soil moisture availability. Upon rehydration, the gradual degradation or swelling of the polymer activates microbial metabolism, thereby synchronizing nutrient release with biological nutrient transformation processes. This dual functionality enhances nutrient use efficiency, minimizes leaching and volatilization losses, and improves root–microbe interactions, leading to better root architecture, higher water uptake, and improved crop resilience under water-limited conditions. Consequently, polymer-entrapped microbe–coated fertilizers constitute a promising eco-sustainable tool for improving fertilizer efficiency and soil productivity in arid and semi-arid agroecosystems. Rebelo Romão et al. (2022) showed that seed encapsulation with xerotolerant Actinobacteria formulated in a dry polymer matrix improved maize performance under drought conditions. The formulation maintained high microbial survival during long-term storage and enhanced plant growth and biomass under greenhouse water-deficit conditions. Moreover, treated plants preserved grain yield despite reduced irrigation, indicating that this dry bioformulation is a promising tool for improving water-use efficiency in drought-affected cropping systems. Further, Kaur et al. (2023) demonstrated that a rice bran-gum Arabic-alginate polymer matrix used to encapsulate the phosphor bacterium Myroidgitamensis generated low-cost, slow-release microbeads. These microbeads achieved 94.11% encapsulation efficiency and maintained high bacterial survival for up to six months at 4°C. Additionally, they showed a 99.75% biodegradation rate after 110 days in soil, providing a sustainable carrier for microbial fertilizers in water-limited environments. Compared with liquid biofertilizer, the dried polymer-entrapped formulation significantly increased wheat germination rate, root and shoot length, fresh and dry biomass, and improved potassium and phosphorus uptake in pot-grown seedlings under natural open-field conditions. Per consequent, polymer-entrapped microbe-coated fertilizers can enhance nutrient use efficiency and early growth under limited soil moisture status.

18.3.3.5 Nanofertilizers

Nanofertilizers constitute nutrient formulations in the nanometer size range that aim to increase nutrient use efficiency while reducing the environmental footprint of conventional mineral fertilizers, making them particularly attractive for drought-prone systems. Nano-enabled products typically fall into three broad categories: (1) nanoscale coatings or carrier materials (e.g., nanopolymers, zeolites, or hydroxyapatite matrices) applied to conventional fertilizers to slow dissolution; (2) nanoscale additives that modify soil–fertilizer interactions; and (3) synthesized nanofertilizers in which the nutrient itself is present as or associated with nanoparticles. In water-limited environments, these systems act as smart delivery platforms by combining high specific surface area with controlled release kinetics and strong sorption to carrier matrices, thereby synchronizing nutrient availability with crop demand and minimizing losses *via* leaching, volatilization, and fixation. In the same context, (Astaneh et al., 2021) demonstrated that soil application with nano-chelated nitrogen at 41 kg N ha^{-1} combined with conventional urea significantly improved wheat physiological traits under flowering-stage droughty period. This treatment also reduced environmental N losses compared with sole urea fertilization, highlighting the potential of soil-applied nano-N in drought-prone systems.

Rashmi and Prakash (2023) reported that soil application with nanophosphorus fertilizers based on hydroxyapatite or nano rock phosphate, together with a 25% reduction in the conventional P dose, significantly improved maize growth and yield in the Central Dry Zone of Karnataka. In addition, combined soil and foliar application of hydroxyapatite nano-P (NP1) achieved higher cob length, kernel number, and grain and stover yields than the full recommended P dose from single superphosphate alone. The same authors also reported that treated soil with foliar application of hydroxyapatite nanophosphorus allowed a 25% reduction in conventional P while maintaining or increasing maize yield in a dry zone, showing that nano-P can sustain productivity with less fertilizer in water-limited systems. Astaneh et al. (2021) showed that replacing part of soil-applied urea with nano-chelated nitrogen in wheat under flowering-stage droughty period increased relative water content, reduced electrolyte leakage, and improved grain N, P, and K at lower total N rates, indicating better drought resilience and nutrient use efficiency than conventional urea alone.

Furthermore, Reddy et al. (2025) showed that integrating nanofertilizer in the form of nano-DAP into phosphorus management for spring wheat allows substantial reductions in soil-applied P without sacrificing yield or nutrient uptake under semi-arid conditions. In a 2-year field experiment at New Delhi, treatments combining 75% of the recommended soil P dose with two foliar sprays of nano-DAP (5 mL L^{-1}) produced grain yields (4.25–6.04 t ha^{-1}), N uptake (84–125 kg ha^{-1}), and P uptake (20–24 kg ha^{-1}) statistically comparable to those achieved with 100% conventional P. This research study showed that nano-DAP used as a foliar supplement to reduce basal P can maintain or improve productivity and profitability, offering a more resource-efficient and environmentally sustainable nutrient management strategy well suited to regions where both water and mineral P are limiting.

18.3.4 Use of rhizosphere microbes

Drought-prone agroecosystems present two linked constraints: limited water availability and impaired nutrient acquisition. Rhizosphere microbes, including free-living and symbiotic bacteria and fungi that colonize the root zone, are shown to mitigate these constraints by enhancing nutrient solubilization or biological fixation, improving root growth, and regulating plant water relations. Use of targeted microbial inoculants can improve nutrient use efficiency and crop resilience in water-limited systems, reduce fertilizer inputs, and lower environmental degradation. Recent reviews consolidate the mechanisms and practical outcomes associated with nitrogen-fixing bacteria, phosphate-solubilizing bacteria, and arbuscular mycorrhizal fungi (AMF) to mitigate nutrient limitations in drought-prone regions. In dryland cropping systems, nitrogen is commonly the primary limiting nutrient. Biological nitrogen fixation (BNF) mediated by rhizosphere bacteria represents a promising strategy for enhancing nitrogen availability and decreasing dependence on mineral fertilizers.

18.3.4.1 Nitrogen-fixing bacteria

The root-nodule symbiosis between Rhizobium bacteria and legumes enables substantial biological nitrogen fixation. In drought-prone systems, integrating inoculated legumes enhances nitrogen availability for subsequent cereal crops through residual nitrogen contributions and improved soil structure.

Working on nodulated soybean plants inoculated with *Bradyrhizobium diazoefficiens* and exposed to progressive water deficit, Rubia et al. (2025) illustrate how rhizobial symbiosis sustains biological nitrogen fixation and root–nodule carbon sink strength, highlighting the potential of *Rhizobium*-based strategies for nutrient management in drought-prone regions. Moreover, *Rhizobium* strain QHCD11, isolated from faba bean nodules in arid regions of the Qinghai-Tibet Plateau, exhibited superior drought tolerance and significantly improved host plant resilience to water deficit through multiple mechanisms (Li et al., 2023): enhancing root nodule formation, increasing chlorophyll content, improving relative water content, stimulating proline accumulation, and elevating superoxide dismutase activity under drought conditions compared with uninoculated seedlings.

Another important group of nitrogen-fixing bacteria is *Azospirillum*. These microorganisms live in free association close to plant roots (in the rhizosphere) without forming nodules. In the review paper of Sun et al. (2025), it is indicated that the combined capacity of Azospirillum spp. for biological nitrogen fixation, phytohormone production, and enhancement of mineral and water uptake makes these bacteria particularly suitable for enhancing crop performance under drought stress.

Azotobacter, on the other hand, comprises species that function as free-living, nitrogen-fixing plant growth-promoting rhizobacteria PGPR. They fix atmospheric nitrogen independently, without the need for a specific plant host, making them compatible with a wide range of nonleguminous crops, including cereals like maize. Their utility is enhanced in drought-prone environments when native, stress-adapted strains are used. For instance, Shirinbayan et al. (2019) isolated and screened *Azotobacter* strains from semi-arid regions of Iran. The study identified strains such as *Azotobacter salinestris* and *A. chroococcum* that exhibited tolerance to osmotic stress and possessed multiple plant-growth-promoting traits. In greenhouse trials under drought stress (40%–80% field capacity), inoculation with these selected strains significantly increased maize shoot dry weight, plant height, chlorophyll content, and the concentration of key nutrients like nitrogen, phosphorus, and iron if compared to noninoculated controls. The authors concluded that these stress-adapted *Azotobacter* strains could help alleviate drought stress in crops and present a promising biofertilizer strategy for arid and semi-arid agriculture.

18.3.4.2 Phosphate-solubilizing microorganisms

Phosphorus (P) is a vital plant nutrient, yet its bioavailability in soils is frequently constrained by its tendency to form insoluble compounds, such as calcium or iron phosphates. Drought stress further aggravates this limitation by lowering soil water content, which restricts P diffusion and suppresses microbial activity, thereby restricting plant uptake of this essential element (Adhikari, 2024). Phosphorus-solubilizing microorganisms (PSMs), encompassing bacterial genera such as Bacillus and Pseudomonas as well as fungal species like Aspergillus and Penicillium, can mitigate this constraint by secreting organic acids and enzymes that convert fixed P into plant-available forms. Saadouli et al. (2021) demonstrated that the phosphate-solubilizing bacterium (PSB), *Pantoea agglomerans* strain V8R67, significantly enhanced phosphorus availability in semi-arid soil, increasing soluble phosphorus by 69% after 15 days of inoculation. The findings underscore the potential of *P. agglomerans* as an eco-friendly biofertilizer to improve phosphorus nutrition and support soil health in drought-prone agricultural systems. Lopez et al. (2025) showed that peanut-associated PSB exhibited the highest root dry weight and nodule number compared with uninoculated plants under drought and phosphorus deficiency. Referring to these authors, drought-tolerant PSB have significant potential as biofertilizers for enhancing peanut resilience and productivity under concurrent drought and phosphorus stress, supporting sustainable

Table 18.2 Soluble Phosphorus (mg kg^{-1}) in two agricultural regions "Mediterranean (Spain) and semi-arid (Morocco)" with different drought level and phosphorus-solubilizing microorganisms.

Region	Drought level	Control	*Bacillus subtilis*	*Pseudomonas fluorescens*
Mediterranean	Control	8.5	10.2	11.0
	Moderate	6.8	8.5	9.1
	Severe	5.2	6.2	6.5
Semi-arid	Control	7.8	9.5	10.5
	Moderate	6.0	7.5	8.1
	Severe	4.8	5.8	6.3

Source: Adhikari, R., 2024. Microbial phosphorus solubilization under drought conditions. Journal of Soil and Future Research. 5, 45–47.

agriculture in contrasting environments. Adhikari (2024) demonstrated that drought-tolerant phosphorus-solubilizing microorganisms (PSMs), including *Bacillus* and *Pseudomonas* species, increased soluble P levels by 20%–35% under water-limited conditions relative to noninoculated controls. While drought significantly reduced soil enzyme activities and microbial biomass, PSM inoculation partially alleviated these declines (Table 18.2). These results underscore the role of PSMs in enhancing P availability and reinforcing crop resilience under water stress, offering valuable perspectives for sustainable agriculture in drought-prone areas. In conclusion, integrating PSMs into biofertilizers can reduce the need for chemical phosphorus fertilizers, advancing more sustainable agricultural practices in drought-affected regions.

18.3.4.3 Arbuscular mycorrhizal fungi

Arbuscular mycorrhizal fungi (AMF) represent an eco-friendly asset in organic agriculture. These fungi colonize root cortical cells and extend extraradical hyphae, thereby expanding the soil–root interface. Through high-affinity phosphate (P) and ammonium transporters, AMF can supply up to 90% of a plant's phosphorus and 25% of its nitrogen. Studies have documented these symbiotic associations across a wide range of plants, including angiosperms, bryophytes, pteridophytes, and gymnosperms, highlighting the diverse benefits they provide. These benefits extend beyond enhanced mineral nutrition to include improved drought resistance via better water uptake and overall ecosystem stability (Parihar et al., 2020; Guo et al., 2022). Zhou et al. (2025) evaluated the capacity of arbuscular mycorrhizal fungi (AMF) to enhance drought tolerance in olive cultivars (Arbequina and Koroneiki) under simulated arid conditions. The AMF species *Rhizophagus intraradices* and *Funneliformis mosseae* were found to restore photosynthesis, improve the maximum quantum efficiency of photosystem II (Fv/Fm), and enhance leaf water potential. They also increased the accumulation of osmoregulatory compounds such as soluble sugars, starch, and proline, while reducing lipid peroxidation under drought stress in both Koroneiki and Arbequina cultivars. The study highlights that pairing specific AMF strains with olive cultivars provides a scalable, chemical-free strategy to stabilize olive productivity in the arid climate of southwest China, advancing climate-smart agriculture in drought-prone regions. Chandrasekaran (2022) conducted a meta-analysis of published research to evaluate the efficacy of arbuscular mycorrhizal fungi (AMF) in alleviating drought stress. The overall analysis indicated that AMF-inoculated plants exhibited 49% greater growth promotion compared to nonmycorrhizal plants under drought conditions. Biomass analysis revealed significant increases in root dry weight (49%), shoot dry weight (54%), and total dry weight (58%). Root morphological traits were also improved, with root length increasing by 37%, root surface area by 31%, and root volume by 65%. Nutrient uptake analysis showed substantial enhancements in phosphorus (P) uptake (86%), nitrogen (N) uptake (35%), and potassium (K) uptake (46%). These results revealed that AMF improve significantly P uptake ($P < .001$) and root volume ($P < .001$) under drought stress, suggesting a linear relationship between root length and P acquisition. This meta-analysis supports the broader application of AMF for enhanced plant productivity, stress resilience, and sustainable agriculture, particularly in nutrient management within drought-prone regions. By improving nutrient uptake efficiency and reducing dependency on chemical fertilizers, AMF offers a nature-based solution to sustain crop nutrition and soil health under water-limited conditions and advance climate-resilient farming systems.

18.4 Conclusion

Effective nutrient management in drought-prone regions requires moving beyond conventional fertilization paradigms toward integrated, site-specific strategies that explicitly address the interactive constraints imposed by water scarcity on nutrient cycling and crop uptake. The evidence reviewed demonstrates that chronic nutrient imbalances, accelerated erosion, declining soil biological function, and nutrient fixation under alkaline- or saline- conditions collectively reduce fertilizer use efficiency and threaten long-term agricultural sustainability across global drylands. However, adaptive management practices offer substantial potential to overcome these constraints through complementary mechanisms.

Organic amendments, including green manure, farmyard manure, compost, and biochar, provide multiple benefits by gradually releasing nutrients in synchrony with crop demand, increasing soil organic matter and cation exchange capacity, enhancing water retention capacity critical for drought resilience, and supporting diverse microbial communities essential for nutrient transformations. Conservation agriculture practices encompassing reduced tillage, residue retention, and diversified crop rotations simultaneously reduce erosion losses, improve soil structure and infiltration, enhance biological nitrogen fixation through legume integration, and stabilize nutrient cycling across the soil profile. Precision nutrient management technologies including soil-test-based fertilization, 4 R stewardship principles, slow/controlled-release formulations, and nanofertilizers enable targeted nutrient delivery that minimizes losses through leaching, volatilization, and runoff while maintaining or improving crop productivity at reduced input levels. Rhizosphere microbial interventions utilizing nitrogen-fixing bacteria, phosphate-solubilizing microorganisms, and arbuscular mycorrhizal fungi enhance nutrient bioavailability through biological nitrogen fixation, mineral solubilization, and expanded root–soil interface, while also improving plant water relations and stress tolerance (Fig. 18.2).

Implementing these integrated nutrient management strategies requires recognizing that no single intervention provides a universal solution across the diverse biophysical and socioeconomic contexts characterizing global drylands. Rather, optimal approaches will combine complementary practices adapted to local soil properties, climate patterns, crop species, water availability, and farmer resources. Future research priorities should emphasize long-term field experiments evaluating synergistic effects of combined interventions, development of decision-support tools incorporating soil–crop–climate interactions, breeding programs selecting crop varieties with enhanced nutrient acquisition efficiency under drought, and participatory approaches ensuring that recommended practices align with farmer knowledge, preferences, and constraints. Advancing climate-resilient agriculture in drought-prone regions ultimately depends on translating scientific knowledge into accessible, economically viable, and environmentally sustainable nutrient management systems that enhance food security while preserving the natural resource base for future generations.

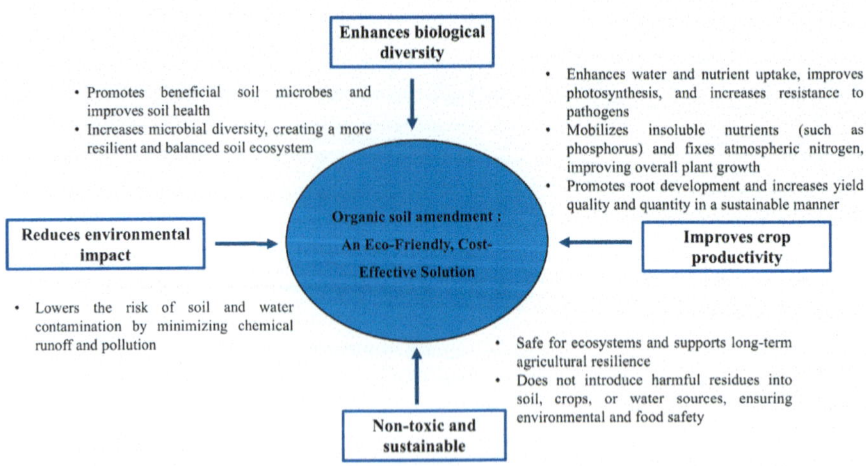

FIGURE 18.2

Soil health comparison of conservation agriculture systems compared to conventional tillage.

18.5 Future considerations

Recent advances in nutrient management research for drought-prone regions have revealed complex interactions between water availability, soil processes, and plant nutrient acquisition. However, numerous critical questions remain unresolved that require further investigation to elucidate the precise mechanisms governing nutrient cycling and fertilizer use efficiency under water-limited conditions. The principal research gaps that demand more comprehensive investigation include:

- Role of emerging technologies and unexplored nutrient formulations, including artificial intelligence–driven decision support systems, nanofertilizer delivery platforms, and polymer-entrapped microbial inoculants, with mechanisms of action under field conditions still poorly characterized.
- Impact of multi-scale interactions between water–nutrient dynamics, soil microbial community structure, and soil physicochemical properties, particularly how drought-induced shifts in microbial functional guilds influence nitrogen mineralization, phosphorus solubilization, and micronutrient chelation across different soil types.
- Effects of extreme climate change on integrated nutrient management strategies, including the influence of temperature variability, altered precipitation intensity, and prolonged dry spells on nutrient release from organic amendments and slow-release fertilizers.
- Socioeconomic and behavioral determinants governing farmer adoption of conservation agriculture, precision nutrient management, and organic amendment strategies, particularly in terms of economic viability, knowledge transfer, and long-term investment decisions in resource-constrained dryland communities.
- Genetic and molecular basis of nutrient use efficiency under drought stress, focusing on crop varieties with enhanced root architecture, mycorrhizal responsiveness, and stress signaling pathways that optimize nutrient acquisition while maintaining yield stability.

Advancing climate-resilient agriculture in drought-prone regions ultimately depends on translating these insights into accessible, economically viable, and environmentally sustainable nutrient management systems that enhance food security and preserve natural resources for future generations.

References

Adeleke, K.A., Atoloye, I.A., Creech, J.E., Dai, X., Reeve, J.R., 2021. Nutritive and non-nutritive effects of compost on organic dryland wheat in Utah. Agronomy Journal 113 (4), 3518–3531. Available at https://doi.org/10.1002/agj2.20698.

Adhikari, R., 2024. Microbial phosphorus solubilization under drought conditions. Journal of Soil and Future Research 5, 45–47.

Ahmad, S., Abbas, G., Tariq, M., Fatima, Z., Wahab, A.A., Ahmed, M., et al., 2025. Nitrogen nutrition for cotton in a semi-arid environment. Journal of Agricultural Science 163 (1), 27–41. Available at https://doi.org/10.1017/S0021859624000613.

Amirahmadi, E., Ghorbani, M., Moudrý, J., Bernas, J., Mukosha, C.E., Hoang, T.N., 2024. Environmental assessment of dryland and irrigated winter wheat cultivation under compost fertilization strategies. Plants 13 (4). Available at https://doi.org/10.3390/plants13040509.

Astaneh, N., Bazrafshan, F., Zare, M., Amiri, B., Bahrani, A., 2021. Nano-fertilizer prevents environmental pollution and improves physiological traits of wheat grown under drought stress conditions. Scientia Agropecuaria 12 (1), 41–47. Available at https://doi.org/10.17268/sci.agropecu.2021.005.

Ayankojo, I.T., Morgan, K.T., 2021. Optimizing tomato growth and productivity using nitrogen and irrigation application timing. Agronomy 11 (10), 1968. Available at https://doi.org/10.3390/agronomy11101968.

Bekchanova, M., Campion, L., Bruns, S., Kuppens, T., Lehmann, J., Jozefczak, M., et al., 2024. Biochar improves the nutrient cycle in sandy-textured soils and increases crop yield: a systematic review. Environmental Evidence 13 (1). Available at https://doi.org/10.1186/s13750-024-00326-5.

Bogati, K., Walczak, M., 2022. The impact of drought stress on soil microbial community, enzyme activities and plants. Agronomy 12 (1), 189. Available at https://doi.org/10.3390/agronomy12010189.

Carneiro, M.P., Gabetto, F.P., Costa, F.M., Meneghini, V., de Sousa, J.E.S., Pereira, G.A.G., et al., 2025. Biochar-mediated improvements in soil health in drylands: a bibliometric review of the potential and mechanisms for combating desertification and climate change. Journal of Arid Environments 232, 105490. Available at https://doi.org/10.1016/j.jaridenv.2025.105490.

Chandrasekaran, M., 2022. Arbuscular mycorrhizal fungi mediated enhanced biomass, root morphological traits and nutrient uptake under drought stress: a meta-analysis. Journal of Fungi 8 (7). Available at https://doi.org/10.3390/jof8070660.

Chen, G., Wang, Q., Li, H., He, J., Lu, C., Gong, S., et al., 2022. Research on the effect of conservation tillage mode on the suppression of near-surface dust in farmland. Agriculture 12 (5), 703. Available at https://doi.org/10.3390/agriculture12050703.

Chen, Q., Chang, L., Han, F., Khan, K.S., Chai, Y., Chai, S., et al., 2025. Straw strip mulching boosts potato yields by enhancing soil moisture and water use efficiency. Frontiers in Plant Science 16. Available at https://doi.org/10.3389/fpls.2025.1659236.

Cheng, B., Dai, H.Y., Liu, T.J., Liu, H., Li, J.R., Qiao, R.J., et al., 2024. Mineralization characteristics of soil organic carbon under different herbaceous plant mosaics in semi-arid grasslands. Scientific Reports 14 (1). Available at https://doi.org/10.1038/s41598-024-73018-2.

Choudhary, M., Meena, V.S., Panday, S.C., Mondal, T., Yadav, R.P., Mishra, P.K., et al., 2021. Long-term effects of organic manure and inorganic fertilization on biological soil quality indicators of soybean-wheat rotation in the Indian mid-Himalaya. Applied Soil Ecology 157. Available at https://doi.org/10.1016/j.apsoil.2020.103754.

Choudhary, S., Rajpoot, S.K., Tripathi, A., Choudhary, M., Radha, L., Sen, M., 2025. Diversified cropping systems for improving the crop productivity and soil health of dryland ecosystem. Land Degradation and Development. Available at https://doi.org/10.1002/ldr.70140.

Crews, T.E., Kemp, L., Bowden, J.H., Murrell, E.G., 2022. How the nitrogen economy of a perennial cereal-legume intercrop affects productivity: can synchrony be achieved? Frontiers in Sustainable Food Systems 6. Available at https://doi.org/10.3389/fsufs.2022.755548.

Dargie, S., Girma, T., Chibsa, T., Kassa, S., Boke, S., Abera, A., et al., 2022. Balanced fertilization increases wheat yield response on different soils and agroecological zones in Ethiopia. Experimental Agriculture 58. Available at https://doi.org/10.1017/s0014479722000151.

De Silva, S., Kariyawasam Hetti Gamage, L., Thapa, V.R., 2025. Impact of drought on soil microbial communities. Microorganisms 13 (7), 1625. Available at https://doi.org/10.3390/microorganisms13071625.

Deng, Z., Huang, M., Zhang, W., Wang, G., Huang, X., Liang, G., et al., 2023. Effects of five years conservation tillage for hedging against drought, stabilizing maize yield, and improving soil environment in the drylands of northern China. PLoS One 18 (3), e0282359. Available at https://doi.org/10.1371/journal.pone.0282359.

Dinesh, G.K., Sharma, D.K., Jat, S.L., Venkatramanan, V., Boomiraj, K., Kadam, P., et al., 2024. Residue retention and precision nitrogen management effects on soil physicochemical properties and productivity of maize-wheat-mungbean system in Indo-Gangetic Plains. Frontiers in Sustainable Food Systems 8. Available at https://doi.org/10.3389/fsufs.2024.1259607.

Donovan, T.C., Comas, L.H., Schneekloth, J., Schipanski, M., 2025. Nitrogen and water availability affect soil nitrogen mineralization and maize nitrogen uptake dynamics. Nutrient Cycling in Agroecosystems 130 (3), 387–405. Available at https://doi.org/10.1007/s10705-025-10406-8.

Ghafoor, I., Rahman, M.H.U., Hasnain, M.U., Ikram, R.M., Khan, M.A., Iqbal, R., et al., 2022. Effect of slow-release nitrogenous fertilizers on dry matter accumulation, grain nutritional quality, water productivity and wheat yield under an arid environment. Scientific Reports 12 (1). Available at https://doi.org/10.1038/s41598-022-18867-5.

Guo, X., Wang, P., Wang, X., Li, Y., Ji, B., 2022. Specific plant mycorrhizal responses are linked to mycorrhizal fungal species interactions. Frontiers in Plant Science 13. Available at https://doi.org/10.3389/fpls.2022.930069.

Gupta, N., Singh, Y., Jat, H.S., Singh, L.K., Choudhary, K.M., Sidhu, H.S., et al., 2023. Precise irrigation water and nitrogen management improve water and nitrogen use efficiencies under conservation agriculture in the maize wheat systems. Scientific Reports 13 (1). Available at https://doi.org/10.1038/s41598-023-38953-6.

Gupta, R.K., Sraw, P.K., Kang, J.S., Kaur, J., Sharma, V., Pathania, N., et al., 2024. Interactive effects of long-term management of crop residue and phosphorus fertilization on wheat productivity and soil health in the rice–wheat. Scientific Reports 14 (1). Available at https://doi.org/10.1038/s41598-024-51399-8.

Hu, C., Xia, X., Han, X., Chen, Y., Qiao, Y., Liu, D., et al., 2018. Soil organic carbon sequestration as influenced by long-term manuring and fertilization in the rice-wheat cropping system. Carbon Management 9 (6), 619–629. Available at https://doi.org/10.1080/17583004.2018.1526625.

Idham, I., Pagiu, S., Lasmini, S.A., Nasir, B.H., 2021. Effect of doses of green manure from different sources on growth and yield of maize in Dryland. International Journal of Design and Nature and Ecodynamics 16 (1), 61–67. Available at https://doi.org/10.18280/ijdne.160108.

Jiang, M., Dong, C., Bian, W., Zhang, W., Wang, Y., 2024. Effects of different fertilization practices on maize yield, soil nutrients, soil moisture, and water use efficiency in northern China based on a meta-analysis. Scientific Reports 14 (1). Available at https://doi.org/10.1038/s41598-024-57031-z.

Kaur, R., Kaur, S., Dwibedi, V., Kaur, C., Akhtar, N., Alzahrani, A., 2023. Development and characterization of rice bran-gum Arabic based encapsulated biofertilizer for enhanced shelf life and controlled bacterial release. Frontiers in Microbiology 14. Available at https://doi.org/10.3389/fmicb.2023.1267730.

Kugedera, A.T., Kokerai, L.K., 2024. A review on the effects of mineral fertilizer, manure and water management in improving sorghum grain yields in semi-arid areas. Journal of Plant Nutrition 47 (7), 1175–1188. Available at https://doi.org/10.1080/01904167.2023.2297955.

Li, P., Teng, C., Zhang, J., Liu, Y., Wu, X., He, T., 2023. Characterization of drought stress-mitigating Rhizobium from faba bean (Vicia faba L.) in the Chinese Qinghai-Tibet Plateau. Frontiers in Microbiology 14. Available at https://doi.org/10.3389/fmicb.2023.1212996.

Lopez, A., Anzuay, M.S., Loser, U.A., Taurian, T., Furlan, A.L., 2025. The combined effects of drought stress and phosphorus deficiency on peanut (Arachis hypogaea L.) plants are mitigated by drought tolerant phosphate-solubilizing bacteria. Symbiosis 97 (1), 53–70. Available at https://doi.org/10.1007/s13199-025-01080-z.

Lv, L., Gao, Z., Liao, K., Zhu, Q., Zhu, J., 2023. Impact of conservation tillage on the distribution of soil nutrients with depth. Soil and Tillage Research 225, 105527. Available at https://doi.org/10.1016/j.still.2022.105527.

Madejón, P., Fernández-Boy, E., Morales-Salmerón, L., Navarro-Fernández, C.M., Madejón, E., Domínguez, M.T., 2023. Could conservation tillage increase the resistance to drought in Mediterranean faba bean crops? Agriculture, Ecosystems and Environment 349. Available at https://doi.org/10.1016/j.agee.2023.108449.

Mahmood, M., Wang, J., Mehmood, S., Ahmed, W., Ayyoub, A., Seleiman, M.F., et al., 2025. Influence of drought stress on phosphorus dynamics and maize growth in tropical ecosystems. BMC Plant Biology 25 (1). Available at https://doi.org/10.1186/s12870-025-06092-x.

Meena, S.N., Sharma, S.K., Singh, P., Ram, A., Meena, B.P., Jain, D., et al., 2023. Tillage-based nutrient management practices for sustaining productivity and soil health in the soybean-wheat cropping system in Vertisols of the Indian semi-arid tropics. Frontiers in Sustainable Food Systems 7. Available at https://doi.org/10.3389/fsufs.2023.1234344.

Moreira Cevallos, A.A., Cedeño García, G.A., Guillen Mendoza, S.V., Medranda Vera, E.F., Avellán Cedeño, B.A., León Castro, L.X., 2024. Liquid fertilization and biochar to boost the yield and profitability of *Arachis hypogaea* L. In dryland agroecosystems. Chilean journal of agricultural & animal sciences 40 (2), 357–370. Available at https://doi.org/10.29393/chjaas41-32lisk60032.

Moreno-Jiménez, E., Plaza, C., Saiz, H., Manzano, R., Flagmeier, M., Maestre, F.T., 2019. Aridity and reduced soil micronutrient availability in global drylands. Nature Sustainability 2 (5), 371–377. Available at https://doi.org/10.1038/s41893-019-0262-x.

Mudare, S., Jing, J., Makowski, D., He, X., Liang, Z., Sims, Z., et al., 2025. Crop rotations synergize yield, nutrition, and revenue: a meta-analysis. Nature Communications 16 (1). Available at https://doi.org/10.1038/s41467-025-64567-9.

Nawaz, A., Farooq, M., 2017. Nutrient management in dryland agriculture systems. In: Farooq., M., Siddique, K. (Eds.), Innovations in Dryland Agriculture. Springer

International Publishing, Cham, pp. 115–142. Available at 10.1007/978-3-319-47928-6_5.

Oueriemmi, H., Kidd, P.S., Trasar-Cepeda, C., Rodríguez-Garrido, B., Zoghlami, R.I., Ardhaoui, K., et al., 2021. Evaluation of composted organic wastes and farmyard manure for improving fertility of poor sandy soils in arid regions. Agriculture (Switzerland) 11 (5). Available at https://doi.org/10.3390/agriculture11050415.

Parihar, M., Rakshit, A., Meena, V.S., Gupta, V.K., Rana, K., Choudhary, M., et al., 2020. The potential of arbuscular mycorrhizal fungi in C cycling: a review. Archives of Microbiology 202 (7), 1581–1596. Available at https://doi.org/10.1007/s00203-020-01915-x.

Perera, M.D.A.M., Wijesundara, W.R.A.T.P., Dissanayaka, D.M.S.B., De Silva, S.H.N.P., 2025. Enhancing resilience and productivity of drought-prone cropping systems through conservation agricultural practices. Plant and Soil 518, 891–916. Available at https://doi.org/10.1007/s11104-025-08038-6.

Qi, S., Degen, A., Wang, W., Huang, M., Li, D., Luo, B., et al., 2024. Systemic review for the use of biochar to mitigate soil degradation. GCB Bioenergy 16 (6). Available at https://doi.org/10.1111/gcbb.13147.

Rafique, M.I., Al-Wabel, M.I., Al-Farraj, A.S.F., Ahmad, M., Aouak, T., Al-Swadi, H.A., et al., 2025. Incorporation of biochar and semi-interpenetrating biopolymer to synthesize new slow release fertilizers and their impact on soil moisture and nutrients availability. Scientific Reports 15 (1). Available at https://doi.org/10.1038/s41598-025-90367-8.

Ramos, M.C., 2024. Assessment of nitrogen and phosphorus losses due to erosion in compost-treated and non-treated vineyard soils: effect of rainfall intensity. Environmental Monitoring and Assessment 196, 1120. Available at https://doi.org/10.1007/s10661-024-13269-8.

Rashmi, C.M., Prakash, S.S., 2023. Effect of nano phosphorus fertilizers on growth and yield of maize (Zea mays L.) in central dry zone of Karnataka. Mysore Journal of Agricultural Sciences 57, 286–293.

Rebelo Romão, I., Rodrigues Dos Santos, A.S., Velasco, L., Martínez-Ferri, E., Vilchez, J.I., Manzanera, M., 2022. Seed-encapsulation of desiccation-tolerant microorganisms for the protection of maize from drought: phenotyping effects of a new dry bioformulation. Plants 11 (8). Available at https://doi.org/10.3390/plants11081024.

Reddy, K.S., Shivay, Y.S., Kumar, D., Parida, B.K., Bora, R., Borate, R.B., et al., 2025. Nano DAP augments productivity, phosphorus use efficiency, and profitability of spring wheat. Scientific Reports 15 (1). Available at https://doi.org/10.1038/s41598-025-92364-3.

Rehman, A., Ullah, A., Nadeem, F., Farooq, M., 2019. Sustainable nutrient management. In: Farooq, M., Pisante, M. (Eds.), Innovations in Sustainable Agriculture. Springer International Publishing, Cham, pp. 167–211. Available at 10.1007/978-3-030-23169-9_7.

Ren, H., Lv, H., Xu, Q., Yao, Z., Yao, P., Zhao, N., et al., 2024. Green manure provides growth benefits for soil mesofauna by promoting soil fertility in agroecosystems. Soil and Tillage Research 238, 106006. Available at https://doi.org/10.1016/j.still.2024.106006.

Rubia, M.I., Larrainzar, E., Arrese-Igor, C., 2025. Drought stress modifies the source-sink dynamics of nitrogen-fixing soybean plants prioritizing roots and nodules. Physiologia Plantarum 177 (3). Available at https://doi.org/10.1111/ppl.70276.

Saadouli, I., Mosbah, A., Ferjani, R., Stathopoulou, P., Galiatsatos, I., Asimakis, E., et al., 2021. The impact of the inoculation of phosphate-solubilizing bacteria pantoea agglomerans on phosphorus availability and bacterial community dynamics of a semi-arid soil. Microorganisms 9 (8), 1661. Available at https://doi.org/10.3390/microorganisms9081661.

Samijan, S., Nurwahyuni, E., Minarsih, S., Supriyo, A., Jauhari, S., Hindarwati, Y., et al., 2024. Optimizing sorghum productivity using balanced fertilizers on dryland. Phyton-International Journal of Experimental Botany 93 (7), 1403–1420. Available at https://doi.org/10.32604/phyton.2024.048339.

Santos, J.M.R.D., Pereira, A.P.D.A., Moniz, T.D.C.D., Henrique, G.D.S., Barbosa, F.L.A., Nascimento, Í.V.D., et al., 2025. Biochar as a strategy to mitigate greenhouse gases in degraded drylands of the Brazilian semiarid region: carbon stocks and CO_2 Fluxes. Journal of Plant Nutrition and Soil Science 188 (6), 1014–1026. Available at https://doi.org/10.1002/jpln.70016.

Santoso, U., Akhmad, G., Emmy, S.M., Raihani, W., 2021. Application of livestock manure and edamame harvest waste to improve the chemical properties of acid dry land. International Journal of Biosciences 19 (4), 41–52. Available at http://dx.doi.org/10.12692/ijb/19.4.41-52.

Selvakumar, S., Thiruppathi, M., Immanuel, R.R., Sivaraman, A., Chakravarthy, G.A., 2022. Impact of foliar fertilization of nutrients and biofertilizer on the growth and yield of hybrid cotton (*Gossypium hirsutum L*). Crop Research 57 (4), 275–280. Available at https://doi.org/10.31830/2454-1761.2022.CR-821.

Shakoor, A., Qamar, M., Abbas, S.H., Vaseer, S.G., Waqar, S., Rashid, K., et al., 2024. Response of foliar applied urea on growth, yield and quality of wheat (*Triticum Aestivum L.*) at different growth stages. Biological and Clinical Sciences Research Journal 2024 (1), 920. Available at https://doi.org/10.54112/bcsrj.v2024i1.920.

Shi, J., Zhou, H., Xu, M., Zhang, Q., Li, J., Wang, J., 2023. Fertilization highly increased the water use efficiency of spring maize in dryland of Northern China: a meta-analysis. Agronomy 13 (5), 1331. Available at https://doi.org/10.3390/agronomy13051331.

Shirinbayan, S., Khosravi, H., Malakouti, M.J., 2019. Alleviation of drought stress in maize (Zea mays) by inoculation with *Azotobacter* strains isolated from semi-arid regions. Applied Soil Ecology 133, 138–145. Available at https://doi.org/10.1016/j.apsoil.2018.09.015.

Sileshi, G.W., 2022. Dose-response models to guide site-specific nutrient management and lessons for fertiliser trial design in sub-Saharan Africa. Experimental Agriculture 58. Available at https://doi.org/10.1017/s0014479721000193.

Song, Y., Ma, L., Duan, Q., Xie, H., Dong, X., Zhang, H., et al., 2024. Development of slow-release fertilizers with function of water retention using eco-friendly starch hydrogels. Molecules (Basel, Switzerland) 29 (20), 4835. Available at https://doi.org/10.3390/molecules29204835.

Sun, R., He, H., Jing, Y., Leng, S., Yang, G., Yihe, L., et al., 2024. Global wind erosion reduction driven by changing climate and land use. Earth's Future 12 (10). Available at https://doi.org/10.1029/2024EF004930.

Sun, W., Shahrajabian, M.H., Wang, N., 2025. A study of the different strains of the genus *Azospirillum* spp. on increasing productivity and stress resilience in plants. Plants 14 (2). Available at https://doi.org/10.3390/plants14020267.

Tarolli, P., Luo, J., Park, E., Barcaccia, G., Masin, R., 2024. Soil salinization in agriculture: mitigation and adaptation strategies combining nature-based solutions and bioengineering. iScience 27 (2), 108830. Available at https://doi.org/10.1016/j.isci.2024.108830.

Wang, F., Wang, Y., Lyu, H., Fan, Z., Hu, F., He, W., et al., 2023. No-tillage mulch with leguminous green manure retention reduces soil evaporation and increases yield and water productivity of maize. Agricultural Water Management 290, 108573. Available at https://doi.org/10.1016/j.agwat.2023.108573.

Wang, J., Sun, L., Sun, Y., Yang, S., Qin, Q., Xue, Y., 2025. Long-term biochar-based fertilizer substitution promotes carbon, nitrogen, and phosphorus acquisition enzymes in dryland soils by affecting soil properties and regulating bacterial community. Applied Soil Ecology 206. Available at https://doi.org/10.1016/j.apsoil.2024.105801.

Xie, L., Du, Y., Tang, L., He, X., Ming, W., Rong, X., et al., 2025. Polymer-coated urea effects on yield and gaseous loss of nitrogen for planting Chinese cabbage. PLoS One 20 (11), e0337171. Available at https://doi.org/10.1371/journal.pone.0337171.

Xu, Q., Fu, H., Zhu, B., Hussain, H.A., Zhang, K., Tian, X., et al., 2021. Potassium improves drought stress tolerance in plants by affecting root morphology, root exudates and microbial diversity. Metabolites 11 (3), 1–17. Available at https://doi.org/10.3390/metabo11030131.

Xue, S., Li, X., Huang, D., Zhang, D., Gao, Y., Wang, X., 2025. Green manure enhances soil organic carbon sequestration while increasing the risk of soil inorganic carbon loss in calcareous soils. Geoderma 461, 117467. Available at https://doi.org/10.1016/j.geoderma.2025.117467.

Yang, X., Leys, J., Zhang, M., Gray, J.M., 2023. Estimating nutrient transport associated with water and wind erosion across New South Wales, Australia. Geoderma 430, 116345. Available at https://doi.org/10.1016/j.geoderma.2023.116345.

Yang, X., Xiong, J., Du, T., Ju, X., Gan, Y., Li, S., et al., 2024. Diversifying crop rotation increases food production, reduces net greenhouse gas emissions and improves soil health. Nature Communications 15 (1). Available at https://doi.org/10.1038/s41467-023-44464-9.

Zhang, H., Peng, J., Zhao, C., 2024a. Wind speed and vegetation coverage in turn dominated wind erosion change with increasing aridity in Africa. Earth's Future 12 (6). Available at https://doi.org/10.1029/2024EF004468.

Zhang, H., Cheng, Y., Zhong, Y., Ni, J., Wei, R., Chen, W., 2024b. Roles of biochar properties in their water-holding capacity and bound water evaporation: quantitative importance and controlling mechanism. Biochar 6, 30.

Zhang, Y., Hou, K., Qian, H., Gao, Y., Fang, Y., Xiao, S., et al., 2022. Characterization of soil salinization and its driving factors in a typical irrigation area of Northwest China. Science of The Total Environment 837, 155808. Available at https://doi.org/10.1016/j.scitotenv.2022.155808.

Zhao, Y., Tian, Y., Li, X., Song, M., Fang, X., Jiang, Y., et al., 2022. Nitrogen fixation and transfer between legumes and cereals under various cropping regimes. Rhizosphere 22, 100546. Available at https://doi.org/10.1016/j.rhisph.2022.100546.

Zhou, J., Deng, Y., Li, J., Xu, Z., Wang, B., Xu, X., et al., 2025. Genotype-specific synergy between arbuscular mycorrhizal fungi and olive cultivars enhances drought resilience in China's olive belt. Agronomy 15 (11), 2568. Available at https://doi.org/10.3390/agronomy15112568.

CHAPTER

Salinity stress and nutrient fertilization

19

Noureddine Chaachouay[1,2], Jamaa Habchaoui[2,3] and Azamal Husen[2,4]

[1]*Agri-Food and Health Laboratory (AFHL), ESEF of Berrechid, Hassan First University, Settat, Morocco,*

[2]*Department of Biotechnology, Graphic Era (Deemed to be University), Dehradun, Uttarakhand, India,*

[3]*Laboratory of Botany, Biotechnology and Plant Protection, Department of Biology, Faculty of Sciences, Ibn Tofail University, Kenitra, Morocco,*

[4]*Wolaita Sodo University, Wolaita Sodo, South Ethiopia Regional State, Ethiopia*

List of abbreviations

HAK	high-affinity K^+ transporter
KT	K^+ transporter
KUP	K^+ uptake transporter
ROS	reactive oxygen species
SOS	salt overly sensitive

19.1 Introduction

Salinization of soil is a very important issue of the world, which threatens food security and the sustainability of agriculture (Kumar and Sharma, 2020). Saline soils have detrimental effects on crop yields, ecosystem productivity on over 800 million hectares of land worldwide, mainly osmotic stress and interference with nutrient homeostasis as well as toxic ions and affect the growth of plants (El Sabagh et al., 2020). The ensuing crisis infiltrates the lives of farmers, food-production chains, and socioeconomic stability, especially in arid and semi-arid lands where irrigation activities and poor drainage networks are routine in adding salinity (Zahran, 2010). These effects indicate that urgent sustainable control measures are essential to alleviate soil salinization and hence preserve agricultural yield.

Depending on its origin and spatial distribution, salinization is divided into two categories, primary and secondary (Zhang et al., 2022). The main salinization occurs naturally through the following processes: parent rock weathering, intrusion

Mineral Nutrients to Mitigate Abiotic Stress in Crops. DOI: https://doi.org/10.1016/B978-0-443-44712-9.00019-0

of sea water, salt deposition in the arid climates due to high levels of evaporation (Okur and Örçen, 2020; Salama et al., 1999). Secondary salinization, in its turn, is mainly anthropogenic and is caused by irrigation with saline or not-most-optimal quality of water, insufficient drainage, and alterations of the territory (Zhang et al., 2023). Human activities increase soil salinity and accelerate erosion of arable lands at critical rates. At this pace, arable land is projected to decline by 50% by 2050. (Hossain et al., 2020; Kebede, 2023).

One of the counter paradoxes of soil salinization is the presence of deficiencies in the form of nutrients in an environment of sediment abundance of ions (Chhabra, 2021). Despite the deposition of salts in the soil, salinity often triggers the occurrence of imbalance and lack of vital minerals like potassium, calcium, and nitrate (Yadav et al., 2012). High dosages of sodium and chloride ions disrupt the uptake of nutrients, osmotic balance, and disturb physiological processes of the plants (Murtaza et al., 2023). Plants are therefore faced with a multifaceted nutritional problem in which salt toxicity and lack of minerals coexist to negatively affect both plant growth and yield in spite of the occurrence of high soluble ions (Kabir and Díaz-Pérez, 2025).

Discussing the scope and consequences of soil salinization, along with its causes and the nutrient paradox behind it, is critical to creating the most effective mitigation and adaptation strategies. Salt tolerance and nutrient acquisition can be improved by incorporating soil and water management, incorporating improved fertilizer management, and exploiting the association between plants and soil and water microbes (Chaachouay, 2026; Bargaz et al., 2018). The mentioned strategies are fundamental in maintaining agricultural output and sustaining the health of ecosystems in the face of a growing salinity pressure on a global scale. In this respect, the aim of the chapter is to outline the adverse effect of salinity on nutrient levels and their accessibility by plants, and to estimate the effectiveness of strategic fertilization methods as a feasible way to reduce salt stress and strengthen the tolerance.

19.2 The physiological and biochemical basis of salinity stress

Salinity stress on plants occurs in two phases, which are described by the two-phase model that has long been established (Wang et al., 2021). The initial stage is the osmotic stress, which takes place within seconds, and when subjected to high levels of salt in the soil solution, the water potential reduces abruptly. It restricts the capacity of the plant to take in water, leading to the physiological drought of the plant even though water is present and the hindrances in cell expansion and reduction in growth in the young tissues (Ilyas et al., 2021). Plants respond to this osmotic stress by early signaling, which leads to stomatal closing and osmotic readjustment to save water, which is an immediate defense mechanism against this hyperosmotic environment (Yu et al., 2024).

Ion-specific toxicity is the second stage, which occurs in the long-term because of sodium (Na^+) and chloride (Cl^-) ions building up in tissues of the plants. Abundant Na^+ and Cl^- interfere with cell ionic balance by substituting important nutrients like potassium and calcium and rupturing cell structures, such as membranes and enzymes (Yadav et al., 2024). This ionic disequilibrium affects metabolism, resulting in chlorosis, necrosis, and lowered photosynthetic capability. The extent of ion toxicity is determined by the ability of the plants to have ion compartmentalization and exclusion processes that tend to reduce the salt levels in the cytoplasm (Imtiaz et al., 2023).

Osmotic stress and ionic stress act in collaboration and result in an oxidative stress through excessive generation of reactive oxygen species (ROS) including hydrogen peroxide and superoxide radicals. The buildup of ROS leads to assault on lipids, proteins, and nucleic acids, which further amplifies the impairment of cell functioning (Jomova et al., 2023). Plants neutralize oxidative stress with the help of antioxidant defense mechanisms that employ enzymes such as superoxide dismutase and catalase that trap ROS. A combination of the osmotic stress, ionic stress, and oxidative stress allows physiological functions to be largely compromised, such as photosynthesis, stomatal, and protein synthesis, to restrict growth and productivity with salinity stress (Chaachouay et al., 2024; Choudhary et al., 2023). Overall, salinity stress affects plants in a two-stage process consisting of a preoccurrence of an osmotic shock that lowers water content and ion toxicity caused by salt expansion, which leads to oxidative injury. It is important to know these coordinated physiological and biochemical activities to be able to create ways of increasing the salt tolerance of crops and maintaining agriculture in salty soils (Shabaan et al., 2022).

19.3 Salinity-induced alterations in nutrient dynamics

Salt stress can manipulate the plant nutrient processes in complex processes, but by far, the inhibition of nutrient uptake and translocation in the plant system is the most important phenomenon (Awais et al., 2023). Overload with Na ions and Cl ions interferes with vital elements such as potassium (K^+), calcium (Ca^{2+}), magnesium (Mg^{2+}), nitrate (NO_3^-), phosphate ($H_2PO_4^-$), and sulfate (SO_4^{2-}) and will thus disrupt their absorption and assimilation (Bhatla and Kathpalia, 2023; Chhabra, 2021). Competition at the root surface of ionic antagonism, where salt ions supersede the main carrier proteins and binding sites use the nutrients, thus leading to these deficiencies in the face of the large quantity of ions in the soil solution (Fig. 19.1). An example is the replacement of K^+ by Na^+ in plant cells, which disrupts enzyme reactions and osmotic pressure.

Salt-induced changes in the functions of the roots and the chemistry of the rhizosphere worsen the impairment of nutrient acquisition (Liu et al., 2021). High levels of salinity cause changes in soil pH and changes in the ion exchange capacity, which in turn affect the availability of nutrients and their solubility (Syed et al., 2021).

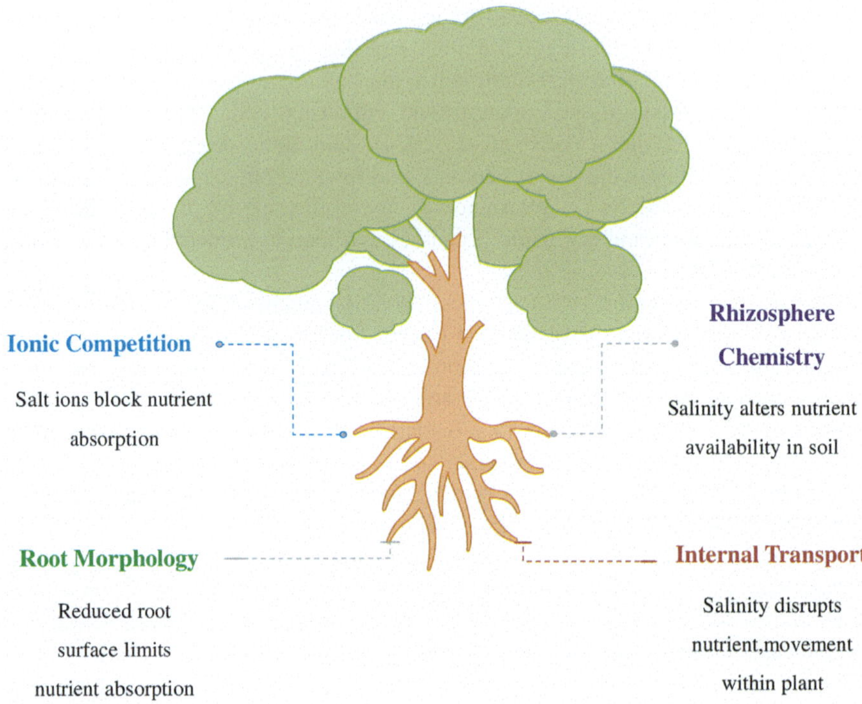

FIGURE. 19.1

Navigating nutrient limitation under saline conditions.

Alteration of the morphology of roots secondary to salt, including the reduction of root surface area and density of root hair, also undermine the ability of the plant to utilize its ability to take up nutrients effectively (Farooq et al., 2024). These physical and chemical changes in the root environment inhibit active and passive methods of transportation that are critical in the process of maintaining homeostasis of nutrition.

In the plant, internal nutrient transport and distribution are disturbed by salinity (Husen, 2024). The imbalances caused by ionic disturbances interfere with nutrient translocation via the xylem and the phloem processes and distort the supply of vital components to growing tissues (Singh, 2024). Na^+ and Cl^- build up, usually to death levels in leaves and reproductive organs. These perturbations reduce physiological functions, such as photosynthesis, protein synthesis, and growth, and therefore enhance the negative impact of salt stress and worsen the solution of plant health and effectiveness (Balasubramaniam et al., 2023).

All of the above are collectively expected to cause salinity-induced changes in nutrient processes (antagonist action at the root level, changes in rhizosphere chemistry, and disruption of internal transport mechanisms). An in-depth study of those mechanisms is essential to the development of the measures that can promote the reduction of nutrient imbalance in plants and enhance their tolerance to stress caused by salinity, which includes the production of genotypes resistant to salinity, the optimization of fertilization, and the improvement of soil management (EL Sabagh et al., 2021).

19.4 The protective role of essential nutrients in salinity tolerance

The essential nutrients play an important role in the development of plant tolerance to salinity stress, and each of them results in different mechanisms that maintain physiological and biochemical balance during stress (Chaachouay and Husen, 2025; Balasubramaniam et al., 2023). K^+ is the major counterion to Na^+ and maintains a K^+/Na^+ homeostasis, without which enzymatic activity and osmotic homeostasis are unable to occur (Raddatz, 2021). K^+ controls the rate of transpiration and photosynthesis when exposed to salt by modulating stomatal dynamics. Higher intracellular K^+/Na^+ ratio enables sodium toxicity tolerance in that it ensures ionic balance and preserves metabolic process (Boussora et al., 2024). The role of calcium ions is to act as an ideal example of secondary messengers in salt-stress signaling pathways, which helps stabilize cellular membranes and walls. Ca^{2+} mediates the process of defensive response development, thus making the plants sense and react to high levels of salinity. At the same time, Ca^{2+} maintains membrane integrity by reducing sodium permeability and strengthens cell-wall integrity, therefore, alleviating the harm caused by salt. This Ca^{2+} is a membrane-bound action of Ca^{2+} that plays a significant role in cellular and whole-plant salt tolerance development (Mansour, 2023).

Nitrogen has a complicated impact on the ability to tolerate salinity and relies on the form of nitrogen through nutrition, such as nitrate (NO_3^-) or ammonium (NH_4^+). Mostly, the supply of nitrates increases the salt tolerance by encouraging the uptake of nitrogen and osmo-protectants production of proline and glycine betaine that stabilize the osmotic pressure (Kumari et al., 2022). These compatible solutes can both protect cell structures against degradation caused by salt and oxidative stress. Phosphorus plays a crucial role in metabolism of energy, signal transduction, and phosphate levels in alkaline-saline soils could be lower than those found in normal soils, but sufficient phosphate to stimulate the generation of ATP and mechanism of signal transduction to cope with stress are required (Phour and Sindhu, 2023).

Micronutrients such as zinc, copper, and manganese are cofactors of antioxidant enzymes that counteract the reactive oxygen species produced during salt stress and, thus, decrease oxidative damage (Kesawat et al., 2023). Silicon is not an essential element, but it promotes salt tolerance through strengthening of cells, constrained sodium absorption, and improved plant water condition (Souri et al., 2021). These positive implications give importance to the need to incorporate all the required and even potentially positive factors into crop management that is intended to address salinity stress. Accordingly, the overall activity of crucial nutrients and other positive factors incorporates an inherent system of biochemical and physiological defense mechanisms that forms the basis of the resilience of plants to the complex challenges of salinity (Singh et al., 2024). An explanation of these functions will be used to develop specific practices in nutrient management that aim to improve crop salt tolerance and yield.

19.5 Strategic fertilization management under saline conditions

The strategic fertilization in saline conditions is crucial to the maximization of nutrient availability and alleviation of salt-related harm, and a range of methods based on different approaches have been developed to meet these problematic conditions (Fig. 19.2). Nutrient renewal to the soil is still the accepted method; nevertheless, foliar renewal is accepted as an effective additional method that can allow the usage of the nutrient by foliage and avoid the salty zone of the roots (Wahid et al., 2025). Foliar fertilization is especially beneficial in the provision of micronutrients, and a rapid remedy of deficiencies at phases of crucial growth when root absorption is blocked by salinity.

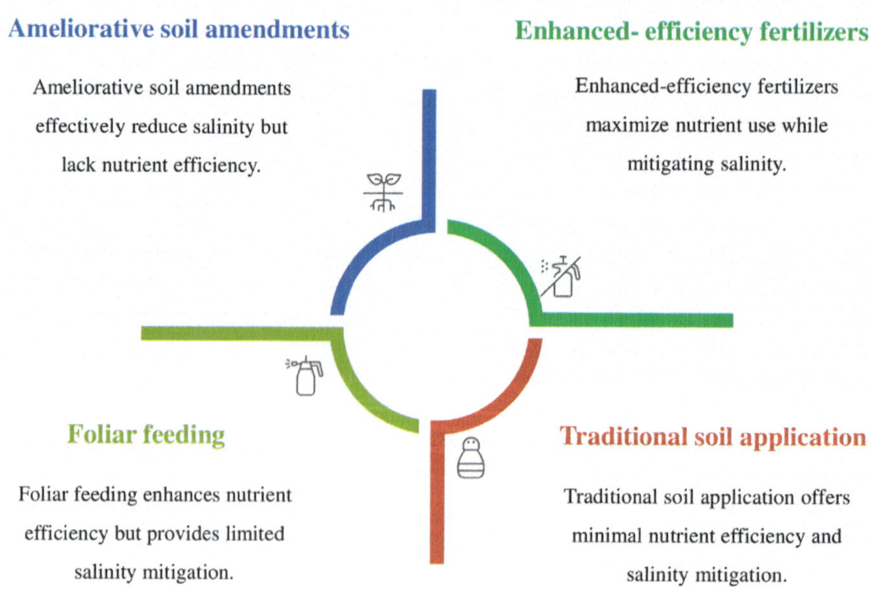

FIGURE 19.2

Strategic fertilization management in saline conditions.

Fertilizer represents another essential component that allows timing and splitting of nutrient levels in the crop to align with nutrient demand to improve nutrient use efficiency and mitigate losses in saline environments (Sood et al., 2025). The fixation of nutrients and erosion in the sodic soils can be reduced by isolating a dose of fertilizer and making them in a series of applications that are less frequent but more frequent applications. Sustainable fertilization practices with saline soils are also based on the high-efficiency fertilizers that are controlled-release and stabilized formulations, which slow nutrient release, reduce volatilization and salt accumulation, further enhancing efficiency.

Gypsum ($CaSO_4$), elemental sulfur, and organic amendments are important for soil amelioration, as they help reduce salinity and improve nutrient dynamics. Gypsum supplies calcium to replace sodium on exchange sites and promote aggregate stability, while sulfur acidifies alkaline soils and increases nutrient solubility. (Abdul Qadir et al., 2022). The addition of organic matter amounts to increase moisture retention of soils, increase microbial activities, and therefore provides greater penetration of roots and better uptake of nutrients. The fact that important crops like tomato, wheat, and rice are responsive to integrated approaches to their fertilization requirements supports this assertion since the overall effect of fertilization approaches is not only to raise crop yield but also to enhance resilience to salinity as crops in salt-contaminated agricultural systems (Brempong and Addo-Danso, 2022). These are the measures that optimize the fertilization process, timing, modern fertilizers, and amendments to the soils so that they form a comprehensive management system of fertilization process designed to the reduction of the effects of salinity and also the maintenance of the production of crops. The application of this requires an all-encompassing appreciation of the nutrient requirements of crops and soils salinity variations, and the environmental factors, hence guiding both applied and scientific efforts to optimize the efficiency of fertilizers and increase food supply in salt-impacted areas.

19.6 Molecular and genetic insights into nutrient–salinity interactions

Salt Overly Sensitive (SOS) pathway and NHX antiporters serve an important role in the salt tolerance of plants by responding to Na + homeostasis (Xie et al., 2022). SOS pathway consists of the plasma membrane Na^+/H^+ antiporter SOS1, serine/threonine protein kinase SOS2, and calcium-sensing protein SOS3 and SCaBP8. SOS pathway includes SOS1, which is a plasma-membrane antiporter of Na^+/H^+, and SOS2, a serine/threonine protein kinase, and SOS3, SCaBP8, which are calcium sensors. Saline stress causes effects of SOS2 activating SOS3 and activation of SOS1, thus causing the extrusion of Na ions in the cytoplasm into the soil, thus alleviating toxic Na ion excesses (Gupta et al., 2021). At the same time, vacuolar NHX antiporters move Na + out of the cytoplasm into vacuoles, trapping it there and helping to maintain cellular ionic balance. It involves a coordinated action of these elements to maintain low cytosolic Na + levels, a requirement for cell viability in saline (Fig. 19.3).

Regarding K^+ intake and regulation, plants use the HAK/KUP/KT family of transporters, which guarantee high-affinity absorption of K^+. Saline stress leads to the upregulation of these transporters to maintain a good K^+/Na^+ ratio, which is considered vital to enzyme activities, osmotic adaptation, and stomatal control. The fact that these transporters are highly specific to K^+ allows them to uptake K^+ effectively, even in cases where the availability of K^+ is limited by competition with Na^+ externally (Wang et al., 2023). By regulating these transporters, plants are able to maintain vital physiological functions even in salty conditions when they are in Na^+ antagonism.

The designing of nutrient-efficient, salt-tolerant/cultivar crops is simultaneously occurring through the means of biotechnology that is under exploitation of the molecular knowledge of nutrients–salinity interactions. Strategies of genetic engineering, such as overexpressing elements of the SOS pathway and NHX antiporters, are used to aid in Na^+ exclusion and sequestration (Khan et al., 2015). The use of alleles that provide elevated nutrient uptake and salt tolerance is

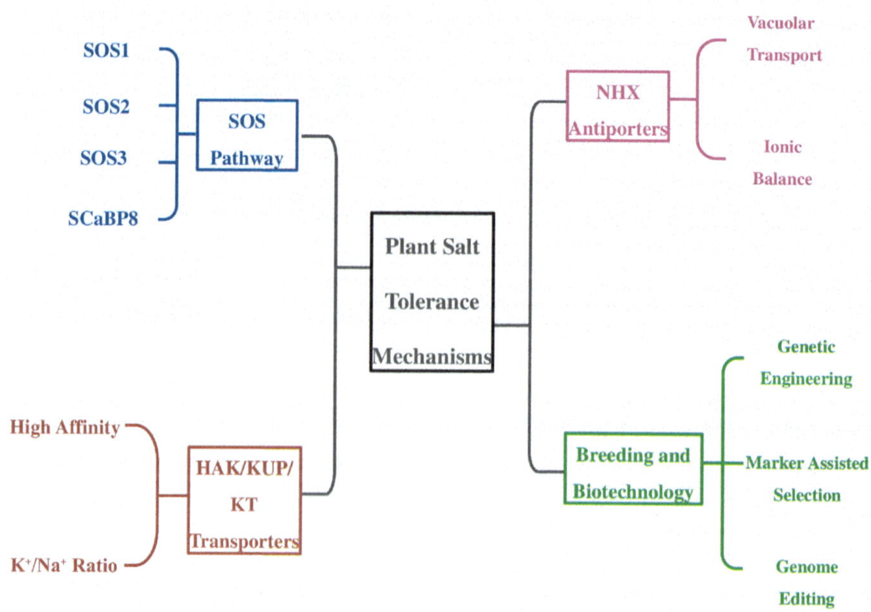

FIGURE 19.3

Salinity-triggered regulation of nutrient-related molecular mechanisms.

aided by the technology of marker-assisted selection and genome editing. Such integrative strategies have the capacity to produce the varieties that remain in homeostasis and produce productive harvests on the salinized soils, thus helping to overcome global food-security issues of increased salinity levels in agricultural land.

19.7 Conclusion and future research directions

It is of great significance to carry out nutrient management in integrated salinity-control systems, for enhancing cross-salinity adaptability, relieving the harmful effect of saline environment, and ensuring agricultural production in saline area. Management of nutrients enhances efficacy of nutrient combination, osmotic regulation, and alleviation of physiological harm caused by salt, and sustains vegetable growth and crop yield in the presence of salinity. Therefore, soil amendments, timing and methods of applying fertilizers, or the introduction of auxiliary microorganisms, including potassium solubilizing bacteria, are promising measures that together increase the health and nutrients of the soil, suitable for the complex problems of salinity. The significant knowledge gaps exist in the development of salt tolerance and nutrient regulation, especially the involvement of root microbiome in nutrient uptake under salt-stress conditions is becoming a key determinant, as salt-tolerant plant-growth-promoting microbes have the capacity to increase the nutrient uptake and loss by regulating stress-hormonal interactions and reinforcing antioxidant defense mechanisms. A more comprehensive repertoire of this microbial interaction can have prospects of on-the-fly salinity alleviation interventions. At the same time, precision farming technologies using soil and plant sensors can be used to optimize the use of fertilizer in saline soils and minimize environmental impact by supporting real-time monitoring and controlling the levels of nutrients on specific sites. Climate change, which also happens alongside salinity stress due to drought and high CO_2 levels, only exacerbates the situation as it increases the effect of the plants to nutrient and water stress. Such interactions should be carefully studied so that to work out adaptive tactics, including the breeding of cultivars with improved nutrient use efficiency, the use of molecular and gene-editing technology, and the combination of environmental sensing data with the microbiome. The multidisciplinary investigations in these areas would generate a more viable and resilient agroecosystem that would be able to handle the increasing menace of salinity and climatic constraints. To conclude, more nutrient management development would focus on large-scale management of soil, vegetation, the biosphere, and technology, and active investigation of root microbiomes, precision agriculture, and climate interactions. The efforts will form a backbone toward the design of robust cropping systems, hence, protecting food security, and environmental sustainability in a setting that is facing increasing pressures and growing demands at a rate that is exponentially expanding, especially in soil salinization and climate change.

References

Abdul Qadir, A., Murtaza, G., Zia-ur-Rehman, M., Waraich, E.A., 2022. Application of gypsum or sulfuric acid improves physiological traits and nutritional status of rice in calcareous saline-sodic soils. Journal of Soil Science and Plant Nutrition 22 (2), 1846–1858. Available at https://doi.org/10.1007/s42729-022-00776-1.

Awais, M., Rafiq, M., Shabbir, R., Aslam, M., 2023. Salinity stress effects on nutrient uptake in plants and its influence on plant growth efficiency. Trends in Animal and Plant Sciences 1, 64–72.

Balasubramaniam, T., Shen, G., Esmaeili, N., Zhang, H., 2023. Plants' response mechanisms to salinity stress. Plants 12 (12), 2253. Available at https://doi.org/10.3390/plants12122253.

Bargaz, A., Lyamlouli, K., Chtouki, M., Zeroual, Y., Dhiba, D., 2018. Soil microbial resources for improving fertilizers efficiency in an integrated plant nutrient management system. Frontiers in Microbiology 9. Available at https://doi.org/10.3389/fmicb.2018.01606.

Bhatla, S.C., Kathpalia, R., 2023. Essential and functional mineral elements. In: Bhatla, S.C., Lal, M.A. (Eds.), Plant Physiology, Development and Metabolism. Springer Nature, Singapore, pp. 25–49. Available at https://doi.org/10.1007/978-981-99-5736-1_2.

Boussora, F., Triki, T., Bennani, L., Bagues, M., Ben Ali, S., Ferchichi, A., et al., 2024. Mineral accumulation, relative water content and gas exchange are the main physiological regulating mechanisms to cope with salt stress in barley. Scientific Reports 14 (1). Available at https://doi.org/10.1038/s41598-024-65967-5.

Brempong, M.B., Addo-Danso, A., 2022. Improving soil fertility with organic fertilizers. In: Turan, M., Yildirim, E. (Eds.), New Generation of Organic Fertilizers. IntechOpen. Available at 10.5772/intechopen.103944.

Chaachouay, N., 2026. Carbon nanotube in plant stress alleviation: Role in photosynthesis, oxidative stress reduction, and nutrient uptake. In: Abdel Latif, A.A.H. (Eds.), Nanotechnology in Plant Sciences, Nanoparticles in Abiotic and Biotic Stress Alleviation: Mechanisms and Sustainable Plant Solutions. Spinger, Cham, pp. 83–99. Available at 10.1007/978-3-032-24219-8_4.

Chaachouay, N., Husen, A., 2025. Osmolytes and their crosstalk mechanism with plant hormones for higher plant productivity and survival under adverse situations. In: Husen, A. (Eds.), Roles of Osmolytes in Changing Environment. Academic Press, London, pp. 13–43. Available at 10.1016/b978-0-443-26549-5.00016-3.

Chaachouay, N., Azeroual, A., Bencherki, B., Douira, A., Zidane, L., 2024. Use of melatonin in plants' growth and productions. In: Husen, A. (Eds.), Biostimulants in Plant Protection and Performance. Elsevier, Amsterdam, pp. 107–115. Available at 10.1016/b978-0-443-15884-1.00004-x.

Chhabra, R., 2021. Salt-affected Soils and Marginal Waters: Global Perspectives and Sustainable Management. Springer, Cham, pp. 1–714. Available at https://doi.org/10.1007/978-3-030-78435-5.

Choudhary, S., Wani, K.I., Naeem, M., Khan, M.M.A., Aftab, T., 2023. Cellular responses, osmotic adjustments, and role of osmolytes in providing salt stress resilience in higher plants: polyamines and nitric oxide crosstalk. Journal of Plant Growth Regulation 42 (2), 539–553. Available at https://doi.org/10.1007/s00344-022-10584-7.

Farooq, M., Rafique, S., Zahra, N., Rehman, A., Siddique, K., 2024. Root system architecture and salt stress responses in cereal crops. Journal of Agronomy and Crop Science 210 (6). Available at https://doi.org/10.1111/jac.12776.

Gupta, A., Shaw, B.P., Sahu, B.B., 2021. Post-translational regulation of the membrane transporters contributing to salt tolerance in plants. Functional Plant Biology 48 (12), 1199–1212. Available at https://doi.org/10.1071/FP21153.

Hossain, A., Krupnik, T.J., Timsina, J., Mahboob, M.G., Chaki, A.K., Farooq, M., et al., 2020. Agricultural land degradation: processes and problems undermining future food security. In: Fahad, S., Hasanuzzaman, M., Alam, M., Ullah, H., Saeed, M., Ali Khan, I., Adnan, M. (Eds.), Environment, Climate, Plant and Vegetation Growth. Springer International Publishing, pp. 17–61. Available at https://doi.org/10.1007/978-3-030-49732-3_2.

Husen, A., 2024. Essential Minerals in Plant-Soil System (Coordination, Signalling, and Interactionunder Adverse Situations). Elsevier Inc, Cambridge, MA.

Ilyas, M., Nisar, M., Khan, N., Hazrat, A., Khan, A.H., Hayat, K., et al., 2021. Drought tolerance strategies in plants: a mechanistic approach. Journal of Plant Growth Regulation 40 (3), 926–944. Available at https://doi.org/10.1007/s00344-020-10174-5.

Imtiaz, H., Mir, A.R., Corpas, F.J., Hayat, S., 2023. Impact of potassium starvation on the uptake, transportation, photosynthesis, and abiotic stress tolerance. Plant Growth Regulation 99 (3), 429–448. Available at https://doi.org/10.1007/s10725-022-00925-7.

Jomova, K., Raptova, R., Alomar, S.Y., Alwasel, S.H., Nepovimova, E., Kuca, K., et al., 2023. Reactive oxygen species, toxicity, oxidative stress, and antioxidants: chronic diseases and aging. Archives of Toxicology 97 (10), 2499–2574. Available at https://doi.org/10.1007/s00204-023-03562-9.

Kabir, M.Y., Díaz-Pérez, J.C., 2025. Calcium route in the plant and blossom-end rot incidence. Horticulturae 11 (7). Available at https://doi.org/10.3390/horticulturae11070807.

Kebede, F., 2023. Status, drivers, and suggested management scenarios of salt-affected soils in Africa. In: Choukr-Allah, R., Ragab, R. (Eds.), Biosaline Agriculture as a Climate Change Adaptation for Food Security. Springer, Cham, pp. 259–284. Available at https://doi.org/10.1007/978-3-031-24279-3_13.

Kesawat, M.S., Satheesh, N., Kherawat, B.S., Kumar, A., Kim, H.U., Chung, S.M., et al., 2023. Regulation of reactive oxygen species during salt stress in plants and their crosstalk with other signaling molecules—current perspectives and future directions. Plants 12 (4). Available at https://doi.org/10.3390/plants12040864.

Khan, M.S., Ahmad, D., Khan, M.A., 2015. Trends in genetic engineering of plants with (Na+/H+) antiporters for salt stress tolerance. Biotechnology and Biotechnological Equipment 29 (5), 815–825. Available at https://doi.org/10.1080/13102818.2015.1060868.

Kumari, R., Bhatnagar, S., Kalra, C., 2022. Nitrogen assimilation in plants. Advances in Plant Nitrogen Metabolism. CRC Press, pp. 38–54.

Kumar, P., Sharma, P.K., 2020. Soil salinity and food security in India. Frontiers in Sustainable Food Systems 4. Available at https://doi.org/10.3389/fsufs.2020.533781.

Liu, H.-Q., Lu, X.-B., Li, Z.-H., Tian, C.-Y., Song, J., 2021. The role of root-associated microbes in growth stimulation of plants under saline conditions. Land Degradation & Development 32 (13), 3471–3486. Available at https://doi.org/10.1002/ldr.3955.

Mansour, M.M.F., 2023. Role of vacuolar membrane transport systems in plant salinity tolerance. Journal of Plant Growth Regulation 42, 1364–1401. Available at https://doi.org/10.1007/s00344-022-10655-9.

Murtaza, G., Qadir, A.A., Farooqi, Z.U.R., Murtaza, B., 2023. Nutrient depletion in salt-affected soils and yield reduction. Soil Constraints and Productivity. CRC Press, pp. 239–254.

Okur, B., Örçen, N., 2020. Soil salinization and climate change. Climate Change and Soil Interactions. Elsevier, pp. 331–350.

Phour, M., Sindhu, S.S., 2023. Soil salinity and climate change: microbiome-based strategies for mitigation of salt stress to sustainable agriculture. In: Parray, J.A. (Ed.), Climate Change and Microbiome Dynamics. Springer International Publishing, pp. 191–243. Available at https://doi.org/10.1007/978-3-031-21079-2_13.

Raddatz, N., Morales de los Ríos, L., 2021. Nutrient Use-Efficiency in Plants: An Integrative Approach.

El Sabagh, A., Hossain, A., Barutçular, C., Iqbal, M.A., Islam, M.S., Fahad, S., et al., 2020. Consequences of salinity stress on the quality of crops and its mitigation strategies for sustainable crop production: an outlook of arid and semi-arid regions. In: Fahad, S. (Ed.), Environment, Climate, Plant and Vegetation Growth. Springer, Cham, pp. 503–533. Available at https://doi.org/10.1007/978-3-030-49732-3_20.

EL Sabagh, A., Islam, M.S., Skalicky, M., Ali Raza, M., Singh, K., Anwar Hossain, M., et al., 2021. Salinity stress in wheat (*Triticum aestivum* L.) in the changing climate: adaptation and management strategies. Frontiers in Agronomy 3. Available at https://doi.org/10.3389/fagro.2021.661932.

Salama, R.B., Otto, C.J., Fitzpatrick, R.W., 1999. Contributions of groundwater conditions to soil and water salinization. Hydrogeology Journal 7 (1), 46–64. Available at https://doi.org/10.1007/s100400050179.

Shabaan, M., Asghar, H.N., Zahir, Z.A., Zhang, X., Sardar, M.F., Li, H., 2022. Salt-tolerant PGPR confer salt tolerance to maize through enhanced soil biological health, enzymatic activities, nutrient uptake and antioxidant defense. Frontiers in Microbiology 13. Available at https://doi.org/10.3389/fmicb.2022.901865.

Singh, A., Rajput, V.D., Agrawal, S., Ghazaryan, K., Minkina, T., Al Tawaha, A.R.M., et al., 2024. Nanoparticles mediated salt stress resilience: a holistic exploration of physiological, biochemical, and nano-omics approaches. Reviews of Environmental Contamination and Toxicology 262 (1). Available at https://doi.org/10.1007/s44169-024-00070-4.

Singh, A.K., 2024. Impacts of Minerals on the Plant's Growth and Metabolism. Addition Publishing House.

Sood, T., Kapoor, S., Kaur, J., Hussain, N., Sood, S., 2025. Fertigation: a paradigm shift in nutrient delivery for sustainable agriculture. In: Hussain, N., Hung, C.-Y., Wang, L. (Eds.), Agricultural Nutrient Pollution and Climate Change. Springer Nature, Switzerland, pp. 135–164. Available at https://doi.org/10.1007/978-3-031-80912-5_5.

Souri, Z., Khanna, K., Karimi, N., Ahmad, P., 2021. Silicon and plants: current knowledge and future prospects. Journal of Plant Growth Regulation 40 (3), 906–925. Available at https://doi.org/10.1007/s00344-020-10172-7.

Syed, A., Sarwar, G., Shah, S.H., Muhammad, S., 2021. Soil salinity research in 21st century in pakistan: its impact on availability of plant nutrients, growth and yield of crops. Communications in Soil Science and Plant Analysis 52 (3), 183–200. Available at https://doi.org/10.1080/00103624.2020.1854294.

Wahid, N., Wahab, S., Iqbal, S., Kanwal, K.M.N., Hayat, K., Karishma, K.A., et al., 2025. Biostimulants as a sustainable strategy to augment salinity tolerance in oilseed crops. In: Abdel Latef, A.A.H. (Ed.), Oilseed Crops Under Abiotic Stress. Springer Nature, Singapore, pp. 1–50. Available at https://doi.org/10.1007/978-981-96-8346-8_1.

Wang, L., Liu, J., Yu, X., Wang, T., Feng, R., 2021. A simplified model for the phase composition curve of saline soils considering the second phase transition. Water Resources Research 57 (1). Available at https://doi.org/10.1029/2020WR028556.

Wang, Y., Zhang, Y., Wei, Y., Meng, J., Zhong, C., Fan, C., 2023. Characterization of HAK protein family in *Casuarina equisetifolia* and the positive regulatory role of CeqHAK6 and CeqHAK11 genes in response to salt tolerance. Frontiers in Plant Science 13. Available at https://doi.org/10.3389/fpls.2022.1084337.

Xie, Q., Zhou, Y., Jiang, X., 2022. Structure, function, and regulation of the plasma membrane Na+/H+ antiporter salt overly sensitive 1 in plants. Frontiers in Plant Science 13. Available at https://doi.org/10.3389/fpls.2022.866265.

Yadav, R.S., Meena, S.C., Patel, S.I., Patel, K.I., Akhtar, M.S., Yadav, B.K., et al., 2012. Bioavailability of soil P for plant nutrition. In: Lichtfouse, E. (Ed.), Farming for Food and Water Security. Springer Netherlands., pp. 177–200. Available at https://doi.org/10.1007/978-94-007-4500-1_8 Vol. 10.

Yadav, S., Yadav, J., Kumar, S., Singh, P., 2024. Metabolism of macro-elements (calcium, magnesium, sodium, potassium, chloride and phosphorus) and associated disorders. In: Singh, R.L., Singh, P., Pathak, N. (Eds.), Clinical Applications of Biomolecules in Disease Diagnosis. Springer Nature, Singapore, pp. 177–203. Available at https://doi.org/10.1007/978-981-97-4723-8_8.

Yu, B., Chao, D.-Y., Zhao, Y., 2024. How plants sense and respond to osmotic stress. Journal of Integrative Plant Biology 66 (3), 394–423. Available at https://doi.org/10.1111/jipb.13622.

Zahran, M.A., 2010. Afro-Asian mediterranean coastal lands. In: Gilbert, F. (Ed.), Climate—Vegetation. Springer Netherlands., pp. 1–103. Available at https://doi.org/10.1007/978-90-481-8595-5_1 Vol. 4.

Zhang, Y., Hou, K., Qian, H., Gao, Y., Fang, Y., Tang, S., et al., 2023. Natural-human driving factors of groundwater salinization in a long-term irrigation area. Environmental Research 220, 115178. Available at https://doi.org/10.1016/j.envres.2022.115178.

Zhang, Y., Hou, K., Qian, H., Gao, Y., Fang, Y., Xiao, S., et al., 2022. Characterization of soil salinization and its driving factors in a typical irrigation area of Northwest China. Science of The Total Environment 837, 155808. Available at https://doi.org/10.1016/j.scitotenv.2022.155808.

CHAPTER

Mineral nutrients and root architecture

20

Libia Iris Trejo-Téllez[1], Víctor Hugo Carbajal-Vázquez[2] and Fernando Carlos Gómez-Merino[2]

[1]*Department of Soil Science-Plant Nutrition, College of Postgraduates in Agricultural Sciences, Montecillo, Texcoco, State of Mexico, Mexico,*

[2]*Department of Plant Physiology, College of Postgraduates in Agricultural Sciences, Montecillo, Texcoco, State of Mexico, Mexico*

Abbreviations

2,4-D	2,4-dichlorophenoxyacetic acid;
3-MA	3-methyladenine;
^{13}C-NMR	^{13}C nuclear magnetic resonance;
ABA	abscisic acid;
ABCB1	ATP-binding cassette subfamily B member 1;
ACA8	autoinhibited Ca^{2+}-ATPase 8;
ACC	1-aminocyclopropane-1-carboxylate;
ACO	ACC oxidase;
AFB3	auxin F-box 3;
AKT1	Arabidopsis K^+ transporter 1;
AMT1; 3	ammonium transporter 1;3;
ANR1	Arabidopsis nitrate regulated 1;
APS1	ATP sulfurylase 1;
ATPS	ATP sulfurylase;
ATAF1	Arabidopsis transcription activation factor 1;
ATAF2	Arabidopsis transcription activation factor 2;
AUX	auxin;
AUX1	auxin resistant 1;
AVG	aminoethoxyvinylglycine;
BAPTA	1,2 bis(o-aminophenoxy)ethane-N,N,N′,N′-tetraacetic acid;
BES1	BRI1-EMS-SUPPRESSOR 1. Transcription factor acting in brassinosteroid signaling; identified as a suppressor of ethyl methanesulfonate (EMS) induced mutations in BRI1 (BRASSINOSTEROID INSENSITIVE 1);
BR	brassinostroids;
bHLH038 and bHLH39	basic helix–loop–helix transcription factors;
BR6ox1* and *BR6ox2	brassinosteroid-6-oxidase 1 and 2: *caprice*: *cpc*;
CBLs	calcineurin B-like proteins;
CCC	cation-chloride cotransporter;

Mineral Nutrients to Mitigate Abiotic Stress in Crops. DOI: https://doi.org/10.1016/B978-0-443-44712-9.00021-9

CEPR1	CEP receptor 1;
CEPs	C-end-encoded peptides;
CIPKs	CBL-interacting protein kinases;
CKX2 protein	cytokinin oxidase/dehydrogenase 2;
CLE	CLAVATA3/embryo surrounding region-related;
CLV1	CLAVATA 1;
CLV3	CLAVATA 3;
CNGC1	cyclic nucleotide-gated channel 1;
CNGC15s	cyclic nucleotide-gated channel 15s;
CPK	calcium-dependent protein kinases;
CTK	cytokinin;
CTr	copper transporter;
CuONP	copper oxide nanoparticles;
CW	cell wall;
DELLA	Asp–Glu–Leu–Leu–Ala motif;
DMI1	does not make infections 1;
DPI	diphenyleneiodonium;
DTX33	detoxification 33;
DTX35	detoxification 35;
ECA1	endoplasmic reticulum Ca^{2+}-ATPase 1;
EIN3	ethylene insensitive 3;
ER	endoplasmic reticulum;
ERS1	ethylene response sensor 1;
ET	ethylene;
FIT	FER-like iron deficiency-induced transcription factor;
FRO2	ferric reduction oxidase 2;
GA	gibberellin;
GSH	reduced glutathione;
HAK5	high-affinity K^+ transporter 5;
HOLT	homolog of legume too much love;
HKT	high-affinity K^+/Na^+ transporter family;
HRS1	hypersensitivity to low phosphate-elicited primary root shortening 1;
IAA	indole-3-acetic acid;
IBA	indole-3-butyric acid;
IRT1	iron-regulated transporter 1;
JA	jasmonic acid;
LAXs	like AUX1;
LPR1	low phosphate root 1;
LPR2	low phosphate root 2;
LR	lateral roots;
MADS-box	an acronym derived from MCM1, AGAMOUS, DEFICIENS, and SRF;
MGT	magnesium transporter;
MIZ1	MIZU-KUSSEI 1;
Mn-SOD	manganese superoxide dismutase;
MOT1	molybdate transporter 1;

MV	methyl viologen;
NAA	1-naphthaleneacetic acid;
NAC	transcription factor family named after;
NAM	no apical meristem,
ATAF	*Arabidopsis* transcription activation factor
CUC	CUP-shaped cotyledon;
NBIP1	NRT1.1B-interacting protein 1;
NCL1	Na^+/Ca^{2+} ion exchanger 1;
NiONPs	nickel oxide nanoparticles;
NLP3	NIN-like protein 3; named after nodule inception (NIN);
NO	nitric oxide;
NPA	N-1-naphthylphthalamic acid;
NPF6.3/NRT1.1	dual-affinity nitrate transporter and sensor belonging to the NRT1/PTR (peptide transporter) family;
NR	nitrate reductase;
NRAMP	natural resistance-associated macrophage;
NTR	nitrate transporter;
NRT1.1B	rice homolog of the nitrate transporter/sensor NRT1.1;
NRT1.5/NPF7.3	nitrate transporter 1.5/nitrate/peptide family 7.3;
OAS-TL3	O-acetylserine (thiol) lyase 3;
PCaP1	plasma membrane–associated Ca^{2+}-binding protein 1;
PDR2	phosphate deficiency response 2;
Pi	phosphate;
PIN2	PIN-formed 2;
PINs	PIN-formed;
PIP	plasma membrane intrinsic proteins;
PLT1	PLETHORA1;
PLT2	PLETHORA2;
PMS	phenazine methosulfalte;
POTs	potassium transporters;
PR	primary root;
PUB9	PLANT U-BOX 9, an E3 ubiquitin ligase;
raz	requires additional Zn;
RG-II	rhamnogalacturonan II;
RGF/GLV/CLEL	root growth factor/GOLVEN/CLE-like;
RHD3	root hair defective 3;
RNS	reactive nitrogen species;
ROS	reactive oxygen species;
RPL13aC	ribosomal protein large subunit 13a C;
RSA	root system architecture;
RSL4	root hair defective 6-like 4;
rth3	root hairless 3;
SCR	scarecrow;
SDK6/ARK2	S-domain receptor kinase 6 (also known as Arabidopsis receptor kinase 2);
SHR	short root;
SNF	symbiotic nitrogen fixation;

SOS1	salt overly sensitive 1, also known as *NHX7*;
SPX4	SPX domain-containing protein 4; the SPX domain is named after SYG1, PHO81, and XPR1;
SST1	sulfate transporter 1;
STOP1-ALMT1	*STOP1*, sensitive to proton toxicity 1; *ALMT1*, aluminum-activated malate transporter 1;
TAA1	tryptophan aminotransferase of Arabidopsis 1;
TAR	tryptophan aminotransferase related;
TAR2	tryptophan aminotransferase related 2;
TCA	tricarboxylic acids;
TfMn	manganese translocation factor;
TIR1	transport inhibitor response 1;
TML	too much love;
SULTR1; 1 and SULTR1; 2	high-affinity sulfate transporters;
U-box/ARMADILLO repeat (ARM)	combination of a U-box E3 ubiquitin ligase domain and ARMADILLO repeat (ARM) protein–protein interaction domains;
werwolf	*wer*;
WRKY71	WRKY transcription factor 71. Member of the WRKY family of transcription factors, characterized by the conserved WRKY amino acid motif and a zinc-finger DNA-binding domain;
WT	wild type;
***YUCs* (*YUCCA* genes)**	genes encoding flavin monooxygenases involved in auxin biosynthesis; named after the yucca-like phenotype (elongated hypocotyls and narrow leaves) observed in *Arabidopsis* gain-of-function mutants.

20.1 Introduction

Plants are capable of absorbing virtually all elements that occur in nature; however, only 17 of them are essential. Of these, C, H, and O are supplied from air and water and are normally not considered limiting for the life cycle of a plant. The remaining 14 elements are supplied through the soil solution or fertilizers and, according to their concentration in plant tissue, are classified as macronutrients: N, P, K, Ca, Mg, and S (≥ 500 mg kg^{-1} dry matter); or micronutrients: Cl, Fe, Mn, Zn, Cu, B, Co, and Ni (< 500 mg kg^{-1} dry matter) (Kirkby, 2023). Potentially, any macronutrient or micronutrient may affect root architecture, although the metabolic, physiological, and molecular maps of such effects for each element and their interactions with one another are still under construction (Motte et al., 2019).

Roots are fundamental for anchoring terrestrial plants in the soil and for the efficient uptake of water and nutrients. In addition, roots enable plants to establish relationships with soil flora and fauna by taking advantage of symbiotic associations, and they may serve as reservoirs for the storage of organic and mineral reserves (Lombardi et al., 2021). The root system is composed of embryonic roots and postembryonic roots. Root architecture, or the overall spatial arrangement of roots, is an important factor that determines the efficiency with which plants can access the resources provided by the soil or the growth medium (Aibara and Miwa, 2014; Shahzad and Amtmann, 2017).

From a plant biology perspective, the term *root architecture* refers to different aspects of the form of root systems. In order to clarify the term and distinguish it from other related concepts, Lynch (1995) proposed a conceptual distinction among root morphology, topology, distribution, and architecture. Root morphology alludes to the superficial characteristics of an individual root axis as an organ, considering epidermal features such as root cap, root hairs, root diameter, arrangement of emergence of secondary roots, root axis undulations, and zone of cortical senescence. Anatomical characteristics of a root linked to cellular and tissular conformations are considered at a deeper level and are not habitually included as part of root architectural traits. In contrast, root topology describes the hierarchical arrangement of the plant root system, depicting how individual roots branch and connect to one another. Taking mathematical models as a guide, topology does not change when a figure is deformed; in roots, the branching pattern is maintained even if axes bend or rotate, which enables detailed analyses of the topology even after excavation. Beyond root orientation, root distribution defines the spatial arrangement, positional gradient, or grid (structured sampling pattern) used to study density, length, or biomass of roots in the soil, as a function of traits such as position among neighboring plants, distance from the stem, and soil depth, among others. Taking into consideration morphological, topological, geometrical, and distributional traits of roots, the term root system architecture defines the spatial distribution and arrangement of plant roots in the soil, focusing mainly on the entire root system. Accordingly, root branching (i.e., topology, geometry) and root shape (i.e., depth, spread) determine the root architecture of a plant, which is a vital attribute for nutrient and water uptake, anchorage to soil, and tolerance or resistance to environmental stresses. Both the genetic makeup of the plant and its interactions with the environment are determinants of the final phenotype of root architecture.

Because of the strong effect of the interaction between genetics and environment on shaping root architecture, this is an intricate attribute that exhibits considerable divergence both among and within genotypes, with soil attributes such as aeration, impedance, bulk density, drainage, pH, contents of nutrients, water, and contaminants substantially affecting its final configuration.

According to Yen (1987), Fig. 20.1 depicts the five prevailing branching arrangements of plant root system architecture. The captivating variety and ability of plants to change their root phenotype in response to different environmental cues constitute major methodological challenges and, at the same time, represent fascinating aspects of plant functional morphology.

As stated by Yen (1987), the H-type pattern refers to horizontal root systems in which roots extend evenly and expand broadly. In this type, the main root allocation (>80% of biomass) occurs within medium extents of up to 60 cm. R-type corresponds to right or straight root systems characterized by diagonally growing

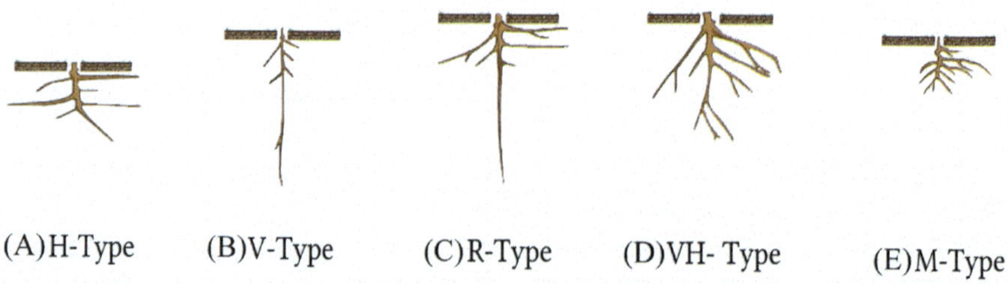

FIGURE 20.1

The five types of root branching defining root system architecture according to Yen (1987) and Wang et al. (2025). (A) H-type (horizontal) root system; (B) V-type (vertical) root system; (C) R-type (right or oblique) root system; (D) VH-type (vertical-horizontal) root system; and (E) M-type (massive) root system. VH types confer stabilization in inclined surfaces, as well as resistance to lodging. H and M types are appropriate for soil strengthening, contributing to avoiding water and soil particles runoff and the risk of erosion. V types are also optimal to confer wind resistance.

roots with widespreadness; occasionally, lateral roots are present. The depths that these root systems reach are vast, with just about 20% of their extension occurring in the top 60 cm of soil. Root systems exhibiting the VH pattern display a rigid taproot and various solid laterals branching copiously at low angles. They reach moderate to profound depths, with ca. 80% of the root material found in the top 60 cm of soil. In vertical root systems, denoted as V-type, there are almost erect roots with a slight extension and a small number of lateral roots that reach moderate to highly keen depths. Finally, the M-type corresponds to enormous root systems in which roots divide and subdivide, growing in multiple orientations. For agronomic practices, it has to be taken into consideration that about 80% of the root develops within the top 30 cm of topsoil (due to the richness in oxygen, water, and nutrients), albeit the whole root system may reach longer depths.

The architecture of the root system is an essential attribute for plant growth and crop yield, especially in fields with extremely low or high stratification of water and nutrients. This architecture comprises a set of root phenotypes that regulate the temporal and spatial allocation of roots within the intricate soil solution and, therefore their capacity to exploit resources. The plasticity and diversity of root architecture, as reflected in diverse root phenotypes, render higher efficiency of roots when plants are established in challenging environments with restricted availability of nutrients and water, among others.

Root phenotypes and phenotypic states that enhance exploration of deeper soil layers improve the acquisition of water and nutrients in mobile ionic forms such as nitrate or sulfate. These ionic species are often available in deeper soil layers during the growing season as a result of leaching and surface desiccation. In contrast, the availability of low-mobility nutrients, such as phosphate and potassium, is generally greater in surface soil layers.

Root system architecture is a determining feature of plant productivity, since it shapes growth, nutrient uptake, and stress responses. Attributes such as number, diameter, branching, and angles of roots represent some of the elements of root architecture. Indeed, root number, as well as the density and length of lateral root branching, directly influence plant performance under conditions of low nitrogen availability (Postma et al., 2014).

Root architecture displays tremendous phenotypic plasticity, since it is influenced by the genetic background of the plant and the environmental conditions where it thrives or survives. Thus, different species or ecotypes may develop diverse forms of root architecture depending on prevailing soil conditions. Even within the same genotype, different spatial arrangements of the root system may occur throughout the life cycle. Fluctuations in oxygen, water, and nutrients, in addition to stress factors of physical (i.e., soil mechanical impedance or compaction), chemical (i.e., contaminants, xenobiotics), or biological (i.e., pathogenic microorganisms, insect pests) nature, all impact the rate at which the root system develops and grows.

Since plants are sessile, they are uncapable of moving or relocate seeking more favorable environments to thrive, grow, and develop. Thence, plants have evolved sophisticated mechanisms of cellular signaling that involve receptors, transduction, and responses that perceive and transduce environmental signals. Significant

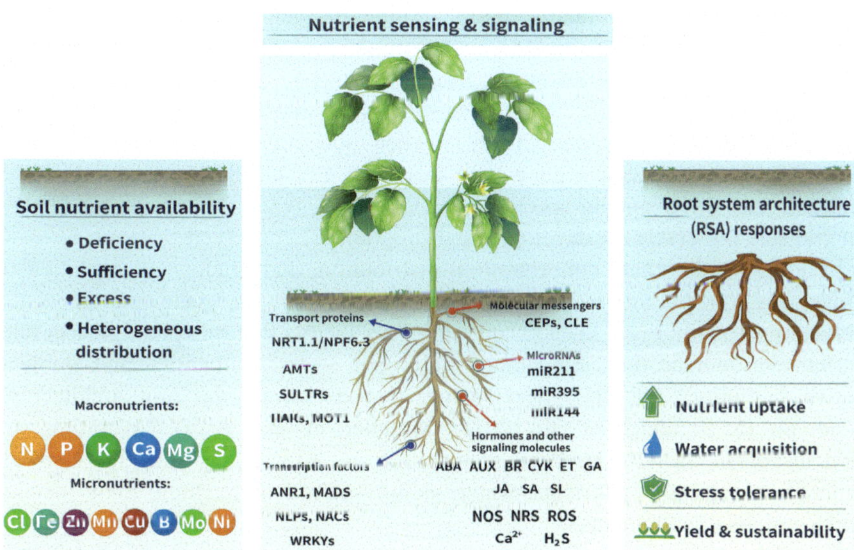

FIGURE 20.2

Conceptual overview of nutrient availability, sensing and signaling pathways, and their integrated effects on root system architecture (RSA) and plant performance to enable sustainable agriculture.

advances have been made in understanding these mechanisms that control root architecture responses to the deficiency or excess of individual nutrients; however, the interactions among signals from different nutrients and the benefits of root architecture responses under specific conditions remain unresolved tasks.

In this chapter, we analyze and discuss some of the most recent and outstanding advances in research on the impact of nutrients on root architecture. In addition, we approach new avenues related to plant mineral nutrients as essential components that shape root architecture for an adaptive sustainable agriculture. Ultimately, as depicted in Fig. 20.2, soil nutrient availability renders nutrient-sensing and signaling processes that in the end determine the dynamics of root system architecture.

20.2 Endogenous and exogenous factors affecting root architecture

The phenotype of a root system results from intricate interactions between the genotype or genetic constitution of the plant and the environment. The plant's genetic makeup encompasses the total of genes that regulate all vital processes related to root development, growth and branching, whereas environmental cues (i.e., availability of water and nutrients as well as abiotic stresses such as water deficit, saline soils, and contaminants) determine the manner how the root system reacts and adjusts its metabolism to cope with these challenges or stimulants, and thus leverage the available resources, thrive, or survive.

Endogenous, genetically determined programs that define root architecture differ among plant species, particularly between monocotyledons and dicotyledons. Therefore a root architecture phenotype shaped as a result of a specific nutritional condition may differ among species, notwithstanding when the signaling pathway that controls the divergent developmental programs is conserved.

In addition, the exact concentration, distribution, and gradient of a particular nutrient; the concentrations of other nutrients involved in synergistic or antagonistic interactions; the life cycle of the plant; and other factors that rule plant growth rate and nutrient demand (i.e., temperature, light, soil water availability, and ambient relative humidity) are all critical factors that determine the root architecture phenotype (Kellermeier et al., 2014; Mounier et al., 2014). According to Młodzińska-Michta (2023), root architecture is affected by factors such as limitation or excess of nutrients, water, salinity, temperature, light, toxic metals, and hazardous contaminants.

Given that root architecture can vary due to different factors, both genetic and environmental, it is evident that agricultural yield or the growth and development of the root system of a given species or cultivar cannot be extrapolated to other genotypes or management conditions. If, on the other hand, studies are used to identify nutrient sensors and signals, the environmental conditions are less critical, and the results can be generalized.

In the following sections, the existing scientific knowledge on the effects of macronutrients and micronutrients on root architecture is described.

20.3 Effect of macronutrients on root architecture

Macronutrients profoundly shape plant root architecture by triggering specific growth changes. For instance, low N often enhances lateral root development to explore the soil, whereas high N availability slows growth; low P promotes deeper roots and increased root hair formation to improve acquisition; and overall nutrient balance dictates root depth versus lateral spread, thereby affecting water and nutrient foraging strategies. Roots adapt by modifying elongation, branching, and root hair density to optimize nutrient uptake.

20.3.1 Nitrogen

Nitrogen (N) is the most abundant element in the Earth's atmosphere, albeit its molecular form is extremely stable and thus inert and unavailable for use by plants, since plants lack the necessary biochemical and molecular machinery to directly use atmospheric N. Consequently, N becomes a *limiting nutrient in crop production.* Once becoming available by biological fixation, abiotic fixation, or industrial synthesis, plants can leverage N primarily as nitrate (NO_3^-), ammonium (NH_4^+), or both.

Nitrogen availability plays a pivotal role in shaping root architecture. It may prompt plants to develop enhanced root hairs, more lateral roots, or more extensive exploration area in order to optimize forage and uptake. N deficiencies render more extensive root systems, with longer and deeper individual roots that can forage a larger soil volume and contribute to N use efficiency. Instead, N sufficiency may render penalties of root growth and prompt a tighter root structure. These responses are mediated through intricate hormonal and molecular signals that involve nitrate transporters (NRT1 and NRT2) and other regulatory molecules (Mohd-Radzman et al., 2013). Furthermore, ammonium may also promote root variations in the architecture and anatomy of roots, associated with modified suberin and lignin levels, which may drastically change hydraulic conductivity and solute permeability in rice (*Oryza sativa*) roots (Ranathunge et al., 2016). Indeed, NO_3^- plays a critical role as part of an elaborated signaling system that involves a wide array of transcription factors that control root system architecture, crucial for plant development and nutrient acquisition (Asim et al., 2020).

In *Arabidopsis thaliana* plants grown under a constant and homogeneous nutrient supply, the dose–response curve for NO_3^- and lateral root (LR) length exhibits a bell-shaped pattern. Instead, primary root (PR) length is regularly hampered as NO_3^- availability is reduced within the same concentration range (Araya et al., 2014, 2016). The effect of nitrate on lateral root initiation can display contrasting responses. Depending on its availability and distribution in the plant growth medium, nitrate may exert both stimulant and inhibitory effects on lateral root development and growth. In *Arabidopsis* plants exposed to a NO_3^--enriched medium, the number of lateral roots within the nitrate-rich zone did not change; however, the average elongation rate of lateral roots was increased, although lateral root development was also inhibited. Thus root branching can be adjusted by antagonizing signals derived from the plant's internal N status and the external supply of this element (Zhang et al., 1999). Purely positive effects have also been reported. Localized and heterogeneous nitrate supply stimulates lateral root elongation in nitrate-rich environments (Mounier et al., 2014).

In addition to its nitrate transport activity, NPF6.3/NRT1.1 (NPF, NITRATE/PEPTIDE FAMILY; NRT, NITRATE TRANSPORTER) acts as a nitrate sensor capable of promoting physiological responses related to root system architecture and expression levels of numerous genes involved in NO_3^- signaling (Noguero and Lacombe, 2016). NRT1.1 alters its affinity at the plasma membrane through reactions that rely on NO_3^- availability and operates as a NO_3^- sensor acting upstream of low-nitrate-induced transcriptional responses (Ho et al., 2009). It also functions as a basipetal AUX transporter in the tips of developing lateral roots, and then modulates local and systemic lateral root responses to low NO_3^- availability (Krouk et al., 2010; Mounier et al., 2014). Under conditions of NO_3^- deficiencies, AUX transport and inhibition of lateral root development are mediated by a process of phosphorylation of NRT1.1 at the T101 residue (Bouguyon et al., 2015). Additionally, NO_3^- may regulate root development by posttranscriptional regulation of the NRT1.1/NPF6.3 transporter/sensor, a transceptor that operates as both a dual-affinity nitrate transporter and a nitrate sensor (Bouguyon et al., 2016). Furthermore, in response to NO_3^-, NRT1.1 fairly regulates AUX biosynthesis and transport to direct root branching (Maghiaoui et al., 2020). Moreover, low-nitrogen-induced lateral root growth in *A. thaliana* depends on the function of the AUX biosynthesis gene *TAR2* (TRYPTOPHAN AMINOTRANSFERASE RELATED 2) (Ma et al., 2014).

Applications of localized NO_3^- activate NRT1.1-mediated signaling that triggers the transcription factor ANR1 (ARABIDOPSIS NITRATE REGULATED 1), a member of the MADS-box family, which induces lateral root development (Remans et al., 2006). In *Arabidopsis* plants subjected to low NO_3^- availability, the ectopic expression of *ANR1* rendered longer lateral roots, without affecting lateral root density or primary root length compared to the wild type (WT) (Gan et al., 2012). In the rice proteome, 30 ANR1-like MADS-type transcription factors have been found (Zhou et al., 2025b); among them, *OsMADS23*, *OsMADS27a*, *OsMADS27b*, and *OsMADS57* are targets of the microRNA *miR144-a*, which is conserved in monocotyledons but missing in *Arabidopsis* (Yan et al., 2014).

In rice, the ectopic expression of *miR144-a* inhibits the transcriptional activation of *MADS*-box transcription factors and hampers lateral root induction when localized NO_3^- is supplied, which indicates that one or more *ANR1*-like genes in rice may display a similar function as that of *ANR1* in *Arabidopsis*. In rice, a homolog of *Arabidopsis NRT1*.1, *NRT1*.1B, has been found to be responsive to N signaling (Hu et al., 2015). Once NRT1.1B binds to NO_3^-, it recruits the E3 ubiquitin ligase NBIP1 (NRT1.1B-INTERACTING PROTEIN 1), prompting the ubiquitination and break down of the protein SPX4 (SPX DOMAIN-CONTAINING PROTEIN 4), which results in the release of the master transcription factor NLP3 (NIN-LIKE PROTEIN 3) from the SPX4-NLP3 complex, thus prompting its translocation to the nucleus (Hu et al., 2019). Likewise, the ZmNRT1.1B-ZmNLP3.1 complex regulates NO_3^- signal perception and transduction in maize roots (Cao et al., 2024).

In the legume *Lotus japonicus*, the microRNA *miR2111* modulates shoot control of root branching. This microRNA systemically directs lateral root induction and the N response through its root target gene *TML* (TOO MUCH LOVE); together, *miR2111* and *TML* control symbiotic infections with N-fixing bacteria. Systemic regulation of lateral root induction by *miR2111* and *TML/HOLT* (*HOLT*, HOMOLOG OF LEGUME *TML*) is conserved in *A. thaliana*, which follows an ecological strategy distinct from that of legumes. Consequently, the miR2111–TML/HOLT regulon arises as a fundamental and conserved factor in the adaptive modulation of root architecture in dicotyledons (Sexauer et al., 2023).

The regulation of root system architecture in response to N supply is also affected by C-terminally encoded peptides (CEPs), key communicators between roots and shoots in plants. In addition, *CEP* genes participate in NO_3^- assimilation under N-limiting conditions (Tabata et al., 2014). CEPs operate as long-distance root to shoot signals generated under the N-starvation condition and result in the induction of *NRT2*.1 (Ohkubo et al., 2017). CEPs interaction with CEPR, also known as XIP1 (XYLEM INTERMIXED WITH PHLOEM1) (Bryan et al., 2012), induces the expression of *CEPD1* and *CEPD2* in shoots, which function as rootward signals required for the induction of *NRT2*.1 (Tabata et al., 2014). This systemic signaling network allows plants to optimize nutrient foraging and maintain overall nitrogen homeostasis. CEP signaling boosted by its interaction with CEP RECEPTOR 1 (CEPR1) modulates systemic N demand signaling, legume nodulation, and root system architecture (Taleski et al., 2024). Moreover, CEPs have been categorized as novel phytohormones that participate in the modulation of plant growth, development, and mechanisms of tolerance or resistance to environmental cues, while functioning as pivotal nodes in nutrient uptake and translocation, root development, and rhizosphere interactions (Aggarwal et al., 2020; Taleski et al., 2024).

Another group of peptides regulating root development in response to N is represented by CEP peptides, CLE peptides (CLAVATA3/EMBRYO SURROUNDING REGION-RELATED) (Okamoto et al., 2015). These peptides are encountered in xylem sap (Okamoto et al., 2003, 2015) and participate in adjusting root development when N availability fluctuates (Araya et al., 2014). Low nitrate levels induce the transcriptional activation of genes encoding CLE peptides, whereas high nitrate levels repress them. The overexpression of *CLE1* to *CLE7* genes in *Arabidopsis* inhibits lateral root development under uniform low-nitrate conditions (Araya et al., 2014).

The molecular interaction between NO_3^- and AUX signaling pathways triggers the initiation of lateral root emergence, and involves NRT1.1-dependent nitrate accumulation, which induces signaling mediated by AUXIN F-BOX 3 (AFB3) (Forde, 2014), as well as the action of a NAC (NAM–ATAF–CUC) domain-containing transcription factor, NAC4, which acts downstream of AFB3 (Vidal et al., 2013, 2014). Indeed, several NAC transcription factors promote root growth in some plant species, including *A. thaliana* (He et al., 2005; Hu et al., 2019; Wu et al., 2019; Xu et al., 2022), soybean (*Glycine max*) (Hao et al., 2011; Quach et al., 2014; Mazarei et al., 2025), rice (*Oryza sativa*) (Redillas et al., 2012; Jeong et al., 2013; Lee et al., 2017; Moreno-Alvarado et al., 2017; Escobar-Sepúlveda et al., 2017), and peanut (*Arachis hypogaea*) (Li et al., 2021), among others.

Exogenous applications of CEP to **barrelclover** (*Medicago truncatula*) and *Arabidopsis* plants promote PO_4^{3-} and SO_4^{2-} uptake, in addition to NO_3^- uptake (Roy et al., 2022). In the wild rice *Oryza longistaminata*, CEPs may participate in the process of ammonium uptake and assimilation when the availability of NH_4^+ is not homogeneous (Kawai et al., 2022). Furthermore, CEP signaling may also induce root-to-shoot NO_3^- translocation (Lin et al., 2008) as demonstrated by CEP hormone inducing *NRT1.5* upregulation (Delay et al., 2019), and *cepr1* mutant plants showing *NRT1.5* downregulation (Tabata et al., 2014; Chapman et al., 2019).

However, the precise signaling pathway that mediates the response to NH_4^+ is still under investigation. It is well established that NH_4^+ prompts changes in cytosolic pH, transcriptional activation of genes, and posttranslational protein modifications, with a consequent apoplastic acidification, coordinated NH_4^+ uptake, improved NH_4^+ assimilation, modified concentrations of reactive oxygen species (ROS) and phytohormones, and a reshaping of root system architecture (Liu and von Wirén, 2017).

Ammonium signaling alters root architecture by impairing primary root elongation while inducing lateral root branching, a process that involves ammonium transporters acting as both nutrient transporters and signaling components. Apoplastic acidification, NH_4^+ concentrations, and phytohormone crosstalk, especially AUX and cytokinin (CTK), represent factors that significantly affect NH_4^+ signaling. Loss of function and knockout mutants with modified transporter function, such as the *amt1;3* mutant of AMT1;3 (AMMONIUM TRANSPORTER 1;3), display decreased NH_4^+-induced lateral root branching, which suggests a pivotal role for specific transporters in this regard (Jia et al., 2020).

Hence, both nitrate and ammonium influence root growth and architecture in plants.

20.3.2 Phosphorus

Phosphorus (P), as phosphate (Pi), is the available form of P that plants take up, translocate, and assimilate. Pi deficiency drastically impairs primary root growth, prompts lateral root growth, and increases the number of root hairs (Bouain et al., 2016). In *Arabidopsis* Col-0, *LPR1* (LOW PHOSPHATE ROOT 1) and *PDR2* (PHOSPHATE DEFICIENCY RESPONSE 2) constitute two central components involved in sensing phosphate deprivation and triggering the repression of root growth (Hoehenwarter et al., 2016). *LPR1* and its homolog *LPR2* encode ferroxidases, whereas *PDR2* encodes a P5-type ATPase that operates in the endoplasmic reticulum (ER) and is necessary for appropriate expression of SCR (SCARECROW), a crucial regulator of root shaping, as well as for stem-cell preservation in Pi-starved roots, thereby uncovering a connection between root shaping and the adaptation of meristem activity to Pi availability (Ticconi et al., 2009).

Phosphate deprivation–induced impairment of primary root growth is substantially decreased in the single mutants *lpr1* and *lpr2*, and the response is absolutely inhibited in *lpr1–lpr2* double mutant plants. Conversely, *pdr2* mutants show a hypersensitive short-root phenotype in Pi-starved plants (Svistoonoff et al., 2007; Müller et al., 2015). In phosphate-deprived WT plants, cell division in the root meristems of primary roots is augmented, while the quiescent center appears disrupted (Cederholm and Benfey, 2015). Apoplastic callose deposition is also observed in phosphate-starved plants, potentially as a result of an enhanced accumulation of ROS in the stem cell niche and the root apical meristem. When comparing the *pdr2* single mutant versus *lpr1–lpr2* double mutant, this phenotype is more evident in the former (Müller et al., 2015). The cell-to-cell movement of SHR (SHORT ROOT), a pivotal transcription factor controlling radial root shaping, is hindered by callose deposition, which demonstrates the importance of maintaining root apical meristem. In the *lpr1–lpr2–pdr2* triple mutant, neither callose deposition nor a hypersensitive short-root phenotype under Pi-limiting conditions is observed, demonstrating that *LPR1* functions downstream of *PDR2* within the same signaling route (Ticconi et al., 2009; Müller et al., 2015). Root meristematic activity repressed under Pi-deficient conditions is anteceded by reduced cell expansion in the root elongation zone, a process that again depends on *LPR1*. A functional STOP1-ALMT1 module (STOP1, SENSITIVE TO PROTON TOXICITY 1; ALMT1, ALUMINUM-ACTIVATED MALATE TRANSPORTER 1) operates upstream of *LPR1* to regulate this early response (Balzergue et al., 2017; Mora-Macías et al., 2017; Sadhukhan et al., 2021).

Another crucial element in the detection and modulation of root architecture under conditions of low Pi availability is represented by signaling peptides (Lay and Takahashi, 2018; Hastwell et al., 2024). *Arabidopsis* mutant plants for *RGF2*, a member of the RGF/GLV/CLEL (ROOT GROWTH FACTOR/GOLVEN/CLE-like) peptide family, exhibit a hypersensitive short-root phenotype under Pi-limiting conditions (Cederholm and Benfey, 2015), demonstrating that *RGF2* preserves longitudinal growth of the primary root under phosphate deprivation. In addition, *RGF2* induces root branching in the epidermis, cortex, and endodermis. Conversely, *RGF1*, another member of the RGF family, downregulates root cell divisions in these tissues (Cederholm and Benfey, 2015). Hence, different RGF peptides display distinct functions in regulating root architecture under Pi deficiency. Interestingly, there is an engaging convergence between the *PDR2–LPR1* pathway and GOLVEN (GLV)/CLE-like peptides in the modulation of root development. Their interaction has been found in terms of nutrient signal sensing and root meristem activity. Genes encoding GLV/RGF peptides are generally well-conserved throughout the land plant kingdom, which highlights their importance in root architecture shape in relation to foraging N and Pi. In barrelclover, the expression of five out of 15 *GLV/RGF* genes was remarkably induced during nodule organogenesis, and their response to AUX and N-depletion was differential. In response to the synthetic GOLVEN10 peptide, roots augmented the number of cells within the root cortical cell layers, though cell length did not change, thus resulting in root elongation (Roy et al., 2024).

The functional interaction between the S-domain receptor kinase 1–6 (SDK6/ARK2) and the E3 ubiquitin ligase AtPUB9 (PLANT U-BOX 9), which contains a U-box/ARMADILLO repeat (ARM), regulates lateral root development in Pi-starved plants (Deb et al., 2014). The *Arabidopsisark2–pub9* double mutant shows a drastic reduction in phosphate starvation–induced stimulation of lateral root growth, which also exhibits reduced AUX accumulation in the primary root as well as in lateral root tips compared with the WT. With the interaction of ARK2, the PUB9 protein localizes to autophagosomes, suggesting a possible function in autophagy during the elicitation of lateral root formation under Pi-starvation. Indeed, when Pi-starved roots were received the application of 3-methyladenine (3-MA), an inhibitor of autophagy, no stimulation of lateral root formation was observed (Deb et al., 2014).

As a consequence of prolonging the elongation period or enhancing the growth rate, Pi-deprived plants develop longer root hairs (Xue et al., 2021). During these events, ethylene (ET) induces a reorganization of root system architecture (Song and Liu, 2015). For instance, Song et al. (2016) reported that the *Arabidopsis* mutant *hps5* exhibits constitutive ET responsiveness and enhanced sensitivity to Pi deprivation owing to a mutation in the ET receptor ERS1 (ETHYLENE RESPONSE SENSOR 1). This molecular alteration provokes an integral ET response as a result of increased accumulation of the EIN3 (ETHYLENE INSENSITIVE 3) protein, a crucial transcription factor that takes part in the ET signaling pathway. Notably, EIN3, a key nuclear transcription factor overexpressed under Pi-limiting conditions, activates direct targets of RSL4 (ROOT HAIR DEFECTIVE 6-LIKE 4), another transcription factor that regulates root hair induction.

In Pi-deprived plants, root architecture is reshaped accordingly, enabling the uptake of Pi to the fullest extent. In this process, various Pi-dependent regulatory mechanisms critical interactions among different phytohormones and Fe have been observed (Crombez et al., 2019). Furthermore, Pi deprivation triggers symbiotic and recycling of internal Pi. Interestingly, in Pi-deprived environments, lateral root division is enhanced by symbiotic relationships with arbuscular mycorrhizal fungi, but the specific molecular machinery behind the regulation of root branching in response to fungi symbiosis needs to be further elucidated to enable the establishment of a deeper insight of this phenomenon and eventually improve crop productivity and tolerance to environmental stressors under Pi-limiting conditions (Chen et al., 2023; Sato and Miura, 2011).

20.3.3 Potassium

Potassium (K) is a vital nutrient for plants; just after N and P, K represents the third major essential element in plants (Soumare et al., 2023). Although in the past K had been eclipsed in root research in part by N and P, the prevalent surge in K-related research synchronizes with an expanding consciousness of its critical roles in plant growth, metabolism, stress tolerance, and efficient nutrient management (Sustr et al., 2019). Indeed, root system architecture as well as root hair production have evolved adaptive adjustments to cope with K-limiting conditions, thus enhancing K forage (Wang and Wu, 2010, 2013).

There exists a plethora of proteins involved in K^+ transport with precise functions in plants. Metabolic and physiological processes related to AUX homeostasis, signal perception and transduction, membrane trafficking, phloem transport, and cell expansion have been associated with K^+ transport, which ranks K^+ as a pivotal regulator for root growth. Potassium displays moderate mobility within the soil solution, with recent research demonstrating both systemic and localized effects on growth and stress responses (Sustr et al., 2019).

In rice genotypes displaying different mechanisms for K use efficiency, Jia et al. (2008) demonstrated divergent responses in root morphology under K-limiting stress. When applying 5 mg K L^{-1} (low-K condition), root growth was hampered in all genotypes, whereas the application of 10 mg L^{-1} (moderate-K condition), root length of the L-efficient genotypes was enhanced. In terms of fine root induction (<0.2 mm in diameter), deficient and moderate K levels triggered enhanced production of this sort of roots in K-efficient rice genotypes as compared to the inefficient ones. Potassium-limiting condition was better tolerated high-efficient genotypes, which reduced root membrane damage and adjustes root architecture to thrive under low K conditions.

Under K deficiency (0.01 mM KCl), *Arabidopsis* seedlings display a pronounced decrease in lateral root growth, and this response was genotype-dependent. Metabolic exchanges between the main root and lateral root elongation revealed the existence of two strategies leading to morphological adjustments to low K. In strategy I, Col-0 plants kept primary root growth but reduced lateral root elongation. In strategy II, Catania-1 plants impaired principal root elongation to promote lateral root division (Kellermeier et al., 2013).

In two maize genetically uniform lines, 90-21-3 and D937, which are tolerant and sensitive to K starvation, respectively, K deficiency inhibited the growth of both lines, with more pronounced effects in shoots than in roots. Total root length and root surface area, particularly those of fine roots, were drastically impaired under K-limiting conditions. Compared with D937, line 90-21-3 showed only slight reductions in root length and surface area under K deficiency, along with a substantial promotion of root activity (Du et al., 2017).

In maize exposed to water-stress, K nutrition (17 or 30 mg kg^{-1} vs. 5 mg kg^{-1}) increased primary root biomass, though the diameter of lateral roots was not affected by K under water limiting conditios (Jordan-Meille et al., 2018). Interestingly, in rice, the potassium transporter OsHAK5 (HIGH-AFFINITY K^+ TRANSPORTER 5) modifies rice architecture via ATP-dependent transmembrane AUX fluxes. Its overexpression increased root hair length, regardless of the K^+ status, thus contributing to the maintenance of root architecture by adjusting cellular chemiosmotic gradients and regulating ATP-dependent polar AUX transport for lateral root and root-hair growth (Yang et al., 2020).

In a high-throughput profiling of root system alterations triggered by K starvation, (Yawson et al., 2020) reported that the progressive effect of K deficiency rendered a substantial reduction (ca. 25.5%) in the main attributed describing the root system. The magnitude of the abatement was ca. 38 % in root biomass and 23 % in root length.

In rice roots, the traits width-to-depth ratio, maximum width, and projected area were significantly decreased in response to K starvation under water deficit. When increasing the dose of K, root top and bottom angles were retained, while root steepness was avoided under mild water deficit. Nonetheless, high K availability did not improve seedling growth (Patel et al., 2021).

The NRT1.5/NPF7.3 (NITRATE TRANSPORTER 1.5/NITRATE/PEPTIDE FAMILY 7.3) transporter is a NO_3^-/H^+ and K^+/H^+ cotransporter and is involved in the translocation of NO_3^- and K^+ from roots to shoots. When K becomes a limiting nutrient, NRT1.5/NPF7.3 inhibits primary root growth by modulating the content of the PIN2 (PIN-FORMED 2) protein and AUX. In the *nrt1.5* and *lks2* mutants, K starvation results in shorter primary roots, broad meristematic regions and elongation zones of primary roots, and increased cellular activity in the meristem region compared with WT plants, thus unveiling the participation of NRT1.5 in low-K–induced impairment of primary root growth. Additionally, external administration of AUX (IAA, indole-3-acetic acid), AUX analogs (NAA, 1-naphthaleneacetic acid; 2,4-D, 2,4-dichlorophenoxyacetic acid), or an AUX precursor (IBA, indole-3-butyric acid) promoted primary root growth in WT plants and in the NRT1.5 COM complementation line. Moreover, NPA (N-1-naphthylphthalamic acid) supply hampered the primary root growth of the *nrt1.5* and *lks2* mutants. As compared to WT plants, *nrt1.5* plants accumulated higher levels of AUX in the root tip, suggesting that NRT1.5 controls root growth suppression by modulating AUX allocation, thus revealing that NRT1.5 hinders primary root growth by adjusting AUX concentrations in the root tip by degrading PIN2 (Wang et al., 2022).

The *Arabidopsis* gene *RPL13aC* (RIBOSOMAL PROTEIN LARGE SUBUNIT 13a C) encodes a specific ribosomal protein that is component of the cytosolic 60 S large ribosomal subunit. The main role of RPL13aC in the ribosome is protein synthesis, though it also contributes significantly to shaping the root system architecture, by regulating root slanting and coiling through its effect on K accumulation in shoots (Ma et al., 2023).

The performance of the cultivars Xushu-32 (low-K-tolerant) and Ningzishu-1 (low-K-sensitive) of sweetpotato (*Ipomoea batatas*) under low-K^+ conditions (0.1 mM) revealed a significant inhibition of root length, surface area, number of forks, and number of crossings. Notably, the cultivar Xushu-32 (low-K-tolerant) developed greater lateral root length, density, and surface area as compared to Ningzishu-1 (low-K-sensitive). This behavior was associated with higher IAA contents in roots and the induction of AUX signaling pathway genes *IbIAA4* and *IbIAA8* (Liu et al., 2023).

In wheat (*Triticum aestivum*), the application of 0.01 mM K^+ (low-K condition) hampered root biomass by 42%-64%, hindering root morphology by decreasing root length, surface area, and tip number as compared to the application of 6 mM K^+ (sufficient-K condition). Main K^+ transporters [HAK, HKT (HIGH-AFFINITY K^+/Na^+ TRANSPORTER family). Significant induction of AKT1 (ARABIDOPSIS K^+ TRANSPORTER 1)] and Ca^{2+} signaling components (CBLs, CALCINEURIN B-LIKE proteins; CIPKs, CBL-INTERACTING PROTEIN KINASES) were observed, indicating central roles of roles in K^+ acquisition, root system shape, and stress adaptation (Liu et al., 2025).

20.3.4 Calcium

Calcium (Ca^{2+}) acts as a universal second messenger, relaying signals to trigger diverse responses including root architecture remodeling to optimize water and nutrient foraging.

The arbuscular mycorrhizal endosymbiosis in barrelclover triggers a flux of nuclear Ca^{2+} signals, essential for the establishment of coordinated functions of two membrane-localized ion channels: N-fixing and P-delivering DMI1 (DOES NOT MAKE INFECTIONS 1) and the complex of cyclic nucleotide-gated channels CNGC15s (CYCLIC NUCLEOTIDE-GATED CHANNEL 15s). This Ca^{2+} signaling cascade boosts primary root growth, meristem development, and AUX homeostasis (Leitão et al., 2019).

Calcium chloride gradients affect the expression of the AUX efflux carrier PIN2, which in turn regulates root architecture and lateral root initiation. In *Arabidopsis*, the abundance of the PIN2 protein has a direct effect on root curvature in response to $CaCl_2$ gradients. At low $CaCl_2$ concentrations, PIN2 subsequent accumulation leads to a straight root architecture. In contrast, increasing $CaCl_2$ concentrations results in PIN2 destabilization and subsequent degradation, which in turn gives rise to a wavy root phenotype (Nisar et al., 2020).

According to Zhang et al. (2020), calcium regulates root morphogenesis mainly by mediating or modulating phytohormone and stress signaling pathways, particularly those involving AUX, CTK, brassinosteroids (BR), abscisic acid (ABA), ET, and ROS, as well as by regulating CW reformation. Furthermore, in sucrose-deprived *Arabidopsis* plants, ACA8 (AUTOINHIBITED Ca^{2+}-ATPase 8) mediates Ca^{2+} signaling and controls primary root development (Zhang et al., 2014a). By regulating the activity of PLETHORA1 (PLT1) and PLETHORA2 (PLT2), two essential proteins determining root development, Ca^{2+} signaling may affect primary root growth. In addition, AtDMI1, a regulator of nuclear Ca^{2+} signatures, down-regulates the expression of PLT1 and PLT2 during primary root growth in *Arabidopsis* (Leitão et al., 2019).

Populations of *Arabidopsis lyrata* exhibit variation in early lateral root emergence in response to Ca and Ni (Veatch-Blohm et al., 2023). Lateral root formation and taproot length were decreased in all populations analyzed in response to Ni compared with Ca, with the smallest contraction observed in populations adapted to serpentine soils characterized by high Ni and low Ca levels. All populations showed similar frequencies of root exploration as affected by Ca.

Calcium signaling regulates root hydrotropism (Liang et al., 2025a), which leads roots to expand toward more saturated soils. Depending on the water potential of the soil, long-distance Ca^{2+} signals are transmitted from the root cap to the elongation zone via the phloem to promote root bending toward areas with higher water availability (Shkolnik et al., 2018). A significant decrease in root hydrotropic response has been observed in mutants of the plasma membrane–associated cation-binding protein-encoding gene, *PCaP1* (PLASMA MEMBRANE–ASSOCIATED Ca^{2+}-BINDING PROTEIN 1) (Tanaka-Takada et al., 2019). The overexpression of *MIZ1* (MIZU-KUSSEI 1), whose encoded protein is a critical positive regulator of the root's ability to grow toward water, enables this phenotype to be restored. Diverse Ca-dependent protein kinases (CPKs), including CPK4, CPK5, CPK6, and CPK11, can phosphorylate the protein MIZ1. This phosphorylation mitigates the repressive effect of MIZ1 on the ER-localized Ca^{2+}-ATPase ECA1 (ENDOPLASMIC RETICULUM Ca^{2+}-ATPase 1), with a consequent change of cytoplasmic Ca^{2+} influx and outflow (Ju et al., 2025). An intracellular phosphoregulatory module composed of CPK and ECA1 integrates cytosolic Ca^{2+} flux with ABA homeostasis to regulate root growth under osmotic stress. Disruption of this module aggravates root growth hypersensitivity to osmotic stress. It was demonstrated that CPKs and ECA1 synergically operate phosphoregulatory hub regulating Ca^{2+} signaling and root growth (Liang et al., 2025b).

20.3.5 Magnesium

Magnesium (Mg) markedly modifies root architecture, since its deficiency triggers robust adaptations to forage nutrients, including an increased root hair growth to stimulate nutrient uptake. In turn, adequate levels of Mg support overall root biomass and function by enabling energy metabolism, carbohydrate transport, and nutrient cycling, influencing root hair density, lateral root development, and the exudation of organic acids to adapt to soil conditions. Modifications of root architecture triggered by Mg unbalance include growing more hairs, changing lateral root length and number, and ET, nitric oxide (NO), and AUX signaling.

In rice cv. Wuyunjing 7, the supply of low Mg (0.01 mM) with high K (6 mM) significantly enhanced root growth (by 29.2%) as compared to the control (1 mM

Mg and 1.0 mM K). The dry biomass of roots improved by 12.1% in comparison to the control, with a consequent rise in the root-to-shoot ratio. Other root traits, including extension, surface area, volume, and diameter, substantially augmented at Mg-limiting conditions, as compared to the control. The evident gain in root to shoot ratio may be robust mechanism to cope with low-Mg stress during the early growth of rice plants (Ding and Xu, 2011).

In *Arabidopsis*, root hair development decreases progressively with increasing Mg supply, a response associated with the emergence of new trichoblast files and with the probability of trichoblasts forming root hairs (Niu et al., 2014b). Tip-focused ROS and cytosolic Ca^{2+} concentrations during root hair elongation are enhanced under low Mg conditions but decrease under high Mg supply. Under low Mg, application of the ROS inhibitor DPI (diphenyleneiodonium) or the Ca^{2+} chelator BAPTA [1,2-bis(o-aminophenoxy)ethane-N,N,N′,N′-tetraacetic acid] blocks the induced development of root hairs. Conversely, under high Mg conditions, treatment with the electron acceptors PMS (phenazine methosulfate) or MV (methyl viologen), as well as with $CaCl_2$, restores root hair development. In addition, Mg supply does not change root hair growth in the *rhd2-1* mutant, which exhibits reduced levels of ROS and cytosolic Ca^{2+}. These outcomes demonstrate a pronounced effect of Mg supply on root hair development in *Arabidopsis*, mediated by Ca^{2+} and ROS signaling pathways that regulate root hair elongation and the expression of root hair morphogenetic genes. Consistently, Mg starvation in *Arabidopsis* does not change the transcriptional activation of most genes putatively regulating Mg transport. Primary root elongation and lateral root formation are not affected by low Mg but are inhibited by high Mg. Furthermore, root hair traits including emergence, density, and length that are regulated by Ca^{2+} and ROS-dependent signaling, are also stimulated by low Mg supply, but reduced by high Mg treatments (Niu et al., 2014a).

According to Liu et al. (2018b), Mg starvation prompts the emergence and growth of root hairs, enabling plant roots to forage more Mg under Mg-deprived conditions. Several chemicals, signaling molecules, including ET, NO, AUX, ROS, and Ca^{2+}, modulate root hair development triggered by Mg starvation.

In roots, Mg deprivation increases the concentrations of AUX, which in turn modulates the emergence and growth of root hairs (Liu et al., 2018a). Additionally, the root hair increase promoted by Mg starvation is potentially modulated by a relatively particular AUX transport pathway. Root hair development prompted by Mg deprivation was impaired in the AUX transport mutants *pin1-1*, *aux1–7* and *pin2*, but not *inpin3–5*, *pin4-3*, and *pin7-2*. Accordingly, Mg limitation also induces the transcriptional activity of *AUX1* (AUXIN RESISTANT 1), *PIN1* and *PIN2* genes, but does not affect that of *PIN3*, *PIN4* and *PIN7* (Liu et al., 2017). In addition, ET and NO production are induced in the roots of Mg-starved plants; these signaling molecules, in turn, modulate root hair morphogenesis (Liu et al., 2017). Thus both AUX and ET/NO are involved in modulating root hair growth under Mg-limiting conditions. In *Arabidopsis*, ET operates upstream of AUX in regulating root cell elongation and lateral root growth. NO functions downstream of AUX to prompt root development, while both ET and NO act upstream of AUX in controlling root hair morphogenesis (Liu et al., 2018b). The enzymatic activities of nitrate reductase and NO synthase-like proteins to stimulate NO production are both enhanced in response to ET. In turn, NO increases ET concentrations by promoting the enzymatic activity of 1-aminocyclopropane-1-carboxylate (ACC) oxidase and ACC synthase. Both ET and NO prompt AUX concentration by inducing the expression of *AUX1*, *PIN1*, and *PIN2* genes and proteins in the roots and function upstream of AUX to induce root hair morphogenesis in Mg-limiting conditions (Liu et al., 2017). When Mg supply is increased, root hair growth may be affected since Ca^{2+} gradient is disrupted in the root hair cells. Indeed, (Niu et al., 2014a) reported that increased Mg supply hampers the growth of root hair by decreasing the concentration gradient of Ca^{2+} in the root hair tip. Furthermore, it has been reported that adding Ca^{2+} to high Mg nutrient solution enhanced cytosolic Ca^{2+} concentration in the tip of root hairs and rescued root hair growth (Ogawa et al., 2025).

In potato (*Solanum tuberosum*) plants exposed to severe Mg deficiency, Mg retranslocation is impaired, resulting in disrupted photoassimilate partitioning and, ultimately, reduced root growth. An early interruption of Mg retranslocation during initial plant development may serve as an indicator of growth reductions at later developmental stages (Koch et al., 2020).

In tomato (*Solanum lycopersicum*) plants subjected to an adaptation period under internal Mg limitations, the expression of six MAGNESIUM TRANSPORTER (MGT) genes (two expressed in shoots and four in roots) was upregulated in Mg-deficient plants, although Mg deficiency resulted in a root system that was approximately 50% smaller. Auxin concentrations decreased by about 23% in Mg-deprived roots, albeit a stimulated accumulation of gibberellin (GA), CTK, and ABA. Consistent with this reduction in AUX levels, there was an overall transcriptional downregulation of thirteen genes involved in AUX biosynthesis (*TAR*, TRYPTOPHAN AMINOTRANSFERASE RELATED; *YUCs*, YUCCA genes encoding flavin monooxygenases involved in AUX biosynthesis), transport *LAXs* (LIKE AUX1;*PINs*, PIN-FORMED), and signaling (*IAAs*, INDOLE-3-ACETIC ACID–INDUCIBLE proteins; *ARFs*, AUXIN RESPONSE FACTORS) (Ishfaq et al., 2021).

In tomato and cucumber (*Cucumis sativus* L.), increasing Mg levels enhanced root growth and nutrient accumulation up to maximum values at 2.0 mM, after which both parameters declined. Excessive Mg supply promoted the exudation of phenolic acids by roots and hindered plant uptake of K and Ca, despite having no effect on Mg absorption; cucumber was more tolerant to Mg excess than tomato (Qu et al., 2023).

In tea (*Camellia sinensis*) plants, an adequate Mg supply increases the diameter of metaxylem cells in root tips, rendering greater average root length and diameter. A severe deprivation of Mg supply leads to the accumulation of photosynthates in stems and new shoots by impairing their translocation to roots. Specifically, severe Mg deficiency inhibits photosynthate transport to roots, causing their accumulation in tea plant stems, whereas moderate Mg deprivation hampers transport to new shoots, leading to accumulation in mature leaves (Chen et al., 2024).

Exogenous applications of Mg peanut significantly enhanced tolerance to salt stress by regulating gene expression of *CNGC1* (CYCLIC NUCLEOTIDE-GATED CHANNEL 1), *NCL1* (Na^{+}/Ca^{2+} ION EXCHANGER 1), and *SOS1* (SALT OVERLY SENSITIVE 1, also known as *NHX7*). Concomitantly, the corresponding encoded proteins effectively decreased cellular Na accumulation. Additionally, the exogenous application of Mg modulated the gene expression of *ACAs* and *POTs* (POTASSIUM TRANSPORTERS) and maintains cellular Ca^{2+} and K^{+} homeostasis, with a consequent reestablishment of root growth after salt stress (Wang et al., 2024a).

In a three-dimensional root quantification system, soybean plants exposed to Mg deficiency exhibited reductions in different root traits, including length, surface area, centroid, root tip count, volume, maximum and minimum root width, depth, and the maximum width-to-maximum depth ratio as compared to the control. In contrast, Mg deficiency had only minor effects on root solidity, bushiness, and root volume distribution (Liu et al., 2024).

20.3.6 Sulfur

Sulfur (S) availability profoundly impacts plant root architecture. In general, S-starvation renders increased primary root elongation and lateral root density. While excess S can also enhance lateral root growth, specific responses may differ. Sulfur-triggered modifications in root architecture may involve complex hormonal signaling networks, particularly AUX-related pathways, and nutrient-sensing mechanisms that adjust root traits for optimal nutrient foraging, with a concurrent impact on the overall plant growth and yield.

In *A. thaliana*, low sulfate availability promotes primary root elongation. Root cysteine, glutathione (GSH), and IAA levels are directly dependent on external sulfate concentrations within the physiological range, ultimately affecting root system architecture. Low sulfate concentrations also downregulate the genes involved in AUX production and translocation and elevate the accumulation of *PLT1* and *PLT2*. Sulfate level affects the primary root elongation by modulating the internal AUX levels and root stem cell niche maintenance (Zhao et al., 2014).

Sulfur-starved seedlings of sugar beet (*Beta vulgaris* L. ssp. *vulgaris* var. *altissima*) displayed a drastic increment in the number of root tips, in the miR395 expression in leaves but not in roots, and in the transcriptional activation of *APS1* (ATP SULFURYLASE 1)gene (Stevanato et al., 2015). Similarly, in Chinese kale (*Brassica oleracea* var. *alboglabra*), S starvation upregulates the expression of *miR395*, which targets the gene *BoATPS* (ATP SULFURYLASE) to regulate S supply and modify S assimilation pathways. In S-deprived plants, levels of primary sulfur metabolites including tryptophan, isoleucine, and GSH decline substantially, whereas valine content increases markedly. Moreover, levels of secondary sulfur metabolites such as glucosinolates, are consistently decreased. It is worth noting that S-deprived plants show longer roots, which may be associated with increased valine and decreased isoleucine levels, as valine boosts root growth whereas isoleucine hampers it (Yang et al., 2025).

In wheat plants, S-deficiency rendered enhanced root growth at Z39 stage (14% more length and 16% more tips) as compared to S sufficiency. Sulfur absorption was greater in S-sufficient treatments at Z39 and Z51 stages (Carciochi et al., 2017).

In different legumes, once nodulated, sulfur applications improve symbiotic nitrogen fixation, while sulfur deficiency decreases nodulation, SNF, and nodule metabolism. These inhibitory outcomes may result from impaired nitrogenase biosynthesis and activity, accumulation of nitrogen-rich amino acids, and declines in leghemoglobin, ferredoxin, ATP, and glucose levels in nodules. In the Japanese lotus (*Lotus japonicus*), the protein SST1 (SULFATE TRANSPORTER 1) is crucial for symbiotic N fixation in root nodules. Accordingly, this protein transports sulfate from the cytoplasm of the root cell to the intracellular rhizobia, where sulfate is crucial for protein and cofactor biosynthesis, including nitrogenase biosynthesis (Krusell et al., 2005). Within this symbiotic relationship, the GSH produced in the bacteroids and host cells is essential for nodule activity, while sulfur assimilation in the whole plant is reprogrammed during symbiosis (Becana et al., 2018; Siegl et al., 2024).

In *Arabidopsis*, the signaling module integrated by CLV3 (CLAVATA 3) peptide, a member of CLE family, and CLV1 (CLAVATA 1) leucine-rich repeat receptor kinase, regulates lateral root development in response to altered S supply. Wild-type seedlings treated with low-S supply displayed reduced lateral root density, which was recovered when sulfate was resupplied. Conversely, the *clv1* mutant shows a higher daily increase rate of lateral root density than the WT when S is deficient; this increase is reduced to WT levels when sulfate is applied, which indicates that *CLV1* mediates a signal that inhibits lateral root development when S is a limiting nutrient. Transcript levels of *CLE2* and *CLE3* decrease under S starvation and are subject to *CLV1*-mediated feedback regulation, indicating that CLE peptide signal levels are modified during the process of lateral root development (Dong et al., 2019).

Root hairs may support nutrient uptake, albeit with some variations among nutrients. In the model plant *Arabidopsis thaliana*, the proteins SULTR1;1 and SULTR1;2 are two high-affinity sulfate transporters active in roots. The enhanced expression of these two transporters in S-starved plants prompts S acquisition. Mutants with more [*werwolf* (*wer*)] and fewer [*caprice* (*cpc*)] root hairs, resulting from higher and lower expression of *SULTR1;1*, respectively, exhibited similar sulfate uptake rates across plant lines under both S-sufficiency and S-deficiency. In plants exposed to S-limiting conditions, the expression levels of *SULTR1;1* and *SULTR1;2* exhibited a negative correlation with the number of root hairs. Thus both S-deficiency-induced *SULTR* expression and S uptake rates relied on the root hair number that the genotype could develop. Moreover, primary root length and root hair number also displayed a negative correlation, together with a greater number of root hairs in S-deprived plants, as compared with plants subjected to S sufficiency. These results demonstrate that under both S-sufficient and S-deficient conditions, sulfate acquisition is influenced primarily by root biomass rather than by the number of root hairs (Kimura et al., 2019).

In *Arabidopsis*, Khan et al. (2010) examined the role of sulfite reductase (SiR) in assimilatory reduction of inorganic sulfate to sulfide. When the enzymatic activity of sulfite reductase was decreased, plants displayed growth retardation and severe disruption of S, N, and C metabolism. T-DNA insertion lines (*sir1-1* and *sir1–2*) displayed reduced levels of *SiR* transcript, protein, and enzymatic activity compared with the wild type, while growth was retarded and stunted and root length considerably reduced. In addition, shoot or root of the *sir1-1* mutant was combined with the corresponding WT organ using micrografting to generate homo and heteromeric chimeric plants. Interestingly, the alteration of S assimilation in shoots affected sulfate uptake at the root level, the composition of primary sulfur compounds and root growth, but not *vice versa* (Forieri et al., 2021).

In soybean, the S mutant *M18* was developed from the WT cultivar JN18, outstanding for its magnified root system, higher protein content in seeds, and increased levels of methionine, an S-containing amino acid. In this line, the gene *OAS-TL3* (O-ACETYLSERINE (THIOL) LYASE 3) is overexpressed in roots, which is essential in S metabolism. The corresponding encoded enzyme is determinant in the synthesis of cysteine, another important S-containing amino acid. GSH concentrations in lines overexpressing *OAS-TL3* were directly correlated with diverse root attributes including length, volume, projected area, and surface area, as well as with the number of total root tips, bifurcations, and crossings. Hence, *OAS-TL3* enhanced soybean root growth and activity, as well as methionine content in seeds through the OAS-TL–Cys–GSH pathway (Zhang et al., 2023).

Hydrogen sulfide (H_2S) is a gaseous signaling molecule that modulates seed germination, root development, stress responses, stomatal movement, and senescence. The phytohormone CTK is a key negative regulator of root system architecture and is antagonized by H_2S through targeting critical enzymes involved in CTK metabolism. Hydrogen sulfide induces persulfidation of cysteine-62 in the CKX2 protein (CYTOKININ OXIDASE/DEHYDROGENASE 2), enhancing CKX2 activity and accelerating CTK degradation, thus reducing CTK levels in root tips (Carrillo et al., 2025; Li, 2025). This reduction in CTK disrupts asymmetric cell elongation in the root elongation zone, evidenced by increased lateral root angles and shortened root hairs (Wang et al., 2024b). Treatment with low concentrations of sodium hydrosulfide (NaHS, an H_2S donor) decreases CTK content and enhances lateral root density. H_2S-mediated impairment of primary root growth and induction of lateral and adventitious root formation are in full agreement with the known effects of AUX on root development. Furthermore, low-concentration NaHS treatment enhances polar AUX transport, promoting AUX accumulation at lateral root initiation sites, which stimulates lateral root primordium formation and elongation while suppressing primary root growth (Jia et al., 2015). Under waterlogging and salinity stresses, H_2S interacts with ET and ABA to form a complex regulatory network. During waterlogging, H_2S modulates the activity of ACO (ACC OXIDASE), a rate-limiting enzyme in ET biosynthesis, reducing ACC levels and preventing ET overaccumulation–induced root senescence. At the same time, H_2S activates ABA signaling to promote aerenchyma formation, thus enhancing root ventilation. Under salt stress, H_2S, in coordination with ABA, induces the expression of ABA-responsive genes and ion channels, maintaining Na^+/K^+ homeostasis and strengthening root stress tolerance (Jiang et al., 2019; Li, 2025).

20.4 Effect of micronutrients on root architecture

Plant micronutrients (i.e., Cl, Fe, Mn, B, Zn, Cu, Ni, and Mo) are crucial for shaping root architecture by altering primary root growth, lateral root formation, and root hair development, depending on their availability in the growth media. These elements display signaling functions that boost specific root responses to optimize nutrient foraging, often through hormone-mediated pathways (e.g., AUX and ET), thus allowing root systems remodeling to leverage nutrient patches in the soil. Micronutrient deficiencies, heterogeneous or excessive availability may have profound impact on root morphology and spatial distribution, thereby affecting overall plant productivity.

20.4.1 Chlorine

Chlorine (Cl) is absorbed by plants as chloride ion (Cl^-). It is an essential element that influences root architecture. Plants exhibiting Cl^- deficiencies develop deeper and wider roots, whereas excessive Cl^- levels can be toxic and may interfere with NO_3^- uptake. In order to forage for scarce Cl^- or cope with chloride excess, plants adjust their root system architecture, which affects water and nutrient capture as well as overall plant fitness. Optimal Cl^- uptake generally occurs near neutral pH. As a micronutrient, optimal Cl^- concentrations contribute to enhanced elongation of root cells, improved plant water relations, and increased water and nutrient use efficiency (Colmenero-Flores et al., 2019).

Cl^- homeostasis is a crucial process in plant cells. For instance, when the genes *DTX33* (DETOXIFICATION 33) and *DTX35* (DETOXIFICATION 35) codifying vacuolar channels involved in Cl^- compartmentalization during cell expansion are disrupted, plants develop shorter root hairs (Zhang et al., 2017). Moreover, different lines of diverse plant species harboring mutations of genes codifying cation-chloride cotransporter (CCC) proteins, which modulate Cl^- homeostasis, exhibit critical disruptions in the elongation of different cell types such as the elongating root cells (Chen et al., 2019).

In tobacco (*Nicotiana tabacum* L.) var. Havana, Franco-Navarro et al. (2021) demonstrated that chloride nutrition supplied within a beneficial range (1–5 mM) improved drought resistance by enhancing water-deficit avoidance and tolerance mechanisms. In particular, Cl^- increased cell size, improved osmoregulation, and decreased water consumption without hindering photosynthetic efficiency, resulting in higher overall water-use efficiency. Importantly, Cl^- also increased root biomass under drought stress. Furthermore, Lucas et al. (2021) reported that tobacco and tomato plants treated with 5 mM chloride salts exhibited longer primary roots and shorter lateral roots during early vegetative development. These root architecture changes allow Cl-fertilized crops to develop longer roots to reach deeper zones in the soil solution where more water and nutrients may be available, thus helping to withstand drought periods that are increasingly more frequent as a consequence of global climate change.

In the salinity sensitive *Prunus* hybrid 'Nemaguard', the exposure to Cl^- solution compoed of 20 mM NaCl, 3.3 mM $MgCl_2$ and 3.3 mM $CaCl_2$, with a total of 33.2 mM Cl^- per liter and the EC of the Cl^- solution measured as 3.3 dS m^{-1}, resulted in a significant decrease of root tissue density and a negative relationship between root biomass and salt treatment duration, with an increase in root surface area without visible changes of average root diameter (Zhou et al., 2024).

20.4.2 Boron

Boron (B) is a key micronutrient shaping root architecture. It significantly affects CW formation, membrane integrity, and transport of nutrients and sugars. Similar to other nutrients, B-deprived plants exhibit stunted root growth, whereas excess B can be toxic and impair root development. Under limited B availability, plants adapt by modifying root system architecture, often through enhanced lateral root elongation and improved root hair development to improve uptake. Notably, B availability is tightly regulated by soil pH and Ca uptake, thereby influencing root growth and function.

In *Arabidopsis* seedlings, low B supply alters root system architecture, thereby reducing primary root growth and enhancing root hair length and number. Just 4 h after treatment applications, B-limiting conditions impaired primary root cell elongation and hindered growth, through mechanisms mediated by ET and AUX. Furthermore, there is evidence of crosstalk among ET, AUX, ABA, Ca^{2+} sensors, and ROS, in root responses to B deficiency (González-Fontes et al., 2016).

In B-limiting conditions, plant root development, N uptake, and the integrity and function of CW can be impaired, with a concomitant reduced crop productivity. Notably, more than 90% of the B present in plants is localized in the CW, highlighting its significance in root system growth and development. Boron is an integral structural component of rhamnogalacturonan II (RG-II), where it acts as a cross-linking element within the CW matrix. Thus B fosters membrane integrity, appropriate function of the indole-3-acetic acid (IAA) pathway, and modulation of CW synthesis (Srivastava et al., 2025).

Trifoliate orange [*Poncirus trifoliata* (L.) Raf.] grown under B-limiting conditions, root growth and development are impaired and altered by interactions mediated by the auxin IAA. Long-term B deprivation (≈0.25 mg L^{-1}) inhibits root elongation, enlarges root tips, and induces acute necrosis accompanied by a serious reduction of soluble B and IAA contents in seedling root tips. The expression of

IAA biosynthetic genes (*TAA1* (TRYPTOPHAN AMINOTRANSFERASE OF ARABIDOPSIS 1); *TAR2* (TRYPTOPHAN AMINOTRANSFERASE RELATED 2); *YUC3*; and *YUC8*) is significantly enhanced in just 3 h after treatment and subsequently markedly decreased. In contrast, the expression of rootward IAA transport genes (*AUX1*, *PIN1*, and *PIN4*) decreases drastically under B starvation, whereas the expression of shootward IAA transport genes (*LAX1*; *ABCB1* (ATP-BINDING CASSETTE SUBFAMILY B MEMBER 1); and*PIN3*) increases significantly (Li et al., 2016a). Furthermore, B-deficient roots of trifoliate orange exhibit pronounced enlargement of root tips and reduced CW B and uronic acid contents in pectin compared with B-adequate roots. Cell wall thickness is 1.82-fold greater in B-deprived plants than in control roots, and disrupted protein structure along with accumulation of waxes and cellulose in the CW is evident under B starvation. ^{13}C nuclear magnetic resonance (^{13}C-NMR) spectra indicate that B deprivation alters the organic carbon structure of the CW and increases the contents of amino acids, cellulose, phenols, and lignin. Swelling and weakened structural integrity of the CW, resulting from alterations in the pectin network and overall CW composition and structure under B deficiency, are likely major causes of rapid growth interruption and markedly increased root tips in trifoliate orange roots under B-limiting conditions (Wu et al., 2017). Moreover, Riaz et al. (2018) reported that B starvation drastically hindered primary and lateral root elongation and induced leaf chlorosis, with a concomitant accumulation of ROS, membrane peroxidation, and root cell damage. Boron starvation decreased GSH concentrations and impaired the activity of major antioxidant enzymes. Furthermore, B-deficient plants exhibited increased accumulation of Ca^{2+} modified subcellular cell wall structure with disorganized homogalacturonan epitopes.

Arabidopsis plants grown under B-limiting conditions display disrupted BR biosynthesis and signaling, thereby affecting root growth, with an evident coregulation between BR-regulated genes and B-deficiency-responsive genes. BR receptor mutants *bri1–119* and *bri1–301* were more insensitive to B-limiting conditions, whereas overexpression of the dominant ***bes1-D*** and ***pBZR1:bzr1-D*** was insensitive to low-B stress. Under B-deficient conditions, exogenous application of 24-epibrassinolide rescues root growth inhibition, while treatment with the BR biosynthesis inhibitor brassinazole exacerbates this hampering effect. The nuclear-localized signal of BES1 (BRI1-EMS-SUPPRESSOR 1) displayed an evident reduction under B-deficient conditions in comparison to B-sufficient conditions. Ultimately, deficiency of B hampers brassinolide accumulation, thus suppressing BR signaling and root elongation, likely by downregulating the expression of *BR6ox1* and *BR6ox2* (BRASSINOSTEROID-6-OXIDASE 1 and 2) (Zhang et al., 2021).

In sweetpotato, changes in B availability modify root architecture traits at the initiation of storage root formation in three cultivars (i.e., Beauregard, Murasaki, and Okinawa). Although cultivar-specific responses to B deficiency were recorded, reductions in main root length and lateral root length occurred in all three cultivars evaluated (Villordon and Gregorie, 2021).

In maize, plants strikingly did not show any change upon severe B starvation under *in terra* growth conditions with either inhibited root growth or compensatory root foraging behavior, which is atypical for most plant species. In root hairless *rth3 (ROOT HAIRLESS 3)* mutant plants, leaf B concentrations were significantly lower both in B-deficient or B-sufficient conditions compared to the WT. Nonetheless, the *rth3* mutant did not show larger total root length, greater number of fine roots, nor enhanced expression of B uptake transporters as compensatory adaptations (Bienert et al., 2023). Conversely, in a previous report in maize, the number of lateral roots was reduced under B-deficient conditions (0.0 mM $\mathbf{H_3BO_3}$) compared to B-sufficient (0.05 mM $\mathbf{H_3BO_3}$) or B-excessive conditions (0.5 mM $\mathbf{H_3BO_3}$) (Wilder et al., 2022). Similar effects were recorded when testing the putative B deficiency mimic phenylboronic acid in maize (Housh et al., 2020).

In different plant species including rapeseed, maize, rice, and *Arabidopsis*, B deficiency renders prompt aberrations in primary and lateral root elongation, including complete loss of the apical meristem, and alterations in lateral root number (Chu et al., 2025). Moreover, swollen and necrotic root tips composed of enlarged cells are typical features of B-deficient roots.

20.4.3 Iron

Iron (Fe) is a critical micronutrient that profoundly shapes root system architecture. It acts as a signal that modulates primary root growth, lateral root branching, and root hair formation, enabling plants to adapt to both Fe deficiency and Fe excess, with AUX and ET, eventually mediating these responses.

In P-deprived *Arabidopsis* plants, primary root elongation is hindered as a consequence of Fe toxicity. Plants exposed to low Fe concentrations under P-deficient conditions, primary root elongation is restored without increasing P availability or changing the transcriptional activation of P starvation-regulated genes. These results indicate that manipulating Fe availability may represent a valuable strategy to improve plant tolerance to P deficiency (Ward et al., 2008). Furthermore, localized Fe supply in horizontally separated agar plates doubles lateral root length without affecting lateral root number. In the Fe uptake–defective mutant *iron-regulated transporter 1* (*irt1*), lateral root development is drastically hindered; however, this requisite for IRT1 can be bypassed by Fe supply to shoots, suggesting that symplastic Fe triggers local lateral root elongation. Fe-prompted initiation of lateral root primordia and root cell elongation depends on the rootward AUX stream and is followed by increased activity of the AUX reporter DR5–β-glucuronidase in lateral root apices. The essential function of the AUX transporter *AUX1* in Fe-induced lateral root elongation is supported by Fe-responsive *AUX1* promoter activity in lateral root apices and by the inability of the *aux1-T* mutant to increase lateral roots into Fe-enriched agar patches (Giehl et al., 2012). Excess Fe directly inhibits primary root growth and hampers lateral root formation by operating at the tip of the growing primary root. This response may contribute to adjust root system architecture by limiting excessive Fe uptake in rhizosphere regions facing severe toxicity, while maintaining the uptake of other nutrients in less stressed parts of the root system. Notably, repression of primary root growth and lateral root formation in plants facing excessive Fe is attenuated by adding K^+, underscoring the potential of nutrient management strategies to ameliorate root system architecture and alleviate Fe toxicity symptoms (Li et al., 2016b).

Tomato plants treated with inhibitors of AUX polar transport and of ET biosynthesis (NPA, N-1-naphthylphthalamic acid; and AVG, aminoethoxyvinylglycine, respectively) grown by the split-root technique that allows Fe heterogeneous distribution exhibited a systemic action of AUX on *FERRIC REDUCTASE OXIDASE 1* (*SlFRO1*) expression, while ET signaling was effective not only locally but also systemically. Physiological adjustments to this diversified Fe supply are regulated by the integration of AUX and ET (Celletti et al., 2020).

Fe nutrition is drastically affected by phytohormones in both a direct and indirect manner. Both ET and AUX may modulate alterations in root system architecture to thrive in Fe-limiting environments. AUX mediate responses to Fe deficiency and uptake by optimizing the growth of secondary roots and root hair proliferation, whereas ET alters rhizospheric adjustments by controlling the synthesis of root exudates particularly phytosiderophores. CKs and ABA mediate Fe starvation by changing the transcriptional activity of genes that regulate AUX biosynthesis to shape a favorable root system architecture and stress tolerance strategy. GA may modulate the allocation of the DELLA (Asp–Glu–Leu–Leu–Ala motif) growth repressors in roots and thus regulate the Fe uptake machinery under Fe starvation by inducing the expression of *bHLH038* and *bHLH39* (basic helix–loop–helix transcription factors), which control IRT1 and FRO2 (FERRIC REDUCTION OXIDASE 2). Furthermore, ABA, BR, CKs, and jasmonic acid (JA) may negatively regulate Fe acquisition and uptake by modifying the transcriptional expression of the *FIT* (FER-LIKE IRON DEFICIENCY-INDUCED TRANSCRIPTION FACTOR) gene. Adaptations in root attributes such as the induction of an advantageous root system architecture or changes in the rhizosphere promote a higher availability and uptake of Fe under Fe-starvation (Divte et al., 2021).

In *A. thaliana*, an alteration in the CW-localized ferroxidase LPR2 (LOW PHOSPHATE ROOT 2) or a decrease in Fe supply readily attenuates the impairment of primary root growth when NH_4^+ is supplied as the main N source. Compared with NO_3^-, NH_4^+ leads to excessive Fe content in the phloem apoplast, in a process that relies on LPR2. Such a disrupted Fe content renders a vast callose deposition in the phloem derived from an oxidative burst, which hampers the function of the phloem. NH_4^+ hinders primary root development by prompting an insufficient allocation of sucrose to the growth zone (Liu et al., 2022b).

Sultana et al. (2023) explored 100 wheat varieties to examine whether the architecture influences Fe concentration in grains. A broad genetic diversity was found among indigenous wheat genotypes analyzed. Especially, Fe concentration in grains and root system architecture displayed a wide variation, which represents a valuable information to be exploited for the generation of new varieties through breeding programs, taking into consideration root system architecture attributes aimed at stimulating Fe biofortification in wheat.

Iron helps rice plants adapt to toxic arsenic (As) by altering root system architecture through the transcription factor *OsWRKY71* (WRKY TRANSCRIPTION FACTOR 71), a mechanism that may involve GA signaling. This is a clear evidence that underscores the key role of Fe in alleviating the detrimental effects of As stress on rice (Mirza and Gupta, 2024).

Rice plants have evolved sophisticated mechanisms to cope with Fe toxicity. In Fe-tolerant plants, such mechanisms involve marked modifications of root architecture including reduced root length and increased root thickness, which restrict Fe uptake. Moreover, other adaptive responses include enhanced lignin deposition for structural reinforcement and strengthened antioxidant systems. In contrast, Fe-sensitive genotypes exhibit pronounced root damage, with reduced root length, surface area, and volume, leading to impaired root performance, leaf bronzing, and reduced yield. By exploring 16 distinct rice varieties, Sonu et al. (2023) found that Fe toxicity rendered root system impairments, especially in root length, surface area, and root volume, which represent critical factors enabling signal perception of Fe stress. Interestingly, genotypes displaying better tolerance mechanisms to Fe stress displayed healthier roots than the sensitive ones. One of the greatest mechanisms for tolerance among genotypes was Fe exclusion, by decreasing Fe^{2+} uptake.

20.4.4 Manganese

Manganese (Mn) is an essential element that functions as an activator or cofactor of hundreds of enzymes in plants. Since it participates in lignin biosynthesis, its application may contribute to improve mechanical strength to roots. Both Mn deficiency and Mn toxicity adversely affect root growth and function. Adequate Mn supply supports healthy root development, whereas excess Mn can disrupt AUX transport and signaling, resulting in stunted and damaged roots, with the degree of tolerance varying among plant species and genotypes (Alejandro et al., 2020).

In roots, Mn deficiency exacerbates endodermal suberization and disturbs ion homeostasis, thereby limiting hydraulic conductance (Chen et al., 2019; Mimoun et al., 2024). In Mn-deprived plants, reduced chlorophyll content, impaired photosynthetic efficiency, and decreased overall plant vitality are commonly observed (Ben Maachia and Namsi, 2025).

In *Arabidopsis*, a decline in manganese superoxide dismutase (Mn-SOD) led to impaired root growth and affected tricarboxylic acid (TCA) cycle flux and mitochondrial redox homeostasis. Nevertheless, the adjustment of plant metabolism with TCA cycle impairment little effected overall respiratory rates (Morgan et al., 2008). In barley (*Hordeum vulgare*) roots, the suberization of endodermal cells was modified by Mn starvation, while the magnitude of Mn deprivation ultimately determined whether suberization was risen or dropped. Moderate Mn limitation enhanced the length of the unsuberized zone close to the root tip and augmented the distance from the root tip at which the fully suberized zone developed. Conversely, a pronounced Mn starvation stimulated suberization closer to the root tip. When plants were resupplied with Mn, suberization was similar to that observed on plants with sufficient Mn supply. The decreased suberization in moderately Mn-starved plants induced radial Mn transport across the endodermis at a greater distance from the root tip. Less suberin prompted inwards radial transport of Ca^{2+} and Na^{+}, but reduced the K^{+} concentration in the stele. During dramatic Mn starvation, Mn uptake was directed toward the root tip. With more suberin, the inward radial transport of Ca^{2+} and Na^{+} is diminished, whereas that of K^{+} is enhanced (Chen et al., 2019).

In *Arabidopsis*, Mn starvation triggered a general modification in the arrangement and attributes of the root epidermal cells (Wei Yang et al., 2008). Part of the additional hairs formed in response to Mn starvation were located in atrichoblast positions, which is an indication of a postembryonic reprogramming of the cell fate acquired during embryogenesis. Plants grown under a light intensity of >50 μmol m^{-2} s^{-1} with sufficiency of Mn displayed inhibition of root hair elongation. Instead, Mn-deficient plants exhibited enhanced root hair development. Mn starvation did not affect the expression of genes involved in epidermal cell fate specification, which suggests that the mechanism regulating the differentiation of rhizodermal cells during embryogenesis was bypassed under Mn-starved conditions. This observation was evidenced by a partial rescue of the hairless *cpc* mutant by Mn deprivation. In addition to a reduction in Mn concentration in plant tissues, Mn starvation elevated Fe concentration. This induction was linked to a reduction of transcript levels of the iron transporter *IRT1*, suggesting a more efficient transport of Fe when Mn is lacking. The Fe transporter IRT1 displays low substrate specificity and transports Mn^{2+} too, which may constitute another route for Mn^{2+} uptake in root cells. This mechanism has been identified and characterized in *Arabidopsis* (Vert et al., 2002), rice (Bughio et al., 2002), and barley (Pedas et al., 2008).

In soybeana, high Mn supply leads to thickening of root diameter without detectable alterations in the cells of the cortex, epidermis, exodermis, or endodermis. Anatomical changes associated with nutritional stress are expressed mainly as extensive disorganization of root cell cytoplasm and increased vacuolation under high Mn doses. It has been observed that the intensity of root changes in response to Mn fluctuations diverges among genotypes, which indicates the existence of genotype-specific tolerance mechanisms to different mannitudes of Mn-limiting or excessive conditions (Junior et al., 2009).

A comprehensive transcriptomic and proteomic analysis conducted by Rodríguez-Celma et al. (2016) revealed widespread changes in gene and protein expression in response to Mn starvation in roots of *Arabidopsis*. The approach detected 22,385 transcripts, while 11,606 proteins were unambiguously determined. Interestingly, just 10% of the changes in the abundance of proteins was linked to transcription, denoting a large impact of protein-level regulation. Differentially expressed genes encompassed a broad range of biological functions, including mRNA maturation, translation, and transport, as well as primary and secondary metabolic processes. In addition, steady-state levels of various major glucosinolates were substantially modified under Mn-deprived conditions, in both roots and leaves, which may represent a compensatory response to a rise in pathogen susceptibility linked to Mn-deficient conditions.

It has been observed that Mn toxicity hinders primary root elongation by reducing the meristematic cell division potential in *Arabidopsis*. In addition, Mn toxicity decreases AUX levels in root tips by impairing IAA biosynthesis and downregulating the expression of the AUX efflux carriers PIN4 and PIN7. Loss-of-function *pin4* and *pin7* mutants exhibit less suppression of root growth than Col-0 WT seedlings. Thus the detrimental effect of Mn toxicity on primary root growth is mediated through alterations in AUX biosynthesis and the expression of the AUX efflux transporters PIN4 and PIN7 (Zhao et al., 2017).

20.4.5 Zinc

Plants often adapt deeper or denser root systems in low-zinc (Zn) soils; however, excess Zn can be toxic, altering root architecture and requiring specific transport mechanisms for efficient Zn allocation to shoots and grains.

In barrelclover, excess Zn increased Zn accumulation in both roots and shoots of WT plants and the Zn-accumulating mutant *raz* (which requires additional Zn). However, *raz* plants exhibited lower Zn concentrations in roots than WT plants. Roots of *raz* plants suffering excess Zn showed reduced root volume, surface area, and total length compared with WT plants, and also developed fewer lateral roots than WT plants. Excess Zn increased root diameter in both genotypes. The Mn translocation factor (TfMn), calculated as the shoot-to-root Mn ratio ([Mn] shoot/[Mn] root), enhanced response to Zn supply in both *raz* and WT plants, though the increase in *raz* plants was more visible. Overall, TfMn values were higher in *raz* plants but were decreased in both genotypes upon Zn applications. Neither Zn applications nor plant genotypes affected nodulation. In addition, the expression of the *miR166* was induced upon excess Zn conditions in WT plants (Lewis et al., 2012).

Bochicchio et al. (2015) developed a screening strategy using Petri dishes with a gradient of distances between germinating seeds and a metal-contaminated medium to study modifications in root architecture and morphology of *A. thaliana* exposed to Cu and Zn at sub-toxic concentrations. In the presence of Cu, primary root length did not change substantially compared with the control, whereas Zn applied alone increased primary root length by 45% relative to the control. In metal-treated seedlings, total root system length augmented as a consequence of a higher number of lateral roots. Seedlings located closer to agar sectors containing metals exhibited pronounced root curvature and greater root branching than those farther from the metal source. This performance, in addition to the observed induction of root diameter in metal-treated seedlings, represents a compensatory growth response, with thicker roots possibly functioning as a boundary to protect roots from metal toxicity.

The response of nine cowpea (*Vigna unguiculata*) genotypes to two Zn levels (0.00 and 0.05 mg L^{-1}) was evaluated in a pot experiment. The results showed that plants supplied with Zn (+Zn) developed greater root length, diameter, volume, and number of forks compared with Zn-deficient plants (–Zn). However, genotype-dependent differences were observed in the number of root tips and total root biomass (Bhatt et al., 2019). In soybean, another legume, (Yusefi-Tanha et al., 2022) demonstrated that both Zn oxide nanoparticles **(ZnONPs; around 38 nm) and Zn^{2+} ions at optimal doses (~200 mg kg^{-1}) significantly improved soybean root growth, architecture, and Zn uptake, leading to better biomass and yield. The order of Zn bioaccumulation was: root > seed > leaf > stem. Nevertheless**, higher Zn concentrations showed toxicity, reducing root length and volume.

At low concentrations (70 ppm), Zn induces morphological and physiological adaptations of the root system, whereas at high concentrations (560 ppm) it exerts toxic effects on plants. In rapeseed (*Brassica napus* subsp. *napus*), growth-inhibiting Zn treatments result in elevated protein tyrosine nitration due to imbalances in ROS and reactive nitrogen species (RNS) homeostasis, although the overall nitration pattern remains similar to that of the control. This nitro-oxidative stress is accompanied by pronounced changes in CW composition and by reduced cell proliferation and viability, caused by high Zn uptake and disrupted microelement homeostasis in root tips. By contrast, under conditions that promote positive root growth responses, reorganization of the tyrosine nitration pattern is observed, with no substantial alterations in ROS and RNS balance. Under these conditions, decreases in meristematic zone viability and proliferation are less drastic, possibly as a consequence of lower Zn uptake (Feigl et al., 2019).

Zn application rates ≤11.4 kg ha^{-1} substantially raised root dry weight, root length density, and root surface area within the 0–30 cm soil layer in winter wheat, whereas higher Zn application rates diminished these root parameters. Both root growth and availability of soil Zn drastically affected Zn uptake, translocation, and remobilization to grain (Liu et al., 2019). Furthermore, genetic variability in wheat root system architecture significantly affects Zn uptake and grain Zn accumulation. Varieties with more robust root systems usually display enhanced Zn acquisition and higher Zn concentrations in the grain, underscoring root system architecture as a key target trait for Zn biofortification in breeding programs (Noor et al., 2024).

Zinc deficiency in millets (seven species representing different millet types) induces pronounced changes in root system architecture, including increased root hair length and density, as adaptive responses to enhance Zn uptake. These morphological alterations are triggered by differential regulation of genes such as *TIR1* (TRANSPORT INHIBITOR RESPONSE 1) and *RHD3* (ROOT HAIR DEFECTIVE 3), which are associated with root development and nutrient acquisition. Under Zn-sufficient conditions, root growth and architecture remain largely unaltered, and development proceeds normally (Ajeesh Krishna et al., 2023).

Just recently, Coşkun et al. (2025) demonstrated that grafting is a useful technique to attenuate Zn stress in cucumber by enhancing nutrient uptake, protecting metabolic and hormonal balance, and reinforcing antioxidative defense mechanisms. From a total of five rootstock genotypes tested, the *Cucurbita maxima* × *C. moschata* hybrids TZ148 and Maximus drastically enhanced physiological and biochemical resilience under both Zn-deficient and Zn-toxic conditions. These hybrids exhibit strong root systems and adaptability to various abiotic stress.

20.4.6 Copper

Copper (Cu) is an essential micronutrient that markedly influences root system architecture, both in deficient and excessive conditions. Although Cu deficiency has not been fully addressed, it is well documented that Cu excess results in phytotoxicity reflected in reduced primary root elongation, enhanced lateral root production that results in a denser and more fibrous root system, increased root diameter, and pronounced changes in root cell structure, including enhanced lignification.

In Rhodes grass (*Chloris gayana* Knuth.), Cu (<1 μM) leads to toxic effects that drastically hamper roots growth, causing cuticle damage, reduced root hair development, root thickening and cracking, and drastic deformation of the root system. Root damage began at approximately 0.2 μM Cu, disrupting root structure and function, while higher Cu levels induced severe morphological alterations and growth inhibition (Sheldon and Menzies, 2005).

In *bluegum eucalyptus* (*Eucalyptus globulus*), the application of chemical root pruning with 25 g $CuCO_3 \cdot Cu(OH)_2$ L^{-1} improved root morphology in treated plants as compared to the control, though statistical differences were not detected among treatments tested. In the experiments carried out in the field, plants exposed to Cu did not display improved traits as compared to control plants four years after planting, and the modifications in root morphology recorded in the nursery did not alter plant survival or growth. Hence, the use of chemical root pruning with Cu aimed at improving stability of *eucalyptus* is not recommended, at least not for the first four years after outplanting (Fernández et al., 2007).

In pot trials that lasted three years, rice plants exposed to toxic levels of Cu displayed toxicity symptoms that relied on Cu concentrations applied. The application of 100 mg Cu kg^{-1} soil was the start point to observe the first symptoms of root inhibition, whereas 300–500 mg Cu kg^{-1} resulted in nearly 50% of root dry weight was lost in the soil. Root growth was almost completely arrested at 800–1000 mg Cu kg^{-1}. Interactions between rice growth stage and soil Cu concentration influenced the degree of Cu toxicity to roots. The application of 400 mg Cu kg^{-1} significantly hindered total root length per plant and average root system length, with the inhibitory effects intensifying as the duration of Cu stress increased. Root length in Cu-stressed plants did not differ among the tested genotypes. Soil Cu treatment increased the number of cell layers and the thickness of the epidermal CW at both root bases and root tips. Compared with the control, epidermal thickness increased by 75% at the root base and by 27% at the root tip under Cu stress. In addition, cortical cells became larger and distorted, and xylem vessels were deformed and reduced in number relative to the control (Xu et al., 2008).

In *Arabidopsis*, the exposure to concentrations up to 5 μM Cu^{2+} in hydroponics impaired root biomass more drastically than shoot biomass, with the highest Cu accumulations observed in roots. In agar plates, the application of up to 50 μM Cu^{2+} decreased primary root growth and enhanced the density of short lateral roots, with local Cu^{2+} toxicity observed throughout the whole root system. Root tips exposed to Cu^{2+} displayed altered AUX and CTK accumulations and mitotic activity. Harmful Cu^{2+} concentrations caused at 50 μM also triggered higher root apical meristem death, which in turn rendered in a pronounced reorganization of the root system architecture (Lequeux et al., 2010).

In Korean bush-clover [*Kummerowia stipulacea* (Maxim.) Makino], a metallophyte species, plants originating from an ancient Cu mine exhibited higher root-to-shoot biomass allocation and a more complex root system architecture than plants from noncontaminated sites when exposed to Cu stress. Acid invertase was pivotal in modulating biomass allocation, and the associated physiological responses favored increased root-to-shoot biomass ratios. These traits are considered advantageous during adaptive evolution in Cu-enriched mine soils (Zhang et al., 2014b).

Scots pine (*Pinus sylvestris*) seedlings established under Cu-limiting conditions (1.2 nM) exhibited no visible disruption of growth. When exposed to excess Cu (0.32, 1.0, 2.5, 5.0, or 10 μM $CuSO_4$), the main root lost dominance, while lateral root development exhibited a drastic impairment, revealing a deficient adaptive reorganization of the root system architecture under excess Cu stress (Ivanov et al., 2016).

The responses of two grapevine rootstocks, Fercal and 196.17 (both *Vitis vinifera* × *Vitis berlandieri*) to Cu toxicity were tested under monocropping and oat intercropping systems (Marastoni et al., 2019). Root morphology was dependent on both rootstock genotype and cropping system. The expression patterns of Cu transporter genes, including *VvCTr* (COPPER TRANSPORTER), *VvNRAMP* (NATURAL RESISTANCE-ASSOCIATED MACROPHAGE PROTEIN), and *VvIRT1*, varied between rootstocks, with stronger induction generally observed in Fercal plants. Gene expression profiles in intercropped grapevines indicated rootstock-specific response mechanisms that depended on Cu availability. When Fercal plants were exposed to low Cu concentrations, root system growth and the expression of transporter genes were both enhanced, whereas 196.17 plants showed enhanced apoplastic accumulation of divalent cations together with enhanced transporter expression. Under high Cu concentrations, Fercal plants enhanced the activity of all divalent cation transporters as well as the release of root exudates. In a similar approach, Morsch et al. (2022) assessed root structural traits (e.g., enhanced branching and larger root systems) in three grapevine genotypes, IAC 572 [(*Vitis riparia* × *V. rupestris*) × *V. caribaea*], Magnolia (*V. rotundifolia*), and Paulsen 1103 (*V. berlandieri* × *V. rupestris*), to identify Cu tolerance mechanisms. Magnolia was identified as a Cu-tolerant genotype that maintained enhanced root development, whereas P application mitigated Cu toxicity by reducing Cu accumulation in roots and improving plant growth. Hence, adaptations of root system architecture and P management are key factors for alleviating Cu stress in grapevine.

Soybean plants established in copper oxide nanoparticle (CuONP)-amended soil showed that smaller CuONPs (25 nm) drastically enhanced seed Cu concentration, thus improving nutritional value, but at the same time inhibited root growth and altered root system architecture, revealing a clear trade-off between nutritional enhancement and root development. Cu bioaccumulation was the highest in roots, followed by leaves, stems, and seeds, with smaller CuONPs resulting in greater Cu accumulation in seeds. The study suggests that CuONPs have the potential to enhance the nutritional profile of soybean; however, it also emphasizes that possible toxicity risks for human consumers of Cu-biofortified seeds require further investigation before widespread agricultural application can be recommended (Yusefi-Tanha et al., 2020).

The Brazilian guajuvira (*Cordia americana*) exhibits a high tolerance to excess Cu in soils and growth media. The species uses specific tolerance mechanisms to manage the metal without compromising its growth or biomass production. Lovato et al. (2024) demonstrated that root morphology, photosynthetic variables, chlorophyll *a* fluorescence, and photosynthetic pigment contents were not negatively affected by Cu addition. The mitigation of the toxic effects was attributed to a preferential accumulation of the metal in its root system, thereby protecting shoots and maintaining photosynthetic activity relatively stable. Consequently, shoot and root growth, as well as overall biomass production, remained unaffected across increasing Cu concentrations. Therefore the *Cordia americana* species has the potential to be used in areas contaminated with Cu.

Chinese privet (*Ligustrum lucidum*) seedlings exposed to 100 g $Cu(OH)_2$ L^{-1} maintained growth linked to the development of more active roots. Untreated plants exhibited five root orders and did not develop roots with diameters >2.0 mm, whereas plants exposed to Cu produced eight root orders, roots with diameters of 2.0–3.5 mm, and a greater quantity of first-order and finest roots. The accumulation of Cu in roots hindered Cu redistribution to shoots, while restoration of Cu concentrations to normal levels in vegetative organs after transplanting fostered Cu homeostasis and reestablished nutrient balance (Zhou et al., 2025a).

20.4.7 Nickel

Nickel (Ni) status may significantly influence root system architecture, with deficiencies diminishing the root system with dead fibrous roots, impacting overall growth and functionality, while excessive Ni applications induce root gravitropic impairments and localized inhibition of root elongation (Lešková et al., 2020). Some plants have evolved elaborate mechanisms to thrive under elevated Ni availability by developing metallophilic root systems that actively grow toward and thrive in soil patches with high concentrations of Ni. This trait is especially pronounced in Ni hyperaccumulator plants, including species of the genera *Alyssum* and *Noccaea* (Brassicaceae), in which specialized root architectural responses contribute to efficient Ni uptake and tolerance (Rosatto et al., 2021).

Maize seedlings, two days old, were established in nutrient solutions containing 15, 20, 25, and 35 μM $Ni(NO_3)_2$ and analyzed after 2 and 7 days of incubation. Nickel was detected in all root tissues, and its concentration augmented both with exposure time and along the longitudinal axis from the root tip to the root base. Regardless of root region or tissue type, Ni accumulation in protoplasts exceeded that in the CW. Nickel penetrated the endodermal barrier and accumulated to the highest concentrations in the endodermis and pericycle. Accumulation of Ni in the pericycle restricted root branching. Nickel had no effect on final cell length; thus disruption of root growth resulted primarily from inhibited cell division rather than reduced cell elongation. Maize plants maintained Ni concentrations below toxic levels in shoots, indicating that this species displays a Ni excluder behavior (Seregin et al., 2003).

In wheat, Ni supply notably impaired root development, both in deficient or excess Ni conditions. Low Ni concentrations induced relatively minor morphological changes, with just the formation of additional cell layers in the exodermis. However, high Ni supply dramatically hindered root growth, cell division, and cell differentiation, ultimately leading to root necrosis. These inhibitory effects are mainly associated with disruption of cell cycle progression, including inhibition of DNA synthesis, and with the induction of oxidative stress through H_2O_2 accumulation. In addition, Ni interferes with AUX transport and CW-related processes, resulting in the arrest of meristematic activity and a shift of the differentiation zones closer to the root tip (Demchenko et al., 2005).

Barley root hair development differs markedly among growth media such as soil, sand, and water. Root hairs grown in water cultures were generally longer but showed greater sensitivity to Ni toxicity. Instead, soil offered a buffering capacity that attenuated Ni toxicity, although it limited root hair development compared with water culture. Root morphology contrasted according to the substrates, with total root surface area following the order soil > sand > water, mainly due to differences in root hair development, whereas root surface area per unit root length (reflecting root hair density) displayed the order water > sand > soil. The greater development of root hairs in water culture was a result of increased sensitivity to Ni toxicity, in contrast to root elongation, which was more pronounced in soil. Soil chemical properties such as pH and cation exchange capacity are key factors modulating Ni toxicity, making soil a less toxic growth environment for barley roots than aqueous systems (Lin et al., 2016).

In *Arabidopsis*, Ni toxicity severely hinders root growth and alters gravitropic responses by disrupting CW formation and hampering PIN2-mediated AUX transport. Ni stress decreases cell elongation and disrupts AUX allocations by reducing shootward AUX transport and lowering PIN2 protein concentrations. Ni toxicity also affects cytoskeleton dynamics by altering microtubule organization necessary for optimal root growth and gravitropic responses. Furthermore, ROS production and transcriptional activation of genes involved in CW biosynthesis are also altered in the event of Ni toxicity, leading to root gravitropism alterations and shortened roots, ultimately hindering normal root development (Lešková et al., 2020).

In tomato, the application of 100 μM melatonin promoted seedling tolerance to Ni toxicity (50 μM) by enhancing nutrient uptake, root system architecture, photosynthetic performance, and antioxidant capacity, and then attenuating growth inhibition and limiting Ni accumulation at deleterious levels. Melatonin acts as a protective agent, optimizing nutrient balance and protecting cells from oxidative stress caused by Ni, making plants healthier under stress (Altaf et al., 2021).

In a population of six southern cattail (*Typha domingensis*) ecotypes, Ni toxicity significantly reduced root fresh and dry weights as treatment levels increased. Ionic homeostasis (Na^+, K^+, Cl^-, and N) and contents of photosynthetic pigment were impaired across all ecotypes in response to Ni stress (Akhter et al., 2021).

Nickel hyperaccumulator species such as *Alyssoides utriculata* and *Noccaea caerulescens*, have developed intricate mechanisms to tolerate high Ni concentrations, including efficient translocation and sequestration of Ni in shoots with minimal detrimental effects on root growth and function (Roccotiello et al., 2015; Seregin et al., 2022). In contrast, nonhyperaccumulator (excluder) species, including *Alyssum montanum* and *Thlaspi arvense*, moderate Ni translocation, maintaining it mainly in root tissues. This response is promoted by an enhanced root growth or surface area to immobilize Ni and limit its translocation to shoots. Overall, hyperaccumulator species maintain root functionality and biomass to sustain nutrient uptake under high Ni conditions, whereas excluder species rely on root-based sequestration strategies that can lead to growth inhibition, highlighting contrasting adaptive strategies to cope with Ni toxicity (Rosatto et al., 2021).

In an experiment with sweetpotato exposed to different levels of Ni, Ni supply significantly altered root growth traits, depending on the concentrations tested. For instance, plants exposed to 7.5 and 15 mg Ni L^{-1} showed stimulation of root growth, with the greatest enhancement observed at 15 mg L^{-1}. Compared with the control, treatment with 15 mg Ni L^{-1} increased root length (54.6%), root volume (32.8%), surface area (44.1%), average root diameter (9.8%), projected area (43.2%), number of root tips (69.1%), crossings (17.1%), forks (110.1%), and root length per unit volume (53.7%). Conversely, higher Ni concentrations (30 and 60 mg L^{-1}) provoked drastic decreases in root morphological traits, with the most severe effects observed at 60 mg L^{-1}. Relative to the control, maximum reductions at this concentration were 43.8% for root length, 18.9% for total root volume, 38.1% for average diameter, 33.1% for projected area, 32.6% for surface area, 78.5% for root tips, 53.8% for forks, and 44.3% for root length per unit volume. These results indicate that sweetpotato can cope with up to 15 mg L^{-1} by impairing oxidative stress, whereas increases in osmolyte levels, antioxidant capacity, and enzyme activities are insufficient to counteract the detrimental effects of higher Ni toxicity (Kumar et al., 2022).

Nickel oxide nanoparticles (NiONPs) induce strong defense responses in the Ni-hyperaccumulator *Odontarrhena lesbiaca* (Brassicaceae), leading to pronounced modifications in root structure and signaling. NiONPs rendered thickening of root CW by enhancing deposition of lignin, suberin, and pectin, as well as anatomical alterations such as cortex expansion. In parallel, NiONPs trigger intricate nitrosative signaling responses, thereby altering NO and peroxynitrite levels. Different tolerance strategies were observed among ecotypes analyzed: some promote NO signaling, whereas others restrict stress by controlling protein nitration. Together, these cellular and biochemical changes attenuate excessive NiONP entry into root tissues and mitigate toxicity, underscoring the diverse adaptive mechanisms that enable *O. lesbiaca* to thrive and survive under these conditions (Kondak et al., 2024).

20.4.8 Molybdemum

Molybdenum (Mo) may notably reshape plant root architecture. Traits such as root length, volume, surface area, forks per root, tips per root, and root activity may be improved in response to optimal Mo supply. In legumes, Mo is particularly crucial for N fixation by boosting rhizobia activity, increasing nodule number, enhancing specialized enzymatic activity, and improving the overall N assimilation.

In wheat exposed to drought stress, Mo may substantially enhance water uptake by prompting deeper and more extensive root systems, inducing aquaporin gene expression, particularly that of *TaPIP* (PLASMA MEMBRANE INTRINSIC PROTEINS), and maintaining nutrient balance (K^+, Ca^{2+}, and Mg^{2+}). These adjustments enable plant to survive under water stress conditions and improve water-use efficiency. Molybdenum also triggers antioxidant activity and modulates root hydraulic conductivity, thus improving tolerance to water-limited conditions (Wu et al., 2019). Further, Mo stimulates winter wheat root growth by increasing NO production, which induces nitrate reductase (NR) activity, enhances nitrogen use efficiency, and stimulates nitrate transporter genes (*NRTs*) involved in N uptake, particularly when NO_3^- or mixed nitrogen sources are supplied. This Mo-induced NO signaling pathway ameliorates oxidative stress, promotes N assimilation, and boosts increases in root biomass, root length, and overall root vigor, confirming the pivotal role of Mo in modulating root development under both optimal and stress conditions (Imran et al., 2021).

A sand culture system was established with grapevine seedlings exposed to different N sources (NO_3^-, NH_4NO_3, and NH_4^+) in order to test the effect of Mo supply (0.0 or 1.0 μM) on N uptake. It was observed that the different N forms and Mo application profoundly altered biomass accumulation, as well as root architecture and activity. Total N content in response to Mo was treated in NH_4NO_3 supplied plants, followed by NO_3^- and finally NH_4^+. Mo and N enhanced molybdate transporter genes *VvMOT1* and *VvNRT1.1* expression synergistically. The application of Mo modified the utilization of the different N sources tested, with NH_4NO_3 showing the highest effect, while NH_4^+ showed the least. Furthermore, N-content and N-use efficiency were the highest in plants treated with Mo supplied with NH_4NO_3 source, demonstrating the synergistic effects of Mo with the coapplication of NO_3^- and NH_4^+ sources (Liu et al., 2022a). Both soybean (Vatansever et al., 2016) and barrelclover (Gil-Díez et al., 2019) genomes have members of the MOT1 (MOLYBDATE TRANSPORTER 1) family. MtMOT1.2 is necessary for molybdate delivery to the endodermis cells of the roots, while MtMOT1.3 contributes to the translocation of molybdate into nodule cells, where bacterial transporters take up the delivered molybdate to synthesize the essential nitrogenase FeMo cofactor (Tejada-Jiménez et al., 2017). In another legume, *Lotus japonicus*, LjMOT1 is localized in the plasma membrane and might be responsible for molybdate uptake from the soil and its translocation inside the plant, instead of Mo delivery to the nodules (Gao et al., 2016). In *Arabidopsis*, Weber et al. (2024) have also reported the presence of MOT genes, with special focus on the mechanisms of molybdate distribution and homeostasis. Just recently, Awasthi et al. (2025) found that Mo supply together with *Rhizobium* inoculation substantially enhanced plant height, root length, total nodules, effective nodules, fresh weight of root, dry weight of root, root volume, as well as yield attributes of the legume mung bean (*Vigna radiata*).

20.5 Combined effects of nutrients on root architecture

Whether plants grow in natural ecosystems or in agricultural systems, they are often exposed to multiple nutrient deficiencies simultaneously. The significance of integrative nutrient effects on root system architecture was first systematically addressed by Kellermeier et al. (2014), who applied multiple combinations of deficiency and sufficiency of N, P, K, and S, in addition to low and high light conditions, to *A. thaliana* plants. Several types of interactions among nutrients (antagonistic, synergistic, or neutral) were observed when combined nutrient limitations were compared with single-nutrient ones. In particular, S was found to exert significant interactive effects on root system architecture, whereas K and N enhanced lateral root division under single nutrient limitations but not under combined limiting conditions (Gruber et al., 2013; Kellermeier et al., 2014). These findings highlight that conventional single-nutrient deficiency experiments reveal only a limited portion of the complex nutrient signaling network underlying root architecture under diverse soil conditions. Large-scale experiments employing multidimensional experimental designs would therefore be required to allow accurate quantification of root architecture across multiple concentration ratios of all nutrients.

In *Arabidopsis*, Medici et al. (2015) investigated the perception and crosstalk between phosphate and nitrate signaling and reported that a substantial component of phosphate starvation–induced primary root inhibition is regulated by the transcription factor HRS1 (HYPERSENSITIVITY TO LOW PHOSPHATE-ELICITED PRIMARY ROOT SHORTENING 1) and its homolog HHO1, only in the presence of NO_3^- in the growth medium. The genes *HRS1* and *HHO1* are activated in response to early NO_3^- deficiency in reactions dependent on NRT1.1, whereas the HRS1 protein is stabilized by Pi deprivation. Hence, the transcriptional and posttranscriptional control of HRS1 by NO_3^- and Pi, respectively, facilitates an essential response of roots to two distinct nutritional signals: nitrogen and phosphorus.

It has also been demonstrated a crosstalk between Pi and micronutrients such as Fe and Zn. For example, Pi-deprivation–induced impairment of primary root growth depends on iron availability (Strieder et al., 2017), which is likely related to the significance of Fe^{3+} in the PDR2–LPR1 signaling module (Müller et al., 2015). The LPR1 protein is localized to both the ER and the CW, but its main function is primarily defined as a CW-localized ferroxidase. *LPR1*-dependent apoplastic accumulation of Fe^{3+} at the root tip may likely trigger ROS production. Within this model, ROS accumulation renders meristem-specific callose production, thus limiting cell-to-cell movement of *SHR1*, thereby hindering primary root development.

In cassava (*Manihot esculenta*), Ca and Zn cooperate to enhance root architecture, Ca and Zn content in the edible storage root, and foraging (Busener et al., 2020). In serpentine soils, characterized by low Ca availability and toxic Ni concentrations, crosstalk between Ca and Ni alters early lateral root development in *Arabidopsis lyrata*. This response is mediated by disruptions in Ca signaling and AUX allocation, thereby shaping root system architecture under the extreme conditions imposed by serpentine soils (Veatch-Blohm et al., 2023).

20.6 Conclusions and prospects

Nutrient signal perception and transduction remain a pivotal active field of research influencing root growth and root system architecture. As a phenotypic outcome resulting from the genetic constitution of the plant and its interaction with the environment, the final root system architecture of a plant species brings about excellent opportunities to identify key components of nutrient signaling pathways. In recent years, major advances have arisen particularly from: first, the integration of transcriptional, translational, redox-based, and CW-mediated regulatory processes; second, the discovery of peptide signals involved in local and systemic nutrient signaling; third, the identification of molecular hubs underlying the interactive effects of different nutrients; and finally, the correlation of root system architecture with nutrient-use efficiency and yield in crop genotypes. A concerted effort is now required to better control and monitor the nutrient profile surrounding the root and to systematically manipulate multiple nutrients, as well as other environmental factors that determine plant growth and development. When combined with genetics, particularly the exploration of natural variation, this system-level approach could generate the knowledge needed to build models capable of predicting root system architecture (RSA) for any genotype under any given combination of nutrients. Linking RSA with nutrient uptake, growth, and yield represents the next critical step. It is conceivable that nutrient-use efficiency under moderate fertilization regimes could be improved by desensitizing plants to nutrient deficiency signals. Exploring this and other crop improvement strategies requires a precise understanding of how nutrient signaling networks are interconnected.

Root form or architecture can influence various plant functions such as nutrient and water acquisition, anchorage, and interactions with other species. It is a trait that is rigorously controlled through plant breeding to improve water-use efficiency and stress tolerance.

Root system architecture is a fundamental determinant of plant productivity, particularly challenging in environments characterized by low or high stratification of water and nutrients. This architecture is composed of a set of root phenotypes that define the temporal and spatial distribution of roots within the heterogeneous soil matrix and, consequently, their capacity to acquire both mobile and immobile resources. In many regions of the world, agricultural production is constrained by water and nutrient availability. Understanding how different root architectural traits can be modified to cope with those challenges will enable more efficient resource acquisition by roots under conditions of limited water and nutrient availability.

Root phenotypes and phenotypic states that enhance exploration of deeper soil layers improve the acquisition of mobile soil components, such as nitrate and water, across a wide range of environments. These mobile resources are typically available in deeper soil layers during the growing season as a result of leaching and surface drying. Conversely, the availability of relatively immobile resources, such as phosphate and potassium, is generally higher in surface soil layers.

Studies of root traits have long been constrained by the difficulty of excavating root systems for phenotyping. Nevertheless, root system architecture, including root branching patterns, number, diameter, and growth angles across different root classes, has a strong impact on plant growth. For example, the number of roots within a specific class, as well as the density and length of lateral root branching, directly affect plant performance under conditions of low nitrogen availability.

The advent of novel technologies including advanced imaging, noninvasive sensors, as well as robotics and artificial intelligence (IA) is enabling high-throughput analysis of root architecture, growth, and function, moving beyond destructive methods to capture dynamic root traits in soil, crucial for improving crop yields and resilience. In addition, genome-wide association studies will become even more prominent for deciphering the genetic diversity underlying root system architecture responses to nutrient signals and for identifying useful natural alleles. Building on these advances, favorable alleles can be exploited through root tissue- or cell type-specific expression and allele-specific modification in crops, taking advantage of genome editing to improve nutrient and water use efficiency.

Plants have evolved intricate systems to dynamically regulate root system architecture and transporter expression in response to nutrient availability. One major remaining challenge is to determine how plants sense external and internal nutrient status as the initial step in responding to a specific nutrient. In particular, it remains to be clarified how sensing occurs (via local sensors and/or systemic signals), where it takes place (in the external environment, within cells, or in specific organelles), and what is perceived (the mineral ion itself or derived metabolites).

Spatial and temporal heterogeneity of nutrient availability under field condition, including nutrient deficiencies and excess, represents another challenging field of research for analyzing the physiological benefits and likely trade-offs of these nutrient-dependent control mechanisms. The incorporation of artificial intelligence, computational and mathematical modeling with experimental data will facilitate the explanation of how plant system architecture is optimized to leverage the resources, thrive, or survive under particular environments. In addition, a profound comprehension of nutrient-dependent responses will enable fundamental questions in plant science to be addressed, especially regarding root system architecture remodeling. Overall, unveiling novel mechanisms of nutrient transport and signaling affecting root growth and structure will provide critical insights into plant adaptive strategies under changing environmental conditions, which are being exacerbated by global climate change. Ultimately, the application of the new discoveries in agriculture will be crucial to mitigate climate impacts and build sustainable, inclusive, and resilient agriculture, ensuring long-term environmental and human well-being.

References

Aggarwal, S., Kumar, A., Jain, M., Sudan, J., Singh, K., Kumari, S., et al., 2020. C-terminally encoded peptides (CEPs) are potential mediators of abiotic stress response in plants. Physiology and Molecular Biology of Plants 26 (10), 2019–2033. Available at https://doi.org/10.1007/s12298-020-00881-4.

Aibara, I., Miwa, K., 2014. Strategies for optimization of mineral nutrient transport in plants: multilevel regulation of nutrient-dependent dynamics of root architecture and transporter activity. Plant and Cell Physiology 55 (12), 2027–2036. Available at https://doi.org/10.1093/pcp/pcu156.

Ajeesh Krishna, T.P., Maharajan, T., Antony Ceasar, S., Ignacimuthu, S., 2023. Zinc supply influences the root-specific traits with the expression of root architecture modulating genes in millets. Journal of Soil Science and Plant Nutrition 23 (4), 5527–5541. Available at https://doi.org/10.1007/s42729-023-01419-9.

Akhter, N., Aqeel, M., Hameed, M., Sakit Alhaithloul, H.A., Alghanem, S.M., Shahnaz, M.M., et al., 2021. Foliar architecture and physio-biochemical plasticity determines survival of *Typha* domingensis pers. Ecotypes in nickel and salt affected soil. Environmental Pollution 286. Available at https://doi.org/10.1016/j.envpol.2021.117316.

Alejandro, S., Höller, S., Meier, B., Peiter, E., 2020. Manganese in plants: from acquisition to subcellular allocation. Frontiers in Plant Science 11. Available at https://doi.org/10.3389/fpls.2020.00300.

Altaf, M.A., Shahid, R., Ren, M.X., Altaf, M.M., Jahan, M.S., Khan, L.U., 2021. Melatonin mitigates nickel toxicity by improving nutrient uptake fluxes, root architecture system, photosynthesis, and antioxidant potential in tomato seedling. Journal of Soil Science and Plant Nutrition 21 (3), 1842–1855. Available at https://doi.org/10.1007/s42729-021-00484-2.

Araya, T., Miyamoto, M., Wibowo, J., Suzuki, A., Kojima, S., Tsuchiya, Y.N., et al., 2014. CLE-CLAVATA1 peptide-receptor signaling module regulates the expansion of plant root systems in a nitrogen-dependent manner. Proceedings of the National Academy of Sciences of the United States of America 111 (5), 2029–2034. Available at https://doi.org/10.1073/pnas.1319953111.

Araya, T., Kubo, T., von Wirén, N., Takahashi, H., 2016. Statistical modeling of nitrogen-dependent modulation of root system architecture in *Arabidopsis thaliana*. Journal of Integrative Plant Biology 58, 254–265. Available at https://doi.org/10.1111/jipb.12433.

Asim, M., Ullah, Z., Xu, F., An, L., Aluko, O.O., Wang, Q., et al., 2020. Nitrate signaling, functions, and regulation of root system architecture: insights from *Arabidopsis thaliana*. Genes 11 (6), 633. Available at https://doi.org/10.3390/genes11060633.

Awasthi, L.K., Pandey, L., Chalise, D., 2025. Effect of Rhizobium and molybdenum on growth, root nodulation and yield of mung bean (*Vigna radiata* L.) in Tikapur, Kailali. Journal of Tikapur Multiple Campus 8, 141–163. Available at https://doi.org/10.3126/jotmc.v8i8.75977.

Balzergue, C., Dartevelle, T., Godon, C., Laugier, E., Meisrimler, C., Teulon, J.M., et al., 2017. Low phosphate activates STOP1-ALMT1 to rapidly inhibit root cell elongation. Nature Communications 8. Available at https://doi.org/10.1038/ncomms15300.

Becana, M., Wienkoop, S., Matamoros, M.A., 2018. Sulfur transport and metabolism in legume root nodules. Frontiers in Plant Science 9. Available at https://doi.org/10.3389/fpls.2018.01434.

Ben Maachia, S., Namsi, A., 2025. Manganese deficiency, soil chemistry, and root dysfunction drive physiological and metabolic changes in date palm under field conditions. Agronomy 15 (11), 2490. Available at https://doi.org/10.3390/agronomy15112490.

Bhatt, S.C., Rawat, D., Srivastava, P.C., 2019. Effect of Zn application on root growth parameters and shoot dry matter content of some cowpea genotypes. International Journal of Current Microbiology and Applied Sciences 8 (4), 1338–1349. Available at https://doi.org/10.20546/ijcmas.2019.804.156.

Bienert, M.D., Junker, A., Melzer, M., Altmann, T., von Wirén, N., Bienert, G.P., 2023. Boron deficiency responses in maize (*Zea mays* L.) roots. Journal of Plant Nutrition and Soil Science. Available at https://doi.org/10.1002/jpln.202300173.

Bochicchio, R., Sofo, A., Terzano, R., Gattullo, C.E., Amato, M., Scopa, A., 2015. Root architecture and morphometric analysis of *Arabidopsis thaliana* grown in Cd/Cu/Zn-gradient agar dishes: A new screening technique for studying plant response to metals. Plant Physiology and Biochemistry 91, 20–27. Available at https://doi.org/10.1016/j.plaphy.2015.03.010.

Bouain, N., Doumas, P., Rouached, H., 2016. Recent advances in understanding the molecular mechanisms regulating the root system response to phosphate deficiency in *Arabidopsis*. Current Genomics 17 (4), 308–314. Available at https://doi.org/10.2174/1389202917666160331201812.

Bouguyon, E., Brun, F., Meynard, D., Kubeš, M., Pervent, M., Leran, S., et al., 2015. Multiple mechanisms of nitrate sensing by *Arabidopsis* nitrate transceptor NRT1.1. Nature Plants 1 (3). Available at https://doi.org/10.1038/nplants.2015.15.

Bouguyon, E., Perrine-Walker, F., Pervent, M., Rochette, J., Cuesta, C., Benkova, E., et al., 2016. Nitrate controls root development through posttranscriptional regulation of the NRT1.1/NPF6.3 transporter/sensor. Plant Physiology 172 (2), 1237–1248. Available at https://doi.org/10.1104/pp.16.01047.

Bryan, A.C., Obaidi, A., Wierzba, M., Tax, F.E., 2012. XYLEM INTERMIXED WITH PHLOEM1, a leucine-rich repeat receptor-like kinase required for stem growth and vascular development in *Arabidopsis thaliana*. Planta 235 (1), 111–122. Available at https://doi.org/10.1007/s00425-011-1489-6.

Bughio, N., Yamaguchi, H., Nishizawa, N.K., Nakanishi, H., 2002. S. Mori, cloning an iron-regulated metal transporter from rice. Journal of Experimental Botany 53 (374), 1677–1682. Available at https://doi.org/10.1093/jxb/erf004.

Busener, N., Kengkanna, J., Saengwilai, P.J., Bucksch, A., 2020. Image-based root phenotyping links root architecture to micronutrient concentration in cassava. Plants People Planet 2 (6), 678–687. Available at https://doi.org/10.1002/ppp3.10130.

Cao, H., Liu, Z., Guo, J., Jia, Z., Shi, Y., Kang, K., et al., 2024. ZmNRT1.1B (ZmNPF6.6) determines nitrogen use efficiency via regulation of nitrate transport and signalling in maize. Plant Biotechnology Journal 22 (2), 316–329. Available at https://doi.org/10.1111/pbi.14185.

Carciochi, W.D., Divito, G.A., Fernández, L.A., Echeverría, H.E., 2017. Sulfur affects root growth and improves nitrogen recovery and internal efficiency in wheat. Journal of Plant Nutrition 40 (9), 1231–1242. Available at https://doi.org/10.1080/01904167.2016.1187740.

Carrillo, R., Moreno, I., Romero, L.C., Aroca, A., Gotor, C., 2025. Hydrogen sulfide-induced barley resilience to drought and salinity through protein persulfidation. Plant Physiology and Biochemistry 221, 109644. Available at https://doi.org/10.1016/j.plaphy.2025.109644.

Cederholm, H.M., Benfey, P.N., 2015. Distinct sensitivities to phosphate deprivation suggest that RGF peptides play disparate roles in *Arabidopsis thaliana* root development. New Phytologist 207 (3), 683–691. Available at https://doi.org/10.1111/nph.13405.

Celletti, S., Pii, Y., Valentinuzzi, F., Tiziani, R., Fontanella, M.C., Beone, G.M., et al., 2020. Physiological responses to fe deficiency in split-root tomato plants: possible roles of auxin and ethylene? Agronomy 10 (7). Available at https://doi.org/10.3390/agronomy10071000.

Chapman, K., Taleski, M., Ogilvie, H.A., Imin, N., Djordjevic, M.A., 2019. CEP-CEPR1 signalling inhibits the sucrose-dependent enhancement of lateral root growth. Journal of Experimental Botany 70 (15), 3955–3967. Available at https://doi.org/10.1093/jxb/erz207.

Chen, A., Husted, S., Salt, D.E., Schjoerring, J.K., Persson, D.P., 2019. The intensity of manganese deficiency strongly affects root endodermal suberization and ion homeostasis. Plant Physiology 181 (2), 729–742. Available at https://doi.org/10.1104/pp.19.00507.

Chen, Z., Wang, L., Cardoso, J.A., Zhu, S., Liu, G., Rao, I.M., et al., 2023. Improving phosphorus acquisition efficiency through modification of root growth responses to phosphate starvation in legumes. Frontiers in Plant Science 14. Available at https://doi.org/10.3389/fpls.2023.1094157.

Chen, L., Liu, M., Cai, Y., Wu, L., Zhang, Q., 2024. Magnesium deficiency differentiated effects on the roots growth and shoot metabolism by regulating the distribution of photosynthetic products in tea plants. Scientia Horticulturae 338, 113748. Available at https://doi.org/10.1016/j.scienta.2024.113748.

Chu, L., Schäfer, C.C., Matthes, M.S., 2025. Molecular mechanisms affected by boron deficiency in root and shoot meristems of plants. Journal of Experimental Botany 76 (7), 1866–1878. Available at https://doi.org/10.1093/jxb/eraf036.

Colmenero-Flores, J.M., Franco-Navarro, J.D., Cubero-Font, P., Peinado-Torrubia, P., Rosales, M.A., 2019. Chloride as a beneficial macronutrient in higher plants: new roles and regulation. International Journal of Molecular Sciences 20 (19). Available at https://doi.org/10.3390/ijms20194686.

Coşkun, Ö.F., Aydın, A., Toprak, S., Başak, H., Turan, M., 2025. Enhancing cucumber tolerance to zinc stress through grafting: a comprehensive study on biochemical, hormonal, and metabolic responses. BMC Plant Biology 25 (1). Available at https://doi.org/10.1186/s12870-025-07287-y.

Crombez, H., Motte, H., Beeckman, T., 2019. Tackling plant phosphate starvation by the roots. Developmental Cell 48 (5), 599–615. Available at https://doi.org/10.1016/j.devcel.2019.01.002.

Deb, S., Sankaranarayanan, S., Wewala, G., Widdup, E.E., Samuel, M.A., 2014. The S-domain receptor kinase AtARK2 and the U-box/ARMrepeat-containing E3 ubiquitin ligase AtPUB9 module mediates lateral root development under phosphate starvation in Arabidopsis. Plant Physiology 165, 1647–1656. Available at https://doi.org/10.1104/pp.114.244376.

Delay, C., Chapman, K., Taleski, M., Wang, Y., Tyagi, S., Xiong, Y., et al., 2019. CEP3 levels affect starvation-related growth responses of the primary root. Journal of Experimental Botany 70 (18), 4763–4774. Available at https://doi.org/10.1093/jxb/erz270.

Demchenko, N.P., Kalimova, I.B., Demchenko, K.N., 2005. Effect of nickel on growth, proliferation, and differentiation of root cells in Triticum aestivum seedlings. Russian Journal of Plant Physiology 52 (2), 220–228. Available at https://doi.org/10.1007/s11183-005-0034-5.

Ding, Y., Xu, G., 2011. Low magnesium with high potassium supply changes sugar partitioning and root growth pattern prior to visible magnesium deficiency in leaves of rice (*Oryza sativa* L.). American Journal of Plant Sciences 02 (04), 601–608. Available at https://doi.org/10.4236/ajps.2011.24071.

Divte, P.R., Yadav, P., Pawar, A.B., Sharma, V., Anand, A., Pandey, R., et al., 2021. Crop response to iron deficiency is guided by cross-talk between phytohormones and their regulation of the root system architecture. Agricultural Research 10 (3), 347–360. Available at https://doi.org/10.1007/s40003-020-00532-w.

Dong, W., Wang, Y., Takahashi, H., 2019. CLE-CLAVATA1 signaling pathway modulates lateral root development under sulfur deficiency. Plants 8 (4), 103. Available at https://doi.org/10.3390/plants8040103.

Du, Q., Zhao, X., Jiang, C., Wang, X., Han, Y., Wang, J., et al., 2017. Effect of potassium deficiency on root growth and nutrient uptake in maize (*Zea mays* L). Agricultural Sciences 08 (11), 1263–1277. Available at https://doi.org/10.4236/as.2017.811091.

Escobar-Sepúlveda, H.F., Trejo Téllez, L.I., García-Morales, S., Gómez-Merino, F.C., 2017. Expression patterns and promoter analyses of aluminum-responsive NAC genes suggest a possible growth regulation of rice mediated by aluminum, hormones and NAC transcription factors. PLoS One 12 (10). Available at https://doi.org/10.1371/journal.pone.0186084.

Feigl, G., Molnár, Á., Szőllősi, R., Ördög, A., Törőcsik, K., Oláh, D., et al., 2019. Zinc-induced root architectural changes of rhizotron-grown B. napus correlate with a differential nitro-oxidative response. Nitric Oxide: Biology and Chemistry / Official Journal of the Nitric Oxide Society 90, 55–65. Available at https://doi.org/10.1016/j.niox.2019.06.003.

Fernández, M., Tejero, J.R., Pérez, I., Soria, F., Ruiz, F., López, G., 2007. Effect of copper coating nursery containers on plant growth and root morphology of *Eucalyptus globulus* Labill. cuttings and seedlings. Silva Lusit 15, 215–227.

Forde, B.G., 2014. Nitrogen signalling pathways shaping root system architecture: an update. Current Opinion in Plant Biology 21, 30–36. Available at https://doi.org/10.1016/j.pbi.2014.06.004.

Forieri, I., Aref, R., Wirtz, M., Hell, R., 2021. Micrografting provides evidence for systemic regulation of sulfur metabolism between shoot and root. Plants 10 (8), 1729. Available at https://doi.org/10.3390/plants10081729.

Franco-Navarro, J.D., Díaz-Rueda, P., Rivero-Núñez, C.M., Brumós, J., Rubio-Casal, A.E., De Cires, A., et al., 2021. Chloride nutrition improves drought resistance by enhancing water deficit avoidance and tolerance mechanisms. Journal of Experimental Botany 72 (14), 5246–5261. Available at https://doi.org/10.1093/jxb/erab143.

Gan, Y., Bernreiter, A., Filleur, S., Abram, B., Forde, B.G., 2012. Overexpressing the ANR1 MADS-box gene in transgenic plants provides new insights into its role in the nitrate regulation of root development. Plant and Cell Physiology 53 (6), 1003–1016. Available at https://doi.org/10.1093/pcp/pcs050.

Gao, J.S., Wu, F.F., Shen, Z.L., Meng, Y., Cai, Y.P., Lin, Y., 2016. A putative molybdate transporter LjMOT1 is required for molybdenum transport in *Lotus japonicus*. Physiologia Plantarum 158 (3), 331–340. Available at https://doi.org/10.1111/ppl.12489.

Giehl, R.F.H., Lima, J.E., von Wirén, N., 2012. Localized iron supply triggers lateral root elongation in *Arabidopsis* by altering the AUX1-mediated auxin distribution. The Plant Cell 24 (1), 33–49. Available at https://doi.org/10.1105/tpc.111.092973.

Gil-Díez, P., Tejada-Jiménez, M., León-Mediavilla, J., Wen, J., Mysore, K.S., Imperial, J., et al., 2019. MtMOT1.2 is responsible for molybdate supply to *Medicago truncatula* nodules. Plant Cell and Environment 42 (1), 310–320. Available at https://doi.org/10.1111/pce.13388.

González-Fontes, A., Herrera-Rodríguez, M.B., Martín-Rejano, E.M., Navarro-Gochicoa, M.T., Rexach, J., Camacho-Cristóbal, J.J., 2016. Root responses to boron deficiency mediated by ethylene. Frontiers in Plant Science 6. Available at https://doi.org/10.3389/fpls.2015.01103 2016.

Gruber, B.D., Giehl, R.F.H., Friedel, S., von Wirén, N., 2013. Plasticity of the *Arabidopsis* root system under nutrient deficiencies. Plant Physiology 163 (1), 161–179. Available at https://doi.org/10.1104/pp.113.218453.

Hao, Y.J., Wei, W., Song, Q.X., Chen, H.W., Zhang, Y.Q., Wang, F., et al., 2011. Soybean NAC transcription factors promote abiotic stress tolerance and lateral root formation in transgenic plants. Plant Journal 68 (2), 302–313. Available at https://doi.org/10.1111/j.1365-313X.2011.04687.x.

Hastwell, A.H., Chu, X., Liu, Y., Ferguson, B.J., 2024. The parallel narrative of RGF/GLV/CLEL peptide signalling. Trends in Plant Science 29 (12), 1342–1355. Available at https://doi.org/10.1016/j.tplants.2024.07.014.

He, X.J., Mu, R.L., Cao, W.H., Zhang, Z.G., Zhang, J.S., Chen, S.Y., 2005. AtNAC2, a transcription factor downstream of ethylene and auxin signaling pathways, is involved in salt stress response and lateral root development. Plant Journal 44 (6), 903–916. Available at https://doi.org/10.1111/j.1365-313X.2005.02575.x.

Ho, C.H., Lin, S.H., Hu, H.C., Tsay, Y.F., 2009. CHL1 functions as a nitrate sensor in plants. Cell 138 (6), 1184–1194. Available at https://doi.org/10.1016/j.cell.2009.07.004.

Hoehenwarter, W., Mönchgesang, S., Neumann, S., Majovsky, P., Abel, S., Müller, J., 2016. Comparative expression profiling reveals a role of the root apoplast in local phosphate response. BMC Plant Biology 16 (1). Available at https://doi.org/10.1186/s12870-016-0790-8.

Housh, A.B., Matthes, M.S., Gerheart, A., Wilder, S.L., Kil, K.E., Schueller, M., et al., 2020. Assessment of a18f-phenylboronic acid radiotracer for imaging boron in maize. International Journal of Molecular Sciences 21 (3). Available at https://doi.org/10.3390/ijms21030976.

Hu, B., Wang, W., Ou, S., Tang, J., Li, H., Che, R., et al., 2015. Variation in NRT1.1B contributes to nitrate-use divergence between rice subspecies. Nature Genetics 47 (7), 834–838. Available at https://doi.org/10.1038/ng.3337.

Hu, B., Jiang, Z., Wang, W., Qiu, Y., Zhang, Z., Liu, Y., et al., 2019. Nitrate–NRT1.1B–SPX4 cascade integrates nitrogen and phosphorus signalling networks in plants. Nature Plants 5 (4), 401–413. Available at https://doi.org/10.1038/s41477-019-0384-1.

Imran, M., Sun, X., Hussain, S., Rana, M.S., Saleem, M.H., Riaz, M., et al., 2021. Molybdenum supply increases root system growth of winter wheat by enhancing nitric oxide accumulation and expression of NRT genes. Plant and Soil 459 (1-2), 235–248. Available at https://doi.org/10.1007/s11104-020-04765-0.

Ishfaq, M., Zhong, Y., Wang, Y., Li, X., 2021. Magnesium limitation leads to transcriptional down-tuning of auxin synthesis, transport, and signaling in the tomato root. Frontiers in Plant Science 12. Available at https://doi.org/10.3389/fpls.2021.802399.

Ivanov, Y.V., Kartashov, A.V., Ivanova, A.I., Savochkin, Y.V., Kuznetsov, V.V., 2016. Effects of copper deficiency and copper toxicity on organogenesis and some physiological and biochemical responses of Scots pine (*Pinus sylvestris* L.) seedlings grown in hydroculture. Environmental Science and Pollution Research 23 (17), 17332–17344. Available at https://doi.org/10.1007/s11356-016-6929-1.

Jeong, J.S., Kim, Y.S., Redillas, M.C.F.R., Jang, G., Jung, H., Bang, S.W., et al., 2013. OsNAC5 overexpression enlarges root diameter in rice plants leading to enhanced drought tolerance and increased grain yield in the field. Plant Biotechnology Journal 11 (1), 101–114. Available at https://doi.org/10.1111/pbi.12011.

Jia, Y.B., Yang, X.E., Feng, Y., Jilani, G., 2008. Differential response of root morphology to potassium deficient stress among rice genotypes varying in potassium efficiency. Journal of Zhejiang University. Science. B 9 (5), 427–434. Available at https://doi.org/10.1631/jzus.B0710636.

Jia, H., Hu, Y., Fan, T., Li, J., 2015. Hydrogen sulfide modulates actin-dependent auxin transport via regulating ABPs results in changing of root development in Arabidopsis. Scientific Reports 5 (1). Available at https://doi.org/10.1038/srep08251.

Jia, Z., von Wirén, N., Xu, G., 2020. Signaling pathways underlying nitrogen-dependent changes in root system architecture: from model to crop species. Journal of Experimental Botany 71 (15), 4393–4404. Available at https://doi.org/10.1093/jxb/eraa033.

Jiang, J.L., Tian, Y., Li, L., Yu, M., Hou, R.P., Ren, X.M., 2019. H_2S alleviates salinity stress in cucumber by maintaining the Na+/K+ balance and regulating H_2S metabolism and oxidative stress response. Frontiers in Plant Science 10. Available at https://doi.org/10.3389/fpls.2019.00678.

Jordan-Meille, L., Martineau, E., Bornot, Y., Lavres, J., Abreu-Junior, C.H., Domec, J.-C., 2018. How does water-stressed corn respond to potassium nutrition? A shoot-root scale approach study under controlled conditions. Agriculture 8 (11), 180. Available at https://doi.org/10.3390/agriculture8110180.

Ju, C., Javed, L., Fang, Y., Zhao, Y., Cao, C., Deng, Y., et al., 2025. *Arabidopsis* calcium-dependent protein kinases 4/5/6/11 negatively regulate hydrotropism via phosphorylation of MIZU-KUSSEI1. The Plant Cell 37 (1). Available at https://doi.org/10.1093/plcell/koae279.

Junior, J.L., Malavolta, E., Nogueir, N.D.L., Moraes, M.F., Reis, A.R., Rossi, M.L., et al., 2009. Changes in anatomy and root cell ultrastructure of soybean genotypes under manganese stress. Revista Brasileira de Ciencia do Solo 33 (2), 395–403. Available at https://doi.org/10.1590/s0100-06832009000200017.

Kawai, M., Tabata, R., Ohashi, M., Honda, H., Kamiya, T., Kojima, M., et al., 2022. Regulation of ammonium acquisition and use in *Oryza longistaminata* ramets under nitrogen source heterogeneity. Plant Physiology 188 (4), 2364–2376. Available at https://doi.org/10.1093/plphys/kiac025.

Kellermeier, F., Chardon, F., Amtmann, A., 2013. Natural variation of *Arabidopsis* root architecture reveals complementing adaptive strategies to potassium starvation. Plant Physiology 161 (3), 1421–1432. Available at https://doi.org/10.1104/pp.112.211144.

Kellermeier, F., Armengaud, P., Seditas, T.J., Danku, J., Salt, D.E., Amtmann, A., 2014. Analysis of the root system architecture of *Arabidopsis* provides a quantitative readout of crosstalk between nutritional signals. The Plant Cell 26 (4), 1480–1496. Available at https://doi.org/10.1105/tpc.113.122101.

Khan, M.S., Haas, F.H., Samami, A.A., Gholami, A.M., Bauer, A., Fellenberg, K., et al., 2010. Sulfite reductase defines a newly discovered bottleneck for assimilatory sulfate reduction and is essential for growth and development in *Arabidopsis thaliana*. The Plant Cell 22 (4), 1216–1231. Available at https://doi.org/10.1105/tpc.110.074088.

Kimura, Y., Ushiwatari, T., Suyama, A., Tominaga-Wada, R., Wada, T., Maruyama-Nakashita, A., 2019. Contribution of root hair development to sulfate uptake in *Arabidopsis*. Plants 8 (4), 106. Available at https://doi.org/10.3390/plants8040106.

Kirkby, E.A., 2023. Introduction, definition, and classification of nutrients. Marschner's Mineral Nutrition of Plants. Elsevier, United Kingdom, pp. 3–9. Available at http://doi.org/10.1016/B978-0-12-819773-8.00016-2.

Koch, M., Winkelmann, M.K., Hasler, M., Pawelzik, E., Naumann, M., 2020. Root growth in light of changing magnesium distribution and transport between source and sink tissues in potato (*Solanum tuberosum* L). Scientific Reports 10 (1). Available at https://doi.org/10.1038/s41598-020-65896-z.

Kondak, S., Janovszky, P., Szőllősi, R., Molnár, Á., Oláh, D., Adedokun, O.P., et al., 2024. Nickel oxide nanoparticles induce cell wall modifications, root anatomical changes, and nitrosative signaling in ecotypes of Ni hyperaccumulator *Odontarrhena lesbiaca*. Environmental Pollution 341, 122874. Available at https://doi.org/10.1016/j.envpol.2023.122874.

Krouk, G., Lacombe, B., Bielach, A., Perrine-Walker, F., Malinska, K., Mounier, E., et al., 2010. Nitrate-regulated auxin transport by NRT1.1 defines a mechanism for nutrient sensing in plants. Developmental Cell 18 (6), 927–937. Available at https://doi.org/10.1016/j.devcel.2010.05.008.

Krusell, L., Krause, K., Ott, T., Desbrosses, G., Krämer, U., Sato, S., et al., 2005. The sulfate transporter SST1 is crucial for symbiotic nitrogen fixation in *Lotus japonicus* root nodules. The Plant Cell 17 (5), 1625–1636. Available at https://doi.org/10.1105/tpc.104.030106.

Kumar, S., Wang, M., Liu, Y., Fahad, S., Qayyum, A., Jadoon, S.A., et al., 2022. Nickel toxicity alters growth patterns and induces oxidative stress response in sweetpotato. Frontiers in Plant Science 13. Available at https://doi.org/10.3389/fpls.2022.1054924.

Lay, K.S., Takahashi, H., 2018. Nutrient-responsive small signaling peptides and their influence on the root system architecture. International Journal of Molecular Sciences 19 (12), 3927. Available at https://doi.org/10.3390/ijms19123927.

Lee, D.K., Chung, P.J., Jeong, J.S., Jang, G., Bang, S.W., Jung, H., et al., 2017. The rice OsNAC6 transcription factor orchestrates multiple molecular mechanisms involving root structural adaptions and nicotianamine biosynthesis for drought tolerance. Plant Biotechnology Journal 15 (6), 754–764. Available at https://doi.org/10.1111/pbi.12673.

Leitão, N., Dangeville, P., Carter, R., Charpentier, M., 2019. Nuclear calcium signatures are associated with root development. Nature Communications 10 (1). Available at https://doi.org/10.1038/s41467-019-12845-8.

Lequeux, H., Hermans, C., Lutts, S., Verbruggen, N., 2010. Response to copper excess in *Arabidopsis thaliana:* impact on the root system architecture, hormone distribution, lignin accumulation and mineral profile. Plant Physiology and Biochemistry 48 (8), 673–682. Available at https://doi.org/10.1016/j.plaphy.2010.05.005.

Lešková, A., Zvarík, M., Araya, T., Giehl, R.F.H., 2020. Nickel toxicity targets cell wall-related processes and PIN2-mediated auxin transport to inhibit root elongation and gravitropic responses in *Arabidopsis*. Plant and Cell Physiology 61 (3), 519–535. Available at https://doi.org/10.1093/pcp/pcz217.

Lewis, R.W., Tang, G., McNear, D.H., 2012. Morphological and genetic changes induced by excess Zn in roots of *Medicago truncatula* A17 and a Zn accumulating mutant. BMC research notes 5, 657. Available at https://doi.org/10.1186/1756-0500-5-657.

Li, G., Kronzucker, H.J., Shi, W., 2016a. Root developmental adaptation to Fe toxicity: mechanisms and management. Plant Signaling & Behavior 11 (1), e1117722. Available at https://doi.org/10.1080/15592324.2015.1117722.

Li, Q., Liu, Y., Pan, Z., Xie, S., Peng, S.-ang, 2016b. Boron deficiency alters root growth and development and interacts with auxin metabolism by influencing the expression of auxin synthesis and transport genes. Biotechnology & Biotechnological Equipment 30 (4), 661–668. Available at https://doi.org/10.1080/13102818.2016.1166985.

Li, P., Peng, Z., Xu, P., Tang, G., Ma, C., Zhu, J., et al., 2021. Genome-wide identification of NAC transcription factors and their functional prediction of abiotic stress response in peanut. Frontiers in Genetics 12. Available at https://doi.org/10.3389/fgene.2021.630292.

Li, J.-S., 2025. Novel regulatory mechanisms of plant root architecture: roles of hydrogen sulfide and phytohormones. Life Research 8 (2), 12. Available at https://doi.org/10.53388/lr20250012.

Liang, H., Wang, L., Gong, F., Chang, J., 2025a. Functions of plant hormones and calcium signaling in regulating root hydrotropism. Journal of Plant Physiology 308, 154490. Available at https://doi.org/10.1016/j.jplph.2025.154490.

Liang, X., Zhou, Y., Xu, W., Liang, J., 2025b. An intracellular CPK-ECA1 phosphoregulatory circuit couples calcium signatures to ABA homeostasis for plant osmosensivity. Science Advances 11 (40). Available at https://doi.org/10.1126/sciadv.adz2428.

Lin, S.H., Kuo, H.F., Canivenc, G., Lin, C.S., Lepetit, M., Hsu, P.K., et al., 2008. Mutation of the *Arabidopsis* NRT1.5 nitrate transporter causes defective root-to-shoot nitrate transport. The Plant Cell 20 (9), 2514–2528. Available at https://doi.org/10.1105/tpc.108.060244.

Lin, Y., Allen, H.E., Di Toro, D.M., 2016. Barley root hair growth and morphology in soil, sand, and water solution media and relationship with nickel toxicity. Environmental Toxicology and Chemistry 35 (8), 2125–2133. Available at https://doi.org/10.1002/etc.3389.

Liu, M., Liu, X.X., He, X.L., Liu, L.J., Wu, H., Tang, C.X., et al., 2017. Ethylene and nitric oxide interact to regulate the magnesium deficiency-induced root hair development in *Arabidopsis*. New Phytologist 213 (3), 1242–1256. Available at https://doi.org/10.1111/nph.14259.

Liu, Y., von Wirén, N., 2017. Ammonium as a signal for physiological and morphological responses in plants. Journal of Experimental Botany 68 (10), 2581–2592. Available at https://doi.org/10.1093/jxb/erx086.

Liu, M., Bi, J., Jin, C., 2018a. Developmental responses of root hairs to Mg deficiency. Plant Signaling & Behavior 13 (9), e1500068. Available at https://doi.org/10.1080/15592324.2018.1500068.

Liu, M., Zhang, H., Fang, X., Zhang, Y., Jin, C., 2018b. Auxin acts downstream of ethylene and nitric oxide to regulate magnesium deficiency-induced root hair development in *Arabidopsis Thaliana*. Plant and Cell Physiology 59 (7), 1452–1465. Available at https://doi.org/10.1093/pcp/pcy078.

Liu, D.Y., Liu, Y.M., Zhang, W., Chen, X.P., Zou, C.Q., 2019. Zinc uptake, translocation, and remobilization in winter wheat as affected by soil application of zn fertilizer. Frontiers in Plant Science 10. Available at https://doi.org/10.3389/fpls.2019.00426.

Liu, L., An, M.M., Li, X.J., Han, Z., Li, S.X., Li, B., 2022a. Molybdenum-induced effects on nitrogen absorption and utilization under different nitrogen sources in *Vitis vinifera*. Journal of Plant Interactions 17 (1), 756–765. Available at https://doi.org/10.1080/17429145.2022.2089752.

Liu, X.X., Zhang, H.H., Zhu, Q.Y., Ye, J.Y., Zhu, Y.X., Jing, X.T., et al., 2022b. Phloem iron remodels root development in response to ammonium as the major nitrogen source. Nature Communications 13 (1). Available at https://doi.org/10.1038/s41467-022-28261-4.

Liu, M., Zhang, Q., Jin, R., Zhao, P., Zhu, X., Wang, J., et al., 2023. The role of IAA in regulating root architecture of sweetpotato (*Ipomoea batatas* [L.] Lam) in response to potassium deficiency stress. Plants 12 (9), 1779. Available at https://doi.org/10.3390/plants12091779.

Liu, Z., Li, S., Mao, J., Yan, H., Liang, C., Lu, X., et al., 2024. Effect of magnesium nutrition on growth and root system architecture traits of soybean seedlings. Journal of South China Agricultural University 45 (3), 321–328. Available at https://doi.org/10.7671/j.issn.1001-411X.202311009.

Liu, D., Jiao, Q., Liu, H., Fan, L., Yu, P., Jiang, G.Y., et al., 2025. Wheat roots adapt to potassium deficiency through coordinated antioxidant, hormonal, and metabolic reprogramming: insight from morpho-physiological, biochemical, and transcriptomic analysis. Plant Stress 18. Available at https://doi.org/10.1016/j.stress.2025.101109.

Lombardi, M., De Gara, L., Loreto, F., 2021. Determinants of root system architecture for future-ready, stress-resilient crops. Physiologia Plantarum 172 (4), 2090–2097. Available at https://doi.org/10.1111/ppl.13439.

Lovato, L.D.R., Aguilar, M., Rosa, M.V.M., da, F.C.R., Peixoto, Wertonge, T.W., et al., 2024. Tolerance of Cordia americana plants exposed to excess copper. Ciênc. Florest. 34, e85568. Available at https://doi.org/10.5902/1980509885568.

Lucas, M., Peinado-Turrubia, P., Moreno-Racero, F.J., García-Rodríguez, A.F., Colmenero-Flores, J.M., Rosales, M.A., 2021. Chloride nutrition affects root architecture during early vegetative development in tobacco and tomato plants. XXIV Meeting of the Spanish Society of Plant Biology and XVII Spanish-Portuguese Congress on Plant Biology. pp. 381-382.

Lynch, J., 1995. Root architecture and plant productivity. Plant Physiology 109 (1), 7–13. Available at https://doi.org/10.1104/pp.109.1.7.

Ma, D., Fukuda, H., Sotta, N., Fujiwara, T., 2023. *Arabidopsis thaliana* RPL13aC affects root system architecture through shoot potassium accumulation. The Plant Journal 116 (2), 497–509. Available at https://doi.org/10.1111/tpj.16386.

Ma, W., Li, J., Qu, B., He, X., Zhao, X., Li, B., et al., 2014. Auxin biosynthetic gene TAR 2 is involved in low nitrogen-mediated reprogramming of root architecture in *A rabidopsis*. The Plant Journal 78 (1), 70–79. Available at https://doi.org/10.1111/tpj.12448.

Maghiaoui, A., Bouguyon, E., Cuesta, C., Perrine-Walker, F., Alcon, C., Krouk, G., et al., 2020. The *Arabidopsis* NRT1.1 transceptor coordinately controls auxin biosynthesis and transport to regulate root branching in response to nitrate. Journal of Experimental Botany 71 (15), 4480–4494. Available at https://doi.org/10.1093/jxb/eraa242.

Marastoni, L., Sandri, M., Pii, Y., Valentinuzzi, F., Cesco, S., Mimmo, T., 2019. Morphological root responses and molecular regulation of cation transporters are differently affected by copper toxicity and cropping system depending on the grapevine rootstock genotype. Frontiers in Plant Science 10. Available at https://doi.org/10.3389/fpls.2019.00946.

Mazarei, M., Coffey, N., Shipp, S.E.A., Stewart, C.N., Hewezi, T., 2025. Increased root growth and seed yield in transgenic soybean overexpressing NAC genes GmNAC19 and GmGRAB1. Plant Cell Reports 44 (7). Available at https://doi.org/10.1007/s00299-025-03550-6.

Medici, A., Marshall-Colon, A., Ronzier, E., Szponarski, W., Wang, R., Gojon, A., et al., 2015. AtNIGT1/HRS1 integrates nitrate and phosphate signals at the *Arabidopsis* root tip. Nature Communications 6. Available at https://doi.org/10.1038/ncomms7274.

Mimoun, A., Rey, H., Jourdan, C., Banamar, H., Yakoubi, F., Babou, F., et al., 2024. Moderate salinity stimulates root plasticity and growth parameters of date palm seedlings (*Phoenix dactylifera* L). Rhizosphere 30, 100876. Available at https://doi.org/10.1016/j.rhisph.2024.100876.

Mirza, Z., Gupta, M., 2024. Iron reprogrammes the root system architecture by regulating OsWRKY71 in arsenic-stressed rice (*Oryza sativa* L). Plant Molecular Biology 114 (1). Available at https://doi.org/10.1007/s11103-024-01420-5.

Młodzińska-Michta, E., 2023. Abiotic factors determine the root system architecture – review and update. Acta Societatis Botanicorum Poloniae 92 (1). Available at https://doi.org/10.5586/asbp/168700.

Mohd-Radzman, N.A., Djordjevic, M.A., Imin, N., 2013. Nitrogen modulation of legume root architecture signaling pathways involves phytohormones and small regulatory molecules. Frontiers in Plant Science 4. Available at https://doi.org/10.3389/fpls.2013.00385.

Mora-Macías, J., Ojeda-Rivera, J.O., Gutiérrez-Alanís, D., Yong-Villalobos, L., Oropeza-Aburto, A., Raya-González, J., et al., 2017. Malate-dependent Fe accumulation is a critical checkpoint in the root developmental response to low phosphate. Proceedings of the National Academy of Sciences of the United States of America 114 (17). Available at https://doi.org/10.1073/pnas.1701952114 E3563.

Moreno-Alvarado, M., García-Morales, S., Trejo-Téllez, L.I., Hidalgo-Contreras, J.V., Gómez-Merino, F.C., 2017. Aluminum enhances growth and sugar concentration, alters macronutrient status and regulates the expression of NAC transcription factors in rice. Frontiers in Plant Science 8. Available at https://doi.org/10.3389/fpls.2017.00073.

Morgan, M.J., Lehmann, M., Schwarzländer, M., Baxter, C.J., Sienkiewicz-Porzucek, A., Williams, T.C.R., et al., 2008. Decrease in manganese superoxide dismutase leads to reduced root growth and affects tricarboxylic acid cycle flux and mitochondrial redox homeostasis. Plant Physiology 147 (1), 101–114. Available at https://doi.org/10.1104/pp.107.113613.

Morsch, L., Somavilla, L.M., Trentin, E., Silva, K.R., de Oliveira, J.M.S., Brunetto, G., et al., 2022. Root system structure as a criterion for the selection of grapevine genotypes that are tolerant to excess copper and the ability of phosphorus to mitigate toxicity. Plant Physiology and Biochemistry 171, 147–156. Available at https://doi.org/10.1016/j.plaphy.2021.12.034.

Motte, H., Vanneste, S., Beeckman, T., 2019. Molecular and Environmental Regulation of Root Development. Annual Review of Plant Biology 70, 465–488. Available at https://doi.org/10.1146/annurev-arplant-050718-100423.

Mounier, E., Pervent, M., Ljung, K., Gojon, A., Nacry, P., 2014. Auxin-mediated nitrate signalling by NRT1.1 participates in the adaptive response of *Arabidopsis* root architecture to the spatial heterogeneity of nitrate availability. Plant, Cell and Environment 37 (1), 162–174. Available at https://doi.org/10.1111/pce.12143.

Müller, J., Toev, T., Heisters, M., Teller, J., Moore, K.L., Hause, G., et al., 2015. Iron-dependent callose deposition adjusts root meristem maintenance to phosphate availability. Developmental Cell 33 (2), 216–230. Available at https://doi.org/10.1016/j.devcel.2015.02.007.

Nisar, M., Ali, Z., Ali, A., Aman, R., Park, H.J., Ullah, I., et al., 2020. $CaCl_2$ salt signaling in primary root architecture and lateral root emergence in *Arabidopsis thaliana*. Russian Journal of Plant Physiology 67 (3), 515–520. Available at https://doi.org/10.1134/S1021443720030176.

Niu, Y., Jin, G., Zhang, Y.S., 2014a. Root development under control of magnesium availability. Plant Signaling & Behavior 9 (9), e29720. Available at https://doi.org/10.4161/psb.29720.

Niu, Y., Chai, R., Liu, L., Jin, G., Liu, M., Tang, C., et al., 2014b. Magnesium availability regulates the development of root hairs in *Arabidopsis thaliana* (L.) Heynh. Plant Cell and Environment 37 (12), 2795–2813. Available at https://doi.org/10.1111/pce.12362.

Noguero, M., Lacombe, B., 2016. Transporters involved in root nitrate uptake and sensing by *Arabidopsis*. Frontiers in Plant Science 7. Available at https://doi.org/10.3389/fpls.2016.01391 September.

Noor, M., Kiran, A., Shahbaz, M., Sanaullah, M., Wakeel, A., 2024. Root system architecture associated zinc variability in wheat (*Triticum aestivum* L). Scientific Reports 14 (1). Available at https://doi.org/10.1038/s41598-024-52338-3.

Ogawa, S.T., Zhang, W., Staiger, C.J., Kessler, S.A., 2025. MLO-mediated Ca2+ influx regulates root hair tip growth in *Arabidopsis*. New Phytologist 248 (6), 3024–3039. Available at https://doi.org/10.1111/nph.70378.

Okamoto, S., Suzuki, T., Kawaguchi, M., Higashiyama, T., Matsubayashi, Y., 2015. A comprehensive strategy for identifying long-distance mobile peptides in xylem sap. The Plant Journal 84, 611–620. Available at https://doi.org/10.1111/tpj.13015.

Ohkubo, Y., Tanaka, M., Tabata, R., Ogawa-Ohnishi, M., Matsubayashi, Y., 2017. Shoot-to-root mobile polypeptides involved in systemic regulation of nitrogen acquisition. Nature Plants 3 (4). Available at https://doi.org/10.1038/nplants.2017.29.

Okamoto, M., Vidmar, J.J., Glass, A.D.M., 2003. Regulation of NRT1 and NRT2 gene families of *Arabidopsis thaliana*: responses to nitrate provision. Plant and Cell Physiology 44 (3), 304–317. Available at https://doi.org/10.1093/pcp/pcg036.

Patel, D.S., Kirti, B., Dhiraji, P.P., Vipulkumar, P., Suchismita, J., Ajay, V.N., et al., 2021. Does plant root architecture respond to potassium under water stress? A case from rice seedling root responses. Current Science 120 (6), 1050–1056. Available at https://doi.org/10.18520/cs/v120/i6/1050-1056.

Pedas, P., Ytting, C.K., Fuglsang, A.T., Jahn, T.P., Schjoerring, J.K., Husted, S., 2008. Manganese efficiency in Barley: identification and characterization of the metal ion transporter HvIRT1. Plant Physiology 148 (1), 455–466. Available at https://doi.org/10.1104/pp.108.118851.

Postma, J.A., Dathe, A., Lynch, J.P., 2014. The optimal lateral root branching density for maize depends on nitrogen and phosphorus availability. Plant Physiology 166 (2), 590–602. Available at https://doi.org/10.1104/pp.113.233916.

Qu, S., Li, H., Zhang, X., Gao, J., Ma, R., Ma, L., et al., 2023. Effects of magnesium imbalance on root growth and nutrient absorption in different genotypes of vegetable crops. Plants 12 (20), 3518. Available at https://doi.org/10.3390/plants12203518.

Quach, T.N., Tran, L.S.P., Valliyodan, B., Nguyen, H.T.M., Kumar, R., Neelakandan, A.K., et al., 2014. Functional analysis of water stress-responsive soybean GmNAC003 and GmNAC004 transcription factors in lateral root development in *Arabidopsis*. PLoS One 9 (1). Available at https://doi.org/10.1371/journal.pone.0084886

Ranathunge, K., Schreiber, L., Bi, Y.M., Rothstein, S.J., 2016. Ammonium-induced architectural and anatomical changes with altered suberin and lignin levels significantly change water and solute permeabilities of rice (*Oryza sativa* L.) roots. Planta 243 (1), 231–249. Available at https://doi.org/10.1007/s00425-015-2406-1.

Redillas, M.C.F.R., Jeong, J.S., Kim, Y.S., Jung, H., Bang, S.W., Choi, Y.D., et al., 2012. The overexpression of OsNAC9 alters the root architecture of rice plants enhancing drought resistance and grain yield under field conditions. Plant Biotechnology Journal 10 (7), 792–805. Available at https://doi.org/10.1111/j.1467-7652.2012.00697.x.

Remans, T., Nacry, P., Pervent, M., Filleur, S., Diatloff, E., Mounier, E., et al., 2006. The *Arabidopsis* NRT1.1 transporter participates in the signaling pathway triggering root colonization of nitrate-rich patches. Proceedings of the National Academy of Sciences of the United States of America 103 (50), 19206–19211. Available at https://doi.org/10.1073/pnas.0605275103France.

Riaz, M., Yan, L., Wu, X., Hussain, S., Aziz, O., Jiang, C., 2018. Boron deprivation induced inhibition of root elongation is provoked by oxidative damage, root injuries and changes in cell wall structure. Environmental and Experimental Botany 156, 74–85. Available at https://doi.org/10.1016/j.envexpbot.2018.08.032.

Roccotiello, E., Serrano, H.C., Mariotti, M.G., Branquinho, C., 2015. Nickel phytoremediation potential of the Mediterranean *Alyssoides utriculata* (L.) Medik. Chemosphere 119, 1372–1378. Available at https://doi.org/10.1016/j.chemosphere.2014.02.031.

Rodríguez-Celma, J., Tsai, Y.H., Wen, T.N., Wu, Y.C., Curie, C., Schmidt, W., 2016. Systems-wide analysis of manganese deficiency-induced changes in gene activity of *Arabidopsis* roots. Scientific Reports 6. Available at https://doi.org/10.1038/srep35846.

Rosatto, S., Mariotti, M., Romeo, S., Roccotiello, E., 2021. Root and shoot response to nickel in hyperaccumulator and non-hyperaccumulator species. Plants 10 (3), 508. Available at https://doi.org/10.3390/plants10030508.

Roy, S., Griffiths, M., Torres-Jerez, I., Sanchez, B., Antonelli, E., Jain, D., et al., 2022. Application of synthetic peptide CEP1 increases nutrient uptake rates along plant roots. Frontiers in Plant Science 12. Available at https://doi.org/10.3389/fpls.2021.793145.

Roy, S., Torres-Jerez, I., Zhang, S., Liu, W., Schiessl, K., Jain, D., et al., 2024. The peptide GOLVEN10 alters root development and noduletaxis in *Medicago truncatula*. Plant Journal 118 (3), 607–625. Available at https://doi.org/10.1111/tpj.16626.

Sadhukhan, A., Kobayashi, Y., Iuchi, S., Koyama, H., 2021. Synergistic and antagonistic pleiotropy of STOP1 in stress tolerance. Trends in Plant Science 26 (10), 1014–1022. Available at https://doi.org/10.1016/j.tplants.2021.06.011.

Sato, A., Miura, K., 2011. Root architecture remodeling induced by phosphate starvation. Plant Signaling & Behavior 6, 1122–1126. Available at https://doi.org/10.4161/psb.6.8.15752.

Seregin, I.V., Kozhevnikova, A.D., Kazyumina, E.M., Ivanov, V.B., 2003. Nickel toxicity and distribution in maize roots. Russian Journal of Plant Physiology 50 (5), 711–717. Available at https://doi.org/10.1023/a:1025660712475.

Seregin, I.V., Kozhevnikova, A.D., Schat, H., 2022. Nickel tolerance and accumulation capacities in different populations of the Hyperaccumulator *Noccaea caerulescens*. Russian Journal of Plant Physiology 69 (4). Available at https://doi.org/10.1134/s1021443722040148.

Sexauer, M., Bhasin, H., Schön, M., Roitsch, E., Wall, C., Herzog, U., et al., 2023. A micro RNA mediates shoot control of root branching. Nature Communications 14 (1). Available at https://doi.org/10.1038/s41467-023-43738-6.

Shahzad, Z., Amtmann, A., 2017. Food for thought: how nutrients regulate root system architecture. Current Opinion in Plant Biology 39, 80–87. Available at https://doi.org/10.1016/j.pbi.2017.06.008.

Sheldon, A.R., Menzies, N.W., 2005. The effect of copper toxicity on the growth and root morphology of rhodes grass (*Chloris gayana* Knuth) in resin buffered solution culture. Plant and Soil 278 (1-2), 341–349. Available at https://doi.org/10.1007/s11104-005-8815-3.

Shkolnik, D., Nuriel, R., Bonza, M.C., Costa, A., Fromm, H., 2018. MIZ1 regulates ECA1 to generate a slow, long-distance phloem-transmitted Ca^{2+} signal essential for root water tracking in Arabidopsis. Proceedings of the National Academy of Sciences of the United States of America 115 (31), 8031–8036. Available at https://doi.org/10.1073/pnas.1804130115.

Siegl, A., Afjehi-Sadat, L., Wienkoop, S., 2024. Systemic long-distance sulfur transport and its role in symbiotic root nodule protein turnover. Journal of Plant Physiology 297, 154260. Available at https://doi.org/10.1016/j.jplph.2024.154260.

Song, L., Liu, D., 2015. Ethylene and plant responses to phosphate deficiency. Frontiers in Plant Science 6. Available at https://doi.org/10.3389/fpls.2015.00796.

Song, L., Yu, H., Dong, J., Che, X., Jiao, Y., Liu, D., et al., 2016. The molecular mechanism of ethylene-mediated root hair development induced by phosphate starvation. PLoS Genetics 12 (7), e1006194. Available at https://doi.org/10.1371/journal.pgen.1006194.

Sonu, Nandakumar, S., Singh, V.J., Pandey, R., Gopala Krishnan, S., Bhowmick, P.K., et al., 2023. Implications of tolerance to iron toxicity on root system architecture changes in rice (*Oryza sativa* L). Frontiers in Sustainable Food Systems 7. Available at https://doi.org/10.3389/fsufs.2023.1334487.

Soumare, A., Sarr, D., Diédhiou, A.G., 2023. Potassium sources, microorganisms and plant nutrition: challenges and future research directions. Pedosphere 33 (1), 105–115. Available at https://doi.org/10.1016/j.pedsph.2022.06.025.

Srivastava, A., Thakur, M., Pandey, S., Kumar, C., Sharma, S., Deshmukh, R., et al., 2025. Functions of boron in plant roots: current Insights. South African Journal of Botany 177, 201–210. Available at https://doi.org/10.1016/j.sajb.2024.07.041.

Stevanato, P., Fedito, P., Trebbi, D., Cagnin, M., Saccomani, M., Cacco, G., 2015. Effect of sulfate availability on root traits and microRNA395 expression in sugar beet. Biologia plantarum 59 (3), 491–496. Available at https://doi.org/10.1007/s10535-015-0511-4.

Strieder, M.L., Pinto, K.G., Bertoldi, C., Schneider, A.D.B., Delatorre, C.A., 2017. Response of *Arabidopsis thaliana* root growth to phosphorus and its relation to media chemical composition. Biologia Plantarum 61, 587–594. Available at https://doi.org/10.1007/s10535-017-0713-z.

Sultana, R., Kiran, A., Sanaullah, M., Wakeel, A., 2023. Exploring the linkage between root system architecture and grain iron content in wheat (*Triticum aestivum* L). Frontiers in Sustainable Food Systems 7. Available at https://doi.org/10.3389/fsufs.2023.1156728.

Sustr, M., Soukup, A., Tylova, E., 2019. Potassium in root growth and development. Plants 8 (10), 435. Available at https://doi.org/10.3390/plants8100435.

Svistoonoff, S., Creff, A., Reymond, M., Sigoillot-Claude, C., Ricaud, L., Blanchet, A., et al., 2007. Root tip contact with low-phosphate media reprograms plant root architecture. Nature Genetics 39 (6), 792–796. Available at https://doi.org/10.1038/ng2041.

Tabata, R., Sumida, K., Yoshii, T., Ohyama, K., Shinohara, H., Matsubayashi, Y., 2014. Perception of root-derived peptides by shoot LRR-RKs mediates systemic N-demand signaling. Science (New York, N.Y.) 346 (6207), 343–346. Available at https://doi.org/10.1126/science.1257800.

Taleski, M., Jin, M., Chapman, K., Taylor, K., Winning, C., Frank, M., et al., 2024. CEP hormones at the nexus of nutrient acquisition and allocation, root development, and plant-microbe interactions. Journal of Experimental Botany 75 (2), 538–552. Available at https://doi.org/10.1093/jxb/erad444.

Tanaka-Takada, N., Kobayashi, A., Takahashi, H., Kamiya, T., Kinoshita, T., Maeshima, M., 2019. Plasma membrane-associated Ca^{2+}-binding protein PCaP1 is involved in root hydrotropism of *Arabidopsis thaliana*. Plant and Cell Physiology 60 (6), 1331–1341. Available at https://doi.org/10.1093/pcp/pcz042.

Tejada-Jiménez, M., Gil-Díez, P., León-Mediavilla, J., Wen, J., Mysore, K.S., Imperial, J., et al., 2017. *Medicago truncatula* molybdate transporter type 1 (MtMOT1.3) is a plasma membrane molybdenum transporter required for nitrogenase activity in root nodules under molybdenum deficiency. New Phytologist 216 (4), 1223–1235. Available at https://doi.org/10.1111/nph.14739.

Ticconi, C.A., Lucero, R.D., Sakhonwasee, S., Adamson, A.W., Creff, A., Nussaume, L., et al., 2009. ER-resident proteins PDR2 and LPR1 mediate the developmental response of root meristems to phosphate availability. Proceedings of the National Academy of Sciences of the United States of America 106 (33), 14174–14179. Available at https://doi.org/10.1073/pnas.0901778106.

Vatansever, R., Filiz, E., Ozyigit, I.I., 2016. In silico identification and comparative analysis of molybdenum (Mo) transporter genes in plants. Revista Brasileira de Botanica 39 (1), 87–99. Available at https://doi.org/10.1007/s40415-015-0217-z.

Veatch-Blohm, M.E., Medina, G., Butler, J., 2023. Early lateral root formation in response to calcium and nickel shows variation within disjunct populations of *Arabidopsis* lyrata spp. lyrata. Heliyon 9 (2), e13632. Available at https://doi.org/10.1016/j.heliyon.2023.e13632.

Vert, G., Grotz, N., Dédaldéchamp, F., Gaymard, F., Guerinot, M.L., Briat, J.F., et al., 2002. IRT1, an *Arabidopsis* transporter essential for iron uptake from the soil and for plant growth. The Plant Cell 14 (6), 1223–1233. Available at https://doi.org/10.1105/tpc.001388.

Vidal, E.A., Moyano, T.C., Riveras, E., Contreras-López, O., Gutiérrez, R.A., 2013. Systems approaches map regulatory networks downstream of the auxin receptor AFB3 in the nitrate response of *Arabidopsis thaliana* roots. Proceedings of the National Academy of Sciences of the United States of America 110 (31), 12840–12845. Available at https://doi.org/10.1073/pnas.1310937110.

Vidal, E.A., Álvarez, J.M., Gutiérrez, R.A., 2014. Nitrate regulation of AFB3 and NAC4 gene expression in *Arabidopsis* roots depends on NRT1.1 nitrate transport function. Plant Signaling and Behavior 9. Available at https://doi.org/10.4161/psb.28501.

Villordon, A., Gregorie, J.C., 2021. Variation in boron availability alters root architecture attributes at the onset of storage root formation in three sweetpotato cultivars. HortScience: a Publication of the American Society for Horticultural Science 56 (11), 1423–1429. Available at https://doi.org/10.21273/HORTSCI16134-21.

Wang, Y., Wu, W.H., 2010. Plant sensing and signaling in response to K+-deficiency. Molecular Plant 3 (2), 280–287. Available at https://doi.org/10.1093/mp/ssq006.

Wang, Y., Wu, W.H., 2013. Potassium transport and signaling in higher plants. Annual Review of Plant Biology 64, 451–476. Available at https://doi.org/10.1146/annurev-arplant-050312-120153.

Wang, Y., Wang, R., Zhao, S., Lu, C., Zhu, Z., Li, H., 2022. Transporter NRT1.5/NPF7.3 suppresses primary root growth under low K+ stress by regulating the degradation of PIN-FORMED2. BMC Plant Biology 22 (1). Available at https://doi.org/10.1186/s12870-022-03730-6.

Wang, R., Dong, X., Gao, Y., Hao, F., Zhang, H., Lin, G., 2024a. Molecular mechanism of exogenous magnesium in regulating cation homeostasis in roots of peanut seedlings under salt stress. Agronomy 14 (4), 724. Available at https://doi.org/10.3390/agronomy14040724.

Wang, X., Liu, C., Li, T., Zhou, F., Sun, H., Li, F., et al., 2024b. Hydrogen sulfide antagonizes cytokinin to change root system architecture through persulfidation of CKX2 in *Arabidopsis*. New Phytologist 244 (4), 1377–1390. Available at https://doi.org/10.1111/nph.20122.

Wang, Y., Guo, D., Xu, J., Liu, Z., Wang, L., Chen, C., 2025. Quantitatively analyzed root system architecture influence on root-soil mechanical interactions using 3D printing models. Plant and Soil 513 (2), 2587–2604. Available at https://doi.org/10.1007/s11104-025-07333-6.

Ward, J.T., Lahner, B., Yakubova, E., Salt, D.E., Raghothama, K.G., 2008. The effect of iron on the primary root elongation of *Arabidopsis* during phosphate deficiency. Plant Physiology 147 (3), 1181–1191. Available at https://doi.org/10.1104/pp.108.118562.

Weber, J.N., Minner-Meinen, R., Kaufholdt, D., 2024. The mechanisms of molybdate distribution and homeostasis with special focus on the model plant *Arabidopsis thaliana*. Molecules (Basel, Switzerland) 29 (1). Available at https://doi.org/10.3390/molecules29010040.

Wei Yang, T.J., Perry, P.J., Ciani, S., Pandian, S., Schmidt, W., 2008. Manganese deficiency alters the patterning and development of root hairs in *Arabidopsis*. Journal of Experimental Botany 59 (12), 3453–3464. Available at https://doi.org/10.1093/jxb/ern195.

Wilder, S.L., Scott, S., Waller, S., Powell, A., Benoit, M., Guthrie, J.M., et al., 2022. Carbon-11 radiotracing reveals physiological and metabolic responses of maize grown under different regimes of boron treatment. Plants 11 (3). Available at https://doi.org/10.3390/plants11030241.

Wu, X., Riaz, M., Yan, L., Du, C., Liu, Y., Jiang, C., 2017. Boron deficiency in trifoliate orange induces changes in pectin composition and architecture of components in root cell walls. Frontiers in Plant Science 8. Available at https://doi.org/10.3389/fpls.2017.01882.

Wu, S., Sun, X., Tan, Q., Hu, C., 2019. Molybdenum improves water uptake via extensive root morphology, aquaporin expressions and increased ionic concentrations in wheat under drought stress. Environmental and Experimental Botany 157, 241–249. Available at https://doi.org/10.1016/j.envexpbot.2018.10.013.

Xu, J., Wang, Y., Liu, J., 2008. Copper toxicity on the development and architecture of rice root. 2nd International Conference on Bioinformatics and Biomedical Engineering, Shanghai, China, pp. 4470–4474. https://doi.org/10.1109/ICBBE.2008.611.

Xue, C., Li, W., Shen, R., Lan, P., 2021. PERK13 modulates phosphate deficiency-induced root hair elongation in *Arabidopsis*. Plant Science 312, 111060. Available at https://doi.org/10.1016/j.plantsci.2021.111060.

Xu, P., Ma, W., Hu, J., Cai, W., 2022. The nitrate-inducible NAC transcription factor NAC056 controls nitrate assimilation and promotes lateral root growth in Arabidopsis thaliana. PLOS Genetics 18, e1010090. Available at https://doi.org/10.1371/journal.pgen.1010090.

Yan, Y., Wang, H., Hamera, S., Chen, X., Fang, R., 2014. miR444a has multiple functions in the rice nitrate-signaling pathway. The Plant Journal 78 (1), 44–55. Available at https://doi.org/10.1111/tpj.12446.

Yang, T., Feng, H., Zhang, S., Xiao, H., Hu, Q., Chen, G., et al., 2020. The potassium transporter OsHAK5 alters rice architecture via ATP-dependent transmembrane auxin fluxes. Plant Communications 1 (5), 100052. Available at https://doi.org/10.1016/j.xplc.2020.100052.

Yang, Z., Zhang, L., Li, J., Dong, X., Guan, B., Yu, X., et al., 2025. miR395 regulates sulfur metabolism and root development in Chinese kale under sulfur deficiency. Environmental and Experimental Botany 237, 106204. Available at https://doi.org/10.1016/j.envexpbot.2025.106204.

Yawson, D.O., Amoah, K.K., Asare, P.A., Techie-Menson, J., Afutu, E., Atiah, K., et al., 2020. A meta-analysis of modifications of root system traits of crop plants to potassium

(K) deprivation. Plant Roots. InTech, London, UK. Available at https://doi.org/10.5772/intechopen.95044.

Yen, C.P., 1987. Tree root patterns and erosion control. Proceedings of the International Workshop on Soil Erosion and its Counter-Measures. Bangkok, Thailand, pp. 92–111.

Yusefi-Tanha, E., Fallah, S., Rostamnejadi, A., Pokhrel, L.R., 2022. Responses of soybean (*Glycine max* [L.] Merr.) to zinc oxide nanoparticles: understanding changes in root system architecture, zinc tissue partitioning and soil characteristics. Science of The Total Environment 835, 155348. Available at https://doi.org/10.1016/j.scitotenv.2022.155348.

Yusefi-Tanha, E., Fallah, S., Rostamnejadi, A., Pokhrel, L.R., 2020. Root system architecture, copper uptake and tissue distribution in soybean (*Glycine max* (L.) Merr.) grown in copper oxide nanoparticle (CuONP)-amended soil and implications for human nutrition. Plants 9 (10), 1326. Available at https://doi.org/10.3390/plants9101326.

Zhang, H., Jennings, A., Barlow, P.W., Forde, B.G., 1999. Dual pathways for regulation of root branching by nitrate. Proceedings of the National Academy of Sciences of the United States of America 96 (11), 6529–6534. Available at https://doi.org/10.1073/pnas.96.11.6529.

Zhang, J., Zhang, X., Wang, R., Li, W., 2014a. The plasma membrane-localised Ca^{2+}-ATPase ACA8 plays a role in sucrose signalling involved in early seedling development in *Arabidopsis*. Plant Cell Reports 33 (5), 755–766. Available at https://doi.org/10.1007/s00299-014-1590-y.

Zhang, L., Pan, Y., Lv, W., Xiong, Z.-T., 2014b. Physiological responses of biomass allocation, root architecture, and invertase activity to copper stress in young seedlings from two populations of *Kummerowia stipulacea* (maxim.) Makino. Ecotoxicology and Environmental Safety 104 (1), 278–284. Available at https://doi.org/10.1016/j.ecoenv.2014.03.013.

Zhang, H., Zhao, F.G., Tang, R.J., Yu, Y., Song, J., Wang, Y., et al., 2017. Two tonoplast MATE proteins function as turgor-regulating chloride channels in *Arabidopsis*. Proceedings of the National Academy of Sciences of the United States of America 114 (10). Available at https://doi.org/10.1073/pnas.1616203114 E2036.

Zhang, X.P., Ma, C.X., Sun, L.R., Hao, F.S., 2020. Roles and mechanisms of Ca^{2+} in regulating primary root growth of plants. Plant Signaling and Behavior 15 (5). Available at https://doi.org/10.1080/15592324.2020.1748283.

Zhang, C., He, M., Wang, S., Chu, L., Wang, C., Yang, N., et al., 2021. Boron deficiency-induced root growth inhibition is mediated by brassinosteroid signalling regulation in *Arabidopsis*. The Plant Journal 107 (2), 564–578. Available at https://doi.org/10.1111/tpj.15311.

Zhang, Y., Zhang, H.Z., Lamboro, A., Fu, J.Y., Du, Y.Y., Qu, J., et al., 2023. Enhancement of root sulfur metabolic pathway by overexpression of OAS-TL3 to increase total soybean seed protein content. Molecular Breeding 43 (1). Available at https://doi.org/10.1007/s11032-022-01348-y.

Zhao, Q., Wu, Y., Gao, L., Ma, J., Li, C.-Y., Xiang, C.-B., 2014. Sulfur nutrient availability regulates root elongation by affecting root indole-3-acetic acid levels and the stem cell niche. Journal of Integrative Plant Biology 56 (12), 1151–1163. Available at https://doi.org/10.1111/jipb.12217.

Zhao, J., Wang, W., Zhou, H., Wang, R., Zhang, P., Wang, H., et al., 2017. Manganese toxicity inhibited root growth by disrupting auxin biosynthesis and transport in *Arabidopsis*. Frontiers in Plant Science 8. Available at https://doi.org/10.3389/fpls.2017.00272.

Zhou, S., Walker, R.R., Edwards, E.J., 2024. Fine-root and leaf acquisitive traits decoupled from chloride accumulation in reflecting the differential salinity tolerance among *Prunus* hybrids. Frontiers in Plant Science 15. Available at https://doi.org/10.3389/fpls.2024.1502201.

Zhou, Y., Zhang, Y., Chen, H., Jia, J., Meng, G., Wu, S., et al., 2025a. Root architecture and nutritional status of copper-pruned container-grown *Ligustrum lucidum* seedlings after transplanting. New Forests 56 (3). Available at https://doi.org/10.1007/s11056-025-10098-6.

Zhou, Z.Z., Teng, N.H., Liang, J.L., Zeng, Y.Y., Wang, Y.Z., Zeng, L.Q., et al., 2025b. Identification, characterization, and expression profiling of rice MADS-box transcription factor genes associated with fluroxypyr-meptyl and oxyfluorfen metabolism. Genetica 153 (1). Available at https://doi.org/10.1007/s10709-025-00248-1.

CHAPTER

Beneficial elements in stress management

21

Fernando Carlos Gómez-Merino[1], Eréndira Esmeralda Hernández-Andrade[2] and Libia Iris Trejo-Téllez[2]

[1]*Department of Plant Physiology, College of Postgraduates in Agricultural Sciences, Montecillo, Texcoco, State of Mexico, Mexico,*

[2]*Department of Soil Science-Plant Nutrition, College of Postgraduates in Agricultural Sciences, Montecillo, Texcoco, State of Mexico, Mexico*

Abbreviations

ABA	abscisic acid
ABI5	abscisic acid-insensitive 5
ABC1	gene encoding ATP-binding cassette transporter 1
ALMT1	aluminum-activated malate transporter 1
AMF	arbuscular mycorrhizal fungi
ANAC017	NAC transcription factor 017
ANAC017-XTH	ANAC017-regulated xyloglucan endotransglucosylase/hydrolase
ANAC050	NAC transcription factor 050
ANAC070	NAC transcription factor 070
APase	acid phosphatase
APX	ascorbate peroxidase
AsA	ascorbic acid
AsA-GSH	ascorbate–glutathione cycle
AUX	auxins
Bjgb	*Bradyrhizobium japonicum* globin
CAS	cysteine synthase
CAT	catalase
$Ce(NO_3)_3$	cerium nitrate
CeO_2	cerium dioxide
CH_3OH	methanol
CH_4	methane
Chl *a*	chlorophyll a
Chl *a* + *b*	total chlorophyll
Chl *b*	chlorophyll b
CO_2	carbon dioxide
$CoSO_4$	cobalt sulfate

Mineral Nutrients to Mitigate Abiotic Stress in Crops. DOI: https://doi.org/10.1016/B978-0-443-44712-9.00023-2

CRISPR-Cas9	clustered regularly interspaced short palindromic repeats–CRISPR-associated protein 9
Cys	cysteine
DHAR	dehydroascorbate reductase
DNA	deoxyribonucleic acid
DPPH	2,2-diphenyl-1-picryl-hydrazyl
EC	electrical conductivity
ETR	electron transport rate
Fv/Fm	maximum photochemical efficiency of PSII, or chlorophyll fluorescence efficiency
GA	gibberellins
GABA	γ-aminobutyric acid
GH3	auxin-amido synthetase
GHGs	greenhouse gases
Gly I	glyoxalase I
Gly II	glyoxalase II
GPX	glutathione peroxidase
GR	glutathione reductase
G_s	stomatal conductance
GSH	reduced glutathione
GSH/GSSG	reduced/oxidized glutathione ratio
GSH1	gene encoding glutathione synthetase 1
GSSG	oxidized glutathione
GST	glutathione S-transferase
H_2O_2	hydrogen peroxide
H_2S	hydrogen sulfide
H_4SiO_4	monosilicic acid
$HgCl_2$	mercury(II) chloride
HvLsi1	*Hordeum vulgare* low silicon transporter 1
HvLsi2	*Hordeum vulgare* low silicon transporter 2
I	iodide
IAA	indole-3-acetic acid
Ci	intracellular CO_2 concentration
IO_3	iodate
Ip	isopentenyl adenine
KI	potassium iodide
KIO_3	potassium iodate
La(III)-AA	lanthanum(III)-amino acid chelate
$LaCl_3$	lanthanum(III) chloride
Lsi	low silicon transporter
MCM	methylmalonyl-CoA mutase
MDA	malondialdehyde
MDH	methanol dehydrogenase
MDHAR	monodehydroascorbate reductase
Met	methionine
MS	methionine synthase
NAC	NAC transcription factor

NaCl	sodium chloride
NaF	sodium fluoride
Na_2S	sodium sulfide
Na_2SeO_3	solium selenite
Na_2SeO_4	sodium selenate
NCED	9-cis-epoxy carotenoid dioxygenase gene
$nCeO_2$	cerium dioxide nanoparticles
NO	nitric oxide
NO_2	nitrite
NO_3	nitrate
NOAEL	no observable adverse effect level
N_2	dinitrogen
N_2O	nitrous oxide
TiO_2-NPs	titanium dioxide nanoparticles
OA	organic acids
$O_2\bullet^-$	superoxide radical
PAL	phenylalanine ammonia-lyase
PCS	gene encoding phytochelatin synthase
P_N	net photosynthetic rate
POD	peroxidase
PPO	polyphenol oxidase
Prx	peroxiredoxin
PSII	photosystem II
q_P	photochemical quenching
ΦPSII	PSII photochemistry (effective quantum yield of photosystem II)
REE	rare earth element
REEs	rare earth elements
RNR	ribonucleotide reductase
RNS	reactive nitrogen species
ROS	reactive oxygen species
RSNO	S-nitrosothiol
Rubisco	ribulose-1,5-bisphosphate carboxylase/oxygenase
RWC	relative water content
SA	salicylic acid
SDH	shikimate dehydrogenase
Se-NPs	selenium nanoparticles
$SiO_2\bullet nH2O$	amorphous silica
Slp1	siliplant1 protein
SOD	superoxide dismutase
SPAD	soil plant analysis development
STOP1	sensitive to proton rhizotoxicity 1
SwE	seaweed extract
TCA	tricarboxylic acid cycle
TG	titanium gypsum
TiO_2	titanium dioxide
TM	transmembrane
TM1–2	transmembrane helices 1–2

TM5–6	transmembrane helices 5–6
T_r	net transpiration rate
Trx	thioredoxin
UV-B	ultraviolet-B radiation

21.1 Introduction

Plants require various essential nutrients to perform well and have high crop yields. These nutrients are essential in supporting significant physiological processes in plants, which include germination, root development, flowering, and fruiting. Each nutrient has a unique role to play, and without it, it can have negative effects on plant health and crop production. These chemical elements are thus indispensable for a plant's growth, development, and reproduction. By definition, a chemical element to be considered essential must fulfill three criteria: (1) the plant cannot complete its life cycle without it (criterion of life cycle completion); (2) no other element can substitute for it (criterion of specificity); and (3) it is part of a necessary metabolic process, enzyme function or molecular structure (criterion of direct action) (Kirkby, 2023). Taking into consideration the criteria of essentiality proposed by Arnon and Stout (1939), in the field of classical plant nutrition, there are 17 elements required for most higher plants to complete their normal life cycle, thus fulfilling the principles of essentiality. These 17 essential nutrients are categorized into macronutrients (typically present at > 500 mg kg^{-1} in plant tissues on a dry basis) and micronutrients (regularly present at < 500 mg kg^{-1} in plant tissues on a dry basis). Macronutrients include the basic elements obtained from air and water, and others from the soil: carbon (C), hydrogen (H), oxygen (O), nitrogen (N), phosphorus (P), potassium (K), calcium (Ca), magnesium (Mg), and sulfur (S). Micronutrients may be provided either via precipitation or a soil solution and include chlorine (Cl), iron (Fe), manganese (Mn), zinc (Zn), copper (Cu), boron (B), molybdenum (Mo), and nickel (Ni). Beneficial elements are not essential for plants but may benefit metabolic, physiological, or stress responses, thus conferring better performance under challenging environmental conditions (Brown et al., 2022). Beneficial elements encompass an additional list of 12 elements that boost growth, stress resistance or tolerance, and yield in specific plants, enhancing overall plant health and development. To date, the list of beneficial elements arranged alphabetically by their chemical symbol includes: aluminum (Al), cerium (Ce), cobalt (Co), iodine (I), lanthanum (La), sodium (Na), neodymium (Nd), praseodymium (Pr), selenium (Se), silicon (Si), titanium (Ti) and vanadium (V) (Gómez-Merino and Trejo-Téllez, 2018; Trejo-Téllez and Gómez-Merino, 2023). Moreover, other nonessential elements, such as silver (Ag), arsenic (As), cadmium (Cd), chromium (Cr), fluorine (F), mercury (Hg), lithium (Li), lead (Pb), and tungsten or wolfram (W), may also improve plant performance in terms of growth, development, yield and responses to environmental stressors (Poschenrieder et al., 2013; Arif et al., 2016; Escobar-Sepúlveda et al., 2017), though available evidence is insufficient or even contradictory to make a firm determination and classification as beneficial elements.

It is especially noteworthy that beneficial elements are capable of boosting physiological, biochemical, and molecular adjustments in plant metabolism as a strategy to thrive in challenging environments. In addition, these elements may stimulate biological processes and productive attributes including seed germination and bud emergence, biomass production, flowering and flower development, fruit set, yield, and crop quality in a hormetic manner. Hormesis is the natural phenomenon where low doses of an agent trigger beneficial, adaptive responses, whereas high doses of that agent cause inhibitory, toxic, or harmful effects, showing opposite outcomes depending on concentration (Agathokleous et al., 2020; Husen, 2024). Although hormesis is well documented for lanthanum (Agathokleous et al., 2019), the rest of the beneficial elements need to be further investigated to determine if their application indeed results in biphasic dose–responses in plants, pointing to a hormetic behavior.

Factors influencing the achievement of favorable responses in plants to exposure to beneficial elements include, but are not limited to, dosage or concentration, rate, timing, and application technique, considering the plant's needs. Simultaneously, genotype and phenological stage of plant species, agronomic methods, and environmental conditions represent key factors when deciding which, how, and when a beneficial element should be applied.

During evolution, plants developed two different strategies for coping with environmental stress, namely resistance and tolerance. While resistance relates to a plant's ability to prevent or restrict the damage or impact of a stressor through physical, chemical, or specific genetic machinery, tolerance is the capability of a plant to endure the detrimental effects of a stressor on its fitness (survival, growth, and reproduction) despite the presence of the stress or damage. Notably, beneficial elements can trigger both resistance and tolerance mechanisms in plants to cope with abiotic stressors, resulting in adaptive responses to survive. Consequently, plant growth, development, reproduction, and quality attributes of final products can be improved in crops grown under limiting conditions if they are properly treated with these elements (Trejo-Téllez and Gómez-Merino, 2023). Given that beneficial elements markedly enhance productivity and resilience within the scope of hormesis (Agathokleous et al., 2020; Jalal et al., 2021), it is crucial to calculate the precise range of doses at which these elements display stimulatory effects on plants and establish the no observable adverse effect level (NOAEL) and the stress sensitization and toxicity (Agathokleous, 2025). Additionally, the physiological, metabolic, and genetic pathways behind the positive, neutral, or negative effects of these elements on plant biology are issues of paramount relevance to be investigated in detail. Moreover, the absorption, distribution, metabolization, and storage of beneficial elements in plant cells are processes of great significance for developing innovative applications supporting sustainable agriculture (Vatansever et al., 2016).

This chapter outlines some cutting-edge advancements in research regarding the stimulatory effects of 12 recognized beneficial elements for plants: Al, Ce, Co, I,

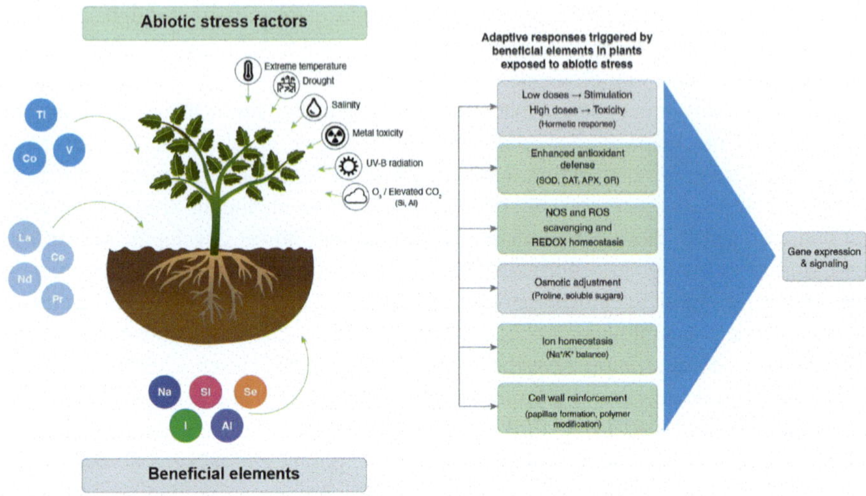

FIGURE 21.1

Beneficial elements enhance plant tolerance and resistance to abiotic stress through physiological, biochemical, and molecular mechanisms.

La, Na, Nd, Pr, Se, Si, Ti, and V. Specifically, we highlight the role of these elements in boosting morpho-physiological, biochemical, and genetic processes to enhance resilience to abiotic stress factors. Fig. 21.1 summarizes the most prominent findings on these elements for plants and their biological significance in driving improved capacities to thrive in challenging environments.

21.2 Physiological and molecular mechanisms underlying the stimulatory effects of beneficial and essential trace elements

21.2.1 Aluminum

Aluminum (Al) is the third most plentiful element overall (after O and Si) and the most abundant metal in the lithosphere, comprising between 8.1% and 8.3% of its mass (Lide, 2017). Applied at optimal concentrations, Al may exert stimulatory effects on plants, depending on various factors including its chemical form in the growth medium, cultivation conditions, and genetic makeup of the plant (Bojórquez-Quintal et al., 2017; Ofoe et al., 2023). In tea plants (*Camellia sinensis*), Al is essential for root growth, development, and nutrient uptake (Fan et al., 2019; Sun et al., 2020). Like most, if not all, beneficial elements, Al may improve plant tolerance to some environmental challenges when applied at low concentrations (Bojórquez-Quintal et al., 2017; Muhammad et al., 2019). Conversely, at high levels, it may result in toxic effects, limiting plant performance and inducing yield penalties of 25%–80% depending on the plant genotype (Sade et al., 2016; Fan et al., 2019). Importantly, ca. 40% of global arable land has high levels of Al, especially in the tropics and subtropics (de Jesus et al., 2017).

The root apoplast accumulates most of the Al absorbed via root hairs from the soil solution or growth medium, accounting for up to 80% of the total Al fraction absorbed (Sade et al., 2016). Since cell walls accumulate up to 90% of this absorbed Al, they become significantly firmer. Once inside the plant cell, Al may move into the symplast, thus inducing changes in different processes including responses to environmental stressors, nutrient uptake, reorganization of subcellular components, and metabolism of the entire plant (Silva et al., 2020a).

In plants that thrive in acidic environments where P availability is low, both $Fe^{2+/3+}$ and Al^{3+} may stabilize the protein Sensitive to Proton rhizotoxicity 1 (STOP1), which gives rise to proton tolerance (Sawaki et al., 2009). As a transcription factor, STOP1 accumulates in the nucleus and induces the expression of the transmembrane protein Aluminum-activated Malate Transporter 1 (ALMT1). This protein plays a critical role in plant nutrition and root protection in acidic, aluminum-toxic soils by excluding toxic Al^{3+} ions from the root apex (Sharma et al., 2016; Godon et al., 2019). By binding to the extracellular surface of ALMT1 in *Arabidopsis*, Al^{3+} ions trigger conformational changes in its transmembrane loops, opening the channel gate for malate efflux to secrete organic acids (OA) for external detoxification (Wang et al., 2022). In rice bean (*Vigna umbellata*), Al signals trigger the endogenous accumulation of abscisic acid (ABA), while ABA-mediated Al responses are driven by the transcription factor ABI5, which induces cell wall adjustments and osmotic balance (Fan et al., 2018). Similarly, in mandarin lima (*Citrus limonia*), aluminum toxicity prompts the transcriptional activation of the 9-*cis*-epoxy carotenoid dioxygenase (*NCED*) gene, boosting ABA biosynthesis in the roots and gas exchange in leaf stomata (Gavassi et al., 2021).

Al may also boost mechanisms of resistance or tolerance against nutrient imbalances and phytotoxicity (Kaur et al., 2016; Ofoe et al., 2023). In tea plants, Al alleviated B-deficiency by up-regulating C and N metabolism, activating antioxidative defense, and enhancing B uptake and transport (Hajiboland et al., 2014). In rice (*Oryza sativa*) seedlings, the application of 200 μM Al attenuated Mn toxicity by adjusting membrane potential and modifying cell membrane binding properties; this dampened Mn toxicity, thereby reducing Mn absorption in root cells and accumulating it in shoots (Wang et al., 2015). In barley (*Hordeum vulgare*), Al triggers antagonistic interactions with Mn and mitigates Mn toxicity (Muhammad et al., 2016). In Indian rhododendron (*Melastoma malabathricum*), Al has been reported to alleviate Fe toxicity (Watanabe et al., 2006), while in the rice relative *Oryza rufipogon* and in maize (*Zea mays*), Al mitigates Fe deficiency chlorosis (Alves et al., 2026). Alleviation of Fe toxicity by Al has also been found in tea and rice cultivars, where it reduces Fe absorption and remobilization, thus inhibiting chlorosis, purpling, stunting, and bronzing in leaves, while decreasing root cell surface charges (Hajiboland et al., 2013; Alia et al., 2015; Alves et al., 2025). In rice, Al alleviated Fe deficiency chlorosis by interfering with Pi homeostasis (Alves et al., 2025). In tea plants, Al attenuates fluoride (F) toxicity by forming Al-F complexes and consequently enhancing plant growth (Yang et al., 2016). Similarly, Al promoted P uptake in citrus plants (*Citrus sinensis*) (Li et al., 2016), while in maize, Al stimulated the transcriptional activation of genes that mitigate Pi deficiency (Maron et al., 2008). In citrus, Al induced the accumulation of low P-responsive proteins in leaves. These proteins play a pivotal role in triggering adaptive responses to Pi-starvation, and include glycerophosphodiester phosphodiesterase (GDPD), phosphoenolpyruvate carboxylase (PEPCase), purple acid phosphatase (PAP), pyrophosphatases (PPase), and ribonucleases (RNase) (Li et al., 2016).

In rice, exposure to 200 μM Al stimulated growth and increased the concentrations of chlorophylls, sugars, and essential elements (P, K, and Mg), with a concomitant impact on the expression of *NAC* transcription factors (Moreno-Alvarado et al., 2017). Coincidently, the *Arabidopsis* ANAC017 transcription factor modulates Al tolerance by modifying the expression of cell wall-remodeling genes (Tao et al., 2022), whereas ANAC070 represses the ANAC017-XTH31 module linked to xyloglucan biosynthesis, thereby participating in the control of Al tolerance (Li et al., 2024). Recent research found that the *Arabidopsis* ANAC050 transcription factor confers aluminum resistance by cooperating with OA exudation and the accumulation of hemicelluloses in the cell wall (Tao et al., 2025).

Considering that Al is the most prevalent metallic element in the lithosphere (up to 8.3% by weight) and becomes more soluble in acid soils as the pH decreases, plants may come into contact with certain amounts of this metal during their life cycle (Pax et al., 2015). Therefore, Al may prompt positive, neutral, or negative effects depending on the plant's genetic makeup, the ionic form of the metal, and the environmental conditions where plants thrive (Bojórquez-Quintal et al., 2017; Liu et al., 2020; Ofoe et al., 2023).

21.2.2 Cerium

Cerium (Ce) is the most abundant rare earth element in the lithosphere, with an average concentration of approximately 66.5 mg kg^{-1}, though it is present in very low concentrations in the oceans (i.e., 1.2×10^{-6} mg L^{-1}) (Lide, 2017; Kotelnikova et al., 2019; NCBI National Center for Biotechnology Information, 2026).

Cerium is an essential cofactor for the key metabolic enzyme methanol dehydrogenase (MDH) in certain methylotrophic bacteria. These bacteria, known as lanthanide-dependent methylotrophs, require rare earth elements (REEs) like Ce to grow on one-carbon compounds such as methanol (CH_3OH) (Daumann, 2019; Skovran et al., 2019). Methylotrophic bacteria are pivotal to plant ecosystems and agricultural sustainability. They promote plant productivity and enhance resilience against environmental stressors by inducing mechanisms like hormone biosynthesis, nutrient solubilization, and N fixation. By acting as biofertilizers, they reduce chemical reliance and improve soil health and fertility (Masuda et al., 2018; Nysanth et al., 2023). They colonize plant roots (rhizosphere) and leaves (phyllosphere), improving nutrient access and helping plants adapt to harsh conditions, offering a green solution for food security (Ameesha et al., 2025). Moreover, methylotrophic bacteria, particularly methanotrophs, play a crucial role in climate change by consuming potent greenhouse gases (GHGs) like methane (CH_4) and CO_2, acting as a natural sink. These versatile microbes can convert GHGs into biomass, offering pathways for carbon sequestration and biotechnological solutions for climate-neutral chemical production, making their study vital for mitigation strategies and sustainable agriculture (Kumar et al., 2018; Ehinmitan et al., 2025). In particular, *Oharaeibacter diazotrophicus*, a methylotrophic and diazotrophic bacterium isolated from the rice rhizosphere, harbors alpha-proteobacterial methylotrophy genes that utilize La^{3+}, Ce^{3+}, Nd^{3+}, and Pr^{3+} as cofactors for the corresponding enzymatic activity (Lv et al., 2017; Lv and Tani, 2018). It is worth mentioning that plant growth significantly influences methanol-utilizing methylotrophs in soil by altering their diversity and boosting their metabolic activity, particularly in the rhizosphere (Macey et al., 2020).

In members of the Fabaceae plant family, such as common bean (*Phaseolus vulgaris*), soybean (*Glycine max*), and chickpea (*Cicer arietinum*), the endosymbiotic bacterium *Bradyrhizobium diazoefficiens* is a crucial component of N fixation and confers resistance against flooding. In soybean nodules, the hemoglobin Bjgb from *B. diazoefficiens* controls nitric oxide (NO) homeostasis to protect symbiotic nitrogen fixation and mitigate hypoxia damage induced by flooding (Salas et al., 2020). Notably, *B. diazoefficiens* can consume methanol as an energy or carbon source to grow when light lanthanides (i.e., La^{3+}, Ce^{3+}, Nd^{3+}, or Pr^{3+}) are abundant, which simultaneously stimulates MDH activity (Wang et al., 2019). In both common bean (Xie et al., 2019) and tomato (*Solanum lycopersicum*) (Singh et al., 2019), not only cerium dioxide nanoparticles ($nCeO_2$) but also Ce^{3+} ions ($CeCl_3$) drive the activation of the enzymatic antioxidant system. Similarly, in wheat seedlings exposed to salt stress (40, 80, 120, or 160 mM NaCl), cerium nitrate [$Ce(NO_3)_3$] enhanced the activity of antioxidant enzymes, including ascorbate peroxidase (APX), catalase (CAT), glutathione peroxidase (GPX), glutathione reductase (GR), and superoxide dismutase (SOD). This treatment also increased the ratios of ascorbate/dehydroascorbate and reduced glutathione (GSH)/oxidized glutathione (GSSG), the content of chlorophylls and carotenoids, net photosynthetic rate (P_N), plant height, plant dry mass, and root length, and significantly reduced the content of hydrogen peroxide (H_2O_2) and that of malondialdehyde (MDA) in chloroplasts, with an overall improved resistance to salt stress (Chen and Shan, 2019).

Although studies testing the ionic form of bulk Ce on the amelioration of plant responses to stress are limited, research into the impact of nanoceria or $nCeO_2$ is increasing (i.e., Preetha et al., 2023; Ukwattage and Zhiyong, 2025; Soleymani et al., 2025), unveiling novel nanoparticle effects. It is important to take into consideration that absorption, transport, and utilization of nanoparticles vary not only among but also within species. Moreover, other biological conditions of the plant such as phenological stages and tissues treated, as well as the soil and general environmental conditions (i.e., soil moisture, temperature, or microbial activity), represent critical factors that regulate the final positive, neutral, or negative effects of the nanoparticles applied (Solanki et al., 2015; Madlala et al., 2024; Demeke et al., 2025). As a consequence of the substantial decrease in size, nanomaterials expand their surface mass ratio, exhibiting free, gradual, and constant movements for longer time (Pozzi et al., 2024), offering greater efficacy for agricultural applications, including enhanced efficiency in nutrient use under adverse environmental conditions exacerbated by the impact of climate change (Yang et al., 2025).

In plants grown under high-salt environments, exogenous supply of $nCeO_2$ can help mitigate the effects of salt stress. In canola (*Brassica napus*) treated with 100 mM NaCl, application of 200 or 1000 mg $nCeO_2$ kg^{-1} dry substrate enhanced growth, biochemical parameters, and photosynthesis (Rossi et al., 2016). When cotton (*Gossypium hirsutum*) seeds imbibed in 200 mM NaCl were treated with $nCeO_2$, significant increases in germination were observed, with further improvements in seedlings including better morphological and physiological characteristics, reduction in reactive oxygen species (ROS), stimulation of antioxidant activity, higher concentrations of Ca^{2+} and Mg^{2+}, and maintenance of ion homeostasis (An et al., 2020). In $nCeO_2$-primed seeds of rapeseed (*Brassica napus* subsp. *napus*) exposed to 200 mM NaCl, the nanoparticles maintained ion homeostasis by effectively modulating the Na^+/K^+ ratio while increasing α-amylase activity, thereby attenuating salt stress (Khan et al., 2021). In *Arabidopsis*, $nCeO_2$ helps attenuate oxidative stress by efficiently neutralizing free radicals and activating the transcriptional expression of K^+ channels, thereby triggering the plant's tolerance mechanisms to cope with salt stress (Wu et al., 2018). In Moldavian dragonhead (*Dracocephalum moldavica) exposed to* 50 and 100 mM NaCl, foliar applications of $nCeO_2$ enhanced leaf and shoot fresh and dry weight, photosynthetic pigments, antioxidant levels of SOD, APX, GPX, and proline content, whereas levels of MDA and H_2O_2, and electrolyte leakage were significantly reduced (Mohammadi et al., 2021). Likewise, in grapevine (*Vitis vinifera*) under 75 mM NaCl, foliar spray of $nCeO_2$ (25, 50, and 100 mg L^{-1}) significantly alleviated the deleterious effects of salt stress by enhancing growth parameters, relative water content (RWC), photosynthetic efficiency, and antioxidant enzyme activities (GPX, APX, SOD) while reducing oxidative damage markers such as proline, MDA, and electrolyte leakage (Gohari et al., 2021). In purslane (*Portulaca oleracea*) treated with 50 mM NaCl, a combination of nanoceria and salicylic acid (SA) improved seed weight as well as the contents of total soluble solids, anthocyanin, chlorophyll *b* (chl *b*), and K, compared to control plants (Hassanpouraghdam et al., 2022).

Nanoceria supply has also been found to boost plant performance under drought stress. In sorghum (*Sorghum bicolor*), foliar spray of $nCeO_2$ attenuated the harmful effects of drought by activating the antioxidant system, decreasing oxidative damage, and improving photosynthesis and pollen fertility, thereby increasing yield compared to control plants (Djanaguiraman et al., 2018). Similarly, in apple (*Malus × domestica*) cv. Red Delicious, $nCeO_2$ alleviated drought stress by adjusting ion balance, scavenging free radicals, activating transcriptional expressions of genes involved in osmotic stress tolerance, and boosting phytohormone cross-talk (Soleymani et al., 2025).

In *Arabidopsis* plants subjected to chilling stress (4°C for 5 days), nanoceria accumulated in chloroplasts, activating the antioxidant system, stimulating photosynthetic parameters, and increasing Rubisco activity compared to stressed plants that did not receive nanoceria. This demonstrates that $nCeO_2$ represents a valuable multitasking tool for alleviating various abiotic environmental stressors, including high-intensity light and extreme temperatures (Wu et al., 2017).

Toxic effects of some chemical elements in plants may also be alleviated by applications of $nCeO_2$. In maize, the supplementation of 1000 ppm $nCeO_2$ in plants treated with 0.5 mg Cd kg^{-1} soil on a dry basis attenuated the deleterious effects of cadmium (Ayub et al., 2023). Finally, in mungbean (*Vigna radiata*) exposed to 50 ppm $HgCl_2$, the combination of $nCeO_2$ with *Bacillus coagulans* and *Escherichia coli* significantly mitigated ROS accumulation induced by $HgCl_2$ and enhanced yield components (Mukherjee et al., 2023).

21.2.3 Cobalt

In the Earth's crust, cobalt (Co) is present at 22.5 mg kg^{-1}, while in seawater its concentration is ca. 2×10^{-5} mg L^{-1} (Lide, 2017); it is therefore considered a rather scarce element in nature (Okamoto and Eltis, 2011).

While not considered a nutrient for all taxa, some plant species benefit considerably from Co applications. However, Co is considered an essential element for some lower plants, including marine algal species (diatoms, chrysophytes, and dinoflagellates), and for vascular plants within the family *Fabaceae* (legumes) (Gál et al., 2008; Hu et al., 2021). In legume nodules, Co is a critical component of cobalamin, a key coenzyme for several specific bacterial enzymes—including ethanolamine ammonia-lyase, glutamate mutase, diol dehydratase, reductive dehalogenases, methionine synthase, methylmalonyl-CoA mutase, and ribonucleotide reductase (class II)—some of which are determinant factors for nitrogen fixation and metabolization. While not an integral part of the nitrogenase enzyme itself, Co is required for the synthesis and activity of enzymes supporting optimal nodule function and nitrogenase efficiency (Chatzistathis, 2018; Gad et al., 2025).

When fenugreek (*Trigonella foenum-graecum*) plants were grown in soils containing toxic levels of cobalt, inoculation with flavonoid-activated *Rhizobium tibeticum* resulted in improved biomass accumulation and higher nodulation rates, with a concomitant increase in the enzymatic activity of both glutamine synthetase and nitrogenase when plants were supplied with 50 mg Co kg^{-1} (Abd-Alla et al., 2014).

In barley grown in saline soil (pH 8.6) supplemented with different Co concentrations (0.0, 7.5, 10.0, 12.5, 15.0, 17.5, and 20.0 ppm), all cobalt levels significantly stimulated growth and yield compared to the control. Increasing Co levels up to 10.0 ppm enhanced growth, dry matter content, grain quality, and yield; furthermore, while N, P, K, Ca, Mg, Mn, Zn, and Cu contents increased, Na and Cl levels decreased (Gad et al., 2011).

Similarly, in maize planted in saline soil (pH 8.6), applying 15, 17.5, or 20 ppm Co substantially increased root and shoot growth parameters, while auxins (AUX), gibberellins (GA), cytokinins (CK), and ABA gradually rose in shoots as cobalt levels in plant growing media were increased. Results also indicated a decrease in SPAD values and Fe content in new leaves when Co levels rose from 0.0 to 25.0 ppm. Application of 20.0 ppm Co produced the highest ear length, diameter, weight, and grain yield values. Under salt stress, cobalt (at all concentrations tested) markedly enhanced nutrient contents (i.e., N, P, K, Ca, Mg, Zn, Cu, and Mn) in grains relative to control plants (Gad and El-Metwally, 2015).

Seed priming with cobalamin also provided significant protection against salinity stress in common bean, mainly by triggering an effective antioxidant system and regulating osmotic balance (Keshavarz and Moghadam, 2017). In cobalamin, the key coenzyme during N fixation in legume nodules, cobalt occupies the center of its complex structure.

In cucumber (*Cucumis sativus*) grown in three saline soils (3.35, 3.85, and 5.40 ds m^{-1}), cobalt application considerably promoted biomass production and fruit yield. Of all levels tested, the application of 15 ppm Co was the most effective, demonstrating that Co prompts protective mechanisms to cope with salt stress (Gad et al., 2018a).

In a field experiment where common bean plants were cultivated under different water regimes (100, 80, and 60%) and irrigated once with 4, 8, 12, 16, and 20 ppm Co, all cobalt concentrations substantially increased growth, yield parameters, and mineral content compared to untreated plants. The 12 ppm Co treatment yielded the highest values and improved water use efficiency (Gad et al., 2018b).

For onion plants (*Allium cepa*) in saline soils (3.3–5.5 dS m^{-1}), the application of 12.5 ppm Co markedly stimulated biomass production, yield, mineral concentration, and bulb quality (Gad et al., 2020).

(Gad et al., 2020). Finally, mungbean seeds soaked with Co (4, 8, 12 mg L^{-1}) were planted under three irrigation regimes (50%, 60%, and 70% of available water). Plants receiving Co at 8–12 mg L^{-1} showed improved plant height, leaf number, leaf area, total root dry weight, and water use efficiency, thereby reducing the effects of water stress (Alak and Al-Sabagh, 2020).

In rice, Co enhanced drought-stress tolerance by attenuating oxidative stress and boosting yield. In particular, Co significantly improved plant height, biomass, water content, tillering, and leaf area. It also increased pigments [chlorophyll *a* (Chl *a*), carotenoids, total chlorophyll], sugars (soluble sugars and total carbohydrates), and

Co content in shoots and roots. Furthermore, Co stimulated the antioxidant system by elevating the concentration of total phenols, flavonoids, proline, and enzymatic ROS scavengers, including CAT, peroxidase (POD), and polyphenol oxidase (PPO), while significantly decreasing chl *b*, MDA, and 2,2-diphenyl-1-picryl-hydrazyl (DPPH). Yield attributes, such as the number of tillers, panicle parameters (number, length, and weight), 100-grain weight, harvest index, and grain nutritive value (sugars, vitamin B, and Co contents), were significantly enhanced by Co (Tourky et al., 2023). Similarly, in split grass (*Pennisetum divisum*) plants exposed to three salinity levels (200, 500, or 1000 mM NaCl), the foliar application of Co (15.0 or 25.0 mg L^{-1}) downregulated the content of the highly reactive molecule methylglyoxal. It also stimulated the activity of GSH, GSSG, GR, Glyoxalase I (Gly I), and Glyoxalase II (Gly II) in roots under critical salt stress compared to the control (Alhammad et al., 2023).

In peanut (*Arachis hypogaea*) cultivated at 60% gross water requirement, the application of Zn to the soil, either alone or combined with foliar Co, significantly increased yield by 89.7% and 191.3%, respectively, compared to the control. Nonetheless, Co treatments alone decreased Fe and Cu uptake, which was restored when Zn was applied. At 80% gross water requirement, the combined application of Zn and Co dramatically enhanced N, P, Mg, and Zn uptake, thereby improving plant performance and yield while decreasing water losses. Under this irrigation treatment, a singular foliar spray of Zn resulted in a 32.4% increase relative to the corresponding control. In combination with Co, water use efficiency increased by 70.0% compared to the control; thus, the detrimental effects of water deficit were attenuated, enabling higher productivity and optimal irrigation management (Elshamly and Nassar, 2024a). In the same crop species exposed to water deficit, plants enhanced the biosynthesis of osmolytes (e.g., proline) and antioxidants, maintaining optimal nutrient balance in response to potassium humate and Co, especially when treated with 3 g L^{-1} potassium humate + 3.25 mg L^{-1} Co under the 75% irrigation strategy (Elshamly et al., 2024b).

In chili pepper (*Capsicum annuum*) plants exposed to drought (field capacity: 40%), the supply of Co (20 mg L^{-1}) in combination with fulvic acid (4 g L^{-1}) and biochar (0.5%) significantly increased plant height (23.29%), plant dry weight (28.85%), fruit length (20.17%), fruit girth (21.41%), and fruit yield (25.13%) compared to control. The combination of fulvic acids, Co, and biochar also enhanced total chlorophyll content, as well as leaf N, P, and K levels compared to the control. In summary, the application of Co with fulvic acids and biochar contributed to alleviating drought stress in chili pepper (Hareem et al., 2024).

In radish (*Raphanus sativus*) grown in a saline soil (pH 8.29; EC 5.09 dS m^{-1}), the application of 20 mg Co L^{-1} (as $CoSO_4$) and proline (0.25 mM) enhanced shoot ($\sim$20%) and root growth ($\sim$23%), as well as plant dry ($\sim$19%), and fresh weight ($\sim$41%) compared to the control. Furthermore, this combination substantially improved plant performance under salinity stress (Inayat et al., 2024).

Similarly, faba bean (*Vicia faba*) plants receiving 75% of the recommended N level and 12 ppm Co showed markedly improved growth parameters, productivity, and nutritional quality of grain. Cotreatment positively influenced root nodulation, nitrogen fixation, and N, P, K, Mn, Zn, and Cu contents. This demonstrates a synergistic interaction between N and Co that promotes sustainable agriculture and plant resilience (Gad et al., 2025).

In summary, when supplied at optimal concentrations, Co can help alleviate various abiotic environmental stressors—including extreme temperatures, water deficit, salinity, nutrient deficiencies, and metal toxicity—using different mechanisms such as enhancing the uptake of essential elements. However, when applied at high toxic levels, Co can induce detrimental effects in plants (Panda et al., 2023).

21.2.4 Iodine

Iodine (I) is a nonmetallic element in the halogen family (Group 17). It ranks 63rd in abundance in the Earth's crust at 0.45–0.55 mg kg^{-1}, while its concentration in the oceans ranges from 0.045 to 0.060 mg L^{-1} (Lide, 2017; Zhang et al., 2018).

In the field of mineral nutrition of higher plants, I is not considered a plant nutrient; however, in marine plants and algae, it is crucial for growth, antioxidant defense, and pigment synthesis, helping them adapt to UV light (Küpper et al., 2008; Paz et al., 2018). For humans, iodine is categorized as an essential trace element; its deficiency in diets is the leading cause of preventable brain damage, intellectual disability, and certain types of cancer worldwide (Rahmani et al., 2024; Liu et al., 2025a). To address this issue, crop biofortification represents a valuable tool to provide a portion, if not the total amount, of the mineral needed (Medrano-Macías et al., 2016). When plants are treated with I for biofortification purposes, they can simultaneously be biostimulated to promote better growth, productivity, and quality of the final products, while enhancing responses to abiotic stress factors (Consentino et al., 2023). Nonetheless, I, like most if not all beneficial elements, exhibits a hormetic dose–response curve, becoming harmful to plants when supplied at higher concentrations compared to other chemical elements (Medrano-Macías et al., 2016; Zhang et al., 2023). Below, we describe the potential role of iodine in modulating plant tolerance mechanisms against environmental abiotic stressors across selected plant species.

Iodine can act as an efficient electron donor and consequently be effective as an antioxidant and scavenger of free radicals. In lettuce (*Lactuca sativa*), IO_3^- attenuated the ionic, osmotic, and oxidative stress caused by the application of 100 mM NaCl. When applying IO_3^-, toxic ion and soluble sugar concentrations increased, while the activities of antioxidant enzymes—such as SOD, APX, dehydroascorbate reductase (DHAR) and GR—were enhanced. Consequently, ROS such as the superoxide radical ($O_2^{\bullet -}$) and H_2O_2 reach minimum levels in NaCl-stressed plants. Thus, IO_3^- can be considered a beneficial element that attenuates salinity damage (Leyva et al., 2011). Similarly, in lettuce exposed to 100 mM NaCl, the application of iodate (20 or 40 μM I) significantly increased biomass production and stimulated the activity of the enzymes shikimate dehydrogenase (SDH), phenylalanine ammonia-lyase (PAL), and polyphenol oxidase (PPO). A concomitant increase in hydroxycinnamic acids, their derivatives and total phenols represented protective mechanisms against salinity (Blasco et al., 2013). Even so, lettuce plants exposed to moderate salinity (50 mM NaCl) and supplemented with iodine (0.09 mg KIO_3 L^{-1}) displayed improved nutritional value in terms of antioxidants, photosynthetic pigments, and anthocyanins, but did not improve their salt tolerance (Maglione et al., 2022), a response that may be ascribed to the low concentration of KIO_3 used. Recently, Shalaby (2025) reported that lettuce plants grown in a saline environment responded to iodine (0, 3, 6, and 9 mg KIO_3 L^{-1}) in a dose-dependent manner. Low concentrations enhanced their antioxidant capacity, decreased toxic compounds, improved nutritional value, maintained physiological balance, and promoted plant growth and yield, though high I concentrations disrupted physiological processes and reduced productivity. Interestingly, plants sprayed with 3 mg KIO_3 L^{-1} exhibited high levels of antioxidant enzymes (CAT, SOD, APX), nonenzymatic antioxidants [ascorbic acid (AsA), proline, and phenols], chlorophyll, and nutrients, whereas the contents of MDA, H_2O_2, and Na decreased, resulting in increased head weight and total yield and reduced nitrate content.

In soybean plants grown under Cd-contaminated conditions, iodine supplied as potassium iodate (KIO_3; 20, 40, 80 μM I) activated the enzymatic antioxidant system (e.g., APX, GR, and SOD), demonstrating the pronounced beneficial effect of I in attenuating the deleterious effects of cadmium (Gupta et al., 2015). Safflower (*Carthamus tinctorius*) plants challenged with dehydration stress (0, -0.4, -0.8, -1.2, -1.6 MPa) displayed improved performance when supplied with iodine (0, 40, 80 μM); this treatment reduced the detrimental effects of the stress (Jafarian et al., 2020). Similarly, potassium iodide (10 and 20 μM L^{-1}) supplied to soybean plants exposed to water deficit enhanced biomass by ca. 40% compared to the control. Iodine also boosted gas exchange by approximately 71% and minimized lipid peroxidation via increased enzymatic antioxidant activity. Nevertheless, in control plants (not exposed to water deficit), iodine concentrations higher than 10 μM negatively affected photosynthetic efficiency, biomass accumulation, and partitioning (Lima et al., 2023). Recently, a natural biostimulant formulation composed of bee honey (15 g L^{-1}), KIO_3 (25 mg L^{-1}), and silymarin (0.2 mg L^{-1}) was tested to evaluate its potential in preventing water deficit stress in faba bean. Plants exposed to water deficit received foliar applications of the biostimulant formulation. These biostimulated plants displayed reduced damage from water deficit through enhanced osmolyte accumulation, antioxidant activity, and recovering photosynthetic activity. The formulation also increased the concentrations of AUX, CK, and GA while reducing ABA, highlighting the potential of the formulation to promote sustainable agriculture (Rady et al., 2025).

In tomatoes grown under greenhouse conditions and subjected to salinity stress (100 mM NaCl), foliar applications of iodine (100 μM KIO_3 every 15 days) increased fruit production by 23%. Higher concentrations of N, Fe, and I were found in the leaves but not in the fruits. However, GPX activity, antioxidant potential, and concentrations of lycopene, Ca, and Mg in fruits decreased in I-treated plants (García-Fuentes et al., 2022). Similarly, in tomato plants treated with 30 or 60 mM NaCl, the application of 0.2–10.0 μM KIO_3 in growth chambers or 50–100 μM KIO_3 in greenhouses increased fruit weight, water content, and yield, while quality traits (i.e., fruit firmness and total soluble solids) were also enhanced in salt-stressed plants treated with iodine (Kiferle et al., 2022). When rocket (*Eruca sativa*) seedlings were grown under salt stress (100 mM NaCl) in a hydroponic system, the application of potassium iodate (50 and 100 μM KIO_3) alleviated salinity damage and supported growth. Specifically, the reduction in fresh and dry biomass weight for both shoots and roots was mitigated by the treatment. Furthermore, iodine application increased chlorophyll content, H_2O_2, MDA, proline, and sucrose levels, as well as antioxidant enzyme activities (Dizkara et al., 2025).

Finally, applying 100 mg L^{-1} to broccoli (*Brassica oleracea* var. *italica*) seedlings exposed to cold stress (20°C/2°C, day/night) significantly decreased leaf injuries (9.05%) by inducing different defense mechanisms, including enhanced APX, CAT, POD, and SOD antioxidant activity, and the accumulation of compatible solutes such as proline, free amino acids, sucrose, and reducing sugars (Quispe et al., 2025).

21.2.5 Lanthanum

With an approximate abundance of 30–39 mg kg^{-1} in the Earth's crust, lanthanum (La) is the third most plentiful lanthanide, after Ce and Nd (Thornton and Burdette, 2019). La concentration in seawater is significantly lower, typically around 3.4×10^{-6} mg L^{-1} (Lide, 2017).

La is the most extensively investigated rare earth element (REE) in plants, inducing characteristic hormetic dose responses that have been largely demonstrated across several species and multiple endpoints. At low concentrations (i.e., ≤ 56 μM), La stimulates growth and physiological processes, whereas it causes toxicity at higher doses. Factors affecting La hormetic dose responses include genotypic variation of the plant species evaluated, pH of the growth media, source and ionic form, duration of the exposure to La, and the number of doses below the no observed adverse effect level (NOAEL), which has been identified at 249 μM (Agathokleous et al., 2018, 2019).

In snow lotus (*Saussurea involucrata*) plants exposed to salinity stress, La^{3+} mitigated plant growth repression, increased leaf water potential and moisture content, and enhanced the contents of soluble protein and proline, while abating MDA content. Moreover, La^{3+} applications significantly stimulated the activity of the antioxidant enzymes APX, CAT, GR, and SOD, attenuated the decomposition of photosynthetic pigments, and boosted the ratio of total chlorophyll to carotenoids in plants exposed to salt toxicity (Xu et al., 2007).

In maize seedlings, pretreatment with lanthanum prior to salt-induced osmotic stress substantially increased chlorophyll and carotenoid contents, diverse photosynthetic parameters (i.e., F_v/F_m, q_P, Φ_{PSII}, and P_N), biomass, and antioxidant activity of APX, CAT, SOD, and GPX, compared with the salt-stressed plants that did not receive La. Furthermore, La significantly decreased MDA and H_2O_2 contents induced by the saline challenge (Liu et al., 2016).

Lanthanum may also attenuate the stress caused by nutrient deficiencies. For instance, in 45-day-old adzuki beans (*Vigna angularis*) exposed to P deficiency (P at 1% or 5% of its original concentration), the application of 0.4 mM $LaCl_3$ markedly ameliorated phosphorus uptake efficiency under critical P deficiency, enhancing leaf area, gas exchange, contents of chlorophylls and carotenoids, APase activity, root morphology, water uptake and overall growth (Lian et al., 2019).

In wheat, La supplementation decreased Cd accumulation, leading to better plant growth. La supply reduced Cd concentration in root cells and xylem sap. Furthermore, in pot experiments, La application also reduced Cd accumulation in all tissues, while the concentration of Cd in grains was reduced by 35.0%. The transcriptional activity of *TaNramp5* and *TaHMA2* was repressed by La supplementation. La is able to effectively control Cd accumulation in wheat. The reduced accumulation of Cd in wheat with La is likely due to two reasons: reduced Cd uptake in wheat due to the down-regulation of TaNramp5 and reduced translocation of Cd from roots to shoots due to the down-regulation of TaHMA2. *TaHMA2* (Yang et al., 2019). In rice, supplying 20 mg L^{-1} $La(NO_3)_3$-amino acid chelates [La (III)-AA] effectively alleviated $CuSO_4$ (50 mg L^{-1}) stress by reducing oxidative stress and increasing chlorophyll content, thus promoting growth. An antagonistic effect between La(III) and Cu(II) was evident. Exogenous La(III)-AA decreased Cu (II) content in leaves, stems, and roots by 55.56%, 59.46%, and 26.29%, respectively, and attenuated Cu(II) damage by maintaining the mesophyll cell ultrastructure. La (III) induced the expression of the auxin-amido synthetase *OsGH3*.4, but inhibited IAA (indole-3-acetic acid) -dependent expansin expression; this regulates cell wall loosening and growth under stress, ultimately protecting cell structures and boosting photosynthesis to confer resistance against Cu stress (Zhong and Chen, 2020).

In tomato plants grown under low light stress (i.e., ~50 μmol m^{-2} s^{-1}), root elongation was reduced by approximately half. Interestingly, supplementing these plants with 20 mM lanthanum considerably alleviated this suppression. Under low light stress, the expression of the genes for phytochrome-interacting factor, which induces AUX synthesis, and several AUX-synthesis related proteins was markedly upregulated by lanthanum. Thus, La stimulates AUX-induced cell elongation and alleviates growth suppression under low light stress conditions (Iguchi et al., 2023).

In forked fern (*Dicranopteris pedata*) subjected to drought, the application of lanthanum (9.5 and 1.0 mM) and yttrium (0.25 and 0.50 mM) increased CAT and POD activities by 99.04% and 81.25%, respectively. It also increased proline, soluble protein, and soluble sugar content by 97.52%, 71.24%, and 61.81%, respectively, compared to the control. Lipid and flavonoid metabolism were also upregulated, enabling the plant to successfully cope with drought stress (Feng et al., 2024).

In soybean plants challenged with salt stress, La application effectively enhanced protein content and the activity of the antioxidant enzymes SOD and POD, while decreasing Na^+ content and the Na^+/K^+ ratio, thus reducing oxidative damage. Additionally, La prevented chloroplast disintegration and mitigated the decline of protein, carbohydrates, and pectin in the leaves. La reduced leaf flavonoid content and synthesis by inhibiting key metabolites in the phenylalanine metabolic pathway under NaCl stress. Overall, these findings suggest that La mitigates cellular damage by modulating the antioxidant system and secondary metabolite production, which are important mechanisms supporting the adaptive response of soybean leaves and enhancing plant tolerance to salinity (Tong et al., 2024).

Similarly, in Chinese licorice (*Glycyrrhiza uralensis*) seedlings established in pots with a sandy loam substrate and treated with 300 mL of a solution containing 160 mM NaCl once every two weeks for a total of five times, applications of $La(NO_3)_3$ (0.75 mM) rendered photosynthetic compensation recovery, activated antioxidant defense, and synergistically improved both biomass and medicinal quality compared to control plants (Shang et al., 2025).

It has been well documented that La induces characteristic hormetic dose–response patterns; plants can tolerate and metabolize relatively high concentrations of this element (i.e., ~50–250 μM) before adverse effects become evident. In terms of crop management, La-induced hormesis is a valuable trait that can activate metabolic pathways to enhance crop growth, productivity, and quality, while promoting intrinsic defense mechanisms against abiotic environmental stresses. Integrating these aspects with comprehensive information on the environmental implications of La application will be critical for future research.

21.2.6 Selenium

In the Earth's crust, selenium (Se) is found at concentrations ranging from 0.05 to 0.5 mg kg^{-1}, thus representing the 30th most abundant metalloid. In the oceans, Se exhibits average concentrations of 2×10^{-4} mg L^{-1} (Tan et al., 2016; Lide, 2017). In the periodic table, Se is localized in group VIa between sulfur (S) and tellurium (Te), displaying chemical attributes analogous to both (Pilbeam et al., 2015).

The essentiality of Se has been proven across the three main branches of life: bacteria, archaea, and eukaryotes (Araie and Shiraiwa, 2016; Dervisi et al., 2023). Nonetheless, the necessity of selenium is not universal among all eukaryotes; while vertebrates and algae are rich in selenoproteins, most higher plants and fungi lost the machinery to synthesize them during evolution, perhaps because Se was not readily available in terrestrial habitats (Zhang and Gladyshev, 2009; Garousi, 2017; Schiavon and Pilon-Smits, 2017; Liu et al., 2022). Se insufficiency impacts a significant global population, affecting 500 million to 1 billion people, primarily due to deficient dietary intake (Shreenath et al., 2023), leading to critical health disorders such as infertility, autoimmune thyroiditis, and carcinogenesis (Bai et al., 2025; Estarriaga-Navarro et al., 2025). Therefore, agronomic biofortification with Se has become a major focus as an efficient approach to increase crop nutritional quality and combat Se-related hidden hunger (Pal et al., 2026). Moreover, genetic biofortification involving breeding or engineering crops may also enhance their ability to take up, transport, and accumulate Se in edible parts (Zhou et al., 2023).

Because of the chemical analogy between Se and S, Se uptake and translocation within plant cells can be mediated by S transporter proteins (SULTR), localized to the plasmalemma; subsequent incorporation into selenocysteine (SeCys) or selenomethionine (SeMet) occurs via the sulfur assimilatory pathway (Pilbeam et al., 2015; Trippe and Pilon-Smits, 2021). Due to this close chemical similarity, a higher risk of Se toxicity is expected when S is not sufficiently available, as organisms may indiscriminately absorb and metabolize toxic Se compounds in place of essential S (White et al., 2004). In fact, besides selenoproteins, oxyproteins and nitroproteins may form during Se excess (Kolbert et al., 2019). Beyond uptake and translocation, remobilization of Se within different plant tissues may also occur (Keskinen et al., 2010; Shinmachi et al., 2010). In addition, foliar applications of Se may be remobilized to other plant organs (Mezey et al., 2025).

Although Se does not fulfill the essentiality principles to be considered a plant nutrient, it can enhance various productive traits in plants and attenuate the impact of complex environmental pressures (Pilbeam et al., 2015; Abdullah et al., 2025).

Cadmium-stressed rapeseed (*Brassica napus*) plants treated with 2 μM Se displayed regulation of the DNA (deoxyribonucleic acid) methylation, Cd sequestration, and attenuation of ROS burst (Filek et al., 2008). Likewise, in sorghum plants challenged with high temperatures (40/30°C daytime maximum/nighttime minimum), the application of Se decreased membrane damage by enhancing antioxidant defense and reducing ROS accumulation (Djanaguiraman et al., 2010). In wheat (*Triticum aestivum*), optimal Se supply (i.e., 0.5–2.0 mg Se kg^{-1} soil) stimulated seedling root growth while decreasing oxidative stress caused by increased levels of UV-B (Yao et al., 2010). Selenium may also activate protection mechanisms against low-temperature stress by increasing the contents of chlorophylls, anthocyanins, flavonoids, and phenolic compounds, as well as POD and CAT enzymatic activity, thereby increasing biomass production (Chu et al., 2010). In ryegrass (*Lolium perenne*), Se attenuated the oxidative stress triggered by toxic levels of aluminum in the growth media by reducing superoxide radicals and subsequently enhancing the enzymatic activity of POD (Cartes et al., 2010). In rapeseed exposed to 100 or 200 mM NaCl, Hasanuzzaman et al. (2011) found that the application of 25 μM Na_2SeO_4 increased AsA and GSH contents, the GSH/GSSG ratio, and the activity of APX, CAT, DHAR, Gly I, Gly II, GPX, GR, glutathione S-transferase (GST), and MDHAR (monodehydroascorbate reductase), ultimately leading to a significant decrease in H_2O_2 and MDA, thus enhancing salt stress tolerance. Similar stimulatory responses of the antioxidant system triggered by Se were observed in rapeseed seedlings exposed to dehydration (Hasanuzzaman and Fujita 2011). In recalcitrant seeds of sugar maple (*Acer saccharum*), Se played a protective role during desiccation (Pukacka et al., 2011). Specifically, seeds treated with sodium selenite (10 mg L^{-1}) for 6 h displayed higher germination capacity, lower electrolyte leakage, and lower MDA content in embryonic axes, while the activities of GPX and GR in embryonic axes were higher compared to untreated seeds. The supply of Se (2.5 or 5.0 μM) to mungbean plants challenged with arsenic (2.5, 5.0, or 10 μM As) had different beneficial effects. When supplied at 5 μM, Se significantly decreased Se uptake; consequently, damage to membranes, chlorophyll, and cellular disruptions were less drastically affected by As-induced stress. Furthermore, As-boosted oxidative damage was lowered with Se application, which could be related to increased concentrations of APX, AsA, CAT, GR, GSH, and SOD. In addition, the detoxification activity triggered by metallothioneins, thiols, and GST was elevated in Se-treated plants (Malik et al., 2012). Selenium may also attenuate As toxicity in rice. Co-application of Se and AUX enhanced growth as well as concentrations of chlorophyll, protein, MDA, Cys, and proline in As-treated seedlings (Pandey and Gupta, 2015). In maize challenged with H_2O_2, Se maintained homeostatic levels of cytosolic Ca^{2+} and pollen viability (Del Pino et al., 2019). Furthermore, extracts of emmer wheatgrass (*Triticum turgidum* spp. *dicoccum*) biofortified with Se mitigated *in vitro* oxidative stress induced by H_2O_2during maize pollen germination (Del Pino et al., 2022).

In common beans, seed priming and a foliar spray of 70 ppm Se-nanoparticles (Se-NPs) improved stomatal conductance (G_s; 14.2%) and net transpiration rate (T_r, 11.7%) in plants subjected to sodium fluoride (NaF) stress. Moreover, Se-NPs enhanced DPPH and AsA activities, total protein, proline, and phenol content, as well as the concentrations of phosphorus, potassium, calcium, magnesium, and zinc (Ahmed et al., 2025).

In safflower challenged with 0.5, 1.5, or 2.5 g kg^{-1} NaCl, the application of 0.01 or 0.02 g kg^{-1} Na_2SeO_4 stimulated seed germination and biomass production by altering anatomical, biochemical, and physiological attributes. Specifically, Se substantially increased stem diameter, cortical parenchyma thickness, and pith diameter under salinity stress. When 0.02 g kg^{-1} Se was applied to plants receiving 2.5 g kg^{-1} NaCl, significant improvements were observed in Na^+, K^+, Chl *a*, Chl *b*, Chl *a* + *b*, Car, soluble carbohydrates, and RWC, whereas H_2O_2 and MDA levels decreased in the shoots (Fatahiyan et al., 2025).

Curly endive (*Cichorium endivia*) plants cultivated under three different irrigation levels (100%, 75%, or 50% of substrate water holding capacity) were treated with *Ecklonia maxima*-based seaweed extract (SwE; Kelpstar®, Italy; 3 mL L^{-1}), Se (8.0 μmol Na_2SeO_4 L^{-1}) or their combination (SwE + Se). The application of SwE, Se, or SwE + Se enhanced growth, productive features, and soluble solid content, while reducing dry matter percentage, proline, and MDA levels compared to untreated control plants (Consentino et al., 2026).

21.2.7 Silicon

Silicon (Si) concentrations are estimated to be 2.82×10^5 mg kg^{-1} in the Earth's crust, while in seawater it is found at 2.2 mg L^{-1} (Lide, 2017). After oxygen (O), Si is the second most abundant element on Earth (Hussain et al., 2025). Notably, the soil Si cycle is of paramount significance for ecosystem functioning; therefore, a deeper understanding of the drivers of Si cycling (e.g., predominant speciation), human disturbances, and the implications for vital soil properties (water storage, nutrient availability, and microaggregate stability) is of fundamental importance (Schaller et al., 2021).

Among the list of beneficial elements, Si is, by and large, the most researched one in plant sciences. It may significantly influence various aspects of plant ontogeny and responses to environmental signals and stress. The complex processes of Si uptake and accumulation in plant cells trigger diverse ecosystem services like soil appropriation, biomass supply, and carbon sequestration (Katz et al., 2021). Importantly, Si plays critical roles in attenuating the effect of adverse environmental conditions by enhancing plant resilience in challenging environments (Ali and Bijay-Singh, 2025). The influence of Si on plant physiology and metabolism involves intricate mechanisms, impacting nutrient uptake, photosynthesis, and hormonal regulation (Pereira et al., 2024).

In sugarcane plants facing water deficit, Si supply promoted better recovery—evidenced by superior sugar, stalk yield, and dry biomass—following both early and late water deficit (Camargo et al., 2017). Indeed, monocots can accumulate as much as 10–100 g Si kg^{-1} in their tissues on a dry weight basis, displaying substantial benefits from Si supply (Tubana et al., 2016). In rice plants exposed to excessive Fe, Si accelerates cell wall POD and phenylpropanoid pathway activity, restricting excess Fe entry into the plant as a defense mechanism (Mehrabanjoubani et al., 2019). In another study with rice plants exposed to Fe toxicity, Si attenuated the impairments of excess Fe on photosynthetic performance by altering leaf diffusive conductance with negligible effects on carbon metabolism (Dos Santos et al., 2019).

In barley plants exposed to toxic levels of chromium (Cr), the supply of 1–2 mM Si significantly reduced the toxic effect of Cr, increasing plant growth as a consequence of enhanced photosynthetic parameters (e.g., SPAD units, C_i, G_s, P_N, and Fv/Fm) while alleviating ultrastructural disorders caused by Cr both in roots and leaves (Ali et al., 2013). Similarly, in cowpea (*Vigna unguiculata*) plants exposed to excess Al, Si application substantially decreased Al accumulation in plant tissues, while improving gas exchange, Fv/Fm, pigments, and antioxidant enzyme concentration, and reducing oxidative compounds (de Jesus et al., 2017). Furthermore, applying a combination of biochar and ash obtained from silicon-rich rice husk burning significantly decreased inorganic As accumulation in rice grain to levels below the maximum inorganic As level in husked rice established by the Codex (i.e., 0.35 mg kg^{-1}) (Leksungnoen et al., 2019). In wheat, both organosilicon and inorganic silicon fertilizers attenuated the toxic effects of Cd and Pb, enhancing photosynthetic parameters and yield while diminishing oxidative damage, as evidenced by lower levels of H_2O_2 and MDA relative to control plants (Huang et al., 2019). In barley grown under soil flooding conditions, monosilicic acid (H_4SiO_4) reduced carbon-driven denitrification emissions and the N_2O:N_2 ratio, while increasing the potential denitrification emission. The presence of H_4SiO_4 in the system provided a more complete denitrification process, favoring the formation of N_2 within the $NO_3^- \rightarrow NO_2^- \rightarrow NO \rightarrow N_2O \rightarrow N_2$ reaction sequence (Włodarczyk et al., 2019). Another study on barley exposed to osmotic stress demonstrated that Si accumulation in shoots is controlled by the silicon transporters HvLsi1 and HvLsi2 (low silicon, Lsi transporters) only when plants face K starvation. Furthermore, Si increased starch concentrations under both osmotic stress and K deprivation and regulated glycolytic and tricarboxylic acid cycle (TCA)-related pathways; this stimulatory effect was likely mediated by ABA homeostasis and the active cytokinin isopentenyl adenine (iP) (Hosseini et al., 2017). Similarly, in soybean under low light stress, Si promoted the transcriptional activity of genes involved in lignin biosynthesis, thereby strengthening the stem and reducing lodging (Hussain et al., 2020). By efficiently regulating both reactive oxygen and nitrogen species (i.e., ROS and RNS) metabolic processes, Si may also promote salt stress resilience in soybean. Specifically, Si application recovered chlorophyll content, transpiration rates, and P_N in salt-stressed plants. Conversely, the transcriptional activity of antioxidant-related genes (*GmCAT1*, *GmCAT2*, and *GmAPX1*) and the expression of S-nitrosothiol (RSNO) and NO-related genes were reduced in salt-stressed plants following Si supplementation. Furthermore, Si applications repressed the transcriptional activity of genes involved in ROS and RNS biosynthesis and metabolism and reduced the concentrations of antioxidant enzymes and radical scavengers, thereby mitigating the detrimental effects of NaCl (Chung et al., 2020). Finally, Gómez-Merino et al. (2020) performed a comprehensive *in silico* analysis of the mechanisms in Si flow from root to shoot in pepper plants, unveiling a potential regulatory link between transcriptional activations of genes and phytohormonal signaling cascades that may be involved in vital processes such as growth, development, and responses to environmental stressors such as wounding, salinity, and dehydration.

Over the past five years, several studies have developed innovative mechanistic models describing the positive impact of Si on plants. For instance, Ahanger et al. (2020) highlighted the pivotal role of Si in enhancing plant stress tolerance by regulating the biosynthetic, metabolic, and molecular restructuring of secondary metabolites (i.e., phenolics, terpenes, and nitrogen-containing compounds). This allows Si to act as a *quasi-essential* agent against abiotic and biotic stresses. Similarly, Gaur et al. (2020) provided a comprehensive analysis of Si's role in plant growth, development, and responses to both biotic (i.e., bacteria, oomycetes, fungi) and abiotic (salinity, water deficit, metal toxicity, and radiation) stress factors. Overall, Si uptake and transport are active processes facilitated by specific transporters—Lsi1, Lsi2, Lsi3, and Lsi6—which are located in the roots, stems, and leaves. Importantly, plants display genetic variation regarding Si uptake and related benefits, with rice exhibiting the highest accumulation levels. Mandlik et al. (2020) performed an in-depth analysis of these beneficial effects of Si in plants, including the molecular mechanisms involved in Si transport from roots via the xylem to shoots, where Lsi transporters enable its unloading into tissues. Once transported, silicon is deposited as silica in epidermal cells, sclerenchyma, vascular tissues, and intercellular spaces, strengthening cell walls and enhancing plant resilience under various stress factors. Within the cell walls, Si is primarily found as amorphous silica ($SiO_2 \bullet_n H_2O$) and covalently bound organosilicon complexes. These forms interact with cell wall components through both covalent bonds and noncovalent interactions to provide structural rigidity and enhance stress resistance. Specifically, covalent linkages (Si–O–C) form between Si and cell wall matrix components (primarily hemicellulose and lignin, and to a lesser extent, pectin and cellulose), which are critical for plant cell wall dynamics and thus plant development and resilience to environmental challenges (Sheng and Chen, 2020).

Si can attenuate the phytotoxic effects of metals and metalloids, whether essential (e.g., Cu, Ni, or Zn) or nonessential (e.g., Al, As, Cd, Cr, Pb, or Sb). The mechanisms of resistance involve: (1) restriction of root uptake and immobilization of these elements in the rhizosphere; (2) sequestration in the root apoplast; (3) complexation and storage within the apoplast, thus limiting symplasmic allocation; and (4) redox homeostasis by boosting antioxidant activity, improving membrane integrity, and regulating transcriptional activity of different Si-responsive genes (Vaculík et al., 2020).

The genetic control of Si uptake and distribution within plant cells was reviewed by Trejo-Téllez et al. (2022). Influx Lsi channels have been implicated in the absorption of Si in root cells, while efflux Lsi channels are responsible for transferring Si to other tissues and organs. Moreover, biosilicification is under the control of Siliplant1 proteins (Slp1). An intricate coordination among these proteins enables Si uptake, mobilization, and accumulation within plant cells. Interestingly, the abundance of these proteins varies across Viridiplantae (i.e., Chlorophyta and Streptophyta), thus influencing the ability of plants to thrive in challenging environments.

In rice—the premier Si accumulator among crops—substantial gains have been made in identifying Si transporters involved in uptake, transport, and distribution. These processes are mediated by Lsi1 (influx, root cells) and Lsi2 (efflux, root cells) during uptake; Lsi3 (pericycle) facilitates xylem loading, while Lsi6 (xylem parenchyma) handles unloading and node distribution (Huang and Ma, 2024). Furthermore, Silicon Efflux Transporter 4 (SIET4) is involved in Si deposition (Mitani-Ueno et al., 2023), and Siliplant1 (Slp1) is implicated in silica precipitation (Zexer et al., 2023), enhancing the plant's ability to cope with different environmental challenges.

Symbiotic relationships between plants and arbuscular mycorrhizal fungi (AMF) stimulate growth and help plants tackle abiotic stress factors such as extreme temperature, drought, salinity, and toxic metals. This occurs through enhanced nutrient and water absorption, improved root architecture, osmolyte buildup, and the increased transcriptional activity of stress-related genes (Samanta et al., 2025). In the legume *Mimosa caesalpiniaefolia*, the co-application of a SwE from *Solieria filiformis* and Si boosted plant growth, with the highest mycorrhizal colonization found at 150 mg Si kg^{-1} with SwE (da Silva et al., 2025).

In their editorial for the Research Topic "New avenues of silicon's role in plant biology: trends and controversies," Mukarram et al. (2025) posit two major functions of Si: the fortification of cell walls to serve as a mechanical barrier and the induction of adaptive mechanisms in response to environmental stress. They contend that the dichotomy between anatomical and physiological modes of Si action represents a rapidly evolving frontier in the field.

Narula and Chaudhry (2025) recently conducted a comprehensive review addressing the critical role of Si in enhancing intricate plant mechanisms to cope with a variety of abiotic stress factors. Their findings underscore the importance of Si for regenerative agricultural practices amidst the global climate crisis and suggest further research pathways to optimize Si application in agricultural systems.

21.2.8 Sodium

Sodium is the sixth most abundant element in the Earth's crust (2.36×10^4 mg kg^{-1}). Furthermore, at 10.76 g L^{-1}, Na is the major ion found in seawater (Lide, 2017; Chen et al., 2019).

Within the discipline of plant science, Na is one of the most extensively studied elements; it has been regarded as a general toxicant, though it may also elicit beneficial effects in various species when supplied at lower levels. In some, such as certain C4 species, it is indeed essential (Kronzucker et al., 2013). The effects of Na as a plant toxicant are well established and were recently reviewed (Garcia-Daga et al., 2025; Kushwaha et al., 2025; Zhang et al., 2025). Regarding the beneficial effects of Na^+, it has been shown that Na^+ can operate as a functional ion partially substituting for K^+ in plants, especially under K-starvation conditions, by regulating cellular osmotic pressure, turgor maintenance, and stomatal function (Thorne and Maathuis, 2022). At moderate levels, Na can enhance the growth, quality, and taste of some horticultural products (Naz et al., 2023; Sarraf et al., 2023; Laclau et al., 2024).

In the Australian succulent coastal plant karkalla (*Carpobrotus rossii*) exposed to cadmium stress, NaCl supply promoted plant growth and reduced shoot cadmium content by up to 87%, resulting from a significant abatement of Cd uptake by roots and its subsequent translocation to the shoot (Cheng et al., 2018).

In Venetian dogbane (*Apocynum venetum*) grown under osmotic stress (−0.2 MPa), the application of 25 mM NaCl substantially enhanced Na^+ accumulation in leaves and contributed to osmotic adjustment. This boosted RWC and photosynthetic activity, thereby enhancing overall plant growth (Cui et al., 2019).

In black locust (*Robinia pseudoacacia*) plants grown in soils contaminated with lead (Pb) and zinc (Zn), sodium sulfide (Na_2S) supply increased shoot biomass and decreased bioavailable Pb and Zn, reducing the Pb and Zn content in *plants*. Furthermore, Na_2S promoted phosphatase activity and increased available P content while stimulating AMF activity and diversity, leading to enhanced metal immobilization in soils (Zhang et al., 2020a).

In the natrophilic sugar beet (*Beta vulgaris*), a supply of 50 mM NaCl was optimal for plant growth. In plants subjected to sorbitol-induced osmotic stress, the addition of 50 mM NaCl also attenuated the detrimental impact of sorbitol on growth and reduced shoot osmotic potential, while enhancing Na^+ uptake and proline levels, and decreasing K^+ accumulation (Wu et al., 2015).

In the euhalophyte striped-stalked saltwort (*Suaeda salsa*), Na^+ and Ca^{2+} significantly increased leaf succulence by boosting leaf water content and enlarging epidermal cells. Under water-deficit conditions, Na^+ and Ca^{2+} mitigated the stress by stimulating succulence and promoting CAT, POD, and SOD enzymatic activity, consequently decreasing oxidative stress (Zhang et al., 2024).

Overall, though sodium is a nonessential element for most higher plants, it is potentially advantageous in diverse challenging contexts—especially under potassium-limiting conditions—and is regarded as a functional element amongst alkali metals. Understanding the processes of absorption, transport, and allocation of Na is crucial for alleviating osmotic, oxidative, and nutrient imbalances triggered by salinity stress, and thus for expanding its use as an inorganic biostimulant. Hence, this lays the groundwork for subsequent investigations to unveil novel mechanisms controlling the fundamental processes of uptake, transport, translocation, and metabolism of Na in plants, and, consequently, to establish the proper dosage and timing of Na to be applied in different crop plants to cope with climate change and support sustainable agriculture.

21.2.9 Titanium

Titanium (Ti) is the ninth most abundant element in the Earth's crust, with an average concentration of ca. 0.56% by mass, equivalent to 5600 mg kg^{-1}, making it the seventh most abundant metal and the fourth most abundant structural metal (after Al, Fe, and Mg). In seawater, its concentration is 1×10^{-3} mg L^{-1} (Lide, 2017; Zheng et al., 2019).

Research consistently shows that Ti applications exert positive effects on plant growth and stress tolerance, which has led to its classification as a beneficial element in plant physiology (Lyu et al., 2017; Singhal et al., 2023). In particular, Ti—frequently applied as titanium dioxide (TiO_2) or titanium dioxide nanoparticles (TiO_2-NPs)—has been extensively studied as a strategy for abiotic stress management in plants. Its effects depend on the dose, chemical form, application method, and plant species (Trela-Makowej et al., 2024).

For example, in soybean exposed to shade conditions (i.e., because of intercropping), some photosynthetic parameters including Fv/Fm, qP, PSII, and electron transport rate (ETR) were enhanced when 2.5 mg Ti per plant were applied, as compared to the control receiving 0 mg Ti per plant (Hussain et al., 2019). Furthermore, the uptake of toxic elements (i.e., As, Cd, and Pb) by rice plants grown in gleyic soils was substantially reduced when they were supplied with 0.15%–0.30% titanium gypsum (TG) (Zhai et al., 2020).

It has been well established that Ti interacts with plant micronutrients such as Fe (Lyu et al., 2017). Both Ti and Fe are transition metals that exhibit similar physical and chemical features (Cigler et al., 2010). If plants are Fe-deprived, Ti may induce the expression of genes involved in Fe uptake. However, if plants are exposed to high concentrations of Ti, imminent competition with Fe will cause drastic metabolic disruptions. Hence, Ti supply is only suggested under Fe-deprived conditions, particularly when soil pH is ≥ 6.5 (Lyu et al., 2017).

The co-application of TiO_2-NPs and biochar to sorghum seedlings grown in antimony (Sb) contaminated soil significantly enhanced the concentrations of both chl *a* and chl *b*, while TiO_2-NPs stimulated the capacity of seedlings to accumulate Sb (Daryabeigi-Zand et al., 2020). Similarly, in rice plants challenged with Cd stress, TiO_2-NPs increased plant growth and photosynthetic pigment contents in leaves while reducing MDA and Cd contents in grains (Zhang et al., 2020b).

In maize, seed priming with 60 ppm TiO_2-NPs prior to exposure to 200 mM NaCl increased various germination, growth, and antioxidant parameters, as well as K^+ concentration and RWC. Conversely, it decreased mean emergence time, sodium concentration, electrolyte leakage, and lipid peroxidation relative to untreated seeds exposed to salt stress (Shah et al., 2021).

In tomato, foliar application of Ti resulted in responses that depended on the cultivation system, phenological stage, and salinity level; in hydroponically grown seedlings, it increased macronutrient concentrations and antioxidant activity under 50 mM NaCl; whereas in plants at the reproductive stage, it did not mitigate the salinity-induced yield reduction, although it did modify fruit quality parameters under moderate salinity (Carbajal-Vázquez et al., 2020; Carbajal-Vázquez et al., 2022). Recently, it was reported that supplying 156.6 and 208.8 μM TiO_2-NPs significantly improved both the vigor index of seeds and growth parameters of tomato seedlings during the early developmental stage, with hormetic dose responses observed (Carbajal-Vázquez et al., 2025).

In *Arabidopsis thaliana*, incorporating TiO_2-NPs into the substrate at a concentration equivalent to 2 mM significantly increased plant survival under prolonged drought stress imposed by complete suspension of irrigation, achieving recovery rates close to 90% compared to less than 15% in untreated plants, demonstrating enhanced drought stress tolerance induced by Ti application (Pérez-Zavala et al., 2022).

Similarly, foliar application of Ti at 50 and 100 mg L^{-1} in split grass reduced ionic imbalance and methylglyoxal accumulation under severe salinity by decreasing Na^+ and Cl^- levels while increasing foliar K^+ and Ca^{2+} concentrations, thereby contributing to the attenuation of stress prompted by high NaCl levels (Alhammad et al., 2023). Furthermore, pretreatment with 200 ppm TiO_2-NPs in safflower attenuated the damage induced by 50 mM NaCl only in the Balci cultivar, as shown by significant reductions of MDA and improvements in root growth parameters, further confirming a genotype-dependent response (Dinler et al., 2024).

Although salinity induces an osmotic stress component, it differs from water deficit and nonionic osmotic stress because it additionally involves ionic toxicity and nutritional imbalance. In micropropagated sweet potato (*Ipomoea batatas*) plantlets, osmotic stress induced by sorbitol (0.22 M) significantly reduced root number, whereas incorporating TiO_2-NPs into the culture medium increased both root number and length. Under *in vitro* water-deficit conditions, applications of 50 and 100 mg TiO_2-NPs L^{-1} increased root number, while the greatest root lengths were observed at 100 and 150 mg TiO_2-NPs L^{-1} (with no significant differences between these doses), resulting in longer and thicker roots advantageous for water and nutrient uptake processes (Ogunnubi et al., 2025).

In mint (*Mentha* × *gracilis*), applying 100 mg TiO_2-NPs L^{-1} improved growth, essential oil yield, and the enzymatic activity of SOD and APX; furthermore, MDA and H_2O_2 levels in plants exposed to 50 and 100 mM NaCl were decreased, evidencing a mitigating effect of TiO_2-NPs against salt toxicity (Morshedloo et al., 2025). In soybean, foliar application of 30 ppm TiO_2-NPs synthesized using aloe (*Aloe vera*) extract efficiently mitigated the negative effects of increasing NaCl concentrations (25–200 mM) by increasing photosynthetic pigment contents, soluble protein levels, and antioxidant activity, while reducing salinity-induced ultrastructural damage compared to plants subjected only to salt stress (Abdelhameed et al., 2025). In pea (*Pisum sativum*) plants subjected to 100 mM NaCl, foliar spraying with 50 and 100 mg TiO_2-NPs L^{-1} reduced membrane damage, H_2O_2 accumulation, and lipid peroxidation; additionally, it improved photosystem II (PSII) efficiency, electron transport, and photosynthetic rate, with more pronounced effects at 100 mg L^{-1} (Yotsova et al., 2025).

Thermal stress, particularly low-temperature stress, has also been mitigated through Ti application. In fenugreek, foliar spraying of TiO_2-NPs at concentrations of 2 and 5 ppm under cold conditions (10 °C) increased photosynthetic pigment content and the activity of antioxidant enzymes while reducing ROS relative to the stress-only control. Under the same cold conditions, TiO_2-NPs application induced the overexpression of genes associated with sterol biosynthesis, including *SEP*, *SQS*, *CAS*, and *SSR*, evidencing a Ti-modulated transcriptional response to suboptimal temperatures (Babaei et al., 2025). Although some studies have focused on stresses other than thermal stress, Ti application has shown a protective effect on key photosynthetic apparatus parameters, such as photosystem efficiency and electron transport. In pea plants subjected to salinity, TiO_2-NPs preserved PSII and PSI functionality, improved electron transport, and contributed to the functional stabilization of photosynthetic components compared to plants challenged with salinity stress (Yotsova et al., 2025).

Furthermore, Ti has shown a relevant role in managing challenges imposed by toxic metals and metalloids. Treatment of rice plants with 25 and 50 mg TiO_2-NPs L^{-1} mitigated the phytotoxic effects of arsenic (50 μM As) by improving plant growth and reducing ROS impairments, lowering the concentrations of H_2O_2, MDA, and methylglyoxal, and decreasing electrolyte leakage compared to plants exposed only to the metalloid. Additionally, TiO_2-NPs enhanced the antioxidant system and induced the glyoxalase cycle. At the molecular level, TiO_2-NPs application induced the overexpression of genes associated with arsenic detoxification and sequestration, including glutathione synthetase 1 (*GSH1*), phytochelatin synthase (*PCS*), and ATP-binding cassette transporter 1 (*ABC1*). This promoted the accumulation of GSH and phytochelatins, facilitating the vacuolar compartmentalization of As (Kiany et al., 2022). Similarly, in daylily (*Hemerocallis* sp.), foliar application of 200 mg TiO_2-NPs L^{-1} reduced the deleterious effects of 5 mM Cd treatment by stimulating photosynthetic activity and triggering the antioxidant system, resulting in enhanced Cd tolerance. At the molecular level, transcriptomic analysis revealed that TiO_2-NPs significantly induced gene expression, with marked overexpression of genes involved in photosynthesis, chlorophyll metabolism, and redox processes, with the involvement of the transcription factors MYB, NAC, and WRKY playing key roles in the attenuation of Cd stress (Liu et al., 2025b). Finally, in *Arabidopsis*, application of TiO_2-NPs at a concentration of 2 mM induced a broad transcriptomic response that enhanced plant resilience against water deficit, salinity, phosphorus deficiency, and *Botrytis cinerea* infection. This response was linked to the activation of intricate defense pathways and hormonal signaling mediated by ABA and salicylic acid, consistent with a Ti-induced priming state (Pérez-Zavala et al., 2022).

Nevertheless, TiO_2-NPs may also prompt detrimental impacts on plants. To cite just two examples, in wheat, TiO_2-NPs hindered photosynthesis at both photochemical and nonphotochemical phases (Dias et al., 2019). Moreover, it was found that TiO_2-NPs cause drastic metabolic dysfunction by breaking down sugar biosynthesis through photosynthetic disturbances. This forces a shift to alternative energy sources, such as amino acids, to feed the TCA cycle, while ROS and sugars act as signals triggering antioxidant defenses. These impacts on carbon/amino acid pathways can significantly compromise plant growth and crop yield (Silva et al., 2020b).

The results presented here suggest that Ti, particularly in the form of TiO_2 and TiO_2 nanoparticles, can modulate plant responses to different types of abiotic stress, including toxic levels of salts, water deficit, thermal stress, and metal and metalloid toxicity. Additionally, it has been documented that TiO_2-NPs application can induce transcriptomic responses associated with plant defense against biotic stress, such as infection by *Botrytis cinerea*, through priming-like mechanisms. However, these effects follow a well-defined hormetic pattern and are highly dependent on the experimental context (dose, chemical form, particle size, application method, and plant species), highlighting the need for careful and standardized evaluation before considering potential agronomic applications (Trela-Makowej et al., 2024).

21.2.10 Vanadium

Vanadium (V) is moderately abundant in the Earth's crust, averaging approximately 1.2×10^2 mg kg^{-1} (0.02%), making it more common than Cu or Zn but highly dispersed. In open-ocean environments, its concentration typically ranges from 1.5 to 2.5 μg L^{-1}, indicating that V is one of the most abundant trace transition metals in the ocean (Lide, 2017; Petranikova et al., 2020).

The importance of V in the functioning of vascular plants is extremely diverse, as this element may exert both positive and harmful effects (Hanus-Fajerska et al., 2021). While its role in alleviating abiotic stress impacts on plants has only recently emerged, research remains relatively limited or scarce compared to other beneficial elements.

Indeed, bioaccumulation of Cu, Hg, and Pb at toxic levels may be prevented by appropriate applications of V (Wang et al., 2013; Akoumianaki-Ioannidou et al., 2016). For example, sweet basil plants supplied with up to 40 mg V L^{-1} displayed enhanced root dry matter accumulation, while Fe and Pb concentrations in leaves and roots significantly decreased (Akoumianaki-Ioannidou et al., 2016). Likewise, V application in cucumber drastically reduced Cd and Pb absorption and accumulation (Osu and Onyema, 2016). In *Arabidopsis thaliana*, V may also exhibit protective effects during oxidative stress (Rojek et al., 2019). Collectively, these data indicate that V can operate as a strong inducer of secondary metabolism and a stress-buffering agent for plant priming against abiotic stress (Hanus-Fajerska et al., 2021), though its application is currently in its early stages.

21.3 Conclusions and perspectives

Climate change and variability, encompassing extreme temperatures, altered precipitation, and increased extreme weather, among other abiotic stressors, are dramatically impacting agriculture. Therefore, immediate attention is required for climate-resilient production systems and effective adaptation strategies. Beneficial elements may aid plants to cope with different extreme conditions by regulating the activity of transporters, improving mineral acquisition, accumulating S-containing metabolites, mitigating ROS and RNS production, and boosting antioxidant defense responses. Notably, beneficial elements may impact the biosynthesis of phytohormones, which directly activate stress tolerance mechanisms to overcome diverse environmental challenges. Furthermore, by supporting vital enzymatic processes, DNA synthesis, and metabolic functions in beneficial microorganisms, some beneficial elements indirectly contribute to attenuating diverse abiotic stress factors.

Aside from their roles in stimulating natural processes to promote plant growth, boost yield, and improve quality, beneficial elements can prompt a plethora of physiological mechanisms aimed at developing tolerance, resistance, and resilience to environmental stresses, acting as plant conditioners or bioactivators that help plants adapt to environmental challenges like drought, salinity, toxicants, and extreme temperatures and radiation.

Currently, the use of beneficial elements is gaining momentum in plant science and notably in sustainable agriculture approaches aimed at developing strategies to tackle the great systemic environmental risks exacerbated by the global climate crisis. Since most, if not all, beneficial elements prompt hormesis, extensive future investigations are needed to define the most favorable dose–response curves for each element.

Emerging technologies are paving the way to further elucidate physiological and metabolic pathways prompted by beneficial elements in plants. These technologies involve high-throughput analysis through AI-driven automation, precision gene editing, advanced imaging, and *in situ* monitoring, utilizing tools like CRISPR-Cas9, nanobots, hyperspectral sensors, remote sensing, genetically encoded biosensors, and automated phenotyping platforms to rapidly assess physiology, map diversity, and understand gene function from single cells to entire landscapes. Ultimately, these elements are opening a new chapter in sustainable agriculture, taking center stage in the worldwide push for green and regenerative agriculture.

References

Abd-Alla, M.H., Bagy, M.K., El-enany, A.W., Bashandy, S.R., 2014. Activation of *Rhizobium tibeticum* with flavonoids enhances nodulation, nitrogen fixation, and growth of fenugreek (*Trigonellafoenum-graecum* L.) grown in cobalt-polluted soil. Archives of Environmental Contamination and Toxicology 66, 303–315. Available at https://doi.org/10.1007/s00244-013-9980-7.

Abdelhameed, R.E., Hegazy, H.S., Abdalla, H., Adarosy, M.H., 2025. Efficacy of green synthesized titanium dioxide nanoparticles in attenuation salt stress in *Glycine max* plants: modulations in metabolic constituents and cell ultrastructure. BMC Plant Biology 25, 221. Available at https://doi.org/10.1186/s12870-025-06194-6.

Abdullah, Wani, K.I., Hayat, K., Naeem, M., Aftab, T., 2025. Multifaceted role of selenium in plant physiology and stress resilience: a review. Plant Science (Shannon, Ireland) 355, 112456. Available at https://doi.org/10.1016/j.plantsci.2025.112456.

Agathokleous, E., 2025. Dosing requirements to untangle hormesis in plant science. Trends in Plant Science 30, 1189–1192. Available at https://doi.org/10.1016/j.tplants.2025.06.011.

Agathokleous, E., Kitao, M., Calabrese, E.J., 2018. The rare earth element (REE) lanthanum (La) induces hormesis in plants. Environmental Pollution (Barking, Essex: 1987) 238, 1044–1047. Available at https://doi.org/10.1016/j.envpol.2018.02.068.

Agathokleous, E., Kitao, M., Calabrese, E.J., 2019. Hormetic dose responses induced by lanthanum in plants. Environmental Pollution (Barking, Essex: 1987) 244, 332–341. Available at https://doi.org/10.1016/j.envpol.2018.10.007.

Agathokleous, E., Kitao, M., Calabrese, E.J., 2020. Hormesis: highly generalizable and beyond laboratory. Trends in Plant Science 25, 1076–1086. Available at https://doi.org/10.1016/j.tplants.2020.05.006.

Ahanger, M.A., Bhat, J.A., Siddiqui, M.H., Rinklebe, J., Ahmad, P., 2020. Silicon and secondary metabolites integration in plants: a significant association in stress tolerance. Journal of Experimental Botany eraa291. Available at https://doi.org/10.1093/jxb/eraa291 2020.

Ahmed, S., Qasim, M., Sardar, R., Yasin, N.A., Umar, I., 2025. Multidimensional role of selenium nanoparticles to promote growth and resilience dynamics of *Phaseolus vulgaris* against sodium fluoride stress. International Journal of Phytoremediation 27, 733–750. Available at https://doi.org/10.1080/15226514.2024.2440110.

Akoumianaki-Ioannidou, A., Barouchas, P.E., Ilia, E., Kyramariou, A., Moustakas, N.K., 2016. Effect of vanadium on dry matter and nutrient concentration in sweet basil (*Ocimum basilicum* L.). Australian Journal of Crop Science 10, 199–206.

Alak, M.K., Al-Sabagh, T.M.H.B., 2020. Role of soaking seeds with cobalt and ascorbic acid in alleviation of mung bean under water stress effect. Plant Arch 20, 253–259.

Alhammad, B.A., Saleem, K., Asghar, M.A., Raza, A., Ullah, A., Farooq, T.H., et al., 2023. Cobalt and titanium alleviate the methylglyoxal-induced oxidative stress in *Pennisetum divisum* seedlings under saline conditions. Metabolites 13, 1162. Available at https://doi.org/10.3390/metabo13111162.

Ali, A.M., Bijay-Singh, 2025. Silicon: a crucial element for enhancing plant resilience in challenging environments. Journal of Plant Nutrition 48, 486–521. Available at https://doi.org/10.1080/01904167.2024.2406479.

Ali, S., Farooq, M.A., Yasmeen, T., Hussain, S., Arif, M.S., Abbas, F., et al., 2013. The influence of silicon on barley growth, photosynthesis and ultra-structure under chromium stress. Ecotoxicology and Environmental Safety 89, 66–72. Available at https://doi.org/10.1016/j.ecoenv.2012.11.015.

Alia, F.J., Shamshuddin, J., Fauziah, C.I., Husni, M.H.A., Panhwar, Q.A., 2015. Effects of aluminum, iron and/or low pH on rice seedlings grown in solution culture. International Journal of Agriculture and Biology 17, 702–710. Available at https://doi.org/10.17957/IJAB/14.0019.

Alves, J.S., Menguer, P.K., Lima-Melo, Y., Fiorentini, V.H.R., Ponte, L.R., Olsson, R.V., et al., 2025. Aluminum alleviates iron deficiency chlorosis by interfering with phosphorus homeostasis in rice (*Oryza sativa* L). Plant Physiology and Biochemistry: PPB / Societe Francaise de Physiologie Vegetale 220, 109427. Available at https://doi.org/10.1016/j.plaphy.2024.109427.

Alves, J.D.S., Lima-Melo, Y., Wairich, A., Sasso, V.M., Nascimento, V.L., Sperotto, R.A., et al., 2026. Aluminum alleviation of iron deficiency chlorosis is conserved in wild rice relative *Oryza rufipogon* and in maize. Plants 15, 159. Available at https://doi.org/10.3390/plants15010159.

Ameesha, A.U., Krishnan, A.A.U., Anu-Rajan, S., Chitra, N., Anith, K.N., Soumya, V.I., et al., 2025. Phyllosphere methylotrophs and climate resilience: the emerging role of PPFMS in mitigating abiotic stresses. Int. J. Adv. Biochem. Res. SP-9, 1776–1784. Available at https://www.doi.org/10.33545/26174693.2025.v9.i10Sv.6089.

An, J., Hu, P., Li, F., Wu, H., Shen, Y., White, J.C., et al., 2020. Emerging investigator series: molecular mechanisms of plant salinity stress tolerance improvement by seed priming with cerium oxide nanoparticles. *Environmental Science: Nano* 7, 2214–2228. Available at https://doi.org/10.1039/D0EN00387E.

Araie, H., Shiraiwa, Y., 2016. Selenium in algae. In: In: Borowitzka, M.A., Beardall, J., Raven, J.A. (Eds.), The Physiology of Microalgae, Developments in Applied Phycology 6. Springer, Cham, pp. 281–288. Available at https://doi.org/10.1007/978-3-319-24945-2_12.

Arif, N., Yadav, V., Singh, S., Singh, S., Ahmad, P., Mishra, R.K., et al., 2016. Influence of high and low levels of plant-beneficial heavy metal ions on plant growth and development. Front. Environ. Sci. 4, 69. Available at https://doi.org/10.3389/fenvs.2016.0006.

Arnon, D.I., Stout, P.R., 1939. The essentiality of certain elements in minute quantity for plants with special reference to copper. Plant Physiology 14, 371–375. Available at https://doi.org/10.1104/pp.14.2.371.

Ayub, M.A., Rehman, M.Z., Ahmad, H.R., Rico, C.M., Abbasi, G.H., Umar, W., et al., 2023. Divergent effects of cerium oxide nanoparticles alone and in combination with cadmium on nutrient acquisition and the growth of maize (*Zea mays*). Frontiers in Plant Science 14, 1151786. Available at https://doi.org/10.3389/fpls.2023.1151786.

Babaei, M.J., Ebrahimi, A., Heidari, P., Azadvari, E., Gharanjik, S., Chaghakaboodi, Z., 2025. Titanium dioxide-mediated regulation of enzymatic and non-enzymatic antioxidants, pigments, and diosgenin content promotes cold stress tolerance in *Trigonella foenum-graecum* L. Scientific Reports 15, 1837. Available at https://doi.org/10.1038/s41598-024-84472-3.

Bai, S., Zhang, M., Tang, S., Li, M., Wu, R., Wan, S., et al., 2025. Effects and impact of selenium on human health, A Review. Molecules (Basel, Switzerland) 30, 50. Available at https://doi.org/10.3390/molecules30010050.

Blasco, B., Leyva, R., Romero, L., Ruiz, J.M., 2013. Iodine effects on phenolic metabolism in lettuce plants under salt stress. Journal of Agricultural and Food Chemistry 61, 2591–2596. Available at https://doi.org/10.1021/jf303917n.

Bojórquez-Quintal, E., Escalante-Magaña, C., Echevarría-Machado, I., Martínez-Estévez, M., 2017. Aluminum, a friend or foe of higher plants in acid soils. Frontiers in Plant Science 8, 1767. Available at https://doi.org/10.3389/fpls.2017.01767.

Brown, P.H., Zhao, F.J., Dobermann, A., 2022. What is a plant nutrient? Changing definition to advance science and innovation in plant nutrition. Plant and Soil 476, 11–23. Available at https://doi.org/10.1007/s11104-021-05171-w.

Camargo, M.S., Bezerra, B.K.L., Vitti, A.C., Silva, M.A., Oliveira, A.L., 2017. Silicon fertilization reduces the deleterious effects of water deficit in sugarcane. J. Soil Sci. Plant Nutr. 17, 99–111. Available at https://doi.org/10.4067/S0718-95162017005000008.

Carbajal-Vázquez, V.H., Gómez-Merino, F.C., Herrera-Corredor, J.A., Contreras-Oliva, A., Alcántar-González, G., Trejo-Téllez, L.I., 2020. Effect of titanium foliar applications on tomato fruits from plants grown under salt stress conditions. Not. Bot. Horti. Agrobo. 48, 924–937. Available at https://doi.org/10.15835/nbha48211904.

Carbajal-Vázquez, V.H., Gómez-Merino, F.C., Alcántar-González, G., Sánchez-García, P., Trejo-Téllez, L.I., 2022. Titanium increases the antioxidant activity and macronutrient concentration in tomato seedlings exposed to salinity in hydroponics. Plants 11, 1036. Available at https://doi.org/10.3390/plants11081036.

Carbajal-Vázquez, V.H., Trejo-Téllez, L.I., Mejía-Méndez, J.L., Salinas-Ruiz, J., Gómez-Merino, F.C., 2025. Biostimulant effects of titanium dioxide nanoparticles on germination and initial growth of tomato: evidence of hormesis. PeerJ 13, e20516. Available at https://doi.org/10.7717/peerj.20516.

Cartes, P., Jara, A.A., Pinilla, L., Rosas, A., Mora, M.L., 2010. Selenium improves the antioxidant ability against aluminium-induced oxidative stress in ryegrass roots. Ann. Appl. Biol.156 297–307. Available at https://doi.org/10.1111/j.1744-7348.2010.00387.x.

Chatzistathis, T., 2018. Physiological importance of manganese, cobalt and nickel and the improvement of their uptake and utilization by plants. In: Hossain, M.A., Kamiya, T., Burritt, D.J., Tran, L.P., Fujiwara, T. (Eds.), Plant Micronutrient Use Efficiency. Academic Press, London, pp. 129–131. Available at https://doi.org/10.1016/B978-0-12-812104-7.00008-3.

Chen, X.D., Shan, C.J., 2019. Cerium nitrate improves salt tolerance of wheat seedlings by regulating the antioxidant capacity of chloroplasts. Biologia Plantarum 63, 323–327. Available at https://doi.org/10.32615/bp.2019.037.

Chen, M., Liu, Q., Wang, S.W., Wang, E., Guo, X., Chou, S.L., 2019. High-abundance and low-cost metal-based cathode materials for sodium-ion batteries: problems, progress, and key technologies. Energy Mater 9, 1803609. Available at https://doi.org/10.1002/aenm.201803609.

Cheng, M., Wang, A., Liu, Z., Gendall, A.R., Rochfort, S., Tang, C., 2018. Sodium chloride decreases cadmium accumulation and changes the response of metabolites to cadmium stress in the halophyte *Carpobrotus rossii*. Annals of Botany 122, 373–385. Available at https://doi.org/10.1093/aob/mcy077.

Chu, J., Yao, X., Zhang, Z., 2010. Responses of wheat seedlings to exogenous selenium supply under cold stress. Biological Trace Element Research 136, 355–363. Available at https://doi.org/10.1007/s12011-009-8542-3.

Chung, Y.S., Kim, K.S., Hamayun, M., Kim, Y., 2020. Silicon confers soybean resistance to salinity stress through regulation of reactive oxygen and reactive nitrogen species. Frontiers in Plant Science 10, 1725. Available at https://doi.org/10.3389/fpls.2019.01725.

Cigler, P., Olejnickova, J., Hruby, M., Csefalvay, L., Peterka, J., Kuzel, S., 2010. Interactions between iron and titanium metabolism in spinach: a chlorophyll fluorescence study in hydropony. Journal of Plant Physiology 167, 1592–1597. Available at https://doi.org/10.1016/j.jplph.2010.06.021.

Consentino, B.B., Vultaggio, L., Iacuzzi, N., La Bella, S., De Pasquale, C., Rouphael, Y., et al., 2023. Iodine biofortification and seaweed extract-based biostimulant supply interactively drive the yield, quality, and functional traits in strawberry fruits. Plants 12, 245. Available at https://doi.org/10.3390/plants12020245.

Consentino, B.B., Mancuso, F., Vultaggio, L., Bellitto, P., Ntatsi, G., Cannata, C., et al., 2026. Selenium biofortification and an *Ecklonia maxima*-based seaweed extract jointly compose curly endive drought stress tolerance in a soilless system. Plants 15, 170. Available at https://doi.org/10.3390/plants15010170.

Cui, Y.N., Xia, Z.R., Ma, Q., Wang, W.Y., Chai, W.W., Wang, S.M., 2019. The synergistic effects of sodium and potassium on the xerophyte *Apocynum venetum* in response to drought stress. Plant Physiology and Biochemistry: PPB / Societe Francaise de Physiologie Vegetale 135, 489–498. Available at https://doi.org/10.1016/j.plaphy.2018.11.011.

da Silva, I.A., de Andrade, J.L.S., Barbosa, F.L.A., Almeida, M.S., Araújo, M.L.H., de Souza, A.J., et al., 2025. Co-application of seaweed extract (*Solieria filiformis*) and silicon: effect on sporulation, mycorrhizal colonization, and initial growth of *Mimosa caesalpiniaefolia*. Microorganisms 13, 1581. Available at https://doi.org/10.3390/microorganisms13071581.

Daryabeigi-Zand, A., Tabrizi, A.M., Heir, A.V., 2020. Co-application of biochar and titanium dioxide nanoparticles to promote remediation of antimony from soil by *Sorghum bicolor*: metal uptake and plant response. Heliyon 6, e04669. Available at https://doi.org/10.1016/j.heliyon.2020.e04669.

Daumann, L.J., 2019. Essential and ubiquitous: the emergence of lanthanide metallobiochemistry. Angewandte Chemie (International Ed. in English) 58, 12795–12802. Available at https://doi.org/10.1002/anie.201904090.

de Jesus, L.R., Batista, B.L., da Silva Lobato, A.K., 2017. Silicon reduces aluminum accumulation and mitigates toxic effects in cowpea plants. Acta Physiologiae Plantarum /

Polish Academy of Sciences, Committee of Plant Physiology Genetics and Breeding 39, 138. Available at https://doi.org/10.1007/s11738-017-2435-4.

Del Pino, A.M., Guiducci, M., D'Amato, R., Di Michele, A., Tosti, G., Datti, A., et al., 2019. Selenium maintains cytosolic Ca^{2+} homeostasis and preserves germination rates of maize pollen under H_2O_2-induced oxidative stress. Scientific Reports 9, 13502. Available at https://doi.org/10.1038/s41598-019-49760-3.

Del Pino, A.M., Falcinelli, B., D'Amato, R., Businelli, D., Benincasa, P., Palmerini, C.A., 2022. *In vitro* oxidative stress threatening maize pollen germination and cytosolic Ca^{2+} can be mitigated by extracts of emmer wheatgrass biofortified with selenium. Plants 11, 859. Available at https://doi.org/10.3390/plants11070859.

Demeke, E.D., Benti, N.E., Terefe, M.G., Anbessa, T.T., Mengistu, W.M., Mekonnen, Y.S., 2025. A comprehensive review on nano-fertilizers: preparation, development, utilization, and prospects for sustainable agriculture in Ethiopia. Nanoscale Advances 7, 2131–2144. Available at https://doi.org/10.1039/D4NA01068J.

Dervisi, I., Valassakis, C., Koletti, A., Kouvelis, V.N., Flemetakis, E., Ouzounis, C.A., et al., 2023. Evolutionary aspects of selenium binding protein (SBP). Journal of Molecular Evolution 91, 471–481. Available at https://doi.org/10.1007/s00239-023-10105-4.

Dias, M.C., Santos, C., Pinto, G., Silva, A.M.S., Silva, S., 2019. Titanium dioxide nanoparticles impaired both photochemical and non-photochemical phases of photosynthesis in wheat. Protoplasma 256, 69–78. Available at https://doi.org/10.1007/s00709-018-1281-6.

Dinler, B.S., Cetinkaya, H., Koc, F.N., Gül, V., Sefaoğlu, F., 2024. Effects of titanium dioxide nanoparticles against salt and heat stress in safflower cultivars. Acta Botanica Brasilica. 38, e20230136. Available at https://doi.org/10.1590/1677-941X-ABB-2023-0136.

Dizkara, H., Yıldırım, E., Ekinci, M., 2025. The effect of iodine on rocket (*Eruca sativa* L.) plant under salt stress. Journal of Agricultural Production 6, 148–156. Available at https://doi.org/10.56430/japro.1746259.

Djanaguiraman, M., Prasad, P.V., Seppanen, M., 2010. Selenium protects sorghum leaves from oxidative damage under high temperature stress by enhancing antioxidant defense system. Plant Physiology and Biochemistry: PPB / Societe Francaise de Physiologie Vegetale 48, 999–1007. Available at https://doi.org/10.1016/j.plaphy.2010.09.009.

Djanaguiraman, M., Nair, R., Giraldo, J.P., Prasad, P.V.V., 2018. Cerium oxide nanoparticles decrease drought-induced oxidative damage in sorghum leading to higher photosynthesis and grain yield. ACS Omega 3, 14406–14416. Available at https://doi.org/10.1021/acsomega.8b01894.

Dos Santos, M.S., Sanglard, L.M.P.V., Martins, S.C.V., Barbosa, M.L., de Melo, D.C., Gonzaga, W.F., et al., 2019. Silicon alleviates the impairments of iron toxicity on the rice photosynthetic performance via alterations in leaf diffusive conductance with minimal impacts on carbon metabolism. Plant Physiology and Biochemistry: PPB/Societe Francaise de Physiologie Vegetale 143, 275–285. Available at https://doi.org/10.1016/j.plaphy.2019.09.011.

Ehinmitan, E., Siamalube, B., Losenge, T., Mamati, E., Juma, P., Ngumi, V., 2025. *Methylobacterium* spp. in sustainable agriculture: strategies for plant stress management and growth promotion. Microbe 8, 100476. Available at https://doi.org/10.1016/j.microb.2025.100476.

Elshamly, A.M.S., Nassar, S.M.A., 2024a. Impacts of cobalt and zinc on improving peanuts nutrient uptake, yield and irrigation water use efficiency under different irrigation levels. Scientific Reports 14, 7188. Available at https://doi.org/10.1038/s41598-024-56898-2.

Elshamly, A.M.S., Parrey, Z.A., Gaafar, A.Z., Siddiqui, M.H., Hussain, S., 2024b. Potassium humate and cobalt enhance peanut tolerance to water stress through regulation of proline, antioxidants, and maintenance of nutrient homeostasis. Scientific Reports 14, 1625. Available at https://doi.org/10.1038/s41598-023-50714-z.

Escobar-Sepúlveda, H.F., Trejo-Téllez, L.I., Pérez-Rodríguez, P., Hidalgo-Contreras, J.V., Gómez-Merino, F.C., 2017. Diacylglycerol kinases are widespread in higher plants and display inducible gene expression in response to beneficial elements, metal, and metalloid ions. *Frontiers in Plant Science* 8, 129. Available at https://doi.org/10.3389/fpls.2017.00129.

Estarriaga-Navarro, S., Goicoechea, N., Plano, D., Sanmartín, C., 2025. Selenium biofortification: integrating one health and sustainability. Journal of the Science of Food and Agriculture 106, 1955–1967. Available at https://doi.org/10.1002/jsfa.70103.

Fan, W., Xu, J.M., Wu, P., Yang, Z.X., Lou, H.Q., Chen, W.W., et al., 2018. Alleviation by abscisic acid of Al toxicity in rice bean is not associated with citrate efflux but depends on ABI5-mediated signal transduction pathways. Journal of Integrative Plant Biology 61, 140–154. Available at https://doi.org/10.1111/jipb.12695.

Fan, K., Wang, M., Gao, Y., Ning, Q., Shi, Y., 2019. Transcriptomic and ionomic analysis provides new insight into the beneficial effect of Al on tea roots' growth and nutrient uptake. Plant Cell Reports 38, 715–729. Available at https://doi.org/10.1007/s00299-019-02401-5.

Fatahiyan, F., Najafi, F., Shirkhani, Z., 2025. Enhancing salt stress tolerance in *Carthamus tinctorius* L. through selenium soil treatment: anatomical, biochemical, and physiological insights. BMC Plant Biology 25, 100. Available at https://doi.org/10.1186/s12870-025-06078-9.

Feng, L., Chen, Z., Wang, H., Chen, Z., Chen, Z., 2024. Rare Earth Elements induce drought tolerance in *Dicranopteris pedata* from ion-adsorbed rare earth mining area in southern China. Sustainability 16, 10619. Available at https://doi.org/10.3390/su162310619.

Filek, M., Keskinen, R., Hartikainen, H., Szarejko, I., Janiak, A., Miszalski, Z., et al., 2008. The protective role of selenium in rape seedlings subjected to cadmium stress. Journal of Plant Physiology 165, 833–844. Available at https://doi.org/10.1016/j.jplph.2007.06.006.

Gad, N., El-Metwally, I.M., 2015. Chemical and physiological response of maize to salinity using cobalt supplement. International Journal of ChemTech Research 8, 45–52.

Gad, N., El-Moez, M.R.A., Kandil, H., 2011. Barley response to salt stress at varied levels of cobalt I. Growth and mineral composition. Journal of Applied Sciences Research 7, 1454–1459.

Gad, N., Abdel Moez, M.R., Fekry Ali, M.E., Abou-Hussein, S.D., 2018a. Increasing salt tolerance in cucumber by using cobalt. Middle East Journal of Applied Sciences 8, 345–354.

Gad, N., Fekry-Ali, M.E., Abbas, M.M., Abdel-Moez, M.R., 2018b. Maximization of drought tolerance of bean plants using cobalt supplementation A-growth, yield and nutritional status. Middle East Journal of Agricultural Research 7, 1819–1826.

Gad, N., Sayed, S., Hassan, N.M.K., 2020. Influence of cobalt on tolerating climatic change (salinity) in onion plant with reference to growth, yield, and mineral composition. Plant Arch 20, 1491–1495.

Gad, N., Elrahman, E.A.A., Ali, M.E.F., Abou-Shlell, M.K., Teiba, I.I., Almutairi, M.H., et al., 2025. Role and importance of cobalt in faba bean through rationalization of its nitrogen fertilization. BMC Plant Biology 25, 872. Available at https://doi.org/10.1186/s12870-025-06596-6.

Gál, J., Hursthouse, A., Tatner, P., Stewart, F., Welton, R., 2008. Cobalt and secondary poisoning in the terrestrial food chain: data review and research gaps to support risk assessment. Environment International 34, 821–838. Available at https://doi.org/10.1016/j.envint.2007.10.006.

Garcia-Daga, S., Roy, S.J., Gilliham, M., 2025. Redefining the role of sodium exclusion within salt tolerance. Trends in Plant Science 30, 137–146. Available at https://doi.org/10.1016/j.tplants.2024.10.002.

García-Fuentes, J.E., Herrera-Castellanos, B.F., Rivas-Martínez, E.N., Narváez-Ortiz, W.A., Benavides-Mendoza, A., Medrano-Macías, J., 2022. Outcomes of foliar iodine application on growth, minerals and antioxidants in tomato plants under salt stress. Folia Horticulturae 34, 27–37. Available at https://doi.org/10.2478/fhort-2022-0003.

Garousi, F., 2017. The essentiality of selenium for humans, animals, and plants, and the role of selenium in plant metabolism and physiology. Acta Universitatis Sapientiae, Alimentaria 10, 75–90. Available at https://doi.org/10.1515/ausal-2017-0005.

Gaur, S., Kumar, J., Kumar, D., Chauhan, D.K., Prasad, S.M., Srivastava, P.K., 2020. Fascinating impact of silicon and silicon transporters in plants: a review. Ecotoxicology and Environmental Safety 202, 110885. Available at https://doi.org/10.1016/j.ecoenv.2020.110885.

Gavassi, M.A., Silva, G.S., Da Silva, C.D.S., Thompson, A.J., Macleod, K., Oliveira, P.M.R., et al., 2021. NCED expression is related to increased ABA biosynthesis and stomatal closure under aluminum stress. Environmental and Experimental Botany 185, 12. Available at https://doi.org/10.1016/j.envexpbot.2021.104404.

Godon, C., Mercier, C., Wang, X., David, P., Richaud, P., Nussaume, L., et al., 2019. Under phosphate starvation conditions, Fe and Al trigger accumulation of the transcription factor STOP1 in the nucleus of *Arabidopsis* root cells. The Plant Journal: For Cell and Molecular Biology 99, 937–949. Available at https://doi.org/10.1111/tpj.14374.

Gohari, G., Zareei, E., Rostami, H., Panahirad, S., Kulak, M., Farhadi, H., et al., 2021. Fotopoulos V. Protective effects of cerium oxide nanoparticles in grapevine (*Vitis vinifera* L.) cv. Flame Seedless under salt stress conditions. Ecotoxicology and Environmental Safety 220, 112402. Available at https://doi.org/10.1016/j.ecoenv.2021.112402.

Gómez-Merino, F.C., Trejo-Téllez, L.I., 2018. The role of beneficial elements in triggering adaptive responses to environmental stressors and improving plant performance. In: Vats, S. (Ed.), Biotic and Abiotic Stress Tolerance in Plants. Springer Natura, Singapore, pp. 137–172. Available at https://doi.org/10.1007/978-981-10-9029-5_6.

Gómez-Merino, F.C., Trejo-Téllez, L.I., García-Jiménez, A., Escobar-Sepúlveda, H.F., Ramírez-Olvera, S.M., 2020. Silicon flow from root to shoot in pepper: a comprehensive *in silico* analysis reveals a potential linkage between gene expression and hormone signaling that stimulates plant growth and metabolism. PeerJ 8, e10053. Available at https://doi.org/10.7717/peerj.10053.

Gupta, N., Bajpai, M.S., Majumdar, R.S., Mishra, P.K., 2015. Response of iodine on antioxidant levels of *Glycine max* L. grown under Cd^{2+} stress. *Advanced Biomedical Research* 9, 40–48. Available at https://doi.org/10.5829/idosi.abr.2015.9.1.9183.

Hajiboland, R., Barceló, J., Poschenrieder, C., Tolrà, R., 2013. Amelioration of iron toxicity: a mechanism for aluminum-induced growth stimulation in tea plants. Journal of Inorganic Biochemistry 128, 183–187. Available at https://doi.org/10.1016/j.jinorgbio.2013.07.007.

Hajiboland, R., Bahrami-Rad, S., Bastani, S., 2014. Aluminum alleviates boron-deficiency induced growth impairment in tea plants. Biologia Plantarum 58, 717–724. Available at https://doi.org/10.1007/s10535-014-0425-6.

Hanus-Fajerska, E., Wiszniewska, A., Kamińska, I., 2021. A dual role of vanadium in environmental systems-beneficial and detrimental effects on terrestrial plants and humans. Plants 10, 1110. Available at https://doi.org/10.3390/plants10061110.

Hareem, M., Danish, S., Obaid, S.A., Ansari, M.J., Datta, R., 2024. Mitigation of drought stress in chili plants (*Capsicum annuum* L.) using mango fruit waste biochar, fulvic acid and cobalt. Scientific Reports 14, 14270. Available at https://doi.org/10.1038/s41598-024-65082-5.

Hasanuzzaman, M., Fujita, M., 2011. Selenium pretreatment upregulates the antioxidant defense and methylglyoxal detoxification system and confers enhanced tolerance to drought stress in rapeseed seedlings. Biological Trace Element Research 143, 1758–1776. Available at https://doi.org/10.1007/s12011-011-8998-9.

Hasanuzzaman, M., Hossain, M.A., Fujita, M., 2011. Selenium-induced up-regulation of the antioxidant defense and methylglyoxal detoxification system reduces salinity-induced damage in rapeseed seedlings. Biological Trace Element Research 143, 1704–1721. Available at https://doi.org/10.1007/s12011-011-8958-4.

Hassanpouraghdam, M.B., Mehrabani, L.V., Bonabian, Z., Aazami, M.A., Rasouli, F., Feldo, M., et al., 2022. Foliar application of cerium oxide-salicylic acid nanoparticles (CeO_2:SA nanoparticles) influences the growth and physiological responses of *Portulaca oleracea* L. under salinity. International Journal of Molecular Sciences 23, 5093. Available at https://doi.org/10.3390/ijms23095093.

Hosseini, S.A., Maillard, A., Hajirezaei, M.R., Ali, N., Schwarzenberg, A., Jamois, F., et al., 2017. Induction of barley silicon transporter *HvLsi1* and *HvLsi2*, increased silicon concentration in the shoot and regulated starch and ABA homeostasis under osmotic stress and concomitant potassium deficiency. Frontiers in Plant Science 8, 1359. Available at https://doi.org/10.3389/fpls.2017.01359.

Hu, X., Wei, X., Ling, J., Chen, J., 2021. Cobalt: an essential micronutrient for plant growth? Frontiers in Plant Science 12, 768523. Available at https://doi.org/10.3389/fpls.2021.768523.

Huang, S., Ma, J.F., 2024. Silicon transport and its "homeostasis" in rice. Quantitative Plant Biology 5, 1–5. Available at https://doi.org/10.1017/qpb.2024.19 e15.

Huang, H., Rizwan, M., Li, M., Song, F., Zhou, S., He, X., et al., 2019. Comparative efficacy of organic and inorganic silicon fertilizers on antioxidant response, Cd/Pb accumulation and health risk assessment in wheat (*Triticum aestivum* L.). Environmental Pollution (Barking, Essex: 1987) 255, 113146. Available at https://doi.org/10.1016/j.envpol.2019.113146.

Husen, A., 2024. Essential Minerals in Plant-Soil System. Coordination, Signalling, and Interaction under Adverse Situations. Elsevier, Cambridge, MA, USA. Available at https://doi.org/10.1016/C2022-0-02437-5.

Hussain, S., Iqbal, N., Brestic, M., Raza, M.A., Pang, T., Langham, D.R., et al., 2019. Changes in morphology, chlorophyll fluorescence performance and Rubisco activity of soybean in response to foliar application of ionic titanium under normal light and shade environment. The Science of the Total Environment 658, 626–637. Available at https://doi.org/10.1016/j.scitotenv.2018.12.182.

Hussain, S., Shuxian, L., Mumtaz, M., Shafiq, I., Iqbal, N., Brestic, M., et al., 2020. Foliar application of silicon improves stem strength under low light stress by regulating lignin biosynthesis genes in soybean (*Glycine max* (L.) Merr.). Journal of Hazardous Materials 401, 123256. Available at https://doi.org/10.1016/j.jhazmat.2020.123256.

Hussain, R., Aslam, S., Sattar, S., Muhammad, N., Qingjun, G., Qadir, A., 2025. Silicon unveiled: exploring its role in beneficial soil elements. In: Saud, S., Fahad, S., Wang, D. (Eds.), Beneficial Elements for Remediation of Heavy Metals in Polluted Soil. Elsevier, Amsterdam, pp. 341–371. Available at https://doi.org/10.1016/B978-0-443-26522-8.00012-X.

Iguchi, S., Tokunaga, T., Kamon, E., Takenaka, Y., Koshimizu, S., Watanabe, M., et al., 2023. Lanthanum supplementation alleviates tomato root growth suppression under low light stress. Plants 12, 2663. Available at https://doi.org/10.3390/plants12142663.

Inayat, H., Mehmood, H., Danish, S., Alharbi, S.A., Ansari, M.J., Datta, R., 2024. Impact of cobalt and proline foliar application for alleviation of salinity stress in radish. BMC Plant Biology 24, 287. Available at https://doi.org/10.1186/s12870-024-04998-6.

Jafarian, S., Sodaeizade, H., Arani, A.M., Hakimzade, M.A., Sohrabizadeh, Z., 2020. The effect of iodine on increasing drought tolerance of (*Carthamus tinctorius* L.) in seed germination and early growth stage. Journal of Environmental Science Studies 5, 2387–2393.

Jalal, A., de Oliveira Jr., J.C., Ribeiro, J.S., Fernandes, G.C., Mariano, G.G., Trindade, V.D.R., et al., 2021. Hormesis in plants: physiological and biochemical responses. Ecotoxicology and Environmental Safety 207, 111225. Available at https://doi.org/10.1016/j.ecoenv.2020.111225.

Katz, O., Puppe, D., Kaczorek, D., Prakash, N.B., Schaller, J., 2021. Silicon in the soil–plant continuum: intricate feedback mechanisms within ecosystems. Plants 10, 652. Available at https://doi.org/10.3390/plants10040652.

Kaur, S., Kaur, N., Siddique, K.H.M., Nayyar, H., 2016. Beneficial elements for agricultural crops and their functional relevance in defence against stresses. Archives of Agronomy and Soil Science 62, 905–920. Available at https://doi.org/10.1080/03650340.2015.1101070.

Keshavarz, H., Moghadam, R.S.G., 2017. Seed priming with cobalamin (vitamin B12) provides significant protection against salinity stress in the common bean. Rhizosphere 3, 143–149. Available at https://doi.org/10.1016/j.rhisph.2017.04.010.

Keskinen, R., Turakainen, M., Hartikainen, H., 2010. Plant availability of soil selenate additions and selenium distribution within wheat and ryegrass. Plant and Soil 333, 301–313. Available at https://doi.org/10.1007/s11104-010-0345-y.

Khan, M.N., Li, Y., Khan, Z., Chen, L., Liu, J., Hu, J., et al., 2021. Nanoceria seed priming enhanced salt tolerance in rapeseed through modulating ROS homeostasis and α-amylase activities. Journal of Nanobiotechnology 19, 276. Available at https://doi.org/10.1186/s12951-021-01026-9.

Kiany, T., Pishkar, L., Sartipnia, N., Iranbakhsh, A., Barzin, G., 2022. Effects of silicon and titanium dioxide nanoparticles on arsenic accumulation, phytochelatin metabolism, and antioxidant system by rice under arsenic toxicity. Environmental Science and Pollution Research 29, 34725–34737. Available at https://doi.org/10.1007/s11356-021-17927-z.

Kiferle, C., Gonzali, S., Beltrami, S., Martinelli, M., Hora, K., Holwerda, H.T., et al., 2022. Improvement in fruit yield and tolerance to salinity of tomato plants fertigated with micronutrient amounts of iodine. Scientific Reports 12, 14655. Available at https://doi.org/10.1038/s41598-022-18301-w.

Kirkby, E.A., 2023. Introduction, definition and classification of nutrients. In: Rengel, Z., Cakmak, I., White, P.J. (Eds.), Marschner´s Mineral Nutrition of Plants. Elsevier–Academic Press, London, pp. 3–9. Available at https://doi.org/10.1016/B978-0-12-384905-2.00001-7.

Kolbert, Z., Molnár, Á., Feigl, G., Van Hoewyk, D., 2019. Plant selenium toxicity: proteome in the crosshairs. Journal of Plant Physiology 232, 291–300. Available at https://doi.org/10.1016/j.jplph.2018.11.003.

Kotelnikova, A., Fastovets, I., Rogova, O., Schulze, L.M., Wong, J.R., Britto, D.T., 2019. Toxicity assay of lanthanum and cerium in solutions and soil. Ecotoxicology and Environmental Safety 167, 20–28. Available at https://doi.org/10.1016/j.ecoenv.2018.09.117.

Kronzucker, H.J., Coskun, D., Schulze, L.M., Wong, J.R., Britto, D.T., 2013. Sodium as nutrient and toxicant. Plant and Soil 369, 1–23. Available at https://doi.org/10.1007/s11104-013-1801-2.

Kumar, M., Saxena, R., Tomar, R.S., Rai, P.K., Paul, D., 2018. Role of methylotrophic bacteria in climate change mitigation. In: Kashyap, P.L., Srivastava, A.K., Tiwari, S.P., Kumar, S. (Eds.), Microbes for Climate Resilient Agriculture. Wiley, Hoboken, pp. 149–164. Available at https://doi.org/10.1002/9781119276050.ch7.

Küpper, F.C., Carpenter, L.J., McFiggans, G.B., Feiters, M.C., 2008. Iodide accumulation provides kelp with an inorganic antioxidant impacting atmospheric chemistry. Proceedings of the National Academy of Sciences of the United States of America 05, 6954–6958. Available at https://doi.org/10.1073/pnas.0709959105.

Kushwaha, P., Verma, R., Singh, A.K., Dixi, P., 2025. Salinity stress and its alleviation in plants: a general review. International Journal of Scientific Research in Science and Technology 12, 481–547. Available at https://doi.org/10.32628/IJSRST2512364.

Laclau, J.P., Gonçalves, J.L.M., Bouillet, J.P., Nouvellon, Y., 2024. Perspectives: can we apply sodium in tropical plantations to reduce the need of potassium fertilizer? Forest Ecology and Management 553, 121643. Available at https://doi.org/10.1016/j.foreco.2023.121643.

Leksungnoen, P., Wisawapipat, W., Ketrot, D., Aramrak, S., Nookabkaew, S., Rangkadilok, N., et al., 2019. Biochar and ash derived from silicon-rich rice husk decrease inorganic arsenic species in rice grain. The Science of the Total Environment 684, 360–370. Available at https://doi.org/10.1016/j.scitotenv.2019.05.247.

Leyva, R., Sánchez-Rodríguez, E., Ríos, J.J., Rubio-Wilhelmi, M.M., Ruiz, R., Blasco, B., 2011. Beneficial effects of exogenous iodine in lettuce plants subjected to salinity stress. Plant Science (Shannon, Ireland) 181, 195–202. Available at https://doi.org/10.1016/j.plantsci.2011.05.007.

Li, H., Yang, L.T., Qi, Y.P., Guo, P., Lu, Y.B., Chen, L.S., 2016. Aluminum toxicity-induced alterations of leaf proteome in two citrus species differing in aluminum tolerance. International Journal of Molecular Sciences 17, 1180. Available at https://doi.org/10.3390/ijms17071180.

Li, S., Yang, J.B., Li, J.Q., Huang, J., Shen, R.F., Zeng, D.L., et al., 2024. A NAC transcription factor represses a module associated with xyloglucan content and regulates aluminum tolerance. Plant Physiology 196, 564–578. Available at https://doi.org/10.1093/plphys/kiae281.

Lian, H., Qin, C., Zhang, C., Yan, M., Li, H., Zhang, S., 2019. Foliar-applied lanthanum chloride increases growth and phosphorus acquisition in phosphorus-limited adzuki bean seedlings. Plant and Soil 442, 385–399. Available at https://doi.org/10.1007/s11104-019-04197-5.

Lide, R.D., 2017. Abundance of elements in the Erth's crust and in the sea. In: Lide, R.D. (Ed.), CRC Handbook of Chemistry and Physics. CRC Press, Boca Raton, pp. 14–17.

Lima, J.S., Andrade, O.V.S., Santos, L.C., Morais, E.G., Martins, G.S., Mutz, Y.S., et al., 2023. Soybean plants exposed to low concentrations of potassium iodide have better tolerance to water deficit through the antioxidant enzymatic system and photosynthesis modulation. Plants 12, 2555. Available at https://doi.org/10.3390/plants12132555.

Liu, R.Q., Xu, X.J., Wang, S., Shan, C.J., 2016. Lanthanum improves salt tolerance of maize seedlings. Photosynthetica 54, 148–151. Available at https://doi.org/10.1007/s11099-015-0157-7.

Liu, Y., Tao, J., Cao, J., Zeng, Y., Li, X., Ma, J., et al., 2020. The beneficial effects of aluminum on the plant growth in *Camellia japonica*. Journal of Soil Science and Plant Nutrition 20, 1799–1809. Available at https://doi.org/10.1007/s42729-020-00251-9.

Liu, X., Liu, J., Liu, C., Zhang, X., Zhao, Z., Xu, J., et al., 2022. Selenium-containing polysaccharides isolated from *Rosa laevigata* Michx fruits exhibit excellent anti-oxidant and neuroprotective activity *in vitro*. International Journal of Biological Macromolecules 209, 1222–1233. Available at https://doi.org/10.1016/j.ijbiomac.2022.04.

Liu, Z., Lin, Y., Liang, L., Li, X., Wang, Z., Cheng, W., 2025a. Global burden of childhood developmental intellectual disability caused by iodine deficiency from 1990 to 2021 and projection to 2040. Frontiers in Public Health 13, 1598949. Available at https://doi.org/10.3389/fpubh.2025.1598949.

Liu, W., Feng, Y., Chen, S., Chu, R., Li, S., Wang, Y., et al., 2025b. TiO_2 nanoparticles improve cadmium toxicity tolerance in *Hemerocallis citrina* Baroni by modulating photosynthetic and antioxidative profile. Plant Cell Reports 44, 105. Available at https://doi.org/10.1007/s00299-025-03502-0.

Lv, H., Tani, A., 2018. Genomic characterization of methylotrophy of *Oharaeibacter diazotrophicus* strain SM30T. Journal of Bioscience and Bioengineering 126, 667–675. Available at https://doi.org/10.1016/j.jbiosc.2018.05.023.

Lv, H., Masuda, S., Fujitani, Y., Sahin, N., Tani, A., 2017. Oharaeibacter diazotrophicus gen. nov., sp. nov., a diazotrophic and facultatively methylotrophic bacterium, isolated from rice rhizosphere. International Journal of Systematic and Evolutionary Microbiology 67, 576–582. Available at https://doi.org/10.1099/ijsem.0.001660.

Lyu, S., Wei, X., Chen, J., Wang, C., Wang, X., et al., 2017. Titanium as a beneficial element for crop production. Frontiers in Plant Science 8, 597. Available at https://doi.org/10.3389/fpls.2017.00597.

Macey, M.C., Pratscher, J., Crombie, A.T., Murrell, J.C., 2020. Impact of plants on the diversity and activity of methylotrophs in soil. Microbiome 8, 31. Available at https://doi.org/10.1186/s40168-020-00801-4.

Madlala, N.C., Khanyile, N., Masenya, A., 2024. Examining the correlation between the inorganic nano-fertilizer physical properties and their impact on crop performance and nutrient uptake efficiency. Nanomaterials 14, 1263. Available at https://doi.org/10.3390/nano14151263.

Maglione, G., Vitale, E., Costanzo, G., Polimeno, F., Arena, C., Vitale, L., 2022. Iodine enhances the nutritional value but not the tolerance of lettuce to NaCl. Horticulturae 8, 662. Available at https://doi.org/10.3390/horticulturae8070662.

Malik, J.A., Goel, S., Kaur, N., Sharma, J.S., Singh, I., Nayyar, H., 2012. Selenium antagonizes the toxic effects of arsenic on mungbean (*Phaseolus aureus* Roxb.) plants by restricting its uptake and enhancing the antioxidative and detoxification mechanisms. Environmental and Experimental Botany 77, 242–248. Available at https://doi.org/10.1016/j.envexpbot.2011.12.001.

Mandlik, R., Thakral, V., Raturi, G., Shinde, S., Nikolić, M., Tripathi, D.K., et al., 2020. Significance of silicon uptake, transport, and deposition in plants. Journal of Experimental Botany 71, 6703–6718. Available at https://doi.org/10.1093/jxb/eraa301.

Maron, L.G., Kirst, M., Mao, C., Milner, M.J., Menossi, M., Kochian, L.V., 2008. Transcriptional profiling of aluminum toxicity and tolerance responses in maize roots. The New Phytologist 179, 116–128. Available at https://doi.org/10.1111/j.1469-8137.2008.02440.x.

Masuda, S., Suzuki, Y., Fujitani, Y., Mitsui, R., Nakagawa, T., Shintani, M., et al., 2018. Lanthanide-dependent regulation of methylotrophy in *Methylobacteriu maquaticum* strain 22A. mSphere 3, e00462-17. Available at https://doi.org/10.1128/mSphere.00462-17.

Medrano-Macías, J., Leija-Martínez, P., González-Morales, S., Juárez-Maldonado, A., Benavides-Mandoza, A., 2016. Use of iodine to biofortify and promote growth and stress tolerance in crops. Frontiers in Plant Science 7, 1146. Available at https://doi.org/10.3389/fpls.2016.01146.

Mehrabanjoubani, P., Abdolzadeh, A., Sadeghipour, H.R., Aghdasi, M., Bagherieh-Najjar, M.B., et al., 2019. Silicon increases cell wall thickening and lignification in rice (*Oryza sativa*) root tip under excess Fe nutrition. Plant Physiology and Biochemistry: PPB / Societe Francaise de Physiologie Vegetale 144, 264–273. Available at https://doi.org/10.1016/j.plaphy.2019.09.047.

Mezey, J., Mezeyová, I., Selnekovič, A., Bajčan, D., 2025. Foliar selenium biofortification in temperate fruit crops: impact on selenium accumulation and nutritional quality of fruits and juices. Beverages 11, 53. Available at https://doi.org/10.3390/beverages11020053.

Mitani-Ueno, N., Yamaji, N., Huang, S., Yoshioka, Y., Miyaji, T., Ma, J.F., 2023. A silicon transporter gene required for healthy growth of rice on land. Nature Communications 14, 6522. Available at https://doi.org/10.1038/s41467-023-42180-y.

Mohammadi, M.H.Z., Panahirad, S., Navai, A., Bahrami, M.K., Kulak, M., Gohari, G., 2021. Cerium oxide nanoparticles (CeO_2-NPs) improve growth parameters and antioxidant defense system in Moldavian balm (*Dracocephalum moldavica* L.) under salinity stress. Plant Stress 1, 100006. Available at https://doi.org/10.1016/j.stress.2021.100006.

Moreno-Alvarado, M., García-Morales, S., Trejo-Téllez, L.I., Hidalgo-Contreras, J.V., Gómez-Merino, F.C., 2017. Aluminum enhances growth and sugar concentration, alters macronutrient status and regulates the expression of NAC transcription factors in rice. Frontiers in Plant Science 8, 73. Available at https://doi.org/10.3389/fpls.2017.00073.

Morshedloo, M.R., Jabbari, N., Azizi, M., Nazari, M., Zahedi, S.M., Viapiana, A., 2025. Enhancing salt tolerance in *Mentha × gracilis* through foliar applications of titanium and nano-titanium. BMC Plant Biology 25, 1088. Available at https://doi.org/10.1186/s12870-025-07056-x.

Muhammad, N., Cai, S., Shah, J.M., Zhang, G., 2016. The combined treatment of Mn and Al alleviates the toxicity of Al or Mn stress alone in barley. Acta Physiologiae Plantarum / Polish Academy of Sciences, Committee of Plant Physiology Genetics and Breeding 38, 1–7. Available at https://doi.org/10.1007/s11738-016-2296-2.

Muhammad, N., Zvobgo, G., Zhang, G.P., 2019. A review: The beneficial effects and possible mechanisms of aluminum on plant growth in acidic soil. Journal of Integrative Agriculture 18, 1518–1528. Available at https://doi.org/10.1016/S2095-3119(18)61991-4.

Mukarram, M., Corpas, F.J., Lux, A., 2025. Editorial: new avenues of silicon's role in plant biology: trends and controversies. Frontiers in Plant Science 16, 1717302. Available at https://doi.org/10.3389/fpls.2025.1717302.

Mukherjee, S., Krishnamoorthy, S.B., Subrayan, R., Goswami, A., Mitra, S., 2023. A brief study on the role of cerium oxide nanoparticles in growth and alleviation of mercury-induced stress in *Vigna radiata* and soil bacteria *Bacillus coagulans*. Environmental Science and Pollution Research 30, 73952–73963. Available at https://doi.org/10.1007/s11356-023-27496-y.

Narula, S., Chaudhry, S., 2025. Emerging significance of silicon in abiotic stress tolerance: a review. Discover Plants 2, 307. Available at https://doi.org/10.1007/s44372-025-00397-9.

Naz, M., Ghani, M.I., Atif, M.J., Raza, M.A., Bouzroud, S., Afzal, M.R., et al., 2023. Sodium and abiotic stress tolerance in plants. In: Pandey, S., Tripathi, D.K., Singh, V.P., Sharma, S., Chauhan, D.K. (Eds.), Beneficial Chemical Elements of Plants: Recent Developments and Future Prospects. Wiley, Boca Raton, pp. 307–330. Available at https://doi.org/10.1002/9781119691419.ch13.

NCBI (National Center for Biotechnology Information), 2026. PubChem Element Summary for Atomic Number 58, Cerium. Available from: https://pubchem.ncbi.nlm.nih.gov/element/Cerium. (accessed 14.02.26.)

Nysanth, N.S., Rajan, S.A., Sivapriya, S.L., Anith, K.N., 2023. Pink pigmented facultative methylotrophs (PPFMs): potential bioinoculants for sustainable crop production. Journal of Pure and Applied Microbiology 17, 660–681. Available at https://doi.org/10.22207/JPAM.17.2.17.

Ofoe, R., Thomas, R.H., Asiedu, S.K., Wang-Pruski, G., Fofana, B., Abbey, L., 2023. Aluminum in plant: benefits, toxicity and tolerance mechanisms. Frontiers in Plant Science 13. Available at https://doi.org/10.3389/fpls.2022.1085998 1085998.

Ogunnubi, O.O., Owonubi, S.J., Malima, N.M., Ojuederie, B., 2025. Titanium dioxide nanoparticle (TiO_2) mitigation of water-deficit stress in micropropagated sweet potato (*Ipomoea batatas*) plantlets. Journal of Crop Science and Biotechnology 28, 239–249. Available at https://doi.org/10.1007/s12892-025-00275-6.

Okamoto, S., Eltis, L.D., 2011. The biological occurrence and trafficking of cobalt. Metallomics 3, 963–970. Available at https://doi.org/10.1039/C1MT00056J.

Osu, C., Onyema, M., 2016. Vanadium inhibition capacity on nutrients and heavy metal uptake by *Cucumis sativus*. Journal of American Science 12, 63–66. Available at https://doi.org/10.7537/marsjas121016.09.

Pal, V., Kaur, S., Sharma, V., Kaur, M., Verma, G., Sheoran, P., et al., 2026. Agronomic biofortification with micronutrients: an efficient approach to increase nutritional quality in crops and combat hidden hunger. Journal of Agriculture and Food Research 25, 102465. Available at https://doi.org/10.1016/j.jafr.2025.102465.

Panda, D., Behera, P.K., Nayak, J.K., 2023. Role of cobalt in plant growth and tolerance against different environmental stress. In: Roychoudhury, A. (Ed.), Biology and Biotechnology of Environmental Stress Tolerance in Plants Volume 2: Trace Elements in Environmental Stress Tolerance. Apple Academic Press, New York, pp. 19–35. Available at https://doi.org/10.1201/9781003346203-11.

Pandey, C., Gupta, M., 2015. Selenium and auxin mitigate arsenic stress in rice (*Oryza sativa* L.) by combining the role of stress indicators, modulators and genotoxicity assays. Journal of Hazardous Materials 287, 384–391. Available at https://doi.org/10.1016/j.jhazmat.2015.01.044.

Pax, F., Blamey, C., Kopittke, P.M., Wehr, J.B., Menzies, N.W., 2015. Aluminum. In: Barker, A.V., Pilbeam, D.J. (Eds.), Handbook of Plant Nutrition. CRC Press, Boca Raton, pp. 567–605. Available at https://doi.org/10.1201/b18458.

Paz, S., Rubio, C., Gutiérrez, A.J., Revert, C., Hardisson, A., 2018. Iodine: an essential trace element. Medical Journal of Clinical Trials and Case Studies 2, 000171. Available at https://doi.org/10.23880/mjccs-16000171.

Pereira, S., Monteiro, A., Moutinho-Pereira, J., Dinis, L.T., 2024. Silicon, an emergent strategy to lighten the effects of (a)biotic stresses on crops: a review. Journal of Agronomy and Crop Science 210, e12762. Available at https://doi.org/10.1111/jac.12762.

Pérez-Zavala, F.G., Atriztán-Hernández, K., Martínez-Irastorza, P., Oropeza-Aburto, A., López Arredondo, D., Herrera-Estrella, L., 2022. Titanium nanoparticles activate a transcriptional response in *Arabidopsis* that enhances tolerance to low phosphate, osmotic stress and pathogen infection. Frontiers in Plant Science 13, 994523. Available at https://doi.org/10.3389/fpls.2022.994523.

Petranikova, M., Tkaczyk, A.H., Bartl, A., Amato, A., Lapkovskis, V., Tunsu, C., 2020. Vanadium sustainability in the context of innovative recycling and sourcing development. Waste Management (New York, N.Y.) 113, 521–544. Available at https://doi.org/10.1016/j.wasman.2020.04.007.

Pilbeam, D.J., Grethead, H.M.R., Drihem, K., 2015. Selenium. In: Barker, A.V., Oilbeam, D.J. (Eds.), Handbook of Plant Nutrition. CRC Press, Boca Raton, pp. 651–679. Available at https://doi.org/10.1201/b18458.

Poschenrieder, C., Cabot, C., Martos, S., Gallego, B., Barceló, J., 2013. Do toxic ions induce hormesis in plants? Plant Science (Shannon, Ireland) 212, 15–25. Available at https://doi.org/10.1016/j.plantsci.2013.07.012.

Pozzi, M., Jonak-Dutta, S., Kuntze, M., Bading, J., Rüßbült, J.S., Fabig, C., et al., 2024. Visualization of the high surface-to-volume ratio of nanomaterials and its consequences. Journal of Chemical Education 101, 3146–3155. Available at https://doi.org/10.1021/acs.jchemed.4c00089.

Preetha, J.S.Y., Sriram, D., Premasudha, P., Pudake, R.N., Arun, M., 2023. Cerium oxide as a nanozyme for plant abiotic stress tolerance: an overview of the mechanisms. Plant Nano Biology 6, 100049. Available at https://doi.org/10.1016/j.plana.2023.100049.

Pukacka, S., Ratajczak, E., Kalemba, E., 2011. The protective role of selenium in recalcitrant *Acer saccharum* L. seeds subjected to desiccation. Journal of Plant Physiology 168, 220–225. Available at https://doi.org/10.1016/j.jplph.2010.07.021.

Quispe, A.P.V., de Morais, E.G., Benevenute, P.A.N., Lima, J.S., dos Santos, L.C., Silva, M.A., et al., 2025. Priming effect with selenium and iodine on broccoli seedlings: activation of biochemical mechanisms to mitigate cold damages. Plant Physiology and Biochemistry: PPB / Societe Francaise de Physiologie Vegetale 223, 109876. Available at https://doi.org/10.1016/j.plaphy.2025.109876.

Rady, M.M., Elkelish, A., Battaglia, M.L., Abdel-Kawy, M.A.A., Mohamed, G.F., Taie, H.A.A., et al., 2025. Natural biostimulant formulations enriched with honey, iodine, and silymarin enhance drought tolerance and yield in faba bean (*Vicia faba* L.) through antioxidant and hormonal regulation. BMC Plant Biology 25, 1719. Available at https://doi.org/10.1186/s12870-025-07794-y.

Rahmani, A., Shirinzadeh-Dastgiri, A., Neamatzadeh, H., 2024. The hidden effects of iodine deficiency on cancer patients: an urgent appeal for awareness campaigns. Asian Pacific Journal of Cancer Prevention: APJCP 25, 2967–2968. Available at https://doi.org/10.31557/APJCP.2024.25.9.2967.

Rojek, J., Kozieradzka-Kiszkurno, M., Kapusta, M., Aksmann, A., Jacewicz, D., Drzeżdżon, J., et al., 2019. The effect of vanadium(IV) complexes on development of *Arabidopsis thaliana* subjected to H_2O_2-induced stress. Functional Plant Biology: FPB 46, 942–961. Available at https://doi.org/10.1071/FP18262.

Rossi, L., Zhang, W., Lombardini, L., Ma, X., 2016. The impact of cerium oxide nanoparticles on the salt stress responses of *Brassica napus* L. Environmental Pollution 219, 28–36. Available at https://doi.org/10.1016/j.envpol.2016.09.060.

Sade, H., Meriga, B., Surapu, V., Gadi, J., Sunita, M.S.L., Suravajhala, P., et al., 2016. Toxicity and tolerance of aluminum in plants: tailoring plants to suit to acid soils. Biometals: An International Journal on the Role of Metal Ions in Biology, Biochemistry, and Medicine 29, 187–210. Available at https://doi.org/10.1007/s10534-016-9910-z.

Salas, Tortosa, A., Hidalgo-García, G., Delgado, A., Bedmar, E.J., Richardson, D.J., et al., 2020. The hemoglobin Bjgb from *Bradyrhizobium diazoefficien* controls NO homeostasis in soybean nodules to protect symbiotic nitrogen fixation. Frontiers in Microbiology 10, 2915. Available at https://doi.org/10.3389/fmicb.2019.02915.

Samanta, I., Ghosh, K., Saikia, R., Savita, Maity, P.J., Chowdhary, G., 2025. Arbuscular mycorrhizal fungi - A natural tool to impart abiotic stress tolerance in plants. Plant Signal. Behaviour. 20, 2525843. Available at https://doi.org/10.1080/15592324.2025.2525843.

Sarraf, M., Janeeshma, E., Arif, N., Farooqi, M.Q.U., Kumar, V., Ansari, N.A., et al., 2023. Understanding the role of beneficial elements in developing plant stress resilience: signalling and crosstalk with phytohormones and microbes. Plant Stress 10, 100224. Available at https://doi.org/10.1016/j.stress.2023.100224.

Sawaki, Y., Iuchi, S., Kobayashi, Y., Ikka, T., Sakurai, N., Fujita, M., et al., 2009. STOP1 regulates multiple genes that protect *Arabidopsis* from proton and aluminum toxicities. Plant Physiology 150, 281–294. Available at https://doi.org/10.1104/pp.108.134700.

Schaller, J., Puppe, D., Kaczorek, D., Ellerbrock, R., Sommer, M., 2021. Silicon cycling in soils revisited. Plants 10, 295. Available at https://doi.org/10.3390/plants10020295.

Schiavon, M., Pilon-Smits, E.A., 2017. The fascinating facets of plant selenium accumulation - biochemistry, physiology, evolution and ecology. The New Phytologist 213, 1582–1596. Available at https://doi.org/10.1111/nph.14378.

Shah, T., Latif, S., Saeed, F., Ali, I., Ullah, S., Alsahli, A.A., et al., 2021. Seed priming with titanium dioxide nanoparticles enhances seed vigor, leaf water status, and antioxidant enzyme activities in maize (*Zea mays* L.) under salinity stress. Journal of King Saud University – Science 33, 101207. Available at https://doi.org/10.1016/j.jksus.2020.10.004.

Shalaby, O.A., 2025. Iodine application induces the antioxidant defense system, alleviates salt stress, reduces nitrate content, and increases the nutritional value of lettuce plants. Functional Plant Biology: FPB 52, FP24273. Available at https://doi.org/10.1071/FP24273.

Shang, M., Jia, T., Ma, M., 2025. Alleviation of salt stress in *Glycyrrhiza uralensis* by lanthanum nitrate: a predictive modeling approach. Frontiers in Plant Science 16, 1659108. Available at https://doi.org/10.3389/fpls.2025.1659108.

Sharma, T., Dreyer, I., Kochian, L., Piñeros, M.A., 2016. The ALMT family of organic acid transporters in plants and their involvement in detoxification and nutrient security. Frontiers in Plant Science 7, 1488. Available at https://doi.org/10.3389/fpls.2016.01488.

Sheng, H., Chen, S., 2020. Plant silicon-cell wall complexes: identification, model of covalent bond formation and biofunction. Plant Physiology and Biochemistry: PPB / Societe Francaise de Physiologie Vegetale 155, 13–19. Available at https://doi.org/10.1016/j.plaphy.2020.07.020.

Shinmachi, F., Buchner, P., Stroud, J.L., Parmar, S., Zhao, F.J., McGrath, S.P., et al., 2010. Influence of sulfur deficiency on the expression of specific sulfate transporters and the distribution of sulfur, selenium, and molybdenum in wheat. Plant Physiology 153, 327–336. Available at https://doi.org/10.1104/pp.110.153759.

Shreenath, A.P., Hashmi, M.F., Dooley, J., 2023. Selenium deficiency. StatPearls Publishing, StatPearls Treasure Island (FL). Available at https://www.ncbi.nlm.nih.gov/books/NBK482260/.

Silva, C.O., Brito, D.S., da Silva, A.A., Rosa, V.R., Santos, M.F.S., de Souza, G.A., et al., 2020a. Differential accumulation of aluminum in root tips of soybean seedlings. Brazilian Journal of Botany 43, 99–107. Available at https://doi.org/10.1007/s40415-020-00593-9.

Silva, S., Ribeiro, T.P., Santos, C., Pinto, D.C.G.A., Silva, A.M.S., 2020b. TiO_2 nanoparticles induced sugar impairments and metabolic pathway shift towards amino acid metabolism in wheat. Journal of Hazardous Materials 399, 122982. Available at https://doi.org/10.1016/j.jhazmat.2020.122982.

Singh, A., Hussain, I., Singh, N.B., Singh, H., 2019. Uptake, translocation and impact of green synthesized nanoceria on growth and antioxidant enzymes activity of *Solanum lycopersicum* L. Ecotoxicology and Environmental Safety 182, 109410. Available at https://doi.org/10.1016/j.ecoenv.2019.109410.

Singhal, R.K., Fahad, S., Kumar, P., Choyal, P., Javed, T., Jinger, D., et al., 2023. Beneficial elements. new players in improving nutrient use efficiency and abiotic stress tolerance. Plant Growth Regulation 100, 237–265. Available at https://doi.org/10.1007/s10725-022-00843-8.

Skovran, E., Raghuraman, C., Martinez-Gomez, N.C., 2019. Lanthanides in methylotrophy. Current Issues in Molecular Biology 33, 101–116. Available at https://doi.org/10.21775/cimb.033.101.

Solanki, P., Bhargava, A., Chhipa, H., Navin Jain, N., Panwar, J., 2015. Nano-fertilizers and their smart delivery system. In, Rai, M., Ribeiro, C., Mattoso, L., Duran, N. (Eds.), Nanotechnologies in Food and Agriculture. Springer, Cham, pp. 81–101. Available at https://doi.org/10.1007/978-3-319-14024-7_4.

Soleymani, S., Piri, S., Aazami, M.A., Salehi, B., 2025. Cerium oxide nanoparticles alleviate drought stress in apple seedlings by regulating ion homeostasis, antioxidant defense, gene expression, and phytohormone balance. Scientific Reports 15, 11805. Available at https://doi.org/10.1038/s41598-025-96250-w.

Sun, L.L., Zhang, M.S., Liu, X.M., Mao, Q.Z., Shi, C., Kochian, L.V., et al., 2020. Aluminium is essential for root growth and development of tea plants (*Camellia sinensis*). Journal of Integrative Plant Biology 62, 984–997. Available at https://doi.org/10.1111/jipb.12942.

Tan, L.C., Nancharaiah, Y.V., van Hullebusch, E.D., Lens, P.N.L., 2016. Selenium: environmental significance, pollution, and biological treatment technologies.

Biotechnology Advances 34, 886–907. Available at https://doi.org/10.1016/j.biotechadv.2016.05.005.

Tao, Y., Wan, J.X., Liu, Y.S., Yang, X.Z., Shen, R.F., Zhu, X.F., 2022. The NAC transcription factor ANAC017 regulates aluminum tolerance by regulating the cell wall-modifying genes. Plant Physiology 189, 2517–2534. Available at https://doi.org/10.1093/plphys/kiac197.

Tao, Y., Li, S., Liu, Y., Guo, R., Chen, C., Huang, J., et al., 2025. ANAC050 confers aluminium resistance by cooperating with secretion of organic acids and accumulation of cell wall hemicelluloses in plants. Pedosphere 35, 338–351. Available at https://doi.org/10.1016/j.pedsph.2024.02.001.

Thorne, S.J., Maathuis, F.J.M., 2022. Reducing potassium deficiency by using sodium fertilisation. Stress Biology 2, 45. Available at https://doi.org/10.1007/s44154-022-00070-1.

Thornton, B.F., Burdette, S.C., 2019. Seekers of the lost lanthanum. Nature Chemistry 11, 188. Available at https://doi.org/10.1038/s41557-018-0208-3.

Tong, K., Yan, L., Riaz, M., Gao, G., Yu, H., Lu, M., et al., 2024. Mechanisms of Lanthanum-mediated mitigation of salt stress in soybean (*Glycine max* L.). Physiologia Plantarum 176, e14452. Available at https://doi.org/10.1111/ppl.14452.

Tourky, S.M.N., Shukry, W.M., Hossain, M.A., Siddiqui, M.H., Pessarakli, M., Elghareeb, E.M., 2023. Cobalt enhanced the drought-stress tolerance of rice (*Oryza sativa* L.) by mitigating the oxidative damage and enhancing yield attributes. South African Journal of Botany 159, 191–207. Available at https://doi.org/10.1016/j.sajb.2023.05.035.

Trejo-Téllez, L.I., Gómez-Merino, F.C., 2023. Editorial: Beneficial elements: novel players in plant biology for innovative crop production, volume II. Frontiers in Plant Science 14, 1303462. Available at https://doi.org/10.3389/fpls.2023.1303462.

Trejo-Téllez, L.I., Gómez-Trejo, L.F., Escobar-Sepúlveda, H.F., Gómez-Merino, F.C., 2022. The genetics of silicon accumulation in plants. In: Etesami, H., Al-Saeedi, A.H., El-Ramady, H., Fujita, M., Pessarakli, M., Hossain, M.A. (Eds.), Silicon and Nano-silicon in Environmental Stress Management and Crop Quality Improvement. Elsevier-Academic Press, London, pp. 67–75. Available at https://doi.org/10.1016/B978-0-323-91225-9.00011-X.

Trela-Makowej, A., Orzechowska, A., Szymańska, R., 2024. Less is more: the hormetic effect of titanium dioxide nanoparticles on plants. The Science of the Total Environment 910, 168669. Available at https://doi.org/10.1016/j.scitotenv.2023.168669.

Trippe, R.C., Pilon-Smits, E.A.H., 2021. Selenium transport and metabolism in plants: phytoremediation and biofortification implications. Journal of Hazardous Materials 404, 124178. Available at https://doi.org/10.1016/j.jhazmat.2020.124178.

Tubana, B.S., Babu, T., Datnoff, L.E., 2016. A review of silicon in soils and plants and its role in US agriculture. History and future perspectives. Soil Science 181, 393–411. Available at https://doi.org/10.1097/SS.0000000000000179.

Ukwattage, N.L., Zhiyong, Z., 2025. Impacts of cerium dioxide nanoparticles on the soil–plant system and their potential agricultural applications. Nanomaterials 15, 950. Available at https://doi.org/10.3390/nano15120950.

Vaculík, M., Lukačová, Z., Bokor, B., Martinka, M., Tripathi, D.K., Alexander Lux, A., 2020. Alleviation mechanisms of metal(loid) stress in plants by silicon: a review. Journal of Experimental Botany 71, 6744–6757. Available at https://doi.org/10.1093/jxb/eraa288.

Vatansever, R., Ozyigit, I.I., Fili, E., 2016. Essential and beneficial trace elements in plants, and their transport in roots: a review. Applied Biochemistry and Biotechnology 181, 464–482. Available at https://doi.org/10.1007/s12010-016-2224-3.

Wang, H., Wang, T., You, L., Zhong, G., Shi, G., 2013. Effects of vanadate supply on plant growth, Cu accumulation, and antioxidant capacities in *Triticum aestivum* L. Environmental Geochemistry and Health 35, 585–592. Available at https://doi.org/10.1007/s10653-013-9541-z.

Wang, W., Zhao, X.Q., Hu, Z.M., Shao, J.F., Che, J., Chen, R.F., et al., 2015. Aluminium alleviates manganese toxicity to rice by decreasing root symplastic Mn uptake and reducing availability to shoots of Mn stored in roots. Annals of Botany 116, 237–246. Available at https://doi.org/10.1093/aob/mcv090.

Wang, L., Suganuma, S., Hibino, A., Mitsui, R., Tani, A., Matsumoto, T., et al., 2019. Lanthanide-dependent methanol dehydrogenase from the legume symbiotic nitrogen-fixing bacterium *Bradyrhizobium diazoefficiens* strain USDA110. Enzyme and Microbial Technology 130, 109371. Available at https://doi.org/10.1016/j.enzmictec.2019.109371.

Wang, J., Yu, X., Ding, Z.J., Zhang, X., Luo, Y., Xu, X., et al., 2022. Structural basis of ALMT1-mediated aluminum resistance in *Arabidopsis*. Cell Research 32, 89–98. Available at https://doi.org/10.1038/s41422-021-00587-6.

Watanabe, T., Jansen, S., Osaki, M., 2006. Al–Fe interactions and growth enhancement in *Melastoma malabathricum* and *Miscanthus sinensis* dominating acid sulphate soils. Plant, Cell & Environment 29, 2124–2132. Available at https://doi.org/10.1111/j.1365-3040.2006.001586.x.

White, P.J., Bowen, H.C., Parmaguru, P., Fritz, M., Spracklen, W.P., Spiby, R.E., et al., 2004. Interactions between selenium and sulphur nutrition in *Arabidopsisthaliana*. Journal of Experimental Botany 55, 1927–1937. Available at https://doi.org/10.1093/jxb/erh192.

Włodarczyk, T., Balakhnina, T., Matichenkov, V., Brzezińska, M., Nosalewicz, M., Szarlip, P., et al., 2019. Effect of silicon on barley growth and N_2O emission under flooding. The Science of the Total Environment 685, 1–9. Available at https://doi.org/10.1016/j.scitotenv.2019.05.410.

Wu, G.Q., Feng, R.J., Liang, N., Yuan, H.J., Sun, W.B., 2015. Sodium chloride stimulates growth and alleviates sorbitol-induced osmotic stress in sugar beet seedlings. Plant Growth Regulation 75, 307–316. Available at https://doi.org/10.1007/s10725-014-9954-4.

Wu, H., Tito, N., Giraldo, J.P., 2017. Anionic cerium oxide nanoparticles protect plant photosynthesis from abiotic stress by scavenging reactive oxygen species. ACS Nano 11, 11283–11297. Available at https://doi.org/10.1021/acsnano.7b05723.

Wu, H., Shabala, L., Shabala, S., Giraldo, J.P., 2018. Hydroxyl radical scavenging by cerium oxide nanoparticles improves *Arabidopsis* salinity tolerance by enhancing leaf mesophyll potassium retention. Environmental Science: Nano 5, 1567. Available at https://doi.org/10.1039/C8EN00323H.

Xie, C., Ma, Y., Yang, J., Zhang, B., Luo, W., Feng, S., et al., 2019. Effects of foliar applications of ceria nanoparticles and $CeCl_3$ on common bean (*Phaseolus vulgaris*). Environmental Pollution (Barking, Essex: 1987) 250, 530–536. Available at https://doi.org/10.1016/j.envpol.2019.04.042.

Xu, C.M., Zhao, B., Wang, X.D., Wang, Y.C., 2007. Lanthanum relieves salinity-induced oxidative stress in *Saussurea involucrata*. Biologia Plantarum 51, 567–570. Available at https://doi.org/10.1007/s10535-007-0124-7.

Yang, Y., Liu, Y., Huang, C.F., De Silva, J., Zhao, F.J., 2016. Aluminium alleviates fluoride toxicity in tea (*Camellia sinensis*). Plant and Soil 402, 179–190. Available at https://doi.org/10.1007/s11104-015-2787-8.

Yang, H., Xu, Z., Liu, R., Xiong, Z., 2019. Lanthanum reduces the cadmium accumulation by suppressing expression of transporter genes involved in cadmium uptake and translocation in wheat. Plant and Soil 441, 235–252. Available at https://doi.org/10.1007/s11104-019-04112-y.

Yang, Y., Ye, C., Zhao, M., Li, J., Zhang, X., Yang, Z., et al., 2025. Nanoparticles in sustainable agriculture: enhancing nutrient use efficiency and abiotic stress resilience under climate change. Plant Stress 17, 100982. Available at https://doi.org/10.1016/j.stress.2025.100982.

Yao, X., Chu, J., Ba, C., 2010. Responses of wheat roots to exogenous selenium supply under enhanced ultraviolet-B. Biological Trace Element Research 137, 244–252. Available at https://doi.org/10.1007/s12011-009-8580-x.

Yotsova, E., Stefanov, M., Rashkov, G., Dobrikova, A., Apostolova, E., 2025. Mitigating salinity stress in pea plants with titanium oxide nanoparticles. International Journal of Plant Biology 16, 34. Available at https://doi.org/10.3390/ijpb16010034.

Zexer, N., Kumar, S., Elbaum, R., 2023. Silica deposition in plants: scaffolding the mineralization. Annals of Botany 131, 897–908. Available at https://doi.org/10.1093/aob/mcad056.

Zhai, W., Dai, Y., Zhao, W., Yuan, H., Qiu, D., Chen, J., et al., 2020. Simultaneous immobilization of the cadmium, lead and arsenic in paddy soils amended with titanium gypsum. Environmental Pollution (Barking, Essex: 1987) 258, 113790. Available at https://doi.org/10.1016/j.envpol.2019.113790.

Zhang, Y., Gladyshev, V.N., 2009. Comparative genomics of trace elements: emerging dynamic view of trace element utilization and function. Chemical Reviews 109, 4828–4861. Available at https://doi.org/10.1021/cr800557s.

Zhang, L., Goncalves, A.A.S., Jiang, B., Jaroniec, M., 2018. Capture of iodide by $BiVO_4$ and Bi_2O_3: an insightful look at the process and its aftermath. ChemSusChem 11, 1486–1493. Available at https://doi.org/10.1002/cssc.201800327.

Zhang, X., Lou, X., Zhang, H., Ren, W., Tang, M., 2020a. Effects of sodium sulfide application on the growth of *Robinia pseudoacacia*, heavy metal immobilization, and soil microbial activity in Pb-Zn polluted soil. Ecotoxicology and Environmental Safety 197, 110563. Available at https://doi.org/10.1016/j.ecoenv.2020.110563.

Zhang, W., Long, J., Geng, J., Li, J., Wei, Z., 2020b. Impact of titanium dioxide nanoparticles on Cd phytotoxicity and bioaccumulation in rice (*Oryza sativa* L.). International Journal of Environmental Research and Public Health 17, 2979. Available at https://doi.org/10.3390/ijerph17092979.

Zhang, Y., Cao, H., Wang, M., Zou, Z., Zhou, P., Wang, X., et al., 2023. A review of iodine in plants with biofortification: Uptake, accumulation, transportation, function, and toxicity. The Science of the Total Environment 878, 163203. Available at https://doi.org/10.1016/j.scitotenv.2023.163203.

Zhang, D., Tian, C., Mai, W., 2024. Exogenous sodium and calcium alleviate drought stress by promoting the succulence of *Suaeda salsa*. Plants 13, 721. Available at https://doi.org/10.3390/plants13050721.

Zhang, H., Yu, C., Zhang, Q., Qiu, Z., Zhang, X., Hou, Y., et al., 2025. Salinity survival: molecular mechanisms and adaptive strategies in plants. Frontiers in Plant Science 16, 1527952. Available at https://doi.org/10.3389/fpls.2025.1527952.

Zheng, C., Ouchi, T., Iizuka, A., Taninouchi, Y.K., Okabe, T.H., 2019. Deoxidation of titanium using Mg as deoxidant in $MgCl_2$-YCl_3 flux. Metallurgical and Materials Transactions B 50, 622–631. Available at https://doi.org/10.1007/s11663-018-1494-2.

Zhong, Y., Chen, J., 2020. Ameliorative effects of lanthanum(III) on copper(II) stressed rice (*Oryza sativa*) and its molecular mechanism revealed by transcriptome profiling. Plant Physiology and Biochemistry: PPB / Societe Francaise de Physiologie Vegetale 152, 184–193. Available at https://doi.org/10.1016/j.plaphy.2020.05.004.

Zhou, B., Cao, H., Wu, Q., Mao, K., Yang, X., Su, J., et al., 2023. Agronomic and genetic strategies to enhance selenium accumulation in crops and their influence on quality. Foods 12 (24), 4442. Available at https://doi.org/10.3390/foods12244442.

SECTION

V

Emerging technologies, innovations, and future directions

CHAPTER

22 Nanofertilizers: a new frontier in stress tolerance

Weria Weisany

Department of Agriculture Science and Engineering, Science and Research Campus, Islamic Azad University, Tehran, Iran

22.1 Introduction

Crop production is increasingly constrained by heat waves, erratic rainfall, chronic salinity, nutrient-poor or metal-contaminated soils, and overlapping biotic pressures. Under these conditions, conventional fertilizers applied as large, infrequent doses often miss the moving target of plant demand. Much of the nutrient is lost to leaching, volatilization, or fixation; the remainder can arrive too early or too late to prevent oxidative bursts, ionic imbalance, membrane damage, or collapse of photosystem efficiency. The result is familiar: yield instability, high input intensity, and widening environmental footprints.

Nanofertilizers (NFs) offer a different operating logic. Instead of treating nutrition as a bulk supply problem, they frame it as spatiotemporal dosing: delivering the right ion (or signal) to the right interface (leaf cuticle, stomata, rhizoplane) at the right time and rate (hours to weeks), synchronized with stress dynamics (Nongbet et al., 2022). By engineering particle size, surface chemistry, and triggerable release, NFs can keep core physiological levers redox poise, ion/water homeostasis, and photochemistry within a safe operating range through drought, salinity, heat, cold snaps, and episodic pathogen pressure (Llopis-Lorente et al., 2018).

In this chapter, we use "nanofertilizer" broadly to include four practical classes: (1) nanoscale nutrient particles (ZnO, Fe_2O_3, SiO_2); (2) nanostructured carriers that host soluble nutrients (mesoporous silica, LDH, nHAP, lignin–chitosan beads); (3) hybrid systems that coload nutrients with biostimulants or small-molecule signals (NO or SA donors); and (4) stimulus-responsive formulations that alter release with moisture, ionic strength, pH, temperature, or ROS. Across these classes, performance rests on two pillars: interface engineering (adhesion, persistence, entry, translocation) and release kinetics (near-zero-order or programmed pulses matched to stress).

Mechanistically, NFs contribute to stress tolerance in three reinforcing ways. First, they stabilize supply, preventing transient deficiencies that destabilize enzymes and membranes; for example, steady micronutrient availability sustains SOD/CAT/APX activities and the D1 repair cycle in PSII (Elbagory et al., 2025). Second, they shift set-points, modestly raising the plant's tolerance to osmotic and ionic shocks by supporting

Mineral Nutrients to Mitigate Abiotic Stress in Crops. DOI: https://doi.org/10.1016/B978-0-443-44712-9.00022-0

K^+/Na^+ selectivity, membrane integrity, and cell-wall strengthening (notably via silica and calcium) (Aqaei et al., 2020; Weisany et al., 2024a). Third, they coordinate signals, either by enabling endogenous hormone/redox networks to function (through cofactor sufficiency) or by delivering micro-pulses of elicitors (chitosan, SA/JA, NO/H_2S) that prime defenses without imposing growth penalties (Varamin et al., 2020).

A credible path from concept to field requires translation across scales. Laboratory efficacy can vanish in farm water with high hardness, in hot, UV-intense canopies, or in soils that fix phosphate and trace metals. Accordingly, this chapter emphasizes real-matrix characterization (DLS/ζ-potential in spray water; release in leaf-wash and soil extracts), compatibility (surfactants, adjuvants, and microbial partners), and stewardship (biodegradability, nontarget safety, residue fate). We align agronomic scheduling with sensing and forecasts SPAD/NDVI, Fv/Fm or NPQ, canopy temperature, leaf ionomics, RWC/osmotic potential, so applications are triggered by plant state rather than calendar habit.

Several knowledge gaps remain and are flagged throughout: standardized field-matrix release tests; dose boundaries that preserve ROS as a signal but prevent damage; long-term soil transformations of carriers; rules for codeployment with AMF/PGPR; and robust cost–benefit under variable climates. Addressing these gaps will ensure that nanofertilizers are judged not by novelty but by repeatable agronomic value: higher nutrient-use efficiency, fewer passes, steadier photosynthesis and water relations, and resilient yields with smaller environmental footprints.

22.2 Mechanisms, formulations, and applications for crop resilience

Nanofertilizers (NFs) couple essential nutrients with engineered carriers to deliver small, precise doses where and when plants need them most. By tuning particle size, surface chemistry, and release behavior, NFs increase nutrient-use efficiency and stabilize plant physiology under drought, salinity, heat, cold, and toxic-ion stress (Abdel-Hakim et al., 2023). The promise of NFs lies in uniting three pillars: mechanisms, formulations, and applications so that biochemical intent, materials design, and field practice reinforce one another rather than operating in isolation.

Mechanistically, NFs first improve the acquisition and retention of nutrients at plant interfaces. On leaves, nanoscale suspensions adhere more strongly to the cuticle, persist longer, and can traverse stomatal apertures or cuticular microchannels; in soils, carriers slow diffusion and leaching while positioning actives in the rhizosphere (Madlala et al., 2024). This localized, controlled supply evens out the peaks and troughs of plant nutrient status that commonly occur during stress, thereby avoiding both transient deficiency and wasteful excess (Guo et al., 2018). The second core mechanism is redox stabilization. Micronutrient NFs (Zn, Fe, Mn, Cu at agronomically safe doses) serve as cofactors for superoxide dismutase, catalase, and peroxidases, sustaining the enzymatic antioxidant network, while certain carriers (silica, chitosan) help limit lipid peroxidation and protect proteins together dampening H_2O_2 and $^{\bullet}O^{2-}$ accumulation and preserving photosystems (Khosropour et al., 2022; Weisany et al., 2023).

NFs also support ionic and osmotic homeostasis, which is central to resilience under drought and salinity. By sustaining K-dependent enzymes and transporters and reinforcing membranes and cell walls (via silica), plants maintain higher relative water content, better stomatal control, and improved water-use efficiency (Weisany et al., 2024a). The alleviation of nutrient bottlenecks feeds forward into phytohormone crosstalk: ABA-driven stomatal behavior, JA/SA signaling for defense, and ethylene dynamics are all indirectly modulated when nutrient status is stabilized (Wang et al., 2025). In parallel, steady availability of Fe, Mg, Mn, and Zn underpins chlorophyll biosynthesis, D1 protein repair, and nonphotochemical quenching, limiting photooxidative damage during heat waves or high light (Zuo, 2025). A further protective layer comes from detoxification: iron, phosphate, and silica systems can immobilize or coprecipitate toxic ions (Al^{3+}, Cd^{2+}) in the root zone, reducing uptake or sequestering them in less reactive forms to lower oxidative burden (Rahman et al., 2025).

The rhizosphere and phyllosphere microbiomes are integral to these effects. When designed thoughtfully, NFs can be neutral to beneficial microbes or even synergize with them. Chitosan and some trace-element blends mildly elicit plant defenses without heavy growth penalties, priming tissues for stress while avoiding antagonism with biocontrol agents (Mukarram et al., 2023). Coapplication with PGPR or AMF is especially potent under water or salt stress, provided doses and timings are compatible and free ions released from the NF do not inhibit the microbial partners (Rojas-Pirela et al., 2024).

Translating mechanisms into performance depends on formulation science. Inorganic nanonutrients (nano-ZnO, Fe_2O_3/Fe_3O_4, SiO_2, nanohydroxyapatite, layered double hydroxides) offer high active content and tunable solubility, while porous hosts (mesoporous silica, MOFs, nanobiochar fractions) provide large surface areas for loading and diffusion-controlled release. Biopolymers such as chitosan, alginate, pectin, or lignosulfonate add biodegradability, foliar adhesion, and responsiveness to pH or enzymes, and carbon-based coatings or humic/fulvic substances improve compatibility with plant surfaces and hard-water spray solutions (Martău et al., 2019). Two levers dominate performance: interfacial properties (size distribution and ζ-potential in the actual application medium, not just deionized water) and stimuli-responsive release tuned to moisture, ionic strength, pH, or ROS so that more nutrients are available precisely when stress intensifies (Guo et al., 2018).

Application strategy is the third pillar and often the difference between promising lab data and real-world gain. Seed nanopriming short, low-dose exposure before sowing can enhance emergence, early vigor, and preparedness for drought or salinity (Shelar et al., 2021). Foliar sprays provide rapid correction and buffering around sensitive phenophases (tillering, flowering); they work best with moderate droplet sizes, early-morning or late-afternoon timing, and adjuvants that improve spreading without destabilizing the suspension (Wen et al., 2021). Soil placement or fertigation builds root-zone reserves for chronic stress; here, pairing application with moisture-monitoring reduces losses and synchronizes release with plant demand. Tank-mix compatibility is non-negotiable: pH, electrolytes, and surfactant type can make or break dispersion stability, so a simple jar test for each batch should be routine (Pardossi et al., 2009).

Stress-specific playbooks help operationalize choices. Under drought, silica-rich carriers combined with low-dose Zn or Fe support membrane integrity, antioxidant capacity, and water relations; timing matters most when applied as a prestress priming followed by a maintenance spray if the event persists (Weisany et al., 2024a; Weisany et al., 2012). In salinity, silica with K and Ca helps maintain K^+/Na^+ balance, and LDH-based systems can buffer anions in the rhizosphere; scheduling near irrigation can blunt sodium spikes (Chhabra, 2022). For heat, Zn-forward foliar NFs before forecasted high-temperature windows bolster ROS detoxification and membrane stability; for cold snaps, Fe/Mn support protects PSII and reduces photoinhibition (Allakhverdiev et al., 2008). In metal-contaminated soils, iron, silica, or phosphate carriers are best delivered to the root zone to immobilize or sequester toxic ions, with foliar top-ups for photosynthetic cofactors as needed. Where biotic and abiotic stresses co-occur, chitosan-based systems can gently prime defenses, provided biocontrol compatibility is validated (Adrees et al., 2015).

22.3 Targeted nutrient delivery to mitigate abiotic and biotic stresses

Targeted nutrient delivery uses engineered carriers to place essential elements precisely at plant interfaces and tissues where they will have the greatest physiological leverage during stress (Yadav, 2025). Instead of relying on bulk diffusion and chance contact, nano-enabled systems control where (leaf cuticle, stomata, root apoplast, rhizoplane), when (prestress priming, peak stress, recovery), and how fast nutrients are supplied (zero-/near-zero-order or stimulus-responsive release) (Yadav, 2025). This spatiotemporal control stabilizes redox homeostasis, sustains ion balances, and preserves photosynthetic function with smaller inputs and fewer off-target losses.

Temporal targeting hinges on release kinetics. Drought and salinity impose fluctuating demands; formulations that respond to moisture, ionic strength, pH, or ROS can meter nutrients when stress is most acute. For example, silica- or chitosan-based matrices that soften in high-ionic or mildly acidic films release Zn/Fe cofactors during stomatal closure and oxidative bursts, sustaining superoxide dismutase, catalase, and peroxidases precisely when ROS spikes would otherwise damage membranes and photosystems (Ortelli et al., 2016; Quitadamo et al., 2021). For chronic stresses (saline irrigation), soil-placed LDHs or nanohydroxyapatite furnish slow, sustained K, Ca, P, or trace elements over weeks, buffering ion ratios and membrane integrity (Zhao et al., 2025).

Under abiotic stresses, targeted delivery addresses distinct bottlenecks. In drought, Si-rich carriers fortify cell walls and reduce cuticular transpiration, while codelivered Zn or Fe stabilizes antioxidant enzymes and the D1 repair cycle, preserving PSII efficiency and lowering canopy temperature (Wang et al., 2021). In salinity, maintaining high K^+/Na^+ ratios is decisive; K-forward formulations paired with Ca/Si support selective transport and membrane integrity, especially when applied around irrigation events to dampen sodium surges (Akter and Oue, 2018). Heat stress elevates ROS and perturbs membranes preheatwave foliar Zn/Si microdoses bolster redox capacity and lipid stability, whereas impending cold snaps benefit from Fe/Mn support to protect electron transport and reduce photoinhibition (Fortunato et al., 2023).

Biotic stress mitigation leverages two complementary routes: nutritional immunity and gentle elicitation. Adequate micronutrients (Zn, Cu, Mn) strengthen structural barriers (lignification, cross-linking) and fuel defense enzymes and phytoalexin synthesis; delivered in nano-enabled, low-dose, slow-release forms, they enhance basal resistance without broad antimicrobial side effects (Weisany et al., 2019a,b). Biopolymer carriers such as chitosan add a calibrated elicitor signal that primes salicylic/jasmonate pathways, raising readiness against fungi and bacteria while avoiding growth penalties typical of strong elicitation (Arabpoor et al., 2021; Soufiani et al., 2025). Silicon further reinforces epidermal and vascular barriers, reducing penetration and movement of many pathogens and some sap-feeders. Crucially, compatibility with beneficial microbiota (PGPR, AMF, biocontrol agents) must be validated so that trace-metal pulses do not suppress allies (Fig. 22.1).

Efficacy depends on plant uptake and translocation after deposition. Nanometric size, moderate hydrophilicity, and appropriate charge favor passage through cuticular micelles and stomata, while ligand shells reduce aggregation in the apoplast and enable xylem/phloem loading (Wu and Li, 2022). Once inside, colocalization matters: Fe/Mn aimed at chloroplast-rich tissues support photochemistry; Si and Ca that accumulate in cell walls and middle lamellae improve mechanical resilience; Zn that reaches young expanding leaves or reproductive tissues can disproportionately protect yield components during heat or drought (Sági-Kazár et al., 2021).

22.4 Engineering plant resilience through precision nanonutrition

Precision nanonutrition reframes fertilization as a spatiotemporal drug-delivery problem for plants: delivering the right ion, at the right place, at the right time, and at the right rate to stabilize vulnerable processes during stress (Kumawat et al., 2025).

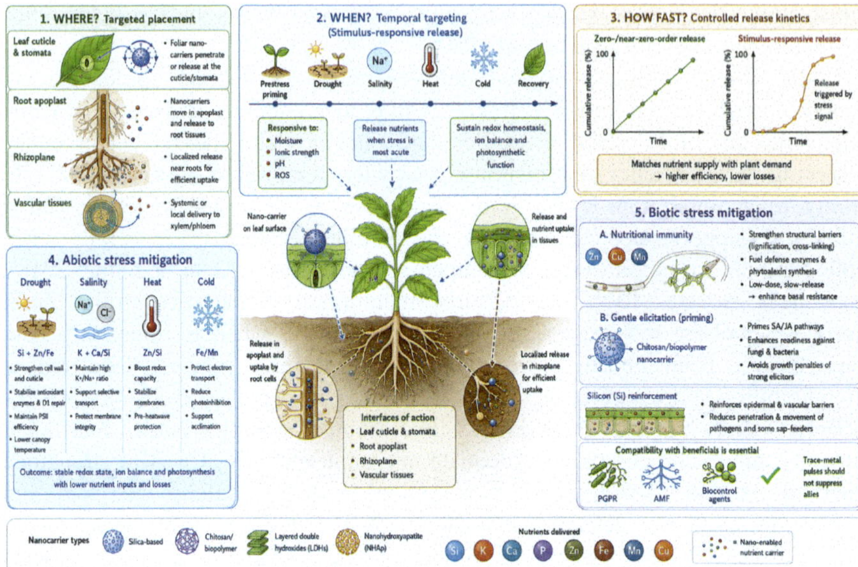

FIGURE 22.1

Targeted nutrient delivery using nano-enabled carriers to alleviate abiotic and biotic stresses in plants through spatially and temporally controlled nutrient release, enhanced stress tolerance, and improved compatibility with beneficial microbiota.

Unlike bulk salts, nano-enabled systems control particle–tissue interfaces (cuticle, stomata, rhizoplane), tune release with environmental cues (moisture, pH, ionic strength, ROS), and codeliver micronutrients with elicitors or osmoprotectants (Shen et al., 2023; Weisany and Khosropour, 2023). The goal is not maximal uptake per se, but homeostatic buffering keeping photosynthesis, redox balance, and ion relations within safe bounds through heat, drought, salinity, cold snaps, and episodic pathogen pressure.

Interface engineering is decisive. Foliar systems favor sub-200 nm hydrodynamically with mildly negative to slightly positive ζ-potentials in the actual spray water to prevent flocculation while maximizing cuticular residence and stomatal entry (Borgatta et al., 2021). Root-zone systems benefit from coatings that bind mucilage and organic matter, creating "nutrient depots" on the rhizoplane that resist leaching and synchronize supply with transpiration cycles. Ligand shells (amino/organic acids, peptides, humics) improve compatibility with hard water, reduce antagonistic precipitation, and can bias tissue translocation (xylem vs phloem) for organ-specific targeting (Zhang et al., 2021).

Resilience emerges when release kinetics match physiological demand. During heat waves, rapid ROS-responsive softening of the matrix can pulse Zn/Fe to support SOD/CAT/APX and D1 repair (Yuan et al., 2025); under salinity, slow

LDH-mediated K/Ca availability buffers membrane selectivity and reduces Na^+ toxicity (Shabala et al., 2006); in drought, Si-rich carriers reduce cuticular water loss and strengthen walls while small Zn doses sustain antioxidant capacity (Manimaran et al., 2025). For cold events, Fe/Mn reaching chloroplasts ahead of the snap lowers photoinhibition and preserves PSII efficiency (Wei et al., 2022). These microdoses prevent the peaks and troughs in tissue concentrations that trigger damage.

22.5 Controlled-release strategies for drought, salinity, and heat tolerance

Controlled-release (CR) nanofertilizers shift nutrient delivery from single, high-amplitude pulses to low, sustained fluxes that track plant demand during stress (Singh et al., 2024). In drought, salinity, and heat, the physiological bottlenecks water status, ion selectivity, and redox stability fluctuate over hours to weeks. Formulations that deliver near-zero-order release or environmentally triggered bursts smooth these fluctuations, preventing the deficiency spikes and toxicity troughs that follow conventional sprays (Yousefi et al., 2024). In practice, performance depends as much on release kinetics in the real matrix (leaf film, rhizosphere solution) as on loading content: a smaller dose released at the right rate outperforms a larger dose released too quickly (Lakshani et al., 2023) (Fig. 22.2).

For drought tolerance, the goal is to sustain antioxidant capacity and fortify cell walls while minimizing transpirational loss. Silica-rich carriers (mesoporous silica, silica-coated minerals, or Si doped biopolymers) provide structural benefits and act as reservoirs for cofactors such as Zn and Fe (Weisany et al., 2024b). Chitosan or alginate shells, mildly cationic in the actual spray water, adhere to the cuticle and thicken the leaf film, extending residence from hours to days (Das et al., 2024). Release that accelerates under higher ionic strength or mild acidity, a proxy for concentrated leaf films during stomatal closure, meters Zn/Fe to support SOD, CAT, and APX exactly when ROS spikes. In the root zone, LDH intercalates or nanohydroxyapatite (nHAP) codeliver Si/K/P with multiday to multiweek diffusion, reducing leaching during infrequent irrigations.

Under salinity, controlled release is used to protect K^+/Na^+ homeostasis and membrane integrity (Cao et al., 2024). Anion- and cation-exchangeable hosts such as LDHs and zeolitic/silicate frameworks sequester competing ions while slowly furnishing K and Ca that stabilize transporters and lipid bilayers (Ebel et al., 1999). Humic- or lignosulfonate-coated nanoparticulates resist flocculation in saline irrigation water and buffer pH, preventing precipitation (El-Ganainy et al., 2022). The most effective profiles supply low, continuous K and Ca to the rhizoplane across weeks, with brief foliar top-ups around irrigation events to blunt Na^+ surges (Zhang et al., 2023). Monitoring K^+/Na^+ ratios in young leaves and leaf osmotic potential guides re-application; growers typically see improved relative water

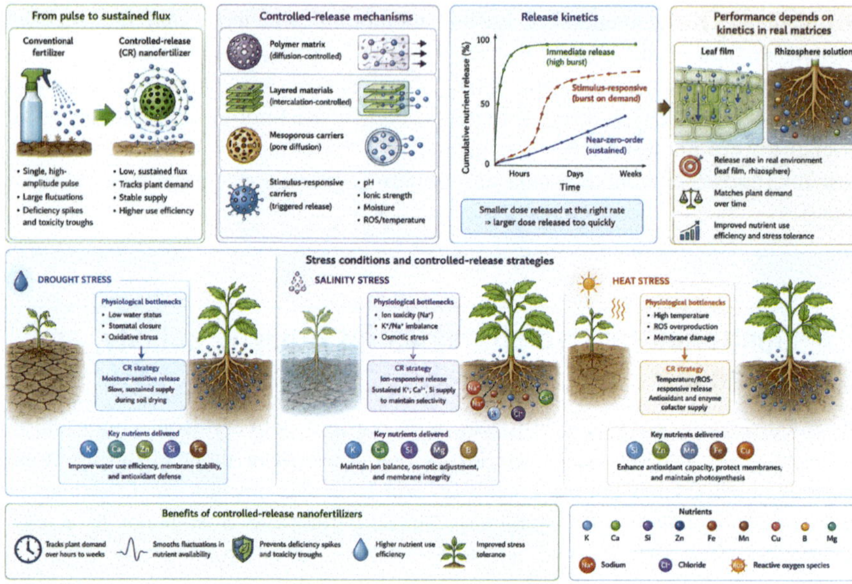

FIGURE 22.2

Controlled-release nanofertilizer strategies for improving drought, salinity, and heat tolerance in plants through sustained and stimulus-responsive nutrient delivery, optimized release kinetics, and enhanced nutrient-use efficiency under stress conditions.

content, cooler canopies, and reduced Na^+ accumulation when release rates are tuned to evapotranspiration cycles rather than calendar intervals.

For heat tolerance, rapid redox destabilization and thylakoid membrane disorder are the primary targets (Fortunato et al., 2023). Foliar CR systems that are ROS-responsive, for example, chitosan or polyurethane networks cross-linked to soften in oxidative microenvironments, can pulse Zn and Mn during midday heat peaks, sustaining D1 repair and nonphotochemical quenching without overloading in the morning or evening (Doğru, 2021). Adding a thin silica overcoat reduces uncontrolled photo-catalysis from any oxide cores and extends persistence under high UV. Because heat waves are acute, the preferred kinetics are "quasi-zero-order for 12–48 h," achieved by sub-200 nm particles with dense, hydrophilic shells that resist rapid wash-off; forecast-triggered sprays 12–24 h ahead of a heat event consistently protect Fv/Fm and maintain transpiration efficiency (Weisany et al., 2024a).

Formulation choices control kinetics. Mesoporous silica (SBA-15/MCM-41) loaded by adsorption gives Higuchi-type diffusion suited to day-scale foliar buffering; polymer-gated pores or lipid bilayers flatten the curve toward zero-order (Weisany et al., 2024b). LDHs intercalated with phosphate or nitrate deliver week-scale release governed by ion exchange, ideal for saline soils. Chitosan/alginate polyelectrolyte complexes offer pH- and salt-responsive gates and excellent leaf adhesion; humic/fulvic shells add chelation and hard-water tolerance (Gierszewska et al., 2018). Carbon dots or lignin nanoparticles can serve as light-stable, wetting-enhancing cocarriers, but doses must be validated to avoid unintended photochemistry (Tavan et al., 2025).

22.6 Modulating redox homeostasis and stress signaling via nanocarriers

Plants walk a tightrope between using reactive oxygen/nitrogen species (ROS/RNS) as signals and preventing them from escalating into damage (Corpas and Palma, 2024). Heat spikes, drought-driven desiccation, and salt-induced ion imbalance intensify electron leakage from photosystems and mitochondria, while biotic encounters trigger apoplastic oxidative bursts via NADPH oxidases (RBOHs) (Sachdev et al., 2021). Precision nanocarriers let us tune this redox landscape: they (1) supply enzyme cofactors at microdoses to maintain antioxidant capacity (Meng et al., 2021), (2) act as catalytic "nanozymes" that quench excess ROS without flattening signaling (Xu et al., 2025), (3) release small-molecule modulators (NO, H_2S, SA/JA) in a stimulus-responsive way to prime defenses (Liu et al., 2022), and (4) stabilize membranes and protein complexes so fewer radicals are generated in the first place (Romero-Ben et al., 2025) (Fig. 22.3).

Cofactor supply that sustains endogenous defenses. Micronutrient nanonutrients Zn, Cu, Mn, Fe delivered in controlled-release matrices keep SOD, CAT, APX, POD, and related peroxidases functioning through stress (Hanif et al., 2024). Sub-200-nm chitosan/silica carriers adhere to the leaf cuticle and meter ions into the apoplast during periods of high ionic strength or mild acidity (a proxy for stomatal closure), preventing the classic crash in enzyme activity seen at stress peaks. Thoughtful iron delivery is critical: sparingly soluble Fe(III) in chelated or LDH-intercalated forms supports chloroplast repair and heme enzymes while minimizing Fenton-type radical formation (Jeong et al., 2008).

Elicitation without overreaction. Biopolymer carriers, especially chitosan/oligo-chitosan, act as gentle elicitors perceived by pattern-recognition receptors, elevating basal defenses (callose deposition, phenylpropanoid enzymes) and tightening redox set-points with minimal yield drag (Soufiani et al., 2025). Carbon dots and lignin nanoparticles can add photoprotective effects and improve foliar spreading, but their doses must be validated to avoid interfering with chloroplast redox signaling (Chowmasundaram et al., 2023).

Lowering ROS generation at the source. Silicon-rich carriers fortify cell walls and thylakoid membranes, reducing ion leak and lipid peroxidation; steady Ca^{2+} availability supports membrane repair and calmodulin-dependent signaling that reins in RBOH activity (Weisany et al., 2024a). Together with improved K^+/Na^+ selectivity under salinity, these structural effects reduce the need for heavy scavenging in the first place.

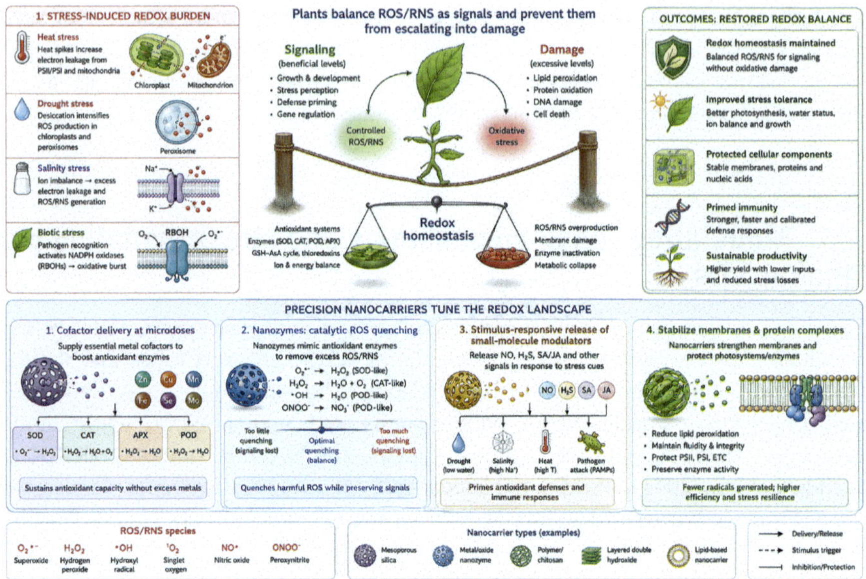

FIGURE 22.3

Schematic illustration of nanocarrier-mediated modulation of redox homeostasis and stress signaling in plants. Precision nanocarriers regulate ROS/RNS balance through controlled cofactor delivery, nanozyme-based ROS scavenging, stimulus-responsive release of signaling molecules, and stabilization of membranes and protein complexes under abiotic and biotic stresses.

Assaying redox and signaling outcomes. Beyond yield, decision-grade readouts include $H_2O_2/O_2\bullet^-$ mapping (DAB/NBT or fluorescent probes), MDA/TBARS and protein carbonyls for oxidative damage, GSH/GSSG and AsA/DHA ratios for redox poise, activities of SOD/CAT/APX/POD, Fv/Fm and NPQ for photochemistry, stomatal conductance/canopy temperature for gas exchange, and qPCR/RT-ddPCR of RBOHD, APX1, CAT2, HSPs, MAPK3/6, WRKYs. Hormone analytics (ABA, SA, JA/JA-Ile, ethylene precursor ACC) verify that nanocarrier signals align with physiological intent.

22.7 Integrating nanofertilizers with biostimulants and microbial consortia

Effective integration starts from complementarity rather than cocktailing. Nanofertilizers excel at placing ions precisely in space and time; biostimulants tune metabolism, water relations, and signaling; microbial consortia expand nutrient supply and build a resilient rhizosphere. When these roles are codesigned, plants

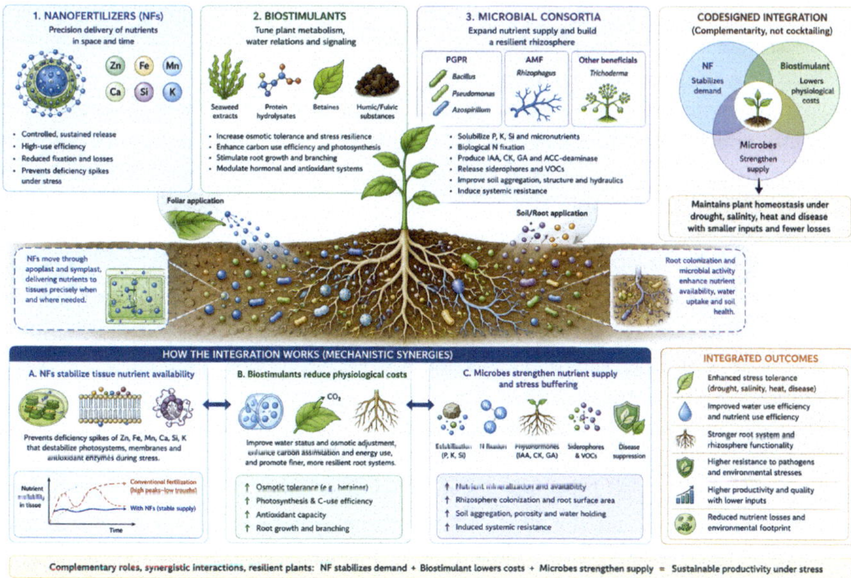

FIGURE 22.4

Schematic illustration of the integrated use of nanofertilizers, biostimulants, and microbial consortia for improving plant stress tolerance. Nanofertilizers stabilize nutrient availability, biostimulants enhance metabolic and osmotic adjustment, and beneficial microbes strengthen nutrient supply, rhizosphere functionality, and stress resilience under adverse environmental conditions.

maintain homeostasis through drought, salinity, heat, and episodic disease pressure with smaller inputs and fewer losses.

Mechanistically, NFs smooth tissue availability of Zn, Fe, Mn, Ca, Si, and K, preventing the deficiency spikes that destabilize photosystems, membranes, and antioxidant enzymes during stress (Mim et al., 2025). Seaweed extracts, protein hydrolysates, betaines, and humic/fulvic substances raise osmotic tolerance, enhance carbon use efficiency, and encourage finer root systems (Ali et al., 2021). Plant-beneficial microbes PGPR such as Bacillus, Pseudomonas, and Azospirillum; AMF like Rhizophagus; and Trichoderma solubilize P/K/Si, fix N, produce IAA/CK/GA and ACC-deaminase, release siderophores and VOCs, and improve soil aggregation and hydraulics (Darakeh et al., 2022). Together, the NF stabilizes demand, the biostimulant lowers physiological costs, and the microbes strengthen supply (Fig. 22.4).

Compatibility is the gatekeeper of synergy. Free ionic pulses especially Cu^{2+} or Zn^{2+} can depress inoculant viability, so trace metals should be chelated or released gradually from carriers (chitosan, alginate, lignosulfonate, humic-coated minerals) (Wei et al., 2021). Physical separation often helps: place the NF on the seed coat and apply microbes as a separate slurry; use layered beads or dual microcapsules that keep living cells away from metal-rich cores; and prefer nonionic wetters over cationic surfactants that harm cells and spores. Aim for stable dispersions in the actual spray water, not just DI water.

Separate applications by route (soil vs foliar) or time (≥48 h) whenever released ion activity approaches inoculant MICs, tank pH/EC drifts outside microbe-tolerant ranges, or strong elicitors (chitosan ≥0.1%) risk over-priming when combined with microbial VOCs. This simple spacing preserves both microbial viability and NF performance. Evaluation should look beyond yield to detect true synergy. Track leaf ionomics (K^+/Na^+, Zn/Fe/Si), chlorophyll fluorescence (Fv/Fm, NPQ), canopy temperature depression, relative water content and osmotic potential, and antioxidant markers. In the rhizosphere, measure root length density, exopolysaccharides, and aggregate stability, and quantify Bacillus/Pseudomonas abundance or AMF colonization (%) by plating or qPCR. In saline or arid systems, monitor leachate EC and Na^+ to document improved ion management.

Stewardship closes the loop. Avoid nano-Ag and high-Cu systems when live inoculants are central; cap trace-metal actives within conservative agronomic ranges; document residue fate and nontarget effects (beneficial microbes, pollinators); and provide clear PPE, reentry, and compatibility guidance (Santás-Miguel et al., 2023). Use life-cycle and partial-budget analyses to confirm that the added complexity pays for itself through fewer passes, lower salt fertilizer rates, and stabilized yields.

22.8 Advances in foliar and soil delivery systems

Modern delivery systems turn nanofertilizers from "better salts" into precision interfaces that cooperate with leaf microfilms and the rhizosphere. On leaves, progress centers on retention, penetration, and timed release. Sub-200 nm suspensions with narrow PDI and mild ζ-potentials (≈ −10 to +25 mV in the actual spray water) resist flocculation and adhere to waxy cuticles; biopolymer or humic shells improve wetting and mitigate bounce, while ROS/ionic-strength–responsive gates meter Zn/Fe/Mn during midday oxidative peaks or under concentrated leaf films when stomata are partially closed (Ding et al., 2023; Pinna and Husted, 2025). Electrospray/ultrasonic droplet generators and air-induction nozzles reduce drift and produce ~200–400 μm droplets that spread without run-off; organosilicone or nonionic adjuvants can help, but cationic quaternary ammonium surfactants are avoided because of phytotoxicity and microbiome sensitivity (Moosavi and Zolghadr, 2023; Wang et al., 2023).

Foliar timing is increasingly forecast-driven. Early-morning or late-afternoon applications overlay the plant's humidity cycle, prolonging the aqueous film that promotes cuticular diffusion and stomatal entry; preevent buffers (12–48 h before heat/dry winds) plus short maintenance passes during prolonged stress outperform one-off high-dose sprays. UAV/tractor variable-rate maps (SPAD/NDVI, canopy temperature) now guide where to intensify or skip passes, cutting nutrient losses and leaf-scorch risk (Tulu et al., 2025; Gibbs et al., 2021).

Cuticle and stomatal entry routes are being targeted explicitly. Ligand shells (amino/organic acids, short peptides) bias adhesion to waxes and facilitate apoplast transport; partially chelated micronutrients traverse cell-wall pores more predictably than free ions (Pinna and Husted, 2025). Peptide tagging and stomata-targeted nanocarriers are emerging to route cargos to specific tissues and stomatal complexes, while near-zero-order release over 24–72 h helps keep chloroplast cofactors steady, preserving Fv/Fm and NPQ under stress (Puangpathumanond et al., 2025).

In soils, innovation focuses on root-zone "depots" synchronized with water movement. Layered Double Hydroxides intercalated with P/K/Ca and nanohydroxyapatite (nHAP) provide week-scale ion exchange without salinity spikes; mesoporous/lignin–chitosan hosts provide trickle-feed trace metals (Roy et al., 2023; Zhang et al., 2024). Banding or in-furrow placement localizes depots at the rhizoplane and limits fixation on high-CEC or calcareous soils. Under saline irrigation, slow K/Ca release helps maintain K^+/Na^+ selectivity, while Si-rich carriers reinforce walls and reduce Na^+ leak (Li et al., 2025; Manimaran et al., 2025). Fertigation of nanosuspensions through drip systems is feasible with shear-stable, anti-settling designs and ~80–120 μm filtration; back-flush routines and biocide-free preservation help prevent emitter fouling (Liu, 2004) (Fig. 22.5).

Soil context matters as much as chemistry. Delivery should be stratified by texture, OM, and pH/EC: sandy soils favor slightly higher loading via frequent small pulses; high-pH calcareous soils benefit from humic-coated or chelated forms to reduce precipitation; sodic fields pair LDH depots with gypsum/acidified irrigation to shift exchange equilibria. Where biochar/compost is used, carriers anchored to colloids can reduce leaching but require release testing in soil extracts to avoid under-delivery (Maaz et al., 2025; Zhang et al., 2024).

Integration with biology is now standard. Coplacement (or staggered timing) with PGPR/AMF boosts nutrient cycling and water relations, provided free-ion activity stays below MICs and living inoculants are physically separated from metal-rich cores (dual capsules, buffered shells) (Khalediyan et al., 2021). Chitosan shells can gently elicit defenses without harming microbes when used at low doses (≈ ≤ 0.1% w/v); in contrast, nano-Ag and high Cu spikes can suppress microbial activity and alter community structure and nutrient cycling, so they are avoided in microbial programs (Zeng et al., 2024).

A deployment checklist keeps systems reproducible: (1) DLS/ζ-potential in farm water after all adjuvants; (2) release profiles in leaf-wash and local soil extracts across pH 5–8 and EC 0.5–3.0 dS m^{-1}; (3) nozzle/droplet verification (target ~200–400 μm; prefer air-induction to cut drift); (4) compatibility no visible flocculation in a 2-h jar test; (5) sensor-triggered scheduling (NDVI↓ > 0.03, canopy–air $\Delta T < -1.5$ °C); and (6) stewardship PPE, reentry, and leachate EC/Na^+ monitoring (Wang et al., 2023; Gibbs et al., 2021).

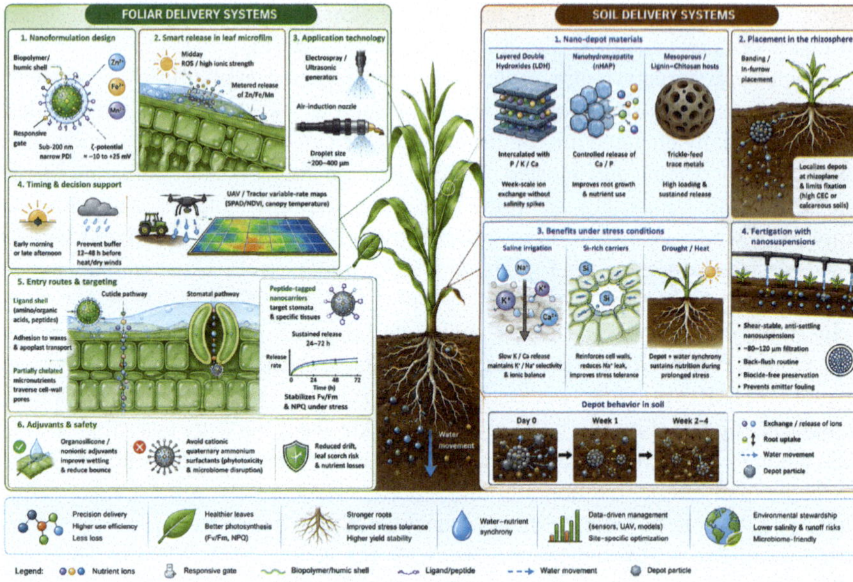

FIGURE 22.5

Advanced foliar and soil delivery systems for nano-enabled fertilizers. The figure illustrates smart foliar application technologies for enhanced retention, penetration, and controlled nutrient release, alongside soil-based nano-depots and fertigation systems designed to improve nutrient-use efficiency, stress tolerance, synchronized water–nutrient transport, and environmental sustainability.

What's next: electrodynamic/electrostatic deposition to steer charged droplets onto target leaves; peptide tags for tissue-specific routing (young leaves, reproductive organs); root-exudate-responsive gates for on-demand release; and closed-loop application where drones read canopy signals and modulate dose in real time (Guo et al., 2024).

22.9 Conclusion

Nanofertilizers shift crop nutrition from bulk supply to spatiotemporal dosing, aligning nutrient delivery with the dynamics of stress. Across the chapter, we showed how mechanistic intent (redox buffering, ion/water homeostasis, photochemical protection) is converted into performance through materials design (carrier choice, interfacial properties, and triggerable release) and disciplined agronomy (route, timing, and sensor-guided scheduling). When these pillars are cooptimized, small, well-timed doses stabilize plant metabolism during drought, salinity, heat, cold snaps, and episodic biotic pressure, delivering resilience with lower inputs and fewer losses.

Evidence converges on three consistent advantages. First, homeostatic buffering: steady microdoses of Zn/Fe/Mn/Cu and Si/Ca keep antioxidant enzymes active, membranes intact, and PSII efficient, avoiding the peaks and troughs that follow conventional pulses. Second, targeting efficiency: foliar systems tuned for cuticle/stomatal entry and soil depots positioned at the rhizoplane reduce leaching and volatilization while meeting tissue demand at critical phenophases. Third, system integration: compatibility with biostimulants and microbial consortia extends benefits to root architecture, osmotic balance, and nutrient cycling, provided free-ion spikes are avoided and applications are separated by route or time when necessary.

Translation from bench to farm depends on real-matrix rigor and stewardship. Formulations must be characterized in farm water (DLS/ζ-potential after adjuvants) and released in leaf-wash and soil extracts, not deionized water. Triggers should match stress signals (moisture for drought, ionic strength/Na^+ for salinity, ROS/temperature for heat) and be calibrated for hours–days (foliar) or days–weeks (root-zone). Safety-by-design favors biodegradable or sparingly soluble carriers, shielding of photocatalytic cores, conservative trace-metal loads, and documented compatibility with PGPR/AMF and biocontrols. Regulatory readiness, worker safety (PPE/reentry), residue fate, and life-cycle impacts are integral, not optional.

Several gaps deserve priority. We need standardized, field-relevant release assays, clearer dose boundaries that preserve "ROS as a signal" while preventing damage, long-term data on soil transformations and cumulative effects, interoperable compatibility tables (tank mixes and inoculant MICs), and multilocation techno-economic analyses that report yield stability alongside input and pass reductions. Closing these gaps will turn promising prototypes into reliable tools across soils, climates, and management systems.

For practitioners, the rule of thumb is simple: design by stress map and phenology, validate in real matrices, and schedule by sensors. In practice, this means preevent priming 12–48 h before heat or dry winds, small maintenance passes only if thresholds persist (Fv/Fm, canopy–air ΔT, K^+/Na^+, RWC), root-zone depots to buffer saline irrigation, and explicit route/time separation when codeploying microbes or strong elicitors. Jar tests in farm water and quick viability checks (≥80% CFU retention after 2 h) prevent costly incompatibilities.

Ultimately, nanofertilizers are not an end in themselves; they are precision interfaces that help keep plants within their safe operating space under volatile climates. When engineered and stewarded as part of an integrated program tolerant germplasm, smart irrigation, microbial partners, and digital decision support they convert environmental variability from a yield threat into a manageable fluctuation, advancing truly climate-smart, resource-efficient agriculture.

References

Abdel-Hakim, S.G., Shehata, A.S.A., Moghannem, S.A., Qadri, M., El-Ghany, M.F.A., Abdeldaym, E.A., et al., 2023. Nanoparticulate fertilizers increase nutrient absorption efficiency and agro-physiological properties of lettuce plant. Agronomy 13 (3). Available at https://doi.org/10.3390/agronomy13030691.

Adrees, M., Ali, S., Rizwan, M., Zia-ur-Rehman, M., Ibrahim, M., Abbas, F., et al., 2015. Mechanisms of silicon-mediated alleviation of heavy metal toxicity in plants: a review. Ecotoxicology and Environmental Safety 119, 186–197. Available at https://doi.org/10.1016/j.ecoenv.2015.05.011.

Akter, M., Oue, H., 2018. Effect of saline irrigation on accumulation of Na+, K+, Ca2+, and Mg2+ ions in rice plants. Agriculture 8 (10), 164. Available at https://doi.org/10.3390/agriculture8100164.

Ali, O., Ramsubhag, A., Jayaraman, J., 2021. Biostimulant properties of seaweed extracts in plants: implications towards sustainable crop production. Plants 10 (3), 531. Available at https://doi.org/10.3390/plants10030531.

Allakhverdiev, S.I., Kreslavski, V.D., Klimov, V.V., Los, D.A., Carpentier, R., Mohanty, P., 2008. Heat stress: an overview of molecular responses in photosynthesis. Photosynthesis Research 98 (1–3), 541–550. Available at https://doi.org/10.1007/s11120-008-9331-0.

Aqaei, P., Weisany, W., Diyanat, M., Razmi, J., Struik, P.C., 2020. Response of maize (*Zea mays* L.) to potassium nano-silica application under drought stress. Journal of Plant Nutrition 43 (9), 1205–1216. Available at https://doi.org/10.1080/01904167.2020.1727508.

Arabpoor, B., Yousefi, S., Weisany, W., Ghasemlou, M., 2021. Multifunctional coating composed of *Eryngium campestre* L. essential oil encapsulated in nano-chitosan to prolong the shelf-life of fresh cherry fruits. Food Hydrocolloids 111, 106394. Available at https://doi.org/10.1016/j.foodhyd.2020.106394.

Borgatta, J.R., Lochbaum, C.A., Elmer, W.H., White, J.C., Pedersen, J.A., Hamers, R.J., 2021. Biomolecular corona formation on CuO nanoparticles in plant xylem fluid. Environmental Science: Nano 8 (4), 1067–1080. Available at https://doi.org/10.1039/d1en00140j.

Cao, X., Sun, L., Wang, W., Zhang, F., 2024. Exogenous calcium application mediates K+ and Na+ homeostasis of different salt-tolerant rapeseed varieties under NaHCO3 stress. Plant Growth Regulation 102 (2), 367–378. Available at https://doi.org/10.1007/s10725-023-01066-1.

Chhabra, R., 2022. Nutrient management in salt-affected soils. Salt-Affected Soils and Marginal Waters: Global Perspectives and Sustainable Management. Springer, pp. 349–429.

Chowmasundaram, Y.A.P., Tan, T.L., Nulit, R., Jusoh, M., Rashid, S.A., 2023. Recent developments, applications and challenges for carbon quantum dots as a photosynthesis enhancer in agriculture. RSC Advances 13 (36), 25093–25117. Available at https://doi.org/10.1039/d3ra01217d.

Corpas, F.J., Palma, J.M., 2024. Reactive oxygen and nitrogen species in plants. Antioxidants 13 (10). Available at https://doi.org/10.3390/antiox13101232.

Darakeh, S.A.S.S., Weisany, W., Tahir, N.A.R., Schenk, P.M., 2022. Physiological and biochemical responses of black cumin to vermicompost and plant biostimulants: arbuscular mycorrhizal and plant growth-promoting rhizobacteria. Industrial Crops and Products 188. Available at https://doi.org/10.1016/j.indcrop.2022.115557.

Das, A., Ghosh, S., Pramanik, N., 2024. Chitosan biopolymer nanocomposites for agriculture applications. Biopolymeric Nanoparticles for Agricultural Applications. Springer, pp. 209–240.

Ding, Y., Zhao, W., Zhu, G., Wang, Q., Zhang, P., Rui, Y., 2023. Recent trends in foliar nanofertilizers: a review. Nanomaterials 13 (21), 2906. Available at https://doi.org/10.3390/nano13212906.

Doğru, A., 2021. Effects of heat stress on photosystem II activity and antioxidant enzymes in two maize cultivars. Planta 253 (4). Available at https://doi.org/10.1007/s00425-021-03611-6.

Ebel, C., Faou, P., Kernel, B., Zaccai, G., 1999. Relative role of anions and cations in the stabilization of halophilic malate dehydrogenase. Biochemistry 38 (28), 9039–9047. Available at https://doi.org/10.1021/bi9900774.

Elbagory, M., Altihani, F.A., El-Nahrawy, S., Shalaby, M., Omara, A.E.D., Singh, J., et al., 2025. Feasibility of Nano-urea and PGPR on Salt Stress Amelioration in Reshmi Amaranth (*Amaranthus tricolor*): stress markers and enzymatic response. Horticulturae 11 (3). Available at https://doi.org/10.3390/horticulturae11030280.

El-Ganainy, S.M., El-Bakery, A.M., Hafez, H.M., Ismail, A.M., El-Abdeen, A.Z., Ata, A.A.E., et al., 2022. Humic acid-coated Fe3O4 nanoparticles confer resistance to *Acremonium* wilt disease and improve physiological and morphological attributes of grain sorghum. Polymers 14 (15). Available at https://doi.org/10.3390/polym14153099.

Fortunato, S., Lasorella, C., Dipierro, N., Vita, F., de Pinto, M.C., 2023. Redox signaling in plant heat stress response. Antioxidants 12 (3). Available at https://doi.org/10.3390/antiox12030605.

Gibbs, J.L., Peters, T.M., Heck, L.P., 2021. Comparison of droplet size, coverage, and drift potential from UAV application methods and ground application methods on row crops. Transactions of the ASABE 64 (3), 819–828. Available at https://doi.org/10.13031/TRANS.14121.

Gierszewska, M., Ostrowska-Czubenko, J., Chrzanowska, E., 2018. pH responsive chitosan/alginate polyelectrolyte complex membranes reinforced by tripolyphosphate. European Polymer Journal 101, 282–290. Available at https://doi.org/10.1016/j.eurpolymj.2018.02.031.

Guo, H., White, J.C., Wang, Z., Xing, B., 2018. Nano-enabled fertilizers to control the release and use efficiency of nutrients. Current Opinion in Environmental Science and Health 6, 77–83. Available at https://doi.org/10.1016/j.coesh.2018.07.009.

Guo, Z., Zhang, J., Chen, L., Wang, Z., Wang, H., Wang, X., 2024. Study on deposition characteristics of the electrostatic sprayer for pesticide application in greenhouse tomato crops. Agriculture 14 (11), 1981. Available at https://doi.org/10.3390/agriculture14111981.

Hanif, S., Javed, R., Cheema, M., Kiani, M.Z., Farooq, S., Zia, M., 2024. Harnessing the potential of zinc oxide nanoparticles and their derivatives as nanofertilizers: trends and perspectives. Plant Nano Biology 10, 100110. Available at https://doi.org/10.1016/j.plana.2024.100110.

Jeong, J., Cohu, C., Kerkeb, L., Pilon, M., Connolly, E.L., Guerinot, M.L., 2008. Chloroplast Fe(III) chelate reductase activity is essential for seedling viability under iron limiting conditions. Proceedings of the National Academy of Sciences of the United States of America 105 (30), 10619–10624. Available at https://doi.org/10.1073/pnas.0708367105.

Khalediyan, N., Weisany, W., Schenk, P.M., 2021. Arbuscular mycorrhizae and rhizobacteria improve growth, nutritional status and essential oil production in

Ocimum basilicum and *Satureja hortensis*. Industrial Crops and Products 160, 113163. Available at https://doi.org/10.1016/j.indcrop.2020.113163.

Khosropour, E., Hakimi, L., Weisany, W., 2022. Role of engineered nanomaterials in biotic stress managements. Engineered Nanomaterials for Sustainable Agricultural Production, Soil Improvement and Stress Management. Elsevier, Iran, pp. 257–272. Available at https://doi.org/10.1016/B978-0-323-91933-3.00001-5.

Kumawat, N., Shah, K., Singh, V., Rana, D., 2025. Nano-engineered fertilizers: enhancing plant nutritional dynamics through precision delivery. Eco-Friendly Nanotechnology: Harnessing Small-Scale Technologies for a Cleaner and Healthier Planet. Deep Science Publishing, pp. 29–44. Available at https://doi.org/10.70593/978-93-49307-12-4_3.

Lakshani, N., Wijerathne, H.S., Sandaruwan, C., Kottegoda, N., Karunarathne, V., 2023. Release kinetic models and release mechanisms of controlled-release and slow-release fertilizers. ACS Agricultural Science & Technology 3 (11), 939–956. Available at https://doi.org/10.1021/acsagscitech.3c00152.

Li, F., Hou, Y., Chen, L., Qiu, Y., 2025. Advances in silica nanoparticles for agricultural applications and biosynthesis. Advanced Biotechnology 3 (2). Available at https://doi.org/10.1007/s44307-025-00067-7.

Liu, Z., 2004. Effects of surfactants on foliar uptake of herbicides – a complex scenario. Colloids and Surfaces B: Biointerfaces 35 (3–4), 149–153. Available at https://doi.org/10.1016/j.colsurfb.2004.02.016.

Liu, C.Y., Chen, H.L., Zhou, H.J., Yu, S.M., Yao, W.H., Wang, N., et al., 2022. Precise delivery of multi-stimulus-responsive nanocarriers based on interchangeable visual guidance. Biomaterials Advances 134. Available at https://doi.org/10.1016/j.msec.2021.112558.

Llopis-Lorente, A., De Luis, B., García-Fernández, A., Jimenez-Falcao, S., Orzáez, M., Sancenón, F., et al., 2018. Hybrid mesoporous nanocarriers act by processing logic tasks: toward the design of nanobots capable of reading information from the environment. ACS Applied Materials and Interfaces 10 (31), 26494–26500. Available at https://doi.org/10.1021/acsami.8b05920.

Maaz, T.M.C., Dobermann, A., Lyons, S.E., Thomson, A.M., 2025. Review of research and innovation on novel fertilizers for crop nutrition. npj Sustainable Agriculture 3 (1). Available at https://doi.org/10.1038/s44264-025-00066-0.

Madlala, N.C., Khanyile, N., Masenya, A., 2024. Examining the correlation between the inorganic nano-fertilizer physical properties and their impact on crop performance and nutrient uptake efficiency. Nanomaterials 14 (15), 1263. Available at https://doi.org/10.3390/nano14151263.

Manimaran, G., Duraisamy, S., Subramanium, T., Rangasamy, A., Alagarsamy, S., James, P., et al., 2025. Silicon-driven approaches to salinity stress tolerance: mechanisms, uptake dynamics, and microbial transformations. Plant Stress 16, 100825. Available at https://doi.org/10.1016/j.stress.2025.100825.

Martău, G.A., Mihai, M., Vodnar, D.C., 2019. The use of chitosan, alginate, and pectin in the biomedical and food sector—biocompatibility, bioadhesiveness, and biodegradability. Polymers 11 (11), 1837. Available at https://doi.org/10.3390/polym11111837.

Meng, J., Lv, Z., Zhang, Y., Wang, Y., Qiao, X., Sun, C., et al., 2021. Precision redox: the key for antioxidant pharmacology. Antioxidants & Redox Signaling 34 (14), 1069–1082. Available at https://doi.org/10.1089/ars.2020.8212.

Mim, J.J., Rahman, S.M.M., Khan, F., Paul, D., Sikder, S., Das, H.P., et al., 2025. Towards smart agriculture through nano-fertilizer-a review. Materials Today Sustainability 30. Available at https://doi.org/10.1016/j.mtsust.2025.101100.

Moosavi, S.S., Zolghadr, A.R., 2023. Effect of quaternary ammonium surfactants on biomembranes using molecular dynamics simulation. RSC Advances 13 (47), 33175–33186. Available at https://doi.org/10.1039/d3ra05030k.

Mukarram, M., Ali, J., Dadkhah-Aghdash, H., Kurjak, D., Kačík, F., Ďurkovič, J., 2023. Chitosan-induced biotic stress tolerance and crosstalk with phytohormones, antioxidants, and other signalling molecules. Frontiers in Plant Science 14. Available at https://doi.org/10.3389/fpls.2023.1217822.

Nongbet, A., Mishra, A.K., Mohanta, Y.K., Mahanta, S., Ray, M.K., Khan, M., et al., 2022. Nanofertilizers: a smart and sustainable attribute to modern agriculture. Plants 11 (19). Available at https://doi.org/10.3390/plants11192587.

Ortelli, S., Poland, C.A., Baldi, G., Costa, A.L., 2016. Silica matrix encapsulation as a strategy to control ROS production while preserving photoreactivity in nano-TiO2. Environmental Science: Nano 3 (3), 602–610. Available at https://doi.org/10.1039/c6en00009f.

Pardossi, A., Incrocci, L., Incrocci, G., Malorgio, F., Battista, P., Bacci, L., et al., 2009. Root zone sensors for irrigation management in intensive agriculture. Sensors 9 (4), 2809–2835. Available at https://doi.org/10.3390/s90402809.

Pinna, A., Husted, S., 2025. Translocation of foliar-applied nanoparticles: a critical review from a plant science perspective. Physiologia Plantarum 177 (5). Available at https://doi.org/10.1111/ppl.70476.

Puangpathumanond, S., Chee, H.L., Sevencan, C., Yang, X., Lau, O.S., Lew, T.T.S., 2025. Stomata-targeted nanocarriers enhance plant defense against pathogen colonization. Nature Communications 16 (1). Available at https://doi.org/10.1038/s41467-025-60112-w.

Quitadamo, F., De Simone, V., Beleggia, R., Trono, D., 2021. Chitosan-induced activation of the antioxidant defense system counteracts the adverse effects of salinity in durum wheat. Plants 10 (7), 1365. Available at https://doi.org/10.3390/plants10071365.

Rahman, M.S., Bhuiyan, M.N.I., Rahman, M., Islam, S., Suchi, P.D., Saha, B.K., et al., 2025. Advancements in nanotechnology for arsenic remediation in agricultural systems: challenges and prospects. Plant Nano Biology 13. Available at https://doi.org/10.1016/j.plana.2025.100169.

Rojas-Pirela, M., Carillo, P., Lárez-Velásquez, C., Romanazzi, G., 2024. Effects of chitosan on plant growth under stress conditions: similarities with plant growth promoting bacteria. Frontiers in Plant Science 15. Available at https://doi.org/10.3389/fpls.2024.1423949.

Romero-Ben, E., Goswami, U., Soto-Cruz, J., Mansoori-Kermani, A., Mishra, D., Martin-Saldaña, S., et al., 2025. Polymer-based nanocarriers to transport therapeutic biomacromolecules across the blood-brain barrier. Acta Biomaterialia 196, 17–49. Available at https://doi.org/10.1016/j.actbio.2025.02.065.

Roy, A.S., de Beer, M., Pillai, S.K., Ray, S.S., 2023. Application of layered double hydroxides as a slow-release phosphate source: a comparison of hydroponic and soil systems. ACS Omega 8 (17), 15017–15030. Available at https://doi.org/10.1021/acsomega.2c07862.

Sachdev, S., Ansari, S.A., Ansari, M.I., Fujita, M., Hasanuzzaman, M., 2021. Abiotic stress and reactive oxygen species: generation, signaling, and defense mechanisms. Antioxidants 10 (2), 1–37. Available at https://doi.org/10.3390/antiox10020277.

Santás-Miguel, V., Arias-Estévez, M., Rodríguez-Seijo, A., Arenas-Lago, D., 2023. Use of metal nanoparticles in agriculture. A review on the effects on plant germination. Environmental Pollution 334, 122222. Available at https://doi.org/10.1016/j.envpol.2023.122222.

Shabala, S., Demidchik, V., Shabala, L., Cuin, T.A., Smith, S.J., Miller, A.J., et al., 2006. Extracellular Ca2+ ameliorates NaCl-induced K+ loss from Arabidopsis root and leaf cells by controlling plasma membrane K +-permeable channels. Plant Physiology 141 (4), 1653–1665. Available at https://doi.org/10.1104/pp.106.082388.

Shelar, A., Singh, A.V., Maharjan, R.S., Laux, P., Luch, A., Gemmati, D., et al., 2021. Sustainable agriculture through multidisciplinary seed nanopriming: prospects of opportunities and challenges. Cells 10 (9). Available at https://doi.org/10.3390/cells10092428.

Shen, M., Liu, S., Jiang, C., Zhang, T., Chen, W., 2023. Recent advances in stimuli-response mechanisms of nano-enabled controlled-release fertilizers and pesticides. Eco-Environment & Health 2 (3), 161–175. Available at https://doi.org/10.1016/j.eehl.2023.07.005.

Singh, M., Goswami, S.P., Ranjitha, G., Sachan, P., Sahu, D.K., Beese, S., Pandey, S.K., 2024. Nanotech for fertilizers and nutrients-improving nutrient use efficiency with nano-enabled fertilizers. Journal of Experimental Agriculture International 46 (5), 220–247. Available at https://doi.org/10.9734/jeai/2024/v46i52372.

Soufiani, S.P., Yousefi, S., Honarvar, M., Weisany, W., Asadi, G., 2025. Polyvinyl alcohol films incorporating cellulose nanocrystals, chitosan, and mesoporous silica nanoparticles for enhanced stability and controlled release of beet extracts. Carbohydrate Polymer Technologies and Applications 10, 100740. Available at https://doi.org/10.1016/j.carpta.2025.100740.

Sági-Kazár, M., Zelenyánszki, H., Müller, B., Cseh, B., Gyuris, B., Farkas, S.Z., et al., 2021. Supraoptimal iron nutrition of *Brassica napus* plants suppresses the iron uptake of chloroplasts by down-regulating chloroplast ferric chelate reductase. Frontiers in Plant Science 12. Available at https://doi.org/10.3389/fpls.2021.658987.

Tavan, M., Yousefian, Z., Bakhtiar, Z., Rahmandoust, M., Mirjalili, M.H., 2025. Carbon quantum dots: multifunctional fluorescent nanomaterials for sustainable advances in biomedicine and agriculture. Industrial Crops and Products 231, 121207. Available at https://doi.org/10.1016/j.indcrop.2025.121207.

Tulu, B.B., Teshome, F., Ampatzidis, Y., Hailegnaw, N.S., Bayabil, H.K., 2025. AgriSenAI: automating UAV thermal and multispectral image processing for precision agriculture. SoftwareX 30. Available at https://doi.org/10.1016/j.softx.2025.102083.

Varamin, J.K., Fanoodi, F., Sinaki, J.M., Rezvan, S., Damavandi, A., 2020. Foliar application of chitosan and nano-magnesium fertilizers influence on seed yield, oil content, photosynthetic pigments, antioxidant enzyme activities of sesame (*Sesamum indicum* L.) under water-limited conditions. Notulae Botanicae Horti Agrobotanici Cluj-Napoca 48 (4), 2228–2243. Available at https://doi.org/10.15835/nbha48411852.

Wang, M., Wang, R., Mur, L.A.J., Ruan, J., Shen, Q., Guo, S., 2021. Functions of silicon in plant drought stress responses. Horticulture Research 8 (1). Available at https://doi.org/10.1038/s41438-021-00681-1.

Wang, S., Li, X., Nuyttens, D., Zhang, L., Liu, Y., Li, X., 2023. Evaluationof compact air-induction flat fan nozzles for herbicide applications: spray drift and biological efficacy. Frontiers in Plant Science 14. Available at https://doi.org/10.3389/fpls.2023.1018626.

Wang, Y.B., Zou, Y.L., Wei, Y.T., Meng, L.S., 2025. Crosstalk between ethylene and JA/ABA/sugar signalling in plants under physiological and stress conditions. Molecular Plant Pathology 26 (3). Available at https://doi.org/10.1111/mpp.70048.

Wei, X., Cao, P., Wang, G., Liu, Y., Song, J., Han, J., 2021. CuO, ZnO, and γ-Fe2O3 nanoparticles modified the underground biomass and rhizosphere microbial community of *Salvia miltiorrhiza* (Bge.) after 165-day exposure. Ecotoxicology and Environmental Safety 217, 112232. Available at https://doi.org/10.1016/j.ecoenv.2021.112232.

Wei, Y., Chen, H., Wang, L., Zhao, Q., Wang, D., Zhang, T., 2022. Cold acclimation alleviates cold stress-induced PSII inhibition and oxidative damage in tobacco leaves. Plant Signaling & Behavior 17 (1). Available at https://doi.org/10.1080/15592324.2021.2013638.

Weisany, W., Khosropour, E., 2023. Chapter 8 - Engineered nanomaterials in crop plants drought stress management. In: Husen, A. (Ed.), Engineered Nanomaterials for Sustainable Agricultural Production, Soil Improvement and Stress Management. Academic Press, pp. 183–204.

Weisany, W., Sohrabi, Y., Heidari, G., Siosemardeh, A., Ghassemi-Golezani, K., 2012. Changes in antioxidant enzymes activity and plant performance by salinity stress and zinc application in soybean (*Glycine max* L.). Plant OMICS 5 (2), 60–67.

Weisany, W., Amini, J., Samadi, S., Hossaini, S., Yousefi, S., Struik, P.C., 2019a. Nano silver-encapsulation of *Thymus daenensis* and *Anethum graveolens* essential oils enhances antifungal potential against strawberry anthracnose. Industrial Crops and Products 141, 111808. Available at https://doi.org/10.1016/j.indcrop.2019.111808.

Weisany, W., Samadi, S., Amini, J., Hossaini, S., Yousefi, S., Maggi, F., 2019b. Enhancement of the antifungal activity of thyme and dill essential oils against *Colletotrichum nymphaeae* by nano-encapsulation with copper NPs. Industrial Crops and Products 132, 213–225. Available at https://doi.org/10.1016/j.indcrop.2019.02.031.

Weisany, W., Razmi, J., Pashang, D., 2023. Improving seed germination and physiological characteristics of maize seedlings under osmotic stress through potassium nano-silicate treatment. Frontiers in Plant Science 14. Available at https://doi.org/10.3389/fpls.2023.1274396.

Weisany, W., Razmi, J., Eshaghabadi, A.H., Pashang, D., 2024a. Silicon Nanoparticles (SiNP): a novel and sustainable strategy for mitigating environmental stresses in plants. Journal of Soil Science and Plant Nutrition 24 (2), 2167–2191. Available at https://doi.org/10.1007/s42729-024-01790-1.

Weisany, W., Yousefi, S., Soufiani, S.P., Pashang, D., McClements, D.J., Ghasemlou, M., 2024b. Mesoporous silica nanoparticles: a versatile platform for encapsulation and delivery of essential oils for food applications. Advances in Colloid and Interface Science 325. Available at https://doi.org/10.1016/j.cis.2024.103116.

Wen, M., Wang, P., Gao, W., Wu, S., Huang, B., 2021. Effects of foliar spraying with different concentrations of selenium fertilizer on the development, nutrient absorption, and quality of citrus fruits. HortScience: A Publication of the American Society for Horticultural Science 56 (11), 1363–1367. Available at https://doi.org/10.21273/hortsci16074-21.

Wu, H., Li, Z., 2022. Nano-enabled agriculture: how do nanoparticles cross barriers in plants? Plant Communications 3 (6), 100346. Available at https://doi.org/10.1016/j.xplc.2022.100346.

Xu, C., Wu, F., Duan, Z., Rajbanshi, B., Qi, Y., Qin, J., et al., 2025. Microneedle-aided nanotherapeutics delivery and nanosensor intervention in advanced tissue regeneration. Journal of Nanobiotechnology 23 (1). Available at https://doi.org/10.1186/s12951-025-03383-1.

Yadav, M., 2025. Nanoparticle-facilitated targeted nutrient delivery in plants: breakthroughs and mechanistic insights. Plant Nano Biology 12, 100156. Available at https://doi.org/10.1016/j.plana.2025.100156.

Yousefi, S., Weisany, W., Hosseini, S.E., Ghasemlou, M., 2024. Mechanisms of nanoencapsulation to boost the antimicrobial efficacy of essential oils: a review. Food Hydrocolloids 150, 109655. Available at https://doi.org/10.1016/j.foodhyd.2023.109655.

Yuan, Z., Li, Y., Sun, M., Yuan, M., Han, Z., Li, X., et al., 2025. Recent progress in ROS-responsive biomaterials for the diagnosis and treatment of cardiovascular diseases. Theranostics 15 (11), 5172–5219. Available at https://doi.org/10.7150/thno.106991.

Zeng, W., Xiang, D., Li, X., Gao, Q., Chen, Y., Wang, K., et al., 2024. Effects of combined inoculation of arbuscular mycorrhizal fungi and plant growth-promoting rhizosphere bacteria on seedling growth and rhizosphere microecology. Frontiers in Microbiology 15. Available at https://doi.org/10.3389/fmicb.2024.1475485.

Zhang, L., Li, X., Zuo, W., Li, S., Sun, G., Wang, W., et al., 2021. Root exuded low-molecular-weight organic acids affected the phenanthrene degrader differently: a multi-omics study. Journal of Hazardous Materials 414, 125367. Available at https://doi.org/10.1016/j.jhazmat.2021.125367.

Zhang, K., Rengel, Z., Zhang, F., White, P.J., Shen, J., 2023. Rhizosphere engineering for sustainable crop production: entropy-based insights. Trends in Plant Science 28 (4), 390–398. Available at https://doi.org/10.1016/j.tplants.2022.11.008.

Zhang, H., Zheng, T., Wang, Y., Li, T., Chi, Q., 2024. Multifaceted impacts of nanoparticles on plant nutrient absorption and soil microbial communities. Frontiers in Plant Science 15. Available at https://doi.org/10.3389/fpls.2024.1497006.

Zhao, Q., Xin, M., Ai, P., Ma, Y., 2025. Effects of continuous saline water irrigation on soil salinization characteristics and dryland jujube tree. Agronomy 15 (8), 1898. Available at https://doi.org/10.3390/agronomy15081898.

Zuo, G., 2025. Non-photochemical quenching (NPQ) in photoprotection: insights into NPQ levels required to avoid photoinactivation and photoinhibition. New Phytologist 246 (5), 1967–1974. Available at https://doi.org/10.1111/nph.70121.

CHAPTER

23 Nutrient biofortification and environmental stress

Lucía Marcial Salvador, Luz Leticia Rivera Solís and Julia Medrano Macías
Plant Physiology Laboratory, Department of Horticulture, Autonomous Agrarian Antonio Narro University, Saltillo, Coahuila, Mexico

Abbreviations

KI	potassium iodide
KIO_3	potassium iodate
CH_2ICOOH	iodoacetic acid
TF	translocation factor
w/v	weight/volume
SOD	superoxide dismutase
CAT	catalase
APX	ascorbate peroxidase
POD	peroxidase
ha	hectares
SSS	shoot silicon-signal
SiNPs	silicon nanoparticles
ROS	reactive oxygen species
H_2O_2	hydrogen peroxide
SAMDC	S-adenosyl-L-methionine decarboxylase.

23.1 Introduction

In recent decades, the environmental conditions that support agricultural production have changed significantly. This is largely due to the population has risen from approximately 2.5 billion inhabitants in 1950 to more than 8 billion today, leading to an enormous expansion of the agricultural frontier toward marginal areas, where the excessive use of water resources, soil degradation, and the emission of gases into the atmosphere have caused critical environmental conditions (Pereira et al., 2019) for the production of the main crops that feed the population, such as maize, wheat and rice (Lyons, 2018).

Mineral Nutrients to Mitigate Abiotic Stress in Crops. DOI: https://doi.org/10.1016/B978-0-443-44712-9.00024-4

Biofortification is a biotechnological tool used mainly to increase the contents of micronutrients and vitamins, such as zinc, iron, copper, selenium, iodine, and recently, silicon, which have been reported to be the most difficult to access for human consumption (Duborská et al., 2022).

However, the use of this technique has also been shown to have many additional benefits, such as modifying plant redox metabolism, increasing the synthesis of molecules with reducing power, increasing tolerance to environmental stress, and generally increasing the homeostasis of the organism. The use of this biotechnological tool has proven to have advantages and disadvantages. Biofortification can be generalized into four groups: traditional (conventional breeding), metabolic engineering, genetic engineering, and nanobiotechnology (Abdalla et al., 2019; Chai et al., 2025; Christian et al., 2022).

In this book chapter, we review published studies from the last 5 years on the four main biofortification strategies using the beneficial elements iodine (I) and silicon (Si). The analysis focuses on their effects on nutritional enhancement and improved tolerance to environmental stress in the major staple crops that supply most of the world's calories: maize, wheat, and rice.

23.1.1 Iodine

23.1.1.1 Iodinebiofortification in crops

Diverse and exhaustive studies on the application of iodine in horticultural species, with the purpose of biofortification and increasing nutritional value, have been carried out (Kiferle et al., 2021; Medrano-Macías et al., 2016; Gonzali et al., 2017).

The results are highly variable, depending on the concentration applied, the chemical form used, and the production system adopted. For example, when iodine is applied in nutrient solution at concentrations on the order of 10^{-6} M of KI in soilless cultures, an increase in biomass has been observed in leafy vegetables (Sánchez-Acosta et al., 2025). The concentration of iodine can be increased to 10^{-5} M (equivalent to 1.3 mg L^{-1}) to obtain biofortified leafy vegetables using different chemical forms, such as I^-, IO_3^-, or iodoacetic acid (CH_2ICOOH^-) (Weng et al., 2008). Other findings have shown that species such as tomato resist higher concentrations (from 1 to 5 × 10^{-3} M of KI and 0.5 to 2 × 10^{-3} M of KIO_3) (Kiferle et al., 2013) once a week for eight weeks from the fruit set of the first cluster of fruits. The results were remarkable: an accumulation of up to 10 mg of iodine per kilogram of fresh fruit weight was achieved, with minimal phytotoxicity.

In strawberry plants, iodine application also increased plant biomass and improved fruit quality (Fuentes et al., 2022). On the other hand, decreases in biomass have been reported in potato crops (Ledwożyw-Smoleń et al., 2023), as well as in carrot (Smoleń et al., 2014) and in nopal (*Opuntia*) (García Osuna et al., 2014). However, in other species where vegetative reserve organs are also harvested, such as onion, iodine does not seem to affect the total weight of the plant (Topcuoğlu et al., 2024). Indeed, in recent years, experiments conducted in

Arabidopsis showed that removing iodine from the nutrient solution reduced plant growth, while restoring it at micromolar levels improved biomass accumulation and induced early flowering. Thus the iodine also regulated the expression of several defense-related genes, indicating a potential protective role against biotic and abiotic stress, the iodine organification was also demonstrated, suggesting a possible role as nutrient in plants (Kiferle et al., 2021).

Agronomic biofortification with iodine in cereals has been rare, especially because of the low accumulation of this element in grains due to its low translocation factor (TF). The main group of researchers on this topic has been led by Cakmak et al. (2017), where the increase in the grain of species such as maize and wheat from 20 to 126 $\mu g\ kg^{-1}$ was achieved; these experiments were carried out in various countries, and this was achieved by foliar applications of inorganic sources (KIO_3 or KI), at high concentrations ranged between 0.01% and 0.05% w/v. The aforementioned experiment has been a global milestone in relation to the iodine biofortification of cereals.

23.1.1.2 *Application of iodine and increased tolerance to environmental stress*

In relation to the improvement in the defense against environmental stress in plants, the broad redox power of iodine is related to its activity. This element has multiple valence states -1, 1, 0, 3, and 5—can act as both an electron donor and an electron scavenger, depending on the system in which it is found. In fact, iodine is considered one of the first inorganic antioxidants used by photosynthetic organisms (Venturi, 2011).

Most of the literature related to biofortification, has been carried out in vegetables, where exogenous applications of iodine increase the activity of antioxidant capacity, and both enzymes such as SOD, CAT, and POX reduce the accumulation of H_2O_2 and MDA (markers of lipid peroxidation), as well as nonenzymatic compounds such as ascorbic acid, glutathione, and total phenolic compounds, which contribute to mitigating oxidative damage induced by adverse factors such as salinity, high temperature, and drought, among others (Blasco et al., 2008; Fuentes et al., 2022; Leyva et al., 2011; Medrano Macías et al., 2021).

The evidence of the beneficial effects of iodine on tolerance to abiotic stress (e.g., salinity, oxidative stress, and drought) is clear but mostly comes from crops with high human consumption(lettuce and tomato) and laboratory studies. An extensive literature review (Medrano-Macías et al., 2016) recommends the following forms of application and concentration according to the desired effect (see Fig. 23.1), which include the activation of the antioxidant system, reduction of peroxides/reactive oxygen species, and, in some cases, better ionic balance (K^+/Na^+).

In cereals, studies on the direct impact on redox metabolism, antioxidant synthesis, and increased tolerance to environmental stress are very scarce. This means a large area of opportunity to encourage the application of this element in

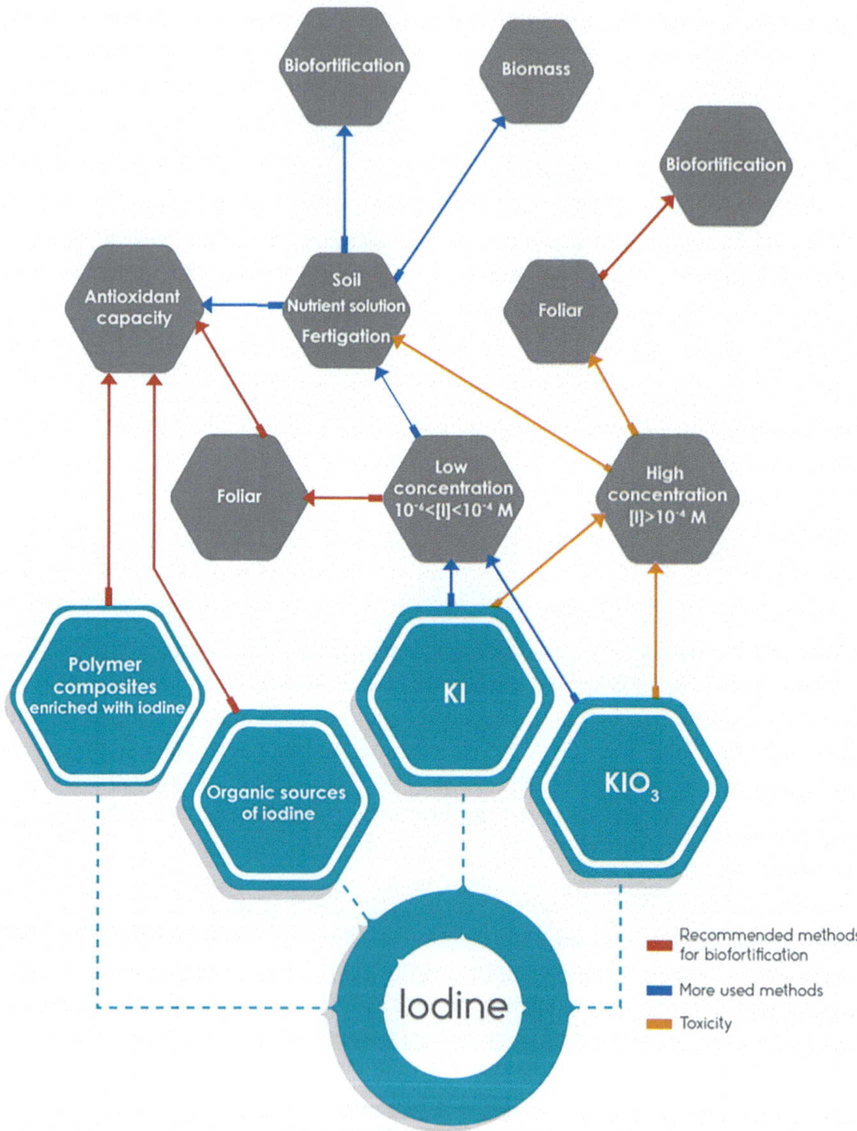

FIGURE 23.1

Recommendation of chemical species and iodine concentrations to achieve biofortification and/or biostimulation in crops.

crops such as wheat, rice, and corn, which would provide a double benefit: biofortification with a direct impact on human nutrition and an improvement in the defense against environmental stress. Table 23.1 shows a compilation of studies conducted in the past 5 years on the application of iodine to wheat, maize, and rice, the doses applied, and the reported beneficial effects. They were divided according to the type of biofortification, and some studies included biofortification accompanied by another beneficial element, such as selenium (Table 23.2).

Table 23.3

23.1.2 Silicon

23.1.2.1 Silicon biofortification in crops

Unlike what happens with iodine, biofortification with silicon has been more widely studied in cereals, especially rice (*Oryza sativa*), a model species because of its high Si accumulation capacity, which is likely due to its high expression of transporters Lsi1 and Lsi2 (Mitani-Ueno et al., 2023; Ramírez-Olvera et al., 2021), which are responsible for the transport of silicic acid (H_4SiO_4) through root membranes. Similarly, consistent studies have been carried out in wheat (*Triticum aestivum*) (Debona et al., 2014), barley (*Hordeum vulgare*) (Ali et al., 2013), maize (*Zea mays*) (Ullah et al., 2025), and sorghum (*Sorghum bicolor*) (Simón et al., 2024), where the application of silicate sources not only increased the content of the element in the grains but also increased resistance to environmental stress, especially against high salinity and drought (Biju et al., 2017). Most studies have been conducted with conventional breeding, including foliar application, fertirrigation, and soil amendments; however, notable advances have been made through the use of metabolic engineering, genetics, and nanotechnology, as shown in Table 23.4 and Table 23.5, respectively.

23.1.3 Conventional breeding applications (foliar/irrigation/amendments)

The application of fertilizers with Si represents an effective practice to increase the concentration of the element in plant tissues and, therefore improve grain yield in crops such as rice. Recent studies have shown that rice yield increases as the application dose of silicon (Si) increases, which is associated with higher concentrations in the flag leaf, straw, and husk, as well as with higher expression of the gene OsLsi6 (Chaiwong et al., 2022).

In maize, Si in the soil solution is absorbed by HvLsi1 or ZmLsi1 transporters, which are polarized in epidermal, hypodermic, and cortical cells, respectively. Unlike in rice roots, since no aerenchyma develops in the roots of maize, Si is transported to the endodermis through the symplastic pathway. The release of Si into the trail is mediated by HvLsi2 or ZmLsi2, respectively, located in the endodermis. This absorption system also results in a high accumulation of Si in the sprouts, but the efficiency is lower than that in rice. The difference in the accumulation of Si could be attributed to the level of expression and polar localization of Lsi1 and Lsi2, the Casparian strip, and the root structure (Mitani-Ueno and Ma, 2021).

Table 23.1 Studies on conventional iodine biofortification.

Crop	Type of stress	Dose/Form of biofortification	Benefit	Key references
Winter wheat	Drought	1.6 mg L^{-1} seed priming + twice application in vegetative of biologically active solution of iodine 80 mg L^{-1}	Higher grain yield under drought stress conditions	Kalenska et al. (2023)
Wheat (*Triticum aestivum* L.)	—	2.3 mM KIO_3 - Oil adjuvant foliar application	Higher concentration of iodine in grains	Magor et al. (2023)
Sweet corn	—	10 μM 5- iodine salycilic acid drench	Higher concentration of iodine in grains	Grzanka et al. (2021)
Sweet corn	—	10 μM KIO_3, KI or iodine salycilic acid + 0.1 μM NH_4VO_3	Higher iodine content in grains and increased growth, reduction in the content of essential macroelements	Grzanka et al. (2020)

Table 23.2 Studies on conventional iodine biofortification combined with selenium in crops such as wheat and maize.

Crop	Type of stress	Dose/form of biofortification	Benefit	Key references
Wheat	—	Foliar application KIO_3 0.04% Na_2SeO_4 0.001%	Increase of I and Se in grains	Ram et al. (2024)
Rice	—	Foliar application KIO_3 0.05% Na_2SeO_4 0.0024%	Increase in iodine content in grains of 12–186 μM kg^{-1} and Se 54–760 μM kg^{-}	Naeem et al. (2022)
Rice	—	Application via foliar application KIO_3 0.05% Na_2SeO_4 0.001%	Increase in iodine content in grains of 11–204 μM kg^{-1} and Se 95–380 μM kg^{-}	Prom-u-thai et al. (2020)

Table 23.3 Studies on nanotechnology and iodine biofortification in wheat and maize.

Crop	Type of stress	Dose/form of biofortification	Benefit	Key references
Wheat var. Zymoyarka	Cold and phytopathogens	Seed priming (1.69 mg L^{-1})	Potentialization of chloroplast sensitivity to oxidative stress, improvement in the antioxidant system	Huliaieva et al. (2025)
Maize seedlings	—	Seed priming 50 mg L^{-1}	Increase in seedling biomass and increase in iodine content	Marcial-Salvador et al. (2024)

Table 23.4 Studies of conventional biofortification of Si and its effects on accumulation and stress tolerance.

Crop	Type of stress	Dose/Form	benefits	Key reference
Maize	Salinity	Si (sodium metasilicate) (0, 2, 4, 6, 8 mM) root application	Na absorption was reduced in 6.69%, 20.7%, 41.12%, and 34.28% in comparasion with control plants The application of 8 mM Si significantly improved antioxidant enzymes such as SOD (50.57%), POD (15.58%), CAT (10.06%), and chlorophyll ratio (21.32%). +	Ullah et al. (2025)
Maize	Salinity	Foliar application 1–2 kg Si ha^{-1}	Improved biomass and yield, water efficiency, antioxidants.	Alayafi et al. (2022)
Maize	Drought/salt stress	Foliar application 0–2 kg Si ha^{-1}	Improved yield, grain weight, and biomass under salinity.	Alayafi et al. (2022)
Maize/ Wheat	Without stress	Ca Mg Silicate were applied at a rate of 1.76 g ha^{-1}. For silicate (equivalent to 185 kg Si ha^{-1}) (soil amendment)	The application of the silicon amendment increased the foliar chlorophyll index, agronomic efficiency, and nitrogen absorption in maize and wheat plants, which favored the development of shoots and roots and resulted in a higher grain yield (5, 2% and 7.6% increase, respectively). In addition, a total accumulation of silicon in shoots, roots and grain of 43 kg Si ha^{-1} in maize and 36 kg Si ha^{-1} in wheat was recorded.	Galindo et al. (2021)
Wheat	Terminal thermal stress	Foliar application of silicon (Si) (2 and 4 mM) in the terminal growth phase, that is, the ear stage of wheat imposed by thermal stress (37 ± 2 °C).	The application of 4 mM Si in the spike stage prominently increased chlorophyll a, b and b, and carotenoids in the leaf, improving the activities of enzymatic antioxidants (catalase, peroxidase, and superoxide dismutase) and osmoprotectants (sugar protein, soluble, and proline) under thermal stress. Foliar application of Si under control conditions improved wheat yield (chlorophyll content, carotenoids, soluble sugar, and proteins and proline).	Mustafa et al. (2021)

(Continued)

Table 23.4 Studies of conventional biofortification of Si and its effects on accumulation and stress tolerance. (*Continued*)

Crop	Type of stress	Dose/Form	benefits	Key reference
Wheat	Osmotic stress/drought	Fertilization with 100 ml of 1.5 Mm sodium metasilicate	Local varieties of wheat showed two groups: high and low Si accumulators. Osmotic stress decreased the Si content in both types of accumulation, while drought increased it.	Thorne et al. (2021)
Wheat	Drought and heat	It was applied at two levels, i.e., Si0=0 mM and Si1=4 mM. On the ground	Treatment with Si significantly increased ($p < 0.001$) the chlorophyll content in all cultivars, regardless of stress treatments, significantly increased ($p \leq 0.05$) the concentrations of all osmolytes estimated in response to stress treatments, physiological improvements (chlorophyll, RWC), antioxidants (POX, CAT, and APX), the concentration of osmolytes increased significantly by > 50%.	Ashfaq et al. (2022)
Wheat	Salinity stress/ Thermal	Si (sodium silicate) and was applied at a rate of 500 mg kg^{-1} of the soil	Significantly improved growth and yield under saline conditions. The exogenous application of Si increased the endogenous level of Si, substantially increased the antioxidant activity, and the accumulation of osmolytes. Improved wheat yield by increasing nutrient absorption, leaf water status, and photosynthetic pigments.	Aouz et al. (2023)
Wheat	Water stress	Foliar application of three forms of silicon (Si) (potassium silicate, calcium silicate and aluminum silicate, each with 0.05% Si)	The different forms of Si reduced Zn and Mn and increased Si compared to the control (without application of Si).	Salem et al. (2022)

Wheat	Salt stress	Foliar application of 0.1 and 0.2% SiO_2	Significant increase in the ICC and the selected gas exchange parameters (P_N, C_i, E, g_s) and chlorophyll fluorescence (PI, RC/ABS, F_V/F_M, Fv/F_0). The doses of Si of 0.1% and 0.2% showed a better mitigating effect compared to the dose of 0.05%. Alteration of DNA methylation in plants as a response to experimental factors. The changes in methylation depended on the dose of silicon.	Tobiasz-Salach et al. (2024)
Rice	Without stress	Silicate (soil amendment) 50–300 kg Si/ha	Higher concentration of Si in flag leaf, husk, and straw; up to +137% total absorption and positive correlation with grain yield and straw weight. The increase in the rate of Si fertilizer significantly increased the level of expression of *OsLsi6*.	Chaiwong et al. (2022)

Table 23.5 Metabolic engineer studies of Si and its effects on accumulation and stress tolerance in rice.

Crop	Experimental condition	Dose/form	Main benefits	Key reference
Rice	Addition to soil	Five fertilization treatments: no application of phosphorus, low phosphorus and normal phosphorus (0, 25, 75 kg·ha^{-1} P_2O_5) (0 P, 25 P, 75 P), along with two levels of silicon (0, 45 kg·ha^{-1} SiO_2), resulting in treatments 0 P, 0PSi, 25 P, 25PSi, and 75 P.	The addition of silicon to the rice crop alleviated the phosphorus limitations by reducing the consumption of ATP in the roots through a decrease in the content of ATPase and P-ATPase. This also minimized the excessive transport of NSC to the roots, thus promoting the growth of the shoots by downregulating *SUT1*, *SWEET11*, *SUS2,* and *CIN2*. In addition to optimizing the root-to-shoot ratio and providing sufficient energy, it increased root volume and upregulated *OsPT2*, *OsPT4,* and *OsPT8*, thus promoting phosphorus absorption	Jiang et al. (2025)
Rice "Suijing 18"	"Changes in plant growth, exchange of photosynthetic gases and oxidative stress of "Suijing 18" rice crops under dry cultivation have been investigated."	The Russian mineral silica Biori was applied as a base fertilizer with an effective silica content of ≥72% at 45 kg · ha^{-1}.	Si improved the photosynthetic capacity, the antioxidant enzymatic activity, reduced the lipid peroxidation of the rice seedlings, promoted the LAI, and promoted the growth of the leaves in rainfed cultivation. – Transcriptomics combined with quasitargeted metabolomics detected 1416 and 520 differentially expressed genes (DEG), 38 and 41 differentially	Jiang et al. (2022)

			accumulated metabolites (DAM) in the leaves and roots of rice, respectively. Among them, 13 DEG participated in the biosynthesis of flavonoids, promoting the accumulation of flavonoids, anthocyanins, and flavonols in the roots and leaves of rice. Meanwhile, 14 DEG participated in photosynthesis, promoting the responses of photosystem I and photosystem II, increasing the abundance of metabolites in the leaves. – 24 AMDs involved in osmoregulatory processes were identified, significantly increasing the amino acids and carbohydrates and their derivatives in the roots.	

Among the silicon pathways supply, foliar application has proven to be particularly effective because it facilitates the rapid absorption of the element through the cuticle and stomata. In this context, the foliar application of silicon promotes an increase in the water content in leaves, improving their morphological characteristics and the conduction capacity of vascular bundles, in addition to increasing the volume of palisade tissue, chlorophyll content, and photosynthetic efficiency (Shen et al., 2022). The application of exogenous silicon can form a silica cuticle layer in the epidermal tissue of the leaf, improving the water status of the tissues in wheat crops (Lamlom et al., 2024). Silicic acid applied foliarly is a better source of silicon than silicates because it is the direct source of silicon available to plants. For most crops, the recommended concentration of silicic acid applied foliarly ranges between 2 and 4 mL per liter of water (Wadas and Kondraciuk, 2025). See Table 23.4.

23.1.4 Metabolic engineering

The nutritional quality of crops has improved through hybrid breeding and biofortification, which is historically defined as the improvement of nutrients through nontransgenic breeding, but it has now been expanded to include strategies of gene editing and metabolic engineering, although its effectiveness has been limited because of the long growth cycles and the drag of genetic linkage and epistatic interactions (Chai et al., 2025). In Table 23.5, some findings about Si application and its effects on alleviating the adverse conditions in rice are presented.

23.1.5 Genetic engineering (CRISPR-Cas)

Gene editing using candidate genes such as the Lsi1 gene in a reference wheat genome could be a possible strategy to induce high absorption of Si (Christian et al., 2022). Recent studies have shown that SIET4 functions as an efflux transporter of Si and is necessary for the adequate deposition of Si in specific tissues and cells because it exports Si from the cells of the leaves to the apoplastic space for final deposition. This is supported by several lines of evidence: SIET4 is polarized in the epidermal cells and the neighboring cells of the bulliform cells of the leaves (Mitani-Ueno et al., 2023). In Table 23.6 are shown some studies on gene editing related to the regulation of absorption and transport of Si and its effect on accumulation and stress tolerance in rice.

23.1.6 Nanotechnology (SiNPs, nanosilica)

Silicon nanoparticles (Si-NPs) have demonstrated considerable potential to enhance crop yield while alleviating the effects of both biotic and abiotic stresses. They can be applied via foliar spraying, directly to the root zone, or through seed priming.

Table 23.6 Genetic engineering studies of Si and its effects on accumulation and stress tolerance in rice.

Crop	Experimental condition	Dose/Form	Main benefits	Key reference
Rice (*Oryza sativa*), Nipponbare variety, along with knockout lines of the SSS gene and some transgenic lines	The work focuses on the regulation of Si transport and how the Shoot-Silicon-Signal (SSS) gene from the "shoot" modulates the uptake in the root.	Silicic acid (Si) of different concentrations: 0, 0.1, 0.2, 0.5, 1.0, 1.5, and 2.0 mM for dose–response tests on vegetative shoots in the nutrient solution.	– In the *SSS* mutants (without the functional SSS gene), the uptake and transport of Si to the leaves were lower than in the wild variety. - Grain production in mutant lines was drastically reduced (to 1/3 of the yield of the wild type) due to insufficient accumulation of Si. – The concentration of Si in the husk of the grains and in the flag leaf was significantly lower in mutants compared to the wild variety. – When they expressed ectopic SSS under the *Lsi1* promoter, the expression of transporters Lsi1 and Lsi2 increased, and Si uptake also improved (up to double that of the wild type under controlled conditions).	Yamaji et al. (2024)
Rice	OsLsi1/OsLsi2 transporters were edited to reduce the accumulation of arsenic with Edit CRISPR, without affecting performance. Arsenite [As(III)], the main form of As in the pore water of rice soil, is inadvertently absorbed by OsLsi1 and OsLsi2, the two key transporters for the absorption of silicon (Si) in the roots of rice. They investigated whether the editing of *OsLsi1* or *OsLsi2* can decrease the accumulation of As in the rice grain without compromising grain yield. They used CRISPR-Cas9 technology to edit the promoter region of *OsLsi1* and the C-terminal coding sequence of *OsLsi1* and *OsLsi2*, and generated a total of 27 mutants.	The absorption and accumulation of Si and As were evaluated both in short-term hydroponic experiments and in a rice field.	The removal of 1.2–2 kb of the *OsLsi1* promoter markedly suppressed the expression of *OsLsi1* in the roots and the uptake of Si and did not affect the uptake of As (III) or the concentration of As in grain. Some of the mutants of the coding sequence *OsLsi1* and *OsLsi2* showed large decreases in the absorption of Si and As (III), as well as large decreases in the accumulation of Si in rice husks. However, only *OsLsi2* mutants showed significant decreases (up to 63%) in the total concentration of As in grain. The editing of *OsLsi2* mainly affected the accumulation of inorganic As in the rice grain with little effect on the accumulation of dimethylarsenate (DMA). The grain yields of the *OsLsi2* mutants were comparable to those of the wild type. The editing of *OsLsi2* provided a promising way to reduce the accumulation of As in the rice grain without compromising grain yield(Xu et al., 2024).	Xu et al. (2024)

(*Continued*)

Table 23.6 Genetic engineering studies of Si and its effects on accumulation and stress tolerance in rice. (*Continued*)

Crop	Experimental condition	Dose/Form	Main benefits	Key reference
Wild-type rice (WT, cv. Nipponbare)	Functionally characterized: SIET4, shares 49% identity with Lsi2. To understand the role of this gene, they generated knockout lines using the CRISPR/Cas9 technique. They used two mutants; siet4-1 and siet4-2, which had a 1-bp insertion in the first and second exon, respectively	Seedlings (16 days old) of WT and two mutants were cultured in a Kimura B solution of medium concentration containing 1 mM Si as silicic acid.	The elimination of the SIET4 gene caused the death of rice in the presence of Si, both in soil and in nutrient solution, although it did not affect its growth in the absence of this element. The lack of SIET4 prevented the export of Si to the apoplast, causing its abnormal accumulation in the mesophyll cells instead of the epidermal cells, which altered the metabolism and led to the death of the plant. The transcriptomic analysis showed that, after 24 h of Si supply, more than a thousand genes were differentially expressed in the mutant with respect to the WT, many associated with responses to biotic and abiotic stress, indicating an unbalanced and dysfunctional growth of rice.	Mitani-Ueno et al. (2023)

When applied foliarly, Si-NPs can penetrate leaf tissues and be translocated to different plant organs through the cuticle or stomata (Wang et al., 2022).

As an environmentally friendly alternative to conventional chemical fertilizers, herbicides, and pesticides, Si-NPs stimulate seed germination, plant growth, biomass accumulation, and nutrient uptake due to their small size, high surface area, and efficient cellular penetration. Furthermore, these nanoparticles mitigate salt stress by modulating gene expression, thereby activating antioxidant enzymes such as superoxide dismutase (SOD), catalase (CAT), and ascorbate peroxidase (APX), which collectively reduce reactive oxygen species (ROS) accumulation.

An optimal concentration of 10 mg L^{-1} SiO_2 nanoparticles was reported to improve relative water content, photosynthetic efficiency, and overall growth parameters, while simultaneously reducing membrane damage and lipid peroxidation (Daler et al., 2024). Additionally, treatments with higher nanosilica concentrations (100–300 mg L^{-1}) have also been shown to enhance salinity stress tolerance (Siddiqui et al., 2020). Table 23.7 shows some advances in Si application and its effect on stress tolerance.

23.1.6.1 Application of silicon and increased tolerance to environmental stress

Silicon applied exogenously is deposited in cell walls and creates a mechanical barrier against external factors (Diana et al., 2024). Additionally, it plays an important role in the regulation of water management in plants, as Si improves the activity of aquaporins by upregulating the expression of PIP genes and relieving the inhibition of aquaporin activity induced by ROS. In addition, an increase in the accumulation of soluble sugars and/or amino acids in the xylem sap through osmoregulation has been observed. The accumulation of osmolytes in the xylem sap increases the osmotic motive force. Consequently, silicon can alter the growth of the roots and increase the root/shoot ratio, which, together with increasing the activity of the aquaporins and the osmotic motive force, contributes to improving the hydraulic conductance of the roots (Saja-Garbarz et al., 2024). Similarly, Si reduces foliar transpiration through cuticular deposition, strengthening resistance to water stress (Vandegeer et al., 2021).

A wide range of genes were modulated in response to abiotic stress factors after Si supplementation (Fig. 23.2). Mechanism by which silicon regulates abiotic stress. (A) Silicon is absorbed through anion transporters. (B) Drought and salinity stress induce osmotic stress and osmotic imbalance. Si improves the synthesis of osmoprotectants such as proline, glycine, and betaine. (C) Tension induces the formation of ROS. It increases the expression of CAT, SOD, and POD to protect against oxidative damage. (D-F) Si supplementation results in the transcriptional regulation of genes related to photosynthesis, such as *PsaH*, which encodes essential polypeptide subunits of the dimer of photosystem I (PSI), the gene *PsbY* (*Os08g02530*), which encodes the polyprotein component of photosystem II, and the *PetC* gene, which encodes the binding polypeptide to the Fe-S Rieske center

Table 23.7 Studies on nanotechnology and silicon biofortification in wheat, maize, and rice.

Crop	Type of stress	Dose/form	Main benefits	Key reference
Wheat	Salt stress	Foliar application with 0 (distilled water), nanochitosan (NC) (0.05% w/v), nanosilicon (NS) (0.05% w/v) and NC-NS (nanochitosan encapsulated in nanosilicon) (0.1% w/v).	NC-NS was the most effective treatment to reduce the accumulation of H_2O_2 in the leaves and roots of plants under salt stress. The SOD activity of plants treated with salt stress and NC, NS, and NC-NS exhibited the same trends. Both salt stress and the application of NC, NS, and NC-NS significantly improved the activity of CAT and APX enzymes in leaves and roots, and their highest activity was observed in the application of NC-NS in the presence of NaCl. In parallel with phenols, flavonoids and anthocyanins, the application of NC, NS, and NC-NS triggered the biosynthesis of other organic antioxidants, such as proline, amino acids, and α-tocopherol.	Hajihashemi and Kazemi (2022)
Maize	Salt stress	10 ppm Zn and 90 ppm Si, applied in nano form (ZnO and SiO_2) and conventional form ($ZnSO_4$ and K_2SiO_3); treatments with and without salt stress (100 mM NaCl), in plants grown with Hoagland solution.	Nano-Si shows a prominent role in the modulation of OP and the limitation of Na accumulation along with a greater accumulation of Zn and Si in plants. Nano-Si: increased chlorophyll content (SPAD) +90% in control and +203% under salt stress.	Shoukat et al. (2024)
Maize	Salt stress	Abiogenic sources of Si, including silicic acid (SA), sodium silicate (Na-Si), potassium silicate (K-Si), and silicon nanoparticles (NPs-Si) were applied in saline-sodic soils.	The application of NPs-Si and K-Si improved plant growth by increasing the availability of nutrients such as P and potassium (K), physiological attributes, and by reducing salt stress and cationic ratios in maize after rotation. of maize crops.	Rizwan et al. (2023)

Rice	Salt stress	Foliar application of SiNP 20 mg L^{-1}	The exogenous SiNPs alleviated the toxic effects of salt stress and promoted the growth of N-22 and Super-Bas, the chlorophyll content (16% and 13%), the carotenoids (15% and 11%), the total protein content (21% and 18%), and the activities of antioxidant enzymes. Real-time quantitative PCR expression analysis showed that SiNPs relieved plants of oxidative bursts by triggering the expression of HKT genes.	Ijaz et al. (2023)
Rice	Water stress	Nanosilicon (at a ratio of 0.4 g/L) as silicon nanodioxide particle size of 20 to 30 nm with a purity of 99.5%, compared to an untreated control.	The foliar application of nanosilicon substantially mitigated the harmful effects of drought stress and significantly improved photosynthetic pigments (it improved chlorophyll a, chlorophyll b, and carotenoids by 21.1, 17.6, and 9.5%), antioxidant enzymes, growth parameters, and traits that contribute to yield compared to the untreated stress control.	Abd-El-Aty et al. (2024)

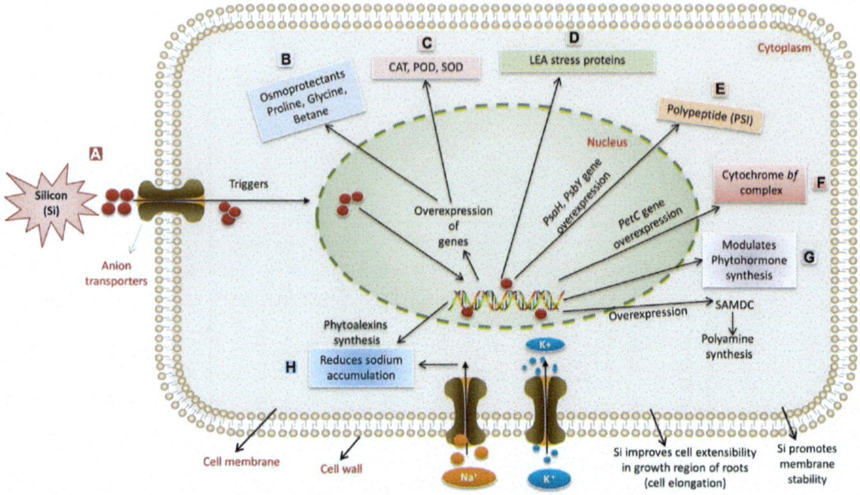

FIGURE 23.2

Representation of the mechanism by which silicon regulates abiotic stress (Mir et al., 2022). From absorption (A), gene regulations (B-G), and phytoalexins synthesis (H).

of the cytochrome bf complex. (G) Si modulates the synthesis of plant growth regulators to relieve stress. It induces the expression of the S-adenosyl-L-methionine decarboxylase (SAMDC) gene, which encodes essential enzymes responsible for synthesizing polyamines. (H) Salinity stress negatively affects plant growth by increasing the accumulation of ions to toxic levels. Owing to a greater influx of NaCl, plants experience more severe oxidative and ionic stress. The accumulation of sodium ions to toxic levels triggers the production of ROS, which, in turn, severely damages cellular components and accelerates senescence in mature leaves, reducing growth and metabolism. The administration of Si improves the absorption of K+, which, in turn, stimulates H+-ATPase enzymes in the plasmatic membrane to overcome salt stress (Mir et al., 2022).

23.2 Conclusions

This chapter reviewed the current state of knowledge on biofortification with iodine and silicon in the three major staple crops that feed the majority of the global population: wheat, rice, and maize. Four types of biofortification were analyzed: conventional breeding, metabolic engineering, genetic engineering, and nanobiotechnology. Tables show the tested doses and the benefits that have been obtained, emphasizing the increase in tolerance to several types of environmental stress.

Conventional breeding for biofortification with iodine is usually carried out foliarly by nutrient solution or seed priming, and no information on biofortification in rice cultivation has available conducted in the past 5 years. With respect to nanobiofortification with iodine, the main form of application is seed priming in wheat and maize. With respect to biofortification with silicon, the bulk of the information is collected from the wheat crop, mainly by foliar and soil application. With respect to this element, several findings have been made in biofortification through metabolic engineering and genetic engineering in rice cultivation. Finally, nanobiofortification was performed by foliar application in wheat and maize, without information on rice.

23.3 Future considerations

Crop biofortification is a dynamic and constantly evolving strategy that seeks to respond to current challenges, including changes in environmental conditions, population growth, and the need to produce plant-based foods with high nutritional quality. According to the findings presented in this chapter, it is essential to continue expanding research and developing alternatives to improve the production of rice, wheat, and maize, both in terms of their tolerance to adverse environmental conditions and in terms of their content of beneficial elements such as silicon and iodine. Advanced biofortification strategies, supported by metabolic, genetic engineering, and nanotechnology, have the potential to offer these benefits and contribute significantly to future food security.

References

Abdalla, N., Ragab, M.I., Fári, M., El-Ramady, H., Alshaal, T., Elhawat, N., et al., 2019. Nanobiotechnology for plants: needs and risks. Environment, Biodiversity and Soil Security 2, 155–174. Available at https://doi.org/10.21608/jenvbs.2018.6711.1041.

Abd-El-Aty, M.S., Kamara, M.M., Elgamal, W.H., Mesbah, M.I., Abomarzoka, E.S.A., Alwutayd, K.M., et al., 2024. Exogenous application of nano-silicon, potassium sulfate, or proline enhances physiological parameters, antioxidant enzyme activities, and agronomic traits of diverse rice genotypes under water deficit conditions. Heliyon 10 (5). Available at https://doi.org/10.1016/j.heliyon.2024.e26077.

Alayafi, A.H., Al-Solaimani, S.G.M., Abd El-Wahed, M.H., Alghabari, F.M., Sabagh, A.E., 2022. Silicon supplementation enhances productivity, water use efficiency and salinity tolerance in maize. Frontiers in Plant Science 13. Available at https://doi.org/10.3389/fpls.2022.953451.

Ali, S., Farooq, M.A., Yasmeen, T., Hussain, S., Arif, M.S., Abbas, F., et al., 2013. The influence of silicon on barley growth, photosynthesis and ultra-structure under chromium stress. Ecotoxicology and Environmental Safety 89, 66–72. Available at https://doi.org/10.1016/j.ecoenv.2012.11.015.

Aouz, A., Khan, I., Chattha, M.B., Ahmad, S., Ali, M., Ali, I., et al., 2023. Silicon induces heat and salinity tolerance in wheat by increasing antioxidant activities, photosynthetic activity, nutrient homeostasis, and osmo-protectant synthesis. Plants 12 (14), 2606. Available at https://doi.org/10.3390/plants12142606.

Ashfaq, W., Fuentes, S., Brodie, G., Gupta, D., 2022. The role of silicon in regulating physiological and biochemical mechanisms of contrasting bread wheat cultivars under terminal drought and heat stress environments. Frontiers in Plant Science 13. Available at https://doi.org/10.3389/fpls.2022.955490.

Biju, S., Fuentes, S., Gupta, D., 2017. Silicon improves seed germination and alleviates drought stress in lentil crops by regulating osmolytes, hydrolytic enzymes and antioxidant defense system. Plant Physiology and Biochemistry 119, 250–264. Available at https://doi.org/10.1016/j.plaphy.2017.09.001.

Blasco, B., Rios, J.J., Cervilla, L.M., Sánchez-Rodrigez, E., Ruiz, J.M., Romero, L., 2008. Iodine biofortification and antioxidant capacity of lettuce: potential benefits for cultivation and human health. Annals of Applied Biology 152 (3), 289–299. Available at https://doi.org/10.1111/j.1744-7348.2008.00217.x.

Cakmak, I., Prom-u-thai, C., Guilherme, L.R.G., Rashid, A., Hora, K.H., Yazici, A., et al., 2017. Iodine biofortification of wheat, rice and maize through fertilizer strategy. Plant and Soil 418 (1-2), 319–335. Available at https://doi.org/10.1007/s11104-017-3295-9.

Chai, N., Xu, J., Zhang, R., Li, G., Wen, J., Su, L., et al., 2025. Synthetic metabolic engineering of functional crops: boosting nutrition and human health. The Crop Journal 14 (1), 8–21. Available at https://doi.org/10.1016/j.cj.2025.06.003.

Chaiwong, N., Pusadee, T., Jamjod, S., Prom-u-thai, C., 2022. Silicon application promotes productivity, silicon accumulation and upregulates silicon transporter gene expression in rice. Plants 11 (7), 989. Available at https://doi.org/10.3390/plants11070989.

Christian, M.M., Shimelis, H., Laing, M.D., Tsilo, T.J., Mathew, I., 2022. Breeding for silicon-use efficiency, protein content and drought tolerance in bread wheat (*Triticum aestivum* L.): a review. Acta Agriculturae Scandinavica Section B: Soil and Plant Science 72 (1), 17–29. Available at https://doi.org/10.1080/09064710.2021.1984564.

Daler, S., Kaya, O., Canturk, S., Korkmaz, N., Kılıç, T., Karadağ, A., Hatterman-Valenti, H., 2024. Silicon nanoparticles (SiO2 NPs) boost drought tolerance in grapevines by enhancing some morphological, physiological, and biochemical traits. Plant Molecular Biology Reporter 43, 1057–1075. Available at 10.1007/s11105-024-01520-y.

Debona, D., Rodrigues, F.A., Rios, J.A., Nascimento, K.J.T., Silva, L.C., 2014. The effect of silicon on antioxidant metabolism of wheat leaves infected by P yricularia oryzae. Plant Pathology 63 (3), 581–589. Available at https://doi.org/10.1111/ppa.12119.

Diana, S.G., Kamila, G.J., Ewa, K., Małgorzata, K.K., Monika, T., Emilia, G., et al., 2024. The effect of silicon supplementation and drought stress on the deposition of callose and chemical components in the cell walls of the Brassica napus roots. BMC Plant Biology 24, 1249. Available at https://doi.org/10.1186/s12870-024-05967-9.

Duborská, E., Šebesta, M., Matulová, M., Zvěřina, O., Urík, M., 2022. Current strategies for selenium and iodine biofortification in crop plants. Nutrients 14 (22), 4717. Available at https://doi.org/10.3390/nu14224717.

Fuentes, J.E.G., Castellanos, B.F.H., Martínez, E.N.R., Ortiz, W.A.N., Mendoza, A.B., Macías, J.M., 2022. Outcomes of foliar iodine application on growth, minerals and antioxidants in tomato plants under salt stress. Folia Horticulturae 34 (1), 27–37. Available at https://doi.org/10.2478/fhort-2022-0003.

Galindo, F.S., Pagliari, P.H., Rodrigues, W.L., Fernandes, G.C., Boleta, E.H.M., Santini, J.M.K., et al., 2021. Silicon amendment enhances agronomic efficiency of nitrogen fertilization in maize and wheat crops under tropical conditions. Plants 10 (7), 1329. Available at https://doi.org/10.3390/plants10071329.

García Osuna, H.T., Mendoza, A.B., Morales, C.R., Rubio, E.M., Star, J.V., Ruvalcaba, R.M., 2014. Iodine application increased ascorbic acid content and modified the vascular tissue in Opuntia Ficus-Indica L. Pakistan Journal of Botany 46 (1), 127–134.

Gonzali, S., Kiferle, C., Perata, P., 2017. Iodine biofortification of crops: agronomic biofortification, metabolic engineering and iodine bioavailability. Current Opinion in Biotechnology 44, 16–26. Available at https://doi.org/10.1016/j.copbio.2016.10.004.

Grzanka, M., Smolen, S., Kováčik, P., 2020. Effect of vanadium on the uptake and distribution of organic and inorganic forms of iodine in sweetcorn plants during early-stage development. Agronomy 10 (11), 1666. Available at https://doi.org/10.3390/agronomy10111666.

Grzanka, M., Smoleń, S., Skoczylas, Ł., Grzanka, D., 2021. Biofortification of sweetcorn with iodine: interaction of organic and inorganic forms of iodine combined with vanadium. Agronomy 11 (9), 1720. Available at https://doi.org/10.3390/agronomy11091720.

Hajihashemi, S., Kazemi, S., 2022. The potential of foliar application of nano-chitosan-encapsulated nano-silicon donor in amelioration the adverse effect of salinity in the wheat plant. BMC Plant Biology 22, 148. Available at https://doi.org/10.1186/s12870-022-03531-x.

Huliaieva, H., Tokovenko, I., Kharchuk, M., Bohdan, M., 2025. Plant nano biology the influence of iodine nano citrates on juvenile wheat plants at phytopathogenic infection and cooling stress. Plant Nano Biology. 13, 100175.

Ijaz, U., Ahmed, T., Rizwan, M., Noman, M., Shah, A.A., Azeem, F., et al., 2023. Rice straw based silicon nanoparticles improve morphological and nutrient profile of rice plants under salinity stress by triggering physiological and genetic repair mechanisms. Plant Physiology and Biochemistry 201. Available at https://doi.org/10.1016/j.plaphy.2023.107788.

Jiang, H., Song, Z., Su, Q.W., Wei, Z.H., Li, W.C., Jiang, Z.X., et al., 2022. Transcriptomic and metabolomic reveals silicon enhances adaptation of rice under dry cultivation by improving flavonoid biosynthesis, osmoregulation, and photosynthesis. Frontiers in Plant Science 13. Available at https://doi.org/10.3389/fpls.2022.967537.

Jiang, H., Li, W., Jiang, Z., Li, Y., Shen, X., Nuo, M., et al., 2025. Silicon enhanced phosphorus uptake in rice under dry cultivation through root organic acid secretion and energy distribution in low phosphorus conditions. Frontiers in Plant Science 16. Available at https://doi.org/10.3389/fpls.2025.1544893.

Kalenska, S., Falko, G., Antal, T., Hordyna, O., Fediv, R., 2023. Iodine-containing preparations in grain growing technologies. Plant and Soil Science 14 (2). Available at https://doi.org/10.31548/plant2.2023.33.

Kiferle, C., Gonzali, S., Holwerda, H.T., Ibaceta, R.R., Perata, P., 2013. Tomato fruits: a good target for iodine biofortification. Frontiers in Plant Science 4. Available at https://doi.org/10.3389/fpls.2013.00205.

Kiferle, C., Martinelli, M., Salzano, A.M., Gonzali, S., Beltrami, S., Salvadori, P.A., et al., 2021. Evidences for a nutritional role of iodine in plants. Frontiers in Plant Science 12. Available at https://doi.org/10.3389/fpls.2021.616868.

Lamlom, S.F., Abdelghany, A.M., Ren, H., Ali, H.M., Usman, M., Shaghaleh, H., et al., 2024. Revitalizing maize growth and yield in water-limited environments through silicon and zinc foliar applications. Heliyon 10 (15). Available at https://doi.org/10.1016/j.heliyon.2024.e35118.

Ledwożyw-Smoleń, I., Pitala, J., Smoleń, S., Liszka-Skoczylas, M., Kováčik, P., 2023. Iodine Biofortification of Dandelion Plants (Taraxacum officinale F.H. Wiggers Coll.) with the Use of Inorganic and Organic Iodine Compounds. Molecules (Basel, Switzerland) 28 (15), 5638. Available at https://doi.org/10.3390/molecules28155638.

Leyva, R., Sánchez-Rodríguez, E., Ríos, J.J., Rubio-Wilhelmi, M.M., Romero, L., Ruiz, J.M., et al., 2011. Beneficial effects of exogenous iodine in lettuce plants subjected to salinity stress. Plant Science 181 (2), 195–202. Available at https://doi.org/10.1016/j.plantsci.2011.05.007.

Lyons, G., 2018. Biofortification of cereals with foliar selenium and iodine could reduce hypothyroidism. Frontiers in Plant Science 9. Available at https://doi.org/10.3389/fpls.2018.00730.

Marcial-Salvador, L., Medrano-Macías, J., Enríquez-Medrano, F.J., Benavides-Mendoza, A., 2024. Synthesis and characterization of submicron iodinated polymer particles with biostimulant effect on the germination and vigor of corn seedlings. Revista Chapingo, Serie Horticultura 30 (3), 47–60. Available at https://doi.org/10.5154/r.rchsh.2024.04.004.

Magor, E., Wilson, M.D., Wong, H., Cresswell, T., Sánchez-Palacios, J., Bell, R.W., Penrose, B., 2023. Selected adjuvants increase the efficacy of foliar biofortification of iodine in bread wheat (Triticum aestivum L.) grain. Frontiers in Plant Science 14. Available at 10.3389/fpls.2023.1246945.

Medrano Macías, J., López Caltzontzit, M.G., Rivas Martínez, E.N., Narváez Ortiz, W.A., Benavides Mendoza, A., Martínez Lagunes, P., 2021. Enhancement to salt stress tolerance in strawberry plants by iodine products application. Agronomy 11 (3), 602. Available at https://doi.org/10.3390/agronomy11030602.

Medrano-Macías, J., Leija-Martínez, P., González-Morales, S., Juárez-Maldonado, A., Benavides-Mendoza, A., 2016. Use of iodine to biofortify and promote growth and stress tolerance in crops. Frontiers in Plant Science 7 (2016). Available at https://doi.org/10.3389/fpls.2016.01146.

Mir, R.A., Bhat, B.A., Yousuf, H., Islam, S.T., Raza, A., Rizvi, M.A., et al., 2022. Multidimensional role of silicon to activate resilient plant growth and to mitigate abiotic stress. Frontiers in Plant Science 13. Available at https://doi.org/10.3389/fpls.2022.819658.

Mitani-Ueno, N., Ma, J.F., 2021. Linking transport system of silicon with its accumulation in different plant species. Soil Science and Plant Nutrition 67 (1), 10–17. Available at https://doi.org/10.1080/00380768.2020.1845972.

Mitani-Ueno, N., Yamaji, N., Huang, S., Yoshioka, Y., Miyaji, T., Ma, J.F., 2023. A silicon transporter gene required for healthy growth of rice on land. Nature Communications 14 (1). Available at https://doi.org/10.1038/s41467-023-42180-y.

Mustafa, T., Sattar, A., Sher, A., Ul-Allah, S., Ijaz, M., Irfan, M., et al., 2021. Exogenous application of silicon improves the performance of wheat under terminal heat stress by triggering physio-biochemical mechanisms. Scientific Reports 11 (1). Available at https://doi.org/10.1038/s41598-021-02594-4.

Naeem, A., Aslam, M., Ahmad, M., Asif, M., Yazici, M.A., Cakmak, I., et al., 2022. Biofortification of diverse basmati rice cultivars with iodine, selenium, and zinc by individual and cocktail spray of micronutrients. Agronomy 12 (1). Available at https://doi.org/10.3390/agronomy12010049.

Pereira, C., Dias, M.I., Petropoulos, S.A., Plexida, S., Chrysargyris, A., Tzortzakis, N., et al., 2019. The effects of biostimulants, biofertilizers and water-stress on nutritional value and chemical composition of two spinach genotypes (Spinacia oleracea L.). Molecules (Basel, Switzerland) 24 (24). Available at https://doi.org/10.3390/molecules24244494.

Prom-u-thai, C., Rashid, A., Ram, H., Zou, C., Guilherme, L.R.G., Corguinha, A.P.B., et al., 2020. Simultaneous biofortification of rice with zinc, iodine, iron and selenium through foliar treatment of a micronutrient cocktail in five countries. Frontiers in Plant Science 11. Available at https://doi.org/10.3389/fpls.2020.589835.

Ram, H., Naeem, A., Rashid, A., Kaur, C., Ashraf, M.Y., Malik, S.S., et al., 2024. Agronomic biofortification of genetically biofortified wheat genotypes with zinc, selenium, iodine, and iron under field conditions. Frontiers in Plant Science 15. Available at https://doi.org/10.3389/fpls.2024.1455901.

Ramírez-Olvera, S.M., Trejo-Téllez, L.I., Gómez-Merino, F.C., Ruíz-Posadas, L.D.M., Alcántar-González, E.G., Saucedo-Veloz, C., 2021. Silicon stimulates plant growth and metabolism in rice plants under conventional and osmotic stress conditions. Plants 10 (4). Available at https://doi.org/10.3390/plants10040777.

Rizwan, A., Zia-ur-Rehman, M., Rizwan, M., Usman, M., Anayatullah, S., Areej, Alharby, H.F., et al., 2023. Effects of silicon nanoparticles and conventional Si amendments on growth and nutrient accumulation by maize (Zea mays L.) grown in saline-sodic soil. Environmental Research 227. Available at https://doi.org/10.1016/j.envres.2023.115740.

Saja-Garbarz, D., Libik-Konieczny, M., Janowiak, F., 2024. Silicon improves root functioning and water management as well as alleviates oxidative stress in oilseed rape under drought conditions. Frontiers in Plant Science 15. Available at https://doi.org/10.3389/fpls.2024.1359747.

Salem, E.M.M., Kenawey, M.K.M., Saudy, H.S., Mubarak, M., 2022. Influence of silicon forms on nutrients accumulation and grain yield of wheat under water deficit conditions. Gesunde Pflanzen 74 (3), 539–548. Available at https://doi.org/10.1007/s10343-022-00629-y.

Sánchez-Acosta, J., Galindo-Guzmán, A.P., Fortis-Hernández, M., Preciado-Rangel, P., Sánchez-Estrada, A., Estrada-Arellano, E., 2025. Iodine biofortification impacts the nutraceutical compounds content and agronomic characteristics of lettuce (*Lactuca sativa* L.). Notulae Botanicae Horti Agrobotanici Cluj-Napoca, 53 (2). Available at https://doi.org/10.15835/nbha53214225.

Shen, Z., Cheng, X., Li, X., Deng, X., Dong, X., Wang, S., et al., 2022. Effects of silicon application on leaf structure and physiological characteristics of *Glycyrrhiza uralensis* Fisch. and *Glycyrrhiza inflata* Bat. under salt treatment. BMC Plant Biology 22 (1). Available at https://doi.org/10.1186/s12870-022-03783-7.

Shoukat, A., Pitann, B., Hossain, M.S., Saqib, Z.A., Nawaz, A., Mühling, K.H., 2024. Zinc and silicon fertilizers in conventional and nano-forms: mitigating salinity effects in maize (*Zea mays* L. Journal of Plant Nutrition and Soil Science 187 (5), 678–689. Available at https://doi.org/10.1002/jpln.202300267.

Siddiqui, H., Ahmed, K.B.M., Sami, F., Hayat, S., 2020. Silicon nanoparticles and plants: Current knowledge and future perspectives. Sustainable Agriculture Reviews 41, 129–142. Available at 10.1007/978-3-030-33996-8_7.

Simón, L.A.Y., Viciedo, D.O., Carbonari, C.A., Duke, S.O., de Carvalho, L.B., 2024. Silicon treatment on sorghum plants prior to glyphosate spraying: effects on growth, nutrition, and metabolism. AgriEngineering 6 (4), 3538–3552. Available at https://doi.org/10.3390/agriengineering6040201.

Smoleń, S., Sady, W., Ledwozyw-Smoleń, I., Strzetelski, P., Liszka-Skoczylas, M., Rozek, S., 2014. Quality of fresh and stored carrots depending on iodine and nitrogen fertilization. Food Chemistry 159, 316–322. Available at https://doi.org/10.1016/j.foodchem.2014.03.024.

Thorne, S.J., Hartley, S.E., Maathuis, F.J.M., 2021. The effect of silicon on osmotic and drought stress tolerance in wheat landraces. Plants 10 (4). Available at https://doi.org/10.3390/plants10040814.

Tobiasz-Salach, R., Stadnik, B., Mazurek, M., Buczek, J., Leszczyńska, D., 2024. Foliar application of silicon influences the physiological and epigenetic responses of wheat grown under salt stress. International Journal of Molecular Sciences 25 (24), 13297. Available at https://doi.org/10.3390/ijms252413297.

Topcuoğlu, B., Ulukapı, K., Nasırcılar, A.G., 2024. Iodine Applications in Onion Cultivation, pp. 23–26. https://eares.org/siteadmin/upload/5701C0924135.pdf

Ullah, M.S., Mahmood, A., Alawadi, H.F.N., Seleiman, M.F., Khan, B.A., Javaid, M.M., et al., 2025. Silicon-mediated modulation of maize growth, metabolic responses, and antioxidant mechanisms under saline conditions. BMC Plant Biology 25 (1). Available at https://doi.org/10.1186/s12870-024-06013-4.

Vandegeer, R.K., Zhao, C., Cibils-Stewart, X., Wuhrer, R., Hall, C.R., Hartley, S.E., et al., 2021. Silicon deposition on guard cells increases stomatal sensitivity as mediated by K+ efflux and consequently reduces stomatal conductance. Physiologia Plantarum 171 (3), 358–370. Available at https://doi.org/10.1111/ppl.13202.

Venturi, S., 2011. Evolutionary significance of iodine. Current Chemical Biology 5 (3), 155–162. Available at https://doi.org/10.2174/187231311796765012.

Wadas, W., Kondraciuk, T., 2025. The role of foliar-applied silicon in improving the growth and productivity of early potatoes. Agriculture 15 (5), 556. Available at https://doi.org/10.3390/agriculture15050556.

Wang, L., Ning, C., Pan, T., Cai, K., 2022. Role of silica nanoparticles in abiotic and biotic stress tolerance in plants: A review. International Journal of Molecular Sciences 23, 1947. Available at 10.3390/ijms23041947.

Weng, H.X., Yan, A.L., Hong, C.L., Xie, L.L., Qin, Y.C., Cheng, C.Q., 2008. Uptake of different species of iodine by water spinach and its effect to growth. Biological Trace Element Research 124 (2), 184–194. Available at https://doi.org/10.1007/s12011-008-8137-4.

Xu, X., Sun, S.K., Zhang, W., Tang, Z., Zhao, F.J., 2024. Editing silicon transporter genes to reduce arsenic accumulation in rice. Environmental Science and Technology 58 (4), 1976–1985. Available at https://doi.org/10.1021/acs.est.3c10763.

Yamaji, N., Mitani-Ueno, N., Fujii, T., Shinya, T., Shao, J.F., Watanuki, S., et al., 2024. Shoot-silicon-signal protein to regulate root silicon uptake in rice. Nature Communications 15 (1). Available at https://doi.org/10.1038/s41467-024-55322-7.

CHAPTER

Diagnostic tools: mineral nutrition and stress detection

24

Maryam Zakavi[1] and Shahnoush Nayeri[2]

[1]*Department of Cellular and Molecular Biology, Faculty of Life Sciences and Biotechnology, Shahid Beheshti University, Tehran, Iran,*

[2]*Muller Lab, Department of Biology, Pennsylvania State University, University Park, PA, United States*

Abbreviations

AI	artificial intelligence
ABA	abscisic acid
CNL	critical nutrient level
DSS	decision support system
EC	electrical conductivity
GIS	geographic information system
ICP-MS	inductively coupled plasma mass spectrometry
ML	machine learning
NDVI	normalized difference vegetation index
NDRE	normalized difference red edge
NIR	near-infrared
PAM	pulse-amplitude modulation
PAR	photosynthetically active radiation
PRI	photochemical reflectance index
PSII	photosystem II
SPAD	soil plant analysis development
UAV	unmanned aerial vehicle
VRA	variable rate application
WBI	water band index

24.1 Introduction

Today, the urgent need for modern agriculture is to ensure global food security for a projected 9.1 billion people in 2050. It requires a 60%–70% increase in crop production, a task greatly hampered because of a large "yield gap" mainly caused by environmental

Mineral Nutrients to Mitigate Abiotic Stress in Crops. DOI: https://doi.org/10.1016/B978-0-443-44712-9.00026-8

stresses (Simane et al., 2025) because plants as immobile organisms have to deal with numerous stresses every day. These stresses include biotic as well as abiotic factors. And out of those stresses given to their immobility, abiotic stresses such as drought, salinity, high and low temperature stresses, micronutrient deficiencies, and metal poisoning have been identified to pose the severest limitations to the sustainable growth and advancement of agriculture. It causes more than 50% yield losses per year in key crops (Rahim et al., 2025; Zakavi et al., 2022; Zakavi and Nayeri, 2025). Moreover, the effect of climate change tends to worsen or modify the above-mentioned factors like precipitation patterns (cooling or warming effect depending on the season) and humidity (Zhang et al., 2013); apart from this, due to combinative effect of stresses among themselves, there is a trade-off situation often developed that makes them more sensitive to other stresses. This trade-off situation makes them vulnerable to another stress (Monaghan et al., 2009); hence, there is a greater economic loss in their yearly yield production. It prevents them to meet the target of sustainable food security aims to achieve global food security (SDG-2) (Zero Hunger) (OECD/FAO, 2017).

Traditionally, the evaluation and mitigation of plant stress have remained subject to a reactive model that relied on expert visual interpretation of typical disease symptoms or pathogen indicative signs (Bock et al., 2010); although such conventional methods have their uses, they remain highly subjective to human error and, more importantly, have remained to be done only after extensive damage has been inflicted to the affected plants or crops, generally beyond the critical threshold of sustainable intervention (Mahlein, 2016). Actually, the widespread reliance on pesticide and fertilizer-based control measures has given expression to a generalized and reactive type (Pereira et al., 2015) that can culminate in serious crop yield losses, as well as commensurate risks of increased environmental damage potential to nontarget harmed recipients through residues (Kariyanna et al., 2024). It is imperative that the cornerstones of sustainable management of agriculture remain transformed away from such reactive approaches to one that is proactive and seeks to identify, detect, or quantify the constituent parts of abiotic stresses before their symptomatic expression to prevent serious nonreversible losses (Galieni et al., 2021); such a "critical window" requires the systematic incorporation of specialized diagnostic technology that could sensitively reveal physiological alterations to plant health to allow early and accurate intervention (Ma et al., 2025).

There is a natural and interlinked process between mineral nutrition and plant stress. Minerals not only act as vital ingredients in plant composition but also as vital regulators of physiological processes and mechanisms of plant stress responses (Marschner, 2011). At the same time, there can be direct disruption of mineral nutrition in stressed plants. Abiotic stresses such as drought, salinity, and high or low pH can induce mineral deficiencies in plants because of their effects on soil nutrients as well as their uptake and transport in plants (Hu and Schmidhalter, 2005). On the other hand, mineral deficiencies or toxicities can also induce serious abiotic stresses to plants. This can trigger a chain of undesirable physiological changes (Hänsch and Mendel, 2009). This creates a vicious cycle wherein stress can induce mineral deficiencies or toxicities. These mineral deficiencies or toxicities can then greatly increase the susceptibility of the plant to the original stress as well as to other stresses (Cakmak, 2005). Thus the nutritional status of a plant is a very sensitive indicator of plant health and of the abiotic stresses to which it is exposed. Leveraging this linkage is the key to predictive agronomy. The conceptual relationship between nutrient status, stress susceptibility, and diagnostic tools is illustrated in Fig. 24.1.

This chapter will focus on bringing forth a detailed insight into the latest technological advancement in diagnosis methods, utilizing the status of plant nutrition to proactively identify and quantify abiotic stresses. This shall take us beyond the destructive methods of chlorophyll or plant nutrition tests that take place in harvests and involve intense manual processing. Rather, we will emphasize the role of the integration of advanced nondestructive methods such as hyperspectral imaging, thermal infrared sensing, fluorescence detection, and nanosensors through platforms such as unmanned aerial vehicles (UAVs) or satellites. Moreover, we will examine the current role of machine learning (ML) and artificial intelligence (AI) in pursuing a paradigmatic change in this area, with their potential to automatically analyze large datasets. These tools have the potential to precisely identify targeted nutrition deficiencies and biotic stresses through the interpretation of complex spectral or physiological patterns to predict their occurrence even before they can be perceived with naked eyes and can diagnose targeted deficiencies or stresses. Thus closing the gap between physiology and technology is crucial to ensure the precision-agricultural mindset itself is critical to ensure the future food security strategy is geared toward sustainability.

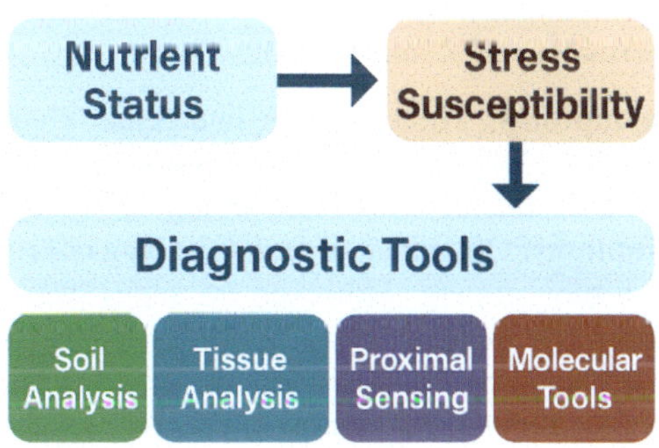

FIGURE 24.1

Linking mineral nutrition to abiotic stress detection.Conceptual framework showing how nutrient status influences stress susceptibility and how complementary diagnostic tools (soil analysis, tissue analysis, proximal sensing, molecular tools) provide convergent evidence for precise interventions

24.2 Foundational concepts: linking nutrient status to stress phenotypes

Nutritional status is one of the principal factors that determine physiological health as well as the potential to endure environmental stresses. Recognizing the direct relationship between nutritional status and patterns of environmental stresses is the first aspect of effective crop management (Bukhari et al., 2019). This section lays out the primary tenets of analyzing nutritional status as it relates to environmental stresses.

24.2.1 Nutrient deficiency and toxicity symptoms: a visual primer

The first signs of nutrient deficiencies appear as visual or morphological alterations to plants. To diagnose them correctly, one must first differentiate between typical signs of deficiencies associated with individual macronutrients and typical signs of stress.

24.2.1.1 Classic deficiency/toxicity signs versus generalized stress

Deficiencies of essential nutrients usually exhibit definite patterns that can be correlated with the elemental mobility within the plant. For mobile micronutrients (e.g., nitrogen, phosphorus, potassium, and magnesium), deficiencies can be supplied to newer tissues when they are scarce (Bhatla and Kathpalia, 2023). As a result, symptoms of Nitrogen deficiencies (e.g., chlorosis and necrosis) are usually localized to the older leaves (Wani et al., 2022). Immobile micronutrients like Calcium, Boron, or Iron have limited mobility and can cause deficiencies to develop in the newer tissues (Bhatla and Kathpalia, 2023).

Classic deficiency symptoms: These include interveinal chlorosis (for Mg, Fe, Mn), uniform chlorosis (for N and S), purplish coloring of leaves (for P in certain plants), and leaf margin death (for K) (Pandey et al., 2024).
Classic toxicity sSymptoms: Often include leaf scorching or bronzing, specific leaf spotting, and stunted root growth. However, in many cases, these particular symptoms are preceded by, or accompany, general signs of stress that involve universal chlorosis or yellowing and necrosis, which is the death of tissues, and stunting, a general and common plant response to a wide range of both biotic and abiotic stresses (Mduma et al., 2015).

24.2.1.2 The symptom overlap challenge

A further diagnostic complication involves the frequent overlap in visual symptoms caused between different nutrient disorders and other stress factors (Biousse and Newman, 2016). For example, interveinal chlorosis is an important symptom for both magnesium deficiency (a mobile nutrient, showing in older leaves) and iron deficiency (an immobile nutrient, showing in newer leaves) (Pandey et al., 2024). General stunting and chlorosis may be caused by nitrogen deficiencies, waterlogging, root pathogen damage, or drought stress (Mduma et al., 2015). This immediately makes visual diagnosis alone an unreliable tool and calls for more precise, analytical diagnostic methods that will help pinpoint the exact stressor in question (Biousse and Newman, 2016).

24.2.2 The "critical nutrient level" and the "stress threshold"

To move beyond subjective visual assessment, agronomists rely on two quantitative physiological concepts that define the plant's nutrient status.

24.2.2.1 *The critical nutrient level*

This refers to the least amount of nutrients in a specific part of the plant, a frequently recently fully matured leaf, at a particular growth stage that relates to maximum growth (e.g., 90%–95% of potential yield) (Lambers et al., 2019). Once the concentration of nutrients in tissue drops below this critical level, growth and yield suffer, and deficiency symptoms may appear. The CNL indicates the threshold between adequate nutrition and deficient nutrition (Abbas et al., 2021).

24.2.2.2 *The stress threshold*

While the CNL marks the point of yield decline, the Stress Threshold is a broader concept. It refers to the point at which a nutrient imbalance, either a deficiency below the CNL or a toxicity above an upper threshold, starts to induce significant physiological dysfunction. Beyond this threshold, primary metabolic pathways such as photosynthesis and respiration are impaired, hormonal signaling is disrupted, and cellular homeostasis is lost. Crucially, crossing this stress threshold makes the plant markedly more susceptible to secondary stressors such as pests, diseases, and extreme weather. In contrast, proactive nutrient management aims to maintain plant tissue concentrations within a sufficient range, comfortably above the deficiency CNL and below the toxicity threshold, to maximize both yield and stress resilience (Choi, 2024; Pandey et al., 2024) (A).

24.2.3 The hidden hunger: latent deficiencies and the consequences for stress resilience

Perhaps the most insidious form of nutrient stress is "hidden hunger." This condition occurs when a plant's nutrient levels are suboptimal-below what is required for robust health and defense-but still above the Critical Nutrient Level where clear visual symptoms or acute yield loss become apparent (Mangrauthia et al., 2024; Nijs and Impens, 1997).

During this latent phase, the plant may look quite healthy to the naked eye, but important physiological processes have already been compromised. Activities of various enzymes can be lowered, protein synthesis rate can be reduced, and biosynthesis of defense-related compounds may be affected. A subclinical deficiency insidiously whittles away at the stress tolerance of the plant, its capability for resistance and recovery to the environmental stresses.

For instance, a plant with latent hunger for potassium may have impaired osmotic regulation and weaker cell walls, making it more susceptible to drought stress and fungal pathogen invasion (Shabala and Pottosin, 2014). Similarly, a plant with latent zinc deficiency may have impaired auxin synthesis, leading to weaker growth that cannot compete effectively under other stresses (Wang et al., 2023). Identifying and correcting such latent deficiencies through precise tissue analysis and advanced sensing technologies is thus a key strategy in building durable and resilient cropping systems that enable intervention long before irreversible damage or yield loss has taken place.

24.3 Conventional diagnostic methods: strengths and limitations

Conventional methods are the historical and usually indispensable basis for plant nutritional and stress diagnosis. Although lacking in real-time capability compared to modern technologies, they provide critical, validated data for informed decision-making (Munson, 2018).

24.3.1 Soil analysis: the preplanting forecast

Analysis of soil is basically done prior to planting; it is a predictive tool used for assessing the inherent fertility and chemical environment of the soil, allowing for corrective amendments.

24.3.1.1 Methodology: chemical extraction and analysis

Soil samples are collected from the root zone, dried, and sieved. Specific chemical extractants, such as Mehlich-3, DTPA, Olsen's bicarbonate, among others, are applied to dissolve and simulate the "plant-available" fraction of nutrients. The resulting solution is analyzed by techniques such as ICP-OES, which quantifies the nutrient concentrations.

24.3.1.2 What it tells us: nutrient availability, pH, salinity, cec

This analysis serves to forecast the nutrient supply availability for crop uptake. It measures soil pH directly, which dictates nutrient solubility and availability. It further assesses salinity through electrical conductivity and cation exchange capacity that shows the potential of the soil to retain and supply cationic nutrients such as K^+, Ca^{2+}, and Mg^{2+} (Munson, 2018).

24.3.1.3 *Limitations: does not reflect plant uptake, root activity, or in-season stress*

The major limitation is that it provides a measure of potential availability, not actual uptake. It cannot take into account the factors that affect root function, including such things as soil compaction, pathogens, or moisture stress. A soil test can return with adequate levels of a nutrient, yet a plant might still be experiencing a deficiency due to poor root health or unfavorable environmental conditions. It is a static picture of the soil before the growing season and cannot detect in-season nutrient deficiencies or toxicities as they develop.

24.3.2 Tissue analysis: the most reliable method for determining nutrient concentrations

Tissue analysis, which verifies visual suspicions with quantitative data, offers a conclusive diagnosis of a plant's internal nutritional status when symptoms appear.

24.3.2.1 *Techniques: laboratory analysis (e.g., ICP-MS), sampling protocols (key tissues, timing)*

According to the report, proper sampling is essential and entails gathering the appropriate plant part (such as freshly matured leaves or petioles) at a particular growth stage (such as early flowering). After being cleaned of impurities, samples are dried, ground, and digested. Advanced tools such as inductively coupled plasma mass spectrometry (ICP-MS), which offer incredibly low detection limits for a variety of elements, are used to analyze the digestate (Jones and Case, 2018).

24.3.2.2 *What it tells us: the precise amount of nutrients and the conclusive identification of toxicities or deficiencies*

The true nutrient concentration in the plant at the time of sampling is revealed by tissue analysis. A conclusive diagnosis of toxicity, sufficiency, or deficiency is made by comparing the results to predetermined "sufficiency ranges" for the crop and growth stage (Mills and Jones, 1996). It may indicate "hidden hunger," a state in which a plant lacks a nutrient but does not yet exhibit outward signs of the problem.

24.3.2.3 *Constraints: time-lag, destructive, and only offers a snapshot rather than a dynamic image*

Because plant tissue must be removed, the process is destructive. The plant's condition might have gotten worse in the days or weeks that pass between sampling, lab analysis, and results. Most significantly, it only offers a single moment in time and is unable to record the dynamic variations in nutrient flux brought about by management techniques or environmental stressors (Marschner, 2011). Yield potential may already be jeopardized by the time a deficiency is confirmed.

24.3.3 The oldest tool in the box: visual scouting

The quickest and easiest diagnostic technique is visual scouting, which uses the human eye to recognize biotic and abiotic stresses from plant appearance.

24.3.3.1 *Practicality and the importance of experience*

It is extremely useful for identifying issues early. Based on the pattern, color, and location of symptoms on the plant and in the field, an experienced agronomist can frequently differentiate between disease symptoms, herbicide damage, pest damage, and nutrient deficiencies (Jones and Case, 2018). For instance, iron deficiency causes interveinal chlorosis on younger leaves, whereas nitrogen deficiency usually causes uniform yellowing (chlorosis) of older leaves (Marschner, 2011).

24.3.3.2 *Major limitation: the onset of symptoms suggests that yield loss has already taken place*

The plant uses visual symptoms as a last resort, which is its biggest disadvantage. Long before necrosis or the degradation of chlorophyll occurs, physiological processes are compromised. The plant has already experienced decreased photosynthesis, stunted growth, and irreversible loss of yield potential by the time a deficiency or stress becomes clearly visible. Furthermore, without confirmatory tissue analysis, it is challenging to make a confident diagnosis because many nutrient deficiencies have similar symptoms (such as interveinal chlorosis) (Mduma, et al., 2015).

24.4 Advanced proximal sensing tools: real-time, in-field diagnostics

A paradigm change from reactive to proactive crop management is represented by sophisticated proximal sensing technologies. These instruments provide real-time data that can be utilized for precision agriculture applications by directly and nondestructively measuring the physiological status of the plant. This enables prompt interventions before a significant loss in yield occurs (Laveglia et al., 2024).

24.4.1 Fundamentals of plant radiometry and spectroscopy

Spectroscopy, the study of the relationship between light and matter, is the basis for most optical sensing instruments (Cavaco et al., 2022). Representative proximal sensing instruments and their operating principles are shown in Fig. 24.2.

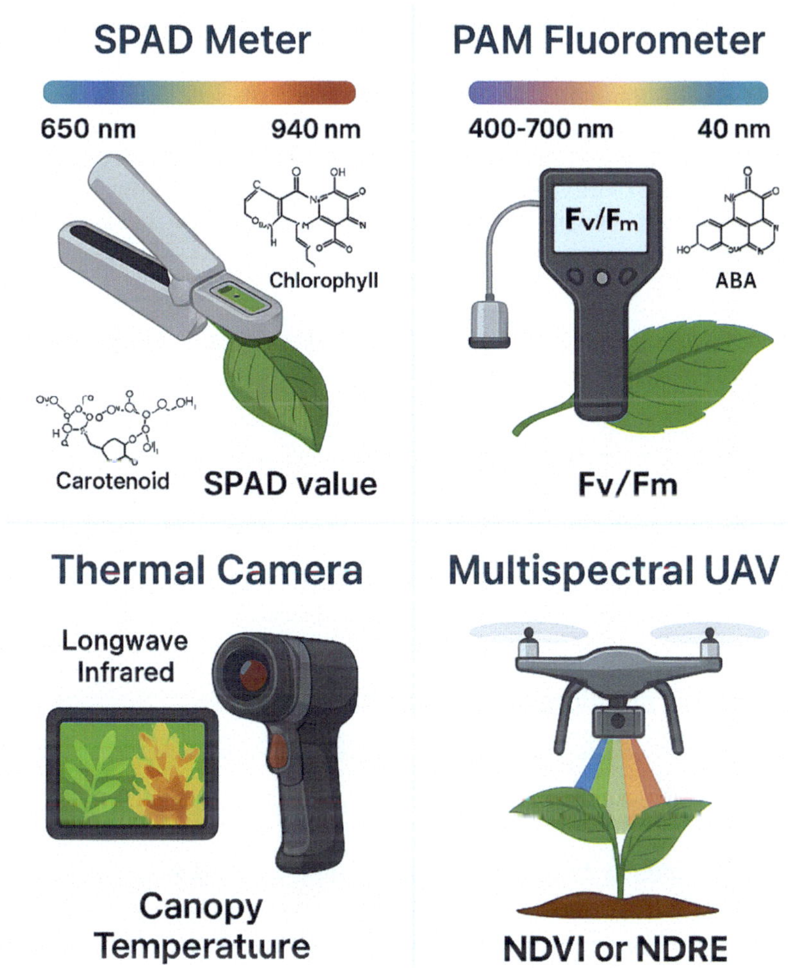

FIGURE 24.2

Advanced proximal sensors for real-time diagnosis. Panels illustrate: (i) SPAD chlorophyll meter operating at 650 nm (red) and 940 nm (IR) to produce SPAD values; molecular insets indicate chlorophyll and carotenoids. (ii) PAM fluorometer measuring Fv/Fm and related parameters to assess PSII efficiency; ABA indicated as a drought-stress biomarker. (iii) Thermal IR camera (8–14 μm) mapping canopy temperature and CWSI. (iv) Multispectral UAV capturing NDVI/NDRE maps for field-scale zonation.

24.4.1.1 *Light's absorption, reflection, and fluorescence in relation to plant tissues*

A leaf can absorb, transmit, or reflect light when it strikes it. The chemical and physical characteristics of the leaf tissues dictate which wavelengths are absorbed, reflected, or emitted (Ustin and Jacquemoud, 2020). For instance, healthy leaves reflect near-infrared (NIR, 700–1300 nm) light because of scattering within the mesophyll cells and strongly absorb photosynthetically active radiation (PAR, 400–700 nm) for photosynthesis.

24.4.1.2 *Connection to plant physiology: water content, nitrogen status, cell structure, and pigments (chlorophyll)*

Plant physiology is quantitatively related to this light interaction. Reflectance in the blue and red bands serves as a stand-in for chlorophyll content, which is frequently associated with nitrogen status because chlorophyll pigments absorb strongly in these regions (Boonupara et al., 2024; Nayeri et al., 2023). Reflectance at these wavelengths indicates the presence of water because water absorbs strongly in the short-wave infrared (SWIR), such as at 970 and 1450 nm. Leaf biomass and internal structure have an impact on the NIR reflectance plateau.

24.4.2 Meters of chlorophyll (like SPAD)

One of the most popular proximal sensors for nitrogen management is the chlorophyll meter.

24.4.2.1 *The idea is to use leaf transmittance as a stand-in for nitrogen and chlorophyll levels*

LEDs are used by handheld meters such as the SPAD-502 to emit light in the infrared (940 nm), where absorption is minimal, and red (650 nm), where chlorophyll absorption is high. A built-in algorithm determines the unitless value (SPAD value) based on the meter's measurement of the transmittance of these two wavelengths through the leaf. This value is highly correlated with the amount of chlorophyll in the leaf and, consequently, with the nitrogen status (Nayeri et al., 2023).

24.4.2.2 *Uses: quick N management; association with salinity and drought stress*

Decisions about side dressings and the evaluation of nitrogen status during the season are the main uses. Additionally, a decrease in SPAD values can be used as an indirect indicator of abiotic stressors like drought and salinity, which hinder chlorophyll synthesis and speed up degradation (Shah et al., 2017).

24.4.3 Reflectance indices at the leaf and canopy levels

Vegetation indices are mathematical combinations of reflectance at important wavelengths that are used to quantify particular plant characteristics (Table 24.1).

24.4.3.1 *The normalized difference vegetation index for chlorophyll and biomass*

(NIR – Red) / (NIR + Red) is the normalized difference vegetation index (NDVI). The most traditional index, it is influenced by fractional vegetation cover, green biomass, and chlorophyll content. It does, however, saturate in high biomass, dense canopies (Xu et al., 2021).

24.4.3.2 *Particular indicators include the water band index, normalized difference red edge, and photochemical reflectance index for light and water stress*

Normalized difference red edge (NDRE): This method works better for mid-to-late season crops because it uses the red-edge region (~700–750 nm), which is highly sensitive to chlorophyll and does not saturate as readily as NDVI.
Water band index: R_{970} / R_{900}. A weak band of water absorption is visible in the reflectance at 970 nm. Early identification of a water deficit is made possible by an increase in this ratio, which denotes a decrease in leaf water content.
Photochemical reflectance index: (R_{531} - R_{570}) / (R_{531} + R_{570}). It is connected to the photoprotective mechanism of the xanthophyll cycle pigments' epoxidation state. Variations in PRI can indicate early light, heat, or drought stress and show how efficiently light is used (Zarco-Tejada et al., 2013).

24.4.3.3 *Handheld spectroradiometers versus multispectral sensors*

Spectroradiometers are high-resolution devices that measure reflectance across hundreds of contiguous narrow bands, offering a complete spectral signature for in-depth analysis and the creation of new indices. This is how handheld spectro-radiometers differ from multispectral sensors (Castroesau et al., 2006).
Multispectral sensors are typically simpler, measuring reflectance at a few predefined, broad wavelengths (e.g., blue, green, red, red-edge, NIR). They are more practical, cost-effective, and commonly deployed on drones, tractors, and satellites (Castroesau et al., 2006).

24.4.4 Chlorophyll fluorescence (e.g., PAM fluorometry)

Chlorophyll fluorescence provides a direct window into the photosynthetic machinery, specifically photosystem II (PSII) (Lichtenthaler et al., 2005).

Table 24.1 Common vegetation indices used in plant stress diagnosis, with formulas, physiological targets, typical use cases, caveats, and recommended pairing with other sensors.

Index	Formula	Target/Sensitivity	Use cases	Caveats	Pairing notes
NDVI	(NIR – Red) / (NIR + Red)	Chlorophyll, green biomass, fractional vegetation cover	Seasonal vigor monitoring; growth assessment	Saturates at high LAI; sensitive to soil background	Combine with NDRE for dense canopies; pair with thermal data
NDRE	(NIR – Red-edge) / (NIR + Red-edge)	Chlorophyll in high-biomass canopies; less prone to saturation	Mid- to late-season nitrogen status; dense canopy monitoring	Sun-angle and sensor-dependent	Calibrate with SPAD or tissue N data
PRI	(R531 – R570) / (R531 + R570)	Xanthophyll cycle activity; light-use efficiency	Early detection of light, heat, or drought stress	Directional reflectance effects; requires narrow bands	Use with thermal (Tc – Ta) and fluorescence (Fv/Fm) for integrated diagnosis
WBI	R970 / R900	Leaf water content (weak water-absorption band)	Early water-deficit detection	Sensitive to leaf structure and species	Pair with CWSI for robust water-stress inference

Abbreviations: *LAI*, leaf area index; *NIR*, near-infrared; *R*xxx, reflectance at wavelength *xxx* nm; *Tc* – *Ta*, canopy-to-air temperature difference.

24.4.4.1 Principle: examining photosystem II's efficiency

Chlorophyll fluorescence is the term for the tiny percentage of light that is reemitted as longer-wavelength red and far-red light after being absorbed by chlorophyll. The photosynthetic efficiency of PSII has an inverse relationship with the emission's intensity and kinetics. The maximum quantum yield of PSII can be calculated using pulse–amplitude–modulation (PAM) fluorometry, which uses a measuring beam to quantify the baseline fluorescence (F_o) and a saturating pulse to measure the maximum fluorescence (Fm): $Fv/Fm=(Fm - F_o)/Fm$.

24.4.4.2 Uses: direct assessment of photosynthetic efficiency in the presence of oxidative stress, heat, cold, and drought

The Fv/Fm ratio of a healthy, dark-adapted plant is usually ~0.83. A decrease in Fv/Fm results from any stress that harms the photosynthetic apparatus, such as photoinhibition from intense heat or light, membrane damage from cold, or metabolic disruption from drought. Because of this, it is an extremely sensitive and straightforward diagnostic tool for identifying stress before outward signs show up.

24.4.5 Thermal imaging

The balance of energy inputs and outputs, mainly transpirational cooling, controls the temperature of plant canopies, which is measured by thermal imaging (Pineda et al., 2021).

24.4.5.1 The measurement of canopy temperature as a proxy for stomatal conductance is the fundamental idea

Transpiration lowers the leaf's temperature to near or below the air temperature when the stomata are open. Under water stress, stomata close to conserve water, transpiration decreases, and evaporative cooling decreases, raising the temperature of the leaves. This temperature differential is detected by an infrared thermal camera, which produces a visual map of canopy stress (Mayanja et al., 2024).

24.4.5.2 Uses: early identification of root health and water deficit (drought stress)

Because it can detect water-stressed areas in a field before wilting is apparent, thermal imaging is an effective tool for irrigation scheduling. Additionally, it can show root health issues like soil compaction or root rot, which restrict water uptake and appear as "hotspots" in a thermal image (Sofi et al., 2019).

24.5 Remote sensing: expanding to the field and terrain

With the use of remote sensing technologies, crop health and stress can be tracked on large spatial scales, providing a comprehensive, field-level perspective as opposed to single-plant measurements. These tools enable precision agriculture by detecting in-field variability that is frequently invisible to the human eye through the deployment of sensors on elevated platforms (Mulder et al., 2011). The workflow for integrating UAV and satellite imagery into GIS-based management zones is depicted in Fig. 24.3.

24.5.1 Platforms: satellites, airplanes, and unmanned aerial vehicles

A hierarchy of remote sensing capabilities is created by the platform selection, which also determines the scale, resolution, and flexibility of data collection.

UAVs (drones): Unaffected by cloud cover, UAVs (drones) provide extremely high spatial resolution (1–10 cm/pixel) and great flexibility for on-demand data collection. These devises are perfect for managing particular management zones and conducting in-depth field-level scouting (Gonzalez et al., 2018).

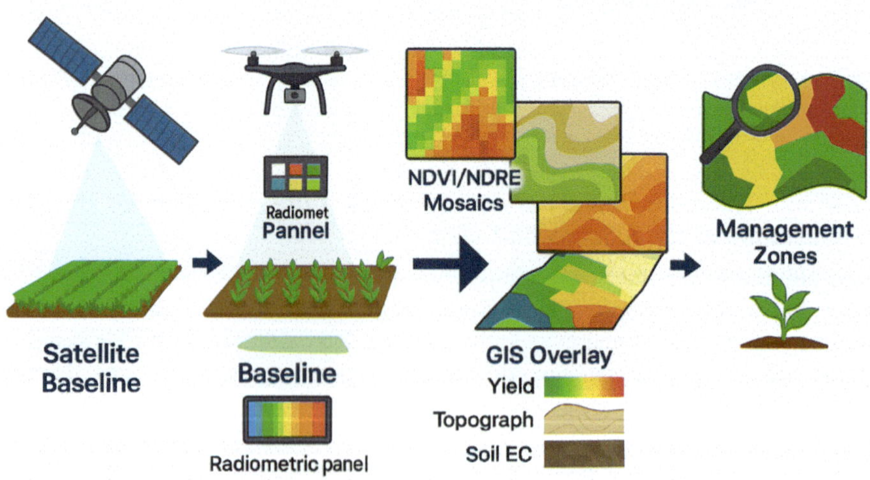

FIGURE 24.3

Field-level stress mapping using UAV and satellite.Workflow: satellite baseline and UAV multispectral flight with radiometric panel → NDVI/NDRE mosaics → overlay with soil EC, topography, and prior yield maps in GIS → delineation of management zones for targeted scouting and stratified sampling.

Aircraft (manned aerial vehicles): Compared to UAVs, aircraft (manned aerial vehicles) can provide higher-resolution imagery (10 cm to 1 m/pixel) over a wider area. Though more expensive and less flexible than UAVs, they are frequently used for regional surveys or when reliable data over a wide, continuous area is required (Shakhatreh et al., 2019).
Satellites: Provide worldwide coverage at different spatial resolutions (10 m/pixel for Sentinel-2, 30 cm/pixel for commercial satellites such as WorldView). Although revisit times and cloud cover may limit them, their primary strength is consistent, repeatable temporal coverage, which allows for long-term trend monitoring (Cuypers et al., 2023).

24.5.2 Temporal and spatial resolution: selecting the appropriate instruments

A crucial trade-off between resolution (spatial detail) and revisit time (frequency of new data collection) must be made when choosing a remote sensing tool (Schacher et al., 2023).

High spatial, low temporal resolution: Although high-resolution commercial satellites, like Planet and Maxar, offer stunning detail, they might only record data once a week or less. This works well for postseason analysis or base map creation.
Moderate space, high temporal resolution: Sentinel-2 (5-day revisit) and Landsat (8-day revisit) are two public satellites that offer free, regular imagery with a resolution of 10–30 m that is appropriate for tracking seasonal dynamics over large farms and landscapes.
Extremely high spatial, on-demand temporal resolution: UAVs bridge the crucial gap for intensive, season-long management. At a resolution that can identify individual plants, they can be used several times a week to track quick changes, like the expansion of a water-stressed area or the reaction to fertilizer application (Schacher et al., 2023).

24.5.3 Combining topographic and soil maps

When remote sensing is combined with other geospatial data to identify the reasons behind observed variability, its full potential is revealed (Avtar et al., 2019; Dritsas and Trigka, 2025; Van der Meer, 2012).

Soil electrical conductivity (ECa) maps: Find trends in organic matter, salinity, and soil texture that have a direct impact on the soil's ability to retain water and the availability of nutrients.
Soil nutrient maps: Offer ground-truthed information on nutrient levels derived from grid or zone soil sampling.

Topographic maps: Information about elevation, slope, and aspect can be used to explain temperature variations throughout a field, drainage patterns, and erosion risk.
Data fusion: An agronomist can make more accurate management decisions by utilizing Geographic Information Systems (GIS) to overlay soil EC and elevation maps with a vegetation index (such as NDVI) from a drone. This allows them to determine whether poor plant health is caused by a waterlogged depression (associated with low elevation) or a nutrient-poor sandy area (correlated with high ECa).

24.6 Cutting-edge technologies and emerging tools

Beyond outward signs and even physiological reactions, the field of plant diagnostics is expanding to the molecular and ionic level (Coatsworth et al., 2023). Future hyper-sensitive, early-warning systems that can identify stress before it affects plant growth and yield are promised by these cutting-edge technologies:

24.6.1 Sensor-powered "plant wearables"

Plant wearables, which are miniature sensors intended to be affixed to or integrated into plants for ongoing, real-time monitoring of their internal chemical environment, were inspired by wearable technology in human health (Banerji et al., 2024).

24.6.1.1 In situ nanosensors for monitoring stress metabolites (H_2O_2, abscisic acid) and ions (K^+, Ca^{2+}, Na^+) in the sap in real time

These sensors make use of functionalized nanomaterials, like graphene or carbon nanotubes, which alter their optical or electrical characteristics when a particular target molecule is present. When a sensor binds to abscisic acid (ABA), a crucial hormone involved in drought stress, it can be made to fluoresce. These sensors can wirelessly transmit data and provide a live feed of plant stress status when embedded in the vascular tissue or leaf mesophyll. Unlike any remote or proximal sensing method, this enables the detection of a potassium (K^+) dip or a sodium (Na^+) surge within minutes of the onset of stress (Rana et al., 2024; Zakavi and Nayeri, 2025).

24.6.2 Ionomics: a diagnostic method based on high throughput

Ionomics offers a comprehensive, macro-scale perspective on the elemental makeup of plants, going beyond a few essential nutrients to a system-level comprehension (Zhang et al., 2023).

24.6.2.1 Principle: quantification of a plant's entire elemental profile

Ionomics simultaneously measures the concentration of nearly every element in a tissue sample, including macronutrients (N, P, K), micronutrients (Fe, Zn, Cu), and nonessential elements (Na, As, Cd), using high-throughput analytical methods such as Inductively Coupled Plasma Mass Spectrometry (ICP-MS) (Wu and Becker, 2012).

24.6.2.2 Use: producing "elemental fingerprints" for particular abiotic stressors

Plant ion homeostasis is disrupted by stress in distinctive ways. For example, a distinct fingerprint of low K and Ca, high Na, and high Cl is produced by salinity stress. A characteristic low P and frequently an accumulation of other elements, such as Zn and Mn, are signs of phosphorus deficiency. These distinct ionomic signatures can be used to train machine learning models to diagnose complex or multiple concurrent stresses by creating sizable databases (Voica et al., 2025).

24.6.3 "Omics" and molecular tools as early-warning systems

The changes in gene expression, protein synthesis, and metabolite production that make up the plant's initial response to stress are the earliest indicators of stress that these instruments can identify (Misra et al., 2019; Zhan, 2025).

24.6.3.1 Gene expression markers: detecting the overexpression of genes that respond to stress prior to visual symptoms

The transcript levels of particular genes known to be induced by stress can be measured using methods such as RNA-Seq or quantitative PCR (qPCR). For instance, a molecular indicator of salinity stress is the fast upregulation of the sodium/proton antiporter NHX1 gene, whereas a marker of drought stress is the induction of AREB/ABF transcription factors. This offers a diagnosis at the genetic programming level of the plant (Wang et al., 2008).

24.6.3.2 Proteomic and metabolomic profiles: biomarkers for combined stressors and nutrient deficiencies

Proteomics: This is the extensive study of proteins. Certain proteins, such as enzymes involved in alternative metabolic pathways or stress protection, may become more abundant (upregulated) in response to a particular nutrient deficiency or stress.

Metabolomics: The thorough examination of all small-molecule metabolites, such as amino acids, organic acids, sugars, and antioxidants, is known as metabolomics. The last downstream result of gene expression, the metabolome, most quickly captures the physiological condition of the plant. A highly sensitive and specific diagnosis of the type and severity of stress can be made using a particular "metabolite fingerprint"—for example, the buildup of proline and particular sugars during drought, or of particular amino acids during nitrogen deficiency (Moshood et al., 2025).

24.7 Integrated diagnostic frameworks: from data to decision

A new problem brought about by the widespread use of diagnostic tools is information overload and the possibility of misdiagnosis. To move from merely identifying an issue to recommending a specific solution, an integrated framework is necessary to combine various data streams into a logical, actionable understanding of crop status (Peng et al., 2011).

24.7.1 The "Toolbox" method: why no one tool is enough

A comprehensive picture cannot be obtained from a single diagnostic technique. Because each tool has its own advantages and disadvantages, using them together is not only advantageous but also required for a precise diagnosis. A sequential diagnostic workflow combining soil analysis, sensing tools, and decision support systems is summarized in Fig. 24.4.

24.7.1.1 Integrating tissue analysis (confirmation), proximal sensing (verification), remote sensing (zoning), and soil analysis (baseline)

Soil analysis: The basic background of the nutrients that were available at the beginning of the season is provided by soil analysis.

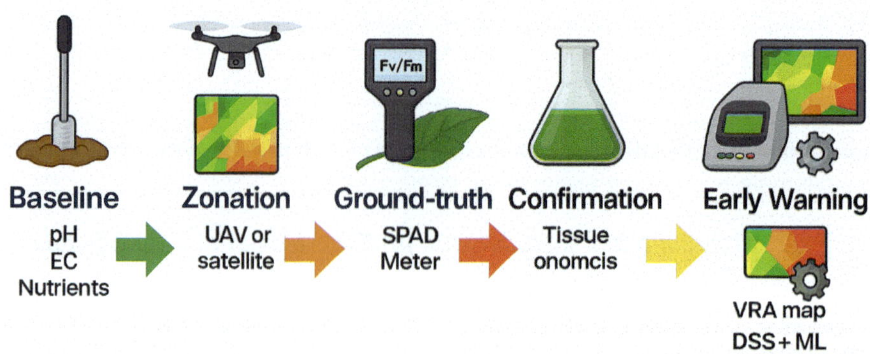

FIGURE 24.4

Integrated diagnostic workflow—from data to decision.Sequential band-layout: Baseline (soil pH, EC, nutrients) → Zonation (UAV/satellite indices) → Ground-truth (SPAD, PAM, thermal) → Confirmation (tissue analysis, ionomics) → Early warning (qPCR/RNA-Seq, metabolomics) → Prescription (DSS + ML) generating variable-rate application maps for irrigation and fertilization.

Remote sensing (e.g., satellite/UAV): Remote sensing (e.g., satellite/UAV) highlights areas of concern (zonation) by identifying the spatial pattern of crop stress throughout the field.
Proximal sensing (e.g., chlorophyll meter, fluorometer): The physiological condition of plants within the designated zones is confirmed by ground-truthing the remote sensing data in the field using proximal sensing (such as a chlorophyll meter or fluorometer).
Tissue analysis: To differentiate between a real deficiency and a stress-induced mimic, tissue analysis provides the conclusive, quantitative validation of the actual nutrient concentrations (Bongiovanni and Lowenberg-Deboer, 2004).

Instead of depending on a single, possibly deceptive piece of data, this methodical, multi-tool approach guarantees that management decisions are supported by convergent evidence.

24.7.2 Interpretation and fusion of data

To produce understandable and practical information, the raw data from multiple tools must be combined.

24.7.2.1 *Establishing management zones through the use of geostatistics and geographic information system*

The main platform for data fusion is a geographic information system (GIS). Overlapping layers include soil ECa maps, yield maps from prior seasons, multitemporal NDVI imagery from drones, and topographic data. Stable, contiguous management zones that represent regions of the field with comparable constraints and yield potential are then defined using geostatistical techniques such as k-means clustering. This substitutes a zoned approach for consistent whole-field management (Ali and Ibrahim, 2020).

24.7.2.2 *The function of machine learning and algorithms in distinguishing between stress resulting from a deficiency (such as a lack of N) and stress causing a deficiency (such as drought restricting N uptake)*

This is a crucial diagnostic problem. To identify intricate patterns, multi-sensor data can be used to train machine learning (ML) models. For example, based on thermal data, a model may discover that a low NDVI (poor vigor), a high crop water stress index (CWSI), and a moderately low NDRE (related to chlorophyll) are all highly predictive of "drought-induced nitrogen deficiency." On the other hand, a "primary nitrogen deficiency" may be indicated by a low NDVI, low CWSI, and very low NDRE. As a result, the proper intervention—fertilization vs. irrigation—can be determined through causal inference (Azimi et al., 2021).

24.7.3 Decision support systems for accurate stress reduction

Converting the integrated diagnosis into a precise, implementable management action is the last stage. (Table 24.2).

24.7.3.1 Converting test results into variable-rate irrigation and nutrient recommendations

A contemporary Decision Support System (DSS) utilizes agronomic models to produce a Variable-Rate Application (VRA) map from the interpreted data, such as a map of nitrogen status zones created from fused sensor data and calibrated with tissue samples. According to reports, this prescription file instructs farm equipment to apply the appropriate input (such as fertilizer or water) at the appropriate rate, location, and time. For instance, for a side-dress nitrogen application, a DSS can produce a VRA map that applies more fertilizer in areas with primary N deficiency and less in areas with secondary drought-induced deficiencies, where irrigation should be addressed first. Similarly, it can produce a variable-rate irrigation prescription to optimize water use and alleviate related nutrient uptake issues by addressing the specific water deficits detected by thermal and reflectance sensors (Shtienberg, 2013).

24.8 Challenges in advanced diagnostics

Even though there are more diagnostic tools available than ever before, there are still many obstacles in the way of their widespread use. For precision plant health management to reach its full potential in global agricultural systems, these issues must be resolved and future development must be guided (Table 24.3).

24.8.1 Technical and logistical difficulties: required expertise, data management, calibration, and cost

Advanced diagnostics implementation is a complex socio-technical issue rather than just a technical one.

Cost: UAVs, high-resolution sensors, and the analytical expenses associated with 'omics technologies are expensive and frequently out of reach for typical growers who lack measurable, obvious returns.
Calibration: There are differences in the correlation between sensor readings (such as NDVI and SPAD) and actual nutrient status. It requires thorough and continuous local calibration to be accurate because it varies by crop species, cultivar, growth stage, and environmental conditions.
Data management: One of the biggest challenges is managing the volume, velocity, and variety of data produced by both proximal and remote sensors, or "big

Table 24.2 Proximal sensing tools for monitoring plant nutrient status and stress, including principle of operation, primary outputs, calibration requirements, advantages, limitations, and approximate cost range.

Instrument/ Tool	Principle	Key bands/ Spectral range	Primary outputs	Calibration needs	Advantages	Limitations	Typical cost (USD)
SPAD Chlorophyll Meter	Leaf transmittance as proxy for chlorophyll and nitrogen	Red (~650 nm); IR (~940 nm)	SPAD value (unitless)	Crop- and growth-stage-specific calibration to leaf N/ chlorophyll	Fast, nondestructive, handheld, simple operation	Affected by leaf thickness, species, and environmental conditions	500–3000
PAM Fluorometer	PSII efficiency via chlorophyll fluorescence kinetics	Excitation: blue; Emission: red/far-red	Fv/Fm, ΦPSII, qP, NPQ	Dark adaptation; strict measurement protocol	Direct indicator of photosynthetic stress; high sensitivity	Sensitive to ambient light and handling conditions	4000–20,000
Thermal IR Camera	Canopy temperature as proxy for stomatal conductance and water stress	8–14 μm (thermal infrared)	Canopy temperature (Tc), CWSI	Emissivity settings; reference surfaces; meteorological data	Early detection of water deficit; useful for irrigation scheduling	Confounded by wind, humidity, and mixed pixels	1500–30,000
Multispectral UAV Camera	Imaging at discrete reflectance bands for vegetation indices	Blue, green, red, red-edge, NIR	NDVI, NDRE, PRI maps	Radiometric panel; sun-angle-corrected flight plan	Field-scale zonation; repeatable mapping; high spatial resolution	Saturates in dense canopy; sun-angle effects	2000–25,000

Abbreviations: *CWSI*, crop water stress index; *NIR*, near-infrared; *NPQ*, nonphotochemical quenching; *PRI*, photochemical reflectance index; *PSII*, photosystem II.

Table 24.3 Integrated diagnostic workflow for precision stress management, showing sequential tools, data products, diagnostic goals, and resulting actionable decisions.

Step	Tool(s)	Data product	Diagnostic goal	Actionable decision
Baseline	Soil analysis (pH, EC, nutrients)	Soil fertility, EC, and elevation layers	Establish pre-plant constraints and variability	Preplant amendments; cultivar selection
Zonation	UAV multispectral; satellite imagery	Management zones; vigor maps	Locate spatial variability and stress hotspots	Targeted scouting; stratified sampling plan
Ground-truth	SPAD; PAM fluorometer; thermal IR	Leaf N proxy; PSII efficiency; CWSI	Confirm physiological stress and differentiate stress types	Prioritize irrigation vs. fertilization; identify timing of intervention
Confirmation	Tissue analysis; ionomics	Elemental concentrations; ionomic profiles	Distinguish deficiency from uptake-limited stress	Variable-rate nutrient application; adjust timing and form of supplements
Early-warning	qPCR/RNA-Seq; metabolomics	Stress-responsive gene or metabolite markers	Detect pre-symptomatic stress states	Preventive foliar application; adjusted irrigation scheduling
Prescription	Decision support system (DSS) + ML	Variable-rate application (VRA) maps; irrigation prescriptions	Translate diagnostic insights into spatially explicit actions	Apply inputs (water, nutrients) by zone and growth stage

Abbreviations: *EC*, electrical conductivity; *CWSI*, crop water stress index; *ML*, machine learning.

data." The task of storing, processing, and interpreting multiple data streams from various platforms and dates can be too much for farmers and agronomists to handle. **Required expertise:** To use these tools effectively, a new set of skills combining data science, geospatial analysis, and agronomy is needed. One major obstacle to adoption is the lack of such cross-disciplinary expertise.

24.9 Future considerations

The transition from general vegetation indices to extremely specific diagnostic models is a major area of research. While current indices such as the NDVI are sensitive to overall "greenness," they are unable to accurately differentiate between conditions like nitrogen deficiency and the early stages of a fungal disease. Building thorough spectral libraries that list the hyperspectral reflectance signatures (400–2500 nm) linked to biotic/abiotic stresses, toxicities, and nutrient deficiencies for every major crop is the way of the future. These libraries will teach advanced machine learning models to recognize and measure stress in addition to detecting it. For example, a model that employs particular absorption characteristics in the red-edge and short-wave infrared could be created to accurately identify and distinguish between potassium and magnesium deficiencies in soybeans. Research into inexpensive smartphone-based sensors (such as utilizing the camera on a phone as a basic spectrophotometer) and the creation of open-source, low-cost unmanned aerial vehicles (UAVs) can significantly lower entry barriers. Technology ownership might not be feasible for smallholders. Instead, these tools can be made available through service-based models wherein nearby cooperatives or agri-service businesses provide soil/tissue analysis and drone scouting as a service. Advanced diagnostics Decision Support Systems (DSS) need to be transformed from complicated software into mobile-first, user-friendly apps that offer precise, directive guidance (e.g., "Apply 50 kg N/ha in the red zone") without necessitating data scientists. AI-powered crop management systems that are autonomous and capable of real-time stress diagnosis and treatment are the promising approach in this case. A closed-loop, self-governing crop management system is the ideal scenario. To provide a constant flow of data to a cloud-based AI platform, this future system would combine a network of in-field sensors (including "plant-wearables"), drones, and satellites. By combining this real-time data with historical weather, soil, and yield models, the AI would serve as a central "crop doctor." In addition to diagnosing the particular stress (for example, "water deficit in Zone B is beginning to induce potassium deficiency"), it would also recommend and carry out a solution. Before any obvious symptoms show up, this could entail sending an autonomous rover to apply a localized foliar K fertilizer to the afflicted plants and automatically activating variable-rate irrigation systems to address the water deficit . This signifies a change from precision farming to agro-ecosystem management that is completely autonomous and predictive.

24.10 Conclusion

Advanced diagnostic technology is a revolution from reactive to predictive approaches for crop management. By using mineral nutrition as a biomarker and through the integration of proximate, remote, and molecular sensing platforms, plant stress can be anticipated before occurring and causing unyielding losses. The convergence of multiple sensor inputs and machine learning capabilities allows for accurate and targeted actions pointing to particular stresses for optimal and sustainable resource allocation. Despite limitations with regard to high cost, accurate calibration, and appropriate know-how, forthcoming innovations in plant wearables, ionomics research, and artificial intelligence-driven decision tools offer potential scalable alternatives for adoption. Finally, "closing the divide between plant physiology and technology is crucial for attaining global food security under fluctuating climatic changes."

References

Abbas, S., Amna, Javed, M.T., Ali, Q., Azeem, M., Ali, S., 2021. Nutrient deficiency stress and relation with plant growth and development. Engineering Tolerance in Crop Plants Against Abiotic Stress. CRC Press, Boca Raton, pp. 239–262. Available at 10.1201/9781003160717-12.

Ali, A.M., Ibrahim, S.M., 2020. Establishment of soil management zones using multivariate analysis and GIS. Communications in Soil Science and Plant Analysis 51 (19), 2491–2500. Available at https://doi.org/10.1080/00103624.2020.1836207.

Avtar, R., Sahu, N., Aggarwal, A.K., Chakraborty, S., Kharrazi, A., Yunus, A.P., et al., 2019. Exploring renewable energy resources using remote sensing and GIS-A review. Resources 8 (3). Available at https://doi.org/10.3390/resources8030149.

Azimi, S., Kaur, T., Gandhi, T.K., 2021. A deep learning approach to measure stress level in plants due to nitrogen deficiency. Measurement 173, 108650. Available at https://doi.org/10.1016/j.measurement.2020.108650.

Banerji, S., Hõrak, H., Torop, J., Huynh, T.-P., 2024. Unravelling the secrets of plants: emerging wearable sensors for plants signals and physiology. Advanced Sensor Research 3 (3). Available at https://doi.org/10.1002/adsr.202300023.

Bhatla, S.C., Kathpalia, R., 2023. Essential and functional mineral elements. In: Bhatla, S.C., Lal, M.A. (Eds.), Plant Physiology, Development and Metabolism. Springer, Singapore, pp. 25–49. Available at https://doi.org/10.1007/978-981-99-5736-1_2.

Biousse, V., Newman, N.J., 2016. Diagnosis and clinical features of common optic neuropathies. The Lancet Neurology 15 (13), 1355–1367. Available at https://doi.org/10.1016/S1474-4422(16)30237-X.

Bock, C.H., Poole, G.H., Parker, P.E., Gottwald, T.R., 2010. Plant disease severity estimated visually, by digital photography and image analysis, and by hyperspectral imaging. Critical Reviews in Plant Sciences 29 (2), 59–107. Available at https://doi.org/10.1080/07352681003617285.

Bongiovanni, R., Lowenberg-Deboer, J., 2004. Precision agriculture and sustainability. Precision Agriculture 5, 359–387. Available at https://doi.org/10.1023/B:PRAG.0000040806.39604.aa.

Boonupara, T., Udomkun, P., Kajitvichyanukul, P., 2024. Quantitative analysis of atrazine impact on UAV-derived multispectral indices and correlated plant pigment alterations: a heatmap approach. Agronomy 14 (4), 814. Available at https://doi.org/10.3390/agronomy14040814.

Bukhari, S.A.H., Peerzada, A.M., Javed, M.H., Dawood, M., Hussain, N., Ahmad, S., 2019. Growth and development dynamics in agronomic crops under environmental stress. In: Hasanuzzaman, M. (Ed.), Agronomic Crops: Volume 1: Production Technologies. Springer, Singapore, pp. 83–114. Available at https://doi.org/10.1007/978-981-32-9151-5_6.

Cakmak, I., 2005. The role of potassium in alleviating detrimental effects of abiotic stresses in plants. Journal of Plant Nutrition and Soil Science 168 (4), 521–530. Available at https://doi.org/10.1002/jpln.200420485.

Castroesau, K., Sanchezazofeifa, G., Rivard, B., 2006. Comparison of spectral indices obtained using multiple spectroradiometers. Remote Sensing of Environment 103 (3), 276–288. Available at https://doi.org/10.1016/j.rse.2005.01.019.

Cavaco, A.M., Utkin, A.B., Marques da Silva, J., Guerra, R., 2022. Making sense of light: the use of optical spectroscopy techniques in plant sciences and agriculture. Applied Sciences 12 (3), 997. Available at https://doi.org/10.3390/app12030997.

Choi, H.W., 2024. From the photosynthesis to hormone biosynthesis in plants. Plant Pathology Journal 40 (2), 99–105. Available at https://doi.org/10.5423/PPJ.RW.01.2024.0006.

Coatsworth, P., Gonzalez-Macia, L., Collins, A.S.P., Bozkurt, T., Güder, F., 2023. Continuous monitoring of chemical signals in plants under stress. Nature Reviews Chemistry 7 (1), 7–25. Available at https://doi.org/10.1038/s41570-022-00443-0.

Cuypers, S., Nascetti, A., Vergauwen, M., 2023. Land use and land cover mapping with VHR and multi-temporal sentinel-2 imagery. Remote Sensing 15 (10), 2501. Available at https://doi.org/10.3390/rs15102501.

Dritsas, E., Trigka, M., 2025. Remote sensing and geospatial analysis in the big data era: a survey. Remote Sensing 17 (3), 550. Available at https://doi.org/10.3390/rs17030550.

Galieni, A., D'Ascenzo, N., Stagnari, F., Pagnani, G., Xie, Q., Pisante, M., 2021. Past and future of plant stress detection: an overview from remote sensing to positron emission tomography. Frontiers in Plant Science 11. Available at https://doi.org/10.3389/fpls.2020.609155.

Gonzalez, F., Mcfadyen, A., Puig, E., 2018. Advances in unmanned aerial systems and payload technologies for precision agriculture. Advances in Agricultural Machinery and Technologies. CRC Press, pp. 33–155.

Hänsch, R., Mendel, R.R., 2009. Physiological functions of mineral micronutrients (Cu, Zn, Mn, Fe, Ni, Mo, B, Cl). Current Opinion in Plant Biology 12 (3), 259–266. Available at https://doi.org/10.1016/j.pbi.2009.05.006.

Hu, Y., Schmidhalter, U., 2005. Drought and salinity: a comparison of their effects on mineral nutrition of plants. Journal of Plant Nutrition and Soil Science 168 (4), 541–549. Available at https://doi.org/10.1002/jpln.200420516.

Jones, J.B., Case, V.W., 2018. Sampling, handling, and analyzing plant tissue samples. Soil Testing and Plant Analysis. Wiley, United States, pp. 389–427. Available at 10.2136/sssabookser3.3ed.c15.

Kariyanna, B., Senthil-Nathan, S., Vasantha-Srinivasan, P., Subba Reddy, B.V., Krishnaiah, A., Meenakshi, N.H., et al., 2024. Comprehensive insights into pesticide residue dynamics: unraveling impact and management. Chemical and Biological Technologies in Agriculture 11 (1). Available at https://doi.org/10.1186/s40538-024-00708-4.

Lambers, H., Oliveira, R.S., Pons, T.L., 2019. Growth and allocation. Plant Physiological Ecology. Springer, New York, NY, pp. 385–449. Available at https://doi.org/10.1007/978-0-387-78341-3_10.

Laveglia, S., Altieri, G., Genovese, F., Matera, A., Di Renzo, G.C., 2024. Advances in sustainable crop management: integrating precision agriculture and proximal sensing. AgriEngineering 6 (3), 3084–3120. Available at https://doi.org/10.3390/agriengineering6030177.

Lichtenthaler, H.K., Buschmann, C., Knapp, M., 2005. How to correctly determine the different chlorophyll fluorescence parameters and the chlorophyll fluorescence decrease ratio RFd of leaves with the PAM fluorometer. Photosynthetica 43 (3), 379–393. Available at https://doi.org/10.1007/s11099-005-0062-6.

Ma, G., Javidan, S.M., Ampatzidis, Y., Zhang, Z., 2025. A novel hybrid technique for detecting and classifying hyperspectral images of tomato fungal diseases based on deep feature extraction and manhattan distance. Sensors 25 (14), 4285. Available at https://doi.org/10.3390/s25144285.

Mahlein, A.K., 2016. Present and future trends in plant disease detection. Plant Disease 100, 1–11.

Mangrauthia, S.K., Molla, K.A., Sundaram, R.M., Chinnusamy, V., Bansal, K.C., 2024. Genomics and genome editing for crop improvement. In: Bansal, K.C., Lakra, W.S., Pathak, H. (Eds.), Transformation of Agri-Food Systems. Springer Nature, Singapore, pp. 297–322. Available at 10.1007/978–981-99–8014-7_22.

Marschner, H., 2011. Marschner's Mineral Nutrition of Higher Plants. Academic press.

Mayanja, I.K., Diepenbrock, C.H., Vadez, V., Lei, T., Bailey, B.N., 2024. Practical considerations and limitations of using leaf and canopy temperature measurements as a stomatal conductance proxy: sensitivity across environmental conditions, scale, and sample size. Plant Phenomics 6. Available at https://doi.org/10.34133/plantphenomics.0169.

Mduma, H.S., Mkindi, M., Karani, A., Ngiha, K., Kalonga, J., Mohamed, Y., et al., 2015. Major signs and symptoms caused by biotic and abiotic agents on plants in the tropical Africa. International Journal of Scientific Research 6, 750–759.

Mills, H.A., Jones, J.B., 1996. Plant Analysis Handbook II. A Practical Sampling, Preparation, Analysis, and Interpretation Guide. Micro-Macro Publishing, Athens.

Misra, B.B., Langefeld, C., Olivier, M., Cox, L.A., 2019. Integrated omics: tools, advances and future approaches. Journal of Molecular Endocrinology 62 (1), R21. Available at https://doi.org/10.1530/JME-18–0055.

Monaghan, P., Metcalfe, N.B., Torres, R., 2009. Oxidative stress as a mediator of life history trade-offs: mechanisms, measurements and interpretation. Ecology Letters 12 (1), 75–92. Available at https://doi.org/10.1111/j.1461-0248.2008.01258.x.

Moshood, A.Y., Abdulraheem, M.I., Li, L., Zhang, Y., Raghavan, V., Hu, J., 2025. Deciphering nutrient stress in plants: integrative insight from metabolomics and proteomics. Functional and Integrative Genomics 25 (1). Available at https://doi.org/10.1007/s10142-025-01551-y.

Mulder, V.L., de Bruin, S., Schaepman, M.E., Mayr, T.R., 2011. The use of remote sensing in soil and terrain mapping — a review. Geoderma 162 (1-2), 1–19. Available at https://doi.org/10.1016/j.geoderma.2010.12.018.

Munson, R.D., 2018. Soil fertility, fertilizers, and plant nutrition. In: Kilmer, V.J. (Ed.), Handbook of Soils and Climate in Agriculture. CRC Press, New York, pp. 269–294. Available at 10.1201/9781351073073.

Nayeri, S., Dolatyari, M., Mouladoost, N., Nayeri, S., Zarghami, A., Mirtagioglu, H., et al., 2023. Ag/ZnO core–shell NPs boost photosynthesis and growth rate in wheat seedlings under simulated full sun spectrum. Scientific Reports 13 (1). Available at https://doi.org/10.1038/s41598-023-41575-7.

Nijs, I.V.A.N., Impens, I.V.A.N., 1997. An analysis of the balance between root and shoot activity in *Lolium perenne* cv. Melvina. Effects of CO_2 concentration and air temperature. New Phytologist 135 (1), 81–91. Available at https://doi.org/10.1046/j.1469-8137.1997.00616.x.

OECD/FAO, 2017. OECD-FAO Agricultural Outlook 2017-2026. OECD Publishing, Paris. Available at https://doi.org/10.1787/agr_outlook-2017-en.

Pandey, A., Hunt, J., Murray, J., Maddern, K., Wang, X., Tang, C., et al., 2024. A comparison of nitrogen fertiliser decision making systems to profitably close grain yield gaps in semi-arid environments. Field Crops Research 318, 109576. Available at https://doi.org/10.1016/j.fcr.2024.109576.

Pandey, P., Singh, S.P., Kumar, U., Gupta, S.P., 2024. Role, nutrient deficiencies and toxicity symptoms of essential plant nutrients. In: Rathinasamy, A. (Ed.), Manure, Fertilizers and Soil Fertility Management. Scientific Publishers, pp. 120–138.

Peng, Y., Zhang, Y., Tang, Y., Li, S., 2011. An incident information management framework based on data integration, data mining, and multi-criteria decision making. Decision Support Systems 51 (2), 316–327. Available at https://doi.org/10.1016/j.dss.2010.11.025.

Pereira, E.I., Giroto, A.S., Bortolin, A., Yamamoto, C.F., Marconcini, J.M., De, A.C., et al., 2015. Perspectives in nanocomposites for the slow and controlled release of agrochemicals: fertilizers and pesticides. In: Rai, M., Ribeiro, C., Mattoso, L., Duran, N. (Eds.), Nanotechnologies in Food and Agriculture. Springer International Publishing, Cham, pp. 241–265. Available at https://doi.org/10.1007/978-3-319-14024-7_11.

Pineda, M., Barón, M., Pérez-Bueno, M.L., 2021. Thermal imaging for plant stress detection and phenotyping. Remote Sensing 13 (1), 1–21. Available at https://doi.org/10.3390/rs13010068.

Rahim, H.U., Ali, W., Uddin, M., Ahmad, S., Khan, K., Bibi, H., et al., 2025. Abiotic stresses in soils, their effects on plants, and mitigation strategies: a literature review. Chemistry and Ecology 41 (4), 552–585. Available at https://doi.org/10.1080/02757540.2024.2439830.

Rana, K., Tripathi, R.M., Pudake, R.N., 2024. Advances in the fabrication of nanosensors for management of abiotic stress in crop plants. In: Al-Khayri, J.M., Alnaddaf, L.M., Jain, S.M., Penna, S. (Eds.), Innovative Methods in Horticultural Crop Improvement. Advances in Plant Breeding Strategies, vol 2. Springer, Cham, pp. 75–97. Available at https://doi.org/10.1007/978-3-031-61095-0_4.

Schacher, A., Roger, E., Williams, K.J., Stenson, M.P., Sparrow, B., Lacey, J., 2023. Use-Specific considerations for optimising data quality trade-offs in citizen science: recommendations from a targeted literature review to improve the usability and utility for the calibration and validation of remotely sensed products. Remote Sensing 15 (5), 1407. Available at https://doi.org/10.3390/rs15051407.

Shabala, S., Pottosin, I., 2014. Regulation of potassium transport in plants under hostile conditions: implications for abiotic and biotic stress tolerance. Physiologia Plantarum 151 (3), 257–279. Available at https://doi.org/10.1111/ppl.12165.

Shah, S.H., Houborg, R., McCabe, M.F., 2017. Response of chlorophyll, carotenoid and SPAD-502 measurement to salinity and nutrient stress in wheat (*Triticum aestivum* L). Agronomy 7 (3), 61. Available at https://doi.org/10.3390/agronomy7030061.

Shakhatreh, H., Sawalmeh, A.H., Al-Fuqaha, A., Dou, Z., Almaita, E., Khalil, I., et al., 2019. Unmanned aerial vehicles (UAVs): a survey on civil applications and key research challenges. IEEE Access 7, 48572–48634. Available at https://doi.org/10.1109/ACCESS.2019.2909530.

Shtienberg, D., 2013. Will decision-support systems be widely used for the management of plant diseases? Annual Review of Phytopathology 51 (1), 1–16. Available at https://doi.org/10.1146/annurev-phyto-082712-102244.

Simane, B., Kapwata, T., Naidoo, N., Cissé, G., Wright, C.Y., Berhane, K., 2025. Ensuring Africa's food security by 2050: the role of population growth, climate-resilient strategies, and putative pathways to resilience. Foods 14 (2), 262. Available at https://doi.org/10.3390/foods14020262.

Sofi, P.A., Ara, A., Gull, M., Rehman, K., 2019. Canopy temperature depression as an effective physiological trait for drought screening. In: Ondrasek, G. (Ed.), Drought-Detection and Solutions. IntechOpen. Available at 10.5772/intechopen.85966.

Ustin, S.L., Jacquemoud, S., 2020. How the optical properties of leaves modify the absorption and scattering of energy and enhance leaf functionality. In: Cavender-Bares, J., Gamon, J.A., Townsend, P.A. (Eds.), Remote Sensing of Plant Biodiversity. Springer, Cham, pp. 349–384. Available at https://doi.org/10.1007/978-3-030-33157-3_14.

Van der Meer, F., 2012. Remote-sensing image analysis and geostatistics. International Journal of Remote Sensing 33 (18), 5644–5676. Available at https://doi.org/10.1080/01431161.2012.666363.

Voica, C., Feher, I., Puscas, R., Iordache, A.M., Cristea, G., 2025. Elemental and Isotopic fingerprints of potatoes. Foods 14 (14), 2440. Available at https://doi.org/10.3390/foods14142440.

Wang, S.-S., Kamphuis, W., Huitinga, I., Zhou, J.-N., Swaab, D.F., 2008. Gene expression analysis in the human hypothalamus in depression by laser microdissection and real-time PCR: the presence of multiple receptor imbalances. Molecular Psychiatry 13 (8), 786–799. Available at https://doi.org/10.1038/mp.2008.38.

Wang, Z., Wang, Y., Du, Q., Yan, P., Yu, B., Li, W.X., et al., 2023. The auxin signaling pathway contributes to phosphorus-mediated zinc homeostasis in maize. BMC Plant Biology 23 (1). Available at https://doi.org/10.1186/s12870-023-04039-8.

Wani, B.A., Rashid, S., Rashid, K., Javid, H., Magray, J.A., Ul Qadir, R., et al., 2022. Nitrogen deficiency in plants. Advances in Plant Nitrogen Metabolism. CRC Press, pp. 28–37.

Wu, B., Becker, J.S., 2012. Imaging techniques for elements and element species in plant science. Metallomics 4 (5), 403. Available at https://doi.org/10.1039/c2mt00002d.

Xu, D., Wang, C., Chen, J., Shen, M., Shen, B., Yan, R., et al., 2021. The superiority of the normalized difference phenology index (NDPI) for estimating grassland aboveground fresh biomass. Remote Sensing of Environment 264, 112578. Available at https://doi.org/10.1016/j.rse.2021.112578.

Zakavi, M., Nayeri, S., 2025. Nanosensors, climate change and agricultural sectors. In: Husen, A. (Ed.), Nanobiosensors for Agricultural and Other Related Sectors. Springer, Singapore, pp. 49–63. Available at https://doi.org/10.1007/978-981-96-6499-3_3.

Zakavi, M., Askari, H., Shahrooei, M., 2022. Maize growth response to different Bacillus strains isolated from a salt-marshland area under salinity stress. BMC Plant Biology 22, 367. Available at https://doi.org/10.1186/s12870-022-03702-w.

Zarco-Tejada, P.J., González-Dugo, V., Williams, L.E., Suárez, L., Berni, J.A.J., Goldhamer, D., et al., 2013. A PRI-based water stress index combining structural and chlorophyll effects: assessment using diurnal narrow-band airborne imagery and the CWSI thermal index. Remote Sensing of Environment 138, 38–50. Available at https://doi.org/10.1016/j.rse.2013.07.024.

Zhan, A., 2025. Multi-omics-driven adaptive management of biological invasions: toward a proactive, predictive, and integrative framework. Biological Diversity 2 (4), 163–192. Available at https://doi.org/10.1002/bod2.70015.

Zhang, Q., Li, J., Singh, V.P., Xiao, M., 2013. Spatio-temporal relations between temperature and precipitation regimes: implications for temperature-induced changes in the hydrological cycle. Global and Planetary Change 111, 57–76. Available at https://doi.org/10.1016/j.gloplacha.2013.08.012.

Zhang, Y., Huang, B., Jin, J., Xiao, Y., Ying, H., 2023. Recent advances in the application of ionomics in metabolic diseases. Frontiers in Nutrition 9. Available at https://doi.org/10.3389/fnut.2022.1111933.

CHAPTER

Breeding strategies for nutrient uptake and stress tolerance

25

Zeynep Görkem Doğaroğlu and Yağmur Uysal

Department of Environmental Engineering, Mersin University, Mersin, Turkiye

Abbreviations

AMF	arbuscular mycorrhizal fungi
APX	ascorbate peroxidase
BR	brassinosteroid
CAT	batalase
CRISPR	clustered regularly interspaced short palindromic repeats
crRNA	CRISPR RNA
Cys2His2	cysteine2-histidine2
GEBVs	genomic estimated breeding values
GR	glutathione reductase
GS	genomic selection
GWAS	Genome-Wide Association Study
MAS	marker-assisted selection
NGS	next-generation sequencing
NUE	nutrient use efficiency
PGPR	plant growth-promoting rhizobacteria
QTL	quantitative trait loci
rhizoSMASH	rhizosphere-competence-related cataboliSM Analysis Shell
ROS	reactive oxygen species
SDG	sustainable development goal
SNPs	single-nucleotide polymorphisms
SOD	superoxide dismutase
SSRs	simple sequence repeats
SynComs	synthetic microbial communities
TALENs	transcription activator-like effector nucleases
tracrRNA	trans-activating CRISPR RNA
ZFNs	zinc finger nucleases

Mineral Nutrients to Mitigate Abiotic Stress in Crops. DOI: https://doi.org/10.1016/B978-0-443-44712-9.00025-6

25.1 Introduction

Global agriculture faces numerous challenges, including drought and excessive fertilizer use, which, exacerbated by climate change, lead to soil salinization and nutrient depletion, reducing crop productivity and nutritional quality (Arif et al., 2025). Disruptions in the production of sufficient and safe food have direct and indirect consequences on multiple interconnected systems, ranging from sustainable economic development and the depletion of natural resources to natural disaster management and sociopolitical governance. Therefore, it is essential to establish effective production strategies by considering multiple factors, including optimized management of agricultural processes, efficient resource utilization, and the application of appropriate technologies (Chowdhuri and Pal, 2025). While chemical fertilizers can temporarily boost yield, they compromise quality and cause long-term environmental harm. Reduced precipitation, floods, and rising temperatures exacerbate soil fertility loss, threaten arable land, decrease biodiversity, and reduce yields. Therefore, ensuring production security under climate change and improving plant stress tolerance are essential for sustainable agriculture (García-Pérez et al., 2025). Recent studies focus on enhancing nutrient uptake and stress-response mechanisms to mitigate agricultural challenges (Arif et al., 2025; Teker Yildiz et al., 2024). Although current systems can meet global food demand, strategic interventions are needed for long-term sustainability, food safety, and environmental protection (Li et al., 2025; Vos and Bellù, 2018).

In line with Sustainable Development Goal 2 (SDG 2), which aims to end hunger, ensure food security, and improve nutrition, as well as promote sustainable and climate-resilient agriculture, all United Nations Member States have adopted this approach. Consequently, in the context of climate change and yield stagnation driven by production costs, modern sustainable agricultural practices inevitably prioritize accelerating production and developing climate-resilient, nutrient-rich crops. As the world population is projected to exceed 9.5 billion by 2050, current production strategies are unlikely to meet future food demand with traditional breeding methods (Kabade et al., 2025). Therefore, systematic efforts are required to advance novel agricultural methods and breeding technologies, reduce costs, enhance genetic diversity, and ensure the efficient use of resources (Thudi et al., 2025). Traditional cultivation methods face inherent limitations in crop yield, diversity, and resilience due to restricted arable land, soil degradation, fluctuating climate conditions, and limited genetic resources.

Plant breeding began with Mendel's discovery of trait inheritance, and modern systematic methods, including genetic hybridization, have accelerated breeding progress (Priyadharsini et al., 2025; Sun et al., 2024; Thudi et al., 2025). Traditional techniques such as hybridization, pedigree breeding, heterosis, mass selection, and backcrossing have substantially improved yields in staple crops like maize, wheat, and rice. Advances in genetic engineering and omics-based approaches have revealed molecular and cellular mechanisms underlying growth, development, and stress responses (Li et al., 2025). While conventional breeding enhances tolerance to biotic and abiotic stresses, it is time-consuming and limited (Sun et al., 2024). Contemporary strategies focus on improving both plant nutrition and stress resilience. Next-generation sequencing (NGS) facilitates decoding crop genomes and mitigating climate-induced stress, supporting breeding advancements (Kabade et al., 2025). Recent approaches include AI-based data monitoring and climate modeling (Li et al., 2025; Su et al., 2025; Wu et al., 2025), the development of green chemistry-derived plant nutrients (Abdulfatah et al., 2025; Motlhalamme et al., 2023), nano-fertilizers (Hamimed et al., 2025; Sharma et al., 2024), photobiological systems (Kabade et al., 2025), and interventions to enhance stress tolerance (Koley and Mondal, 2024; Shoukat et al., 2025).

Plant nutrition strategies encompass a broad spectrum of factors and must be thoroughly evaluated. Sustainable, modern, and ecologically sound alternatives to traditional methods are essential. Consequently, improving crop performance under adverse conditions requires the integration of both conventional and molecular breeding approaches. This section provides a comprehensive review of cultivation strategies that enhance nutrient uptake and stress tolerance, as well as the corresponding applications in contemporary plant breeding practices.

25.2 Importance of nutrient efficiency and stress tolerance in crops

Rising global populations and climate change are degrading agricultural soils. While plants serve as a primary food source for most living organisms, they themselves require essential nutrients to grow and develop, drawing these from soil, water, and air. Plants, as primary food sources, require sixteen essential nutrients for growth and productivity. Carbon (C), hydrogen (H_2), and oxygen (O_2) are primary nutrients from soil and air, while boron (B), calcium (Ca), chlorine (Cl), nitrogen (N), potassium (K), phosphorus (P), sulfur (S), magnesium (Mg), iron (Fe), manganese (Mn), molybdenum (Mo), copper (Cu), and zinc (Zn) are primarily soil-derived. Nutrient bioavailability depends on their chemical forms, and efficient utilization also requires adequate light, water, and suitable temperatures (Silva and Uchida, 2000). These nutrients are classified as macro- (C, H_2, N, O, P, S, Ca, K, Mg) and micronutrients (B, Cl, Fe, Cu, Mn, Zn, Mo), with macroelements needed in larger amounts (1–150 g/kg DW) and micronutrients in trace amounts (0.1 100 mg/kg DW) (Khan et al., 2023). Optimal concentrations are crucial, as deficiencies or excesses impair growth and yield. Therefore, nutrient quantity, combinations, and application rates relative to growth stage are key determinants of crop productivity (Namitha et al., 2025). Each macro- and micronutrient has a distinct functional role, and both deficiencies and excesses produce specific physiological and morphological symptoms.

Understanding the biochemical interactions in the rhizosphere that mediate nutrient uptake and influence plant growth, development, and productivity is complex yet critical (Gupta et al., 2025). Molecular and cellular processes between soil microorganisms, plant roots, and soil pH determine which nutrients are absorbed, by what pathways, and in what quantities. Plant root architecture is the most important element affecting the development of above-ground organs, nutrient uptake (especially N and P), yield, and even resistance to biotic and abiotic stresses. Studies have shown that optimized nitrogen application in maize improves root morphology, increases gibberellic acid levels, root dry weight, root length, and overall plant growth compared to non-fertilized controls (Li et al., 2022; Zhao et al., 2025). In *Grevillea robusta*, phosphorus (P) deficiency was identified as the primary determinant of cluster root formation (Senevirathna et al., 2024).

Agricultural soils serve as the largest reservoirs of soil organic carbon and hold immense potential for sustainable and secure food production. Soil organic carbon accumulation is closely linked to the presence and diversity of soil microorganisms (Bhattacharyya et al., 2021). Biotic and abiotic processes, mediated by microbial activity, enhance soil organic carbon content, reduce water loss, and trigger diverse plant defense mechanisms against stress (Gupta et al., 2025). Climate change amplifies the frequency and intensity of stress factors interacting with inherent soil conditions, rendering plant defense mechanisms highly complex and tissue-specific (Gupta et al., 2025). Exposure to extreme temperatures, salinity, heavy metals, and other stresses disrupts biochemical pathways, leading to the accumulation of reactive oxygen species (ROS) (García-Pérez et al., 2025; Rahman et al., 2024). ROS accumulation causes oxidative stress, damaging membranes, proteins, and DNA, inactivating enzymes, and impairing physiological processes. Unlike animals, plants cannot escape stress and must employ intrinsic defense mechanisms, often through metabolite production at the cellular level. While these mechanisms can provide adaptation to single stress factors, combined stress scenarios frequently overwhelm plant defenses. Cellular communication under abiotic stress, particularly ROS signaling at the plasma membrane and intracellular metabolite balance, is a critical focus of current research (Choudhary et al., 2024). Phytohormones, protein kinases, and gasotransmitters act synergistically with ROS signals to maintain cellular homeostasis and regulate root and shoot development in response to environmental changes (Singh et al., 2024).

Nutrient uptake mechanisms are fundamentally classified as active (energy-dependent, mediated by root transport proteins, with root surface area being critical) and passive (energy-independent, driven by concentration gradients, diffusion, and osmosis). Active and passive transport processes enable efficient nutrient acquisition and distribution. Additionally, plant root systems release a spectrum of low- and high-molecular-weight metabolites into the rhizosphere through rhizodeposition, thereby promoting nutrient solubilization, chelation, and overall bioavailability. These exudates simultaneously function as primary carbon substrates and metabolic energy sources that regulate the composition, activity, and trophic dynamics of rhizosphere microbial consortia (Patra et al., 2021). Within these microbial assemblages, plant growth-promoting rhizobacteria (PGPR) constitute a key functional guild responsible for mediating nutrient acquisition pathways, modulating plant physiological performance, and enhancing host resilience to pathogenic pressures and abiotic stressors (Gupta et al., 2025).

Modern breeding and molecular approaches increasingly leverage omics, genetic, and molecular techniques to evaluate nutrient use efficiency (NUE) for the development of high-yielding varieties. These advanced approaches offer more precise and effective alternatives to conventional breeding but remain incompletely resolved. Continued research seeks to expand knowledge on plant nutrition and identify new opportunities to enhance productivity. Optimizing nutrient use efficiency is critical to maximize the yield potential of limited fertile soils, ensure safe food access, and achieve productive cultivation even in marginal soils (Adhikari et al., 2022).

25.3 Recent strategies in crop breeding

Rapid changes in global climate models are increasingly exposing agricultural systems to a combination of abiotic and biotic constraints. In response, climate-smart cultivation has emerged as an integrative framework that expands traditional breeding objectives while prioritizing resilience to environmental stress, inter-seasonal yield stability, and enhanced water and NUE, simultaneously strengthening the adaptive potential of crop populations (Tabasum et al., 2025). Advances in modern genomics, including genomic selection, high-resolution genotyping platforms, environmental modeling that links genomic datasets with environmental variables, and polygenic prediction tools, now provide breeders with unprecedented capacity to predict and exploit genotype × environment (G×E) interactions, thus, the increase and development of climate-resistant varieties is accelerated, and support is provided for sustainable agriculture (Vikram et al., 2024). Furthermore, emerging genome editing techniques, such as CRISPR/Cas-based primary editing (e.g., Csy4 or Cas6f), enable precise modification of regulatory elements involved in stress signaling, facilitating more efficient management of carbon and resource allocation under heat or drought stress (Lu et al., 2025). When multi-location field evaluations and adaptive performance are considered in depth, climate-smart breeding has fostered a strategic perspective in plant breeding aimed at ensuring food security under changing climatic conditions and promoting sustainable agricultural productivity. Recent research has focused on modern plant breeding strategies, as summarized in Fig. 25.1, highlighting approaches that remain under active development.

25.3.1 Nanoparticle-mediated strategies

Nanotechnological materials, now integrated into nearly all aspects of modern life, have also gained significant attention in agriculture for their roles in yield improvement and plant health. Nanomaterials—with adjustable size, shape, surface charge, functionalization capacity, and strong antimicrobial properties, particularly

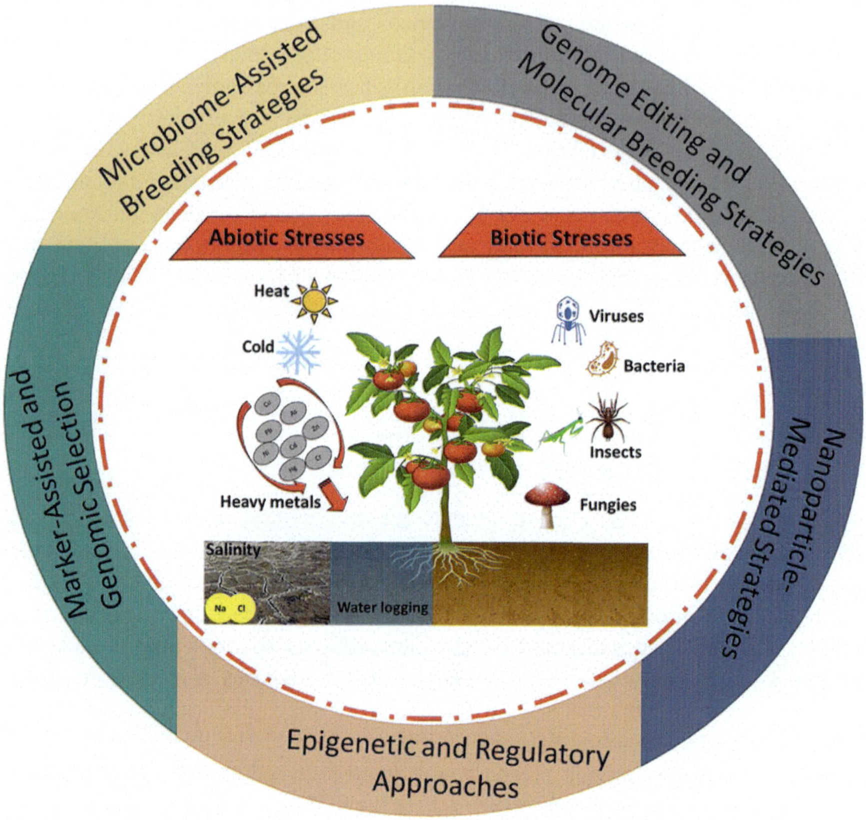

FIGURE 25.1

Modern plant breeding strategies applied against abiotic and biotic stress factors.

ZnO-NPs, AgNPs, SeNPs, SiO_2-NPs, TiO_2-NPs, and FeNPs—are widely explored in applications such as nano-fertilizers and nano-pesticides. However, their safety and long-term ecological effects remain insufficiently understood, despite numerous studies noting short-term improvements in growth and stress resilience (Meléndez-Mori et al., 2025; Uysal and Doğaroğlu, 2025). Nanoparticles can be applied as nano-fertilizers through seed treatment (nanopriming), foliar spray, or root/soil exposure. Seeds are the first structures facing environmental stress, and nano-priming aims to alleviate these stresses, maintain seed quality, and support strong germination and early development. Its efficiency depends on nanoparticle dose, size, synthesis method (green vs. chemical), and plant species (Khan et al., 2025). Reports indicate both positive and negative outcomes. For example, Ochoa-Chaparro et al. (2025) found that chitosan-stabilized ZnO-NPs, SiO_2-NPs, and their combinations enhanced drought tolerance and germination in jalapeño pepper (*Capsicum annuum* L.), with ZnO (100 mg/L) + SiO_2 (10 mg/L) improving seedling viability and antioxidant capacity under severe water deficit. Similarly, AgNP priming increased drought, salinity, and cold tolerance in Chinese cabbage (Zhuang et al., 2025) while also enhancing pathogen resistance via MAPK activation, phytohormone signaling, and detoxification pathways. SeNPs at a concentration of 50 mg/L enhanced maize seed vigor and antioxidant potential while simultaneously suppressing *Fusarium graminearum* infection, thereby supporting sustainable agricultural practices and contributing to improved food security (Dong et al., 2025). FeNP priming also increased salinity tolerance in barley, improving activity of antioxidant enzymes (APX, SOD, GR, CAT) and some agronomic traits (Alqudah et al., 2025).

Beyond seed priming, nanomaterials may also be delivered through foliar or soil/root applications. Foliar treatments are particularly advantageous, as they improve nutrient uptake while enhancing tolerance to biotic and abiotic stresses. Research primarily emphasizes ZnO-NPs, FeNPs, FeO-NPs, and SiO_2-NPs, zinc for its multifunctionality, iron for salinity tolerance, and silicon for reducing drought damage. Meléndez-Mori et al. (2025) showed that foliar application of ZnO-NPs (50 mg/L) increased proline and CAT activity while lowering MDA in salinity-stressed coffee plants, although higher doses (100 mg/L) offered limited photoprotection. Likewise, foliar FeNPs improved barley salinity tolerance by sustaining antioxidant enzyme activity (SOD, CAT, GR, APX) and stabilizing gene expression (PP2C, phosphotransferases), ultimately enhancing yield traits (Alqudah et al., 2025).

The use of nanoparticles in soil as nano-fertilizers has been extensively studied, similar to foliar applications. While most research emphasizes dose-response relationships, fewer studies address how sowing time and geographical location influence nanoparticle efficacy. Ali et al. (2025) demonstrated that the optimal application of ZnO NPs in lentil (*Lens culinaris* L.) was at sowing, with 2.5 mg/L proving most effective in the region with lower native soil zinc. ZnO-NPs have also shown benefits under drought and salinity stress. In cotton, reproductive-stage drought reduced yield, yet both bulk $ZnSO_4$ and nano-ZnO improved antioxidant capacity and reduced membrane damage, with ZnO-NPs specifically enhancing growth, water retention, and biomass. Furthermore, ZnO-NPs supported improved nutrient uptake and accumulation (Anik et al., 2025).

In summary, current literature demonstrates that nano-fertilizers can enhance stress tolerance and yield under key abiotic stresses (salinity, drought, etc.) across multiple crops. Foliar application, in particular, also confers protection against plant pathogens, highlighting the potential of nanoparticles as effective bio-agents in modern sustainable agriculture (Ding et al., 2025; Shoukat et al., 2025; Singh et al., 2025).

25.3.2 Genome editing and molecular breeding strategies

Plant breeding programs have adopted novel genome editing technologies alongside technological advancements. Traditional approaches, such as zinc finger nucleases (ZFNs) and transcription activator-like effector nucleases (TALENs), are effective but often complex, costly, and labor-intensive. In contrast, CRISPR/Cas-based methods developed over the past two decades have become increasingly popular in plant breeding due to their high efficiency, precision, affordability, speed, and capacity for multiplexed gene editing (Fan et al., 2025).

ZFN technology combines Cys2His2 zinc finger domains, responsible for specific DNA recognition, with the FokI endonuclease to create targeted double-strand breaks, enabling gene knockout, replacement, or insertion for the development of more resilient and high-yielding crops (Kanwal et al., 2024). However, its design complexity, high cost, and lengthy processing remain major limitations. TALEN technology, derived from TALE proteins of plant-pathogenic bacteria, similarly induces targeted DNA cleavage through a FokI catalytic domain, with TALE proteins directing sequence-specific binding. TALENs are valued for their precision, flexibility, and lower cost relative to ZFNs, though their large molecular size complicates delivery and production (Zubair et al., 2025). Compared to ZFN and TALEN systems, it is simpler, faster, and more economical, while also enabling multiplex editing. Due to these advantages, CRISPR has become a leading tool for engineering plants resistant to biotic stresses.

Organisms employ diverse defense mechanisms to withstand biotic and abiotic stresses, reprogramming gene expression and protein function to establish adaptive immunity. Understanding this reprogramming, particularly the function of Cas9, has marked a turning point in genome editing technologies (Sutula et al., 2025). CRISPR/Cas9 systems have emerged as highly versatile and efficient platforms for precise genome modification using the evolved adaptive immune mechanism of prokaryotes (bacteria and archaea) (Sutula et al., 2025; Ton et al., 2025). CRISPR-Cas9 enables precise gene modification, though off-target risks remain, and supports rapid breeding, improved stress tolerance, and enhanced plant adaptability. Numerous studies report its success in engineering crops resistant to heat, cold, drought, and salinity stress factors (Amombo et al., 2025; Keerthana et al., 2025; Wang et al., 2025a,b). In this system, crRNA and tracrRNA form a guide RNA that directs Cas9 to the target locus, where Cas9 cleaves DNA like genetic scissors, followed by endogenous repair that facilitates the intended modification (Wang et al., 2025a,b). While Cas9 is most commonly used, alternative nucleases such as Cas12a and Cas13, promising particularly for phytopathogen resistance, are also under investigation (Nerkar et al., 2022; Ton et al., 2025). Additionally, the prime editing technique, which operates through a mechanism similar to CRISPR/Cas, has emerged as a newly discovered and promising genome editing approach that is still under development (Amombo et al., 2025).

CRISPR/Cas9 genome editing systems have been reported to enhance NUE, freezing tolerance, and pathogen resistance, particularly in crops grown in arid and semi-arid regions. It is expected to improve long-term food and nutrition security while strengthening agricultural resilience (Amombo et al., 2025). In alfalfa (*Medicago sativa*), a key forage crop, CRISPR/Cas9 has increased nitrogen use efficiency, frost tolerance, and pathogen resistance, though precautions are needed to prevent gene flow to wild relatives (Fan et al., 2025). Similarly, in tomato, where the genetic base is narrow, CRISPR-based genome editing offers significant potential for improving tolerance to both biotic and abiotic stresses (Keerthana et al., 2025).

Therefore, the CRISPR/Cas technique allows researchers to precisely manipulate plant DNA and control genetic modifications, enabling the development of varieties with enhanced plant productivity, crop yield, pathogen suppression, tolerance to abiotic stresses, and nutritional value. However, researchers also emphasize that plants using the CRISPR/Cas technique must be cultivated with care to avoid disrupting the ecological balance. It is crucial to establish boundaries between cultivated plants and their wild counterparts.

25.3.3 Microbiome-assisted breeding strategies

Microbiome-assisted breeding leverages microorganisms in the rhizosphere, endosphere, or phyllosphere (including epiphytes) to enhance plant growth, nutrient acquisition, yield, and stress resistance. These interactions occur naturally, but increasing biotic and abiotic stresses can disrupt them. Yield losses in staple crops like rice, maize, and wheat have prompted efforts to optimize these natural plant-microbiome relationships, aiming to improve NUE while reducing chemical fertilizer use. Characterizing host plants and stress conditions is crucial because nutrient uptake is influenced by climatic and environmental biotic and abiotic factors such as temperature, sunlight, moisture, soil properties, and the microbial community present. Roots are the primary nutrient uptake sites, with rhizosphere processes playing a direct or indirect role. Arbuscular mycorrhizal fungi (AMF), symbiotic with nearly all higher plant roots, enhance nutrient and water uptake and strengthen plant defenses according to genetic diversity. By supplying carbohydrates to AMF, plants benefit from improved biofortification, stress tolerance, and long-term sustainability (Dwivedi et al., 2025; Singh et al., 2025a, b). Table 25.1 presents the most common microbial species found on the seed surface, as well as in the plant rhizosphere and endosphere, along with their functional roles in the plant.

Plants regulate nutrient uptake, health, and stress tolerance through microorganisms in the rhizosphere, phyllosphere, and endosphere, which contribute to nitrogen fixation, hormone production, and host gene expression. Microbiome engineering increasingly explores human-mediated interventions to optimize these interactions, enhancing productivity, reducing chemical inputs, and mitigating stress effects (Misu et al., 2025). Key microbial genes, termed “M genes,” provide opportunities to shape microbial communities and improve crop resilience (Cernava, 2024; Misu et al., 2025). Additionally, plant breeding has incorporated “R genes” for stress tolerance and “S genes” for environmental susceptibility; while these genes primarily target pathogen resistance, they offer short-term adaptive solutions (Cernava, 2024).

Nearly all plants interact with diverse microorganisms, with root architecture and exudates shaping the rhizosphere microbiome and influencing nutrient uptake and immunity. Key nutrient-related genes include NRT1.1 and PHT1, while defense genes include MYB72, WAK2, and WRKY (Dwivedi et al., 2025). As plants

Table 25.1 The most common microorganism species and their functional roles in plant breeding.

Microorganism/Group	Functional role in plant breeding	Key references
Bacillus spp.	Biological management and control that improve plant performance and nutrient cycling	Hernández-García et al. (2025), Kusstatscher et al. (2021), Zhao et al. (2025)
Burkholderia spp.	Nitrogen fixation and healthy plant growth promotion	dos Santos et al. (2021), Kusstatscher et al. (2021), Misu et al. (2025)
Paenibacillus spp.	Abundant in most rhizospheric soils. Antifungal activity and plant protection	Bziuk et al. (2021), Mendes et al. (2018)
Pseudomonas spp.	Growth promotion, biocontrol and pathogen suppression, and resource competition	Hu et al. (2016), Kusstatscher et al. (2021), Lazcano et al. (2021)
Pantoea spp.	Core seed microbiome, plant growth promotion, and disease protection to various plant species	Duchateau et al. (2024), Hernández-García et al. (2025), Michl et al. (2024)
Enterobacteriaceae	Early plant growth stages, the rhizosphere, and seed microbiome assembly	Bziuk et al. (2021), Hernández-García et al. (2025), Kusstatscher et al. (2021)
Actinobacteria (e.g., *Streptomyces*, *Frankia*)	An environmentally friendly alternative that contributes to higher crop yields, nitrogen fixation, and effective biotic stress management	Kariyanna and Panda (2025), Marco et al. (2022), Misu et al. (2025)
Rhizobium, *Bradyrhizobium*	Nitrogen fixation, symbiosis with legumes, and mechanisms that promote additional growth, alleviating the negative impact of stress-induced ethylene production on nodulation	dos Santos et al. (2025), Lazcano et al. (2021), Misu et al. (2025)
Fungal groups (e.g., Trichoderma, Ascomycota, Basidiomycota, Glomeromycota)	Disease suppression in the endosphere and rhizosphere, mycorrhizal symbiosis, and enhanced stress tolerance	Gopal and Gupta (2016), Li et al. (2024), Misu et al. (2025)

develop, rhizosphere microbial communities evolve. Kusstatscher et al. (2021) analyzed seeds and rhizosphere microbiota of *Cucurbita pepo* L., identifying beneficial bacterial families such as *Enterobacteriaceae*, *Bacillaceae*, *Burkholderiaceae*, and *Pseudomonadaceae*. Bacteria predominated in juvenile seeds, whereas the rhizosphere was enriched with fungi, highlighting seeds' potential for vertical microbial transmission and targeted breeding applications. Building on this, recent studies focus on Seed-Derived Synthetic Microbial Communities (SynComs), assembling bacterial isolates into consortia tailored to host requirements under laboratory conditions (Charron et al., 2025; Gastélum et al., 2025; Herath Dissanayakalage et al., 2025; Durán et al., 2025). Additionally, rhizoSMASH studies have recently gained attention, which aim to identify bacterial catabolic genes regulating the rhizosphere microbiota and integrate them with machine learning approaches (Li et al., 2025).

25.3.4 Marker-assisted and genomic selection

Marker-assisted and genomic selection (GS) has been a focus of plant breeding research since the early 2000s. Marker-assisted selection (MAS) relies on the identification of individuals harboring specific DNA-based markers and is increasingly transforming contemporary breeding strategies by incorporating multi-omics datasets. Through quantitative trait loci (QTL) mapping, genomic regions associated with target traits, including single-nucleotide polymorphisms (SNPs) and simple sequence repeats (SSRs), can be identified and subsequently used to inform selection decisions. The process typically involves phenotyping, QTL mapping, marker validation, and targeted selection.

In genomic selection, a reference population that has been both phenotypically characterized and genotyped provides comprehensive genome-wide SNP datasets. Machine learning–based predictive models, including GBLUP and BayesB, are then trained on these data to estimate genomic breeding values (GEBVs), thereby enabling the informed selection of genetically superior candidates within breeding programs (Alemu et al., 2024). Omics-based approaches, whether low-cost or initially expensive but long-term beneficial, enhance stress tolerance and product quality (Gao and Li, 2025; Rajendra, 2025). MAS and GS have been successfully applied to key crops such as tomato (Rajendra, 2025), potato (Zhang et al., 2025), soybean (Wang et al., 2024), rice (Ma et al., 2025; Mamidi et al., 2026), Japanese plum (Ballesta et al., 2025), sweet pepper (Martina et al., 2025), and maize, improving product quality, disease resistance, and tolerance to abiotic stresses such as drought. For instance, MAS enhanced potato resistance to late blight, viral diseases, and drought (Zhang et al., 2025) and improved soybean tolerance to drought, salinity, extreme temperatures, seed yield, and quality (Gao and Li, 2025; Wang et al., 2024). In rice, resequencing datasets were used to create RiceAtlas, mapping 3131 QTLs across 6044 cultivars under diverse environments, facilitating precise molecular breeding (Ma et al., 2025). Similarly, qCT7, a QTL linked to stem thickness and lodging resistance, was identified for targeted breeding (Mamidi et al., 2026). In *Brassica rapa* subsp. chinensis, leaf morphology genes (BrcLL1-A and BrcLL1-B) were characterized, highlighting SNP roles in stress tolerance (Song et al., 2026). Some studies also discuss integrating MAS with CRISPR to further improve nutrient uptake and stress resilience (Wang et al., 2024).

25.3.5 Epigenetic and regulatory approaches

Plants, as sessile organisms, face diverse environmental stresses throughout their life but possess efficient defense mechanisms. Phenotypic plasticity, allowing rapid adaptation to stress, is a key strategy involving gene expression changes regulated by epigenetic modifications (Singh et al., 2024). These breeding strategies exploit changes occurring in plant chromatin in response to environmental conditions. Epigenetic breeding leverages such chromatin changes to enhance yield and stress tolerance without altering the genome, often via DNA methylation or histone modifications, enabling subsequent generations to “remember” prior stress exposures (Kisvarga et al., 2025). This approach allows regulation of gene activity and understanding of plant responses to biotic and abiotic stresses (Singh et al., 2024; Kaya et al., 2024). For example, DNA methylation through the abscisic acid (ABA) pathway in citrus and barley, as well as RNA-directed DNA methylation (RdDM) via siRNA, enhances drought tolerance (Kaya et al., 2024). In canola, salt stress decreased DNA methylation but increased histone modifications, highlighting their regulatory role (Kong et al., 2025). In tea, epigenetic mechanisms, particularly m6A methylation and miRNAs, influence secondary metabolite synthesis and stress responses (Liao et al., 2025). Wang et al. (2025a,b) combined RNA-seq, ATAC-seq, and GWAS to identify regulatory elements controlling selenium (Se) tolerance in maize, pinpointing ZmGSR2, ZmDOF21, ZmWRKY48, ZmbZIP123, and ZmKNOX6 as key factors in Se metabolism and detoxification.

Epigenetic modifications play essential roles in fertilization, product development, fruit formation, and seed germination (Kisvarga et al., 2025). However, abiotic stress responses can vary between plants, generations, and even genomic regions, showing hypo- or hypermethylation patterns. Epigenetic breeding must therefore address challenges such as identifying epigenetic QTLs (epiQTLs), estimating unexplained heritability, understanding grafting effects, and elucidating mediators of epigenetic changes. For example, Quan et al. (2025) identified rice QTLs/genes (PHS10.1, Hd3a, SD1, D3, Ghd7.1, GH3.5) affecting plant height and grain yield under salinity stress. Integrating beneficial alleles from these loci under both normal and saline conditions can balance growth and stress resilience. Shi et al. (2025) studied brassinosteroid (BR) biosynthesis genes in *C. cathayensis*, including hydrophilic (CcCPD, CcDWF4) and hydrophobic (CcDET2) proteins, which regulate growth and stress responses under salinity, drought, and cold. These genes offer potential targets for molecular breeding of woody plants with enhanced salt tolerance.

25.4 Conclusion

Modern breeding technologies are advancing climate adaptation strategies for plants facing increasing climatic and pathogen pressures, offering opportunities to improve yield and productivity under adverse conditions. Nutrient uptake, essential for plant growth, is central to these approaches, which also enhance resilience to biotic and abiotic stresses. Currently, Marker-Assisted Selection (MAS), QTL mapping, and Genomic Selection (GS) are widely used to improve nutrient use efficiency and stress tolerance. Emerging strategies, including genome editing (e.g., CRISPR/Cas), microbiome-Assisted, and molecular breeding approaches, show promise for enhancing multi-stress tolerance and targeting nutrient-regulating genes, with growing research support. While nanotechnological and epigenetic approaches hold great potential, they remain in development, requiring further research, regulatory frameworks, and commercial application. For instance, epigenetic regulatory networks governing stress responses are not yet fully elucidated and represent an area for future advancement.

25.5 Future considerations

The development of modern breeding approaches requires careful consideration of environment-genotype-management interactions, alongside regulatory, biosafety, and societal factors. While each strategy has unique strengths, its limitations can be mitigated through AI-assisted integration and synergistic combinations. For example, AI driven multiomics analyses can reduce off-target effects in CRISPR/Cas systems, and incorporating epigenetic data into CRISPR design further enhances precision. Addressing these challenges demands not only technological advances but also accurate assessment of climate-induced stresses, support for sustainable agriculture, and coordinated regional and global research efforts. Continuity and innovation are essential to achieve these goals and ensure food security.

References

Abdulfatah, H.F., Abdulrahman, M.F., Naji, E.F., 2025. Green synthesis of iron nanoparticles to promote seed germination of *Zea mays* under salinity condition. Heliyon 11, e41823. Available at https://doi.org/10.1016/j.heliyon.2025.e41823.

Adhikari, S., Anuragi, H., Chandra, K., Tarte, S.H., Dhaka, S.R., Jatav, H.S., et al., 2022. Molecular basis of plant nutrient use efficiency - concepts and challenges for its improvement. Sustainable Plant Nutrition: Molecular Interventions and Advancements for Crop Improvement. Academic Press, pp. 107–151. Available at https://doi.org/10.1016/B978-0-443-18675-2.00001-8.

Alemu, A., Åstrand, J., Montesinos-López, O.A., Isidro y Sánchez, J., Fernández-Gónzalez, J., Tadesse, W., et al., 2024. Genomic selection in plant breeding: key factors shaping two decades of progress. Molecular Plant 17 (4), 552–578. Available at https://doi.org/10.1016/j.molp.2024.03.007.

Ali, M.A., Kordrostami, M., Javar, F.M., Rahimi, M., 2025. Effects of planting date and zinc oxide nano fertilizer on growth and yield of lentil (*Lens culinaris* L.) in two agroecological zones. BMC Plant Biology 25 (1). Available at https://doi.org/10.1186/s12870-025-07187-1.

Alqudah, A.M., Safhi, F.A., Thabet, S.G., 2025. Genetic insights into the synergistic effects of nano-iron on yield, quality, and antioxidant defense in barley under salt stress. Molecular Genetics and Genomics 300 (1). Available at https://doi.org/10.1007/s00438-025-02254-6.

Amombo, E., Mumm, R.H., Nilahyane, A., Hirich, A., Hijri, M., Ntui, V.O., 2025. Enhancing crop nutrition in arid and semiarid regions of africa through genome editing using CRISPR/Cas. Food and Energy Security 14 (3). Available at https://doi.org/10.1002/fes3.70068.

Anik, T.R., Mostofa, M.G., Rahman, M.M., Keya, S.S., Van Ha, C., Khan, M.A.R., et al., 2025. Comparative effects of $ZnSO_4$ and ZnO-NPs in improving cotton growth and yield under drought stress at early reproductive stage. Plant Science 359. Available at https://doi.org/10.1016/j.plantsci.2025.112589.

Arif, M., Ilyas, M., Adnan, M., Kalsoom, R., Ren, M., Xu, R., et al., 2025. Molecular mechanisms and breeding strategies for enhancing wheat resilience to environmental stresses: the role of heat shock proteins and implications for food security. International Journal of Biological Macromolecules 308 (Pt 3), 142468. Available at https://doi.org/10.1016/j.ijbiomac.2025.142468.

Ballesta, P., Fiol, A., Ahumada, S., Osorio, M., Ibañez, J., Fresnedo-Ramírez, J., et al., 2025. Genomic prediction of phenological and fruit-quality traits in a multi-family Japanese plum breeding population. Horticultural Plant Journal. Available at https://doi.org/10.1016/j.hpj.2025.04.014.

Bhattacharyya, R., Rabbi, S.M.F., Zhang, Y., Young, I.M., Jones, A.R., Dennis, P.G., et al., 2021. Soil organic carbon is significantly associated with the pore geometry, microbial diversity and enzyme activity of the macro-aggregates under different land uses. Science of the Total Environment 778. Available at https://doi.org/10.1016/j.scitotenv.2021.146286.

Bziuk, N., Maccario, L., Straube, B., Wehner, G., Sørensen, S.J., Schikora, A., et al., 2021. The treasure inside barley seeds: microbial diversity and plant beneficial bacteria. Environmental Microbiomes 16 (1). Available at https://doi.org/10.1186/s40793-021-00389-8.

Cernava, T., 2024. Coming of age for microbiome gene breeding in plants. Nature Communications 15 (1). Available at https://doi.org/10.1038/s41467-024-50700-7.

Charron, R., Lemée, P., Huguet, A., Minlong, O., Boulanger, M., Houée, P., et al., 2025. Strain-dependent emergence of aminoglycoside resistance in *Escherichia coli* biofilms. Biofilm 9, 100273. Available at https://doi.org/10.1016/j.bioflm.2025.100273.

Choudhary, A., Sharma, S., Kaur, H., Sharma, N., Gadewar, M.M., Mehta, S., et al., 2024. Plant system, abiotic stress resilience, reactive oxygen species, and coordination of engineered nanomaterials: a review. South African Journal of Botany 171, 45–59. Available at https://doi.org/10.1016/j.sajb.2024.05.053.

Chowdhuri, I., Pal, S.C., 2025. Challenges and potential pathways towards sustainable agriculture crop production: a systematic review to achieve sustainable development goals (SDGs). Soil and Tillage Research 248, 106442. Available at https://doi.org/10.1016/j.still.2024.106442.

Ding, M., Xiao, C., Li, S., Liu, J., Kong, Y., Zhang, C., et al., 2025. Effects of iron, zinc, and silicon nanoparticles on morpho-physiological growth, yield, and quality of wheat (*Triticum aestivum* L.) under drought stress. Journal of Hazardous Materials Advances 19, 100789. Available at https://doi.org/10.1016/j.hazadv.2025.100789.

Dong, Q., Huang, T., Zhou, C., Wan, X., He, X., Miao, P., et al., 2025. Nano-priming with selenium nanoparticles reprograms seed germination, antioxidant defense, and phenylpropanoid metabolism to enhance Fusarium graminearum resistance in maize seedlings. Journal of Advanced Research 81, 13–25. Available at https://doi.org/10.1016/j.jare.2025.05.057.

dos Santos, I.B., Pereira, A.Pd.A., de Souza, A.J., Cardoso, E.J.B.N., da Silva, F.G., Oliveira, J.T.C., et al., 2021. Selection and characterization of *Burkholderia* spp. for their plant-growth promoting effects and influence on maize seed germination. Frontiers in Soil Science 1. Available at https://doi.org/10.3389/fsoil.2021.805094.

dos Santos, M.D.T., Favero, V.O., Beatriz, A., Leite, C.A.B., da Costa, G., et al., 2025. Harnessing rhizobial inoculation for sustainable nitrogen management in mung bean (Vigna radiata L.). Plants 14 (23), 3695. Available at https://doi.org/10.3390/plants14233695.

Duchateau, S., Crouzet, J., Dorey, S., Aziz, A., 2024. The plant-associated Pantoea spp. as biocontrol agents: mechanisms and diversity of bacteria-produced metabolites as a prospective tool for plant protection. Biological Control 188. Available at https://doi.org/10.1016/j.biocontrol.2024.105441.

Durán, P., Vailleau, F., Roux, F., 2025. Building microbial synthetic communities: get inspired by the design of synthetic plant communities. New Phytologist 246 (2), 402–405. Available at https://doi.org/10.1111/nph.70011.

Dwivedi, S.L., Vetukuri, R.R., Kelbessa, B.G., Gepts, P., Heslop-Harrison, P., Araujo, A.S.F., et al., 2025. Exploitation of rhizosphere microbiome biodiversity in plant breeding. Trends in Plant Science 30 (9), 1033–1045. Available at https://doi.org/10.1016/j.tplants.2025.04.004.

Fan, S., Jia, L., Wu, J., Zhao, Y., 2025. Harnessing the potential of CRISPR/cas in targeted alfalfa improvement for stress resilience. International Journal of Molecular Sciences 26 (7), 3311. Available at https://doi.org/10.3390/ijms26073311.

Gao, H., Li, H., 2025. Marker-assisted selection (MAS) in soybean breeding. Molecular Plant Breeding. 16 (1), 35–43. Available at https://doi.org/10.5376/mpb.2025.16.0004.

García-Pérez, P., Losada-Barreiro, S., Bravo-Díaz, C., Gallego, P.P., Lucini, L., 2025. Plant oxidative stress and specialized metabolites: a holistic physicochemical and metabolomics perspective. Plant Physiology and Biochemistry 227, 110093. Available at https://doi.org/10.1016/j.plaphy.2025.110093.

Gastélum, G., Gómez-Gil, B., Olmedo-Álvarez, G., Rocha, J., 2025. Harnessing emergent properties of microbial consortia for agriculture: assembly of the Xilonen SynCom. Biofilm 9, 100284. Available at https://doi.org/10.1016/j.bioflm.2025.100284.

Gopal, M., Gupta, A., 2016. Microbiome selection could spur next-generation plant breeding strategies. Frontiers in Microbiology 7. Available at https://doi.org/10.3389/fmicb.2016.01971.

Gupta, D., Dev, V., Garg, S., Bhandari, G., Raverkar, K.P., Punetha, H., et al., 2025. Biochemistry behind stress-induced modulations of plant growth promoting rhizobacteria. Physiological and Molecular Plant Pathology 139, 102752. Available at https://doi.org/10.1016/j.pmpp.2025.102752.

Hamimed, S., Bousdira, F., Harkas, M., 2025. Novel development of sustainable nanofertilizers from olive pomace waste. Biocatalysis and Agricultural Biotechnology 66, 103616. Available at https://doi.org/10.1016/j.bcab.2025.103616.

Herath Dissanayakalage, S.S., Kaur, J., Li, T., Dimech, A.M., Sawbridge, T.I., 2025. Seed-derived synthetic microbial communities (SynComs) from medicago wild relatives modulate early plant microbiome assembly and phenotypic traits in lucerne (*Medicago sativa* L.). Microorganisms 13 (9). Available at https://doi.org/10.3390/microorganisms13092114.

Hernández-García, J.A., Bernal, J.S., Antony-Babu, S., Villa-Tanaca, L., Hernández-Rodríguez, C., De-la-Vega-Camarillo, E., 2025. Teosinte-derived SynCom and precision biofertilization modulate the maize microbiome, enhancing growth, yield, and soil functionality in a Mexican field. Frontiers in Microbiology 16. Available at https://doi.org/10.3389/fmicb.2025.1534327.

Hu, J., Wei, Z., Friman, V.P., Gu, S.H., Wang, X.F., Eisenhauer, N., et al., 2016. Probiotic diversity enhances rhizosphere microbiome function and plant disease suppression. MBio 7 (6). Available at https://doi.org/10.1128/mBio.01790–16.

Kabade, P.G., Kumar, S., Kohli, A., Singh, U.M., Sinha, P., Singh, V.K., 2025. Speed breeding 3.0: mainstreaming light-driven plant breeding for sustainable genetic gains. Trends in Biotechnology 43 (10), 2462–2478. Available at https://doi.org/10.1016/j.tibtech.2025.04.011.

Kanwal, F., Deabel, R.A., Alqahtani, A., Mahmoud, E.H.M., Naeem, M., Ovais, M., et al., 2024. Evolution of genome editing technologies. CRISPRized Horticulture Crops: Genome Modified Plants and Microbes in Food and Agriculture. Elsevier, Pakistan, pp. 21–36. Available at http://doi.org/10.1016/B978-0-443–13229-2.00009-0.

Kariyanna, B., Panda, S., 2025. Unveiling the advanced application of actinomycetes in biotic stress management for sustainable agriculture. Discover Life 55 (1). Available at https://doi.org/10.1007/s11084-025-09704-6.

Kaya, C., Uğurlar, F., Adamakis, I.D.S., 2024. Epigenetic modifications of hormonal signaling pathways in plant drought response and tolerance for sustainable food security. International Journal of Molecular Sciences 25 (15). Available at https://doi.org/10.3390/ijms25158229.

Keerthana, R., Rakshana, P., Salunkhe, S.R., Sakthi, A.R., Kokiladevi, E., Saraswathi, T., et al., 2025. CRISPR-Cas9 mediated enhancement of abiotic stress resilience in tomato: a comprehensive review of target genes. Molecular Biology Reports 52 (1). Available at https://doi.org/10.1007/s11033-025-10634-9.

Khan, M.I.R., Nazir, F., Maheshwari, C., Chopra, P., Chhillar, H., Sreenivasulu, N., 2023. Mineral nutrients in plants under changing environments: a road to future food and nutrition security. Plant Genome 16 (4). Available at https://doi.org/10.1002/tpg2.20362.

Khan, M.N., Li, Y., Mu, Y., Sultan, H., Baloch, A., Din, I., et al., 2025. Recent advances in nano-enabled plant salt tolerance: methods of application, risk assessment, opportunities and future prospects. Journal of Integrative Agriculture 24 (5). Available at https://doi.org/10.1016/j.jia.2024.05.028.

Kisvarga, S., Horotán, K., Orlóci, L., 2025. The role and possibilities of epigenetics in ornamental plant breeding. Plant Molecular Biology Reporter 43 (3), 941–953. Available at https://doi.org/10.1007/s11105-025-01545-x.

Koley, R., Mondal, N.K., 2024. Synthesis of calcium-based nanofertilizer and its efficacy towards reduction of oxidative stress and fluoride uptake in rice (*Oryza sativa* L). Plant Nano Biology 9, 100087. Available at https://doi.org/10.1016/j.plana.2024.100087.

Kong, F., Xiao, Y., Li, J., Wang, J., 2025. Insights into the salt stress response of allotetraploid Brassica napus from epigenetic modification and expression changes. Plant Stress 18, 101016. Available at https://doi.org/10.1016/j.stress.2025.101016.

Kusstatscher, P., Adam, E., Wicaksono, W.A., Bernhart, M., Olimi, E., Müller, H., et al., 2021. Microbiome-assisted breeding to understand cultivar-dependent assembly in cucurbita pepo. Frontiers in Plant Science 12. Available at https://doi.org/10.3389/fpls.2021.642027.

Lazcano, C., Boyd, E., Holmes, G., Hewavitharana, S., Pasulka, A., Ivors, K., 2021. The rhizosphere microbiome plays a role in the resistance to soil-borne pathogens and nutrient uptake of strawberry cultivars under field conditions. Scientific Reports 11 (1). Available at https://doi.org/10.1038/s41598-021-82768-2.

Li, J., Hou, L., Zhang, G., Cheng, L., Liu, Y., 2024. Comparative analysis of rhizosphere and endosphere fungal communities in healthy and diseased faba bean plants. Journal of Fungi 10 (1), 84. Available at https://doi.org/10.3390/jof10010084.

Li, X., Wang, R., Zhou, B., Wang, X., Wang, J., Zhao, M., et al., 2022. Characterization of root morphology and anatomical structure of spring maize under varying n application rates and their effects on yield. Agronomy 12 (11), 2671. Available at https://doi.org/10.3390/agronomy12112671.

Li, Y., Sun, M., Raaijmakers, J.M., Mommer, L., Zhang, F., Song, C., et al., 2025. Predicting rhizosphere-competence-related catabolic gene clusters in plant-associated bacteria with rhizoSMASH. Nature Communications 16 (1). Available at https://doi.org/10.1038/s41467-025-63526-8.

Liao, X., Tang, M., Lei, W., Lin, H., Jin, S., Wang, P., et al., 2025. Epigenetic mechanisms of heterosis in tea plants: insights into m6A RNA methylation and miRNA-mediated regulation. Horticultural Plant Journal. Available at https://doi.org/10.1016/j.hpj.2024.11.005.

Lu, Y., Qiao, D., Wang, J., Sun, W., Cao, Z., Lu, M., et al., 2025. Two highly efficient prime editing systems based on the Csy4 CRISPR endonuclease. Plant Biotechnology Journal. Available at https://doi.org/10.1111/pbi.70337.

Ma, X., Wang, H., Yan, S., Zhou, C., Zhou, K., Zhang, Q., et al., 2025. Large-scale genomic and phenomic analyses of modern cultivars empower future rice breeding design. Molecular Plant 18 (4), 651–668. Available at https://doi.org/10.1016/j.molp.2025.03.007.

Mamidi, A., Krishna, L., Badri, J., Barbadikar, K.M., Durbha, S.R., Prasad, S.S., et al., 2026. Navigating rice culm resilience: high-throughput quantitative trait locus mapping in indica and tropical japonica derived population. Plant Science 362. Available at https://doi.org/10.1016/j.plantsci.2025.112756.

Marco, S., Loredana, M., Riccardo, V., Raffaella, B., Walter, C., Luca, N., 2022. Microbe-assisted crop improvement: a sustainable weapon to restore holobiont functionality and resilience. Horticulture Research 9. Available at https://doi.org/10.1093/hr/uhac160.

Martina, M., Gianoglio, N., Barchi, L., Gaccione, L., Milanesi, C., Acquadro, A., et al., 2025. Towards genomic-assisted breeding in sweet pepper: fruit morphology and selective sweep analysis in MAGIC founders. Scientia Horticulturae 350, 114352. Available at https://doi.org/10.1016/j.scienta.2025.114352.

Meléndez-Mori, J.B., Lapiz-Culqui, Y.K., Huaman-Huaman, E., Zuta Puscan, M., Oliva Cruz, M., 2025. Can zinc oxide nanoparticles alleviate the adverse effects of salinity stress in *Coffea arabica*? Agronomy 15 (5). Available at https://doi.org/10.3390/agronomy15051239.

Mendes, L.W., Mendes, R., Raaijmakers, J.M., Tsai, S.M., 2018. Breeding for soil-borne pathogen resistance impacts active rhizosphere microbiome of common bean. ISME Journal 12 (12), 3038–3042. Available at https://doi.org/10.1038/s41396-018-0234-6.

Michl, K., David, C., Dumont, B., Mårtensson, L.M.D., Rasche, F., Berg, G., et al., 2024. Determining the footprint of breeding in the seed microbiome of a perennial cereal. Environmental Microbiome 19 (1). Available at https://doi.org/10.1186/s40793-024-00584-3.

Misu, I.J., Kayess, M.O., Siddiqui, M.N., Gupta, D.R., Islam, M.N., Islam, T., 2025. Microbiome engineering for sustainable rice production: strategies for biofertilization, stress tolerance, and climate resilience. Microorganisms 13 (2). Available at https://doi.org/10.3390/microorganisms13020233.

Motlhalamme, T., Mohamed, H., Kaningini, A.G., More, G.K., Thema, F.T., Mohale, K.C., Maaza, M., 2023. Bio-synthesized calcium carbonate ($CaCO_3$) nanoparticles: Their antifungal properties and application as nanofertilizer on *Lycopersicon esculentum* growth and gas exchange measurements. Plant Nano Biology 6, 100050. Available at https://doi.org/10.1016/j.plana.2023.100050.

Namitha, V.V., Raj, S.K., Jacob, D., Pillai, P.S., Radhakrishnan, N.V., Paramesha, V., et al., 2025. Impact of crop combination and nutrient management on productivity, profitability, and soil health in a coconut-based multistoried cropping system in Kerala, India. Frontiers in Sustainable Food Systems 9. Available at https://doi.org/10.3389/fsufs.2025.1605962.

Nerkar, G., Devarumath, S., Purankar, M., Kumar, A., Valarmathi, R., Devarumath, R., et al., 2022. Advances in crop breeding through precision genome editing. Frontiers in Genetics 13. Available at https://doi.org/10.3389/fgene.2022.880195.

Ochoa-Chaparro, E.H., Patiño-Cruz, J.J., Anchondo-Páez, J.C., Pérez-Álvarez, S., Chávez-Mendoza, C., Castruita-Esparza, L.U., et al., 2025. Seed nanopriming with ZnO and SiO_2 enhances germination, seedling vigor, and antioxidant defense under drought stress. Plants 14 (11). Available at https://doi.org/10.3390/plants14111726.

Patra, A., Sharma, V.K., Chattopadhyay, A., Mohapatra, K.K., Rakshit, A., 2021. Soil Science: Fundamentals to Recent Advances. Springer, Singapore, pp. 85–126. Available at https://doi.org/10.1007/978-981-16-0917-6_6.

Priyadharsini, R., Rajadurai, K.R., Beaulah, A., Mohan, M.M., Venkatesan, K., Rajesh, S., 2025. A review on crop improvement strategies and breeding methods in ornamental annuals. Horticultural Science 52 (1), 15–32. Available at https://doi.org/10.17221/31/2024-HORTSCI.

Quan, R., Wang, J., Qin, H., Chen, L., Xiao, D., Zhao, Z., et al., 2025. Improving grain yield and salt tolerance by optimizing plant height with beneficial haplotypes in rice (*Oryza sativa*). Journal of Advanced Research 76, 17–32. Available at https://doi.org/10.1016/j.jare.2024.12.007.

Rajendra, V., 2025. Omics Approaches for Tomato Yield and Quality Trait Improvement. Springer, Singapore, pp. 273–305. Available at https://doi.org/10.1007/978-981-96-3890-1_13.

Rahman, M.M., Ghosh, P.K., Akter, M., Al Noor, M.M., Rahman, M.A., Keya, S.S., et al., 2024. Green vanguards: harnessing the power of plant antioxidants, signal catalysts, and genetic engineering to combat reactive oxygen species under multiple abiotic stresses. Plant Stress 13. Available at https://doi.org/10.1016/j.stress.2024.100547.

Senevirathna, A.G.U.N., Wickramasinghe, K.G.G.M., Dissanayaka, D.M.S.B., De Silva, S.H.N.P., 2024. Root growth dynamics, nutrient uptake and use efficiency of *Grevillea robusta* grown under nitrogen and phosphorus deficiency. Journal of Plant Nutrition 47 (17), 2773–2786. Available at https://doi.org/10.1080/01904167.2024.2369065.

Sharma, G., Saharan, V., Pal, A., Sharma, S.S., 2024). Chitosan nanofertilizer to strengthen sink strength and provide resistance against PFSR (post flowering stalk rot) disease in maize. Biocatalysis and Agricultural Biotechnology 60, 103303. Available at https://doi.org/10.1016/j.bcab.2024.103303.

Shi, D., Lin, H., Xu, Y., Ge, Y., Teng, Y., Huang, C., et al., 2025. Brassinosteroid synthesis genes CcCPD/CcDWF4/CcDET2 in *Carya cathayensis* enhance resilience to salt stress through multi-faceted regulatory mechanisms. Plant Physiology and Biochemistry 229 (Pt C), 110552. Available at https://doi.org/10.1016/j.plaphy.2025.110552.

Shoukat, A., Saqib, Z.A., Nawaz, A., Amir, K.Z., Ahmad, I., Hamza, A., et al., 2025. Nanofertilizers benefited maize to cope oxidative stress under saline environment. Plant Nano Biology 11. Available at https://doi.org/10.1016/j.plana.2025.100141.

Silva, J., Uchida, R., 2000. Essential Nutrients for Plant Growth: Nutrient Functions and Deficiency Symptoms. Plant Nutrient Management in Hawaii's Soils, Approaches for Tropical and Subtropical Agriculture, College of Tropical Agriculture and Human Resources. University of Hawaii at Manoa, Honolulu, pp. 31–55.

Singh, A., Bol, R., Lovynska, V., Singh, R.K., Sousa, J.R., Ghazaryan, K., 2025a. Application of nanoparticles for salinity stress management and biofortification in wheat: a review of dual approaches and insights. Frontiers in Plant Science 16. Available at https://doi.org/10.3389/fpls.2025.1592866.

Singh, G., Raigar, O.P., Kaur, S., Bishnoi, R., Mondal, K., et al., 2025b. Advances in genomics-assisted breeding strategies for enhancing nutrient uptake and use efficiency in cereals: a pathway toward sustainable agriculture. Plant Stress 18. Available at https://doi.org/10.1016/j.stress.2025.101002.

Singh, V.K., Ahmed, S., Saini, D.K., Gahlaut, V., Chauhan, S., Khandare, K., et al., 2024. Manipulating epigenetic diversity in crop plants: techniques, challenges and opportunities. Biochimica et Biophysica Acta - General Subjects 1868 (2). Available at https://doi.org/10.1016/j.bbagen.2023.130544.

Song, Z., Yu, Z., Li, H., Sun, Q., Liu, H., Hui, M., 2026. Identification and functional characterization of BrcLL1, a candidate gene controlling leaf lobing in non-heading Chinese cabbage (*Brassica rapa* subsp. chinensis). Plant Science 362, 112785. Available at https://doi.org/10.1016/j.plantsci.2025.112785.

Su, H., Foster, T., Hogeboom, R.J., Luna-Gonzalez, D.V., Mialyk, O., Willaarts, B., Wang, Y., Krol, M.S., 2025. Nutrient production, water consumption, and stresses of large-scale versus small-scale agriculture: A global comparative analysis based on a gridded crop model. Global Food Security 45, 100844. Available at 10.1016/j.gfs.2025.100844.

Sun, L., Lai, M., Ghouri, F., Nawaz, M.A., Ali, F., Baloch, F.S., et al., 2024. Modern plant breeding techniques in crop improvement and genetic diversity: from molecular markers and gene editing to artificial intelligence—a critical review. Plants 13 (19), 2676. Available at https://doi.org/10.3390/plants13192676.

Sutula, M., Tussipkan, D., Kali, B., Manabayeva, S., 2025. Molecular mechanisms underlying defense responses of potato (*Solanum tuberosum* L.) to environmental stress and CRISPR/cas-mediated engineering of stress tolerance. Plants 14 (13), 1983. Available at https://doi.org/10.3390/plants14131983.

Tabasum, S., Nazir, G., Hussain, K., Gautam, A., Mushtaq, F., et al., 2025. Climate smart breeding of vegetable crops: a review. Plant Archives 25, 3298–3307. Available at 10.51470/PLANTARCHIVES.2025.v25.supplement-1.450.

Teker Yildiz, M., Günay, E., Acar, O., 2024. Physiological and biochemical effects of thermo-priming on wheat (*Triticum aestivum* L) under drought and heat stresses. Düzce Üniversitesi Bilim ve Teknoloji Dergisi 12 (1), 375–389. Available at https://doi.org/10.29130/dubited.1213671.

Thudi, M., Mascher, M., Jayakodi, M., 2025. Pangenome charts the genomic path for wheat improvement. Trends in Plant Science 30 (7), 687–689. Available at https://doi.org/10.1016/j.tplants.2025.03.002.

Ton, L.B., Qayyum, Z., Amas, J., Thomas, W.J.W., Edwards, D., Batley, J., et al., 2025. Applications of CRISPR/Cas tools in improving stress tolerance in Brassica crops. Frontiers in Plant Science 16. Available at https://doi.org/10.3389/fpls.2025.1616526.

Uysal, Y., Doğaroğlu, Z.G., 2025. Citrus peel–mediated ZnO nanoparticles: green synthesis, plant responses, and antibacterial activity. Biocatalysis and Agricultural Biotechnology 70. Available at https://doi.org/10.1016/j.bcab.2025.103854.

Vikram, P., Sehgal, D., Prasad, M., Singh, S., Raman, H., 2024. Editorial: harnessing crop biodiversity and genomics assisted pre-breeding approaches for next generation climate-smart varieties, volume II. Frontiers in Plant Science 15. Available at https://doi.org/10.3389/fpls.2024.1387016.

Vos, R., Bellù, L.G., 2018. Chapter 2 - Global trends and challenges to food and agriculture into the 21st century. In: Campanhola, C., Pandey, S. (Eds.), Sustainable Food and Agriculture: An Integrated Approach. Academic Press, pp. 11–30. Available at https://doi.org/10.1016/B978-0-12-812134-4.00002-9.

Wang, R., Zhang, H., Liu, Z., Lu, Y., Lan, S., et al., 2025a. Uncovering the transcriptional regulatory network underlying selenium tolerance in maize seedlings. Journal of Hazardous Materials. 494, 138437. Available at https://doi.org/10.1016/j.jhazmat.2025.138437.

Wang, X., Qi, Y., Sun, G., Zhang, S., Li, W., Wang, Y., 2024. Improving soybean breeding efficiency using marker-assisted selection. Molecular Plant Breeding. Available at https://doi.org/10.5376/mpb.2024.15.0025.

Wang, Y., Phelps, A., Godbehere, A., Evans, B., Takizawa, C., et al., 2025. Revolutionizing agriculture with CRISPR technology: applications, challenges, and future perspectives. Biotechnology Journal 20 (9), e70113. Available at https://doi.org/10.1002/biot.70113.

Wu, B., Cao, C., Mosey, S., Daniell, T., Noy, P., Cui, Y., Rose, M., Snape, J., 2025. How does global agricultural research and innovation cooperation influence agricultural R&I system transformation in the South? Evidence from UK-China cooperation. Food Policy 131, 102813. Available at 10.1016/j.foodpol.2025.102813.

Zhang, Y., Zhu, S., Zhou, J., Wang, F., 2025. Review of genetic mapping and marker-assisted selection in potato breeding. Molecular Plant Breeding 16 (1), 55–62. Available at https://doi.org/10.5376/mpb.2025.16.0006.

Zhao, B., Tong, L., Liu, H., Hao, M., Zhang, R., 2025. Optimizing root morphology is a key to improving maize yield under nitrogen reduction and densification cultivation. Field Crops Research 329, 109958. Available at https://doi.org/10.1016/j.fcr.2025.109958.

Zhuang, D., Zhou, X., Zhao, D., Wang, J., Wang, Y., Chen, M., et al., 2025. AgNPs seeds nanopriming enhanced the tolerance of chinese cabbage to diverse abiotic stresses. ACS Agricultural Science & Technology 5 (8), 1722–1734. Available at https://doi.org/10.1021/acsagscitech.5c00376.

Zubair, A., Ali, M., Ahmad, F., Althobaiti, S.A., 2025. Unlocking the role of transcription activator-like effector nuclease (TALENs) and zinc finger nuclease (ZFN) in the treatment of HIV. Molecular Biology Reports 52, 948. Available at https://doi.org/10.1007/s11033-025-10993-3.

Further reading

Doğaroğlu, Z.G., Uysal, Y., Çaylalı, Z., Karakoç, G., 2024. Antibacterial and phytotoxicological properties assessment of *Momordica charantia* extract-based ZnO nanoparticles. Journal of the Science of Food and Agriculture 104 (5), 2851–2861. Available at https://doi.org/10.1002/jsfa.13176.

Kareem, H.A., Li, Y., Saleem, S., Mustafa, A., Azeem, M., et al., 2025. Nano-bio priming of ZnO and *Sinorhizobium meliloti* mitigates salt stress in alfalfa via improved mineral uptake, antioxidant gene expression, and cellular protection. Industrial Crops and Products 234, 121624. Available at https://doi.org/10.1016/j.indcrop.2025.121624.

Paul, B., Montoya, G., 2020. CRISPR-Cas12a: functional overview and applications. Biomedical Journal 43 (1), 8–17. Available at https://doi.org/10.1016/j.bj.2019.10.005.

Rossmann, M., Pérez-Jaramillo, J.E., Kavamura, V.N., Chiaramonte, J.B., Dumack, K., et al., 2020. Multitrophic interactions in the rhizosphere microbiome of wheat: from bacteria and fungi to protists. FEMS Microbiology Ecology 96 (4), 1–14. Available at https://doi.org/10.1093/femsec/fiaa032.

Index

Note: Page numbers followed by "*f*" and "*t*" refer to figures and tables, respectively.

B

F

G

H

I

J

L

M

N

O

P

Printed and bound by CPI Group (UK) Ltd, Croydon, CR0 4YY
22/09/2026
02220733-0019